AF357329

Sustainability in Geotechnics: The Use of Environmentally Friendly Materials

Sustainability in Geotechnics: The Use of Environmentally Friendly Materials

Editor

Castorina Silva Vieira

MDPI • Basel • Beijing • Wuhan • Barcelona • Belgrade • Manchester • Tokyo • Cluj • Tianjin

Editor
Castorina Silva Vieira
University of Porto
Portugal

Editorial Office
MDPI
St. Alban-Anlage 66
4052 Basel, Switzerland

This is a reprint of articles from the Topical Collection published online in the open access journal *Sustainability* (ISSN 2071-1050) (available at: https://www.mdpi.com/journal/sustainability/special_issues/sustainability_geotechnics).

For citation purposes, cite each article independently as indicated on the article page online and as indicated below:

LastName, A.A.; LastName, B.B.; LastName, C.C. Article Title. *Journal Name* **Year**, *Volume Number*, Page Range.

ISBN 978-3-0365-3419-0 (Hbk)
ISBN 978-3-0365-3420-6 (PDF)

Cover image courtesy of Castorina Silva Vieira.

Contents

About the Editor

Castorina Silva Vieira is Assistant Professor at the Civil Engineering Department of the Faculty of Engineering of the University of Porto (FEUP), Portugal, since 2008. She is also an Integrated Member of the research unit CONSTRUCT—Institute of R&D in Structures and Construction—and Director of the Construction Materials Laboratory from the same university. Currently, she is Deputy Secretary of the Portuguese Geotechnical Society, vice-president of the Portuguese Chapter of the International Geosynthetics Society (IGS) and a member of different Technical Committees of the International Society for Soil Mechanics and Geotechnical Engineering (ISSMGE). She is also an Editorial Board Member of the journals Environmental Geotechnics, Sustainability, Advances in Civil Engineering and Frontiers in Built Environment. Among her major research interests are Sustainability in Geotechnical Engineering, Recycling Construction and Demolition (C&D) waste, Geosynthetics and Soil/geosynthetic interaction.

Editorial

Sustainability in Geotechnics through the Use of Environmentally Friendly Materials

Castorina S. Vieira

CONSTRUCT, Faculty of Engineering, University of Porto, R. Dr. Roberto Frias, 4200-465 Porto, Portugal; cvieira@fe.up.pt

Keywords: sustainability in geotechnical engineering; environmentally friendly materials; low carbon materials; recycled materials; bioengineering techniques; geosynthetics; sustainable ground improvement; sustainable ground remediation; geopolymers

Citation: Vieira, C.S. Sustainability in Geotechnics through the Use of Environmentally Friendly Materials. *Sustainability* **2022**, *14*, 1155. https://doi.org/10.3390/su14031155

Received: 14 January 2022
Accepted: 15 January 2022
Published: 20 January 2022

Publisher's Note: MDPI stays neutral with regard to jurisdictional claims in published maps and institutional affiliations.

1. Introduction

The reduction in the exploitation of non-renewable natural resources is nowadays widely recognized as a pressing need for a more sustainable society. Moreover, the increase in waste valorization and reuse of waste materials are undoubtedly important steps forward for environmental sustainability. Geotechnical design being part of typical civil engineering projects can play a major role in the sustainability of the built environment. Thus, a Special Issue was proposed focused on the use of environmentally friendly materials in geotechnical solutions, highlighting the relevance of geotechnics to reduce our carbon footprint. Their main purpose to collect and publish original research papers pointing out the use of sustainable materials in geotechnics has been achieved, through the great interest of the research community and a high number of submissions. This editorial summarizes the papers published during the 2020–2021 biennium, highlighting their main conclusions.

2. Overview of the Special Issue

One of the biggest challenges facing civil engineers is the design and construction of sustainable structures and infrastructures. Geotechnical engineering, as a branch of civil engineering, can significantly contribute to sustainable development in the construction industry, implementing environmentally friendly and cost-effective solutions. The Special Issue "Sustainability in Geotechnics through the Use of Environmentally Friendly Materials" aims to highlight the ability of geotechnical engineering to contribute to a more sustainable society, through the use of materials and techniques with a smaller environmental footprint. The huge potential of this contribution is clear in the diversity of topics covered by the twenty-three papers published in this issue.

Several papers proposed sustainable ground improvement techniques [1–7] and soil reinforcement with alternative materials such as recycled polypropylene fibers [8], recycled tire-derived aggregates [9], recycled polyethylene terephthalate strips [10] or polypropylene waste strips [11].

The use of recycled materials replacing the soils or natural quarry materials in geotechnical works was presented by various researchers [12–16]. Sustainable solutions using geosynthetics to prevent soil contamination [17,18], to treat the sludge generated in different industries and avoid ground contamination [19] and to reinforce alternative filling materials [12,16,20] are also discussed in the Special Issue.

Low-carbon solutions for the stabilization of contaminated soils [21,22] and bioengineering techniques to prevent soil erosion [23] are put forward as relevant contributions to the planet's sustainability.

The relevance and topicality of the theme raised the attention and interest of researchers from different countries around the world. Ninety-one researchers from sixteen

Sustainability **2022**, *14*, 1155. https://doi.org/10.3390/su14031155   https://www.mdpi.com/journal/sustainability

different countries contributed to this Special Issue (Figures 1 and 2). Based on the first author's institution, the greatest number of papers came from Brazil (seven papers), followed by India with three contributions. Australia, China, Portugal and Thailand contributed two papers each and Germany, Korea, Pakistan, Turkey and USA contributed one paper (Figure 1).

Twelve papers arose from collaborations among researchers from different countries. Figure 2 presents the countries of origin of the ninety-one authors (excluding authors with more than one contribution). In addition to the eleven countries identified in Figure 1, there are also contributions from Canada, Ireland, Japan, Saudi Arabia and Vietnam. As with the number of contributions, the greatest number of authors come from Brazil, followed by China (Figure 2).

The following section summarizes the different contributions with emphasis on the most relevant findings, following the order of their publication.

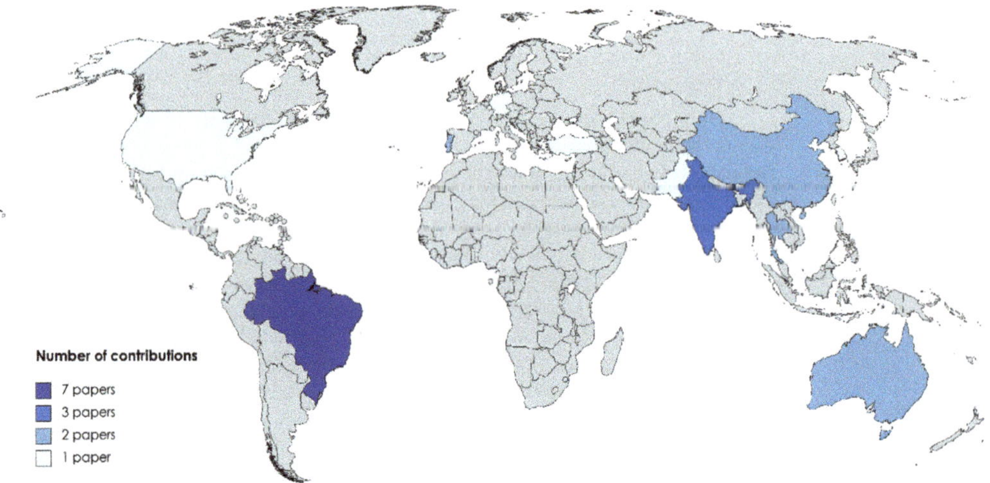

Figure 1. Number of contributions per country (based on first author affiliation).

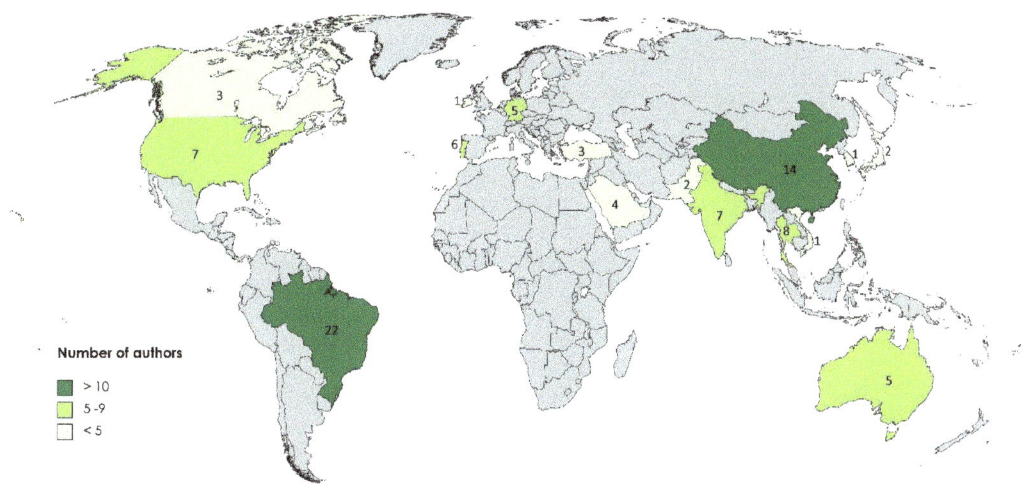

Figure 2. Number of authors from different countries (excluding authors with more than one contribution).

3. Highlights of the Contributions

A laboratory study on a low-carbon cementitious material known as limestone-calcined clay cement (LC^3) for the potential stabilization/solidification of zinc (Zn) and lead (Pb) contaminated soils is presented by Reddy et al. [21]. The authors found that the addition of the LC^3 binder at 8% can improve the compressive strength up to three times compared to untreated Zn- and Pb-contaminated soils. After a 14-day curing period, the pH transformation of acidic to alkaline nature allowed the adsorption of heavy metals in forming various insoluble metal hydroxides. This work shows that limestone-calcined clay cement is a green and sustainable remediation of Zn- and Pb-contaminated soils and the treated soil can be used as a safe and environmentally friendly construction material.

Vieira et al. [16] discuss the pullout behaviour of geogrids embedded in recycled construction and demolition materials, with emphasis on the effects of the specimen's size and pullout displacement rates. The results have shown that the pullout resistance of the geogrids increases with the specimen size and imposed displacement rate. The pullout interaction coefficient, one of the essential parameters in the design of geosynthetic-reinforced structures, exceeded the values typically assumed in the absence of test data (0.5–0.7), which allows us to conclude that the usual practices for conventional backfill materials (soils) are still applicable. The results of this study support the feasibility of using fine-grain recycled construction and demolition materials in the construction of geogrid-reinforced embankments, with obvious benefits in terms of environmental protection and sustainability [16].

The durability of a deposited marine clay treated with cement, copper slag, and hydrated lime was studied in a laboratory by Hanafi et al. [1]. Alternative ground improvement techniques, such as the one presented in this paper, reduce cement usage, and in addition, using waste material, such as copper slag, enables safe disposal of those materials. The authors claim that all proposed mixes resulted in the reduction of embodied energy and CO$_2$ emission.

Ma et al. [2] study the feasibility of using soda residue to produce an alternative material for geotechnical engineering applications. The preparation method of this material in the field is proposed and the mechanical properties for different mixing proportions with fly ash, sand and rubble are investigated. The authors concluded that the subgrade bearing capacity and deformation modulus of this alternative material are higher than those currently recorded with clays and sands.

The addition of lime and/or gypsum to enhance the geotechnical properties of two different types of fly ashes and a multi-criteria decision-making approach to assist practicing engineers in selecting the appropriate mixture for specific civil engineering applications are presented by Moghal et al. [3].

Moghal et al. [22] evaluate the efficacy of enzymatically induced calcite precipitation in the retention of heavy metal in soils. Soils contaminated with cadmium (Cd), nickel (Ni), and lead (Pb) were treated with three types of enzyme solutions. Based on their results, the authors claim that the enzyme-induced calcite precipitation can be an effective alternative in the remediation of soils contaminated with heavy metal ions.

Vianna et al. [23] explore different types of bioengineering techniques as environmentally friendly systems to prevent superficial erosion processes in soil slopes. The efficiency of these techniques was evaluated through high-quality images taken from periodical visits conducted over several months, processed with a computer code. The authors concluded that most of the solutions revealed a deficiency in vegetation establishment and were sensitive to climatological conditions, which has been conditioned by the low fertility and medium acidity of the soil.

Concerning the sustainable treatment of sludge produced in different industries, Aparicio-Ardila et al. [19] present a laboratory study and a statistical analysis on geotextile tube technology used for dewatering of sludge generated at a water treatment plant. The viability of using nonwoven geotextile, a material with better filtration characteristics than the commonly used woven geotextiles, was also studied by [19]. The authors propose to recirculate inside the geotextile tube the effluent collected at the beginning of the dewatering process so that it can be filtered more efficiently and therefore improve its quality. Under the same test conditions, the dewatering performance was better in the bags produced with nonwoven geotextile when compared to those manufactured with woven geotextiles (commonly used).

Lavoie et al. [18] evaluate the performance of two high-density polyethylene (HDPE) geomembranes used, for several years, in a sewage treatment aeration pond and in a municipal landfill leachate pond. The experimental study shows that these geomembranes perform adequately for environmental protection.

Almeida et al. [20] deal with the mechanical damage under repeated loading induced by incinerator bottom ash on three nonwoven geotextiles. The authors concluded that the damage provoked by the incinerator bottom ash on the short-term mechanical behaviour of the geotextiles tends to be lower than the damage induced by the natural aggregates used in their study, which represents good prospects for the use of these alternative materials in contact with geotextiles.

The mechanical properties of compacted lateritic soils reinforced with polypropylene (PP) waste strips cut from recycled plastic packing are discussed by Marçal et al. [11]. Among the many findings of their study, it is worth mentioning that the use of PP waste strips as reinforcements of lateritic soils led to an increase of the unified compressive strength and contributed to the change in soil failure from a brittle to a ductile mode.

Silveira et al. [10] evaluate the effect of recycled polyethylene terephthalate (PET) strips on the mechanical properties of cement-treated lateritic sandy soil. Based on their experimental study the authors concluded the inclusion of recycled PET strips in cement-treated soil can provide a sustainable alternative material with higher strength and ductility.

Safdar et al. [4] report the results of monotonic triaxial drained compression tests performed on sand-cement-fiber mixtures and propose various equations to evaluate the shear modulus and mobilized stress curves at small-strain levels. Their study aims to study the viability of utilizing tsunami waste as ground improvement materials to build sustainable geotechnical infrastructures. The authors claim this research related to increase in stiffness parameters due to the combined effect of cement and fiber additives might be useful for the practicing engineers in the construction of economical and sustainable geotechnical infrastructures.

Using a framework that incorporates resource consumption, environmental and socio-economic concerns, Samuel et al. [5] evaluate the sustainability benefits of a metakaolin-based geopolymer treatment for an expansive soil, when compared with a lime treatment method. Their study shows that the proposed geopolymer, primarily due to metakaolin source material, is a more sustainable alternative to the conventional lime treatment for soil stabilization. In short, the geopolymers can be viable additives or co-additives for the chemical stabilization of problematic expansive soils.

Park [17] evaluates experimental and numerically the permeability characteristics of a geotextile–polynorbornene liner at different oil pollutant contact times and pressure heads. This study confirms the impermeability of the geotextile–polynorbornene liner against oil pollutants and its potential as a solution to prevent pollutant diffusion.

Deng et al. [6] analyze the energy consumption and carbon emissions of Microbial Induced Carbonate Precipitation (MICP), used to strengthen soils and other materials, through the Life Cycle Assessment (LCA) method. The authors found that the current MICP application process consumes less non-renewable resources but has a greater environmental impact. The major environmental impact of MICP techniques is the production of smoke and ash, with other secondary impacts such as global warming, photochemical ozone creation, acidification and eutrophication [6].

A large database of soil/tire-derived aggregates (TDA) compaction tests assembled from the literature was used by Soltani et al. [9] to model the compaction characteristics (optimum moisture content and maximum dry unit weight) of fine-grained soils blended with sand-sized recycled TDA. According to the authors, the proposed empirical models offer a practical procedure towards predicting the compaction characteristics of the soil–TDA blends and can be used for preliminary design assessments and soil–TDA optimization studies.

Udomchai et al. [15] evaluate the interface shear strength between reclaimed asphalt pavement (RAP) and kenaf geogrids (produced with natural kenaf fibers), and assess their viability as environmentally friendly base course materials. A generalized equation was proposed for predicting the interface shear strength, which can be used in the analysis and design of related geotechnical projects including pavement projects, mechanically stabilized earth (MSE) wall design, embankment reinforcement construction and foundation design [15].

Lin et al. [14] analyze the feasibility of utilizing recycled aluminum salt slag (RASS) as a sustainable geomaterial through a comprehensive laboratory test program, focused on geotechnical and environmental engineering tests. Their study has shown that RASS has geotechnical characteristics suitable for their usage as pavement subbase material in road construction and does not pose any environmental and health issues. However, the authors point out that the quality of this aluminum industrial by-product is dependent on the machinery used and manufacturing techniques.

Sukmak et al. [7] examine the feasibility of using garnet residues (GR) as a replacement material in soft clay (SC) prior to cement stabilization, combined with tire-derived aggregates (TDA), to produce an alternative subgrade material. The authors found that GR replacement reduces the specific surface and particle contacts of the SC-GR blends. High amounts of GR led to the reduction in the unconfined compressive strength due to its high water absorption, resulting in insufficient water for cement hydration. Due to the low adhesion property of TDA, unconfined compressive strength and stiffness of cement-stabilized SC-GR reduce with the increase in TDA content. The cement–TDA-stabilized SC–GR at SC:GR of 90:10, cement content of 2% and TDA content of 2% is suggested by [7] as a sustainable subgrade material.

Schwerdt et al. [13] investigate the feasibility of using alternative materials, such as blast furnace slag (BFS), electric furnace slag (EFS), track ballast (TB), and recycled concrete (RC) as filling material in the construction of geogrid-reinforced structures. The laboratory characterization of the materials and a pilot application are presented. From the geotechnical point of view, these alternative materials revealed similar or even better behaviour than natural materials such as gravel. The results of the chemical tests show that only electric furnace slag and track ballast are qualified to be used without restrictions.

The effect of recycled polypropylene (PP) fibers on the shear strength–dilation behaviour of compacted lateritic soils is discussed by Silveira et al. [8]. The authors found that the PP fibers improved the shear strength behaviour of both lateritic soils studied and the shear stress–dilatancy behaviour is affected by the inclusions in the soil mix.

Palmeira et al. [12] present a review on the use of wastes, such as wasted tires, construction and demolition wastes and plastic bottles, as alternative construction materials in geotechnical and geoenvironmental works, giving particular emphasis to their combination with geosynthetics. The authors point out that, despite the benefits to the environment in reusing wastes in geotechnical engineering, it is important to have in mind that some of these wastes will degrade over time or can contain substances that may cause ground contamination. A careful evaluation of such aspects must be carried out before using wastes as construction materials in geotechnical and geoenvironmental works. In this context and when required, geosynthetics such as geomembranes and geocomposite clay liners, if properly specified, can provide efficient barriers for such contaminants [12].

4. Final Remarks

The principle of sustainable development was defined in the report "Our Common Future", also known as the "Brundtland Report", published in 1987 by the United Nations Brundtland Commission, as the "development that meets the needs of the present without compromising the ability of future generations to meet their own needs" [24]. Such a clear, seemingly simple and logical definition of sustainability should be part of everyone's principles and of the different sectors of activity. Due to the huge impact that the construction sector has on the environment, both regarding the consumption of natural resources and energy, and the large volumes of waste produced, this industry has been called upon to change its practices. It is therefore fundamental that researchers and engineers are increasingly encouraged to find solutions that meet sustainable development goals. This Special Issue was intended to bring together different contributions for greater sustainability in geotechnical engineering.

This Special Issue includes several contributions but there is still plenty of room for new contributions on sustainability in geotechnics. For this reason, the issue has been turned into a topical collection and new contributions will continue to be accepted.

Funding: This work was partly prepared within the scope of the research project PTDC/ECI-EGC/30452/2017-POCI-01-0145-FEDER-030452-funded by FEDER funds through COMPETE2020-Programa Operacional Competitividade e Internacionalização (POCI) and by national funds (PIDDAC) through FCT/MCTES.

Data Availability Statement: Not applicable.

Acknowledgments: The Guest Editor would like to thank all the authors and reviewers who contributed to the successful development of this Special Issue.

Conflicts of Interest: The author declares no conflict of interest.

References

1. Hanafi, M.; Ekinci, A.; Aydin, E. Triple-Binder-Stabilized Marine Deposit Clay for Better Sustainability. *Sustainability* **2020**, *12*, 4633. [CrossRef]
2. Ma, J.; Yan, N.; Zhang, M.; Liu, J.; Bai, X.; Wang, Y. Mechanical Characteristics of Soda Residue Soil Incorporating Different Admixture: Reuse of Soda Residue. *Sustainability* **2020**, *12*, 5852. [CrossRef]
3. Moghal, A.A.B.; Rehman, A.U.; Vydehi, K.V.; Umer, U. Sustainable Perspective of Low-Lime Stabilized Fly Ashes for Geotechnical Applications: PROMETHEE-Based Optimization Approach. *Sustainability* **2020**, *12*, 6649. [CrossRef]
4. Safdar, M.; Newson, T.; Schmidt, C.; Sato, K.; Fujikawa, T.; Shah, F. Effect of Fiber and Cement Additives on the Small-Strain Stiffness Behavior of Toyoura Sand. *Sustainability* **2020**, *12*, 10468. [CrossRef]
5. Samuel, R.; Puppala, A.J.; Radovic, M. Sustainability Benefits Assessment of Metakaolin-Based Geopolymer Treatment of High Plasticity Clay. *Sustainability* **2020**, *12*, 10495. [CrossRef]

6. Deng, X.; Li, Y.; Liu, H.; Zhao, Y.; Yang, Y.; Xu, X.; Cheng, X.; Wit, B.d. Examining Energy Consumption and Carbon Emissions of Microbial Induced Carbonate Precipitation Using the Life Cycle Assessment Method. *Sustainability* **2021**, *13*, 4856. [CrossRef]
7. Sukmak, P.; Sukmak, G.; Horpibulsuk, S.; Kassawat, S.; Suddeepong, A.; Arulrajah, A. Improved Mechanical Properties of Cement-Stabilized Soft Clay Using Garnet Residues and Tire-Derived Aggregates for Subgrade Applications. *Sustainability* **2021**, *13*, 11692. [CrossRef]
8. Silveira, M.R.; Rocha, S.A.; Correia, N.d.S.; Rodrigues, R.A.; Giacheti, H.L.; Lodi, P.C. Effect of Polypropylene Fibers on the Shear Strength–Dilation Behavior of Compacted Lateritic Soils. *Sustainability* **2021**, *13*, 12603. [CrossRef]
9. Soltani, A.; Azimi, M.; O'Kelly, B.C. Modeling the Compaction Characteristics of Fine-Grained Soils Blended with Tire-Derived Aggregates. *Sustainability* **2021**, *13*, 7737. [CrossRef]
10. Silveira, M.R.; Lodi, P.C.; Correia, N.d.S.; Rodrigues, R.A.; Giacheti, H.L. Effect of Recycled Polyethylene Terephthalate Strips on the Mechanical Properties of Cement-Treated Lateritic Sandy Soil. *Sustainability* **2020**, *12*, 9801. [CrossRef]
11. Marçal, R.; Lodi, P.C.; Correia, N.d.S.; Giacheti, H.L.; Rodrigues, R.A.; McCartney, J.S. Reinforcing Effect of Polypropylene Waste Strips on Compacted Lateritic Soils. *Sustainability* **2020**, *12*, 9572. [CrossRef]
12. Palmeira, E.M.; Araújo, G.L.S.; Santos, E.C.G. Sustainable Solutions with Geosynthetics and Alternative Construction Materials— A Review. *Sustainability* **2021**, *13*, 12756. [CrossRef]
13. Schwerdt, S.; Mirschel, D.; Hildebrandt, T.; Wilke, M.; Schneider, P. Substitute Building Materials in Geogrid-Reinforced Soil Structures. *Sustainability* **2021**, *13*, 12519. [CrossRef]
14. Lin, Y.; Maghool, F.; Arulrajah, A.; Horpibulsuk, S. Engineering Characteristics and Environmental Risks of Utilizing Recycled Aluminum Salt Slag and Recycled Concrete as a Sustainable Geomaterial. *Sustainability* **2021**, *13*, 10633. [CrossRef]
15. Udomchai, A.; Hoy, M.; Suddeepong, A.; Phuangsombat, A.; Horpibulsuk, S.; Arulrajah, A.; Thanh, N.C. Generalized Interface Shear Strength Equation for Recycled Materials Reinforced with Geogrids. *Sustainability* **2021**, *13*, 9446. [CrossRef]
16. Vieira, C.S.; Pereira, P.; Ferreira, F.; Lopes, M.d.L. Pullout Behaviour of Geogrids Embedded in a Recycled Construction and Demolition Material. Effects of Specimen Size and Displacement Rate. *Sustainability* **2020**, *12*, 3825. [CrossRef]
17. Park, J. Evaluation of Changes in the Permeability Characteristics of a Geotextile–Polynorbornene Liner for the Prevention of Pollutant Diffusion in Oil-Contaminated Soils. *Sustainability* **2021**, *13*, 4797. [CrossRef]
18. Lavoie, F.L.; Valentin, C.A.; Kobelnik, M.; Lins da Silva, J.; Lopes, M.d.L. HDPE Geomembranes for Environmental Protection: Two Case Studies. *Sustainability* **2020**, *12*, 8682. [CrossRef]
19. Aparicio-Ardila, M.A.; Souza, S.T.d.; Silva, J.L.d.; Valentin, C.A.; Dantas, A.D.B. Geotextile Tube Dewatering Performance Assessment: An Experimental Study of Sludge Dewatering Generated at a Water Treatment Plant. *Sustainability* **2020**, *12*, 8129. [CrossRef]
20. Almeida, F.; Carneiro, J.R.; Lopes, M.d.L. Use of Incinerator Bottom Ash as a Recycled Aggregate in Contact with Nonwoven Geotextiles: Evaluation of Mechanical Damage Upon Installation. *Sustainability* **2020**, *12*, 9156. [CrossRef]
21. Reddy, V.A.; Solanki, C.H.; Kumar, S.; Reddy, K.R.; Du, Y.-J. Stabilization/Solidification of Zinc- and Lead-Contaminated Soil Using Limestone Calcined Clay Cement (LC3): An Environmentally Friendly Alternative. *Sustainability* **2020**, *12*, 3725. [CrossRef]
22. Moghal, A.A.B.; Lateef, M.A.; Mohammed, S.A.S.; Lemboye, K.; Chittoori, B.C.S.; Almajed, A. Efficacy of Enzymatically Induced Calcium Carbonate Precipitation in the Retention of Heavy Metal Ions. *Sustainability* **2020**, *12*, 7019. [CrossRef]
23. Vianna, V.F.; Fleury, M.P.; Menezes, G.B.; Coelho, A.T.; Bueno, C.; Lins da Silva, J.; Luz, M.P. Bioengineering Techniques Adopted for Controlling Riverbanks' Superficial Erosion of the Simplício Hydroelectric Power Plant, Brazil. *Sustainability* **2020**, *12*, 7886. [CrossRef]
24. United Nations World Commission on Environment and Development. *Report of the World Commission on Environment and Development: Our Common Future*; Oxford University Press: Oxford, UK, 1987; p. 300.

Article

Triple-Binder-Stabilized Marine Deposit Clay for Better Sustainability

Mohamad Hanafi [1], Abdullah Ekinci [2,* and Ertug Aydin [1]

[1] Civil Engineering, European University of Lefke, North Cyprus, Mersin TR-10, Turkey;
    mhanafi@eul.edu.tr (M.H.); eraydin@eul.edu.tr (E.A.)

[2] Civil Engineering Program, Middle East Technical University, Northern Cyprus Campus, Kalkanli,
    Guzelyurt, North Cyprus, Mersin TR-10, Turkey

* Correspondence: ekincia@metu.edu.tr; Tel.: +90-542-888-1440

Received: 20 March 2020; Accepted: 4 June 2020; Published: 5 June 2020

Abstract: Marine clay deposits are commonly found worldwide. Considering the cost of dumping and related environmental concerns, an alternative solution involving the reuse of soils that have poor conditions is crucial. In this research, the authors examined the strength, microstructure, and wet–dry resistance of triple-binder composites of marine-deposited clays and compiled a corresponding database. In order to evaluate the wetting–drying resistance of the laboratory-produced samples, the accumulated mass loss (ALM) was calculated. The use of slag alone as a binder, at any percentage, increased the ALM up to 2%. However, the use of lime as the third binder seemed to accelerate the chemical reactions associated with the hydration of clay and cementitious material and to enhance the chemical stability, i.e., specimens that included both lime and slag experienced the same ALM as specimens treated with cement only. Scanning electron microscopy analysis confirmed the durability improvements of these clays. The proposed unconfined compressive strength–porosity and accumulated mass loss relationship yielded practical approximation for the fine- and coarse-grained soils blended with up to three binders until 60 days of curing. The laboratory-produced mixes showed reduction of embodied energy and embodied carbon dioxide (eCO_2) emissions for the proposed models.

Keywords: cement; lime; copper slag; strength; durability; microstructure; eCO_2; embodied energy

---

1. Introduction

Marine clay deposits are found worldwide, especially in coastal regions. This research investigated the marine clays disseminated along the Mediterranean and northern coasts of Cyprus Island. Marine clays generate substantial construction problems, mainly due to their low strength and sensitivity against drying/wetting cycles. Rapid development of construction on those formations is comprised mostly of digging and dumping of soil. Environmental pollution created from digging and transportation affects mankind and the ecosystem. Additionally, the huge quantity of excavated soil affects the overall cost of shipping and controlling. Thus, companies try to manage the problems, allowing for environmentally friendly solutions. Considering the structural integrity, utilization, or reuse of untreated marine deposited clay for sub-base construction on highways poses engineering problems. The widely adopted solution is to dispose of the excavated soil into the nearest site or landfill. However, allowing for transportation between the quarrying plant and the excavation area can lead to a large amount of CO_2 emission and increased overall cost. The problems associated with those activities can be minimized via performance optimization of excavated soils, which might reduce the cost and negative effects on the environment. Every project requires different solutions for the management of excavated soil. Previous studies suggested alternative managing strategies for excavated soils, including using them on-site, reusing excavated materials, pre-treating before use in construction, storing them for future consideration, and using them as landfill cover applications [1,2]. Furthermore, Magnusson et al. [3] reported that reusing excavated soil could save as much as 14 kg of carbon dioxide per ton. Additionally, Capobianco et al. [4] stated that the treatment of such soils is more beneficial than digging and dumping.

This study proposes the reuse of copper slag, which is extensively available at the Cyprus Mining Cooperation (CMC) site in the Lefke Region of Cyprus, as a cement replacement. The available slag is left over from the copper mining operations that ended in 1974 and is available in bulk form, having been haphazardly dumped around the Lefke Region. The serious concerns regarding the use of those materials involve heavy metal contamination and their leaching properties. Nevertheless, Alter [5] reported that the leaching values and heavy metal content of copper slag are lower than the levels prescribed by the United States Environmental Protection Agency (USEPA) and Basel Convention. Zain et al. [6] prepared specimens composed of 10% copper slag as a cement replacement. The results showed that the penetration of the trace elements did not exceed the normal rates. Another study revealed that the penetration of heavy metal (copper, nickel, lead, and zinc) ions from copper slag in large volumes was found to be lower compared with the prescribed limits from international authorities [7].

On the other hand, many researchers reported that copper slag does not show pozzolanic properties [8–10]. Moura et al. [11] studied the mechanical behavior of concrete containing 10% copper slag. The authors reported that compressive strength of concrete composed of copper slag had lower strength than reference concrete up to 91 days. However, other researchers mentioned that concrete incorporated with copper slag shows cementitious properties; furthermore, the pozzolanicity of copper slag increases the strength [12–16]. Additionally, another study assessed the pozzolanicity of clay composed of cement and copper slag [17]. The scanning electron microscopy (SEM) results revealed that composites exhibited pozzolanicity. The authors concluded that, at low cement content (30% dry weight of soil), the strength decreased with the copper slag amount. However, with an increase in the testing period and in the cement replacement level, the strength significantly improved with copper slag incorporation.

Few studies have considered the engineering properties of copper slag blended with clay and cement [18]. However, no studies have considered the durability properties of such composites. Durability is defined as the resistance to chemical attack, keeping its stability and integrity over a long period of exposure to a severe environment [19]. The durability of a silty clay–cement combination of composites was studied, where the authors found that the loss in mass of specimens decreased with increasing cement amount [20]. Furthermore, another research demonstrated that the lime addition could be decreased and have a positive effect on the ongoing wet–dry cycles over a long exposure [21]. Consoli et al. [22] and Consoli and Tomasi [23] investigated the porosity/cement and porosity/lime as durability parameters of soil composed of cement to evaluate the durability indices by considering the weight loss after several wet–dry cycles.

Many researchers investigated the lime stabilization of clay [18,19,21,24]. Choquette et al. [24], for example, examined the mineralogy and microstructure of lime-treated Canadian marine clays. Their results revealed that incorporating lime significantly caused the clustering of soil specimens. The flocculated arrangement was preserved by development of cementitious bonds between the particles. The authors correspondingly proposed the incorporation of calcium oxide (lime) into clay soil, which results in the formation of a plate-like morphology. The authors reported that this can increase the bulk volume of the small pores and space available between the clay–cement particles. Additionally, the authors stated that the modification in microstructure as a result of the lime addition agreed well with the mechanical properties of clay. Many researchers analyzed the cement–clay microstructural modification with SEM, reporting a decrease in the deflocculation level with a high amount of cement [25,26].

As specified in many standards around the world, unconfined compressive strength (UCS) is the most quantifying test in construction activities. The Australian earth-building handbook [27] suggests values for designing compressive strength between 0.40 and 0.60 N/mm^2 for rammed earth. The New Zealand Standard for engineering design of earth buildings [28] uses a design compressive strength equal to 0.5 N/mm^2. In Bulletin 5 [29], a safe working compressive stress of 0.25 N/mm^2—rather than an ultimate limit state—for stabilized rammed earth is recommended. Furthermore, according to the recommendations of the US Army Corps of Engineers (USACE) [30], the minimum required unconfined compressive strength for cement-stabilized soil for pavement base course is 500 psi (3400 kPa), and for the subbase, it is 250 psi (1700 kPa). Similarly, Maclean and Lewis [31] reported that the minimum required unconfined compressive strength for base and subbase designs of major roads is 500 psi (3400 kPa), and for minor roads, it is 250 psi (1700 kPa).

Treated soil deteriorates as a result of environmental conditions, such as wet–dry or freeze–thaw cycles and erosion. Under such conditions, the strength and stiffness values of the treated soils are reduced. Bonnot [32] stated that the durability of a treated soil can be studied in terms of loss of mass, expansion, change of strength, or swelling. As the current study was performed in Cyprus, which has a subtropical climate, i.e., Mediterranean and semi-arid-type climate, the wet and dry cycles are governing strength-control mechanisms. The US Army Corps of Engineers (USACE) technical manual [30] states that the maximum permissible mass loss of clay soils after 12 cycles (wet–dry) is 6% of the initial specimen weight for clay stabilization of pavements.

This study aims to evaluate the laboratory-produced triple-binder composites' strength, microstructure, resistance to wet–dry cycles, and sustainability performance. To address this research gap, in the current study, hydrated lime was incorporated with copper slag in cement-stabilized soil as a replacement to facilitate pozzolanic reactions. In addition, a porosity binder index for soil–cement mixes in terms of mass loss was examined for the first time, revealing a correlation between the mass loss and the unconfined compressive strength. Furthermore, embodied energy and embodied carbon dioxide (eCO_2) emissions of each mix were studied in terms of production and transportation of each stabilization product. This research could thereby enable a reduction in the amount of excavated soil through the assessment of the effects of soil disposal on the environment and on CO_2 emission caused by transportation. Additionally, incorporating various amounts of copper slag with cement could potentially decrease the impact of global warming and the amount of accumulated copper slag at the site.

2. Materials and Methods

2.1. Materials

The clay used in this study was collected from the project site, which was located in the Kyrenia District on the northern coast of Cyprus. The samples were obtained from basement excavation of a construction site. Basic characteristic tests, such as sieve analysis, specific gravity, and Atterberg limits, were evaluated based on the international standards in accordance with the corresponding ASTM D2487-17 [33]. The results of these tests are presented in Table 1. The clay was designated as inorganic low-to-medium-plastic clay (CL). The grain size distribution is shown in Figure 1. The soil is composed of clay, silt, and sand with percentages of 49%, 19%, and 32%, respectively. Furthermore, the X-ray fluorescence (XRF) test results showed that this clay is rich in SiO_2, Al_2O_3, and CaO.

Copper slag was collected from an abandoned mine in the Lefke Region in Northern Cyprus. After performing the characterization tests, the slag was classified according to the Unified Soil Classification System USCS as poorly graded sand (SP), and its specific gravity was 3.45. The X-ray spectroscopy analysis allowed the determination of the main components of the slag, i.e., 43.5% ferrous oxide, 32.8% silicon oxide, 8.3% aluminum oxide, 4.0% CaO, and 2.6% SO_3.

Type I cement with a specific gravity of 3.12 and with a Blaine fineness of 289 m^2/kg was used. The chemical composition of the cement is presented in Table 2.

Table 1. Physical properties of marine-deposited clay, hydrated lime, and copper slag.

Properties	Marine Clay	Cement	Hydrated Lime	Copper Slag
Liquid limit (%)	40	-	-	-
Plastic limit (%)	21	-	-	-
Plasticity index (%)	19	-	-	Nonplastic
Specific gravity	2.61	3.12	2.17	3.45
Fine gravel (4.75 < diameter < 20 mm) (%)	0	0	0	0
Coarse sand (2.00 < diameter < 4.75 mm) (%)	2	0	0	10
Medium sand (0.425 < diameter < 2.00 mm) (%)	3	0	0	82
Fine sand (0.075 < diameter < 0.425 mm) (%)	27	0	5	8
Silt (0.002 < diameter < 0.075 mm) (%)	19	90	90	0
Clay (diameter < 0.002 mm) (%)	49	10	5	0
Mean particle diameter (mm)	0.0035	0.015	0.02	0.9
USCS class	CL	ML	ML	SP

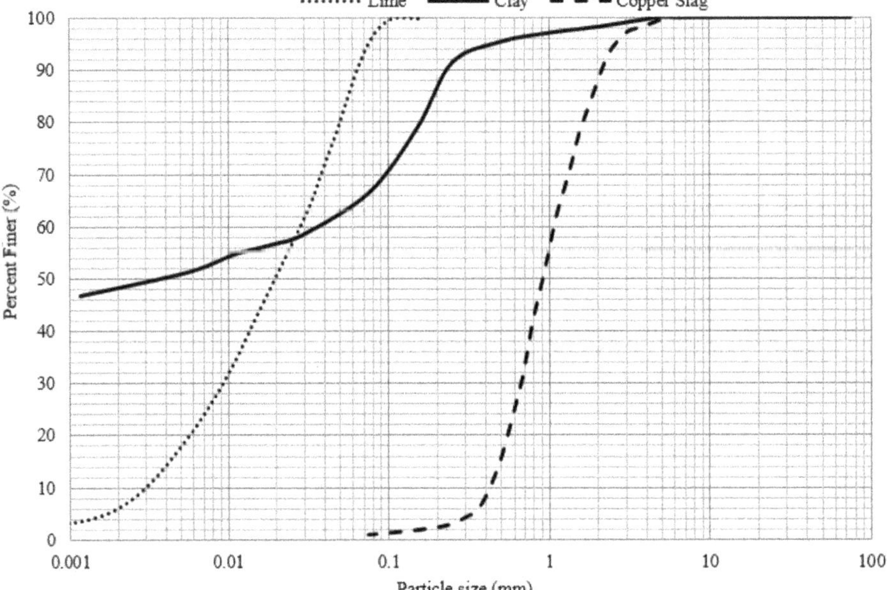

Figure 1. Grain size distribution of the studied clay, copper slag, and hydrated lime.

Table 2. Chemical analysis of portland cement, hydrated lime, and copper slag. (EN197-1).

Compound	Portland Cement (%)	Lime (%)	Copper Slag (%)
SiO_2	21.2	-	32.5
Al_2O_3	5.1	0.38	8.3
Fe_2O_3	2.5	0.3	43.5
CaO	64.7	70.89	4
MgO	0.9	1.95	-
K_2O	0.2	-	-
SO_3	1.5	-	2.6
loss in ignition	2.5	24.59	-

Hydrated lime contains mostly calcium oxide (71% CaO), and was obtained from a local supplier in Cyprus (imported from Turkey). The particle size distribution is shown in Figure 1. The physical properties of all used materials are presented in Table 1.

2.2. Methods

2.2.1. Molding and Curing of Specimens

To investigate the effects of clay treatment, cylindrical specimens of 50 mm diameter and 100 mm height were prepared. First, the amounts of materials were calculated from the targeted dry unit weight. They were then measured and dry-mixed in a tray with a flat-end spatula for at least 5 min to achieve uniformity. After that, water was introduced gradually while the mixing process continued. After ensuring the mixture's homogeneity, it was transferred to a split mold and statically compressed to achieve the required dry density. Upon completion of the mixing and compressing, the prepared specimens were transferred to a curing room in which they were kept for the required curing time [34]. The curing room had a 24 ± 2 °C temperature and a relative humidity of about 95%, according to ASTM C 511 [34]. The preparation data for all specimens are presented in Table 3.

Table 3. Details of molding and curing data.

Name	Soil Type	Cement Contents (%)	Copper Slag Content (%)	Hydrated Lime Content (%)	Molding Dry Unit Weight (kN/m³)	Curing Periods (Days)	Test Type
Clay + Cement (CC)	Marine Deposited Clay	7, 10 and 13	-	-	14.0, 16.0	7, 28, 60	UCS, Wet-Dry Cycles *
Clay + Cement + Slag (CCS)		7 and 10	10%	-	14.0, 16.0	7, 28, 60	UCS, Wet-Dry Cycles *
Clay + Cement + Lime (CCL)		7 and 10	-	5%	14.0, 16.0	7, 28, 60	UCS, Wet-Dry Cycles *
Clay + Cement + Lime + Slag (CCLS)		7, 10 and 13	10%	5%	14.0, 16.0	7, 28, 60	UCS, Wet-Dry Cycles *, SEM **

* Wet–dry cycle done on all tested blends, 1.6 kN/m³ dry unit weights considering 28 days of curing. ** Scanning electron microscopy (SEM) was done on untreated CCLS specimens, 1.6 kN/m³ dry unit weights considering seven, 28, and 60 days of curing.

Blending of the specimens was performed in relation to the relative constant of Portland cement (C), copper slag (CS), and hydrated lime (L). C is the mass of the cement divided by the mass of the dry clay; CS and L are defined as the quotient of mass of cement as a partial replacement for cement.

Porosity was calculated by using a modified version of Equation (1) proposed by Consoli et al. [35], dry unit weight (γ_d), and the weight contents of the marine clay (W_S), Portland cement (W_C), copper slag (W_{CS}), and hydrated lime (W_L). The corresponding unit weights are γ_{ss}, γ_{sc}, γ_{sCS}, and γ_{sL}, respectively.

$$\eta = 100 - 100 \left[\left[\frac{\gamma_d}{total\ mass\ of\ solids} \right] \left[\frac{W_S}{\gamma_{S_S}} + \frac{W_C}{\gamma_{S_C}} + \frac{W_{CS}}{\gamma_{S_{CS}}} + \frac{W_L}{\gamma_{S_L}} \right] \right] \tag{1}$$

Depending on the cement porosity index (η/C_{iv}), a unique relationship was developed in order to predict the behavior of cement-treated soils [36], which only accounted for cement. More recently, Ekinci et al. [18] proposed a more general index X_{iv}, which accounts for all binder contents. In this study, Ekinci et al.'s [18] parameter was modified in an attempt to predict the strength for each mixture, where X_{iv} was calculated from the modified Equation (2), where $V = W/\gamma_s$ is true for all of the used materials.

$$X_{iv} = \frac{V_S + V_C + V_{CS} + V_L}{V} = \frac{\left(\frac{W_S}{\gamma_{S_s}}\right) + \left(\frac{W_C}{\gamma_{sC}}\right) + \left(\frac{W_{CS}}{\gamma_{sCS}}\right) + \left(\frac{W_L}{\gamma_{sL}}\right)}{V} \tag{2}$$

The external exponent of adjusted porosity/binder index $\eta/X_{iv}^{0.32}$ was determined to be the best-fit exponent for all of the blends studied herein and Ekinci et al. [18]. It is in accordance with previous empirical studies on various types of soils that obtained exponents that slightly varied between 0.28 and 0.35 [22,23].

Table 3 provides all of the necessary molding data, including material contents, curing periods, dry unit weights, and the types of tests conducted. The percentages of the cement used are related to the dry weight of the clay; for copper slag and hydrated lime, the percentages are related to the dry weight of the cement. In the triple blend, the copper slag is used as a replacement for cement, but lime is an addition of as much as the dry weight of the cement.

2.2.2. Compressive Strength Test

Strength tests were conducted on specimens after wetting and drying cycles. The tests were conducted according to ASTM C39 [37]. A fully automatic testing machine with 20 kN capacity was used. The failure load was recorded for every specimen, and an average of three specimens was used. Based on the procedure, if the single-specimen compressive strength deviated 10% from the average, the specimen was discarded, and a new specimen was prepared. Thus, the variation of the experimental results was completely eliminated.

2.2.3. Mass Loss by Wet–Dry Cycles

Durability tests were conducted according to ASTM D 559 [38] for durability testing of marine-deposited clays stabilized using various binders. These tests were used to evaluate the mass loss of composites through 12 wetting and drying cycles. Every cycle began with complete immersion of the specimens in water for 5 h; the specimens were then dried in an oven for two days. Subsequently, specimens were brushed with a wire brush using a pre-calibrated controlled load of about 15 N.

2.2.4. Microstructural Investigation

Scanning electron microscopy (SEM) was conducted to evaluate the influences of clay treatment with cement, slag, and lime on the microstructure at seven, 28, and 60 days curing. First, a small dried piece of specimen was attached to the aluminum stubs. Silver paint was applied. After that, the gold coating was applied. Magnifications of SEM images ranged from 3500× to 7500×.

2.2.5. Sustainability Investigation

Sustainability investigation was carried out to evaluate equivalent eCO_2 emission and embodied energy for the production and transport of used materials. It is worth mentioning that eCO_2 and embodied energy calculations were carried out based on their transportation distances to the Middle East Technical University, North Cyprus Campus, as a production site. Table 4 presents the embodied energy values and their equivalent eCO_2 emissions of the materials used in this research.

In Northern Cyprus, most of the raw construction materials are provided from Turkey. The calculations were done in collaboration with the relevant supplier and available scientific resources. Therefore, when considering lime and cement eCO_2 and embodied energy values, the published data from the collaboration of Hammond and Jones [39] and the KASCON Group of Companies (Cyprus-based concrete ready-mix company) were used. Road transportation from the manufacturing place to the port and from the port to the production site was considered by using the European Parliamentary Research Service's (2018) publication for light-duty trucks. Seaway transportation emission and embodied energy values were obtained from the International Maritime Organization [40] and Rossit and Lawson [41]. Furthermore, eCO_2 emissions of the used water were obtained and calculated in collaboration with a governmental water resources agency. Additionally, copper slag values were obtained from Hammond and Jones [39].

Table 4. Equivalent embodied carbon dioxide (eCO$_2$) emission and embodied energy for the production and transport of used materials.

Process	Embodied Energy (MJ/kg)	eCO$_2$ Emission (kg CO$_2$/kg)
Production		
Cement	4.50	0.74
Lime	5.30	0.78
Copper Slag	1.60	0.083
Water	0.0009	0.000155
Transportation		
Cement through road	0.35	0.32
Cement through seaway	0.0162	0.007
Lime through road	0.35	0.32
Lime through seaway	0.0162	0.007
Copper slag through road	0.0702	0.32
Water	0.016	0.32

The soil was assumed to be transported from the same site. Therefore, energy and emissions in excavating, transporting, mixing, and compacting were assumed to be constant for all mixtures and were not included in the total eCO$_2$ and embodied energy calculations. The total embodied energy and eCO$_2$ emissions that each mix will produce, along with the quantities of each material, were calculated and are presented in Table 5. In order to align with strength, durability, and microstructure studies, the mixes were all prepared at 1.6 kN/m^3 density clay + 10% cement (CC), clay + 10% cement + 10% copper slag (CCS), clay + 10% cement + 5% lime (CCL), and clay + 10% cement + 5% lime + 10% copper slag (CCLS).

Table 5. Quantities of each material with eCO$_2$ emission and embodied energy calculations for each mix.

Mix	Quantities (kg/m^3)				Embodied Energy (MJ/m^3)				eCO$_2$ Emission (kg CO$_2$/m^3)			
	CC	CCS	CCL	CCLS	CC	CCS	CCL	CCLS	CC	CCS	CCL	CCLS
Production												
Cement	145.00	131.00	138.00	131.00	652.50	589.50	621.00	589.50	107.30	96.94	102.12	96.94
Lime	0.00	0.00	5.00	5.00	0.00	0.00	26.50	26.50	0.00	0.00	3.90	3.90
Copper Slag	0.00	16.00	0.00	16.00	0.00	25.60	0.00	25.60	0.00	1.33	0.00	1.33
Water	395.00	386.00	392.00	380.00	0.36	0.35	0.35	0.34	0.06	0.06	0.06	0.06
Transportation												
Cement through road	145.00	131.00	138.00	131.00	50.75	45.85	48.30	45.85	46.40	41.92	44.16	41.92
Cement through seaway	145.00	131.00	138.00	131.00	2.35	2.12	2.24	2.12	1.02	0.92	0.97	0.92
Lime through road	0.00	0.00	5.00	5.00	0.00	0.00	1.75	1.75	0.00	0.00	1.60	1.60
Lime through seaway	0.00	0.00	5.00	5.00	0.00	0.00	0.08	0.08	0.00	0.00	0.04	0.04
Copper slag road	0.00	16.00	0.00	16.00	0.00	1.12	0.00	1.12	0.00	5.12	0.00	5.12
Water	395.00	386.00	392.00	380.00	6.32	6.18	6.27	6.08	126.40	123.52	125.44	121.60
Total					712.27	670.72	706.49	698.95	281.18	269.80	278.28	273.42

3. Results and Discussion

3.1. Compressive Strength and Porosity

Figure 2 presents the unconfined compressive strength (q$_u$) of specimens at 1.6 kN/m^3 dry unit weight for all blends considering curing periods of seven, 28, and 60 days. In addition, 1.4 kN/m^3 samples were also tested. However, as the results are, accordingly, similar to those of the 1.6 kN/m^3

specimens, they are not presented in this section. Furthermore, the durability, microstructure, and sustainability assessments were also performed in accordance with 1.6 kN/m^3 dry unit weight specimens; therefore, UCS results are presented accordingly to support the findings. Nevertheless, a table containing all of the results of UCS is presented in Supplementary Materials. Note that a more comprehensive study on unconfined compressive strength of similar mixes with comprehensive statistical analysis has been published by Ekinci et al. [18].

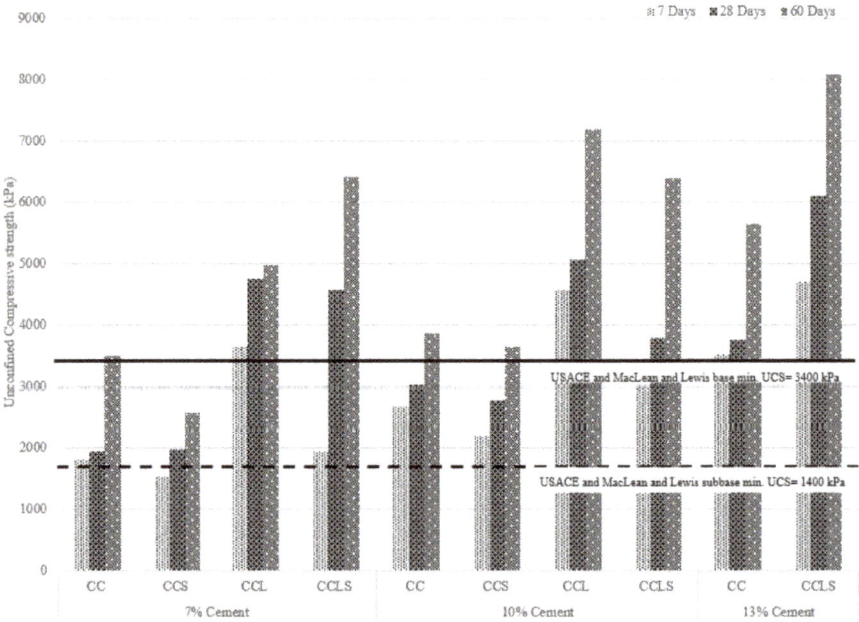

Figure 2. Unconfined compressive strength (UCS; q_u) at 1.6 kN/m^3 dry unit weight for all blends considering curing periods of 7, 28, and 60 days.

Figure 2 reveals that, in all mixes, the increase of cement content and curing duration results in increase of compressive strength. It can be also seen that the CCS mix results in a slight reduction at all cement contents and ages. Adversely, the lime replacement of cement (CCL) mix appears to accelerate the hydration and results in maximizing compressive strength when compared with CC mixes at all cement contents and curing periods. It is also evident that the addition of lime to the CCS mix, which is CCLS, appears to contribute to the pozzolanic reaction; furthermore, as the curing period extends, the copper slag contribution becomes more evident. According to the US Army Corps of Engineers (USACE) technical manual [30] and MacLean and Lewis, the minimum compressive strength requirement for a road subbase construction of a major road can be satisfied via all proposed mixes. Similarly, for base course construction, it can be seen that the CCS mix at 7% cement content is the only mix that fails to satisfy the criteria.

Figure 3 shows the relationship between the adjusted porosity/binder index against variation of cement content and curing period for all composite binders. It is clear that, for each blend, the increase of cement content results in reduction of the adjusted porosity/binder index. As with compression-strength observations, the CCS blend results in a slight reduction in all cement contents. It is interesting to note that the curing period did not affect the adjusted porosity binder/index. Nevertheless, it was only the CCLS specimen out of all other blends that showed reduction in the porosity/binder index as the curing period extended. This observation explains the reason for observed strength gain in CCLS specimens as the samples age. As expected, the lime replacement of cement CCL seems to accelerate the hydration and results in maximized reduction of the adjusted porosity binder index.

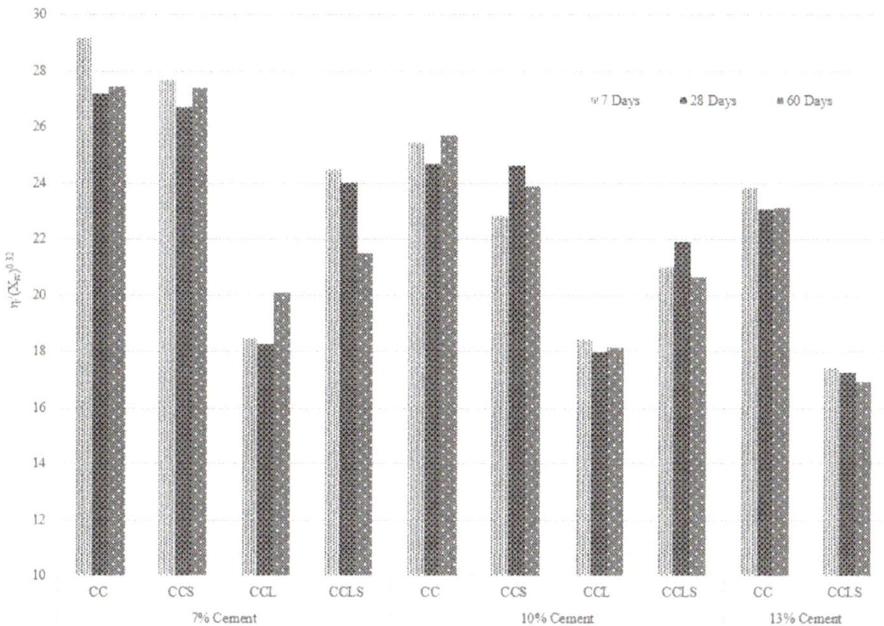

Figure 3. Adjusted porosity/binder index at 1.6 kN/m^3 dry unit weight for all blends considering curing periods of 7, 28, and 60 days.

3.2. Mass Loss by Dry–Wet Cycles

Figure 4A illustrates accumulated mass losses (ALM) of marine-deposited clay soil–cement, soil–cement–copper slag, soil–cement–hydrated lime, and soil–cement–copper slag–hydrated lime blends after 12 wetting and drying cycles. The binder contents, distinct dry unit mass, and curing regime for the laboratory-produced specimens are summarized in Table 3. A sound polynomial fit of ALM versus $\eta/X_{iv}^{0.32}$ after durability tests could be obtained, as shown in Equation (3). Thus, it was observed that the adjusted porosity/binder index can be used to predict durability with up to a triple binder.

$$ALM = 0.031\left(\eta/X_{iv}^{0.32}\right)^2 + 0.864\left(\eta/X_{iv}^{0.32}\right) + 0.867, R^2 = 0.89 \tag{3}$$

The US Army Corps of Engineers (USACE) technical manual [30] states that the maximum permissible mass loss of clay soils after 12 cycles (wet–dry) is 6% of the initial specimen weight for soil stabilization of pavements. In this study, these wetting and drying requirements were satisfied for $\eta/X_{iv}^{0.32}$ of less than about 24. The shaded section in Figure 4A shows the data for the specimens that satisfied this requirement, which are all soil–cement–hydrated lime and soil–cement–copper slag–hydrated lime blends.

Finally, the unconfined compressive strength versus accumulated mass loss after 12 cycles is shown in Figure 4B for the above marine-deposited clays stabilized with clay soil–cement, soil–cement–copper slag, soil–cement–hydrated lime, and soil–cement–copper slag–hydrated lime blends. Unique second-order polynomial relationships with reasonable prediction can be obtained from Equation (4) for such blends.

$$ALM = 131.52(ALM\%)^2 - 2090.9\ (ALM\%) + 9265.5, \quad R^2 = 0.84 \tag{4}$$

Figure 4B also shows that the durability requirement of the USACE [30] for clay soils is only satisfied by the blends with unconfined compressive strengths greater than 1400 kPa. This finding also defined a lower boundary for achieving satisfactory durability of such blends in terms of strength.

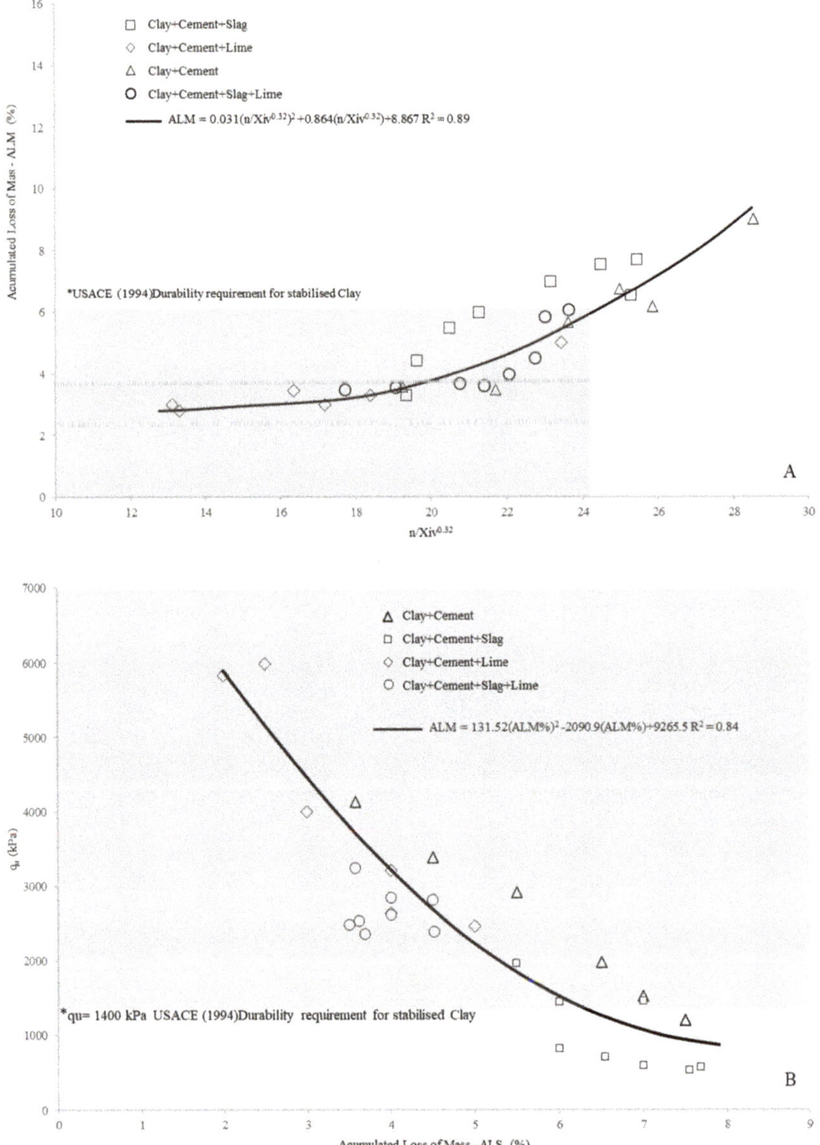

Figure 4. (**A**) Accumulated mass loss (ALM) versus adjusted porosity/binder index, (**B**) ALM considering twelve wet–dry cycles versus unconfined compressive strength (qu); for all tested blends, 1.6 kN/m³ dry unit weights considering 28 days of curing.

3.3. Microstructural Analysis (SEM)

SEM test results performed after compressive strength tests for the composites incorporated with marine-deposited clays at the 7, 28, and 60 days are shown in Figure 5. Those SEM results revealed that no cementitious bonds developed in the untreated clay due to the lack of soil-matrix (bond between the soil, aggregate, and cementitious compounds) cementing bonds. Additionally, as seen in Figure 5A–D, the tested marine-deposited clay is rich in calcium carbonate and contains "hollow-like structures".

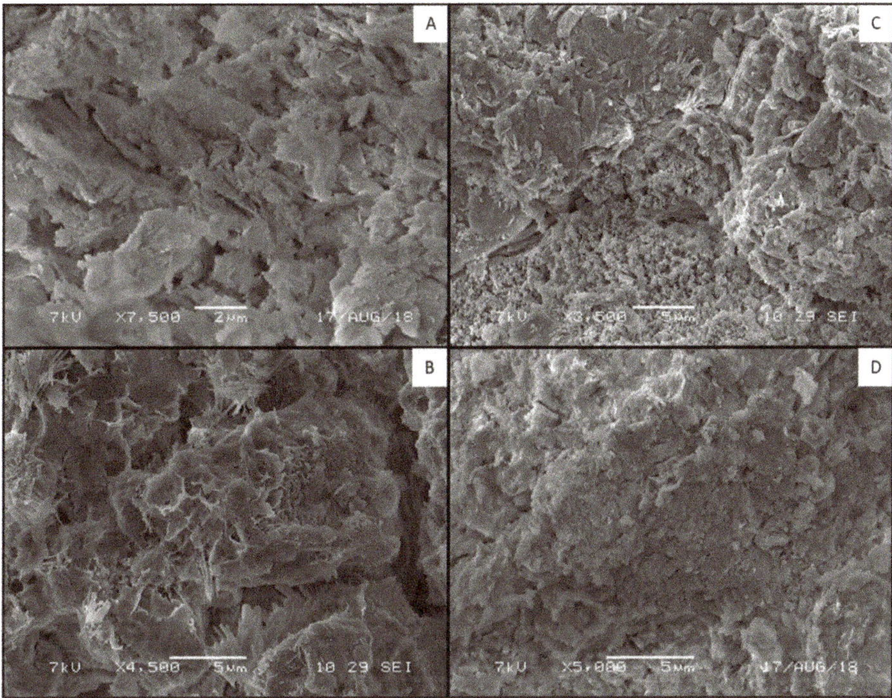

Figure 5. The scanning electron microscopy images, conducted on (**A**) untreated, (**B**) 7 days, (**C**) 28 days and (**D**) 60 days cured cement + copper slag and hydrated lime blend marine-deposited clays.

The clay–cement mixtures were formed from the main chemical reaction of cement and the secondary reactions of the calcium-silicate-hydrate (CSH) product formation during the pozzolanic reaction of soil–cement mixtures. Figure 5B shows the products of hydration, needle-like crystals, and the products that resulted from calcium-silicate hydrates between the soil particles. Based on the SEM micrographs, the products had high aspect ratios. This confirms the formation of CSH needles in the bulk volume as a result of high strength and, thus, reduced ALM. Ettringite is a stable product with needle-like crystals with a hexagonal cross-section, which easily formed from specimens because of the high void ratios. This formation also caused the expansion and ALM increase at later time-points. However, it seems that the ettringite fills the pores in the matrix during the hardening period of seven to 60 days, and the pore space decreases significantly between those curing periods.

The clay compounds, silica, and alumina react with Ca^{2+} and form CSH-calcium aluminate hydrate (CAH) during pozzolanic reactions. These hydrate products grow and harden, thus improving the ALM and strength of the clay-cementitious mixtures over hardening. After the 60 day curing period, the pores were filled with CSH gel, as shown in Figure 5C. Particle aggregation is observable due to the cation exchange reactions caused by introducing lime and slag. This aggregation reduces the "thickness of the double layer" between the clay particles and the attraction between particles, forcing the particles to move closer and initiating the particle aggregation phase. As the curing period increased, the porosity of the specimens was reduced due to cementation, and a further improvement in ALM was observable. It can be assumed that day seven is the beginning of curing and the development of slag–lime reactions and corresponding pore spaces, as illustrated in Figure 4, where the porosities of the specimens cured for seven days are higher, and the porosity decreases with curing and with the addition of slag and lime. This can be seen in Figure 5C,D, which show the completed particle aggregation.

Figure 5D also shows the silica and alumina reaction, where the cementing property is more obvious. This is due to the secondary reaction development after 60 days, indicating a reduction in Portlandite (CH) due to the presence of copper slag and cement because of their reuse and transformation into secondary CSH at this stage of hydration. This feature is not evident in Figure 5B, as the reaction is in its early stage, and CH is more dominant because the CH crystals are absorbed in a later stage, since the copper slag reacts in a later stage. This characteristic is observable in the ALM reduction and strength development as well. Furthermore, hydrated lime addition appears to activate pozzolanic reactions at earlier stages, leading to a reduction in ALM.

3.4. Sustainability Assessment

The environmental assessments of the blends via embodied energy and eCO_2 evaluation are presented in Figures 6 and 7, respectively. Similar to Cabeza et al. [42], evaluation of both figures revealed a clear relationship between embodied energy and CO_2 footprint for primary production. Table 5 shows that greatest contribution to embodied energy and CO_2 emission is due to production and transportation of the cement, followed by the lime.

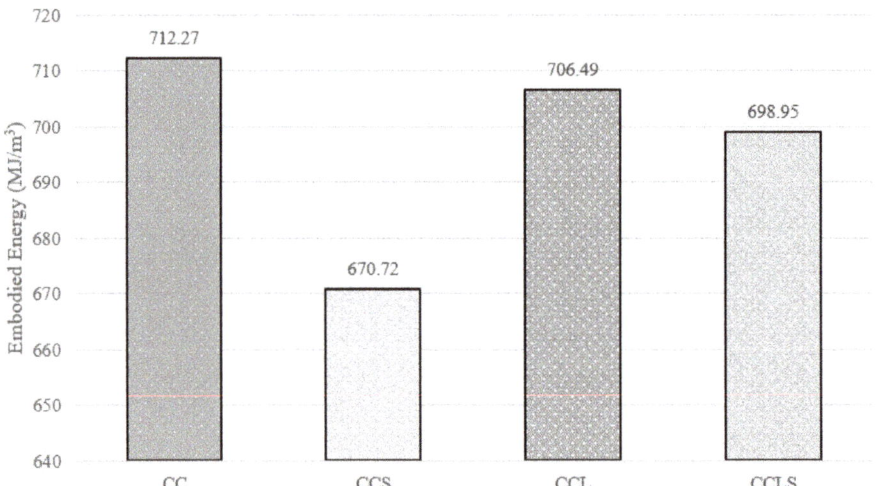

Figure 6. Embodied energy of 10% cement, 10% cement + 10% slag, 10% cement + 5% lime, and 10% cement + 5% lime + 10% slag mixes.

Figures 6 and 7 show that the CC mix, which can qualify as a control mix, produces the highest embodied energy (712.27 MJ/m^3) and eCO_2 emission (281.18 kg CO_2/m^3). Adversely, among all studied mixes, the CCS mix generates the lowest embodied energy (670.72 MJ/m^3) and eCO_2 emission (269.80 kg CO_2/m^3), which proposes at least a 5% reduction of environmental impact. Nevertheless, it is evident in the strength and durability analysis that, at early ages and low cement content, the CCS mixes' performance is degraded and does not satisfy the requirements. Therefore, as an alternative, a lime addition to the cement slag replacement mix (CCLS) was proposed. It can be seen that the CCLS mix generates the second-lowest embodied energy (698.95 MJ/m^3) and eCO_2 emission

(273.42 kg CO_2/m^3), which proposes as much as a 3% reduction of environmental impact when compared with the control mix (CC). Even though the replacement of cement with copper slag greatly reduces the impact, as demonstrated by the replacement of cement with lime (CCL), it contributes positively to strength and durability performance; however, the environmental impact reduction is not as effective, since embodied energy (706.49 MJ/m^3) and eCO$_2$ emission (278.28 kg CO_2/m^3) result in less than a 2% reduction.

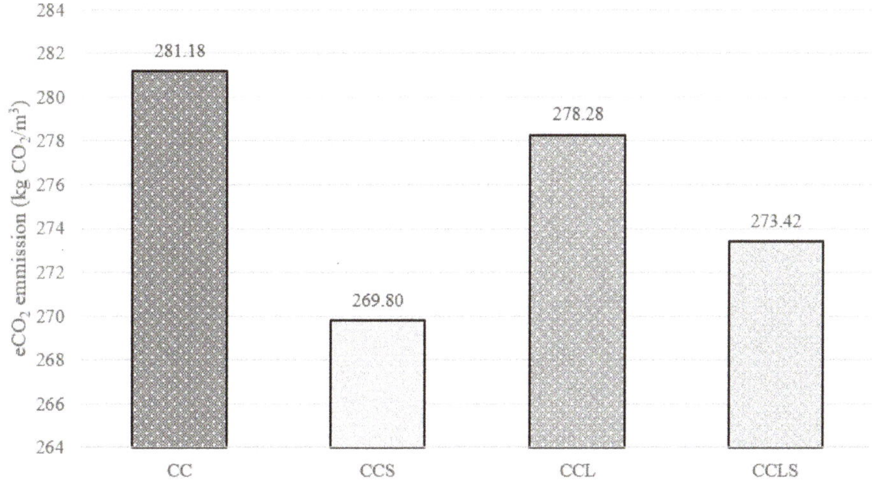

Figure 7. eCO$_2$ emission of 10% cement, 10% cement + 10% slag, 10% cement + 5% lime, and 10% cement + 5% lime + 10% slag mixes.

It is denotative from this study that the proposed binders reduce the consumption of cement and lime by increasing the amount of waste byproducts. The composites could be promising candidates in some major uses such as a base course, subbase course of major roads, rammed earth wall recommended design values, and structural fill. Additionally, they further reduce the environmental impact not just through the production and transport of the process, but also via disposal of waste. Furthermore, in a step ahead, Jiao et al. [43] presented the correlation between embodied energy and cost of individual building components. Therefore, use of this composite will also contribute to cost reduction.

4. Conclusions

This study examined the durability of the deposited marine clay when treated with cement, copper slag, and hydrated lime. The following can be concluded from the study:

- The incorporation of hydrated lime into cement: Slag-treated soils improved the strength and durability performance of the composites and ensured the satisfaction of weight loss and minimum compressive strength according to the requirements of USACE and MacLean and Lewis [31].
- As the curing period increased, the CCLS specimens' porosity declined due to pozzolanic reactions, and further improvement in ALM was observable.
- SEM pictures revealed the formation of "needle-like crystals" with a high aspect ratio between the particles, which resulted from the primary hydration. These crystals are responsible for the improvement in UCS and ALM.
- The incorporation of hydrated lime appeared to accelerate the pozzolanic reactions at earlier stages, resulting in a reduction in ALM.
- Environmental assessment of all proposed mixes resulted in the reduction of embodied energy and eCO$_2$ emission.

- Reusing unsuitable soil and hazardous wastes will reduce environmental and financial impacts. Improving soil with additives will facilitate the use of the available soil on site. In addition to the environmental contribution of cement usage reduction, using waste material, such as copper slag, will enable safe disposal of those harmful materials.

5. Recommendations

In this study, a formula to predict the accumulated loss of mass when adding a triple binder was successfully created. Further research can be conducted to validate this formula when using pozzolanic materials other than copper slag.

Supplementary Materials: The following are available online at http://www.mdpi.com/2071-1050/12/11/4633/s1, Table S1: Unconfined compressive strength results of all performed tests.

Author Contributions: M.H. and A.E. conceived the study and were responsible for scheduling and performing the experimental study. E.A. was responsible for microstructure interpretation. A.E. and M.H. wrote the first draft of the article, and E.A. oversaw and finalized the study. All authors have read and agreed to the published version of the manuscript.

Funding: This research was funded by Office of Research Coordination, Middle East Technical University, Northern Cyprus Campus, grant number FEN-20-YG-4 and The APC was funded by FEN-20-YG-4.

Acknowledgments: The authors greatly appreciate the discussions and help from Pedro Ferreira from University College London and Nilo C. Consoli from Universidade de Federal do Rio Grande do Sul. The authors also thank the graduate students Burak Kın and Dogan Gülaboglu for their support during the laboratory experiments.

Conflicts of Interest: The authors declare no conflict of interest.

References

1. Cabello Eras, J.J.; Gutiérrez, A.S.; Capote, D.H.; Hens, L.; Vandecasteele, C. Improving the environmental performance of an earthwork project using cleaner production strategies. *J. Clean. Prod.* **2013**, *47*, 368–376. [CrossRef]
2. Lafebre, H.; Songonuga, O.; Kathuria, A. Contaminated Soil Management at Cconstruction Sites. *Pract. Period. Hazard. Toxic Radioact. Waste Manag.* **1998**, *2*, 115–119. [CrossRef]
3. Magnusson, S.; Lundberg, K.; Svedberg, B.; Knutsson, S. Sustainable management of excavated soil and rock in urban areas—A literature review. *J. Clean. Prod.* **2015**, *93*, 18–25. [CrossRef]
4. Capobianco, O.; Costa, G.; Baciocchi, R. Assessment of the Environmental Sustainability of a Treatment Aimed at Soil Reuse in a Brownfield Regeneration Context. *J. Ind. Ecol.* **2018**, *22*, 1027–1038. [CrossRef]
5. Alter, H. The composition and environmental hazard of copper slags in the context of the Basel Convention. *Resour. Conserv. Recycl.* **2005**, *43*, 353–360. [CrossRef]
6. Zain, M.F.M.; Islam, M.N.; Radin, S.S.; Yap, S.G. Cement-based solidification for the safe disposal of blasted copper slag. *Cem. Concr. Compos.* **2004**, *26*, 845–851. [CrossRef]
7. Shi, C.; Meyer, C.; Behnood, A. Utilization of copper slag in cement and concrete. *Resour. Conserv. Recycl.* **2008**, *52*, 1115–1120. [CrossRef]
8. Al-Jabri, K.S.; Hisada, M.; Al-Saidy, A.H.; Al-Oraimi, S.K. Performance of high strength concrete made with copper slag as a fine aggregate. *Constr. Build. Mater.* **2009**, *23*, 2132–2140. [CrossRef]
9. Lim, T.T.; Chu, J. Assessment of the use of spent copper slag for land reclamation. *Waste Manag. Res.* **2006**, *24*, 67–73. [CrossRef]
10. Madany, I.M.; Al-Sayed, M.H.; Raveendran, E. Utilization of copper blasting grit waste as a construction material. *Waste Manag.* **1991**, *11*, 35–40. [CrossRef]
11. Moura, W.; Masuero, A.; Molin, D.; Dal Vilela, A. Concrete performance with admixtures of electrical steel slag and copper slag concerning mechanical properties. *Am. Concr. Inst.* **1999**, *186*, 81–100.
12. De Rojas, M.I.S.; Rivera, J.; Frías, M.; Marín, F. Use of recycled copper slag for blended cements. *J. Chem. Technol. Biotechnol.* **2008**, *83*, 209–217. [CrossRef]
13. Al-Jabri, K.S.; Taha, R.A.; Al-Hashmi, A.; Al-Harthy, A.S. Effect of copper slag and cement by-pass dust addition on mechanical properties of concrete. *Constr. Build. Mater.* **2006**, *20*, 322–331. [CrossRef]

14. Pavez, O.; Rojas, F.; Palacious, J.; Nazer, A. Pozzolanic activity of copper slag. In Proceedings of the VI international conference on clean technologies for the mining industry, University of Concepcion, Concepcion, Chile, 18–21 April 2004.

15. Taha, R.; Al-Rawas, A.; Al-Jabri, K.; Al-Harthy, A.; Hassan, H.; Al-Oraimi, S. An overview of waste materials recycling in the Sultanate of Oman. *Resour. Conserv. Recycl.* **2004**, *41*, 293–306. [CrossRef]

16. Mobasher, B.; Devaguptapu, R.; Arino, A.M. Effect of copper slag on the hydration of blended cementitious mixtures. In Proceedings of the Materials Engineering Conference, Washington, DC, USA, 10–14 November 1996.

17. Bharati, S.K.; Chew, S.H. Geotechnical Behavior of Recycled Copper Slag-Cement-Treated Singapore Marine Clay. *Geotech. Geol. Eng.* **2016**, *34*, 835–845. [CrossRef]

18. Ekinci, A.; Scheuermann Filho, H.C.; Consoli, N.C. Copper Slag-Hydrated Lime-Portland Cement Stabilized Marine Deposited Clay. In *Ground Improvement*; Proceedings of the Institution of Civil Engineers: London, UK, 2019. [CrossRef]

19. Dempsey, B.J.; Thompson, M.R. Durability properties of lime-soil Mixtures. *Highw. Res. Board* **1968**, *235*, 61–75.

20. Zhang, Z.; Tao, M. Durability of cement stabilized low plasticity soils. *J. Geotech. Geoenviron. Eng.* **2008**, *134*, 203–213. [CrossRef]

21. Cuisinier, O.; Stoltz, G.; Masrouri, F. Long-term behavior of lime-treated clayey soil exposed to successive drying and wetting. In Proceedings of the Geo-Congress 2014: Geo-characterization and Modeling for Sustainability, Atlanta, GA, USA, 23–26 February 2014.

22. Consoli, N.C.; da Silva, K.; Filho, S.; Rivoire, A.B. Compacted clay-industrial wastes blends: Long term performance under extreme freeze-thaw and wet-dry conditions. *Appl. Clay Sci.* **2017**, *146*, 404–410. [CrossRef]

23. Consoli, N.C.; Tomasi, L.F. The impact of dry unit weight and cement content on the durability of sand-cement blends. *Proc. Inst. Civ. Eng.-Ground Improv.* **2018**, *171*, 96–102. [CrossRef]

24. Choquette, M.; Bérubé, M.A.; Locat, J. Mineralogical and microtextural changes associated with lime stabilization of marine clays from eastern Canada. *Appl. Clay Sci.* **1987**, *2*, 215–232. [CrossRef]

25. Kamruzzaman, A.H.M.; Chew, S.H.; Lee, F.H. Structuration and destructuration behavior of cement-treated Singapore marine clay. *J. Geotech. Geoenviron. Eng.* **2009**, *135*, 573–589. [CrossRef]

26. Chew, S.H.; Kamruzzaman, A.H.M.; Lee, F.H. Physicochemical and engineering behavior of cement treated clays. *J. Geotech. Geoenviron. Eng.* **2004**, *130*, 696–706. [CrossRef]

27. Standards Australia. *The Australian Earth Building Handbook*; Standards Australia: Sydney, Australia, 2002.

28. Standard New Zealand. *Engineering Design of Earth Buildings*; NZS 4297; Standard New Zealand: Wellington, New Zealand, 1998.

29. Middleton, G.F. *Earth Wall Construction*, 4th ed.; CSIRO Division of Building, Construction and Engineering: North Ryde, Australia, 1992.

30. USACE (US Army Corps of Engineers). *Soil Stabilization for Pavements*; USACE: Washington, DC, USA, 1994.

31. MacLean, D.G.; Lewis, W.A. British practice in the design and specification of cement-stabilized bases and sub-bases for roads. *Highw. Res. Rec.* **1963**, *36*, 56–76.

32. Bonnot, J. *Semi-Rigid Pavements*; Association Internationale Permanente des Congrès de la Route: Paris, France, 1991.

33. ASTM. *Standard Practice for Classification of Soils for Engineering Purposes (Unified Soil Classification System)*; D2487-17; ASTM International: West Conshohocken, PA, USA, 2017.

34. ASTM. *Standard Specification for Mixing Rooms, Moist Cabinets, Moist Rooms, and Water Storage Tanks Used in the Testing of Hydraulic Cements and Concretes*; C511; ASTM International: West Conshohocken, PA, USA, 2019.

35. Consoli, N.C.; Filho, H.C.S.; Godoy, V.B.; Rosenbach, C.M.D.C.; Carraro, J.A.H. Durability of rap-industrial waste mixtures under severe climate conditions. *Soils Rocks* **2018**, *41*, 149–156. [CrossRef]

36. Consoli, N.C.; Vaz Ferreira, P.M.; Tang, C.S.; Veloso Marques, S.F.; Festugato, L.; Corte, M.B. A unique relationship determining strength of silty/clayey soils—Portland cement mixes. *Soils Found.* **2016**, *56*, 1082–1088. [CrossRef]

37. ASTM. *Standard Test Method for Compressive Strength of Cylindrical Concrete Specimens*; C39; ASTM International: West Conshohocken, PA, USA, 2020.

38. ASTM. *Standard Test Methods for Wetting and Drying Compacted Soil-Cement Mixtures*; D559; ASTM International: West Conshohocken, PA, USA, 2015.
39. Hammond, G.; Jones, C. *Inventory of Carbon and Energy Version 2.0*; Sustainable Energy Research Team, University of Bath: Bath, UK, 2011.
40. International Maritime Organization. *Third IMO GHG Study 2014 e Executive Summary and Final Report*; International Maritime Organization: London, UK, 2014.
41. Rossit, G.; Lawson, M. Materials Life: The Embodied Energy of Building Materials. *Cannon Des.* **2012**, *1*, 14–35.
42. Cabeza, L.F.; Barreneche, C.; Miró, L.; Morera, J.M.; Bartolí, E.; Inés Fernández, A. Low carbon and low embodied energy materials in buildings: A review. *Renew. Sustain. Energy Rev.* **2013**, *23*, 536–542. [CrossRef]
43. Jiao, Y.; Lloyd, C.R.; Wakes, S.J. The relationship between total embodied energy and cost of commercial buildings. *Energy Build.* **2012**, *52*, 20–27. [CrossRef]

Article

Mechanical Characteristics of Soda Residue Soil Incorporating Different Admixture: Reuse of Soda Residue

Jiaxiao Ma [1], **Nan Yan** [1,2,*], **Mingyi Zhang** [1,2], **Junwei Liu** [1,2], **Xiaoyu Bai** [1,2] **and Yonghong Wang** [1,2]

[1]   Department of Civil Engineering, Qingdao University of Technology, Qingdao 266033, China; majiaxiao1356@163.com (J.M.); zmy58@163.com (M.Z.); zjuljw@126.com (J.L.); baixiaoyu538@163.com (X.B.); hong7986@163.com (Y.W.)

[2]   Cooperative Innovation Center of Engineering Construction and Safety in Shandong Blue Economic Zone, Qingdao University of Technology, Qingdao 266033, China

*   Correspondence: yannan@qut.edu.cn

Received: 2 July 2020; Accepted: 18 July 2020; Published: 21 July 2020

Abstract: Soda residue (SR), a waste by-product of sodium carbonate production, occupies land resources and pollutes the environment seriously. To promote the resource reusing of waste SR, this paper studies the feasibility of utilizing SR for the preparation of soda residue soil (SRS) through laboratory and field tests. The SR and fly ash (FA) were mixed with six different proportions (SR:FA is 1:0, 10:1, 8:1, 6:1, 3:1, 1:1) to prepare SRS, and the optimum water content, maximum dry density, shear strength, and unconfined compression strength of the SRS were measured. The representative SRS (SR:FA is 10:1) was selected to investigate the compression performance and collapsibility. The preparation and filling method of SRS in the field was proposed, and the effects of gravel, sand, and lime on the mechanical properties of SRS were studied through field tests. The results show that the addition of FA contributed to the strength development of SR, and the addition of lime, sand and rubble have a significant effect on the subgrade bearing capacity of SRS. The subgrade bearing capacity and deformation modulus of SRS in field tests is more than 210 kPa and 34.48 MPa, respectively. The results provide experimental basis and reference for the preparation of SRS, the scientific application of SRS in geotechnical engineering to promote sustainable development.

Keywords: soda residue; fly ash; field test; laboratory test; mechanical property

---

1. Introduction

Sodium carbonate is one of the important organic chemical raw materials, widely used in the chemical industry, food industry, metallurgical industry, textile industry, and many other fields [1–3]. The ammonia soda process is the common method for manufacturing sodium carbonate, but this method produces a large amount of soda residue (SR) during production [4,5]. According to statistics [6,7], in China, the production of sodium carbonate using the ammonia soda process exceeds 10 million tons per year, and the resulting SR reaches 3 million tons per year [8]. At present, building a storage yard next to the soda plant and discharging the SR into the storage yard is a common way to dispose of the SR [9–11]. However, this approach has the disadvantage of occupying amounts of land, polluting the local ecology, and failing to achieve sustainable resource and economic development [12,13]. From the remote sensing maps (Figure A1), more than 15 km^2 of land areas near the soda plant were occupied by the storage of soda residue in four stacking fields, and the effective utilization rate of SR is less than 5% [14]. The dry SR powder was dispersed by the wind, and the pollution range is up to 5 km, which can stimulate human eyes and the respiratory tract. In addition, the alkalinity and chloride ion from SR pollute the local soil and water ecological environment through seepage [15]. By monitoring the water quality of the wells around the SR storage yard, the chlorine ion content of one well increased from 86 mg/L to 1490 mg/L within four years, the chloride content of another well increased from 141 mg/L to 3720 mg/L within nine years, and the PH values were all greater than 9 [16]. The scientific treatment of SR has been a troublesome problem.

Therefore, seeking sustainable methods to dispose of the SR has become the main subject of many studies. Previous studies showed that the free chloride ion in SR can react with the fly ash (FA), and the gelling substances generated by the reacting can fill the pores of the SR and reduce the seepage rate of chloride ion and the PH of SR [17]. Moreover, FA can adsorb free chlorine ions in SR and effectively inhibit the dissolution of chlorine ions from SR [18,19]. Mixing SR with FA and other materials can effectively reduce the pollution of the local environment caused by chloride ion seepage from SR. The chemical composition of SR and alinite mineral is very similar, but differing in content of Cao, $CaCI_2$ [20,21]. Hou [22] fired a white cement using SR as raw material. The main component of the cement is alinite minerals, the chemical formula is $Ca_{11}(Si_{0.75}Al_{0.25})O_{18}CI$, which is synthesized by Cao, SiO_2, Al_2O_3, MgO, and $CaCI_2$. The cement has the advantages of the concrete hardening quickly and early strength being high. Kesim et al. [23] used soda sludge (73.5 wt %), clay (26.3 wt %) and iron ore (0.2 wt %) as material to prepare a cement successfully and studied the properties of the cement. The reach of reference [23] showed that the compressive strength of the prepared cement is 26.6 MPa. Some scholars also used SR as the raw material, using different admixtures, different methods to produce different cement and mortar [1,4,8,12]. It is estimated that 7.8 –10.0 million tons of SR have been output annually; less than one percent of the SR is reused scientifically, referenced from the literature [10,14]. It can be seen that the SR consumption of these methods is little, and the application of SR is also well below its discharge.

Fly ash (FA) is a by-product of coal combustion for electricity production, and an average of 0.3 tons of FA output are following the 1.0 ton in the process of coal combustion [24–27]. FA is a corrosive waste, and large amounts of fly ash pollute the local air and environment. The main chemical compositions of FA consist of silicon, aluminum, and iron oxides, and the lime content is less than 10% [25,28]. SR has positive effects on the early strength, good stability, and microstructure of the FA [29], and the mixture of SR and FA as the potential cement and engineering soil alternative for building material. Sun et al. [30] used SR and FA as the main materials to prepare the new non-clinker solidified soil, and investigated the engineering properties of the solidified soil. The research showed that, when the mass percentage of the curing agent is 20%, the mechanical properties of the solidified soil is similar to the solidified soil with 10% composite cement. Zhao et al. [10] developed an alkali-activated fly ash cement that was composed of SR, FA, and NaOH. The results showed that the compressive strength of the mixture is 22.04 MPa, the mixture can be used as an engineering soil. Yan et al. [31] studied the properties of the mixture of SR and FA through laboratory tests. Ji et al. [32] investigated the feasibility of using alkali waste to backfill the waste salt caverns through laboratory model test and found that the compressibility of the SR is very high, the compression modulus is 2.51 MPa. In the subsequent study, FA was used to improve the mechanical properties of the backfilled SR, and the research showed that the mixing of fly ash can enhance the strength and compressibility effectively of the mixture. However, studies on the disposal of SR are not comprehensive and in depth; most scholars used laboratory test to study the mechanical characteristics of soda residue soil (SRS). Moreover, previous studies have not investigated the method and feasibility of field filling of SRS; field tests can better simulate the real stress state of soil, and the test results are more accurate and reliable. It has rarely been reported that sand, rubble, and lime are used as raw materials for preparing the soda residue soil (SRS) through field test, and the feasibility of the field-filling method of SRS has not been investigated in previous studies.

The main objective of this paper was to reuse the SR into a value-added product, as well as study the feasibility of utilizing SR and FA as raw materials to prepare SRS for geotechnical engineering through laboratory and field tests. Five types of SRS with different FA contents were prepared, and the optimum water content, maximum dry density, direct shear strength, and unconfined compression strength were studied. The tests can be used to study the compaction characteristics of SRS with different ratios and the influence of age and mixing proportions on the strength of SRS. The compaction coefficient and collapsibility coefficient of SRS (SR: FA is 10:1) were also investigated. These experiments can be used to study the influence of water content and compaction coefficient on the compressibility characteristics and the water stability of SRS. Moreover, the mechanical properties of SRS with different mixing proportions were studied and probed through a series of field tests. The test results will provide an experimental basis for sustainable resource utilization of SR and FA.

2. Materials and Methods

2.1. Materials

The main raw materials for the test were soda residue (SR) and F class fly ash (FA), which were taken from a soda plant and a power plant in Shandong province of China, respectively. In this study, the chemical composition of SR and FA were obtained using XRF-1800X X-ray fluorescence spectrometer. After drying and grinding the samples, they were passed through a 240 mesh sieve. The samples were dispersed on a plexiglass support, and overlapped particles were manually separated by using a spatula, referenced from the literature [33,34]. The prepared sample was put into the sample table of the spectrometer; the cooling water switch was turned on, and the circulating water pump, and the test was conducted. After the test, we turned off the spectrometer, and we turned off the circulating water pump and the main power supply 15 min later. The chemical composition of SR and FA can be identified by comparing the measured X-ray energy values with the known characteristic X-ray energy values of each element, referenced from the literature [35]. The calibration and test methods refer to the spectrometer: test methods for main performance of energy dispersive X-ray fluorescence spectrometer (GB/T 31364-2015). The chemical composition of SR and FA are shown in Table 1 [34].

Table 1. Chemical compositions of soda residue (SR) and fly ash (FA).

Soda Residue	Percentage (wt %)	Fly Ash	Percentage (wt %)
$CaCO_3$	51.22	SiO_2	51.64
$Mg(OH)_2$	12.78	Al_2O_3	25.17
NaCl	10.87	Fe_2O_3	13.24
$CaSO_4$	9.24	CaO	3.23
Fe_2O_3	5.23	MgO	2.51
$CaCl_2$	4.45	LOI *	3.14
CaO	2.10	Others	1.07
Acid insoluble	4.11	-	-

* LOI: loss on ignition at 1000 °C (wt %).

The main chemical composition of SR is insoluble salts, including calcium carbonate ($CaCO_3$), calcium sulfate ($CaSO_4$), and oxides of iron (Fe_2O_3). The SR includes 67.01% mass percentage of the calcium-containing components, such as calcium carbonate, calcium sulfate, and calcium chloride ($CaCl_2$). It can be seen that the main components of the soil skeleton are present in the SR, which offers the possibility of using SR to make engineering soil.

The main chemical composition of FA consists of oxides of silicon (SiO_2), aluminum (Al_2O_3) and iron (Fe_2O_3), compounds of calcium, and magnesium. The silicon oxide and alumina oxide give FA the properties of volcanic ash [10]. When the FA is mixed with SR, most of the FA is hydrated to form gelling substances, and gelling substances filled in the pores of the mixture, which can improve the strength of the mixture [15].

2.2. Methods

2.2.1. Laboratory Test Methods

In order to investigate the mechanical properties of the SRS, the compaction test, direct shear test, and unconfined compressive strength test were performed with different mix proportions of SRS; the confined compression test and collapsibility test were performed with selected representative proportions.

Compaction test. The SR and FA were dried, crushed, and passed through the 2.5 mm fine screen. Thereafter, the SR and FA were mixed by mass ratio 1:0, 10:1, 8:1, 6:1, 3:1, 1:1, and each ratio prepared 5 samples of different water content. After the mixing and blending, the SRS was loaded into the test tube, and the standard electric compaction apparatus was used to conduct the compaction test according to the Chinese National Standard GB/T 22541-2008. According to the literature [15,30], the distribution range of particle size of SR and FA is 1.0–35.0 μm; light compaction is used to determine the optimum water content and maximum dry density of the SRS. The maximum dry density is the maximum density in the compaction curve. The optimal water content is the water content corresponding to the maximum dry density in the compaction curve. The schematic diagram of the equipment used in the compaction test is shown in Figure 1a.

Direct shear test. According to the results of the compaction test, samples with mass proportions of SR and FA of 1:0, 10:1, 8:1, 6:1, 3:1, and 1:1 were prepared, and two samples of each proportion were prepared for parallel testing. The water content and dry density of the samples were controlled to make it close to the optimal water content and maximum dry density. The mass of SRS required for each sample was calculated from the dry density and volume of the sample, and the samples were compacted by the compact cylinder and hammer. Thereafter, the direct shear test was carried out according to the Chinese National Standard GB/T 50123-2019. The test instrument is the direct shear apparatus, and the test method is quick shear, that is, after applying vertical pressure to the sample, apply horizontal shear stress quickly to the sample to make it shear failure. Cohesion forces is the mutual attraction between soil particles, and internal friction angle is the index of internal friction between soil particles. The values of cohesion forces and internal friction angle are the intercept and inclination of the shear strength line of the direct shear test. The schematic diagram of the equipment used in the direct shear test is shown in Figure 1b.

Unconfined compression strength test. Unconfined compressive strength is the ultimate strength of the sample against axial pressure without lateral pressure. In the test, the pressure when a clear fracture surface appears on the side of the sample is the unconfined compressive strength. The sample preparation method of the unconfined compression strength test is the same as that of the direct shear test, but 6 samples of each proportion were needed for the unconfined compressive strength test. After the samples preparation, the samples were kept in a constant temperature and humidity oven at (20 ± 2) °C and 90% relative humidity for 7 d, 28 d, and 90 d, and the unconfined compressive strength test was performed according to the Chinese National Standard GB/T 21043-2007. The test axial strain rate is 1–3% per minute. The test was completed within 8–10 min. When the dynamometer reached a peak, the test was stopped after 3–5% axial strain. When there is no peak reading, the test should be carried out until the strain reaches 20% axial strain. The schematic diagram of the equipment used in unconfined compression strength test is shown in Figure 1c.

Confined compression test. In order to study the effect of water content and compaction coefficient on the compression performance of the SRS, the representative ratio (SR:FA = 10:1) was selected for the confined compression test. The samples were prepared according to the Chinese National Standard GB/T 50123-2019, two for each sample, and the confined compression test was performed on the samples with different water content and compaction coefficients by high-pressure consolidation apparatus. The schematic diagram of the equipment used in unconfined compression strength test is shown in Figure 1d.

Collapsibility test. In order to study the collapsibility of the SRS, the representative ratio (SR:FA = 10:1) was selected for collapsibility test. The samples were prepared by the cutting ring and tested by the double-line method according to the Chinese National Standard GB 50025-2018. The schematic diagram of the equipment used in the collapsibility test is the same as that in the unconfined compression strength test, as shown in Figure 1d.

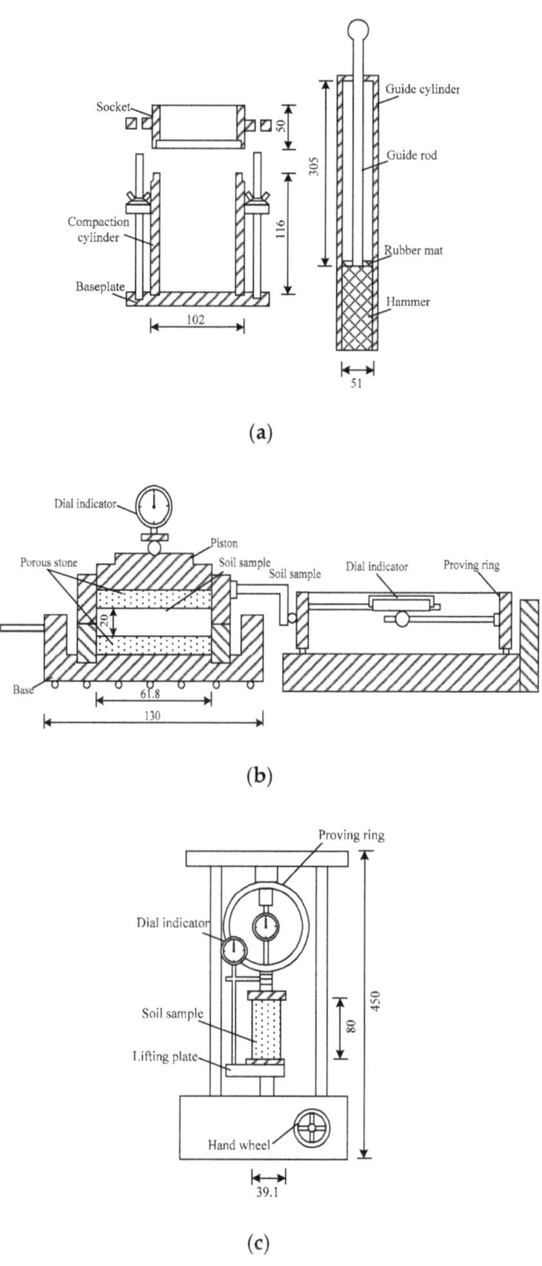

(**a**)

(**b**)

(**c**)

Figure 1. *Cont.*

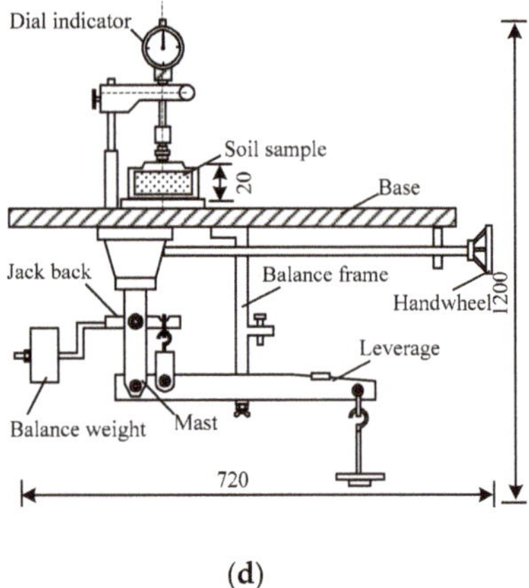

(d)

Figure 1. The schematic diagram of the equipment used in laboratory test: (**a**) Compaction apparatus; (**b**) Strain-controlled direct shear apparatus; (**c**) Unconfined compressive strength test apparatus; (**d**) High-pressure oedometer.

2.2.2. Field Test Methods

Based on the results of the laboratory tests and taking all aspects into consideration, five mixing proportions of the SRS were selected for the field tests. In order to consume as much SR as possible, the proportions of SR to FA were chosen to be 12:1 and 10:1 according to the actual discharge of waste from the soda plants and power plant. For further application of SRS and considering the feasibility of practical engineering, three other formulations of SRS were prepared, referenced from the literature [15,36]. The SRS was prepared using the common building materials such as standard sand (S), II class rubble (R), and II class lime powder (L). The specific mixing proportions of the SRS used in the field test are presented in Table 2.

Table 2. The mixing proportions of the soda residue soil (SRS).

No.	Mixing Proportions *
A1	SR:FA = 12: 1
A2	SR:FA = 10: 1
A3	SR:FA:L = 13:6:1
A4	SR:FA:S:L = 10:5:4:1
A5	SR:FA:R:L = 10:5:4:1

* SR: soda residue, FA: fly ash, S: sand, R: rubble, L: lime powder. The proportions are the mass proportions.

The raw materials of SRS were weighed according to the mixing proportion and mixed the raw materials with JS 500 mixer, as shown in Figure 2a. The mechanical agitation of the mixer can destroy the aggregate skeleton, compound salt skeleton and internal electric field of the SR, promote the dissolution and precipitation of CaCl2, accelerate the dehydration and solidification, and enhance the strength of SRS. When the mixing of the SRS was completed, the SRS was dried out naturally, and when the water content of the SRS was close to the optimum water content determined by the laboratory test, the preparation for SRS filling begins.

(**a**)   (**b**)

Figure 2. The preparation process of soda residue soil (SRS): (**a**) Mixing of raw materials by JS 500 mixer; (**b**) Excavation of test pits.

The test pits were excavated manually with the dimensions of 4.5 m × 1.7 m × 1.3 m (length × width × depth), which are shown in Figure 2b. Then, the SRS A1 to A5 were filled into separate test pits. The filling of SRS using a superficial compaction method. The SRS were filled in five layers, the thickness of each filling layer of the SRS being 30 cm, with frog-rammer tamped 3 times. In order to guarantee the compaction quality, the center of the SRS should be tamped first, and then the surrounding area should be tamped evenly. After compaction of each layer, the micro penetration test was carried out with a WY-4 soil penetration meter produced by the Nanjing soil instrument factory according to the Chinese National Standard GB/T 50123-2019. The blow count of WY-4 soil penetration meter was recorded when the meter penetrated into the SRS at 10 cm. After 25 days of filling, the soil penetration meter was used again to detect the compactness of the surface SRS. When the SRS A1 and SRS A2 were filled, the water content and dry density were determined by taking samples with a cutting ring and calculating the compaction coefficient of the SRS. The micro penetration test points and compaction coefficient test points are shown in Figure 3a.

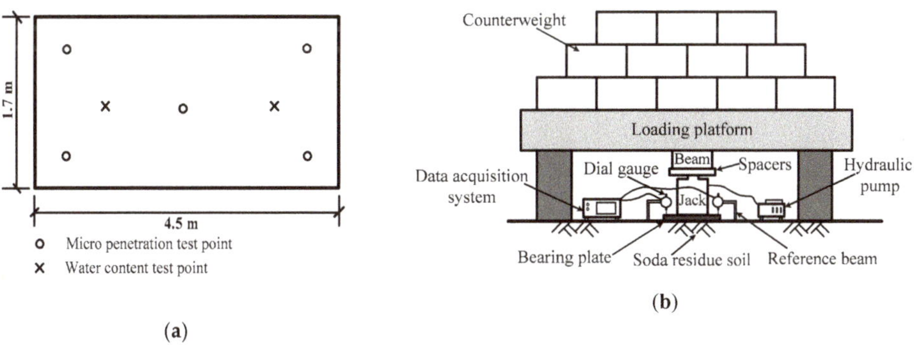

(**a**)   (**b**)

Figure 3. The field test process of soda residue soil: (**a**) The distribution of the test points; (**b**) Schematic diagram of plate load test system.

In order to further study the bearing capacity of the SRS, the plate load test was carried out according to the Chinese National Standard GB 50007-2011. The bearing plate size used in the test was 0.5 m × 0.5 m. The test adopted the method of step-by-step loading and monitored the settlement of the soil through the observation system. The plate load test system is shown in Figure 3b. Three test points were taken for each of the SRS A1 and SRS A2 and were recorded as A1-1, A1-2, A1-3, A2-1, A2-2, and A2-3. One test site was taken for each of A3 alkali cinder, A4 alkali cinder, and A5 alkali cinder, which was recorded as A3, A4, and A5. Due to time constraints, one test site was taken for each of SRS A3, SRS A4, and SRS A5, and recorded as A3, A4, and A5.

3. Laboratory Test Results and Analysis

3.1. Compaction Test Analysis

Compaction test is to determine the relation curve between water content and dry density. The maximum dry density and optimal water content of soil samples can be obtained by the compaction curve, so as to study the compaction characteristics of SRS. The compaction test results are shown in Figure 4. It can be seen that, with the increase in the amount of FA added, the dry density of the SRS is gradually increasing, and the optimal moisture content is gradually decreasing. The maximum dry densities of the SRS with the mixing proportion of 1:1 and 1:0 are 1.00 g/cm^3 and 0.94 g/cm^3, and the optimal water content was 46.0% and 63.1%, respectively. It can be seen that, with the increase of FA addition, the optimal water content of SRS decreases and the maximum dry density increases. This means that the addition of FA can absorb the water in the SR and fill in the voids in the SR agglomerate to increase its compactness. This reaction took place in a short time, increasing the strength of the soda residue soil. The increase of calcium ions and hydroxide ions leads to the following reactions [10,37]:

$$Ca^{2+} + CO_3{}^{2-} = CaCO_3 \downarrow \tag{1}$$

$$Mg^{2+} + 2OH^- = Mg(OH)_2 \downarrow \tag{2}$$

$$Ca^{2+} + SO_4{}^{2-} = CaSO_4 \downarrow \tag{3}$$

From the above equations, it can be seen that the strengthening mechanism of the SRS is as follows: the calcium oxide absorbs water and reduces the water content of the SRS; the ions in the pore water of the SR exchange reaction, resulting in precipitation, and the crystals act as cementation between the particles. Compared with the general engineering soil, the dry density of the SRS is less, and the use of SRS as the backfill soil will produce less additional stress and additional deformation, which is suitable for filling works.

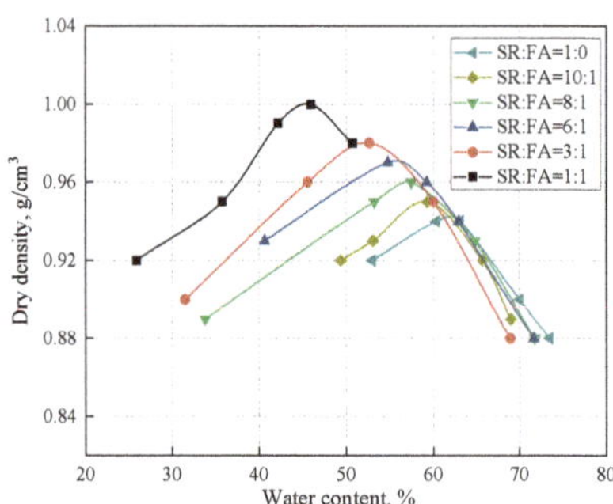

Figure 4. Compaction curve of soda residue soil (SRS) with different proportions.

3.2. Direct Shear Test Analysis

The direct shear test results of the SRS with different proportions are shown in Figure 5. With the increase of FA content, the cohesion forces and internal friction angle of the SRS are improved, but there are some differences in their trends. When the addition of FA is small, the angle of internal friction increases slowly (increase of only 1.6°), and with the increase of the addition of FA, the angle of internal friction increases rapidly (increase of 4.6°). This is due to the fact that when the FA is mixed in small amounts, most of the FA is hydrated to form gelling substances. The hydration reaction of the FA is as follows: when the fly ash and soda residue were mixed, calcium oxide in FA reacts with the water in SR to form calcium hydroxide [37]:

$$CaO + H_2O \rightarrow Ca(OH)_2 \tag{4}$$

Calcium hydroxide reacts with silicon dioxide and aluminum oxide to form hydrate calcium silicate and hydrated calcium aluminate:

$$m\,Ca(OH)_2 + (n-1)H_2O + SiO_2 \rightarrow mCaO \cdot SiO_2 \cdot nH_2O \tag{5}$$

$$mCa(OH)_2 + (n-1)H_2O + Al_2O_3 \rightarrow mCaO \cdot Al_2O_3 \cdot nH_2O \tag{6}$$

The calcium sulfate in the soda residue reacts with these hydrates to form ettringaite:

$$mCaO \cdot Al_2O_3 \cdot nH_2O + CaSO_4 \cdot 2H_2O \rightarrow mCaO \cdot Al_2O_3 \cdot CaSO_4 \cdot (n+2)H_2O \tag{7}$$

The hydrate calcium silicate, hydrated calcium aluminate, and ettringaite are hydraulically setting compositions, which form crystals on the surface of the fly ash vitreous gradually, and play a cementing role on soda residue particles, thus improving the strength of the soda residue soil.

With the increase of FA content in the SRS, due to the limited stimulation effect of SR, the FA filled in the pores of the SR to improve the density and surface mechanical bite force, resulting in the rapid increase of the angle of internal friction.

There are some differences in the trends of cohesion force and internal friction angle with FA addition. With the incremental addition of FA, the internal friction angle was dramatically enhanced, with an increase of 12 kPa for 10:1 SRS, compared with that of 1:0 SRS. With the continuous increase of FA addition, the increase of cohesion was 7 kPa, 5 kpa, and 3 kPa, respectively. This indicated that the hydration reaction between SR and FA produces gelling substances, which can improve the cohesive force of SRS. However, due to the limited stimulation effect of SR on FA, the increase rate of cohesive force decreases.

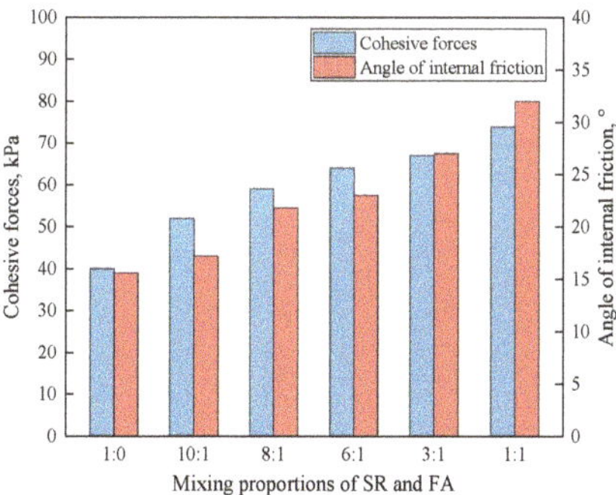

Figure 5. Cohesive forces and angle of internal friction of SRS with different proportions.

3.3. Unconfined Compression Strength Test Analysis

The unconfined compression strength test results of the SRS with different proportions are shown in Figure 6. It can be seen that the SRS compression strength of all ages gradually increases with the increase of FA admixture. The unconfined compression strength of the SRS sample cured for 90 d at the proportion of 1:1 is 21 times that of the SR sample. The unconfined compression strength of the SRS of each proportion cured for 90 days is 1.8 to 2.4 times that of the SRS cured for 7 days. It indicates that the compression strength of the SRS is age-dependent, and the hydration of the FA is a slow process. Liu et al. [38] studied the microstructure of the SRS by scanning electron microscope (SEM) and found that, when the age is early, the SR pores were filled with FA, the FA did not appear to erode and hydrate. With the growth of age, the phenomenon of flocculent gelling material and erosion appeared on the surface of the FA; when curing to 90 days, most of the fly ash was eroded, hydration reaction was sufficient, gelling material filled in the pores, which can explain the SRS with the unconfined compression strength with the age of increasing reasons [39,40].

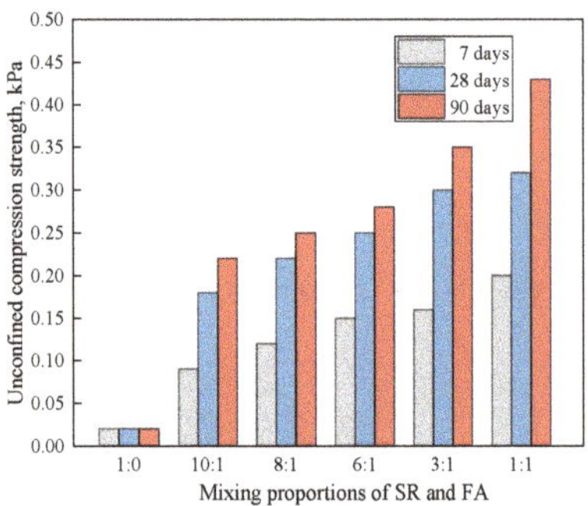

Figure 6. Unconfined compression strength of SRS with different proportions.

3.4. Confined Compression Test Analysis

Samples A1 to A4, B1 to B1 were prepared for confined compression test. The water content and compaction coefficient of the samples are listed in Table 3. The influence of water content (43.2%, 54.1%, 65.2%, 69.9%) and compaction coefficient (0.70, 0.80, 0.90, 0.95) on the coefficient and modulus of compressibility of the SRS were investigated. The confined compression test results are shown in Figure 7. The coefficient and modulus of compressibility of each sample are illustrated in Table 4.

Table 3. Water content and compaction coefficient of each sample.

No.	A1	A2	A3	A4	B1	B2	B3	B4
Water content/%	43.2	54.1	65.2	69.9	65.2	65.2	65.2	65.2
Coefficient of compaction/MPa	0.90	0.90	0.90	0.90	0.70	0.80	0.90	0.95

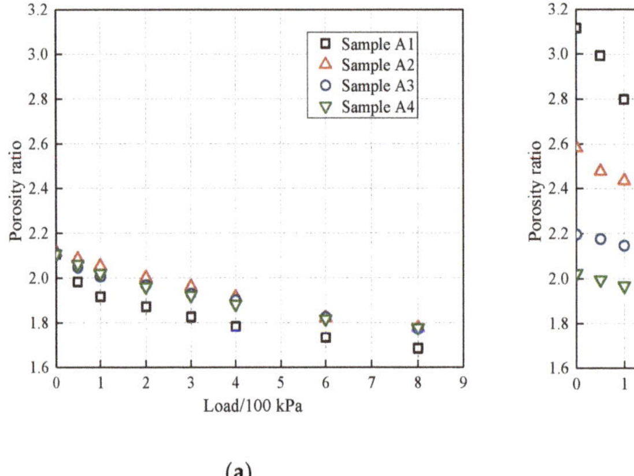

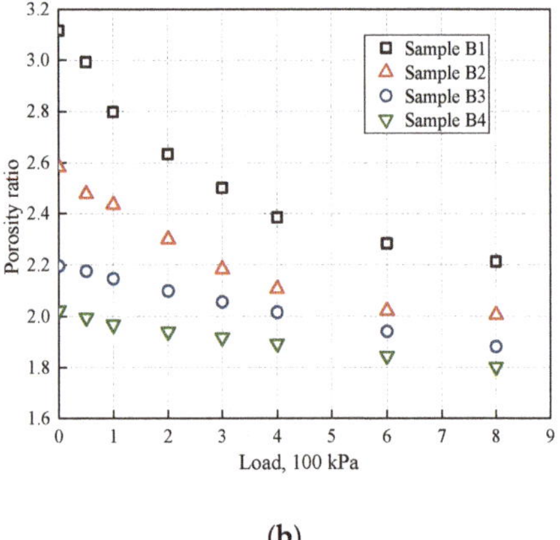

(a)

(b)

Figure 7. Porosity ratio and load curves of samples with different water content and compaction coefficient: (**a**) samples A1 to A4; (**b**) samples B1 to B4.

Table 4. The coefficient and modulus of compressibility of each sample.

No.	A1	A2	A3	A4	B1	B2	B3	B4
Coefficient of compressibility/MPa^{-1}	0.449	0.495	0.388	0.492	1.644	1.347	0.480	0.267
Modulus of compressibility/MPa	6.496	6.619	7.751	6.140	2.310	2.551	6.556	11.016

As shown in Figure 8 and Table 4, it can be seen that, when the compaction coefficient is 0.9, the compressibility coefficient $a_{1\text{-}2}$ of the SRS is 0.1 MPa^{-1} < $a_{1\text{-}2}$ < 0.5 MPa^{-1}, the SRS is of medium compressibility soil. When the compaction coefficient is the same, the water content has less influence on the compressibility of the SRS. The coefficient of compression of SRS A4 with a water content of 69.9% is higher than that of SRS A1 with a water content of 43.2%, the difference between the two is only 0.043 MPa^{-1}. It is indicated that the SRS has good water stability, even in water its compressibility does not change greatly. When the water content is the same, the compaction coefficient of the SRS compression properties has a greater impact. When the compaction coefficient is 0.7 and 0.8, the modulus of compressibility of the SRS B1 and SRS B2 is 2.310 Mpa and 2.551 Mpa, respectively, the SRS is of high compressibility. As the compaction coefficient increases, the modulus of compressibility increases, and the modulus of compressibility of SRS B4 is 11.016 MPa, about 5 times that of SRS B1.

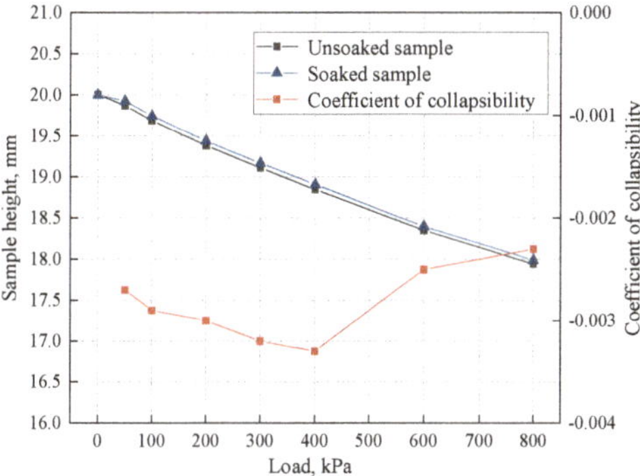

Figure 8. Collapsibility test curve of SRS

As shown in Table 5, the compressibility coefficient $a_{1\text{-}2}$ is usually used to judge the compressibility of soil. When the water content and compaction coefficient are 65.2% and 0.95, respectively, B4 SRS has the best compressibility and compressibility coefficient of 0.267, which is medium-compressibility soil. Therefore, in practical engineering, the medium-compressibility soil can be obtained by controlling the compaction coefficient and water content of the SRS [41,42].

Table 5. The criterion for judging the compressibility of soil.

Compressibility of Soil	Coefficient of Compressibility $a_{1\text{-}2}$/MPa^{-1}
Low-compressibility soil	$a_{1\text{-}2} < 0.1$
Medium-compressibility soil	$0.1 \leq a_{1\text{-}2} < 0.5$
High-compressibility soil	$a_{1\text{-}2} \geq 0.5$

3.5. Collapsibility Test Analysis

The results of the collapsibility test of the SRS are shown in Figure 8. With increasing axial stress, the height of the SRS sample after water immersion is slightly higher than that of the no-water immersion SRS sample; the collapsibility coefficient first decreases and then increases, but is less than 0.015. According to the specification GB 50025-2018, the SRS belongs to non-collapsible soil. This is because the structure of SRS is different from that of the collapsible soil. When the collapsible soil meets water, the cohesion between the soil particles weakens, and the soil structure is destroyed rapidly, resulting in settlement. This indicates that the SRS can maintain a stable soil structure after being exposed to the water and will not collapse. SRS will not settle when it meets water, which will affect the bearing capacity, so it can be used as engineering soil.

4. Field Test Results and Analysis

4.1. Micro Penetration Test Analysis

The results of the micro penetration test are shown in Figure 9. It can be seen that the hammer numbers of A2 SRS are slightly higher than that of A1 SRS. The blow counts of A3 SRS, A4 SRS, and A5 SRS are significantly higher compared to that of A1 SRS and A2 SRS. These results indicated that the strength of SRS can be improved by increasing FA content and adding sand, rubble, and lime to SRS. The hammer numbers of A1 SRS, A2 SRS, A3 SRS, A4 SRS, and A5 SRS increased by 2.62, 1.90, 2.56, 2.08, and 1.81 times, respectively, compared to that of SRS 25 days ago. These results can be attributed to the hydration reaction of SR and FA. As confirmed in the researches of Zhao et al. [1], Liu et al. [38], and Wang et al. [43], the hydration reaction between SR and FA is slow, and the strength of SRS is increased gradually with the increasing of age, which is consistent with the results of the unconfined compression strength test.

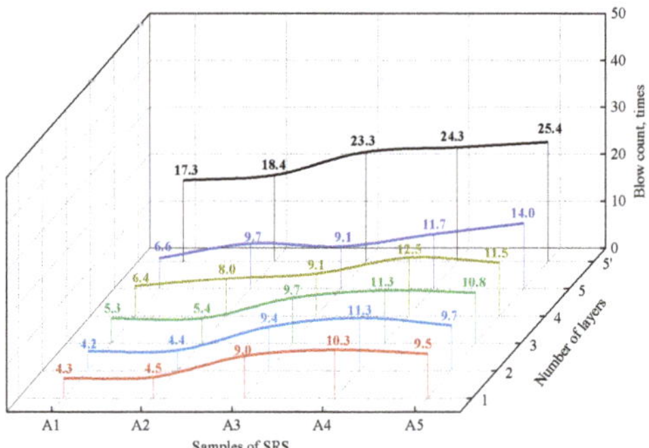

Figure 9. Micro penetration blow count of SRS.

4.2. Compaction Coefficient Test Analysis

Table 6 presents the water content, density, and compaction coefficient of the SRS with different mixing proportions. The formula of compaction coefficient is shown in Formula (8). It can be seen that the compaction coefficient of A1 SRS soil is 0.94 and that of A2 SRS is 0.99. The compaction coefficient of the filling foundation should not be less than 0.90 according to the Chinese National Standard GB 50007-2011. It indicates that the compaction quality of SRS by the filling method used in the test is good, which can improve the soil compactibility, reduce the permeability of soil, and reduce the height of capillary water, so as to prevent the soil-base softening caused by water accumulation. The reason is that mechanical compaction can change the soil structure of soil and improve its strength and stability.

$$K = \frac{\rho_d}{\rho_{dmax}} \tag{8}$$

where K is the compaction coefficient, ρ_d is the dry density, ρ_{dmax} is the maximum dry density.

Table 6. Compaction coefficient of the SRS.

No.	Density /g·cm^{-3}	Water Content /%	Dry Density /g·cm^{-3}	Maximum Dry Density/g·cm^{-3}	Compaction Coefficient	Average *
A1	1.49	71.8	0.88	0.96	0.92	0.94
	1.54	67.1	0.92	0.96	0.96	
A2	1.55	67.5	0.93	0.94	0.99	0.99
	1.51	65.3	0.91	0.93	0.98	

* This is the average of the compaction coefficients.

4.3. Plate Loading Test Analysis

The plate loading test is the most reliable test method for determining the soil parameters such as bearing capacity and deformation modulus. The load-settlement curves for each test point were obtained by monitoring the settlement of the SRS under load. Due to the influence of various factors, there is a certain discrepancy between the measured settlement and the actual settlement, and the measured settlement needs to be corrected by the least squares method, which is calculated by the formulas according to the Chinese National Standard GB 50007-2011:

$$Ns_0 + c_0 \sum p - \sum s' = 0 \tag{9}$$

$$s_0 \sum p + c_0 \sum p^2 - \sum ps' = 0 \tag{10}$$

where N is the load series, s_0 is the corrected settlement, c_0 is the slope of the curve, p is the load, s' is the initial settlement value under different loads.

Combining Formulas (9) and (10), the corrected settlement s_0 and slope c_0 can be calculated:

$$c_0 = \frac{N\sum ps' - \sum p \sum s'}{N \sum p^2 - (\sum p)^2} \tag{11}$$

$$s_0 = \frac{\sum s' \sum p^2 - \sum p \sum ps'}{N \sum p^2 - (\sum p)^2} \tag{12}$$

The measured load-settlement curves and corrected load-settlement curves for each test point are shown in Figure 10. In the test, due to the limitation of the upper limit of loading, the SRS all did not reach the ultimate destruction state, so there is no descending section in the load-settlement curve. Modulus of deformation of the SRS is calculated by the formulas:

$$E_0 = I_0 \left(1 - \mu^2\right) \frac{pd}{s} \tag{13}$$

where E_0 is the modulus of deformation, I_0 is the shape coefficient of the bearing plate, s is the settlement of the bearing plate, d is the length of side of the bearing plate, μ is the Poisson's ratio of the soil. According to the load-settlement curves in Figure 10, the subgrade bearing capacity and modulus of deformation of the SRS with different admixture can be obtained, as shown in Table 6.

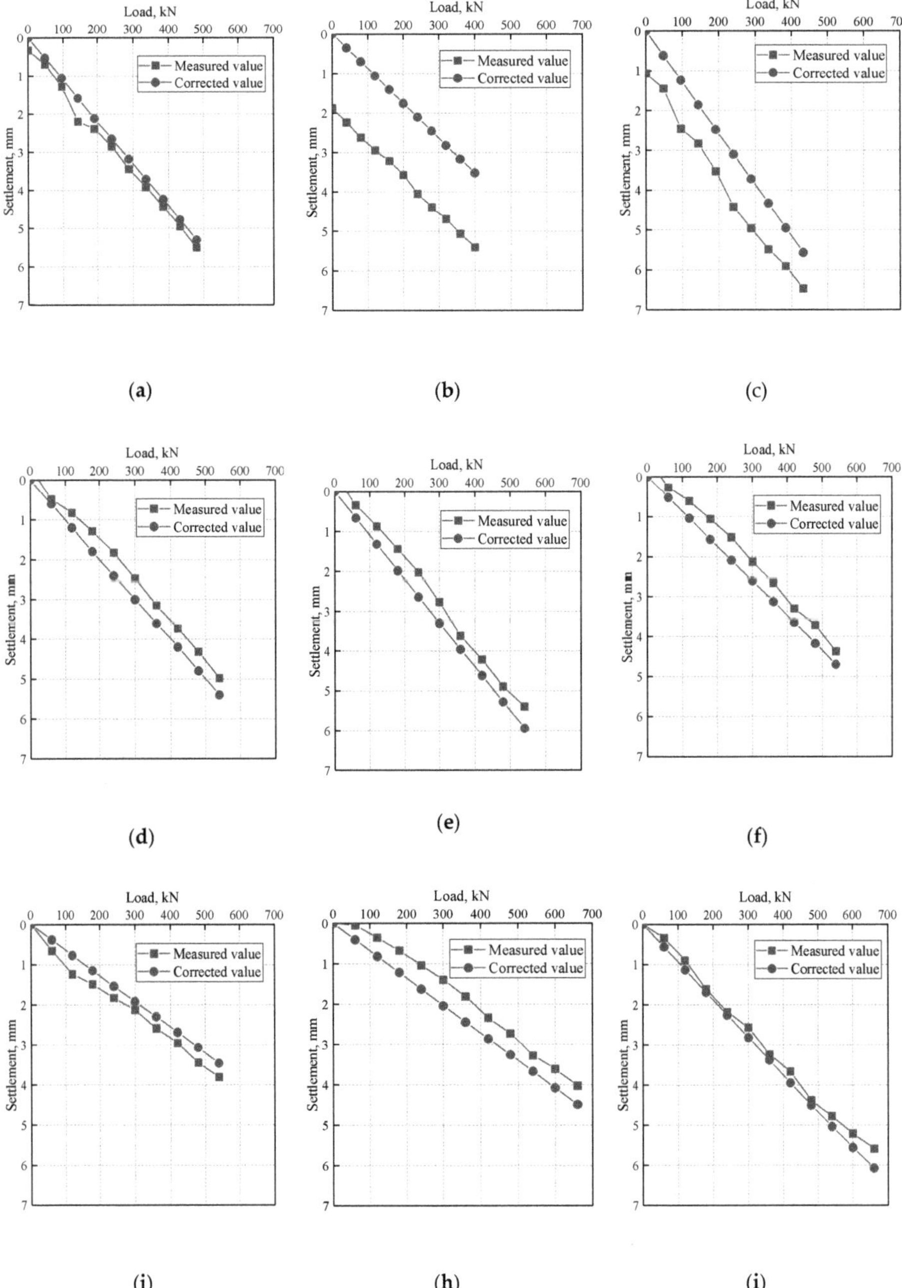

Figure 10. Measured and corrected plate load curves of soda residue soil: (**a**) A1-1 test point; (**b**) A1-2 test point; (**c**) A1-3 test point; (**d**) A2-1 test point; (**e**) A2-2 test point; (**f**) A2-3 test point; (**j**) A3 test point; (**h**) A4 test point; (**i**) A5 test point.

As shown in Figure 10 and Table 7, the Subgrade bearing capacities of A1 SRS, A2 SRS, and A3 SRS are 210 kPa, 220 kPa, and 270 kPa, respectively, and the deformation moduli are 34.48 MPa, 44.68 MPa, and 73.24 MPa, respectively. With the increased FA content, the subgrade bearing capacity and deformation modulus of SRS increased. The strength and compression property of SRS can be improved by adding lime. It is because the water in SRS is absorbed by lime, and the water content of SRS is reduced. Meanwhile, the FA is activated by NaOH to form N-A-S-H gels to fill the pores of solid particles, and the pozzolanic reaction between SR and FA is accelerated to form C-S-H gels to improve the subgrade bearing capacity and deformation modulus of SRS, referenced from the literature [15,38].

Table 7. Subgrade bearing capacity and modulus of deformation of the SRS.

No.	A1-1	A1-2	A1-3	A2-1	A2-2	A2-3	A3	A4	A5
Subgrade bearing capacity/kPa		≥210			≥220		≥270	≥330	≥330
Modulus of deformation/MPa		34.48			44.68		73.24	68.93	50.40

The subgrade bearing capacities of A4 SRS and A5 SRS are more than 330 kPa, and the deformation moduli are 68.93 MPa and 50.40 MPa, respectively. The subgrade bearing capacities of A4 SRS (addition sand) and A5 SRS (addition rubble) are 60 kPa higher than that of A3 SRS, but the deformation moduli of A4 SRS and A5 SRS are 4.31 MPa and 22.84 MPa, which is lower than that of A3 SRS. It can be seen that the addition of sand and rubble had a significant effect on the subgrade bearing capacity of SRS, but a slight influence on modulus of deformation. This is because sand and gravel can serve as the soil skeleton of SRS, significantly improving the strength of the soil [44]. The compressibility of SRS is related to grain composition [45]. When the grain composition of the soil is not good, the rubble (sand) were wrapped by the mixture of SR and FA, the first stage soil skeleton between rubble (sand) cannot be formed. Rubble (sand) becomes the filling material of the second-stage soil skeleton, so the deformation modulus of soil cannot be improved significantly. Therefore, although the strength of A4 SRS and A5 SRS is higher, the deformation modulus is smaller than that of A3 SRS.

The comparison of subgrade bearing capacity and deformation modulus between soda residue soil and general soil are listed in Table 8 [46,47]. It can be seen that the subgrade bearing capacity and deformation modulus of A4 SRS are 330 kPa and 68.93 MPa, respectively, which are obviously higher than clay and sand soil. The maximum subgrade bearing capacity of SRS is 330 kPa (A4 and A5 SRS), which is significantly higher than the test result (180 kPa) of Yan et al. [31], and the maximum modulus of deformation of SRS is 73.24 MPa (A3 SRS). The minimum subgrade bearing capacity and deformation modulus of SRS are 210 kPa and 34.48 MPa, respectively, which are close to that of sand soil. It indicates that the SRS can be used as engineering soil (such as: atrium filling soil, workshop foundation soil, road subgrade backfill soil) in geotechnical engineering. However, SRS should not be used in drinking water sources, cultivated land, and environmentally sensitive areas to avoid secondary pollution of surrounding soil and groundwater, and the effect of cost should be considered in actual filling engineering.

Table 8. Comparison of mechanical indexes between soda residue soil and general soil.

Mechanical Indexes	A4 SRS	Clay	Sand Soil
Subgrade bearing capacity/kPa	>330	120~180	150~220
Modulus of deformation/MPa	68.93	10~22	16~35

5. Conclusions

The objective of this paper was to validate the feasibility of using soda residue as the main raw material for preparation of soda residue soil; the preparation method in the field was proposed, and mechanical properties of soda residue soil with different mixing proportions and materials were investigated. Conclusions could be drawn as follows:

(1) The main chemical composition of SR is insoluble salts, and cohesive forces, angle of internal friction, and unconfined compression strength of SR are 40 kPa, 15.6°, and 0.02 kPa, respectively. The mechanical properties of SR need to be improved.

(2) The addition of FA contributed to the strength development of SR, incorporating about 50% FA makes the admixture possess the highest cohesive forces, angle of internal friction, and unconfined compression strength, which account for 74 kPa, 32°, and 0.43 kPa, respectively. The SRS optimum water content range is 46–63%, and the corresponding dry density is 0.94–1.00 g/cm^3. The SRS has good water stability and will not collapse.

(3) The addition of sand and rubble in SRS has a significant effect on subgrade bearing capacity, but a slight effect on the modulus of deformation. The subgrade bearing capacity and deformation modulus can be improved by adding lime. The subgrade bearing capacity and deformation modulus of SRS in field tests are more than 210 kPa and 34.48 MPa, respectively.

In summary, the research investigated the mechanical properties of SRS with different mixing proportions. However, to further promote the engineering application of SRS, it is necessary to investigate the effect of other admixtures (such as sodium silicate and sodium sulphate), different mixing proportions and the particle size of raw material on the mechanical characteristics of SRS.

Author Contributions: Conceptualization, M.Z. and N.Y.; methodology, N.Y., J.L., and J.M.; data curation, Y.W., J.L., and J.M.; writing—original draft preparation, J.M.; writing—review and editing, X.B., N.Y., and M.Z. All authors have read and agreed to the published version of the manuscript.

Funding: This research was funded by the National Natural Science Foundation of China, grant number 51809146; the Shandong Key Research and Development Program, grant number 2018GSF117008.

Conflicts of Interest: The authors declare no conflict of interest.

Appendix A. Remote Sensing Maps of Stacked Waste Soda Residues in China

Remote sensing maps of stacked waste soda residues in China as shown in Figure A1.

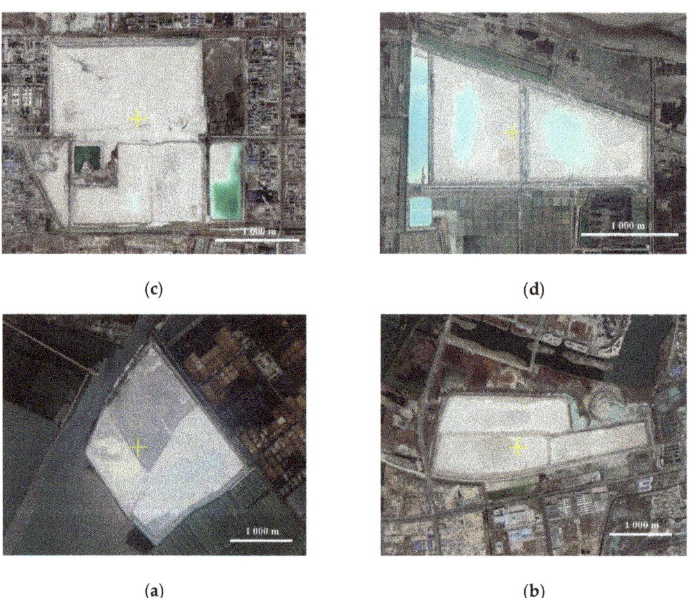

(c)

(d)

(a)

(b)

Figure A1. Remote sensing maps of stacked waste soda residues in China: (**a**) Tangshan; (**b**) Lianyungang; (**c**) Weifang; (**d**) Qingdao.

References

1. Zhao, X.H.; Liu, C.Y.; Wang, L.; Zuo, L.M.; Zhu, Q.; Ma, W. Physical and mechanical properties and micro characteristics of fly ash-based geopolymers incorporating soda residue. *Cem. Concr. Comp.* **2019**, *98*, 125–136. [CrossRef]
2. Liu, R.Z.; Li, J.L.; Wang, Y.W.; Liu, D.W. Flotation separation of pyrite from arsenopyrite using sodium carbonate and sodium humate as depressants. *Colloid. Surface. A* **2020**, *595*, 124669. [CrossRef]
3. Erabi, M.; Goshadrou, A. Bioconversion of Glycyrrhiza glabra residue to ethanol by sodium carbonate pretreatment and separate hydrolysis and fermentation using Mucor hiemalis. *Ind. Crop. Prod.* **2020**, *152*, 112537. [CrossRef]
4. Ucal, G.O.; Mahyar, M.; Tokyay, M. Hydration of alinite cement produced from soda waste sludge. *Construct. Build. Mater.* **2018**, *164*, 178–184. [CrossRef]
5. Zha, F.S.; Pan, D.D.; Xu, L.; Kang, B.; Yang, C.B.; Chu, C.F. Investigations on engineering properties of solidified/stabilized pb-contaminated soil based on alkaline residue. *Adv. Civ. Eng.* **2018**, *8595419*, 1–8. [CrossRef]
6. Yan, C.; Song, X.K.; Zhu, P.; Sun, H.Y.; Li, Y.P.; Zhang, J.F. Experimental study on strength characteristics of soda residue with high water content. *Chin. J. Geotech. Eng.* **2007**, *29*, 1683–1688.
7. Gomes, H.I.; Mayes, W.M.; Rogerson, M.; Stewart, D.I.; Burke, I.T. Alkaline residues and the environment: A review of impacts, management practices and opportunities. *J. Clean. Prod.* **2015**, *112*, 3571–3582. [CrossRef]
8. Wang, B.M.; Wang, L.J.; Mohd Zain, M.F.; Lai, F.C. Development of soda residue concrete expansion agent. *J. Wuhan Univ. Technol. Mater. Sci. Ed.* **2003**, *18*, 79–82.
9. Matthews, D.A.; Eddler, S.W. Decreases in pollutant loading from residual soda ash production waste. *Water Air Soil Poll.* **2003**, *146*, 55–73. [CrossRef]
10. Zhao, X.H.; Liu, C.Y.; Zuo, L.M.; Zhu, Q.; Ma, W.; Liu, Y. Preparation and characterization of press-formed fly ash cement incorporating soda residue. *Mater. Lett.* **2019**, *259*, 126852. [CrossRef]
11. Yang, J.J.; Xie, W.; Zhang, L.; He, C.S.; Bao, G.D. Study on experimental preparation of cement mortar incorporating fly ash-soda residue. *Bull. Chin. Ceram. Soc.* **2010**, *29*, 1211–1216.
12. Kuznetsova, T.V.; Shatov, A.A.; Dryamina, M.A.; Badertdinov, R.N. Use of wastes from soda production to produce nonshrinking oil-well cement. *Russ. J. Appl. Chem.* **2005**, *78*, 698–701. [CrossRef]
13. Zhang, G.; Li, X.; Li, Y.; Wu, T.; Sun, D.; Lu, F. Removal of anionic dyes from aqueous solution by leaching solutions of white mud. *Desalination* **2011**, *274*, 255–261. [CrossRef]
14. Zhao, X.H.; Liu, C.Y.; Zuo, L.M.; Wang, L.; Zhu, Q.; Liu, Y.C.; Zhou, B.Y. Synthesis and characterization of fly ash geopolymer paste for goaf backfill: Reuse of soda residue. *J. Clean. Prod.* **2020**, *260*, 121045. [CrossRef]
15. Zhao, X.H.; Liu, C.Y.; Zuo, L.M.; Wang, L.; Zhu, Q.; Wang, M.K. Investigation into the effect of calcium on the existence form of geopolymerized gel product of fly ash based geopolymers. *Cement Concrete Comp.* **2018**, *103*, 279–292. [CrossRef]
16. Yuan, X.M.; Zhang, H.; Liu, X.M.; Xiong, F. Influence of alkaline residue site on carst groundwater. *Admin. Tech. Environ. Monit.* **2010**, *22*, 36–39.
17. Yu, S.J.; Bi, W.Y. Research on the impact of Cl^- to environment when industrial soda residue is appiled in highway project. *Energy Environ. Prot.* **2007**, *21*, 47–50. [CrossRef]
18. Ohenoja, K.; Pesonen, J.; Yliniemi, J.; Illikainen, M. Utilization of fly ashes from fluidized bed combustion: A review. *Sustainability* **2020**, *12*, 2988. [CrossRef]
19. Czop, M.; Lazniewska-Piekarczyk, B. Evaluation of the leachability of contaminations of fly ash and bottom ash from the combustion of solid municipal waste before and after stabilization process. *Sustainability* **2020**, *11*, 5384. [CrossRef]
20. Ilyukhin, V.V.; Nevsky, N.N.; Bickbau, M.J. Crystal structure of alinite. *Nature* **1977**, *269*, 397–398. [CrossRef]
21. Vaidyanathan, D.; Kapur, P.C.; Singh, B.N. Production and properties of alinite cements from steel plant wastes. *Cement. Concrete. Res.* **1990**, *20*, 15–24.
22. Hou, G.H. Design and research of the mineral composition of white alinite cement. *J. Build. Mater.* **2002**, *5*, 80–83.

23. Kesim, A.G.; Tokyay, M.; Yaman, I.O.; Ozturk, A. Properties of alinite cement produced by using soda sludge. *Adv. Cem. Res.* **2013**, *25*, 104–111. [CrossRef]

24. Sumajouw, D.M.J.; Hardjito, D.; Wallah, S.E.; Rangan, B.V. Fly ash-based geopolymer concrete: Study of slender reinforced columns. *J. Mater. Sci.* **2007**, *42*, 3124–3130. [CrossRef]

25. Moreno, N.; Querol, X.; Andres, J.M.; Stanton, K.; Towler, M.; Nugteren, H.; Janssen-Jurkovicova, M.; Jones, R. Physico-chemical characteristics of European pulverized coal combustion fly ashes. *Fuel* **2005**, *84*, 1351–1363. [CrossRef]

26. Nuccetelli, C.; Pontikes, Y.; Leonardi, F.; Trevisi, R. New perspectives and issues arising from the introduction of (NORM) residues in building materials: A critical assessment on the radiological behaviour. *Constr. Build. Mater.* **2015**, *82*, 323–331. [CrossRef]

27. Cherian, C.; Siddiqua, S. Pulp and paper mill fly ash: A review. *Sustainability* **2019**, *11*, 4394. [CrossRef]

28. Park, S.M.; Khalid, H.R.; Seo, J.H.; Yoon, H.N.; Son, H.M.; Kim, S.H.; Lee, N.K.; Lee, H.K.; Jang, J.G. Pressure-induced geopolymerization in alkali-activated fly ash. *Sustainability* **2018**, *10*, 3538. [CrossRef]

29. Liu, Y.L.; Wang, Y.S.; Fang, G.H.; Alrefaei, Y.; Dong, B.Q.; Xing, F. A preliminary study on capsule-based self-healing grouting materials for grouted splice sleeve connection. *Construct. Build. Mater.* **2018**, *170*, 418–423. [CrossRef]

30. Sun, J.Y.; Gu, X. Engineering properties of the new non-clinker incorporating soda residue solidified soil. *J. Build. Mater.* **2014**, *17*, 1031–1035.

31. Yan, S.W.; Hou, J.F.; Liu, R. Research on geotechnical properties and environmental effect of mixture of soda waste and fly ash. *Rock Soil Mech.* **2006**, *27*, 2305–2308.

32. Ji, G.D.; Yang, C.H.; Liu, W.; Zuo, J.J.; Lei, G.W. An experimental study on the engineering properties of backfilled alkali wastes reinforced by fly ash. *Rock Soil Mech.* **2015**, *36*, 2169–2176.

33. Otsuki, A.; Gonçalves, P.P.; Stieghorst, C.; Révay, Z. Non-destructive characterization of mechanically processed waste printed circuit boards: X-ray fluorescence spectroscopy and prompt gamma activation analysis. *J. Compos. Sci.* **2019**, *3*, 54. [CrossRef]

34. Otsuki, A.; Gonçalves, P.P.; Leroy, E. Selective milling and elemental assay of printed circuit board particles for their recycling purpose. *Metals* **2019**, *9*, 899. [CrossRef]

35. Otsuki, A.; Mensbruge, L.D.L.; King, A.; Serranti, S.; Fiore, L.; Bonofazi, G. Non-destructive characterization of mechanically processed waste printed circuit boards-particle liberation analysis. *Waste Manag.* **2020**, *102*, 510–519. [CrossRef]

36. Han, F.Q.; Zhang, M.Y.; Zhou, Y.Z. Laboratory research on strength and deformation of soda residue soil. *J. Qingdao Univ. Tech.* **2004**, *25*, 20–22.

37. Li, Y.Y.; Yan, S.W.; Zhang, J.Y.; Yin, X.T. Engineering properties and microstructural features of the soda residue. *Chin. J. Geotech. Eng.* **1999**, *21*, 100–103.

38. Liu, C.Y.; Hu, L.; Zhu, Q.; Zhao, X.H.; Liu, Y.C.; Ma, W. Experimental study on performance of fly ash reinforced soda residue padding pad. *J. Hebei Univ. Tech.* **2018**, *47*, 87–93.

39. Ahmadi, S.F.; Eskandari, M. Vibration analysis of a rigid circular disk embedded in a transversely isotropic solid. *J. Eng. Mech.* **2014**, *140*, 04014048. [CrossRef]

40. Eskandari, M.; Samea, P.; Ahmadi, S.F. Axisymmetric time-harmonic response of a surface-stiffened transversely isotropic half-space. *Meccanica* **2017**, *52*, 183–196. [CrossRef]

41. Ahmadi, S.F.; Samea, P.; Eskandari, M. Axisymmetric response of a bi-material full-space reinforced by an interfacial thin film. *Int. J. Solids. Struct.* **2016**, *90*, 251–260. [CrossRef]

42. Wang, Z.Y.; Zhang, N.; Li, Q.; Chen, X.H. Dynamic response of bridge abutment to sand-rubber mixtures backfill under seismic loading conditions. *J. Vibroeng.* **2017**, *19*, 434–446.

43. Wang, Q.; Li, J.J.; Yao, G.; Zhu, X.N.; Hu, S.G.; Qiu, J.; Chen, P.; Lyu, X.J. Characterization of the mechanical properties and microcosmic mechanism of Portland cement prepared with soda residue. *Constr. Build. Mater.* **2020**, *241*, 1–11. [CrossRef]

44. Li, G.X. On soil skeleton and seepage force. *Chin. J. Geotech. Eng.* **2016**, *38*, 1522–1528.

45. San José Martínez, F.; Martín, M.A.; Caniego, F.J.; Tuller, M.; Guber, A.; Pachepsky, Y.; García-Gutiérrez, C. Multifractal analysis of discretized X-ray CT images for the characterization of soil macropore structures. *Geoderma* **2010**, *156*, 32–42. [CrossRef]

46. Kuang, Z.; Zhang, M.Y.; Bai, X.Y. Load-bearing characteristics of fibreglass uplift anchors in weathered rock. *Proc. Inst. Civ. Eng-Geotech. En.* **2020**, *173*, 49–57. [CrossRef]
47. Chen, X.Y.; Zhang, M.Y.; Bai, X.Y. Axial resistance of bored piles socketed into soft rock. *KSCE J. Civ. Eng.* **2019**, *23*, 46–55. [CrossRef]

 *sustainability*

Article

Sustainable Perspective of Low-Lime Stabilized Fly Ashes for Geotechnical Applications: PROMETHEE-Based Optimization Approach

Arif Ali Baig Moghal [1], Ateekh Ur Rehman [2,*], K Venkata Vydehi [1] and Usama Umer [3]

[1] Department of Civil Engineering, National Institute of Technology, Warangal, Telangana 506004, India; baig@nitw.ac.in (A.A.B.M.); vydehi56@student.nitw.ac.in (K.V.V.)

[2] Department of Industrial Engineering, College of Engineering, King Saud University, Riyadh 11421, Saudi Arabia

[3] Advanced Manufacturing Institute, King Saud University, Riyadh 11421, Saudi Arabia; uumer@ksu.edu.sa

* Correspondence: arehman@ksu.edu.sa; Tel.: +966-11-4697177

Received: 11 June 2020; Accepted: 13 August 2020; Published: 17 August 2020

Abstract: In the present scenario of global green environmental and sustainable management, the disposal of large volumes of coal-based ashes (fly ashes) generate significant environmental stress. The aim is to exploit these fly ashes for bulk civil engineering applications to solve societal-environmental issues employing sustainable measures. In this study, the addition of lime and/or gypsum in improving the geotechnical properties (hydraulic conductivity, compressibility, unconfined compression strength, lime leachability, and California bearing ratio) of fly ashes was investigated. To assist the practicing engineers in selecting the right mix of lime and/or gypsum for a given amount of fly ash for a specific application, a multi-criteria approach was adopted. The possible alternatives investigated included untreated fly ash, fly ash treated with lime (1%, 2.5%, 5%, or 10%), and a variation in gypsum dosage (1% or 2.5%) in the presence of lime. Sensitivity analysis was performed to recognize and resolve the conflicting advantages and disadvantages when mixing lime and gypsum. The study revealed that to derive the potential benefits of fly ash, it is essential to combine the lime dosage with gypsum for pavement and liner applications where bulk quantities of fly ash are employed.

Keywords: fly ash; gypsum; lime; liners; pavements; PROMETHEE

---

1. Introduction

Current sustainable energy policies reflect thermal power generation as a major mode of power generation that facilitates industrial development worldwide. A total of 7727.3 Mt of coal was produced worldwide in 2017 [1]. The production of large quantities of sustainable fly ash [2] necessitates adequate disposal facilities, owing to its negative environmental impact [3]. Additionally, the higher disposal costs associated with fly ashes necessitate its recycling for sustainable development. However, the bulk utilization of fly ashes in various civil engineering applications assists in solving sustainable societal and environmental needs, such as liner material for landfills, sub-base material for pavements, backfill material for embankments and retaining walls, substitute material for sand, aggregate, and cement, and other domestic purposes [4–14]. The various engineering properties that facilitate the utilization of fly ashes, especially in geotechnical applications, are elaborated on in Table 1.

Table 1. Summary of the geotechnical properties of fly ashes and fly-ash-amended composites.

Type of Fly Ash and Properties Studied	Potential Sustainable Applications	Reference
Fly ash compacted under controlled test conditions indicated an increase in Φ from 30° to 40° (from DST and Triaxial shear test). Average Cc was 0.2.	Satisfied embankment-fill material criteria.	DiGioia and Nuzzo [15]
Class C fly ash and sand mixtures; 10% bentonite; HC increased by a factor of 1000.	Satisfied liner material criteria for hazardous waste.	Edil et al. [16]
Coal fly ash with lime dust and bentonite (7:2:1 ratio) decreased HC by a factor of 1000, compared to fly ash alone. Migration of heavy metals (Pb, Zn, and Fe) was estimated to be less than 0.10 m (under MSW landfill conditions) over 15 years of service.	Satisfied landfill-barrier material criteria	Nhan et al. [17]
Class C fly ash (20% by weight) with Class F fly ash reduced HC value by a factor of 100. Durability studies revealed HC value increased by a factor between 2 and 3 for wetting and drying cycles.	Satisfied the requirements of liner for waste-containment facilities.	Palmer et al. [6]
The average HC values of Class F fly ash with 30% bentonite reduced by a factor of 10,000. Significant increase in Φ (33° to 42°) and C (36 to 54 kN/m^2) from a Triaxial shear test (CU) was observed.	Satisfied the requirements of liner and cover materials for waste-disposal sites.	Mollamahmutoğlu and Yilmaz [18]
UCS increased by a factor of 2.2 for alkali-activated Class F fly ash. Percentage weight loss decreased by a factor of 5, compared to PCC subjected to acid immersion (at specified curing periods).	Satisfied the requirement of PCC used for construction purpose.	Rostami and Brendley [19]
The HC value of Class F and bottom ashes decreased by a factor of 10 with an increase in Class F fly ash content.	Satisfied the requirements of highway-embankment material.	Kim et al. [8]
The Cc of sedimented class F fly ash deposits was relatively higher (by 3%) when compared to compacted conditions.	Satisfied the requirements of sub-base material and lightweight infrastructure.	Sekhar et al. [10]
With an increase in compaction energy, HC values of Class F fly ash decreased by a factor of 1000. At each compaction energy, a higher CBR value was observed for a water content equal to or slightly less than OMC.	Satisfied the requirements of barrier material for landfill applications.	Zabielska-Adamska [20]
The HC values of Class F fly ashes reduced by a factor of 1000 and the Cc value was reduced by a factor between 6 and 10 due to the addition of lime (at 10%) and gypsum (at 2.5%).	Satisfied the liner material requirements.	Moghal and Sivapullaiah [21]; Moghal and Sivapullaiah [22]
UCS values and the CBR of Class F fly ashes increased by a factor of 20 and 7.5, respectively, due to the addition of lime (at 2.5%) and gypsum (at 2.5%).	Satisfied the requirements of sub-base material in road construction.	Sivapullaiah and Moghal [23]

Note: In above Table 1, Φ: Angle of internal friction; DST: Direct shear test; Cc: Compression index; MSW: Municipal solid waste; C: Cohesion; HC: Hydraulic conductivity; OMC: Optimum moisture content; CU: Consolidated undrained test; PCC: Portland cement concrete; CBR: California bearing ratio; UCS: Unconfined compressive strength.

As most fly ashes produced from harder, older, bituminous, and anthracite coals have very low lime content (Class F type), they require cementing agent(s) in the form of lime, gypsum, or cement to enhance their applicability for various civil engineering applications, as discussed in the above Table 1. The addition of lime will supplement a basic cementing agent to produce pozzolanic compounds over time by dissolving the silica-rich, glassy phases of fly ashes [23–25]. Because lime-fly ash reactions are time-dependent, the addition of gypsum is considered in order to increase the rate at which pozzolanic compounds are formed. When gypsum is added to lime-treated fly ashes, it provides both advantages and disadvantages; it retards the setting time and accelerates the strength [23,26].

The present study focused on the sustainable perspective of lime and/or gypsum addition on the enhancement of the geotechnical properties (hydraulic conductivity (HC), unconfined compression strength (UCS), compressibility characteristics (Cc), lime leachability (LL), and California bearing ratio (CBR)) of two different types of fly ashes. Both lime and gypsum, at varying dosages, were added to enhance the cementation effect. To assist practicing engineers in selecting the appropriate mix of fly ash, lime, and gypsum contents for specific civil engineering applications, a multi-criteria decision-making (MCDM) approach was adopted. An MCDM approach has been successfully applied in civil engineering applications pertaining to hydraulics [27–29], energy [30], solid waste [31,32], transportation [33], green building materials [34], and sustainable development [35].

In the current study, an MCDM approach [36–38] was employed to evaluate the different geotechnical properties of fly ashes, stabilized with lime and gypsum at different dosages, for targeted bulk applications. In the present approach, concepts pertaining to preference flow, sensitivity analyses, and graphical interactive analyses were formed to assist practicing engineers in ranking treatment strategies (i.e., determining the correct dosages of lime and gypsum) to establish the superiority of one treatment strategy over another.

2. Methodology

The adopted methodology was based on the variation of three factors: type of fly ash, lime dosage, and gypsum dosage (Table 2). The combined influence of each factor, at different levels of interaction, was investigated with regard to the response of the resultant geotechnical properties for specific liner and pavement applications.

Table 2. Factors, levels, their objectives, notations, and treatment strategies adopted in the study.

Factor	Levels				Selected Alternative Treatment Strategies for Different Factors and Their Corresponding Levels				
Fly Ash Type	**A**		**B**		**A**	**AL1**	**AL2**	**AL3**	**AL4**
Lime Dosage (%)	L1	L2	L3	L4	AL1G1	AL1G1	AL2G1	AL2G2	AL3G1
Gypsum Dosage (%)	G1		G2		AL3G2	AL4G1	AL4G2	—	—
Objective	Liner Application		Pavement Application		**B**	**BL1**	**BL2**	**BL3**	**BL4**
Response measures (Units)	Liner Application		Pavement Application		BL1G1	BL1G1	BL2G1	BL2G2	BL3G1
UCS (kPa)	Maximize		Maximize		BL3G2	BL4G1	BL4G2	—	—
CBR (%)	Maximize		Maximize						
HC (cm/sec)	Minimize		Maximize						
Cc (kPa)	Minimize		Minimize						
LL (ppm)	Minimize		Minimize						

Note: In Above Table 2, UCS: Unconfined compression strength; CBR: California bearing ratio; Cc: Coefficient of Compressibility; HC: Hydraulic conductivity; LL: Lime leachability; A: Fly ash A; B: Fly ash B; L1, L2, L3, and L4 represent lime dosage at 1%, 2.5%, 5%, and 10%, respectively; G1 and G2 represent gypsum percentage of 1 and 2.5, respectively. For example, AL1G1 represents fly ash type A treated with lime dosage at 1% and gypsum dosage of 1%.

2.1. Materials Used

This study used two low-lime fly ashes, named A and B (Table 3), sourced from thermal power plants in Neyvelli and Muddanur, which are towns in the states of Tamil Nadu and Andhra Pradesh in India, respectively. Analytical-reagent (AR) grade gypsum (CaSO$_4$ 2H$_2$O) and AR grade hydrated lime (Ca(OH)$_2$), supplied by Merck limited, India, were used in the present study. The specific surface area values (SSA) of fly ashes A and B were found to be 9.6 and 8.2 m^2/gm, respectively [25]. Both fly ashes exhibited non-plastic behavior, with fly ash B having greater fines content compared to fly ash A. Mullite and Quartz phases were predominant in both fly ashes.

Table 3. Constituents of selected fly ashes.

Constituent	% Based on Fly Ash Type		Constituent	% Based on Fly Ash Type	
	A	B		A	B
Sodium (Na$_2$O)	0.18	0.19	Calcium (CaO)	9.00	3.62
Potassium (K$_2$O)	0.21	0.27	Ferric (Fe$_2$O$_3$)	16.61	6.28
Titanium (TiO$_2$)	0.26	0.31	Alumina (Al$_2$O$_3$)	18.81	27.65
Magnesium (MgO)	1.41	0.34	Silica (SiO$_2$)	50.97	56.88
Loss on ignition	2.55	4.46			

2.2. Experimental Testing Methodology

UCS tests were conducted in accordance with ASTM D2166 [39] on samples cured for 28 days. Table 4 shows the UCS values for the cured fly ashes A and B. CBR tests were conducted, in accordance with IS 2720 Part 16 [40] and ASTM D1883 [41], on samples cured for 14 days under controlled humidity conditions, and the results are presented in Table 4. Hydraulic conductivity tests were carried out, in accordance with ASTM D5856 [42], on samples cured for 28 days. Compressibility tests were carried out on standard, one-dimensional odometer consolidation tests, as per IS 2720 Part 15 [43] and ASTM D2435 [44]. The compression index values, corresponding to the loading increment from 25 to 50 kPa, are reported in Table 4. The experimental test setup and specimen-testing details are provided in Figure 1. An LL-testing procedure, developed by Moghal and Sivapullaiah [22], was employed in this study. The LL values of cured fly ashes, under 7 days of steady flow conditions, are reported in Table 4. A minimum of three tests were carried out (in triplicates) for each of the studied parameters (hydraulic conductivity; unconfined compressive strength test; California bearing ratio; lime leachability and one-dimensional oedometer fixed-ring consolidation test) as per ASTM standards, and the average values are reported in Table 4.

Table 4. Multi-criteria response measures for each treatment strategy.

Strategy No.	Treatment Strategy	"j" Response Measures				
		UCS (kPa)	CBR (%)	HC (cm/s) $\times 10^{-4}$	Cc (kPa)	LL (ppm)
1	A	173.1	55.5	6.52	0.0099	630
2	AL1	790.3	58.5	1.57	0.0082	790
3	AL2	532	63	1.22	0.0082	860
4	AL3	689.3	79.3	0.69	0.0087	910
5	AL4	1287	80.7	0.54	0.007	1010
6	AL1G1	1796.3	88.1	0.97	0.0055	360
7	AL1G2	2408	120.2	0.61	0.0043	70
8	AL2G1	4500	218	0.096	0.004	440
9	AL2G2	5181	360.6	0.08	0.0038	130
10	AL3G1	3192	276.4	0.02	0.0034	510
11	AL3G2	6842	409.2	0.016	0.0023	210
12	AL4G1	4137.5	128.5	0.0011	0.0049	650
13	AL4G2	6435	180	0.0007	0.0023	370

Table 4. Cont.

Strategy No.	Treatment Strategy	"j" Response Measures				
		UCS (kPa)	CBR (%)	HC (cm/s) $\times 10^{-4}$	Cc (kPa)	LL (ppm)
14	B	391.6	43.7	4.96	0.0156	270
15	BL1	765.2	47.4	3.81	0.0128	390
16	BL2	905	57.9	1.01	0.0129	480
17	BL3	794	87.1	2.01	0.0136	610
18	BL4	1193	72.5	0.91	0.0109	770
19	BL1G1	1112	80.7	0.97	0.0086	260
20	BL1G2	1204.7	117	0.89	0.0066	80
21	BL2G1	2204.3	211.2	0.24	0.0063	270
22	BL2G2	3496.3	320.7	0.1	0.0059	110
23	BL3G1	1950	145.5	0.067	0.0052	330
24	BL3G2	3790.4	238.9	0.028	0.0036	160
25	BL4G1	2642	188.3	0.0021	0.0076	440
26	BL4G2	4987.3	319.2	0.0001	0.0036	180

(a)

(b)

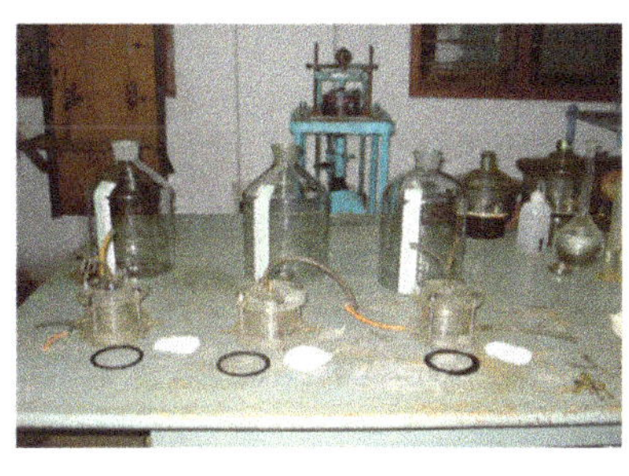

(c)

(d)

Figure 1. Details of the experimental test setup and specimen testing.; (**a**) Unconfined compression strength samples subjected to desiccator curing; (**b**) California bearing ratio test in progress; (**c**) Hydraulic conductivity test setup; (**d**) Consolidation test setup.

3. Experimental Results

The following sections address the fundamental governing mechanism responsible for the enhancement of the targeted properties and the PROMETHEE strategy adopted to choose the ideal mix of fly ash, lime, and gypsum for the selected fly ashes. Pozzolanic reactions of lime (as $Ca(OH)_2$) resulted in the formation of gels, followed by crystallization, namely of calcium silicates, aluminates, and alumina silicates. Precipitation and even distribution of hydration products ($Ca(OH)_2$, C-S-H, etc.) on the surfaces of fly ash contributed to the additional increase in the strength of the stabilized mix. The rate of formation of stabilized compounds, which increases with the curing period, resulted in the reduction of the LL ratio, defined as the ratio of lime leached to the total lime added. It has been demonstrated that the LL decreases with an increase in the lime content. Sulphate ions from gypsum reacted with the alumina phase of fly ash to produce (x CaO. y Al_2O_3. z $CaSO_4$. w H_2O), which enhanced the pozzolanic activity of lime-stabilized fly ashes with higher strength (refer to Table 4). Furthermore, the formation of C-S-H and C-A-S-H gels reduced the LL of fly ashes and increased the unconfined compressive strength and California bearing ratio behavior, as seen in Table 4. Additionally, the resultant denser matrix significantly reduced the hydraulic conductivity values and compression index values (refer to Table 4).

4. MCDM Model: Based on PROMETHEE and GAIA

For the application of the proposed model, the first step was to develop alternative strategies to enhance various geotechnical properties of fly ash. Initially, the untreated fly ashes were adopted for the geotechnical application and were subsequently treated with lime (1%, 2.5%, 5%, and 10%) and a variation in gypsum dosage (1% and 2.5%) in the presence of lime. Thus, a total of 26 ($2 \times 1 + 2 \times 4 + 2 \times 4 \times 2$) alternative strategies were adopted in the present study. For each alternative strategy, three samples were used, and for each sample, five performance measures were obtained. Thus, 390 performance results were obtained and are reported in Table 4. From the results, it is evident that no individual experiment was superior in terms of all five performance measures. The adopted MCDM model was based on PROMETHEE and geometrical analysis for interactive aid (GAIA), proposed by Brans and Mareschal [45]. PROMETHEE was adopted for the partial and complete ranking of the selected alternative treatment strategies; GAIA was used for the sensitivity and comparative analysis of experimental strategies.

4.1. PROMETHEE Partial Ranking

T_a and T_b denote the treatment strategies used to enhance various geotechnical properties for the given fly ash. The effectiveness of the treatment strategies was measured based on multiple performance measures, such as j ϵ performance measures. R_{aj} and R_{bj} are the performance measures for treatment strategies T_a and T_b, respectively. In order to set a partial preference, a generalized preference function P_{abj} was defined for performance measure j, when alternative treatment strategies T_a and T_b were compared; P_{abj} fluctuates between zero and one (refer to Figure 2). The value of P_{abj} can be interpreted as follows:

- $P_{abj} \approx$ one, indicates acceptable preference for strategy T_a over strategy T_b.
- $P_{abj} \approx$ zero, indicates a weak preference for strategy T_a over strategy T_b.
- P_{abj} = zero, indicates no preference for strategy T_a over strategy T_b for a given j performance measure.
- P_{abj} = one, indicates a higher preference for strategy T_a over T_b.

Preference functions (P_{abj}) were adopted [45] based on the response measures (R_{aj} and R_{bj}) and the permissible boundary to derive a full incompatibility (θ) and a full preference ($\varnothing$). The permissible boundary values θ and $\varnothing$ were set by the decision-maker.

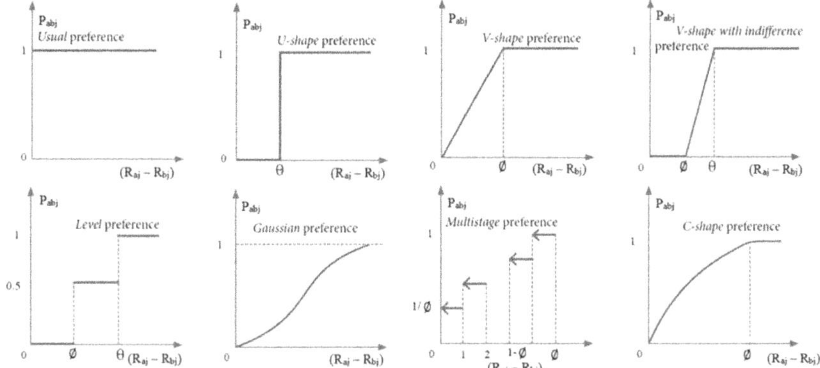

Figure 2. Preference Functions P_{abj}, based on R_{aj}, R_{bj}, θ, and $\varnothing$.

Subsequently, considering all "k" performance measures (j = 1 to k), the preference function value P_{ab} for a pair of treatment strategies T_a and T_b was computed using Equation (1). In Equation (1), W_j is the weight assigned to each performance measure j; P_{abj} is a preference function allocated to a pair of treatment strategies T_a and T_b for each j. P_{ab} fluctuates between zero and one.

$$P_{ab} = \frac{\sum_{j=1}^{k}(W_j \times P_{abj})}{\sum_{j=1}^{k} W_j} \tag{1}$$

However, in practice, more than two treatment strategies exist to enhance various geotechnical properties for the given coal-based fly ash. In this study, 26 treatment strategies were analyzed to find the superiority of a treatment strategy over the others; this is set by estimating two outranking flows (F_+ and F_-), expressed in Equations (2) and (3). Each outranking flow does not result in the same rank for each treatment strategy. Thus, their (F_+ and F_-) intersection is induced to estimate partial ranking.

$$F_{+a} = \frac{1}{n-1}\sum_{1}^{n} P_{ab}\ (b \neq a) \tag{2}$$

$$F_{-a} = \frac{1}{n-1}\sum_{1}^{n} P_{ba}\ (b \neq a) \tag{3}$$

Contrary to P_{ab}, P_{ba} is a preference function to estimate the dominance of treatment strategy T_b over T_a, and n is the number of treatment strategies. For "n" given treatment strategies, each is compared to n-1 other treatment strategies. Thus, the objective in partial ranking is to maximize outranking flows (F_+) and to minimize outranking flows (F_-); these outranking flows express how T_a outranks the other n-1 treatment strategies. By determining positive (F_{+a} and F_{+b}) and negative (F_{-a} and F_{-b}) outranking flows for a pair of treatment strategies T_a and T_b, the dominance relationship can be inferred as shown below:

- If $F_{+a} > F_{+b}$, the outranking relationship is D_{+ab} (i.e., treatment strategy T_a is dominating T_b).
- If $F_{-a} < F_{-b}$, the outranking relationship is D_{-ab} (i.e., treatment strategy T_a is not dominating T_b).
- If $F_{+a} = F_{+b}$, the outranking relationship is ED_{+ab} (i.e., both treatment strategies T_a and T_b equally dominate (ED) the other "n − 2" treatment strategies).
- If $F_{-a} = F_{-b}$, the outranking relationship is ED_{-ab} (i.e., both treatment strategies T_a and T_b are equally dominated (ED) by the other "n − 2" treatment strategies).

Based on the above conditions, three PROMETHEE-based partial rankings were computed as follows to establish the preference relationship between T_a and T_b:

- If (D_{+ab} and D_{-ab})/(D_{+ab} and ED_{-ab})/(ED_{+ab} and D_{-ab}) are true, it indicates that treatment strategy T_a has a higher preference over T_b.
- If ED_{+ab} and ED_{-ab} are true, it indicates that treatment strategy T_a is no different to T_b.
- If (D_{+ab} and D_{-ba})/(D_{+ba} and D_{-ab}) are true, it indicates that treatment strategy T_a is in contrast to T_b. Specifically, on a set of geotechnical performance measures, T_a exhibits the best performance over T_b, which exhibits a low response and vice versa.

In order to determine the best treatment strategy, PROMETHEE partial ranking was extended to PROMETHEE complete ranking.

4.2. PROMETHEE Complete Ranking

For complete ranking, outranking flow F_a is set as the balance between two outranking flows F_+ and F_-, in reference to Equations (2) and (3); F_a is computed using Equation (4). Using PROMETHEE complete ranking, all treatment strategies are comparable.

$$\left\{ \begin{array}{l} F_a = F_{+a} - F_{-a} \\ F_b = F_{+b} - F_{-b} \end{array} \right\} \tag{4}$$

Based on the outranking flows F_a and F_b (in reference to Equation (4)), PROMETHEE complete ranking is established between T_a and T_b as follows:

- If $F_a > F_b$ is true, it indicates that treatment strategy T_a is preferred over treatment strategy T_b.
- If $F_a = F_b$ is true, there is no difference between treatment strategies T_a and T_b.
- If $F_a < F_b$ is true, it indicates that there is no preference of treatment strategy T_a over treatment strategy T_b.

5. Application of PROMETHEE

From above Table 4, it is evident that no treatment strategy exhibits superiority in terms of all five performance measures. Using PROMETHEE, the outranking flows (F_+, F_-, and F) were obtained for each treatment strategy, for two different types of applications (i.e., liner and pavement applications), as presented in the following Table 5. Using F_+, F_-, and F, the treatment strategies ranking and superiority of each treatment strategy are presented in Figures 3–6.

Table 5. Treatment strategies and their ranking based on outranking flows.

Rank	Treatment Strategy Due to Factor Combination	Outranking Flow			Rank	Treatment Strategy Due to Factor Combination	Outranking Flow		
		F	F+	F-			F	F+	F-
1	AL3G2	0.5045	0.5525	0.0480	14	AL4G1	0.0242	0.3123	0.2880
2	AL2G2	0.5044	0.5524	0.0480	15	AL1G1	−0.0118	0.2922	0.3041
3	BL4G2	0.4264	0.5124	0.0860	16	BL1G1	−0.0506	0.2697	0.3203
4	BL2G2	0.4221	0.5102	0.0881	17	BL3	−0.2287	0.1840	0.4127
5	BL3G2	0.3604	0.4804	0.1200	18	AL4	−0.2904	0.1498	0.4401
6	AL4G2	0.2485	0.4245	0.1760	19	BL2	−0.2910	0.1520	0.4430
7	AL1G2	0.2483	0.4243	0.1760	20	BL4	-0.2964	0.1521	0.4484
8	AL2G1	0.2083	0.4004	0.1920	21	B	−0.3209	0.1360	0.4569
9	BL2G1	0.1761	0.3842	0.2081	22	BL1	−0.3286	0.1360	0.4646
10	AL3G1	0.1364	0.3684	0.2320	23	AL1	−0.4137	0.0905	0.5042
11	BL1G2	0.1361	0.3682	0.2321	24	AL3	−0.4194	0.0881	0.5075
12	BL4G1	0.0960	0.3441	0.2482	25	AL2	−0.4561	0.0721	0.5282
13	BL3G1	0.0882	0.3443	0.2561	26	A	−0.4723	0.0641	0.5364

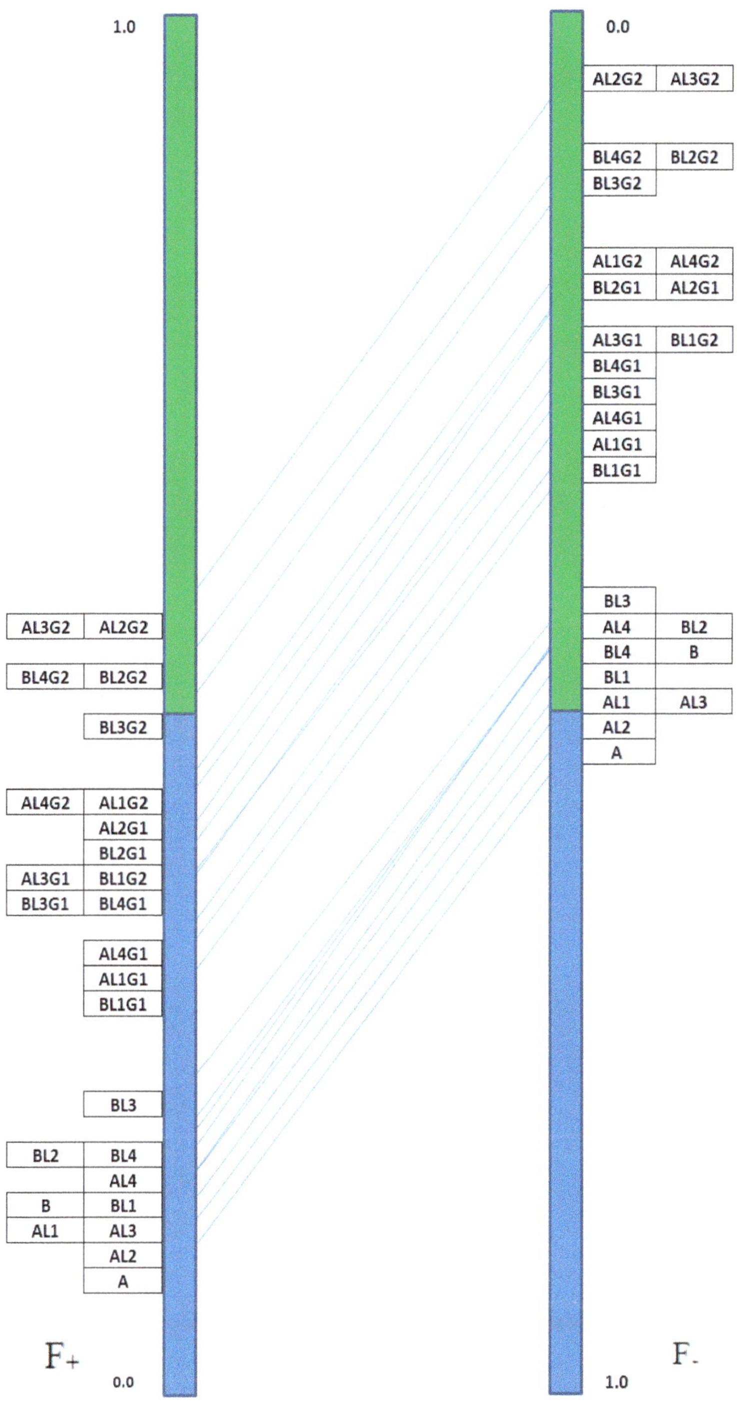

Figure 3. Treatment strategies (partial ranking) based on F_+ and F_-.

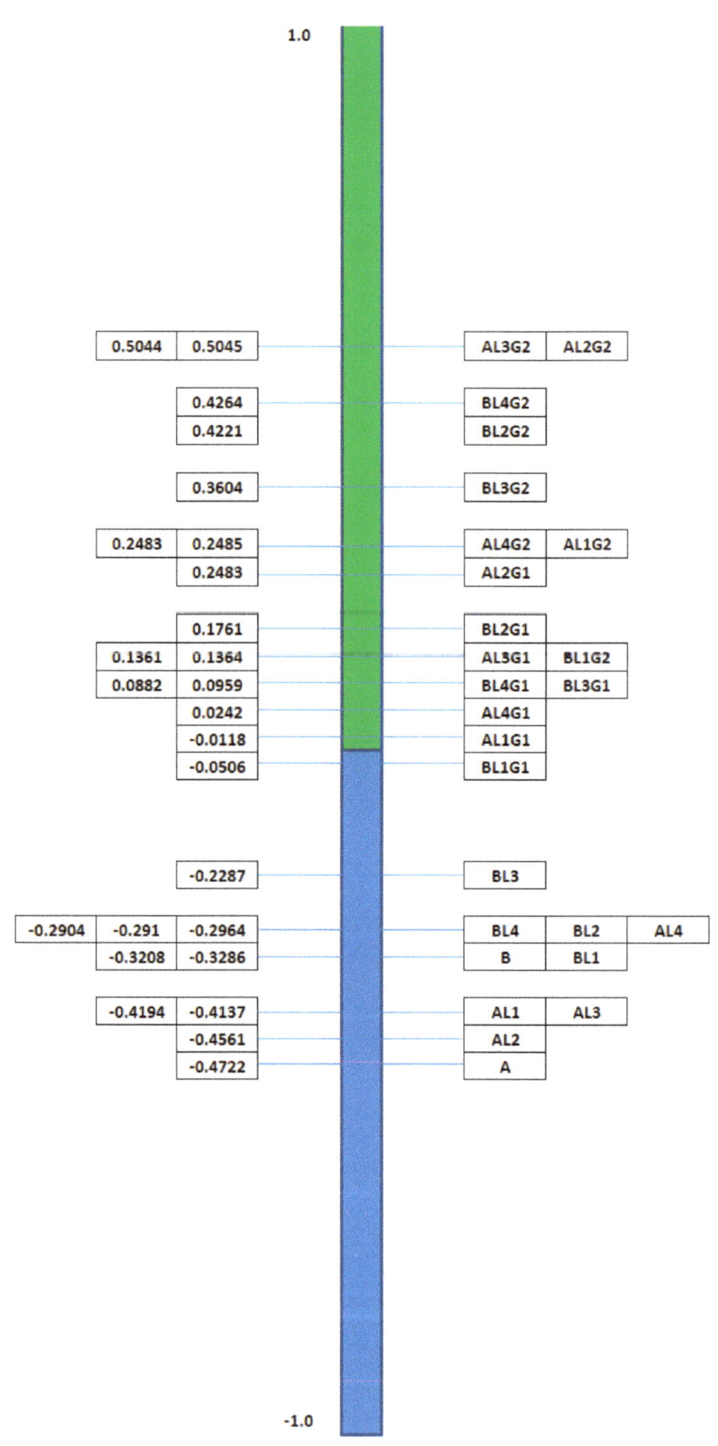

Figure 4. Treatment strategies (final ranking) based on flow F.

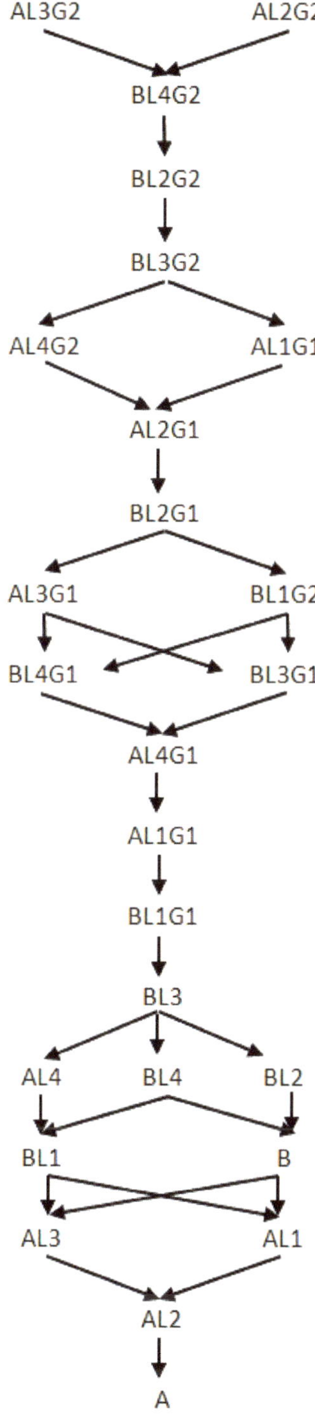

Figure 5. Preference network diagram based on the PROMETHEE approach.

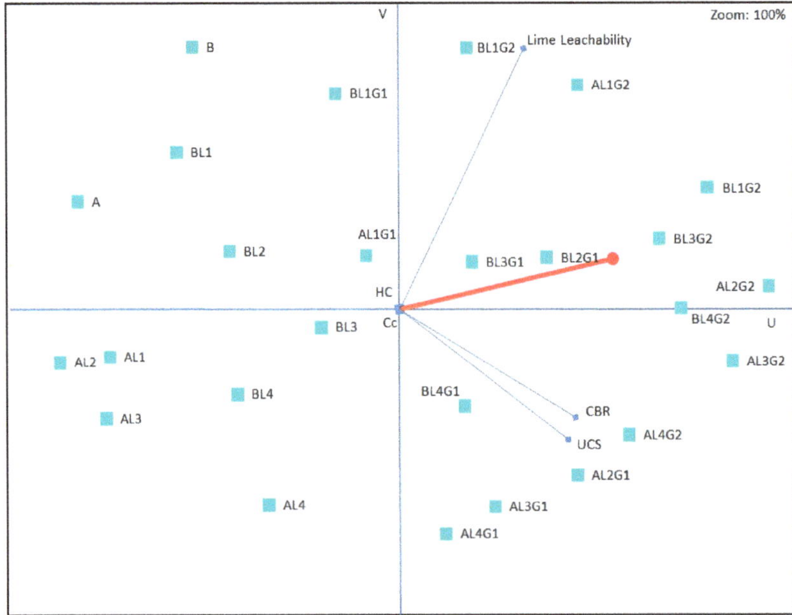

Figure 6. Geometrical analysis based on the interactive aid (GAIA) plane.

In Figure 3, the left-hand column corresponds to the F_+ score and the right-hand column to the F_- score for each treatment strategy. These scores are oriented such that the best are projected upwards. For each treatment strategy, a representative line is drawn from its F_+ to the corresponding F_- score. For any given two treatment strategies, if the representative lines are parallel, the treatment strategy representing the top line is preferred. However, if the two lines intersect, the corresponding treatment strategies are incomparable.

In Figure 3, treatment AL2G2 dominates the other treatments and corresponds to fly ash type A treated with lime (2.5%) and gypsum (2.5%) (Tables 2 and 4). It also reveals that pure fly ash type A underperformed compared to other possible treatments. Similarly, both fly ashes with lime treatment alone underperformed compared to alternative treatments. Generally, the outranking scores (F_+ and F_-) induce two different complete rankings. In order to circumvent this scenario, a complete ranking based on net flow "F" was obtained, as shown in Figure 4. The top half corresponds to F_+ and the bottom half to F_- scores for each treatment.

Similarly, in the above Figure 4, it is evident that AL2G2 and AL3G2 (for example, as calculation case for AL3G2 refer in the Appendix A, to illustrate how to obtain the value of "F") superseded the other treatment strategies, while A and AL2 were the least preferred. Simultaneously, using F_+, F_-, and F, a preference network is drawn (refer to Figure 5) where the treatment strategy and its choice over other treatment strategies are denoted by the arrow →. Figure 5 also shows that treatment strategy AL2G2 was preferred over other treatment strategies, with A and AL2, once again, being least-preferred. It is interesting to note that treatment strategy B was preferred over A, AL1, AL2, and AL3. This is due to the fundamental differences in chemical and mineralogical compositions between the two fly ashes (refer to Table 3). The visual PROMETHEE and GAIA plane represents the conflicts between response measures and highlights the group of treatment strategies of remarkable performance (refer to Figure 6).

The GAIA plane is considered to be a geometrical interactive tool, used to assist decision-makers with sensitivity analyses. Treatment strategies are represented by blue squares in Figure 6, and treatment strategies with no difference appear close, while conflicting strategies are placed in different quadrants.

The various geotechnical properties of fly ash measures, showing equal preference, lean in the same direction in the GAIA plane while conflicting response measures lean in the opposite direction. From Figure 6, it is clear that the fly ash without any additives or with only a lime dosage treatment are scored opposite to the fly ash with both a lime and gypsum treatment. Fly ash type A, treated with a lime dosage of 10% and gypsum dosage of 1%, scored better for response measures UCS and CBR against fly ash type B, which was also treated with a lime dosage of 10% but a gypsum dosage of 2.5%. Similarly, for the lime-leachability response measure, both fly ashes scored better when treated with a 1% lime and 2.5% gypsum dosage. However, for the HC and Cc response measures, only fly ash type A with a 1% lime and gypsum dosage scored better.

Sensitivity analysis was adopted to assign weights to each response measure. To perform sensitivity analyses on the experimental data (refer to Table 4), the Visual PROMETHEE-GAIA walking weights tool was used.

Sensitivity Analysis Using Walking Weights

Sensitivity analysis using walking weights was performed by assigning a set of weights to each response measure, as shown in Table 6. Set 1 represents an equal weight allocation for all response measures (Figure 7). Because fly ashes have the most potential to be used in liners and pavement material applications, a sensitivity analysis was carried out for these two specific applications by assigning due weights. The rationale behind weight allocation is application-specific, as seen in Table 6. For a liner material, the driving factor is "Hydraulic conductivity," and it was assigned a higher weight compared to the other response measures (i.e., UCS, CBR, Cc and LL). Similarly, for the pavement application, the significant factor is the "California bearing ratio," and it was duly assigned higher weightage over the other response measures.

Table 6. Criterion weights allocated for sensitivity analysis.

Criterion Weight		UCS (%)	CBR (%)	HC (%)	Cc (%)	LL (%)
	Objective	Max	Max	Min	Min	Min
For liner applications	Set 1 *	0.2	0.2	0.2	0.2	0.2
	Set 2	0.3	0.0	0.4	0.1	0.2
	Objective	Max	Max	Max	Min	Min
For pavement application	Set 1 *	0.2	0.2	0.2	0.2	0.2
	Set 2	0.2	0.4	0.2	0.05	0.15

Note: * Figures 3–6 represent analyses executed, based on Set 1 weight allocation.

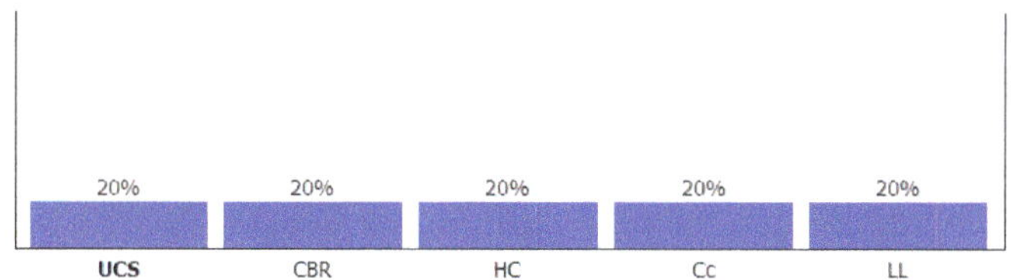

Figure 7. The equal weight allocation for all response measures.

The sensitivity analysis was done relying on the assigned walking weights above (Set 2), using Visual PROMETHEE. The outcome of the sensitivity analysis for liner and pavement applications is presented in the following Figures 8 and 9, respectively. Figure 8a–e represents the analysis corresponding to the ranking of treatment strategies, superiority of each treatment strategy, GAIA plane, and walking weights, respectively, for liner application. Based on the objectives of the response measures, shown in the above Table 6, a maximum of 40% weighting was set to response measure HC and a minimum of 0% to response measure CBR for this set. Hydraulic conductivity was considered a driving prerequisite for liner applications. Therefore, fly ash type A, treated with a 2.5% dosage of lime and 2.5% dosage of gypsum, was preferred over the alternatives (refer to Table 2). With regard to the lime-leachability values, both of the selected fly ashes responded alike to 1% lime content spiked with 2.5% gypsum. At higher lime contents, readily soluble amorphous lime, which is in excess to optimum lime requirements, simply leaches out of the fly ash-lime-gypsum matrix [22]. However, mix AL3G2 showed the maximum possible gain in unconfined compressive strength behavior. Similarly, Figure 9a–e represents the analyses corresponding to the ranking of treatment strategies, the superiority of each treatment strategy, graphical interactive plane, and walking weights, respectively, for pavement applications. In the case of pavement application (refer to Table 6), a maximum of 40% weighting and a minimum of 5% weighting was assigned to the response measures CBR and Cc, respectively. Mix AL3G2 satisfied the requirements for these conditions (refer to Table 2).

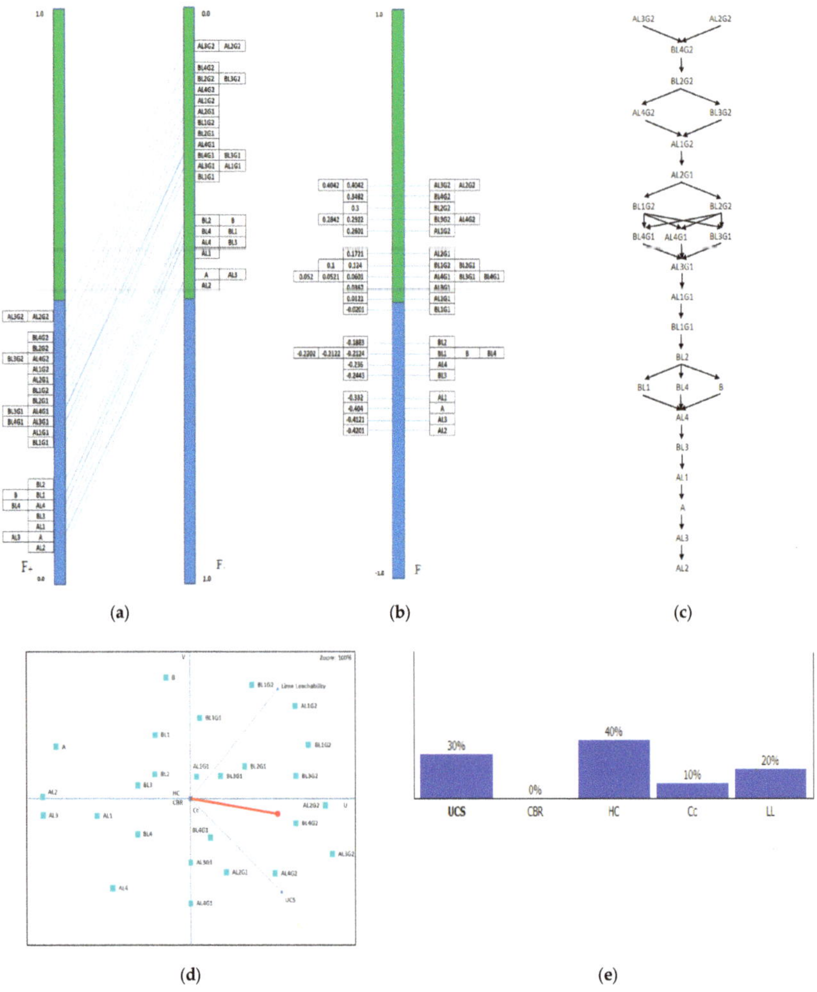

Figure 8. Response measure weight allocation for walking weight Set 2 for the liner application of fly ash: (**a**) Partial ranking of treatment strategies based on F$_+$ and F$_-$; (**b**) Complete ranking of treatment strategies based on outranking flow F; (**c**) Preference network diagram based on the PROMETHEE approach; (**d**) Geometrical analysis for the interactive aid (GAIA) plane; and (**e**) Weight allocation for all response measures (refer to Table 6).

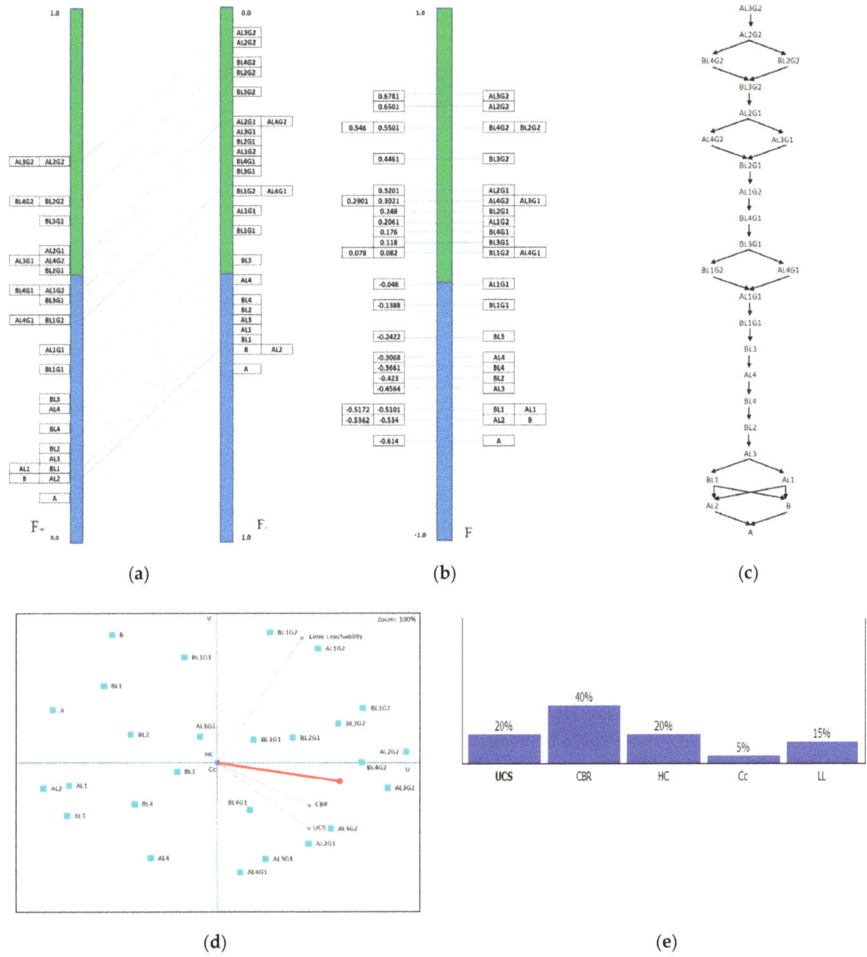

(a)　　　(b)　　　(c)

(d)　　　(e)

Figure 9. Response measure weight allocation for walking weight Set 2 for the pavement application of fly ash: (**a**) Partial ranking of treatment strategies based on F+ and F−; (**b**) Complete ranking of treatment strategies based on outranking flow F; (**c**) Preference network diagram based on the PROMETHEE approach; (**d**) Geometrical analysis for the interactive aid (GAIA) plane; and (**e**) Weight allocation for all response measures (refer to Table 6).

6. Conclusions

The present study investigated the effect of the addition of lime and/or gypsum on improving various geotechnical properties (Cc, HC, UCS, CBR, and LL) of Class F fly ashes. The objective was to assist practicing engineers in the selection of the most effective mix of fly ash type, lime, and gypsum dosage to satisfy various geotechnical properties for specific geotechnical applications. The MCDM model, based on PROMETHEE and GAIA, was adopted to select the best treatment strategy. Multiple treatment strategies were analyzed, based on variations in lime and gypsum dosages, and for each of these alternatives, response measures in the form of geotechnical properties were computed. This approach resulted in obtaining the treatment strategies ranking, superiority of each treatment strategy, and geometrical interactive plane. From these outcomes, it is evident that a group of treatment strategies exhibits similar preferences for given response measures. Thus, sensitivity analysis, based on walking weights, was executed to solve conflicts. In this study, untreated fly ash (A or B) and any fly ash type treated only with lime were found to be the least-preferred options. The proposed MCDM model, based on PROMETHEE and GAIA, works well to assist practicing engineers with identifying the fly ash type, with the appropriate mix of lime and gypsum. The following outcomes were established:

- For liner applications, fly ash type A, treated with a 2.5% dosage of lime and 2.5% dosage of gypsum, is preferred.
- For pavement applications, fly ash type A, treated with a 5% dosage of lime and a 2.5% dosage of gypsum, is preferred.
- When equal walking weights were assigned to all response measures, irrespective of the nature of application, fly ash type A, treated with either a 2.5% or 5% dosage of lime and a 2.5% dosage of gypsum, should be the first choice.

In the eventuality of having fly ash B on-site, it should be treated with a 10% lime and 2.5% gypsum dosage for liner applications. Furthermore, treating it with a 2.5% lime and 2.5% gypsum dosage would meet the pavement application requirements.

Author Contributions: Conceptualization, A.A.B.M. and A.U.R.; methodology, A.A.B.M., A.U.R., and U.U.; formal analysis, A.A.B.M., A.U.R., K.V.V., and U.U.; investigation A.A.B.M., A.U.R., and K.V.V.; resources, A.A.B.M. and A.U.R.; writing—original draft preparation, A.A.B.M. and A.U.R.; writing—review and editing, A.A.B.M., A.U.R., K.V.V., and U.U.; supervision, A.A.B.M. and A.U.R.; funding acquisition, A.U.R. All authors have read and agreed to the published version of the manuscript.

Funding: Deanship of Scientific Research at King Saud University grant number-RG-1439-005.

Acknowledgments: The authors extend their appreciation to the Deanship of Scientific Research at King Saud University for funding this work through research group number RG-1439-005.

Conflicts of Interest: The authors declare no conflict of interest.

Appendix A. Example Calculation for AL3G2

- R_{aj} and R_{bj} are the performance measures for treatment strategies T_a and T_b, respectively;
- W_j is the weight assigned to each performance measure j; and
- P_{abj} is a preference function allocated to a pair of treatment strategies T_a and T_b for each j.

Thus, T_{11} (AL3G2) and T_1 (A) denote the treatment strategies 11 and 1, respectively (refer Table 4); and there are five performance measures (i.e., $j = 5$). Thus, preference function value $P_{11,1}$ for a pair of treatment strategies T_{11} and T_1 is computed using Equation (1).

From Table 4, it is observed that for treatment strategies T_{11} and T_1:

- $R_{11\,1} = 6842$ and $R_{1\,1} = 173.1$ are the performance measures for $j = 1$;
- $R_{11\,2} = 409.2$ and $R_{1\,2} = 55.5$ are the performance measures for $j = 2$;
- $R_{11\,3} = 0.016 \times 10^{-4}$ and $R_{1\,3} = 6.52 \times 10^{-4}$ are the performance measures for $j = 3$;
- $R_{11\,4} = 0.0023$ and $R_{1\,4} = 0.0099$ are the performance measures for $j = 4$;
- $R_{11\,5} = 210$ and $R_{1\,5} = 630$ are the performance measures for $j = 5$, respectively.

V- shape preference function is allocated to the pair of treatment strategies T_{11} and T_1 for each j; and $W_j = 0.2$ equal weight assigned to each j.

The value of $P_{11\,1j}$ can be interpreted as: $P_{11\,1\,1} =$ one; $P_{11\,1\,2} =$ one; $P_{11\,1\,3} =$ zero; $P_{11\,1\,4} \approx$ one $= 0.9905$; and $P_{11\,1\,5} =$ one. Thus, preference function value $P_{11\,1}$ for a pair of treatment strategies T_{11} and T_1 is computed and is equal to 0.7981.

Similarly, $P_{11\,2}$ to $P_{11\,26}$ is used to estimate outranking flow F_+, expressed in Equation (2). Thus, the values are estimated as:

$P_{11\,2} = 0.7464$; $P_{11\,3} = 0.7462$; $P_{11\,4} = 7466$; $P_{11\,5} = 7461$; $P_{11\,6} = 0.7464$; $P_{11\,7} = 0.5461$; $P_{11\,8} = 0.6764$; $P_{11\,9} = 0.3972$; $P_{11\,10} = 0.7548$; $P_{11\,12} = 0.7558$; $P_{11\,13} = 0.5271$; $P_{11\,14} = 0.5279$; $P_{11\,15} = 0.5276$; $P_{11\,16} = 0.5275$; $P_{11\,17} = 0.5285$; $P_{11\,18} = 0.5305$; $P_{11\,19} = 0.3559$; $P_{11\,20} = 0.2462$; $P_{11\,21} = 0.3449$; $P_{11\,22} = 0.2465$; $P_{11\,23} = 0.2469$; $P_{11\,24} = 0.2465$; $P_{11\,25} = 0.7469$; and $P_{11\,26} = 0.5495$.

Therefore, the sum of all P_{ab} (i.e., $P_{11\,1}$ to $P_{11\,26}$) for n = 26 is equal to 13.8125. Thus, the outranking flow F_+, expressed using Equation (2): $F_+ = (1/(26 - 1)) \times 13.8125 = 0.5525$ for treatment strategy T_{11} (AL3G2).

Similarly, $P_{1\,11}$ to $P_{26,\,11}$ estimated to estimate two outranking flow F_-, expressed in Equation (3). The sum of all P_{ba} (i.e., $P_{1\,11}$ to $P_{26\,11}$) for n = 26 is equal to 1.2. Thus, the outranking flow F_- expressed as $F_- = (1/(26 - 1)) \times 1.2 = 0.048$ for treatment strategy T_{11}.

The complete outranking flow F_{11} is set as the balance between two outranking flows F_+ and F_-, in reference to Equations (2) and (3); F_{11} is computed using Equation (4).

Thus, for treatment strategy T_{11}, F_1 is equal to $(0.5525 - 0.048) = 0.5045$ and this value can be seen in Figure 3 assigned to AL3G2.

References

1. British Petroleum. *Statistical Review of World Energy*; Workbook, BP: London, UK, 2018.
2. Cherian, C.; Siddiqua, S. Pulp and Paper Mill Fly Ash: A Review. *Sustainability* **2019**, *11*, 4394. [CrossRef]
3. Ashfaq, M.; Heeralal, M.; Moghal, A.A.B. Characterization studies on coal gangue for sustainable geotechnics. *Innov. Infrastruct. Solut.* **2020**, *5*, 15. [CrossRef]
4. Mehta, P.K. Influence of fly ash characteristics on the strength of portland-fly ash mixtures. *Cem. Concr. Res.* **1985**, *15*, 669–674. [CrossRef]
5. Singh, S.R.; Panda, A.P. Utilization of fly ash in geotechnical construction. In Proceedings of the Indian Geotechnical Conference, Chennai, India, 14–16 December 1996; pp. 547–550.
6. Palmer, B.G.; Edil, T.B.; Benson, C.H. Liners for waste containment constructed with class F and C fly ashes. *J. Hazard. Mater.* **2000**, *76*, 193–216. [CrossRef]
7. Antiohos, S.; Tsimas, S. Activation of fly ash cementitious systems in the presence of quicklime: Part, I. Compressive strength and pozzolanic reaction rate. *Cem. Concr. Res.* **2004**, *34*, 769–779. [CrossRef]
8. Kim, B.; Prezzi, M.; Salgado, R. Geotechnical Properties of Fly and Bottom Ash Mixtures for Use in Highway Embankments. *J. Geotech. Geoenviron. Eng.* **2005**, *131*, 914–924. [CrossRef]
9. Kumar, A.; Walia, B.S.; Mohan, J. Compressive strength of fiber reinforced highly compressible clay. *Constr. Build. Mater.* **2006**, *20*, 1063–1068. [CrossRef]
10. Sekhar, M.R.; Madhav, M.R.; Puppala, A.J.; Ghosh, A. Compressibility and Collapsibility Characteristics of Sedimented Fly Ash Beds. *J. Mater. Civ. Eng.* **2008**, *20*, 401–409. [CrossRef]
11. Sivapullaiah, P.V.; Baig, M.A.A. Gypsum treated fly ash as a liner for waste disposal facilities. *Waste Manag.* **2011**, *31*, 359–369. [CrossRef] [PubMed]
12. Moghal, A.A.B.; Sivapullaiah, P.V. Characterization of Lime and Gypsum Amended Class F Fly Ashes as Liner Materials. *Geo Front.* **2011**, 1162–1171. [CrossRef]
13. Moghal, A.A.B. A state-of-the-art review on the role of fly ashes in geotechnical and geoenvironmental applications. *J. Mater. Civ. Eng.* **2017**, *29*, 04017072. [CrossRef]
14. Xu, P.; Zhao, Q.; Qiu, W.; Xue, Y.; Li, N. Microstructure and Strength of Alkali-Activated Bricks Containing Municipal Solid Waste Incineration (MSWI) Fly Ash Developed as Construction Materials. *Sustainability* **2019**, *11*, 1283. [CrossRef]
15. DiGioia, A.M.; Nuzzo, W.L. Fly Ash as Structural Fill. *J. Power Div.* **1972**, *98*, 77–92.
16. Edil, T.B.; Berthouex, P.M.; Vesperman, K.D. *Fly Ash as a Potential Waste Liner*; ASCE, Geotechnical Practice for Waste Disposal '87, University of Michigan: Ann Arbor, MI, USA, 15–17 June 1987; pp. 447–461.
17. Nhan, C.T.; Graydon, J.W.; Kirk, D.W. Utilizing coal fly ash as a landfill barrier material. *Waste Manag.* **1996**, *16*, 587–595. [CrossRef]
18. Mollamahmutoğlu, M.; Yilmaz, Y. Potential use of fly ash and bentonite mixture as liner or cover at waste disposal areas. *Environ. Earth Sci.* **2001**, *40*, 1316–1324. [CrossRef]
19. Rostami, H.; Brendley, W. Alkali Ash Material: A Novel Fly Ash-Based Cement. *Environ. Sci. Technol.* **2003**, *37*, 3454–3457. [CrossRef]
20. Zabielska-Adamska, K. Fly Ash as a Barrier Material. *Geo Front.* **2012**, 947–956. [CrossRef]
21. Moghal, A.A.B.; Sivapullaiah, P.V. Effect of Pozzolanic Reactivity on Compressibility Characteristics of Stabilised Low Lime Fly Ashes. *Geotech. Geol. Eng.* **2011**, *29*, 665–673. [CrossRef]
22. Moghal, A.; Sivapullaiah, P. Role of lime leachability on the geotechnical behavior of fly ashes. *Int. J. Geotech. Eng.* **2012**, *6*, 43–51. [CrossRef]

23. Sivapullaiah, P.V.; Moghal, A.A.B. CBR and strength behavior of class F fly ashes stabilized with lime and gypsum. *Int. J. Geotech. Eng.* **2011**, *5*, 121–130. [CrossRef]
24. Ghosh, A.; Subbarao, C. Strength Characteristics of Class F Fly Ash Modified with Lime and Gypsum. *J. Geotech. Geoenvironmental Eng.* **2007**, *133*, 757–766. [CrossRef]
25. Moghal, A.A.B. Geotechnical and Physico-Chemical Characterization of Low Lime Fly Ashes. *Adv. Mater. Sci. Eng.* **2013**, *2013*, 1–11. [CrossRef]
26. Sivapullaiah, P.V.; Moghal, A.A.B. Role of Gypsum in the Strength Development of Fly Ashes with Lime. *J. Mater. Civ. Eng.* **2011**, *23*, 197–206. [CrossRef]
27. Zavadskas, E.K.; Antuchevičienė, J.; Kapliński, O. Multi-criteria decision making in civil engineering: Part I—A state-of-the-art survey. *Eng. Struct. Technol.* **2015**, *7*, 103–113. [CrossRef]
28. Zavadskas, E.K.; Antuchevičienė, J.; Kapliński, O. Multi-criteria decision making in civil engineering. Part II—Applications. *Eng. Struct. Technol.* **2015**, *7*, 151–167. [CrossRef]
29. Rousta, B.A.; Araghinejad, S. Development of a Multi Criteria Decision Making Tool for a Water Resources Decision Support System. *Water Resour. Manag.* **2015**, *29*, 5713–5727. [CrossRef]
30. Haralambopoulos, D.A.; Polatidis, H. Renewable energy projects: Structuring a multi-criteria group decision-making framework. *Renew Energy* **2003**, *28*, 961–973. [CrossRef]
31. Fiorucci, P.; Minciardi, R.; Robba, M.; Sacile, R. Solid waste management in urban areas: Development and application of a decision support system. *Resour. Conserv. Recycl.* **2003**, *37*, 301–328. [CrossRef]
32. Vego, G.; Kučar-Dragičević, S.; Koprivanac, N. Application of multi-criteria decision-making on strategic municipal solid waste management in Dalmatia, Croatia. *Waste Manag.* **2008**, *28*, 2192–2201. [CrossRef]
33. Mardani, A.; Zavadskas, E.K.; Khalifah, Z.; Jusoh, A.; Md Nor, K. Multiple criteria decision-making techniques in transportation systems: A systematic review of the state of the art literature. *Transport* **2016**, *31*, 359–385. [CrossRef]
34. Malindu, S.; Chamila, G.; David, L.; Guomin, Z.; Sujeeva, S.; Dennis, W. Sustainable criterion selection framework for green building materials—An optimisation based study of fly-ash Geopolymer concrete. *Sustain. Mater. Technol.* **2020**, *25*, e00178. [CrossRef]
35. Xu, J.; Deng, Y.; Shi, Y.; Huang, Y. A bi-level optimization approach for sustainable development and carbon emissions reduction towards construction materials industry: A case study from China. *Sustain. Cities Soc.* **2020**, *53*, 101828. [CrossRef]
36. Behzadian, M.; Kazemzadeh, R.B.; Albadvi, A.; Aghdasi, M. PROMETHEE: A comprehensive literature review on methodologies and applications. *Eur. J. Oper. Res.* **2010**, 198–215. [CrossRef]
37. Rehman, A.U.; Moghal, A.A.B. The Influence & Optimisation of Treatment Strategy in Enhancing Semi-Arid Soil Geotechnical Properties. *Arab. J. Sci. Eng.* **2018**, *43*, 5129–5141. [CrossRef]
38. Moghal, A.A.B.; Rehman, A.U.; Chittoori, B. Optimizing Fiber Parameters Coupled with Chemical Treatment: Promethee Approach. In Proceedings of the Geotechnical Frontiers 2017: Geotechnical Materials, Modeling and Testing: Selected Papers from Sessions of Geotechnical Frontiers, Orlando, FL, USA, 12–15 March 2017; pp. 30–41. [CrossRef]
39. ASTM D2166 A. *Standard Test Method for Unconfined Compressive Strength of Cohesive Soil*; ASTM: West Conshohocken, PA, USA, 2016.
40. *Indian Standards IS 2720-16: Methods of Test for Soils, Part 16: Laboratory Determination of CBR*; Bureau of Indian Standards: New Delhi, India, 1987.
41. D1883 A. *Standard Test Method for California Bearing Ratio (CBR) of Laboratory-Compacted Soils*; ASTM: West Conshohocken, PA, USA, 2016.
42. D5856 A. *Standard Test Method for Measurement of Hydraulic Conductivity of Porous Material Using a Rigid-Wall, Compaction-Mold Permeameter*; ASTM International: West Conshohocken, PA, USA, 2015.
43. *Indian Standards IS 2720-15: Method of Test for Soils–Determination of Consolidation Properties*; Bureau of Indian Standards: New Delhi, India, 1986.
44. *ASTM D2435 A Standard, Standard Test Methods for One-Dimensional Consolidation Properties of Soils Using Incremental Loading*; ASTM International: West Conshohocken, PA, USA, 2004.
45. Brans, J.P.; Mareschal, B. Promethee Methods. In *Multiple Criteria Decision Analysis: State of the Art Surveys*; Figueira, J., Greco, S., Ehrogott, M., Eds.; Springer: New York, NY, USA, 2005; Volume 200, pp. 163–186.

sustainability

Article

Effect of Fiber and Cement Additives on the Small-Strain Stiffness Behavior of Toyoura Sand

Muhammad Safdar [1,*], Tim Newson [2], Colin Schmidt [3], Kenichi Sato [4], Takuro Fujikawa [4] and Faheem Shah [1]

[1] Earthquake Engineering Center, University of Engineering and Technology Peshawar, Peshawar 25000, Pakistan; fshah_vfciv@uetpeshawar.edu.pk

[2] Department of Civil and Environmental Engineering, Western University, London, ON N6A 3K7, Canada; tnewson@eng.uwo.ca

[3] Thurber Engineering Ltd. 180, 7330 Fisher Street SE, Calgary, AB T2H 2H8, Canada; schmidt.colin5@gmail.com

[4] Department of Civil Engineering, Fukuoka University, 8-19-1 Nanakuma, Jonan, Fukuoka 814-0180, Japan; sato@fukuoka-u.ac.jp (K.S.); takuro-f@fukuoka-u.ac.jp (T.F.)

* Correspondence: drsafdar@uetpeshawar.edu.pk

Received: 23 November 2020; Accepted: 10 December 2020; Published: 15 December 2020

Abstract: The disposal of 2011 Japan earthquake waste has become an important issue in Japan and it is not realistic or economical to send all of these wastes to landfill sites, due to limited space, high costs, and related environmental issues. In sustainable geotechnical applications, mixing of the separated soils from disaster wastes with additives (e.g., cement and fiber) is required to improve their strength and stiffness characteristics. In this study, monotonic triaxial drained compression tests are performed on medium dense specimens of Toyoura sand-cement-fiber mixtures with different percentages of fiber and cement (e.g., 0–3%) additives. The experimental results indicate that behavior of the mixtures is significantly affected by the concentration of fiber and cement additives. Based on a comprehensive set of test results, modifications to the series of equations were developed that can be used to evaluate the shear modulus and mobilized stress curves at small-strain levels. The experimental results and model comparison show that the elastic threshold strain (γ_e), reference strain (γ_r), increases with fiber and cement additives. In addition, the range of curvature parameter, from 0.88 to 1.0, provides a good comparison with the results of small-strain measurements. Overall, the comparison of the results and model shows that the small-strain measurements obtained using local strain transducers fall within the range of model upper and lower bound curves. The results of the unreinforced, fiber, and cemented sand shows a close agreement with the model mean curve, but fiber-reinforced cemented sand shows a good comparison with model upper bound.

Keywords: small-strain stiffness; ground improvement; ground remediation; local strain; triaxial test

---

1. Introduction

The Great East Japan earthquake of 2011 generated a huge quantity of disaster waste and tsunami deposits, which required proper treatment and disposal. To effectively use these waste soils in sustainable geotechnical infrastructures, it is essential to understand the mechanical behavior in their native (pure) or mechanically stabilized form (amended with cement and fiber). The small-strain stiffness of soil plays an important role in the sustainability of many geotechnical problems, such as machine foundations, earthquake ground response analysis, and liquefaction potential evaluations [1–3]. Several techniques have been developed in geotechnical engineering for measuring small-strain stiffness, including resonant columns [4,5], piezoelectric transducers [6–11], and quasi-static loading with high resolution strain measurements [12–15]. A widely used method to measure small-strains is the Hall effect local strain transducer [16]. This type local strain transducer has been employed in various research studies [17–19] to estimate small-strain stiffness moduli. The stress-strain curves observed in the conventional triaxial system are subjected to many errors, especially at small strain range, when the deformations are measured externally. The most common errors observed are seating errors, alignment errors, bedding errors, system compliance, and end restraints. Many researchers [16,20–29] have developed various on-sample strain measuring devices to measure the strains accurately and to compute stiffness at small strain levels. Most of the sophisticated devices reported above are used to estimate small-strains for clean sand specimens.

Laboratory and field testing have shown that the stress-strain behavior of sands can be highly nonlinear, even at stresses well below the peak strength of the material. One of the first comprehensive studies where the parameters that control nonlinear soil behavior were identified was the study by Hardin and Drnevich [30,31]. The empirical equations proposed by Hardin and Drnevich [31] account for the effects of plasticity index, overconsolidation ratio, and confining pressure mainly through adjusting reference strain. The effect of soil type, number of loading cycles, loading frequency, and saturation, amongst other aspects, have also been taken into consideration [11,32,33]. Iwasaki et al. [34] and Kokusho [12] studied the impact of confining pressure, but these studies were limited to observations on clean narrow graded sands tested at low pressures.

Michalowski and Cermak [35] reported that the initial stiffness of a composite material (e.g., sand and fiber) was affected by the different characteristics of the steel and polyamide fibers (e.g., stiffness, roughness, rigidity, size, etc.). Previous research with mixtures of steel fibers and sand [36] indicated that even larger fiber concentrations (e.g., 1.25% by volume) had no adverse effect on the initial stiffness. In addition, steel fiber had a reinforcement effect only slightly higher than less stiff polyamide fiber of the same geometry. It was further concluded that this difference might be attributed to a larger interfacial friction angle of steel fibers compared to polyamide fibers. In addition, it was reported that the strain levels or mobilization resistance for steel fiber (e.g., stiff) is greater than that of polyamide fibers (e.g., flexible) due to their greater stiffness. The literature review on cemented sand shows that natural or artificial cementation increases the small-strain stiffness behavior (G_0) of sands [37–40]. Acar and El-Tahir [37] reported that shear modulus of cemented sands increased with confining stress in the applied range. Conversely, Sharma and Fahey [39] reported that small-strain stiffness (G_0) to be for cemented sands practically independent of the mean stress and dependent on cementation until it was reached a threshold stress corresponding to the onset of major structure degradation. Yun and Santamarina [41] indicated for artificially cemented soils an increase of G_0 with increasing stress after yielding and the values of G_0 remained higher than for the reconstituted soils. Cementation appears to control only G_0 of clays below isotropic or vertical yield stress and the pressure dependency appears to prevail at higher stresses. The latter findings lead to the conclusion that the stiffness of the cemented soils is strongly increased by cementation and independent of confining pressure [40]. Mair [42] and Xu et. al. [43] proposed that the stiffness of a soil is constant below a strain level of 0.001% (e.g., 10^{-5}) and reduces significantly with an increase in strain level (i.e., above 0.001%). In recent decades, researchers have attempted to validate the approximate relationship between stiffness and strain level by employing different instruments, as shown in Figure 1.

Fahey and Carter [44] proposed a hyperbolic model to characterize modulus reduction as a function of shear strength mobilization, as seen in Equation (1). This function requires maximum shear stress (τ_{max}) and an estimate of the small-strain shear modulus (G_0) to be the value of shear modulus at shear strain of 10^{-6} and is assumed to be constant below this value, as well as empirical parameters f and g. Fahey and Carter [44] showed some success fitting this three-parameter model to the data of a wide range of uncemented soils.

$$\frac{G}{G_0} = 1 - f\left(\frac{\tau}{\tau_{max}}\right)^g \tag{1}$$

Darendeli [32] proposed a modified hyperbolic model based on testing of intact sand-gravel sample:

$$\frac{G}{G_0} = \left[\frac{1}{(1 + \gamma/\gamma_r)^a} \right] \tag{2}$$

where a is called the curvature parameter, and γ_r is the reference strain value at which $G/G_0 = 0.50$. This model uses only two parameters, and the reference strain provides an efficient normalization of the shear strain. To better understand the non-linear elastic behavior of sands, and produce a generalized functional relationship, Oztoprak and Bolton [45] conducted a metastudy of the secant shear modulus degradation curves of 454 tests of uncemented sands from the literature. This curve-fitting process led to new interpretations and definitions that enable better predictions of the shear modulus degradation of sands with strain, based on soil classification data.

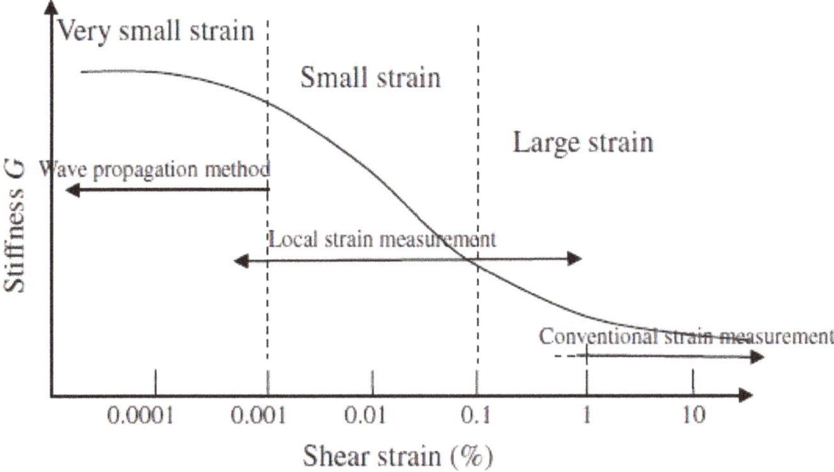

Figure 1. Typical modulus degradation measurement of soil stiffness in laboratory test (after Mair [42]).

In order to enhance the current database on small-strain stiffness behavior and stiffness degradation of amended soils, there is a need to further investigate the effect of fiber and cement additives on the small-strain stiffness (G_0) of sands. In addition, the aim of the work is to develop a rational method and propose few modifications to the series of equations that can be used to evaluate the shear modulus and mobilized stress curves at small-strain levels. Furthermore, the experimental results of amended soils are compared with the Oztoprak and Bolton [45] upper bound and lower bound stiffness degradation models. The modified version of hyperbolic equation for amended soils (e.g., fiber-only, cement-only, and fiber-reinforced cemented sands) leads to a wide range of values for elastic threshold strain (γ_e), the reference strain (γ_r), and the range of curvature parameter (a).

2. Materials and Methods

There has been ongoing long-term collaboration between Western University and Fukuoka University, Japan, with a view to improving those soils, utilizing waste streams and developing industry guidelines for construction. An in-depth coordinated laboratory program of the static and dynamic mechanical effects of various inclusions such as silt, different cementitious additives, and various types of fibers in Toyoura sand has been conducted at both universities over the last seven years. Initial studies on polyvinyl alcohol fiber (PVA) inclusions and Portland cement have been published [46–48] and results from the tests performed confirmed that the addition of polymer fibers and cement improved the liquefaction resistance, undrained shear strength, and stiffness of silty and unreinforced Toyoura sand. Further work on bamboo fibers, gypsum, and cement is currently being conducted in Japan. A comprehensive investigation of the strength and stiffness of these types of materials is vital to support the range of studies being conducted. The current study forms a part of this overall collaborative program with Fukuoka University and addresses this aspect of the work.

2.1. Materials

The three different types of materials e.g., Toyoura sand, polyvinyl alcohol (PVA) fibers, and ordinary Portland cement (OPC) were employed in this study to simulate the use of tsunami wastes in construction of sustainable geotechnical infrastructures. Toyoura sand is a Japanese benchmark sand, which is a well-known laboratory test sand. Toyoura sand has been previously used in a number of investigations and is composed of 75% quartz, 22% feldspar, and 3% magnetite. It can be found primarily in the coastal regions of the Pacific Ocean in Japan [49,50]. The soil has a uniformity coefficient (C_u) of 1.24, a minimum void ratio (e_{min}) of 0.62, and maximum void ratio (e_{max}) of 0.95. Specific gravity test was performed on clean Toyoura sand according to ASTM standard [51] and the specific gravity value of 2.65 was determined. The specific gravity (Gs), like many silicate sands, ranges from 2.64–2.65 for pure Toyoura sand [46,48]. A typical grain size distribution of Toyoura sand is presented in Figure 2a. Toyoura sand is described as having angular to sub-angular particles, is fine grained and poorly graded, which is confirmed by the low coefficient of uniformity and coefficient of curvature, according to the classification of SP by the Unified Soil Classification System (USCS) [46,52,53].

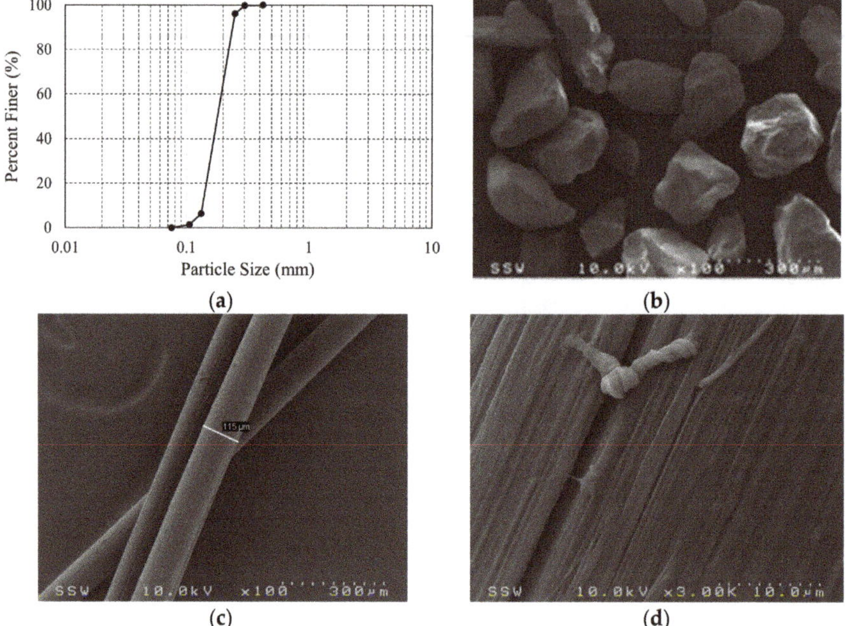

Figure 2. (**a**) Grain size distribution curve for Toyoura sand (**b**) Toyoura sand 100× optical zoom (**c**) PVA fiber 100× optical zoom (**d**) PVA fiber 3000× optical zoom (after Schmidt [50]).

Figure 2b shows a scanning electron microscopic (SEM) image of Toyoura sand and provides an indication of the size, shape, and texture of the particles [46]. The Polyvinyl alcohol (PVA) fibers used in this study shown in Figure 2c have a specific gravity of 1.3. The PVA fibers have a Young's modulus of 28 GPa and a tensile strength of 1200 MPa (Kuraray Cooperation Limited, Tokyo, Japan). Figure 2d shows micro-striations; these striations and filaments give the fibers a surface roughness and, with the existing angularity of the Toyoura sand, might also help in providing the necessary cementitious bonding. Nominal dimensions of the individual fibers are 12 mm long, with a diameter of 0.11 mm. Ordinary Portland Cement Type-I (OPC-I) shipped from Ube-Mitsubishi Cement Corporation in Japan was used as a cementing agent and added as a percent by mass to each specimen. OPC-I has a specific gravity of 3.15 and a composition consisting of approximately 63% tricalcium silicate, 12% di-calcium silicate, 5% tri-calcium aluminate, and 11% tetra-calcium alumino-ferrite [54]. These cementitious and fibrous additives have been previously used to model the monotonic and cyclic properties of amended Toyoura sand [46–48].

2.2. Sample Preparation, Testing Apparatus, Testing Procedure, and Testing Program

The under-compaction moist tamping technique was employed for sample preparation [55]. Cylindrical specimens were formed in five layers with a height of 100 mm and a diameter of 50 mm [56]. Most of the samples were prepared to a target dry density value of ρ_d = 1.49 g/cm^3. This density was selected to replicate a field condition (i.e., medium dense state) for the compacted soil and for comparison with previously published studies [46,47]. Unreinforced, fiber-only, cement-only, and fiber-reinforced cemented Toyoura sand samples were prepared and mixed at 10 percent of water content by dry mass of soil. Figure 2 shows a local strain transducer mounted on a typical sample. All cemented samples were cured for 3 days. Two main reasons for 3 days curing duration used are:

1. The main reason for choosing the shorter curing duration is to speed up the testing process to investigate the effect of cementation on the small-large strain measurements. After three days of curing, an average degree of hydration of 88% is assumed based on empirical data [57].
2. The other reason is to find the lower bound behavior (short term strength and stiffness) of cemented sand. A shorter curing duration provides an initial estimate of strength and stiffness increases. Therefore, the short-term strength and stiffness increases are of vital importance for the design of several geotechnical problems (e.g., machine foundations, embankments etc.). Short curing duration and lower cement content, which are close to the field shallow mixing technique, might help geotechnical engineers in the determination of minimum stiffness and strength of composite materials. In addition, due to the improvements in the strength and stiffness of these amended materials (e.g., despite the short curing times, 0–3% fiber and cement contents), this may be a viable strengthening method for dredged soils, disaster wastes and reclaimed land.

Past research on cemented sands has focused almost exclusively on longer curing durations (e.g., 7–28 days) and higher cement contents (e.g., 0–16%). Overall, sand-cement-fiber composites have been observed to be more effective when specimens are cured for longer durations. These findings are likely to be due to a better contact between the sand-cement-fiber matrix bonding, cement hydration, and improved interaction due to a longer curing period. Limited studies are reported to determine the stiffness and strength of sand-cement-fiber composites for shorter curing duration (e.g., 3 days) and lower cement content (e.g., 0–3% by dry mass of soil). Hence, further laboratory investigations on lower cement content (e.g., 0–3%) and short curing duration is essential in relation to field applications.

Table 1 summarizes the testing program used to evaluate the effect of fiber and cement content on the small-strain shear behavior. A unique test ID is used for the representation of a test i.e., LSM-C0F0M0 represents local strain measurement (LSM) for cement (C) = 0%, fiber (F) = 0% and silt (M) = 0%. A GDS triaxial apparatus was employed to conduct consolidated drained (CD) compression triaxial tests as per accordance to ASTM D7181 [55] to investigate the behavior of unreinforced, fiber-only, cement-only, and fiber-reinforced cemented Toyoura sand specimens. This system is a computer controlled, fully automated advanced GDS Triaxial Testing System (GDSTTS). The GDS Standard Level Pressure/Volume Controllers (STDDPC) allow for pressure measurements to be resolved to 1 kPa, with an accuracy of ±1.5 kPa up to a maximum pressure of 2 MPa. Volume changes can be resolved to 1 mm3 at an accuracy of <0.25% of the current measurement. A 15 kN load balanced internal load cell was installed providing an accuracy of ±1 N [58]. Hall effect local strain transducers were mounted in the middle third of the sample (Figure 3), which is less restrained than the end zones. It is highly desirable that axial deformations are measured locally, if small deformations moduli are to be found. The range, resolution, and accuracy of Hall effect transducer is ±0.3 mm, <0.1 μm, and 0.2% respectively (GDS Instruments). Triaxial tests use external Linear Variable Differential Transformers (LVDTs) to measure large strains (e.g., 0.01–10%). However, these LVDTs measure the global strain applied and not the local strain developed in the triaxial soil sample during shearing. Accurate determination of soil small-strain stiffness is difficult to achieve using global LVDTs attached to the actuator of automated triaxial system in routine laboratory testing. In this study, Hall effect local strain transducers are used to investigate the small-strain stiffness behavior of unreinforced, fiber, cemented, and fiber-reinforced cemented Toyoura sand specimens in triaxial tests.

Figure 3. Hall effect local strain transducer mounted on a typical sample.

Table 1. Testing program for local strain measurements.

Test No.	Test ID	Mean Effective Stress, p′, kPa	Cement Content, %	Fiber Content, %
		Sand Only		
1.	LSM-C0F0M0	100	0	0
		Fiber Only		
2.	LSM-C0F0.5M0	100	0	0.5
3.	LSM-C0F1M0	100	0	1
4.	LSM-C0F2M0	100	0	2
		Cement Only		
5.	LSM-C1F0M0	100	1	0
6.	LSM-C2F0M0	100	2	0
7.	LSM-C3F0M0	100	3	0
8.	LSM-C4F0M0	100	4	0
		Fiber + Cement		
9.	LSM-C3F1M0	100	3	1
10.	LSM-C3F2M0	100	3	2
11.	LSM-C3F3M0	100	3	3
12.	LSM-C2F1M0	100	2	1

All of the specimens were saturated with de-aired water and CO_2 until a B-value of at least 0.96 was reached, before starting the consolidation stage. First, carbon dioxide (CO_2) was slowly flushed through the bottom of the sample for about 30 min to absorb any entrapped air in the voids of specimen with a gradient of pressure for approximately 3 kPa. The top and bottom drainage lines were flushed with de-aired water through back pressure pump at a very slow rate. After flushing the drainage lines, the de-aired water was flushed in the specimen at a very slow rate to fill the voids of specimen and replace CO_2. In addition, the pore water pressure values were also monitored during CO_2 percolation and flushing with water. It was necessary to maintain an effective stress of approximately 3 kPa in order to minimize any sample disturbance. Once the CO_2 percolation and flushing with water was finished, the cell pressure was ramped to 320 kPa and back pressure was ramped to 310 kPa, maintaining an effective stress of 10 kPa. In the next stage, cell pressure was then ramped to 330 kPa (e.g., the back pressure 310 kPa was kept constant and cell pressure starting at 320 kPa was then increased at a rate of 2–3 kPa/minute, till the final target cell pressure of 330 kPa was reached) and pore pressure coefficient B was checked during the saturation stage. Higher B-values were possible in the cemented samples due to the application of higher back pressures (e.g., 320 kPa), short curing duration (e.g., 3 days), and lower cement contents (0–3%). All of specimens for the consolidated drained (CD) tests were isotropically consolidated to the desired mean effective stress (e.g., 100 kPa) under computer control. The consolidation stage was continued until 100% primary consolidation was reached. The rate of axial displacement used to shear all of the specimens was 0.06 mm/min [46,48,59] to eliminate any concerns over rate effects, when comparing the results.

3. Results and Discussion

Typical deviator stress versus shear strain and mobilized stress ($\frac{q}{q_{pk}}$) curves obtained from the Hall effect local strain transducer under drained triaxial shear are shown in Figures 4 and 5. Where, q = deviator stress is the difference between the major and minor principal stresses in a triaxial test, which is equal to the axial load applied to the specimen divided by the cross-sectional area of the specimen, q_{pk} = peak deviator stress, $\frac{q}{q_{pk}}$ = mobilized stress developed with increase in shear strain. It can be seen that the small-strain stiffness (e.g., reference strain range = 0.001–0.05%) reduces and the curve shows slightly flattened response compared to sand only specimen, with the addition of only fibers (Figure 4b). In contrast, the small-large strain stiffness (e.g., reference strain range = 0.1–1.0%) increases by up to 0–137% (Figure 4a); similar results were reported in previous studies [36,60–62] and this behavior is a consequence of the loss of contact between the particles and a reduction in the particle-to-particle friction due to the presence of the fibers [63]. In contrast to the fiber-only specimen, the addition of cement enhances the small-strain stiffness (e.g., reference strain range = 0.01–0.1%) compared to sand only specimen of the Toyoura sand specimens by up to 160–171% (Figure 4b). Furthermore, the small-large strain stiffness (e.g., reference strain range = 0.1–1.0%) increases by up to 175–158% (Figure 4a). Similar results were also reported by Consoli et al. [64] and Schnaid et al. [65]. These results highlight that the weak cementation (e.g., 3 days curing) induced is sufficient to moderately increase the small-strain stiffness and the curve shows relatively stiffer response [66]. In addition, fiber-reinforced cemented sand specimen showed an approximately 145–257% increase in small-strain stiffness (e.g., reference strain range = 0.01–0.1%) behavior compared to the unreinforced specimens (Figure 4b). Furthermore, the small-large strain stiffness (e.g., reference strain range = 0.1–1.0%) increases by up to 260–265% (Figure 4a) and the curve shows significantly stiffer response compared to sand only specimen.

The results reported for fiber-reinforced cemented sand agrees well with previous studies [46–48,67]. The significant increase in small-large strain stiffness (e.g., reference strain range = 0.01–1.0%) of fiber-reinforced cemented sand is attributed to interparticle bonds, particle-to-particle contacts, and fiber-particle friction mechanism.

Figure 5a shows significant reduction in the mobilized stresses (e.g., reference strain range = 0.001–1.0%) and the curves show slightly flattened response compared to sand only specimen, with the addition of only fibers (Figure 5a). In contrast to fiber-only specimens, the mobilized stresses (e.g., reference strain range = 0.05–0.7%) slightly increases (Figure 5b) for cement-only specimens (except for the 1% cement-only specimen). In addition, the fiber-reinforced cemented sand specimen showed limited increases in mobilized stresses (e.g., reference strain range = 0.01–0.1%) compared to the unreinforced specimen (Figure 5c). However, Figure 5c shows moderate increases in mobilized stresses (e.g., reference strain range = 0.1–0.5%) compared to the sand-only specimen.

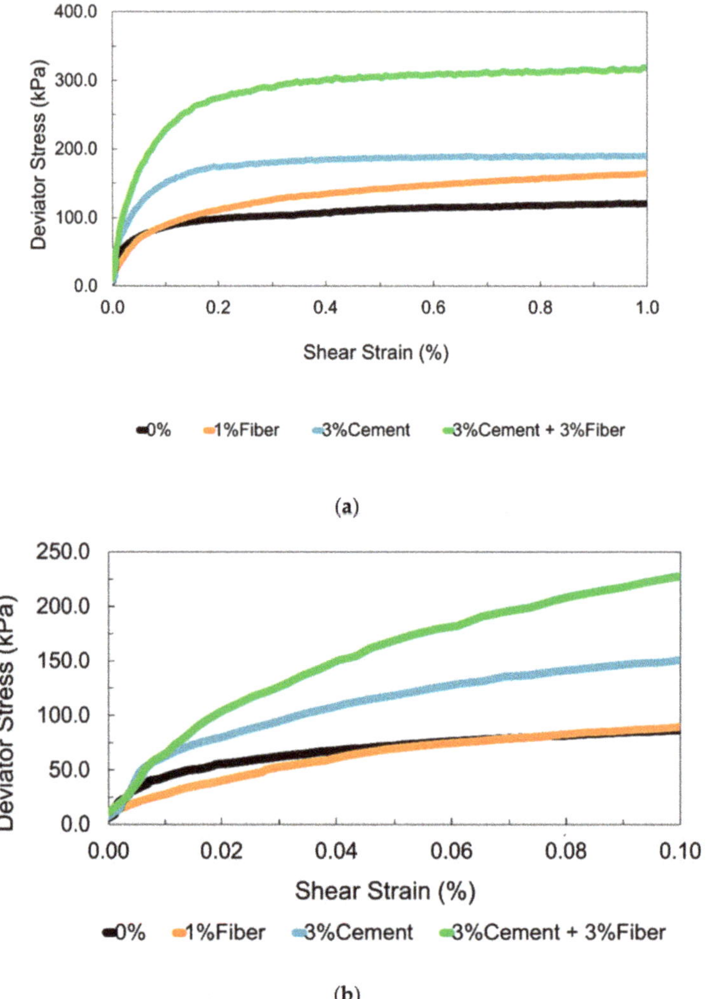

Figure 4. (**a**) Deviatoric stress (q) versus shear strain (ε_q) curves from drained triaxial tests for various Toyoura sand specimens ($\sigma'_c = 100$ kPa) (**b**) Zoomed in until 0.10% shear strain.

Figure 6a,c shows the normalized shear modulus reduction (G/G_i) versus mobilized stress ($\frac{q}{q_{pk}}$) curves. Where, G = shear modulus at any shear strain level and G_i = initial shear modulus. The value of initial shear modulus (G_i) is obtained from the range of local strain measurements (e.g., reference strain, $\gamma_r = 0.00013\%$ to 0.00024%). Fahey and Carter [44] presented similar test results in terms of modulus reduction versus mobilized stress for uncemented granular soils. They proposed a simple hyperbolic relationship for clean sands with a limited range of exponents (0.2–0.4) as shown below:

$$\frac{G}{G_0} = \left[1 - (q/q_{pk}\right]^g \tag{3}$$

where G/G_0 = shear modulus reduction, $\frac{q}{q_{pk}}$ = mobilized stress and g = an exponent to encompass laboratory test data.

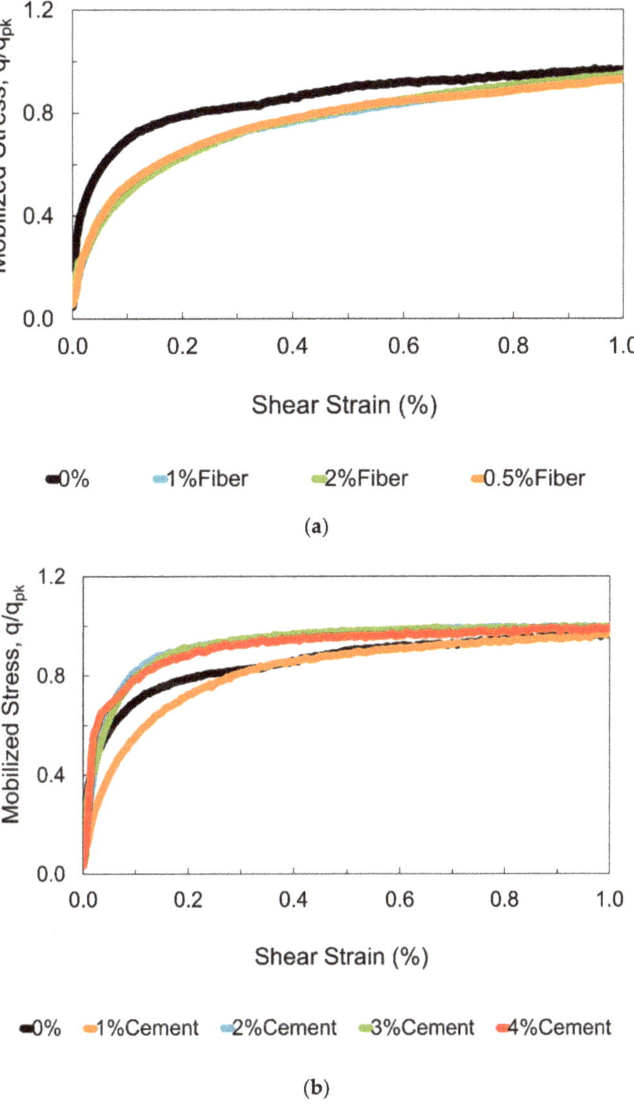

(a)

(b)

Figure 5. *Cont.*

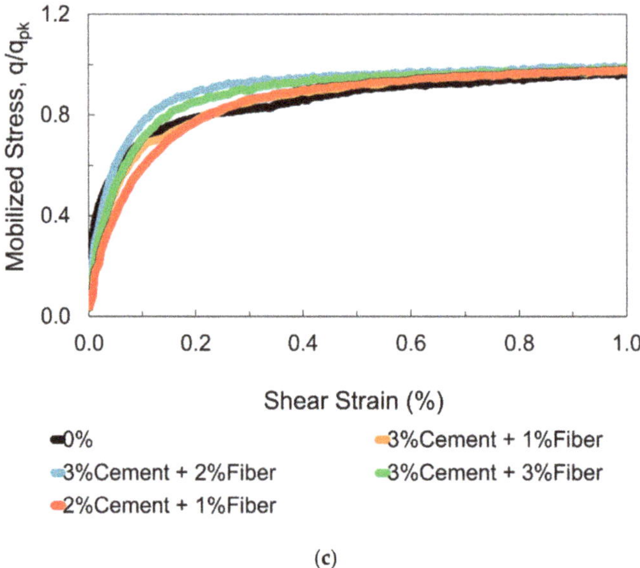

(**c**)

Figure 5. Mobilized stress ($\frac{q}{q_{pk}}$) versus shear strain (ε_q) curves from drained compression tests for various Toyoura sand specimens ($\sigma'_c = 100$ kPa) (**a**) Fiber only (**b**) Cement only (**c**) Cement + Fiber.

For pure Toyoura sand and fiber-only reinforced sand, it can be seen that the results agree well with the hyperbolic model (Equation (3)) employing an exponent value of 0.2–0.3. For purely cemented sand, the results show close agreement adopting values in the range of 0.3–0.4. However, a slightly greater value of exponent (e.g., 0.4–0.6) is required to fit the results of the fiber-reinforced cemented sand. A range of exponent, $g = 0.2$–0.4 was suggested by Fahey and Carter [44] for uncemented granular soils. In contrast, it can be seen that for cemented and fiber-reinforced sands, the range of exponent lies between 0.3 and 0.6, showing a more intense decay of stiffness with straining.

Oztoprak and Bolton [45] proposed a generic relationship for the G/G_0 versus shear strain (ε_q) curves based on a database of 454 tests from the literature. Three curve fitting parameters control the shape of the curve (see Equation (4)): (1) an elastic threshold strain (γ_e), up to which the elastic shear modulus is constant at G_0, and which enables the expression to cover cementation and interlocking effects at small-strains; (2) a reference strain (γ_r), the shear strain at which the secant modulus reduces to $0.5\ G_0$—the two characteristic strains were found to vary with sand type (e.g., uniformity coefficient), state of the soil (e.g., void ratio, relative density), and mean effective stress; and lastly, (3) a curvature parameter (a), which controls the rate of modulus reduction. An average value of curvature parameter, $a = 0.88$, was employed for a database of 379 tests on uncemented sands.

$$\frac{G}{G_0} = \frac{1}{1 + \left[\frac{\gamma - \gamma_e}{\gamma_r}\right]^a} \tag{4}$$

where γ_e = elastic threshold strain, γ_r = reference strain, and a = curvature parameter.

Figures 7–9 show G/G_i versus shear strain (ε_q) curves from similar drained triaxial tests at varying cement (1–4%) and fiber (0.5–3%) contents. Table 2 shows the values of best-fit parameters proposed by Oztoprak and Bolton [45] and for unreinforced and reinforced Toyoura sand.

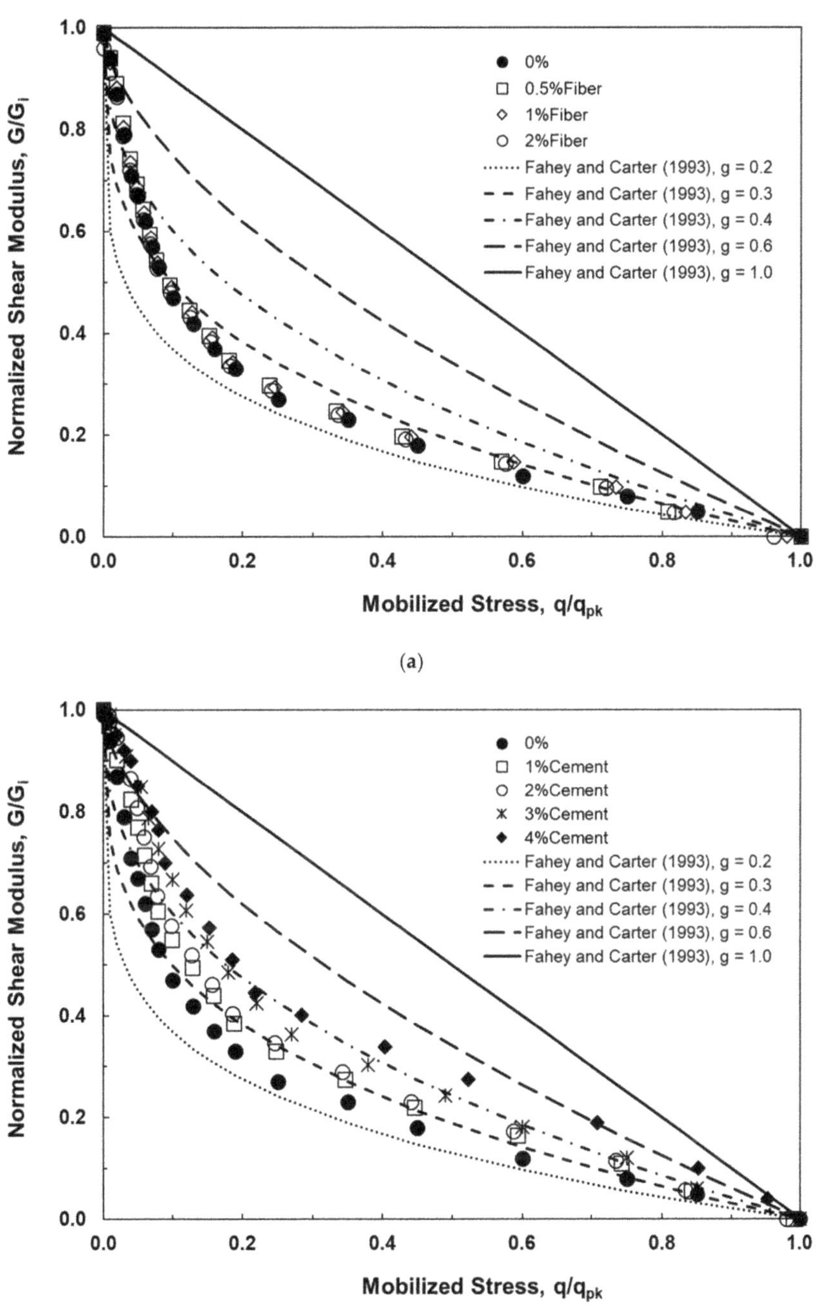

(a)

(b)

Figure 6. *Cont.*

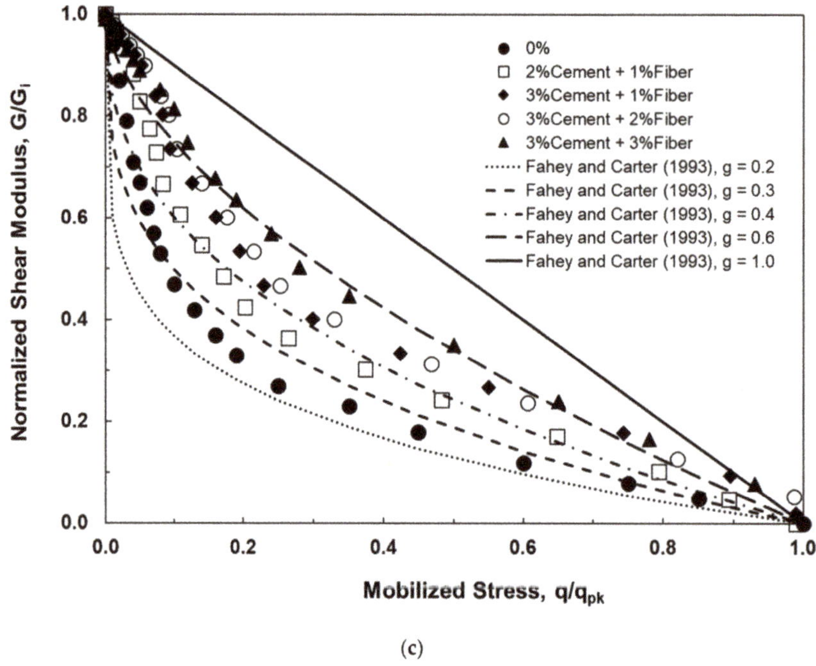

(**c**)

Figure 6. Shear modulus reduction (G/G_i) versus mobilized stress ($\frac{q}{q_{pk}}$) curves from drained compression tests for various Toyoura sand specimens ($\sigma'_c = 100$ kPa). (**a**) Pure Sand, and 0–2% Fibers; (**b**) Pure Sand, and 0–4%; (**c**) Pure Sand, 0–3% Cement and 0–3% Fibers.

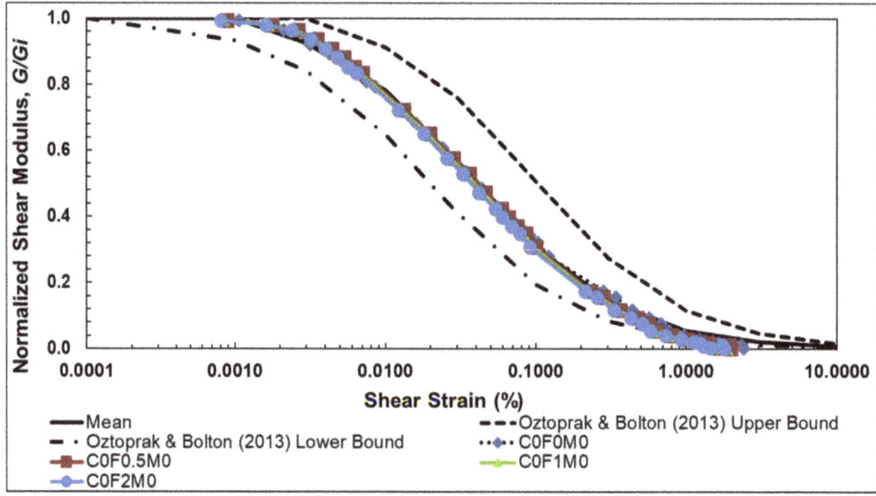

Figure 7. Pure Sand, and 0.5–2% Fibers. G/G_i versus shear strain (ε_q) curves from CD compression tests for unreinforced and fiber-reinforced Toyoura sand specimens consolidated to 100 kPa mean effective stress at different fiber contents (0–2%).

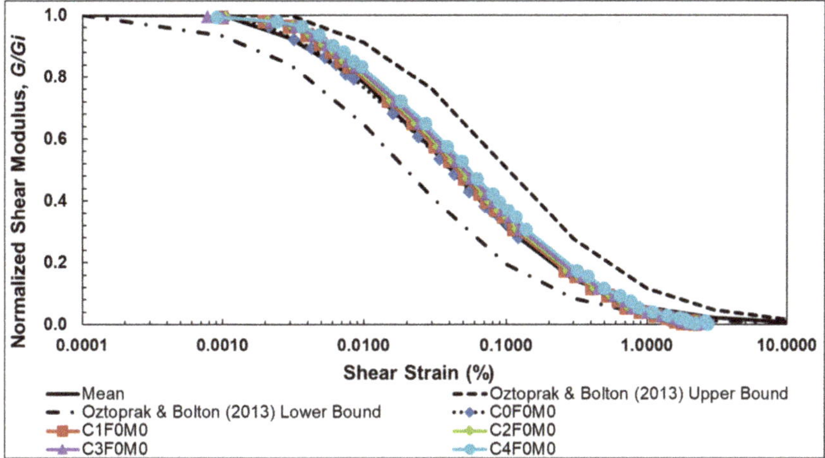

Figure 8. Pure Sand, and 1–4% Cement. G/G_i versus shear strain (ε_q) curves from CD compression tests for unreinforced and cemented Toyoura sand specimens consolidated to 100 kPa mean effective stress at different cement contents (0–4%).

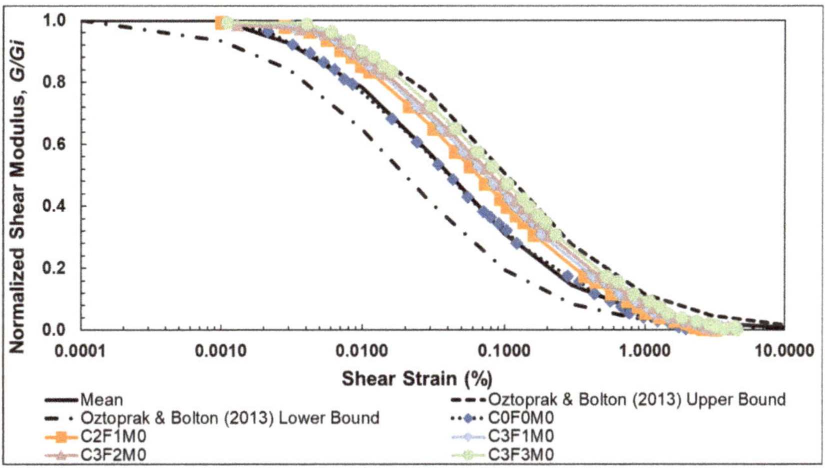

Figure 9. Pure Sand, 2–3% Cement and 1–3% Fibers. G/G_i versus shear strain (ε_q) curves from CD compression tests for unreinforced and fiber-reinforced cemented Toyoura sand specimens consolidated to 100 kPa mean effective stress at different cement (0–3%) and fiber (0–3%) contents.

It can be seen in Figure 7 that the elastic threshold strain (γ_e) ranges from 0.0007% to 0.001% for unreinforced and fiber-reinforced sand. This range slightly increases to 0.0009–0.0014% for cemented sand, shown in Figure 8. For the fiber-reinforced cemented sand shown in Figure 9, the threshold strain increases to a range of 0.0015–0.0022%. The ranges for the reference strain (γ_r), for unreinforced and fiber-reinforced sand (0.039–0.043%), cemented sand (0.048–0.056%), and fiber-reinforced cemented sand (0.065–0.08%) are also shown in Table 2. In addition, it can be seen that the curvature parameter (*a*) for unreinforced and fiber-reinforced sand was 0.88, and 1.0 for cemented and fiber-reinforced cemented sand. The range of curvature parameter from 0.88 to 1.0 provides a good comparison with the results of local strain. Overall, the comparison of the results and model shows that the small-strain results obtained using local strain transducers fall within the range of model upper and lower bound curves. The results of the unreinforced, fiber-reinforced, and cemented sand shows a close agreement with the model mean curve, but fiber-reinforced cemented sand shows a good comparison with model upper bound.

Table 2. Comparison of curve-fitting parameters for unreinforced and reinforced Toyoura Sand with Oztoprak and Bolton [45].

Sample ID	Elastic Threshold Strain, γ_e	Reference Strain, γ_r	Curvature Parameters, a
C0F0M0	0.001	0.043	0.88
C0F0.5M0	0.0008	0.042	0.88
C0F1M0	0.0007	0.040	0.88
C0F2M0	0.0007	0.039	0.88
C1F0M0	0.0009	0.048	1
C2F0M0	0.001	0.050	1
C3F0M0	0.0012	0.052	1
C4F0M0	0.0014	0.056	1
C2F1M0	0.0015	0.065	1
C3F1M0	0.0018	0.074	1
C3F2M0	0.0020	0.076	1
C3F3M0	0.0022	0.080	1
Oztoprak and Bolton [45]			
Lower Bound	0	0.02	0.88
Mean	0.0007	0.044	0.88
Upper Bound	0.003	0.1	0.88

4. Summary and Conclusions

In this study, an effort has been made to utilize the tsunami waste as a ground improvement and remediation technique to build sustainable geotechnical infrastructures. Therefore, a series of local strain measurements were obtained on unreinforced, fiber, cemented, and fiber-reinforced cemented Toyoura sand specimens. It is shown that small-strain stiffness slightly reduces with the addition of fibers. In contrast, addition of cement enhances the small-strain stiffness properties of pure Toyoura sand specimens. The results highlighted that the weak cementation level (e.g., 3 days curing) induced by chemical treatment was sufficient to moderately increase the small-strain stiffness. In addition, fiber-reinforced cemented sand specimens showed increases in small-strain stiffness compared to unreinforced specimens. The fiber used in this study vary in diameter from 110–120 μm, with striation widths of 5 μm to less than 1 μm along the 12 mm length. These micro-striations have small filaments protruding from them, which is likely a result of the extrusion process used in their fabrication. These striations and filaments give the fibers a rough surface and, with the existing angularity of the Toyoura sand, provide an ideal medium for cementitious bonding [46,48,68,69]. Results of the modulus degradation and mobilized stress curves show good agreement with the hyperbolic relation proposed by Fahey and Carter [44]. The comparison of the results with Oztoprak and Bolton [45] model shows that the results of the local strain transducers fall within the range of model upper and lower bound curves. The short curing duration and lower cement content with fiber additive shows close relevance to the field shallow mixing technique. The current research related to increase in stiffness parameters due to the combined effect of cement and fiber additives might be useful for the practicing engineers in the construction of economical and sustainable geotechnical infrastructures. In addition, the results show promising improvements in the strength and stiffness characteristics of fiber-reinforced cemented sand. Hence, much of the debris (such as concrete products, natural and polymeric fibers, and tsunami deposits on the coast of Japan) can be recycled in economical and sustainable geotechnical engineering projects, such as road embankments, recreational park restoration, and agricultural field restoration around the Tokyo Bay region [47,48].

Author Contributions: Conceptualization and methodology, M.S., T.N., K.S., and T.F.; investigation and analysis, M.S. and C.S.; writing—original draft preparation, M.S. and F.S.; writing—review and editing, M.S., T.N., and C.S.; supervision, T.N. All authors have read and agreed to the published version of the manuscript.

Funding: The research project was financially supported by the Western Graduate Research Scholarship at the Department of Civil and Environmental Engineering, Western University, London, Ontario, Canada.

Acknowledgments: The authors would like to acknowledge Miho Nakamichi, Shintaro Koga, and Hiromitsu Shiina for their technical support and for financial support for providing Toyoura Sand for the research project.

Conflicts of Interest: The authors declare no conflict of interest.

References

1. Richart, J.E., Jr.; Hall, J.R., Jr.; Woods, R.O. *Vibrations of Soils and Foundations*, 1st ed.; Prentice Hall: Upper Saddle River, NJ, USA, 1970.
2. Andrus, R.D.; Stokoe, K.H., II. Liquefaction resistance of soils from shear wave velocity. *J. Geotech. Geoenviron. Eng.* **2000**, *126*, 1943–5606. [CrossRef]
3. Yang, J.; Yan, X.R. Site response to multi-directional earthquake loading: A practical procedure. *Soil Dyn. Earthq. Eng.* **2009**, *29*, 710–721. [CrossRef]
4. Hardin, B.O.; Richart, F.E. Elastic wave velocities in granular soils. *J. Soil Mech. Found. Div.* **1963**, *89*, 39–56.
5. Cascante, G.; Santamarina, C.; Yassir, N. Flexural excitation in a standard torsional resonant column device. *Can. Geotech. J.* **1998**, *35*, 478–490. [CrossRef]
6. Brignoli, E.G.M.; Gotti, M.; Stokoe, K.H., II. Measurement of shear waves in laboratory specimens by means of piezoelectric transducers. *Geotech. Test. J.* **1996**, *19*, 384–397. [CrossRef]
7. Nakagawa, K.; Soga, K.; Mitchell, J.K. Observation of Biot compressional wave of the second kind in granular soils. *Géotechnique* **1997**, *47*, 133–147. [CrossRef]
8. Lings, M.L.; Greening, P.D. A novel bender/extender element for soil testing. *Géotechnique* **2001**, *51*, 713–717. [CrossRef]
9. Kumar, J.; Madhusudhan, B.N. Effect of relative density and confining pressure on Poisson ratio from bender–extender element tests. *Géotechnique* **2010**, *60*, 561–567. [CrossRef]
10. Murillo, C.; Sharifipour, M.; Caicedo, B.; Thorel, L.; Dano, C. Elastic parameters of intermediate soils based on bender–extender elements pulse tests. *Soils Found* **2011**, *51*, 637–649. [CrossRef]
11. Ahmad, S. Piezoelectric Device for Measuring Shear Wave Velocity of Soils and Evaluation of Low and High Strain Shear Modulus. Ph.D. Thesis, Western University, London, ON, Canada, 3 February 2016.
12. Kokusho, T. Cyclic triaxial test of dynamic soil properties for wide strain range. *Soils Found.* **1980**, *20*, 45–60. [CrossRef]
13. Hoque, E.; Tatsuoka, F. Anisotropy in elastic deformation of granular materials. *Soils Found.* **1998**, *38*, 163–179. [CrossRef]
14. Ezaoui, A.; Benedetto, H.D. Experimental measurements of the global anisotropic elastic behavior of dry Hostun sand during triaxial tests, and the effect of sample preparation. *Géotechnique* **2009**, *59*, 621–635. [CrossRef]
15. Gu, X.; Yang, J.; Huan, M. Laboratory measurements of small strain properties of dry sands by bender element. *Soils Found.* **2013**, *53*, 735–745. [CrossRef]
16. Clayton, C.R.I.; Khatrush, S.A.A. New device for measuring local axial strains on triaxial specimens. *Géotechnique* **1986**, *36*, 593–597. [CrossRef]
17. Castelblanco, J.M.; Delage, P.; Pereira, J.M.; Cui, Y.J. On-sample water content measurement for a complete local monitoring in triaxial testing of unsaturated soils. *Géotechnique* **2012**, *62*, 597–604. [CrossRef]
18. Jastrzębska, M.; Kowalska, M. Triaxial tests on weak cohesive soils-some practical remarks. *Archit. Civ. Eng. Environ.* **2016**, *9*, 91–103. [CrossRef]
19. Ye, G.L.; Wu, C.J.; Wang, J.F.; Wang, J.H. Influence and countermeasure of specimen misalignment to small-strain behavior of soft marine clay. *Mar. Georesour. Geotechnol.* **2017**, *35*, 170–175. [CrossRef]
20. Brown, S.F.; Snaith, M.S. The measurement of recoverable and irrecoverable deformations in the repeated load triaxial test. *Géotechnique* **1974**, *24*, 225–259. [CrossRef]
21. Cole, D.M. A Technique for measuring radial deformation during repeated load triaxial testing. *Géotechnique* **1978**, *15*, 426–429. [CrossRef]

22. Khan, M.H.; Hoag, D.L.A. Noncontacting transducer for measurement of lateral strains. *Can. Geotech. J.* **1979**, *16*, 409–411. [CrossRef]

23. Brown, S.F.; Austin, G.; Overy, R.F. An instrumented triaxial cell for cyclic loading of clays. *Geotech. Test. J.* **1980**, *3*, 145–152. [CrossRef]

24. Burland, J.B.; Symes, M.A. Simple axial displacement gage for use in the triaxial apparatus. *Géotechnique* **1982**, *32*, 62–65. [CrossRef]

25. Costa-Filho, L.M. Measurement of axial strains in triaxial tests on London clay. *Geotech. Test. J.* **1985**, *8*, 3–13. [CrossRef]

26. Dupas, J.M.; Pecker, A.; Bozetto, P.; Fry, J.J. A 300 mm diameter triaxial cell with a double measuring device. *Adv. Triaxial Test. Soil Rock* **1988**, *977*, 142–1988.

27. Hird, C.C.; Yung, P.C.Y. Discussion on a new Device for measuring local axial strains on triaxial specimens. *Géotechnique* **1987**, *37*, 413–417. [CrossRef]

28. Hird, C.C.; Yung, P.C.Y. The use of proximity transducers for local strain measurements in triaxial tests. *Geotech. Test. J.* **1989**, *12*, 292–296.

29. Gunasekaran, M.; Robinson, R.G. On-sample measurement of strains in triaxial samples using strain gauges. *Indian Geotech. J.* **2008**, *38*, 33–48.

30. Hardin, B.O.; Drnevich, V.P. Shear modulus and damping in soils: Measurement and parameter effects. *J. Soil Mech. Found. Div.* **1972**, *98*, 603–624.

31. Hardin, B.O.; Drnevich, V.P. Shear modulus and damping in soils: Design equations and curves. *J. Geotech. Eng.* **1972**, *98*, 667–692.

32. Darendeli, B.M. Development of a New Family of Normalized Modulus Reduction and Material Damping Curves. Ph.D. Thesis, University of Texas, Austin, TX, USA, 2001.

33. Hurtado, J.; Newson, T. Small-strain behavior of a carbonate clay till underlying a wind turbine shallow foundation by different in-situ and laboratory tests. In Proceedings of the 69th Canadian Geotechnical Conference, Vancouver, BC, Canada, 2–5 October 2016.

34. Iwasaki, T.; Tatsuoka, F.; Takagi, Y. Shear moduli of sands under cyclic torsional shear loading. *Soils Found.* **1978**, *18*, 39–50. [CrossRef]

35. Michalowski, R.L.; Cermak, J. Triaxial compression of sand reinforced with fibers. *J. Geotech. Geoenviron. Eng.* **2003**, *129*, 125–136. [CrossRef]

36. Michalowski, R.L.; Zhao, A. Failure of fiber-reinforced granular soils. *J. Geotech. Eng.* **1996**, *122*, 226–234. [CrossRef]

37. Acar, Y.; El-Tahir, A. Low strain dynamic properties of artificially cemented sand. *J. Geotech. Eng.* **1986**, *112*, 1001–1015. [CrossRef]

38. Saxena, S.K.; Avramidis, A.S.; Reddy, K.S. Dynamic moduli and damping ratios for cemented sands at low strains. *Can. Geotech. J.* **1988**, *25*, 353–368. [CrossRef]

39. Sharma, S.S.; Fahey, M. Deformation characteristic of two cemented calcareous soils. *Can. Geotech. J.* **2004**, *41*, 1139–1151. [CrossRef]

40. Trhlíková, J.; Boháč, J.; Mašín, D. Small-strain behavior of cemented soils. *Géotechnique* **2012**, *62*, 943–947. [CrossRef]

41. Yun, T.S.; Santamarina, J.C. Decementation, softening, and collapse: Changes in small-strain shear stiffness in k_0 loading. *J. Geotech. Geoenviron. Eng.* **2005**, *131*, 350–358. [CrossRef]

42. Mair, R.J. Developments in geotechnical engineering research: Application to tunnels and deep excavations. *Proc. Inst. Civ. Eng. Civ. Eng.* **1993**, *93*, 27–41.

43. Xu, D.; Borana, L.; Yin, J. Measurement of small strain behavior of a local soil by fiber Bragg grating-based local displacement transducers. *Acta Geotech.* **2014**, *9*, 935–943. [CrossRef]

44. Fahey, M.; Carter, J.P. A finite element study of the pressuremeter test in sand using a nonlinear elastic plastic model. *Can. Geotech. J.* **1993**, *30*, 348–362. [CrossRef]

45. Oztoprak, S.; Bolton, M.D. Stiffness of sand through a laboratory test database. *Géotechnique* **2013**, *63*, 54–70. [CrossRef]

46. Schmidt, C.J.R. Static and Dynamic Response of Silty Toyoura Sand with PVA Fiber and Cement Additives. Master's Thesis, Western University, London, ON, Canada, 2015.

47. Nakamichi, M.; Sato, K. A method of suppressing liquefaction using a solidification material and tension stiffeners. In Proceedings of the International Conference on Soil Mechanics and Geotechnical Engineering, Paris, France, 2–6 September 2013.

48. Safdar, M. Monotonic Stress-Strain Behavior of Fiber Reinforced Cemented Toyoura Sand. Ph.D. Thesis, Western University, London, ON, Canada, 2018.

49. Lam, W.K.; Tatsuoka, F. Effects of initial anisotropic fabric and sigma2 on strength and deformation characteristics of sand. *Soils Found.* **1988**, *28*, 89–106. [CrossRef]

50. De, S.; Basudhar, P.K. Steady state strength behavior of Yamuna sand. *Geotech. Geol. Eng.* **2008**, *26*, 237–250.

51. ASTM D854-10. *Standard Test Methods for Specific Gravity of Soil Solids by Water Pycnometer*; ASTM International: West Conshohocken, PA, USA, 2012.

52. Yamaguchi, H.; Kimura, T.; Fuji-I, N. On the influence of progressive failure on the bearing capacity of shallow foundations in dense sand. *Soils Found.* **1976**, *16*, 11–22. [CrossRef]

53. Whitlow, R. *Basic Soil Mechanics*, 4th ed.; Prentice Hall: Upper Saddle River, NJ, USA, 2001.

54. ASTM C150/C150M-12. *Standard Specification for Portland Cement*; ASTM International: West Conshohocken, PA, USA, 2012. [CrossRef]

55. Ladd, R.S. Preparing test specimens using undercompaction. *Geotech. Test. J.* **1978**, *1*, 16–23.

56. ASTM D7181-11. *Method for Consolidated Drained Triaxial Compression Test for Soils*; ASTM International: West Conshohocken, PA, USA, 2011.

57. Shafiq, N.; Nuruddin, M.F. Degree of hydration of OPC and OPC/fly ash paste samples conditioned at different relative humidity. *Int. J. Sustain. Constr. Eng. Technol.* **2010**, *1*, 47–56.

58. Kiss, J.A. Evaluation of Fatigue Response of a Carbonate Clay Till Beneath Wind Turbine Foundation. Master's Thesis, Western University, London, ON, Canada, 2016.

59. Head, K.H. *Manual of Laboratory Testing: Effective Stress Tests*; ELE International Limited: New York, NY, USA, 1986; Volume 3.

60. Gray, D.H.; Al-Refeai, T. Behavior of fabric-versus fiber-reinforced sand. *J. Geotech. Eng.* **1986**, *112*, 804–820. [CrossRef]

61. Diambra, A. Fiber Reinforced Sands: Experiments and Modelling. Ph.D. Thesis, University of Bristol, Bristol, UK, 2010.

62. Din, S.U. Behavior of Fiber Reinforced Cemented Sand at High Pressures. Ph.D. Thesis, University of Nottingham, Nottingham, UK, 2012.

63. Claria, J.J.; Vettorelo, P.V. Mechanical behavior of loose sand reinforced with synthetic fibers. *Soil Mech. Found. Eng.* **2016**, *53*, 11–15. [CrossRef]

64. Consoli, N.C.; Vendruscolo, M.A.; Fonini, A.; Rosa, F.D. Fiber reinforcement effects on sand considering a wide cementation range. *Geotext. Geomembr.* **2009**, *27*, 196–203. [CrossRef]

65. Schnaid, F.; Prietto, P.; Consoli, N. Characterization of cemented sand in triaxial compression. *J. Geotech. Geoenviron. Eng.* **2001**, *127*, 857–868. [CrossRef]

66. Porcino, D.; Marcianò, V.; Granata, R. Static and dynamic properties of a lightly cemented silicate-grouted sand. *Can. Geotech. J.* **2012**, *49*, 1117–1133. [CrossRef]

67. Sadek, S.; Najjar, S.; Abboud, A. Compressive strength of fiber-reinforced lightly-cement stabilized sand. In Proceedings of the 18th International Conference on Soil Mechanics and Geotechnical Engineering, Paris, France, 2–6 September 2013.

68. Toutanji, H.A.; Xu, B.; Lavin, T.; Gilbert, J.A. Properties of polyvinyl alcohol fiber reinforced high performance organic aggregate cementitious material. In Proceedings of the International Congress on Polymers in Concrete, Funcal, Madeira Island, Portugal, 10–12 February 2010; pp. 1–10.

69. Al-Attar, T.S. A quantitative evaluation of bond strength between coarse aggregate and cement mortar in concrete. *Eur. Sci. J.* **2013**, *9*, 22–35.

Publisher's Note: MDPI stays neutral with regard to jurisdictional claims in published maps and institutional affiliations.

Article

Sustainability Benefits Assessment of Metakaolin-Based Geopolymer Treatment of High Plasticity Clay

Rinu Samuel [1], Anand J. Puppala [2,*] and Miladin Radovic [3]

[1] Department of Civil Engineering, University of Texas at Arlington, Arlington, TX 76019, USA; rinu.samuel@mavs.uta.edu

[2] Zachry Department of Civil and Environmental Engineering, Texas A&M University, College Station, TX 77843, USA

[3] Department of Materials Science and Engineering, Texas A&M University, College Station, TX 77843, USA; mradovic@tamu.edu

* Correspondence: anandp@tamu.edu

Received: 10 November 2020; Accepted: 10 December 2020; Published: 15 December 2020

Abstract: Expansive soils are prevalent world over and cause significant hazards and monetary losses due to infrastructure damages caused by their swelling and shrinking behavior. Expansive soils have been conventionally treated using chemical additives such as lime and cement, which are known to significantly improve their strength and volume-change properties. The production of lime and cement is one of the highest contributors of greenhouse gas emissions worldwide, because of their energy-intensive manufacturing processes. Hence, there is a pressing need for sustainable alternative chemical binders. Geopolymers are a relatively new class of aluminosilicate polymers that can be synthesized from industrial by-products at ambient temperatures. Geopolymer-treated soils are known to have comparable strength and stiffness characteristics of lime and cement-treated soils. This study evaluates the sustainability benefits of a metakaolin-based geopolymer treatment for an expansive soil and compares its results with lime treatment. Test results have shown that geopolymers have significantly improved strength, stiffness, and volume-change properties of expansive soils. Increased dosages and curing periods have resulted in further property enhancements. Swell and shrinkage studies also indicated reductions in these strains when compared to control conditions. The sustainability benefits of both geopolymer and lime treatment methods are evaluated using a framework that incorporates resource consumption, environmental, and socio-economic concerns. This study demonstrates geopolymer treatment of expansive soils as a more sustainable alternative for expansive soil treatments, primarily due to metakaolin source material. Overall results indicated that geopolymers can be viable additives or co-additives for chemical stabilization of problematic expansive soils.

Keywords: geopolymer; soil stabilization; expansive soils; sustainability benefits; sustainable ground improvement

---

1. Introduction

Soils that exhibit volume-change upon variation of their moisture content are known as expansive soils. The swelling and shrinking nature of expansive soils is mostly attributed to the proportion of the clay mineral smectite in the soil, as well as the interaction of water with the clay mineral surfaces [1]. The extent of swelling and shrinkage of expansive soils are also dependent on other factors such as soil suction, soil dry unit weight, stress-history, climate, and active zone depth [2]. Expansive soils can prove to be especially hazardous in places with cycles of dry and wet spells resulting in repeated cycles of swelling and shrinkage. The effects of expansive soils are mostly observed near the ground surface where desiccation cracks can be seen in the dry season; further damages caused include pavement distress or failure, differential uplift or settlement of structures, slope and foundation failures, and other damages that compromise the integrity of infrastructures. Expansive soils are present all over the world and are ubiquitous in the south-western United States [3–6]. Millions of dollars are spent each year in the United States alone to fix damages caused to infrastructures by expansive soils [3,7–10]. As such it is important to improve swelling and shrinkage characteristics of expansive soils before proceeding with infrastructure development.

Stabilization of expansive soils has been conventionally performed using chemical additives such as lime and cement, which have proven to significantly improve the strength and lower the volume-change behavior of expansive soils by a series of cationic exchange and pozzolanic reactions between the additive and soil particles [11–13]. These calcium-based conventional chemical additives are known to have durability issues, in addition to having disadvantages in sulfate-rich soils, as certain chemical reactions result in the formation of the mineral ettringite, which causes excessive swelling and volume-change in soils [14,15]. The high demand of lime and cement additives has led to their mass production, which in turn reduces their unit cost, ultimately driving the low-cost production cycle. The cost benefits that lime and cement offer are progressively being overshadowed by their environmental implications. The production of lime and cement are energy-intensive operations that require kilns to be heated between 1000 °C to 1500 °C to process raw materials. A 2018 inventory of the greenhouse gas emissions by the Environmental Protection Agency (EPA) reported that lime and cement production industries produced 97 million metric tons of carbon dioxide (CO_2) from the minerals sector alone [16]. As such, there is an imminent need to focus on sustainable alternatives or co-additives for lime and cement treatment works in pavement geotechnics.

The topic of sustainability is usually met with a lot of apprehension, as it is relatively new and can have a myriad of different interpretations. Nevertheless, Brundtland's Declaration provides a widely recognized commentary which states that "sustainable development is development that meets the needs of the present without compromising the ability of future generations to meet their own needs" [17,18]. The prospect of integrating sustainability into a project is highest during the planning phase and diminishes considerably as the project moves into implementation phases [19]. Since geotechnics is applied in the early stages of a project, it renders an advantage and responsibility to implement sustainable geotechnical practices that positively influence the subsequent phases of infrastructure development. The engineering perspective of sustainability often incorporates cost-efficiency and reasonable control of harmful emissions, in addition to prudent resource consumption [20,21]. Therefore, a comprehensive sustainability approach includes environmental protection, economic development, and social development [18]. Regrettably, environmental impacts have been sidelined for far too long for the sake of monetary benefits, and therefore need to be addressed with more weightage for constructive sustainability with lasting positive impacts.

In recent years, a new class of binder materials known as geopolymers have been hailed as a more sustainable and eco-friendly alternative to lime and cement, due to comparable compressive strength, durability, and low shrinkage properties [22–24]. Geopolymers are aluminosilicate polymers that can be synthesized from industrial by-products, such as metakaolin, fly ash, and clay [25–28], relatively quickly at ambient temperatures thereby having a significantly lower carbon footprint than lime or cement binders [29,30]. Geopolymers consist of extensive three-dimensional structures of covalently bonded aluminosilicates formed by the alkali activation of aluminosilicate rich materials [24]. They are essentially rigid gels that may evolve to form amorphous or crystalline materials under certain temperature and pressure conditions [31]. Chemically, geopolymers can be classified as polysialates and can be represented by their empirical formula as shown in Equation (1) [25].

$$M_n[-(SiO_2)_z - AlO_2]_n \cdot wH_2O \tag{1}$$

where, M is the alkali metal cation (such as Na, K, or Ca), n is the degree of polycondensation, z is the silicon to aluminum (Si:Al) ratio (usually 1, 2, or 3), and w is the molar water amount. Geopolymers are formed in a high pH environment through an alkali-activated polycondensation reaction comprising of five stages—dissolution, speciation equilibrium, gelation, reorganization, and polymerization and hardening [10,24,32]. The synthesis of geopolymers requires an aluminosilicate-rich source (metakaolin, fly ash), an alkali-metal cation source (such as NaOH, KOH, or $Ca(OH)_2$), an additional source of silica (as needed), and water. Predetermined ratios of the components are mixed to form a slurry which on curing form a hardened geopolymer. The transformation of the slurry to form the hardened geopolymer is the result of overlapping polycondensation reactions of the dissolved aluminosilicate species in an aqueous solution resulting in their subsequent polymerization, gelation, and hardening [33]. The gelation of different aluminosilicate species and characteristics of their respective geopolymer formations are dependent on various factors such as concentration of reactive species in solution, raw material type and quality, water content, curing conditions, and time [29,34,35]. Recent studies have shown that metakaolin-based geopolymers significantly improved the strength and volume-change properties of expansive soils [10,33,36,37].

The focus of this study is to assess the sustainability benefits of the metakaolin-based geopolymer used by the authors to successfully treat a high plasticity expansive clayey soil [10,33]. The sustainability benefits of the metakaolin-based geopolymer in this study were evaluated based on the sustainability framework developed at University of Texas at Arlington, which utilizes a weighted multi-criterial evaluation based on resource consumption, environmental impact, and socio-economic impact [18]. Additionally, sustainability benefits were assessed for the conventional lime treatment of the same clay.

2. Materials and Methods

2.1. Materials

A potentially expansive subgrade soil was obtained from Lewisville, Denton county in North Texas for soil treatment work. The soil was subjected to a series of basic geotechnical characterization tests as per the ASTM International (ASTM) standards to better understand its properties, which are summarized in Table 1. The soil was tested for particle size distribution (ASTM D6913-17, D7928-17) and Atterberg limits (ASTM D4318-17, TEX 105-E) and based on these results, the soil was classified as per the Unified Soil Classification System (USCS) method as a high-plasticity clay (CH) with a plasticity index (PI) of 53%, indicating a high swelling potential. Moisture content–dry density relationships (ASTM 44609-94, GR-84-14) were determined to obtain the optimum moisture content (OMC) at which the soil was compacted to its maximum dry density (MDD). The soil was oven-dried, crushed, and subsequently pulverized before being subjected to engineering tests and chemical treatments.

Table 1. Basic geotechnical properties of Lewisville soil.

	Sand Content (%)	Silt Content (%)	Clay Content (%)	LL [1] (%)	PI [2] (%)	MDD [3] (kN/m^3)	OMC [4] (%)	USCS Classification
Lewisville Soil	9.5	37.8	52.4	80	53	15.70	24	High plasticity clay

[1] Liquid limit, [2] plasticity index, [3] maximum dry density, [4] optimum moisture content.

Metakaolin ($2SiO_2 \cdot Al_2O_3$), a highly reactive pozzolanic material was the sole aluminosilicate source of the geopolymer evaluated in this study. Metakaolin is known to have increased compressive strength, reduced permeability, excellent workability, and significantly lower quantities of calcium oxide in comparison to other commonly used aluminosilicate sources like fly ash [38–40], making it an ideal aluminosilicate source for treatment of an expansive soil from a potentially sulfate-rich area in North Texas. Predetermined quantities of metakaolin and an aqueous alkaline activator solution (composed of potassium hydroxide, silica fume, and water) were mixed at room temperature to form a geopolymer slurry with the following mix proportions: $(SiO_2:Al_2O_3) = 4$, $(H_2O:(Al_2O_3 + SiO_2)) = 3$, and $(K_2O:Al_2O_3) = 1$. The workable geopolymer slurry transforms into a hardened geopolymer upon curing at room temperature, with a setting time of 8 days. A simplified schematic of the geopolymer synthesis process is shown below in Figure 1. In this study, geopolymer dosage is presented as the percentage weight of metakaolin (MK) in the geopolymer slurry with respect to the dry weight of soil to be treated. Geopolymer treatment of the high plasticity soil was performed by mixing the appropriate dosage of geopolymer slurry with a predetermined amount of dry soil to form test specimens, and cured at room temperature (22 °C) at 100% relative humidity for different curing periods. Three dosages of 4%, 10%, and 15% metakaolin (MK) were applied to the soil and tested for strength and volume-change properties [33,41].

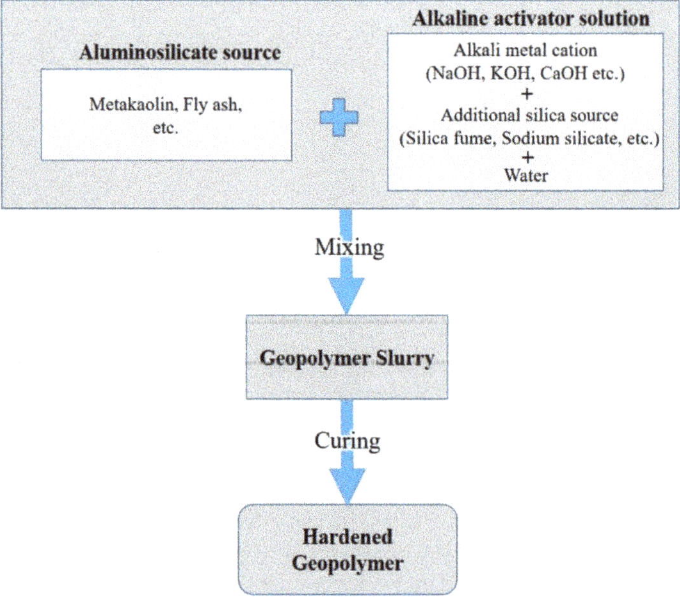

Figure 1. Simplified schematic of geopolymer synthesis [33].

2.2. Engineering Test Methods

Efficiency of geopolymer treatment was assessed by performing unconfined compressive strength (UCS), one-dimensional (1-D) swell, and linear shrinkage tests on control (untreated) and geopolymer-treated soil. A set of three tests were performed for each dosage and curing period of every specimen, to ensure reliability of results. The exact procedures of the test methods are described in detail in [33]. All test specimens were molded in three equivalent lifts at their respective OMCs to 95% of their MDD by static compaction, based on moisture content–dry density relationship tests (ASTM D4609-94) for each soil-geopolymer dosage.

Variations in soil strength before and after geopolymer treatment of CH were evaluated from UCS tests (ASTM D2166-6). UCS is a critical geotechnical parameter and is known as the highest axial compressive strength an unconfined soil mass can bear before failing. These tests were performed by applying strain-controlled uniaxial loads to right-cylindrical soil specimens until they failed. The stress at which the specimen fails is known as the UCS of the soil, which is also the maximum value of the stress-strain curve obtained from the UCS test. UCS tests were performed on control and geopolymer-treated specimens with diameter of approximately 33.3 mm (1.31 inches) and height of 72.4 mm (2.85 inches), with a final height-to-diameter ratio of at least 2. Geopolymer-treated soils were tested for UCS for three different curing periods of 0 (within 2 h of molding), 7, and 28 days. Figure 2 shows a geopolymer-treated soil specimen before and after UCS testing.

(a)   (b)

Figure 2. Unconfined compressive strength (UCS) testing of geopolymer-treated samples: (**a**) intact soil specimen before testing; (**b**) failure crack on soil specimen after testing.

The 1-D swell test quantifies the swelling potential of a soil, by determining its free swell. The free swell of a soil is defined as the percentage swell exhibited by the soil following water absorption at a seating pressure of 1 kPa (0.145 psi). 1-D swell tests were performed for control and geopolymer-treated CH as per ASTM D4546-14e1 using a modified swell test setup, as shown in Figure 3. Compacted soil specimens with approximate diameter of 63.5 mm (2.5 inches) and height of 28.7 mm (1.13 inches) were placed in ring molds to provide radial confinement. These rings were placed in grooved consolidation cells with a top cap load of 1 kPa (0.145 psi), which were subsequently inundated with water. The vertical strain due to water absorption was monitored by a dial gauge placed on the top cap; swell readings were recorded for a period of at least 24 h or until no significant strain change was observed. Geopolymer-treated soils were tested for 1-D swell for three different curing periods of 0 (within 2 h of molding), 3, and 7 days.

Figure 3. Modified one-dimensional (1-D) swell test setup [33].

The shrinkage potential of control and geopolymer-treated CH was determined from linear shrinkage tests (TEX-107-E) using soil passing the 425-μm (No. 40) sieve. Water was mixed with soil to form a slurry comparable to its liquid limit consistency, which was subsequently poured into greased linear shrinkage molds of specific length. The slurry molds were smoothed and oven-dried at 110 °C

(230 °F) until no further change in mass was observed (Figure 4). On cooling, linear shrinkage was determined by measuring the length of oven-dried soil bars as shown in Equation (2):

$$LS = 100 \times (L_W - L_D)/L_W \tag{2}$$

where LS is the linear shrinkage expressed as a percentage; L_W is the length of the wet soil bar, 127 mm (5 inches); and L_D is the length of the dry soil bar, mm (inches). Geopolymer-treated soils were tested for linear shrinkage for three different curing periods of 0 (within 2 h of molding), 3, and 7 days.

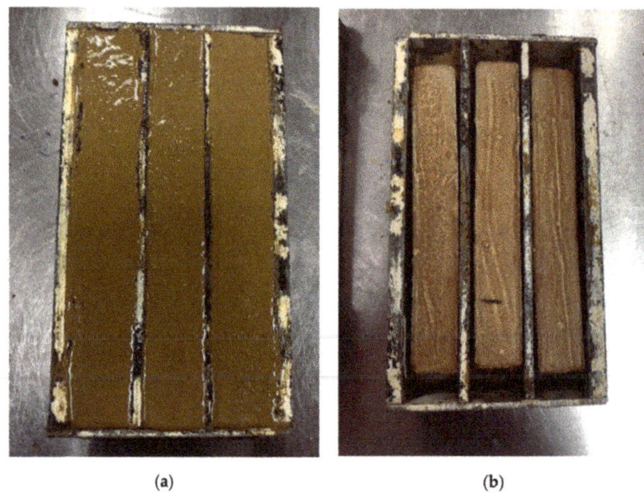

Figure 4. Linear shrinkage testing of geopolymer-treated samples: (**a**) soil slurry in shrinkage mold before oven-drying; (**b**) soil slurry in shrinkage mold after oven-drying.

2.3. Engineering Test Results

Significant improvement was observed in compressive strength of geopolymer-treated CH soil when compared to control CH material, as shown in Figure 5. The 28-day cured strength of CH treated with the lowest (4% MK) geopolymer dosage was observed to be 100% higher than control CH soil with an UCS of 20 psi. The intermediate (10% MK) and highest (15% MK) geopolymer dosages exhibited much higher strength increase by about 225% and 520%, respectively. Immediate strength improvement was observed in treated CH soil with the lower and intermediate dosages of geopolymer, while the highest dosage required a curing period of 7 days for pronounced strength improvement. From UCS tests, the intermediate dosage of 10% MK geopolymer was found to be sufficient for significantly increasing the compressive strength of CH soil.

1-D swell and linear shrinkage tests were conducted to detect changes in volume-change properties of control and geopolymer-treated CH soil. Control CH soil exhibited vertical swell strain and linear shrinkage strain of about 15% and 22%, respectively, indicating a very high swelling and shrinkage potential of this soil. Volume-change test specimens were treated with three geopolymer dosages and cured in air-tight chambers at 100% relative humidity with zero moisture loss. Geopolymer treatment was found to be highly effective for swell and shrinkage tests (as shown in Figure 6). Swelling potential of control CH was reduced by about 80% to 95% within 2 h of geopolymer application. Similarly, shrinkage potential was reduced by about 50% to 90% for all geopolymer dosages within a curing period of 3 days. Immediate swell and shrinkage reduction was observed for all geopolymer dosages within a few hours of geopolymer application.

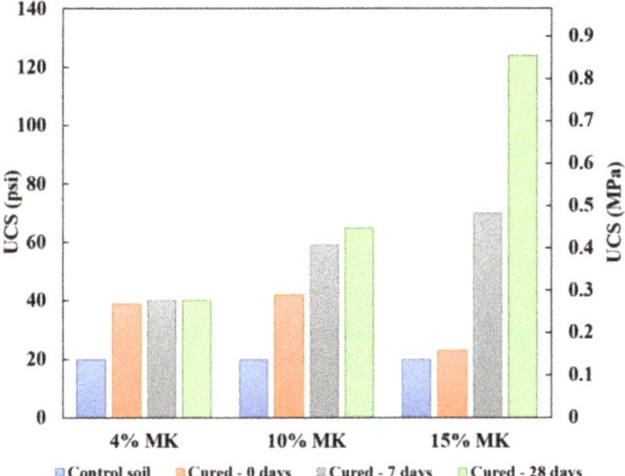

Figure 5. UCS test of control and geopolymer-treated high-plasticity clay (CH). MK—metakaolin.

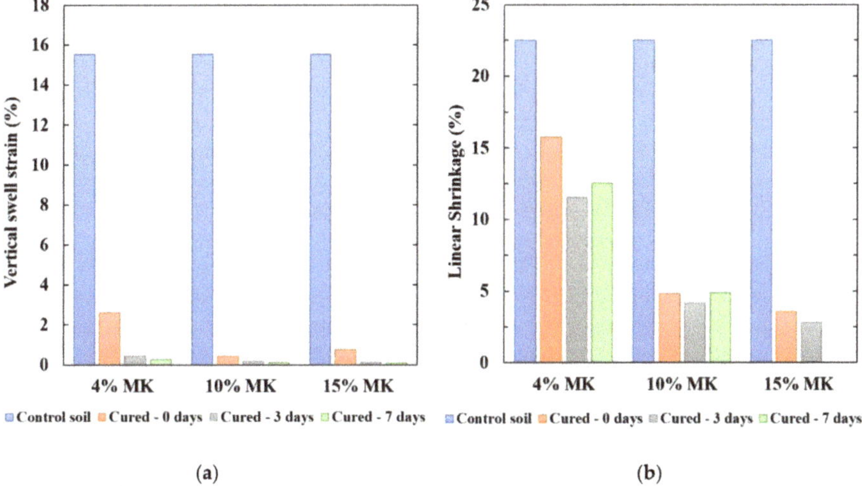

(a)   (b)

Figure 6. Volume-change tests of control and geopolymer-treated CH: (**a**) vertical swell strain plot; (**b**) linear shrinkage plot.

Based on previous tests, a 10% MK dosage of geopolymer was found to be suitable for improvement of the high plasticity Lewisville clay (CH) by increasing UCS and reducing swell and shrinkage strains [33,41]. It should be noted that swell and shrinkage strains reached to near 0% (swell strains) and around 5% (shrinkage strains), respectively. Hence geopolymer with 10% MK is recommended.

3. Sustainability Benefits Assessment Framework

The assessment of sustainability benefits of the metakaolin-based geopolymer in this study was performed by the estimation of the sustainability index (I_{Sus}) as per the framework recently introduced at University of Texas at Arlington [18,42]. The I_{Sus} of a material is proposed to be a function of its resource consumption, environmental impact, and socio-economic impact, and is estimated as shown in Equation (3):

$$I_{Sus} = (W_1 \times I_{Rec}) + (W_2 \times I_{Env}) + (W_3 \times I_{SoEc}) \tag{3}$$

where, I_{Rec} is the resource consumption index, I_{Env} is the environmental impact index, I_{SoEc} is the socio-economic impact index, and W_1, W_2, and W_3 are the weighted values of each associated index.

The weighted values assigned for each index provide an insight into their relevance for a specific project and can be varied based on the executor's judgement. The I_{Sus} can be estimated for different materials for a comparison of the values, with the material with the lowest I_{Sus} being the most sustainable. Life cycle assessment (LCA) is an essential process of obtaining the resource consumption and environmental impact aspects of the sustainability index. The following paragraphs describe the impact factors used in this study to determine the I_{Sus}.

Resource consumption was determined using energy accounting methods during life cycle inventory (LCI) analysis, which is a subset of LCA. The I_{Rec} was estimated using the embodied energy of materials using a "cradle to gate" approach which accounts for the energy expended during the process of production and transportation of materials. The embodied energy of materials used in the study were obtained from the literature [43] and is reported in megajoules (MJ).

Environmental impact assessment is a function of three major components—global warming potential, acidification potential, and eutrophication potential. The global warming potential (GW_P) is an estimate of the impact of raw materials and manufacturing processes on the production of greenhouse gases, which consequentially raises the average global temperature. In this study, the GW_P was represented by the amount of carbon dioxide produced contributing to global warming and is reported in gram equivalent of CO_2 (gCO_2 eq.). The acidification potential (A_P) is the ability of a material to raise the acidity of soils or nearby water bodies by decreasing its pH and is measured in gram equivalent of SO_2 (gSO_2 eq.). Increased A_P usually deposits itself in the form of acid rain, which is known to have harmful effects on living beings as well as infrastructure. The eutrophication potential (E_P) is an indicator of biodiversity and ecological health and is measured in gram equivalent of PO_4^{3-} (gPO_4^{3-} eq.). An increase in E_P or over-nutrification is usually evident in aquatic systems by algal blooms that cause oxygen deficiency, leading to the death of other aerobic organisms, thereby disrupting the biodiversity of adjoining ecosystems.

A cost-benefit analysis is used to evaluate the socio-economic impact index (I_{SoEc}) of different materials that can be used for a project. A life cycle costing (LCC) is used to quantify costs associated with each alternative usually including purchase, construction, operation, maintenance, rehabilitation, and other residual costs [18]. Weighted values are applied to the different categories used to calculate each of the indices, based on their relevance for the project A flowchart of the sustainability benefits assessment framework is shown in Figure 7. This was a pilot study of the sustainability benefits of geopolymers as soil stabilizers and therefore focused on the initial cost of materials.

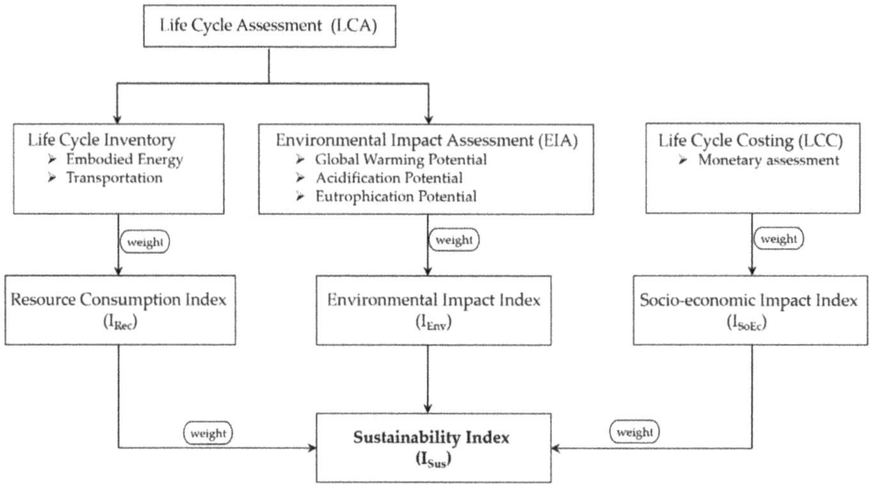

Figure 7. Sustainability assessment framework (adapted from [18]).

4. Comparative Sustainability Benefits

Sustainability benefits were assessed for a laboratory-scale scenario comparing geopolymer and conventional lime treatment of the high-plasticity Lewisville clay (CH). The assessment was performed for dosages of 10% MK for geopolymer treatment and 8% lime for lime treatment of the soil. The appropriate dosage of lime required to stabilize CH was determined based on the Eades and Grim pH test as per TEX 121-E, as well as the soluble sulfate content. In this study, I_{Sus} was evaluated for the primary components of both treatment methods for the same quantity of dry soil (100 kg). As such, this assessment analyzes the sustainability characteristics of metakaolin alone for the geopolymer treatment of soils, as other ingredients (silica fume, KOH) were utilized in lower quantities. Conventional soil treatment of the high plasticity Lewisville clay was performed using commercially available lime. The summary of the treatment methods assessed for sustainability are provided in Table 2:

Table 2. Treatment methods assessed for sustainability.

Treatment ID	A	B
Treatment type	Geopolymer	Lime
Primary component (PC)	Metakaolin	Lime
Soil type	CH	CH
Dry soil (kg)	100	100
Dosage	10% Metakaolin	8% Lime
PC quantity (kg)	10	8

As explained earlier, the I_{Sus} was determined using indices for resource consumption, environmental impact, and socio-economic impact for metakaolin and lime. The embodied energy values for production of metakaolin, as well as its potential for global warming, acidification, and eutrophication were obtained from published literature [43]. The values for embodied energy used during the production of lime, and its acidification and eutrophication potential were obtained from previous studies [44]. The global warming potential of lime was obtained from the Inventory of Carbon and Energy (ICE) database [45]. Additionally, the embodied energy from transportation of materials from source to site was determined from the GREET (Greenhouse gases, Regulated Emissions, and Energy use in Transportation) model [46] developed by Argonne National Laboratory.

The resource consumption of both treatments based on the embodied energy consumed during their production and transportation are summarized in Table 3. The I_{Rec} for the treatment methods were estimated using Equation (4) [18]:

$$I_{Rec} = w_{1a} \times E_{E(material\ 1)} + w_{1b} \times E_{E(material\ 2)} + w_{1c} \times E_{E(transportation)} \tag{4}$$

where, w_{1a}, w_{1b}, w_{1c} are weighted values of each parameter and E_E is the embodied energy. The embodied energy consumed due to transportation of both materials was estimated to be 1.5 MJ/metric ton-km [46], assuming a source to site distance of 80 km (50 miles) covered by a truck. The significantly higher consumption of resources per kg of lime than geopolymer is highlighted by placing higher weighted values on the embodied energy of lime. The resource consumption of Treatment A is observed to be lower than Treatment B with a lower I_{Rec} value of 46.67.

The comparison of environmental impact indices of both treatments is presented in Table 4, where I_{Env} was calculated as per Equation (5) [18]:

$$I_{Env} = w_{2a} \times GW_P + w_{2b} \times A_P + w_{2c} \times E_p \tag{5}$$

where, w_{2a}, w_{2b}, w_{2c} are weighted values of each parameter. The ever-increasing and unimpeded carbon dioxide emissions pose a more imminent concern on a global scale; therefore, higher weightage values were assigned for the GW_P of both treatments than its A_P and E_P. Table 4 shows Treatment A to have a significantly lower I_{Env} value of 38.73 than Treatment B.

Table 3. Calculation of resource consumption index (I_{Rec}).

	Embodied Energy Consumed (MJ)		Per cent Consumption of Embodied Energy (%)		Weights	Weighted Resource Use	
Treatment ID	A [1]	B [2]	A [1]	B [2]		A [1]	B [2]
Geopolymer	25	0	100.00	0.00	0.30	30.00	0.00
Lime	0	63	0.00	100.00	0.40	0.00	40.00
Transportation	1.2	0.96	55.56	44.44	0.30	16.67	13.33
Resource Consumption Index (I_{Rec})						46.67	53.33

[1] Geopolymer treatment, [2] lime treatment.

Table 4. Calculation of environmental impact index (I_{Env}).

	Emission Contribution		Per cent Contribution of Emission (%)		Weights	Weighted Environmental Impact	
Treatment ID	A [1]	B [2]	A [1]	B [2]		A [1]	B [2]
Global Warming Potential (gCO_2 eq.)	3300	6240	34.59	65.41	0.60	20.75	39.25
Acidification Potential (gSO_2 eq.)	3.24	4.95	39.55	60.45	0.20	7.91	12.09
Eutrophication Potential (gPO_4^{3-} eq.)	0.65	0.64	50.30	49.70	0.20	10.06	9.94
Environmental Impact Index (I_{Env})						38.73	61.27

[1] Geopolymer treatment, [2] lime treatment.

The socio-economic impact of the treatments was estimated based on the average unit price of lime obtained from manufacturers and are presented in Table 5. The unit price of the conventional Treatment B was found to be 0.12 USD per kg [47], and the unit price of the novel Treatment A was assumed to be 50% more than the unit price of Treatment B. Note that the actual unit price of Treatment A is significantly higher and it can be attributed to low demand. The higher cost contribution of Treatment A is attributed to the higher unit price of the novel Treatment A compared to the low unit price of the mainstream Treatment B, in addition to the higher quantity of novel material required for soil treatment. The I_{SoEc} was calculated using Equation (6) as [18]:

$$I_{SoEc} = w_3 \times C \tag{6}$$

where $w_3 = 1.0$, and C is the total cost of treatment. Treatment B was found to have a lower socio-economic impact with a lower I_{SoEc} value of 34.78.

Table 5. Calculation of environmental impact index (I_{SoEc}).

	Cost Contribution		Per cent Contribution of Cost (%)		Weights	Weighted Socio-Economic Impact	
Treatment ID	A [1]	B [2]	A [1]	B [2]		A [1]	B [2]
Cost of treatment (USD)	1.80	0.96	65.22	34.78	1.0	65.22	34.78
Socio-Economic Impact Index (ISoEc)						65.22	34.78

[1] Geopolymer treatment, [2] lime treatment.

Finally, the I_{Sus} of both treatments was calculated by adding the weighted values of the three indices as shown in Equation (1) and is summarized in Table 6. For the calculation of I_{Sus}, both the I_{Rec} and I_{Env} were assigned a weight of 40% while, the I_{SoEc} was assigned a lower weight of 20%, since current cost estimates need to be further adjusted based on future supply and demand, in addition to cost being a secondary aspect of this study. According to the sustainability benefits assessment,

Treatment A (geopolymers) is deemed a more sustainable alternative with a lower I_{Sus} value than using Treatment B (lime) for soil improvement. It is important to note that the weighted values applied to each of the indices calculated are left to the discretion of the user. The weighted values will vary significantly for each project and will need to be verified by the user to be reliable and meaningful for the respective application and expected end-goal. The sustainability assessment framework can be used to effectively compare different alternatives for a project, to determine the most sustainable alternative.

Table 6. Calculation of sustainability index (I_{Sus}).

	Index Value		Weights	Weighted Index	
Treatment ID	A [1]	B [2]		A [1]	B [2]
Resource Consumption (I_{Rec})	46.67	53.33	0.40	18.67	21.33
Environmental Impact (I_{Env})	38.73	61.27	0.40	15.49	24.51
Socio-Economic Impact (I_{SoEc})	65.22	34.78	0.20	13.04	6.96
Sustainability Index (I_{Sus})				47.20	52.80

[1] Geopolymer treatment, [2] lime treatment.

Figure 8 shows a graphical representation of the different aspects of the sustainability benefits assessment. It is to be noted that the radar chart does not reflect the unequal individual weights applied to each of the parameters to evaluate the indices [18]. The structural integrity characteristics representative of resiliency, namely probability of fatigue cracking and rutting failure, are also included in this radar chart for a better comparison of both treatments. The hypothetical pavement section and design system used for evaluating the resiliency characteristics have been described in detail in [33]. From Figure 8, the area under geopolymer treatment was estimated to be about 0.64 square units, while it was larger for lime treatment with an estimate of 0.86 square units. Therefore, Treatment A (geopolymer) is confirmed to be a more sustainable alternative than Treatment B (lime) based on the assumptions of the sustainability benefits assessment elaborated in this study.

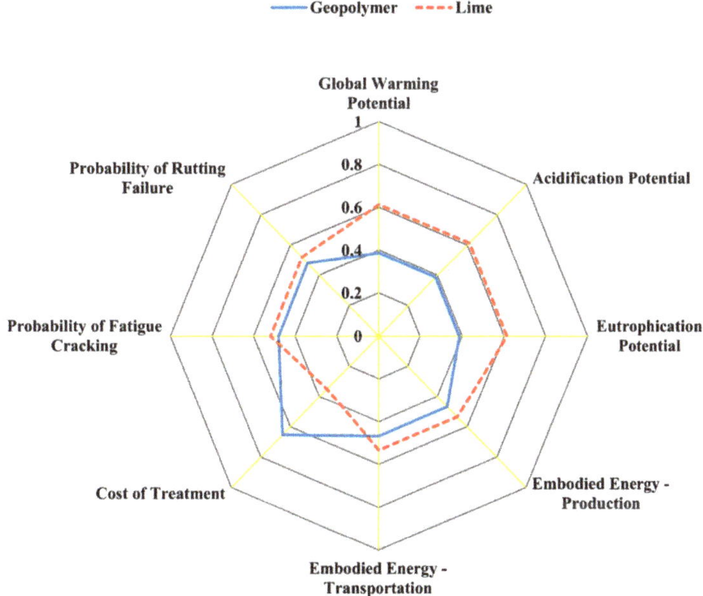

Figure 8. Graphical representation of sustainability benefits assessment [33].

5. Conclusions

Conventional chemical additives such as lime and cement have been used to effectively stabilize expansive soils. The production of lime and cement are major contributors to global greenhouse gas emissions and are therefore not sustainable. Recently, geopolymers have been investigated as alternatives for soil stabilization, due to comparable strength and volume-change properties. This study presents the sustainability benefits assessment of a metakaolin-based geopolymer-treatment for a high plasticity expansive soil, and its comparison with lime-treatment. The sustainability benefits framework used a weighted multi-criteria assessment to estimate a sustainability index (I_{Sus}) for each treatment method. The global warming potential and embodied energy of production of a kilogram of lime were found to be significantly higher than metakaolin, the primary component of the geopolymer. Based on the weightage applied as well as assumptions regarding cost and transportation, the metakaolin-based geopolymer with a lower I_{Sus} was found to be a more sustainable alternative to the conventional lime treatment for soil stabilization.

Author Contributions: The authors confirm contribution to the paper as follows: study conception and design: A.J.P., R.S., and M.R.; data collection: R.S.; analysis and interpretation of results: R.S.; draft manuscript preparation: R.S. All authors have read and agreed to the published version of the manuscript.

Funding: This research was funded by the USDOT's Transportation Consortium for South-Central States (Tran-SET), under research grants 17GTTAM02 and 18CTAMU04.

Acknowledgments: The authors gratefully acknowledge the USDOT's Transportation Consortium for South-Central States (Tran-SET), who funded this study. The first author of this article was the recipient of the USDOT's Dwight D. Eisenhower Transportation Fellowship for the years 2017–2019 and is thankful for the financial support provided. Any findings, conclusions, or recommendations expressed in this material are those of the authors and do not necessarily reflect the views of the funding agency.

Conflicts of Interest: The authors declare no conflict of interest.

References

1. Low, P.F. Structural component of the swelling pressure of clay. *Langmuir* **1987**, *3*, 18–25. [CrossRef]
2. Nelson, J.D.; Chao, K.C.; Overton, D.D.; Nelson, E.J. *Foundation Engineering for Expansive Soils*; John Wiley & Sons: Hoboken, NJ, USA, 2015.
3. Nelson, J.D.; Miller, D.J. *Expansive Soils: Problems and Practice in Foundation and Pavement Engineering*; John Wiley & Sons: Hoboken, NJ, USA, 1992.
4. Olive, W.W.; Chleborad, A.F.; Frahme, C.W.; Schlocker, J.; Schneider, R.R.; Schuster, R.L. *Swelling Clays Map of the Conterminous United States*; USGS: Reston, VA, USA, 1989.
5. Steinberg, M.L. *Geomembranes and the Control of Expansive Soils in Construction*; McGraw-Hill: New York, NY, USA, 1998; ISBN 0070611785.
6. Saride, S.; Puppala, A.J.; Williammee, R. Assessing recycled/secondary materials as pavement bases. Proceedings of the ICE. *Ground Improv.* **2010**, *163*, 3–12. [CrossRef]
7. Puppala, A.; Hoyos, L.; Viyanant, C.; Musenda, C. Fiber and fly ash stabilization methods to treat soft expansive soils. *Soft Ground Technol.* **2001**, *11*, 136–145.
8. Jones, D.E.; Holtz, W.G. Expansive soils—The hidden disaster. *Civ. Eng.* **1973**, *43*, 49–51.
9. Puppala, A.J.; Cerato, A. Heave distress problems in chemically-treated sulfate-laden materials. *Geo-Strata* **2009**, *10*, 28.
10. Samuel, R.; Huang, O.; Banerjee, A.; Puppala, A.; Das, J.; Radovic, M. Case Study: Use of Geopolymers to Evaluate the Swell-Shrink Behavior of Native Clay in North Texas. *Eighth Int. Conf. Case Hist. Geotech. Eng.* **2019**, 167–178. [CrossRef]
11. Little, D.N. *Handbook for Stabilization of Pavement Subgrades and Base Courses with Lime*; Lime Association of Texas: Austin, TX, USA, 1995; ISBN 0840396325.
12. Petry, T.M.; Little, D.N. Review of Stabilization of Clays and Expansive Soils in Pavements and Lightly Loaded Structures—History, Practice, and Future. *J. Mater. Civ. Eng.* **2002**, *14*, 447–460. [CrossRef]

13. Puppala, A.J.; Congress, S.S.C.; Banerjee, A. Research Advancements in Expansive Soil Characterization, Stabilization and Geoinfrastructure Monitoring. In *Frontiers in Geotechnical Engineering*; Springer: Singapore, 2019; pp. 15–29. [CrossRef]

14. Katz, L.; Rauch, A.; Liljestrand, H.; Harmon, J.; Shaw, K.; Albers, H. Mechanisms of soil stabilization with liquid ionic stabilizer. *Transp. Res. Rec.* **2001**, *1757*, 50–57. [CrossRef]

15. Puppala, A.J.; Intharasombat, N.; Vempati, R.K. Experimental Studies on Ettringite-Induced Heaving in Soils. *J. Geotech. Geoenviron. Eng.* **2005**, *131*, 325–337. [CrossRef]

16. U.S. Environmental Protection Agency. Greenhouse Gas Reporting Program (GHGRP)-GHGRP Minerals. 2020. Available online: https://www.epa.gov/ghgreporting/ghgrp-minerals (accessed on 15 August 2019).

17. Brundtland, G.H. Report of the World Commission on Environment and Development: "Our Common Future". *United Nations Gen. Assem.* **1987**. Available online: https://sustainabledevelopment.un.org/content/documents/5987our-common-future.pdf (accessed on 15 August 2019).

18. Das, J.T. *Assessment of Sustainability and Resilience in Transportation Infrastructure Geotechnics*; University of Texas at Arlington: Arlington, TX, USA, 2018.

19. Pantelidou, H.; Nicholson, D.; Gaba, A. Sustainable Geotechnics. *Man. Geotech. Eng.* **2012**. [CrossRef]

20. Graedel, T. Industrial ecology: Definition and implementation. *Ind. Ecol. Glob. Chang.* **1994**. [CrossRef]

21. Kibert, C.J. *Sustainable Construction: Green Building Design and Delivery*; John Wiley & Sons: Hoboken, NJ, USA, 2016.

22. Davidovits, J. *Geopolymer Chemistry and Applications*; Institut Géopolymère, Geopolymer Institute: Saint-Quentin, France, 2008.

23. Provis, J.L.; van Deventer, J.S.J. *Geopolymers: Structures, Processing, Properties and Industrial Applications*; Elsevier: Amsterdam, The Netherlands, 2009.

24. Duxson, P.; Fernández-Jiménez, A.; Provis, J.L.; Lukey, G.C.; Palomo, A.; van Deventer, J.S.J. Geopolymer technology: The current state of the art. *J. Mater. Sci.* **2007**, *42*, 2917–2933. [CrossRef]

25. Davidovits, J. Geopolymers-Inorganic polymeric new materials. *J. Therm. Anal.* **1991**, *37*, 1633–1656. [CrossRef]

26. Van Jaarsveld, J.G.S.; van Deventer, J.S.J.; Lukey, G.C. The effect of composition and temperature on the properties of fly ash-and kaolinite-based geopolymers. *Chem. Eng. J.* **2002**, *89*, 63–73. [CrossRef]

27. Cheng, T.W.; Chiu, J.P. Fire-resistant geopolymer produced by granulated blast furnace 5 slag. *J. Miner. Eng.* **2003**, *16*, 205–210. [CrossRef]

28. Gordon, M.; Bell, J.L.; Kriven, W.M. Comparison of naturally and synthetically-derived potassium-based geopolymers. *Ceram. Trans. Ser.* **2005**, *165*, 95–106.

29. Lizcano, M.; Gonzalez, A.; Basu, S.; Lozano, K.; Radovic, M. Effects of Water Content and Chemical Composition on Structural Properties of Alkaline Activated Metakaolin-Based Geopolymers. *J. Am. Ceram. Soc.* **2012**, *95*, 2169–2177. [CrossRef]

30. Gartner, E. Industrially interesting approaches to "low-CO_2" cements. *Cem. Concr Res.* **2004**, *34*, 1489–1498. [CrossRef]

31. Bell, J.L.; Driemeyer, P.E.; Kriven, W.M. Formation of ceramics from metakaolin-based geopolymers. Part II: K-based geopolymer. *J. Am. Ceram. Soc.* **2009**, *92*, 607–615. [CrossRef]

32. Medri, V.; Fabbri, S.; Dedecek, J.; Sobalik, Z.; Tvaruzkova, Z.; Vaccari, A. Role of the morphology and the dehydroxylation of metakaolins on geopolymerization. *Appl. Clay Sci.* **2010**, *50*, 538–545. [CrossRef]

33. Samuel, R.A. *Synthesis of Metakaolin-based Geopolymer and its Performance as Sole Stabilizer of Expansive Soils*; University of Texas at Arlington: Arlington, TX, USA, 2019.

34. Duxson, P.; Provis, J.L.; Lukey, G.C.; Mallicoat, S.W.; Kriven, W.M.; van Deventer, J.S.J. Understanding the relationship between geopolymer composition, microstructure and mechanical properties. *Colloids Surf. A Physicochem. Eng. Asp.* **2005**, *269*, 47–58. [CrossRef]

35. Steins, P.; Poulesquen, A.; Diat, O.; Frizon, F. Structural evolution during geopolymerization from an early age to consolidated material. *Langmuir* **2012**, *28*, 8502–8510. [CrossRef] [PubMed]

36. Zhang, M.; Zhao, M.; Zhang, G.; Nowak, P.; Coen, A.; Tao, M. Calcium-free geopolymer as a stabilizer for sulfate-rich soils. *Appl. Clay Sci.* **2015**, *108*, 199–207. [CrossRef]

37. Khadka, S.D.; Jayawickrama, P.W.; Senadheera, S. Strength and shrink/swell behavior of highly plastic clay treated with geopolymer. *Transp. Res. Rec.* **2018**, *2672*, 174–184. [CrossRef]

38. Caldrone, M.A.; Gruber, K.A.; Burg, R.G. High reactivity Metakaolin: A new generation mineral admixture for high performance concrete. *Concr. Int.* **1994**, *16*, 11.

39. Gruber, K.A.; Ramlochan, T.; Boddy, A.; Hooton, R.D.; Thomas, M.D.A. Increasing concrete durability with high-reactivity metakaolin. *Cem. Concr. Compos.* **2001**, *23*, 479–484. [CrossRef]

40. McManis, K.L.; Arman, A. Class C fly ash as a full or partial replacement for portland cement or lime. *Transp. Res. Rec.* **1989**, *1219*, 68–81.

41. Samuel, R.; Puppala, A.J.; Banerjee, A.; Huang, O.; Radovic, M.; Chakraborty, S. Improvement of Strength and Volume-Change Properties of Expansive Clays with Geopolymer Treatment. *Transp. Res. Rec..* (under review).

42. Puppala, A.J.; Das, J.T.; Bheemasetti, T.V.; Congress, S.S.C. Sustainability and Resilience in Transportation Infrastructure geotechnics: Integrating Advanced Technologies for better Asset Management. *Geo-Strata* **2018**, *22*, 42–48.

43. Heath, A.; Paine, K.; McManus, M. Minimising the global warming potential of clay based geopolymers. *J. Clean. Prod.* **2014**, *78*, 75–83. [CrossRef]

44. Da Rocha, C.G.; Passuello, A.; Consoli, N.C.; Quiñónez Samaniego, R.A.; Kanazawa, N.M. Life cycle assessment for soil stabilization dosages: A study for the Paraguayan Chaco. *J. Clean. Prod.* **2016**, *139*, 309–318. [CrossRef]

45. Hammond, G.; Jones, C.; Lowrie, F.; Tse, P. *Embodied Carbon: The Inventory of Carbon and Energy (ICE)*; University of Bath and BSRIA: Bracknell, UK, 2011.

46. Wang, M.Q. *Technical Report: GREET 1.5 -Transportation Fuel-Cycle Model: Methodology, Development, Use, and Results*; Argonne National Lab.: Argonne, IL, USA, 1999.

47. Corathers, L.A.; Apodaca, L.E. Lime [Advance Release]. In *Metals and Minerals: U.S. Geological Survey Minerals Yearbook 2016*; U.S. Geological Survey (USGS): Reston, VA, USA, 2016; pp. 43.1–43.13. [CrossRef]

Publisher's Note: MDPI stays neutral with regard to jurisdictional claims in published maps and institutional affiliations.

 *sustainability*

Article

Examining Energy Consumption and Carbon Emissions of Microbial Induced Carbonate Precipitation Using the Life Cycle Assessment Method

Xuejie Deng [1,2,*], Yu Li [1], Hao Liu [1], Yile Zhao [1], Yinchao Yang [2], Xichen Xu [3], Xiaohui Cheng [3] and Benjamin de Wit [4]

[1] School of Energy and Mining Engineering, China University of Mining and Technology, Beijing 100083, China; liyu_cumtb@163.com (Y.L.); liuhaobj@126.com (H.L.); zhaoyile_cumtb@163.com (Y.Z.)
[2] Kailuan (Group) Co. Ltd., Tangshan 063018, China; rcglk@kailuan.com.cn
[3] Department of Civil Engineering, Tsinghua University, Beijing 100084, China; xxc19@mails.tsinghua.edu.cn (X.X.); chengxh@tsinghua.edu.cn (X.C.)
[4] Smith School of Business, Queen's University, Toronto, ON M5V 3K2, Canada; mr.bendewit@gmail.com
* Correspondence: dengxj@cumtb.edu.cn; Tel.: +86-18811352307; Fax: +86-010-62339060

Citation: Deng, X.; Li, Y.; Liu, H.; Zhao, Y.; Yang, Y.; Xu, X.; Cheng, X.; Wit, B.d. Examining Energy Consumption and Carbon Emissions of Microbial Induced Carbonate Precipitation Using the Life Cycle Assessment Method. *Sustainability* **2021**, *13*, 4856. https://doi.org/10.3390/su13094856

Academic Editor: Castorina Silva Vieira

Received: 30 March 2021
Accepted: 22 April 2021
Published: 26 April 2021

Publisher's Note: MDPI stays neutral with regard to jurisdictional claims in published maps and institutional affiliations.

Abstract: Microbial induced carbonate precipitation (MICP) is a new geotechnical engineering technology used to strengthen soils and other materials. Although it is considered to be environmentally friendly, there is a lack of quantitative data and objective evaluation to support conclusions about its environmental impact. In this paper, the energy consumption and carbon emissions of MICP technology are quantitatively analyzed by using the life cycle assessment (LCA) method. The environmental effects of MICP technology are evaluated from the perspectives of resource consumption and environmental impact. The results show that for each tonne of calcium carbonate produced by MICP technology, 1.8 t standard coal is consumed and 3.4 t CO_2 is produced, among which 80.4% of the carbon emissions and 96% of the energy consumption come from raw materials. Comparing using MICP with cement, lime, and sintered brick, the current MICP application process consumes less non-renewable resources but has a greater environmental impact. The major environmental impact that MICP has is the production of smoke and ash, with secondary impacts being global warming, photochemical ozone creation, acidification, and eutrophication. In five potential application scenarios of MICP, including concrete, sintered brick, lime mortar, mine cemented backfill, and foundation reinforcement, the carbon emissions of MICP are 3 to 7 times greater than the emissions of traditional technologies. The energy consumption is 15 to 23 times. Based on the energy consumption and carbon emissions characteristics of MICP technology at the current condition, suggestions are given for the future research of MICP.

Keywords: microbial induced carbonate precipitation; life cycle assessment; energy consumption; carbon emissions

1. Introduction

Microbial induced carbonate precipitation (MICP) is a bio-mineralization process that refers to microorganisms in rock masses or soils generating calcium carbonate mineral crystals. The MICP process is naturally occurring in many circumstances and can be initiated artificially, under specific environmental and nutritional conditions, to take advantage of the good cementing properties created during the process [1,2]. MICP technology has been successfully applied in many areas of engineering, including foundation reinforcement [3,4], cultural relic restoration [5], anti-seepage and anti-leakage controls [6], heavy metal solidification [7–9], mechanical soil improvement [10–12], underground mine waste and backfill optimization [13], in addition to a broad range of prospective applications.

There are four types of MICP technology, including urea hydrolysis, denitrification, iron salt reduction reaction, and sulfate reduction reaction. MICP using urea hydrolysis is the most widely used of the 4 processes, having a reaction process that is relatively simple and easy to control and also one that produces a significant amount of carbonates in a short period. Therefore, MICP using urea hydrolysis has been the mainstream technology for calcium carbonate biomineralization [14]. The principal characteristic of MICP urea hydrolysis is that urea is hydrolyzed to carbonate ions by the catalysis of microbial urease. Then carbonate ions react with free calcium ions in the system to produce calcium carbonate precipitation with gelation properties. The mechanism of MICP can be summarized as Formulas (1)–(3) [4,15], and the reaction mechanism model of MICP with urea hydrolyzed bacteria is shown in Figure 1 [16].

$$CO(NH_2)_2 + 2H_2O \xrightarrow{Bacterial} 2NH_4^+ + CO_3^{2-} \tag{1}$$

$$Ca^{2+} + Cell \rightarrow Cell - Ca^{2+} \tag{2}$$

$$Cell - Ca^{2+} + CO_3^{2-} \rightarrow Cell - CaCO_3 \tag{3}$$

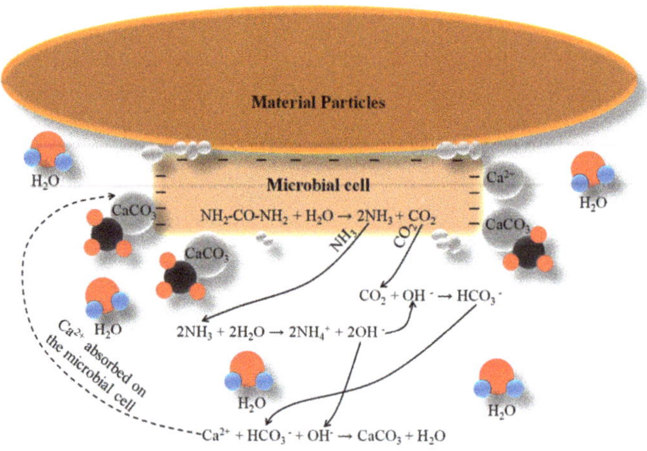

Figure 1. Reaction mechanism model of MICP with urea hydrolyzed bacteria.

Since MICP technology was first introduced, one of its biggest supporting characteristics has been its classification as "environmentally friendly". When comparing MICP to traditional cementation methods, DeJong et al. [17] concluded that MICP provided an alternative and environmentally friendly approach for soil improvement. Nader Hataf [18] used MICP technology to form barriers in waste, water, and foundations, providing an environmentally friendly method for reducing soil permeability of landfill foundations and walls. Li Meng used MICP technology to consolidate heavy metal ions in wastewater to reduce environmental pollution [19]. Gai [20] and Li [21] writes about MICP being used to improve sand and soils and refer to MICP as an environmentally friendly technology in their paper. Adharsh Rajasekar [22] points out that MICP is not 100% environmentally friendly, and the by-products of the reactions can be harmful to human health and local microbiota. To evaluate the environmental effects of MICP scientifically and objectively, energy consumption and carbon emission should be used as indicators for the quantitative analysis.

The consensus viewpoint of current MICP research is that the technology is environment-friendly, but this viewpoint is often reached without providing evaluations that are thoroughly supported by specific, objective, and systematic methods. This contrasts with other industries where research completed on energy consumption and carbon emissions is often supported by evaluation methods, such as life cycle assessment (LCA), which is a quantitative study of energy consumption and carbon emissions [23]. LCA has been applied to various industries, including cement [24], concrete [25], and steel manufacturing, and the strong results achieved support its use for evaluating MICP. Implementing MICP treatment can be applied through different methods, including the grouting method [19,26], the soaking method [27,28], and the mixing method [13], all, of which have relatively low energy consumption. In addition to having lower energy consumption, MICP technology does not directly produce harmful environmental gases, and beyond this, the technology can absorb carbon dioxide from the air by forming carbonate ions to participate in the reaction. However, though there are not direct environmental impacts of applying MICP technology, creating the raw materials used in MICP—including urease bacteria, urea, and calcium—entails significant energy consumption and carbon emissions.

In this paper, LCA is used to study the energy consumption and carbon emissions of MICP technology. From the perspective of resource consumption and environmental impact, the energy consumption and carbon emissions of MICP technology in different application scenarios are quantitatively analyzed, and the environmental impacts of MICP technology are evaluated. The research conclusion of this paper is not to deny the technical advantages of MICP but to point out the future research direction of MCIP based on its energy consumption and carbon emission performance under the current technical level.

2. Materials and Methods

LCA is an assessment method that performs "cradle" to "grave" evaluations, which can be used to thoroughly assess the environmental impacts of technologies [24,29]. LCA includes analyzing the extraction of raw materials, the quantification of energy utilization, the transportation of materials, the production process, and long-term and disposal impacts. As defined in International Organization for Standardization (ISO) 14040:2006 [23], LCA is a compilation and assessment of inputs, outputs, and the potential environmental impacts throughout the life cycle of a product. The LCA method is a systematic multi-stage approach and a reliable assessment tool that is divided into four parts: (a) goal and scope definition; (b) inventory analysis; (c) impact assessment of the process; and (d) analysis of the results.

2.1. Energy Consumption and Carbon Emission Analysis of MICP Technology Based on LCA

(1) Goal and scope

The LCA method is used in this research to study the energy consumption and carbon emissions of MICP technology and evaluate the environmental effects in different applied scenarios using various strength levels.

(2) System boundaries

The core functional product of MICP technology is calcium carbonate ($CaCO_3$). When evaluating the life cycle of MICP, $CaCO_3$ produced by MICP is used as the evaluation index, and the functional unit is set to 1 tonne $CaCO_3$. The system boundaries of MICP technology in the analysis process are shown in Figure 2.

(3) Analysis and assessment

Based on the system boundaries of MICP technology, a quantitative inventory analysis is conducted for raw material consumption, carbon emissions, and energy consumption within the system. On this basis, by calculating environmental potential (EP) and abiotic depletion potential (ADP), the impact of MICP technology on the environment in different applied scenarios is evaluated.

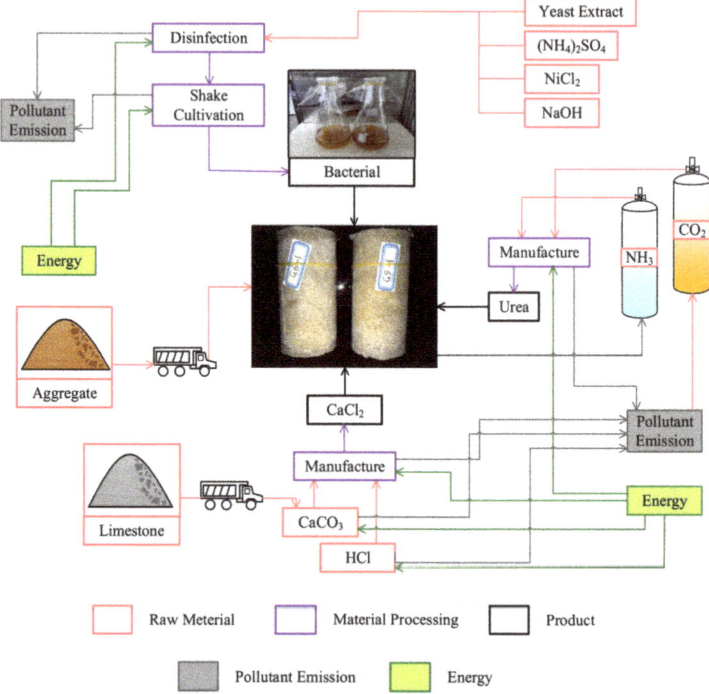

Figure 2. The system boundaries of MICP technology.

2.2. Environmental Impact Assessment Method of MICP
2.2.1. Abiotic Depletion Potential

Natural resources are limited, and as an index of LCA evaluation, abiotic depletion potential (ADP) is a value assigned to a product that is influenced by the amount and type of mineral resources the product consumes in its production. ADP is used to transform the different combinations of consumed resources into a unified reference value that can be used for comparative analyses. The calculation method is shown in Formula (4):

$$ADP = \Sigma[MC(i) \times EF(i)] \tag{4}$$

where *ADP* is the total non-renewable resource potential of the constituent elements used to produce the product; *MC(i)* is the generalized material consumption value of element *i*; *EF(i)* is the equivalent factor of element *i*.

The equivalent factor of ADP is shown in Table 1.

Table 1. The equivalent factor of ADP.

Item	Unified Reference	Category			
		Limestone	Gypsum	Iron Powder	Clay
ADP	kg oil	1.0762	3.0141	3.1283	2.06

2.2.2. Environmental Impact Potential Value

In the analysis of the MICP system and carbon emissions, other pollutant emissions affect the environment. Therefore, the calculated environmental impact potential value refers to the comprehensive index of total emission impacts of various pollutants as seen in all environments of the entire product system, which can be calculated by Formula (5):

$$EP(j) = \Sigma EP(j)_i = \Sigma[Q(j)_i \times EF(j)_i] \tag{5}$$

where $EP(j)$ is the generalized environmental impact potential value of environment j within the production system; $EP(j)_i$ is the environmental impact potential value of the pollutant emission i on the environment j; $Q(j)_i$ is the discharge amount of element i on the environment j; $EF(j)_i$ is the equivalent factor of the environmental impact potential value of element i on the environment j.

The determination of equivalent factors varies with different environmental impacts, and it usually takes one element as a reference to calculate the relative size of other elements.

According to the methods proposed by the International Standardization Organization (ISO), the International Society for Environmental Toxicology (SETAC), and the Technical University of Denmark, the types of environmental impacts considered in this study are global warming (GW) [30], acidification (AC), eutrophication (NE) [31], photochemical ozone creation (POC) [32,33], and smoke and ash (SA). The main environmental impact types and their corresponding equivalent factors are shown in Table 2.

Table 2. Environmental impact types and equivalent factors.

Item	GW (g CO_2·Eq./g)	AC (g SO_2·Eq./g)	NE (g NO_3·Eq./g)	POC (g C_2H_2·Eq./g)	SA (g/g)
CO_2	1				
SO_2		1		0.048	
NO_X	320	0.7	1.35	0.028	
CO	2			0.027	
COD			0.23		
CH_4	25			0.006	
PM					1

The calculated environmental impact potential values are compared after standardization and weighted assessment.

3. Life Cycle Inventory Analysis of MICP

3.1. List of Raw Material Consumption

According to Formulas (1)–(3), 1 mol of $CO(NH_2)_2$ and 1 mol of Ca^{2+} can produce 1 mol of cell $CaCO_3$ under the condition of complete reaction. Thus 0.64 tonnes of urea and 0.4 tonnes of Ca^{2+} are needed to produce 1 tonne of calcium carbonate. The ratio of bacterial solution to nutrient salts (urea and $CaCl_2$) is assumed to 1:10. The raw materials used in the whole reaction process of MICP are bacteria solution, urea, and calcium chloride.

(1)   Bacterial solution

Bacteria and culture medium are needed to prepare the bacterial solution. The bacteria used in this study is *Sporosarcina pasteurii*, which has a highly effective urease activity and is one of the most popular bacteria in MICP studies [20]. Each liter of the culture medium contains 20 g of yeast extract, 10 g of $(NH_4)_2SO_4$, and 10 μmol of $NiCl_2$. The pH of the culture medium is adjusted to a value of about 8.5–9 by using sodium hydroxide solution.

(2)   Urea

The theoretical preparation process of urea is shown in Formula (6):

$$2NH_3 + CO_2 \xrightarrow{\text{high temperature and high pressure}} CO(NH_2)_2 + H_2O \tag{6}$$

The mass ratio of $CO(NH_2)_2$, NH_3, and CO_2 is 64:40:44 under complete reaction conditions, which means 0.4 tonnes of NH_3 and 0.44 tonnes of CO_2 are needed to produce 0.64 tonnes of $CO(NH_2)_2$.

(3) Calcium chloride

The theoretical preparation process of $CaCl_2$ is shown in Formula (7):

$$CaCO_3 + 2HCl \rightarrow CaCl_2 + H_2O + CO_2 \uparrow \tag{7}$$

The mass ratio of $CaCl_2$, $CaCO_3$, and HCl is 1.08:1:0.72 under complete reaction conditions, which means it takes 1 tonne of $CaCO_3$ and 0.72 tonnes of HCl to produce 1.08 tonne of $CaCl_2$.

The list of raw material consumption to produce 1 tonne of cell $CaCO_3$ is shown in Table 3.

Table 3. List of raw material consumption.

Raw Material	NH_3	CO_2	$CaCO_3$	HCl	H_2O *	Yeast Extract	NH_4Cl	$NiCl_2$
Unit Consumption ** kg/t	400	440	1000	720	400	20	10	0.00124

* The H_2O refers to the water consumed to hydrolyze urea, excluding the water in urea solution, calcium chloride solution, and bacterial solution. ** The ratio of materials in the medium solution shown in Table 3 is the optimal ratio during the experiment.

3.2. List of Carbon Emissions

3.2.1. Carbon Emissions of the Bacterial Culture Process

Carbon emissions of the bacterial culture process (C_b) mainly come from the respiration during the bacterial growth and the electrical consumption of equipment used for bacterial cultivation. However, the carbon emission produced by respiration is very small; it is negligible in the calculation. Taking a 1 tonne fermenter as an example, the electrical energy consumed by various instruments during the bacterial cultivation is shown in Table 4.

Table 4. The list of instruments used in the bacterial culture process.

Instruments	Steam Generator	Fermenter	Air Compressor	Air-Drying Machine	Display Panel	Water Production Equipment
Power (kW)	48/24	0.06	15	0.6	0.5	7
Service time (h)	2.5/0.5	16	20	20	0.7	20

The carbon emission of the bacterial culture process can be calculated by Formula (8):

$$C_b = (\Sigma W_i \times t) \times k_1 \tag{8}$$

where W_i is the power of the instrument i, kW; t is the time, h; k_1 is the carbon emissions generated by 1 MJ electricity, 317 g.

According to Formula (8), C_b of 1000 L bacterial solution is 667,955.77 g.

3.2.2. Carbon Emissions of Urea and $CaCl_2$

The energy consumption of urea and calcium chloride (C_{mu} and C_{mca}) mainly includes coal consumption and electricity consumption, resulting from the consumption of raw materials and the production process. The raw materials and energy consumption in the production process of urea and calcium chloride are shown in Table 5 [34].

Table 5. Raw materials and energy consumption of urea and calcium chloride.

Item	Raw Materials (t)		Coal Consumption (kg)	Electricity Consumption (kWh)
Urea (t)	NH_3	0.58	1555.49	1032.57
	CO_2	0.785		
$CaCl_2$ (t)	HCl (31%)	2.33	1000	1593.6138
	Limestone	1.42		

According to Table 5, Formulas (4) and (5), each tonne of urea and calcium chloride produce 0.42 tonnes and 2.79 tonnes of CO_2, respectively. Under complete reaction conditions, MICP technology needs 0.64 tonnes of urea and 1.08 tonnes of calcium chloride to produce 1 tonne of $CaCO_3$, and the carbon emission is 2.74 tonnes.

3.2.3. Carbon Emissions from the Reaction Process of MICP

The entire reaction process of MICP occurs naturally and is mainly dominated by bacteria. In the reaction process, no extra energy is required, and only raw materials are constantly consumed. According to Formulas (1) and (2), all CO_2 generated by urea hydrolysis changes to CO_3^{2-} under the complete reaction condition. This means the carbon emission from the reaction process of MICP (C_p) is 0.

3.2.4. Total Carbon Emissions

Total carbon emissions of MICP (C) include the CO_2 of the bacterial culture process (C_b), the production of raw materials (C_{mu} and C_{mca}), and the reaction process of MICP(C_p):

$$C = C_b + C_{mu} + C_{mca} + C_p \tag{9}$$

According to Formula (9), the total amount of carbon emissions of MICP to generate 1 tonne of $CaCO_3$ is 3399.5 kg.

3.3. Comprehensive Energy Consumption

The comprehensive energy consumption (E) of each tonne of calcium carbonate produced by MICP is obtained by converting the coal consumption and electricity consumption into standard coal, as shown in Formula (10):

$$E = E_C \times k_2 + E_E \times k_3 \tag{10}$$

where E is the comprehensive energy consumption of each tonne of calcium carbonate produced by MICP technology, kg coal equivalent (kgce); E_C is the coal consumed by MICP technology, kg; E_E is the electricity consumed by MICP technology, kWh; k_2 is the standard coal coefficient of raw coal, 0.7143; k_3 is the standard coal coefficient of electricity, 0.1229 kg/kWh.

According to the calculation, 1847.3 kgce of energy is consumed to produce 1 tonne of $CaCO_3$ with MICP technology.

4. Results and Discussion

4.1. LCA of MICP Technology

4.1.1. Carbon Emissions and Energy Consumption Analysis of MICP

The results of carbon emissions and energy consumption inventory analysis of MICP technology are shown in Figure 3.

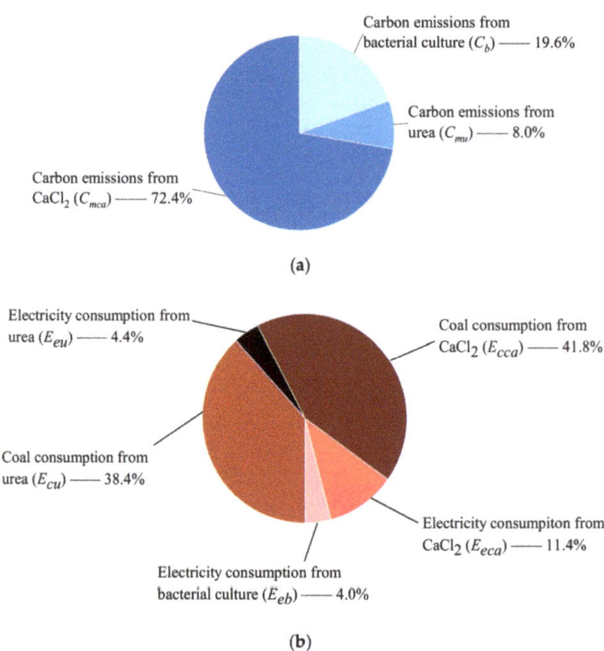

(a)

(b)

Figure 3. Carbon emissions and energy consumption ratio of MICP. (**a**) Carbon emissions; (**b**) energy consumption.

Figure 3 shows that the raw materials of MICP play an important role in carbon emissions and energy consumption. The carbon emissions of MICP mainly come from raw materials, accounting for 80.4% of the total emissions. The rest of the carbon emissions are from the bacterial culture process, accounting for 19.6%. In terms of energy consumption, the coal consumption (E_{cu} and E_{cca}) of MICP accounts for 80.2% of the total energy consumption, and the electricity consumption (E_{eu}, E_{eca}, E_{ep} and E_{eb}) accounts for 19.8%. The coal consumption is all from raw materials, and the electricity consumption is 79.8% from raw materials. The energy consumption of raw materials (E_{cu}, E_{eu}, E_{cca} and E_{eca}) accounted for 96.0% of the total energy consumption. Furthermore, the raw materials are high carbon emission and high-energy consumption materials. The carbon emissions of calcium sources account for 72.4% of the carbon emissions of the entire MICP process. The energy consumption of calcium source and urea account for 53.2% and 42.8% of the total energy consumption, respectively. The energy consumption of the bacterial culture process only accounts for 4.0% of the total energy consumption.

To reduce the energy consumption and carbon emissions of MICP technology, many scholars tried other materials to replace calcium chloride and urea. Chen [35] used the pig urine mainly containing ammonia and urea instead of pure urea. The permeability, porosity, and other mechanical properties of the quartz-sand column were improved obviously by MICP with the pig urine. The energy consumption and carbon emissions were reduced by 43% and 8%, respectively, because of using pig urine. Choi [36] made the calcium source by mixing the eggshell and vinegar at a mass ratio of 1:8. The calcium carbonate content of the eggshell is 94%, and the MICP application effect was good. Cheng [37] used seawater as a calcium source of MICP, and the strength of the sand column increased significantly after repeated treatments with seawater. This research show that energy consumption and carbon emissions can be significantly reduced by using animal waste, eggshell, and seawater in place of industrial urea and calcium sources. Therefore, using organic calcium sources and urea provides a feasible solution to the problems of high-energy consumption and high carbon emissions of MICP raw materials. However, the research in this area is far from mature and more research is needed in the future.

4.1.2. The Relationship between MICP Strength Level and CaCO$_3$ Content

The strength level indicated by the unconfined compressive strength (UCS) of MICP samples directly affects its energy consumption and carbon emissions. The strength level of MICP is related to its calcium carbonate content (CCC). To obtain the quantitative relationship between the UCS and CCC of MICP, the data published by international scholars [10,28,38–45] are summarized and analyzed. The results are shown in Figure 4 and Table 6.

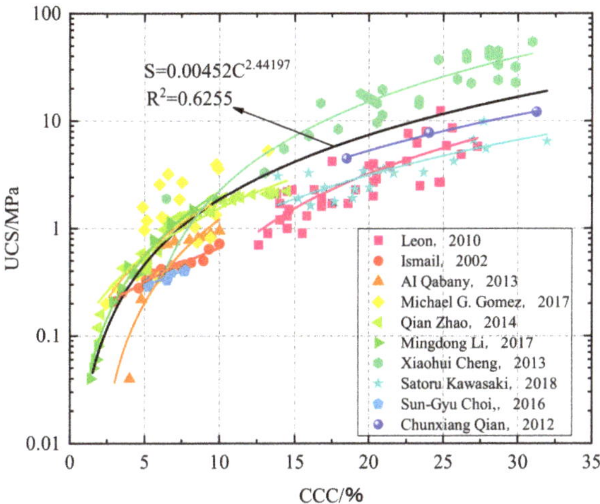

Figure 4. Relationship between calcium carbonate content and strength level of MICP samples. The vertical scale is a logarithmic scale; the horizontal scale is a linear scale.

Table 6. Summarization of scholars on calcium carbonate content and strength level.

Scholar	Bacteria	Calcium Sources	Materials	Methods	Regression Equation	R^2	Reference
Leon	*Sporosarcina pasteurii*	CaCl$_2$	Quarry sand	Grouting	$S = 0.00166C^{2.5419}$	0.5208	[39]
Ismail	*Sporosarcina pasteurii*	CaCl$_2$	RT sand	Grouting	$S = 0.06421C^{1.0020}$	0.9170	[40]
AI Qabany	*Sporosarcina pasteurii*	CaCl$_2$	British sand	Grouting	$S = 0.02342C^{1.6537}$	0.8502	[10]
Michael G. Gomez	*Sporosarcina pasteurii*	-	Silty sand	Grouting	$S = 0.0471C^{1.7643}$	0.4410	[41]
Qian Zhao	*Sporosarcina pasteurii*	CaCl$_2$ 2H$_2$O	Ottawa sand	Soaking	$S = 0.09639C^{1.2163}$	0.9344	[38]
Mingdong Li	*Sporosarcina pasteurii*	CaCl$_2$ 2H$_2$O	Standard sand	Soaking	$S = 0.03592C^{1.7495}$	0.9851	[28]
Xiaohui Cheng	*Sporosarcina pasteurii*	CaCl$_2$	Standard sand	Grouting	$S = 0.00779C^{2.5220}$	0.7312	[42]
Satoru Kawasaki	*Pararhodobacter* sp.	CaCl$_2$	Silty sand	Grouting	$S = 0.00948C^{1.9420}$	0.6349	[43]
Sun-Gyu Choi	*Bacillus* sp.	CaCl$_2$	Ottawa sand	Grouting	$S = 0.0806C^{0.7961}$	0.8587	[44]
Chunxiang Qian	alkalophilic microbes	calcium ion solution	Quartz sand	Grouting	$S = 0.02229C^{1.8322}$	0.9931	[45]

Figure 4 and Table 6 shows that the strength level of MICP technology increases significantly with the increase of CCC, and there is a power function relationship between UCS and CCC. A representative model to describe the relationship between UCS and CCC of MICP samples can be obtained by regression analysis using all the data from these researchers, as shown in Formula (11).

$$S = k_4 C^{k_5} \tag{11}$$

where S is the unconfined compressive strength, MPa; C is the calcium carbonate content,%; k_4 is the coefficient, 0.00452; k_5 is the index, 2.44197.

The quantitative relationship between the strength level and the CCC can provide a basis for the LCA evaluation of energy consumption and carbon emissions of MICP technology in different application scenarios.

Combined with the inventory analysis results, the energy consumption and carbon emissions corresponding to the different strength levels of the MICP are shown in Figure

Figure 5 indicates that the carbon emissions and energy consumption of MICP technology increase rapidly with the increase of the UCS of MICP samples. However, the relationships between strength level and calcium carbonate content of MICP samples are uncertain because of the different MICP materials and methods. The empirical relationship used in this research is an average and representative equation.

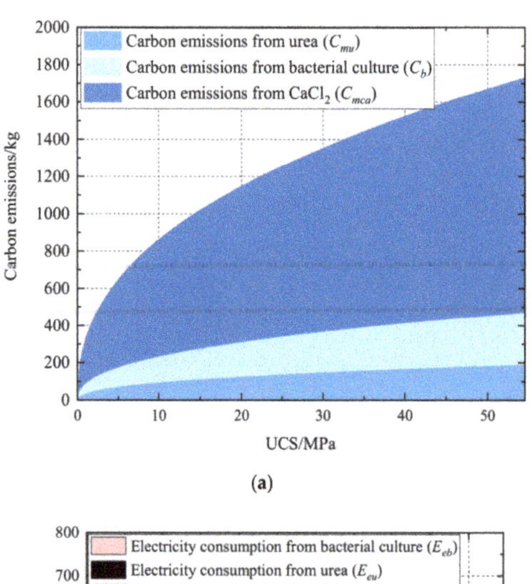

(a)

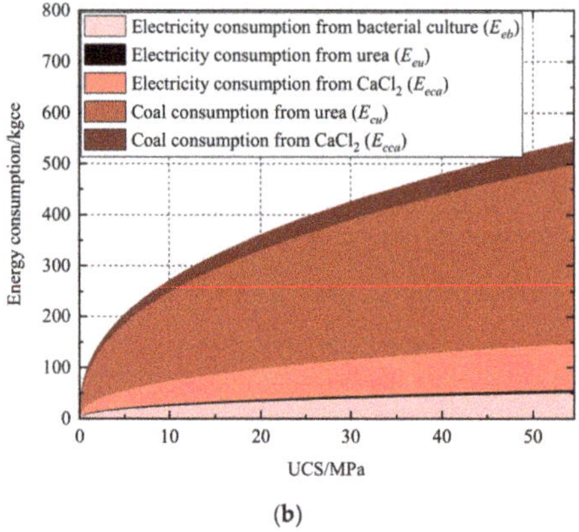

(b)

Figure 5. Carbon emissions and energy consumption versus strength levels of MICP samples. (a) Carbon emissions; (b) energy consumption.

4.2. Carbon Emissions and Energy Consumption of MICP in Different Application Scenarios

4.2.1. Applications of MICP to Replace the Cement Mortar for Concrete

Cement is a common binder, which is widely used in the construction industry [46]. However, using cement has the problem of huge energy consumption and environmental pollution [24]. Replacing cement is one of the important potential applications of MICP technology. In addition, many scholars believe that MICP is a more environmentally friendly technology than cement.

The carbon emissions and energy consumption of cement come from the manufacturing process, transportation, fuels, and electricity [47]. The manufacturing process mainly refers to the calcination process of cement. Because $CaCO_3$, SiO_2, Al_2O_3, and Fe_2O_3 are decomposed in the clinker under high-temperature conditions [48], a large amount of CO_2 is produced in this process. In the transportation process, it is assumed that all raw materials are transported by rail to the factory with a transportation distance of 100 km. The electricity consumption in cement production includes raw material mining, crushing, pre-homogenization, grinding, homogenization, clinker calcination, coal grinding, cement grinding, transportation, etc.

The fuel is mainly coal in the calcination process of cement production. Coal combustion gives off heat but produces many carbon dioxide emissions. The functional unit in this research is set to 1 tonne of cement, and the carbon emissions and energy consumption of cement production are shown in Table 7 [49].

Table 7. List analysis of cement.

Cement Grade (MPa)		52.5
Raw material (kg/t cement)		1656.54
	Manufacture process	835,550
	Transportation	52,545
Carbon emission (g/t cement)	Electricity	150,648
	Fuel(coal)	2814
	Total	1,041,557
	Coal consumption (kJ/t cement)	3,146,042
Energy consumption	Electricity consumption (kJ/t cement)	475,229.4
	Total (kgce/t cement)	123.57

Table 7 shows that the carbon emissions from cement mainly come from the manufacturing process, accounting for 79.9% of the total carbon emissions. Carbon emissions from transportation, electricity and fuel account for 5%, 14.8%, and 0.3% of the total carbon emissions, respectively. The energy consumed in the cement production process is mainly coal, which accounts for 85.7% of the total energy consumption. All fuel is used in the calcination process and the drying process.

The concrete with a strength of 40 MPa (C40) is chosen to compare the energy consumption and carbon emissions of cement mortar and MICP. The C40 concrete needs about 450 kg of cement per unit volume (1 m^3), and its carbon emissions and energy consumption are 468.7 kg and 55.6 kgce, respectively. It can be calculated from Figure 5, to consolidate the same volume of sand and achieve the same strength level by MICP, the carbon emissions and energy consumption is 2107.5 kg and 1145.2 kgce, respectively. The carbon emissions and energy consumption of MICP are 4.5 times and 20.6 times of concrete, respectively. In this application scenario, the carbon emissions and energy consumption of MICP are much higher than cement mortar. MICP is not a more environmentally friendly technology. The main reason is that the raw materials used in MICP, including calcium sources and urea, are all materials with high-energy consumption and high emissions, which harm the LCA environmental evaluation of MICP.

In addition, it is noticed that increasing studies have shown that using MICP technology as an additive in concrete can significantly improve the strength and freeze–thaw resistance of the concrete [50]. In this case, the MICP is used as an admixture rather than a binder, so the environmental impact of MICP should be compared with the traditional admixture in the concrete, not with the cement itself. However, this issue is not discussed due to the limited space of the paper.

4.2.2. Applications of MICP to Replace the Sintered Bricks

Sintered bricks, mainly made of clay, are one of the oldest building materials in the world. It is widely used in the civil and architectural engineering industry with advantages of cheap, durable, fire prevention, heat insulation, noise absorption, and so on [51]. However, the production of sintered bricks needs a huge amount of clay, which destroys cultivated land seriously. According to incomplete statistics, the production of bricks in China damages 467 km^2 of fertile land per year, which is extremely harmful to the environment. MICP technology can be used as a potential alternative to sintering bricks.

For the sintered brick industry, the standard unit brick (SUB) is usually used as the unit to calculate the output of sintered bricks. The volume of a sintered brick is 1,462,800 mm^3 (240 × 115 × 53 mm). The carbon emissions and energy consumption of sintered bricks come from the transportation process, preparation process, fuel combustion, and electricity consumption. The transportation distance is assumed to be 100 km. In the study of this section, the functional unit of sintered brick is set as a SUB. The inventory analysis of the sintered bricks production is shown in Table 8 [52].

Table 8. Inventory analysis of sintered bricks.

Item		Unit	Consumption
Raw material		kg/SUB	4.64
Carbon emission		g/SUB	330.69
Energy consumption	Coal	kg/SUB	0.049
	Electricity	kWh/SUB	0.16
	Total	kgce/SUB	0.055

The strength of sintered brick is 20 MPa. To produce each sintered brick, 330.69 g CO_2 is emitted, and 0.055 kgce is consumed. When bio-bricks with the same size and strength are manufactured by using MICP technology, the calcium carbonate content is about 31.1%, the carbon emission is 2.3 kg per SUB, and the energy consumption is 1.3 kgce per SUB. The carbon emissions and energy consumption of MICP bio-bricks of the same strength level are 7 times and 23 times higher than those of sintered bricks, respectively. Therefore, the MICP technology has a worse impact on the environment than traditional sintered bricks.

However, the clay used to produce sintered bricks is a non-renewable resource, and the environmental impact of the clay consumption is not considered here. Comparing with sintered bricks, MICP consumes less non-renewable resources, and it is discussed in Section 4.3.

4.2.3. Applications of MICP to Replace the Lime Mortar

Lime, mainly composed of calcium oxide, is obtained by calcining natural rocks containing calcium carbonate at an appropriate temperature to decompose carbon dioxide. As one of the building materials, lime mortar is a mixture of lime, sand, and water in a certain proportion. Lime mortar is widely used in masonry and plastering layers, which requires low strength and is not easy to be damp. The use of lime may damage the water and surrounding vegetation, and lime becomes the second-largest source of greenhouse gas emissions after the cement industry [53]. MICP can be used as a potential substitute technology for lime mortar.

Limestone is the raw material of lime; the reaction principle of the lime production is shown in Formula (12):

$$CaCO_3 \xrightarrow{high\ temperature} CaO + CO_2 \uparrow \tag{12}$$

The production of 1 tonne of calcium oxide needs 2 tonnes of limestone containing approximately 50% of calcium oxide and produces 0.79 tonnes of CO_2 simultaneously. The carbon emissions and energy consumption of lime come from manufacturing, transportation, fuel combustion, and electricity consumption. The manufacturing process is the calcining process of limestone, as shown in Equation (12). In the transportation process, it is assumed that all raw materials are transported to the factory by rail, and the transportation distance is 100 km. Fuel combustion and electricity consumption refers to the heat and electricity supply during the process. The functional unit is set to 1 tonne. The carbon emissions and energy consumption of lime produced by different lime kilns are shown in Table 9.

Table 9. Inventory analysis of different lime kilns.

Limekilns	Rotary Kiln	Maliz Shaft Kiln	Sleeve Kiln	Gas-Burning Shaft Kiln	Mechanized Shaft Kiln	Sinopec Shaft Kiln	Other Kilns
Manufacture process/g	790,000	790,000	790,000	790,000	790,000	790,000	790,000
Transportation/g	63,440	63,440	63,440	63,440	63,440	63,440	63,440
Fuel consumption/kg	170	126	140	162	145	140	146
Electricity consumption/kWh	57	44	44	47	26	6	11

Regardless of the lime kiln, the energy consumption and carbon emissions of lime production are very close. The average energy consumption of lime is 149.3 kgce per tonne, and the average carbon emission is about 900 kg per tonne. The carbon emissions produced by lime mainly come from the manufacturing process, accounting for 91.4% of the total emissions. Carbon emissions from transportation, electricity, and fuel accounted for 7.3%, 0.02%, and 1.28% of the total emissions, respectively. The main energy consumed in the lime production process is coal, which accounts for 97% of the total energy consumption, and electricity consumption only accounts for 3%.

The strength of the lime mortar sample is 5 MPa, and the average date is used for calculation. The lime content per unit volume (1 m^3) of lime mortar is 216 kg, the energy consumption is 32.3 kgce, and the carbon emission is 193.8 kg. To achieve the same strength level with MICP, the calcium carbonate content is 17.6%, the energy consumption is 488.8 kgce, and the carbon emission is 899.5 kg. The carbon emissions and energy consumption of MICP are 4.6 times and 15.2 times higher than those of lime mortar, respectively. Therefore, in this application scenario, MICP technology has a worse impact on the environment than lime.

However, it should be noted that lime mortar takes a long time to solidify and has poor durability after solidification, which is seriously affected by the moisture [54]. Conversely, the consolidation speed of MICP is faster, and the durability is better [50]. The performance of carbon emissions and energy consumption simply reflects environmental effects, and it does not indicate technical merits.

4.2.4. Applications of MICP to Replace the Cement for Cemented Backfill

The mining process involves the removal and recovery of economically valuable minerals from the crust of the earth [55]. The underground voids caused by mining activities, which may create serious environmental challenges, are expected to be filled with waste materials by a process known as backfilling technology [56]. The underground backfill can support the ground and dispose of the solid waste. It has significant environmental benefits [57]. Backfilling materials mainly include waste rock, gangue, and fly ash, and binder. The binder for backfilling is mainly cement [58]. MICP technology has the effect of cementing instead of cement, so it can be potentially used in cemented backfill mining.

The ratio of backfill materials used in a coal mine is taken as an example. This backfill material comprises 5% cement, 20% fly ash, 55% gangue, and 20% water. The strength of solidified backfill body is 2 MPa. The 1 m^3 of backfill body is taken as a functional unit, the amount of cement is 155 kg, and the carbon emissions and energy consumption are 161.4 kg and 19.1 kgce, respectively.

If the MICP is used to replace the cement in backfill mining, the calcium carbonate content of the MICP backfill body with a strength of 2 MPa is 12.1%. Therefore, the carbon emission of MICP backfill is 618 kg per cubic meter, and the energy consumption is 335.8 kgce per cubic meter. The carbon emissions and energy consumption of MICP backfill materials are 3.8 times and 17.5 times higher than the traditional cement-based backfill materials, respectively. In the application scenario of backfill mining, MICP has no superiority in terms of life cycle environmental benefits comparing with traditional cemented backfill technology.

4.2.5. Applications of MICP to Replace Cement Grouting for Foundation Reinforcement

Foundation refers to the soil layer within a limited area of the building. The unreinforced natural soil layer is called the natural foundation, which is mostly very week. The strength and deformation properties of the natural foundation usually cannot meet the requirements of construction, so the natural foundation mostly needs to be reinforced. Cement grouting is a common reinforcement method. Its principle is to inject the cement slurry into the natural foundation soil to improve its mechanical properties. As one of the main application directions of MICP, MICP grouting reinforcement technology has the advantages of low grouting pressure, good diffusion performance, and low slurry viscosity.

The strength of the foundation after the cement slurry grouting treatment is about 300 kPa–500 kPa. The proportion of cement slurry for grouting reinforcement should be determined according to the actual situation. Generally, 90 kg of cement is required for a unit volume (1 m^3) of the cement slurry for foundation reinforcement. The carbon emission per unit volume of cement slurry is 93.7 kg, and the energy consumption is 11.1 kgce. The carbon emission and the energy consumption of MICP grouting with the same strength level are 319.7 kg and 173.7 kgce, respectively. The carbon emissions of MICP grouting are 3.4 times of traditional cement grouting reinforcement, and the energy consumption is 15.6 times of cement grouting. Therefore, MICP shows no obvious advantages from the perspective of carbon emission and energy emission in the application scenario of foundation grouting reinforcement.

However, from a technical point of view, MICP grouting has higher diffusion, lower grouting pressure, and better uniformity of foundation reinforcement than traditional cement grouting. Moreover, MICP grouting is not toxic comparing with chemical grouting. Therefore, although the carbon emissions and energy consumption of MICP grouting are relatively high, its technical advantages are obvious; it has a broad development prospect.

4.2.6. Comparison of Various Application Scenarios

To compare energy consumption and carbon emissions of MICP with traditional technology in different application scenarios, the MICP energy consumption index (k_{ME}) and MICP carbon emissions index (k_{MC}) are defined. The k_{ME} is the ratio of the energy consumption of MICP to that of the traditional technology in the current situation. The k_{ME} is the ratio of the carbon emissions of the MICP to that of the traditional technology. When the k_{ME} and k_{MC} are greater than 1, it means that the environmental benefits of MICP are not superior. When the k_{ME} and k_{MC} are equal to 1, it means that the environmental benefits of MICP are equivalent to those of traditional technology. When the k_{ME} and k_{MC} are less than 1, it means that the environmental benefits of MICP are superior in the current situation. The k_{ME} and k_{MC} of MICP in different application scenarios are shown in Figure 6.

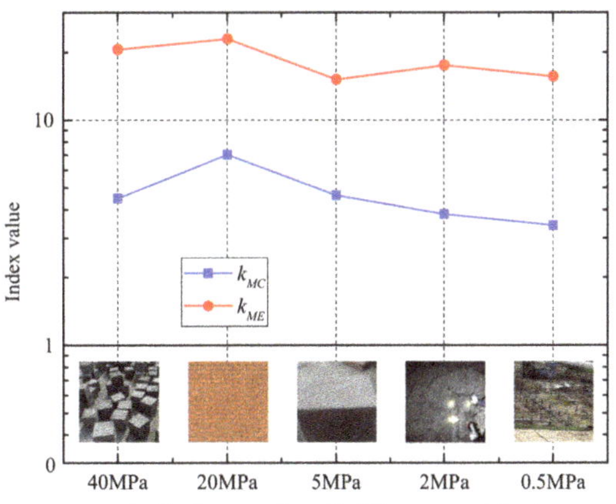

Figure 6. The k_{ME} and k_{MC} of MICP in different applications with different strength levels.

Under the current technical level, from the perspective of LCA, MICP does not show the advantages of environmental benefits but produces more CO_2 and consumes more energy than traditional technologies. In the high-strength scenarios of 40 MPa and 20 MPa, the k_{MC} index is 4.5 and 7, and the k_{ME} index is 20.6 and 23, respectively. In the low-strength scenarios of 5 MPa, 2 MPa and 0.5 MPa, the k_{MC} index is 4.6, 3.8 and 3.4, and the k_{ME} index is 15.2, 17.5 and 15.6, respectively. In general, k_{MC} and k_{ME} values decrease with the decrease of the strength level of application scenarios. The average k_{MC} and k_{ME} are 4.66 and 18.38, respectively. In other words, the carbon emission of MICP is on average 4.66 times of traditional technologies, and the energy consumption is on average 18.38 times of traditional technologies.

In addition, it should be noted that most data of MICP are obtained on the laboratory scale, while the data of traditional technologies is obtained on an industrial scale. The laboratory scale and industrial scale are very different. With developing MICP technology, its energy consumption and carbon emissions will be greatly reduced. Especially after MICP is applied on the industrial scale, the environmental benefits will be more significant.

4.3. Environmental Impact Assessment

4.3.1. Resource Consumption

The five application scenarios involve four kinds of materials: cement, lime, sintered brick, and MICP. The raw materials needed to produce these four materials are shown in Table 10. Abiotic depletion potential (ADP) is used to measure the resource consumption of materials. The functional unit is set to 1 tonne. The ADP is calculated according to Formula (3), and the results are shown in Table 10.

Table 10. The raw materials requirement and ADP value of the four materials.

Item	Raw Materials/kg				ADP
	Limestone	Gypsum	Iron Powder	Clay	kg Oil
Cement	1232.9	50	30		1571.4
Lime	2000				2152.4
Sintered brick				944	1944.64
MICP	1420				1528.2

Table 10 shows that the highest ADP value of the four materials is lime, and the MICP has the lowest ADP value. The ADP values of cement, lime, and sintered bricks are all larger than MICP, which shows that more non-renewable resources are needed to produce these materials. This also shows that MICP technology has advantages over cement, lime, and sintered bricks in terms of resource consumption.

4.3.2. Environmental Impact

In the production process of cement, lime, sintered brick, and MICP, besides carbon dioxide, other emissions are produced. The environmental impact potential value is calculated according to Table 2 and Formula (5). The functional unit is set to 1 tonne. The environmental impact potential values of cement, lime, sintered brick, and MICP on different environmental impacts are shown in Figure 7.

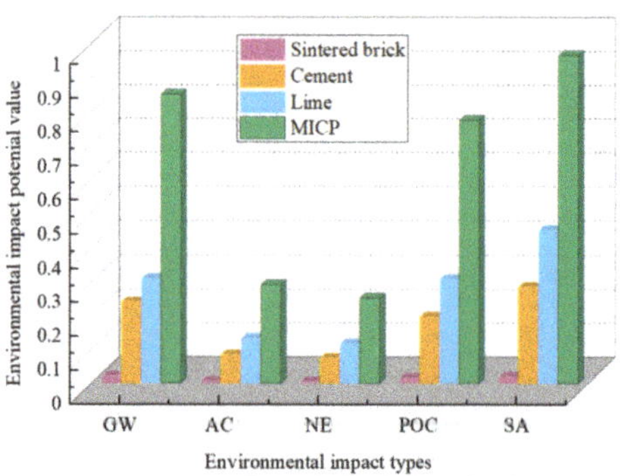

Figure 7. The environmental impact potential value of the four materials on different environmental impacts.

Figure 7 shows that cement, lime, sintered bricks, and MICP have the greatest impact on the environment in smoke and dust (SA), followed by global warming (GW) and photochemical ozone creation (POC), and the least impact is acidification (AC) and eutrophication (NE). The greatest impact of MICP on the environment in SA, which is 4 times that of cement, 2 times that of lime, and 48 times that of sintered bricks. The least environmental impact of MICP is NE, which is 3 times of cement, but 50 times of sintered bricks. Under current technical conditions, MICP has the greatest impact on various environmental impacts, while sintered bricks have the least environmental impacts.

4.4. Limitation and Prospects

4.4.1. Limitation

The energy consumption and carbon emissions studied with LCA can comprehensively and objectively reflect the environmental impact of MICP. However, this research has the following limitations:

(1) In this research, the comparison between MICP and traditional technology is carried out in terms of energy consumption and carbon emissions under the current technological level, and it does not represent the merits of the technology itself;

(2) The MICP data used in this study are all obtained on the laboratory scale, and the small scale is not good for evaluating the life cycle energy consumption and carbon emissions of MICP. Although the research conclusively shows that MICP is not as environmentally friendly as expected under the current technological conditions, it does not deny this technology. MICP has a great potential for environmental benefits with the development on the industrial scale;

(3) Within the life cycle system boundary of MICP technology, the carbon emissions and energy consumption are affected by many factors, such as the respiration of bacteria, the reacting process, the temperature, and the pH. Because the influence of some factors is very small, necessary assumptions and simplifications are made in the research;

(4)   It is assumed that the urea is completely hydrolyzed and reacted, and the carbon emission from the reaction process is zero. However, the obvious irritating odor of ammonia gas can be smelled during the experiment, which indicates that the urea hydrolysis reaction in the MICP process is not complete. Hence, the actual MICP process needs to consume more urea, which means higher energy consumption;

(5)   The treatment method for MICP technology includes grouting, soaking, and mixing. These treatment processes have different technical characteristics, as well as different energy consumption and carbon emissions. However, considering the complexity of the treatment process, it is not covered in the life cycle assessment of MICP in this study.

4.4.2. Prospects

Comparing with traditional cement, lime, and sintered bricks, MICP has technical advantages, such as lower pH, lower viscosity, higher fluidity, lower grouting pressure, better heavy metal ion consolidation performance, and so on. Therefore, although MICP currently does not have an advantage in energy consumption and carbon emissions, it still has great development potential. Based on LCA results of MICP technology at the current technological level, the prospects of MICP research are given as follows:

(1)   Trying diversified raw materials is an effective way to reduce the carbon emissions and energy consumption of MICP. Some organic materials, such as eggshells and livestock urine, have been studied for MICP, and it is worthy of continuing the related research;

(2)   Different bacteria besides *Bacillus* pasteurii can be used for MICP. For example, the carbonic anhydrase mineralizing bacteria can catalyze the hydration reaction of CO_2 and convert it into $CO_{32}-$ to achieve the $CaCO_3$ precipitation. This process can consume CO_2, thereby significantly reducing the carbon emissions of MICP. In addition, the enzyme-induced $CaCO_3$ precipitation is a good path to improve environmental benefits;

(3)   The research of MICP on the industrial scale should be increased. Most of the current research is carried out on the laboratory scale, which is very different from the industrial scale. Usually, the greater scale, the lower unit consumption;

(4)   The application scenarios of MICP should be more diverse. Besides foundation reinforcement, cultural relic restoration, anti-seepage, and anti-leakage, MICP can be applied in more situations to look for better environmental benefits;

(5)   The mechanism of MICP needs to be further studied. A clear understanding of the MICP reaction mechanism is very helpful to optimize the MICP process and technical parameters by which carbon emissions and energy consumption can be reduced.

5. Conclusions

In this paper, the energy consumption and carbon emissions of MICP in different application scenarios are quantitatively analyzed based on LCA. The environmental effects of MICP technology at the current technology level are evaluated. The main conclusions are shown as follows:

(1)   The energy consumption and carbon emissions of MICP are calculated based on LCA. Generating 1 tonne $CaCO_3$ by MICP emits 3399.5 kg of CO_2 and consumes 1847.2 kgce of energy. About 80.4% of carbon emissions and 96% of the energy consumption of MICP are from its raw materials;

(2)   The relationship between the strength level (UCS) and calcium carbonate content (CCC) of MICP is established. The UCS of MICP increases significantly with the increase of the CCC, and a power function relationship is found between the UCS and CCC of MICP samples. Additionally, due to various influencing factors, this relationship is an average and representative equation;

(3)   The abiotic depletion potential (ADP) value of MICP is lower than that of cement, lime, sintered bricks, which indicates that MICP consumes less non-renewable resources

and has advantages in resource consumption. The greatest environmental impact of MICP is smoke and ash, followed by global warming, photochemical ozone creation, acidification, and eutrophication. Environmental impacts of MICP are more serious than cement, lime, and sintered bricks under current technical conditions;

(4) In different application scenarios of concrete, sintered bricks, lime mortar, cemented backfill, and cement grouting foundation reinforcement, the carbon emission of MICP is on average 4.66 times of traditional technologies, and the energy consumption is averagely 18.38 times. It means the environmental benefits of the MICP technology are not superior to those of traditional technology;

(5) Although MICP currently does not have an advantage in energy consumption and carbon emissions, it still has great development potential. Suggestions are given for the future research of MICP based on LCA results.

Author Contributions: Conceptualization, X.C., X.D. and Y.L.; methodology, X.D. and Y.L.; validation, X.D.; formal analysis, Y.L., H.L. and Y.Z.; investigation, Y.L.; resources, X.D.; data curation, X.D. and Y.L.; writing—original draft preparation, X.D. and Y.L.; writing—review and editing, X.C. and B.d.W.; visualization, X.D. and X.X.; supervision, X.D. and Y.Y.; project administration, X.D.; funding acquisition, X.D. All authors have read and agreed to the published version of the manuscript.

Funding: This work was supported by the National Key R&D Program of China (Grant Number 2018YFC0604701), the National Natural Science Foundation of China (Grant Number 51804308, 52034009), the Yue Qi Young Scholar Project (Grant Number 2020QN03), the China Postdoctoral Science Foundation (Grant Number 2020T130269, 2020M670689), and the Postdoctoral Research Project of Hebei Province (Grant Number B2020003029).

Data Availability Statement: Not applicable.

Acknowledgments: The authors would like to thank Hongxian Guo from Tsinghua University and Meng Li from Beijing Technology and Business University for everything they have done for this project. Special thanks to Xi Li for her English writing assistance.

Conflicts of Interest: The authors declare no conflict of interest.

References

1. Dejong, J.T.; Mortensen, B.M.; Martinez, B.C.; Nelson, D.C. Bio-mediated soil improvement. *Ecol. Eng.* **2010**, *36*, 197–210. [CrossRef]
2. Dejong, J.T.; Soga, K.; Kavazanjian, E.; Burns, S.E.; Paassen, L.A.V.; Qabany, A.A.; Aydilek, A.; Bang, S.S.; Burbank, M.; Caslake, L.F.; et al. Biogeochemical processes and geotechnical applications: Progress, opportunities and challenges. *Geotechnique* **2013**, *63*, 287–301. [CrossRef]
3. Chu, J.; Stabnikov, V.; Ivanov, V. Microbially induced calcium carbonate precipitation on surface or in the bulk of soil. *Geomicrobiol. J.* **2012**, *29*, 544–549. [CrossRef]
4. Cheng, L.; Cord-Ruwisch, R.; Shahin, M.A. Cementation of sand soil by microbially induced calcite precipitation at various degrees of saturation. *Can. Geotech. J.* **2013**, *50*, 81–90. [CrossRef]
5. Tittelboom, K.V.; Belie, N.D.; Muynck, W.D.; Verstraete, W. Use of bacteria to repair cracks in concrete. *Cem. Concr. Res.* **2009**, *40*, 157–166. [CrossRef]
6. Zhang, H.; Guo, H.; Li, M.; Cheng, X. Experimental research of microbial-induced clogging in sands. *Ind. Constr.* **2015**, *45*, 139–142.
7. Kang, C.H.; Kwon, Y.J.; So, J.S. Bioremediation of heavy metals by using bacterial mixtures. *Ecol. Eng.* **2016**, *89*, 64–69. [CrossRef]
8. Chen, X. Heavy metal immobilisation and particle cementation of tailings by biomineralisation. *Environ. Geotech.* **2017**, *5*, 107–113. [CrossRef]
9. Yang, J.; Pan, X.; Zhao, C.; Mou, S.; Achal, V.; Misned, F.A.A.; Mortuza, M.G.; Gadd, G.M. Bioimmobilization of heavy metals in acidic copper mine tailings soil. *Geomicrobiol. J.* **2016**, *33*, 261–266. [CrossRef]
10. Qabany, A.A.; Soga, K. Effect of chemical treatment used in MICP on engineering properties of cemented soils. *Geotechnique* **2013**, *63*, 331–339. [CrossRef]
11. Lee, M.L.; Ng, W.S.; Tanaka, Y. Stress-deformation and compressibility responses of bio-mediated residual soils. *Ecol. Eng.* **2013**, *60*, 142–149. [CrossRef]
12. Montoya, B.M.; Dejong, J.T. Stress-strain behavior of sands cemented by microbially induced calcite precipitation. *J. Geotech. Geoenviron. Eng.* **2015**, *141*, 04015019. [CrossRef]

13. Deng, X.; Zongxuan, Y.; Yu, L.; Hao, L.; Jianye, F.; Benjamin, D.W. Experimental study on the mechanical properties of microbial mixed backfill. *Constr. Build. Mater.* **2020**, *265*, 120643. [CrossRef]

14. Muynck, W.D.; Belie, N.D.; Verstraete, W. Microbial carbonate precipitation in construction materials: A review. *Ecol. Eng.* **2009**, *36*, 118–136. [CrossRef]

15. Montoya, B.M. Bio-mediated soil improvement and the effect of cementation on the behavior, improvement, and performance of sand. *Diss. Theses Gradworks* **2012**, *7*, 209–223.

16. Umar, M.; Kassim, K.A.; Chiet, K.T.P. Biological process of soil improvement in civil engineering: A review. *J. Rock Mech. Geotech. Eng.* **2016**, *8*, 767–774. [CrossRef]

17. DeJong, J.T.; Fritzges, M.B.; Nüsslein, K. Microbially induced cementation to control sand response to undrained shear. *J. Geotech. Geoenviron. Eng.* **2006**, *132*, 1381–1392. [CrossRef]

18. Hataf, N.; Baharifard, A. Reducing soil permeability using microbial induced carbonate precipitation (MICP) method: A case study of shiraz landfill soil. *Geomicrobiol. J.* **2020**, *37*, 147–158. [CrossRef]

19. Li, M.; Cheng, X.; Guo, H.; Yang, Z. Biomineralization of carbonate by terrabacter tumescens for heavy metal removal and biogrouting applications. *J. Environ. Eng.* **2015**, *142*, C4015005. [CrossRef]

20. Gai, X.; Sánchez, M. An elastoplastic mechanical constitutive model for microbially mediated cemented soils. *Acta Geotech.* **2019**, *14*, 709–726. [CrossRef]

21. Li, M.; Li, L.; Ogbonnaya, U.; Wen, K.; Tian, A.; Amini, F. Influence of fiber addition on mechanical properties of MICP-treated sand. *J. Mater. Civ. Eng.* **2015**, *28*, 04015166. [CrossRef]

22. Rajasekar, A.; Moy, C.K.S.; Wilkinson, S. MICP and advances towards eco-friendly and economical applications. *IOP Conf. Ser. Earth Environ. Sci.* **2017**, *78*, 012016. [CrossRef]

23. ISO 14040. *Environmental management—Life cycle assessment—Principles and framework*; ISO 14040, 2006.

24. Huntzinger, D.N.; Eatmon, T.D. A life-cycle assessment of Portland cement manufacturing: Comparing the traditional process with alternative technologies. *J. Clean. Prod.* **2009**, *17*, 668–675. [CrossRef]

25. Heede, P.V.D.; Belie, N.D. Environmental impact and life cycle assessment (LCA) of traditional and 'green' concretes: Literature review and theoretical calculations. *Cem. Concr. Compos.* **2012**, *34*, 431–442. [CrossRef]

26. Xu, X.; Guo, H.; Cheng, X.; Li, M. The promotion of magnesium ions on aragonite precipitation in MICP process. *Constr. Build. Mater.* **2020**, *263*, 120057. [CrossRef]

27. Li, L.; Amini, F.; Zhao, Q.; Li, C.; Ogbonnaya, U. Development of a flexible mold for bio-mediated soil materials. *Geotech. Spec. Publ.* **2015**, *12*, 2339–2348.

28. Li, M.; Wen, K.; Li, Y.; Zhu, L. Impact of oxygen availability on microbially induced calcite precipitation (MICP) treatment. *Geomicrobiol. J.* **2017**, *35*, 15–22. [CrossRef]

29. Pryshlakivsky, J.; Searcy, C. Fifteen years of ISO 14040: A review. *J. Clean. Prod.* **2013**, *57*, 115–123. [CrossRef]

30. IPCC. *Guidelines for National Greenhouse Gas Inventories, Reference Manual*; IPCC: Bracknell, UK, 1996.

31. Huijbregts, M.A.J.; Schöpp, W.; Verkuijlen, E.; Heijungs, R.; Reijnders, L. Spatially explicit characterization of acidifying and eutrophying air pollution in life-cycle assessment. *J. Ind. Ecol.* **2010**, *4*, 75–92. [CrossRef]

32. Jenkin, M.E.; Hayman, G.D. Photochemical ozone creation potentials for oxygenated volatile organic compounds: Sensitivity to variations in kinetic and mechanistic parameters. *Atmos. Environ.* **1999**, *33*, 1275–1293. [CrossRef]

33. Derwent, R.G.; Jenkin, M.E.; Saunders, S.M.; Pilling, M.J. Photochemical ozone creation potentials for organic compounds in northwest Europe calculated with a master chemical mechanism. *Atmos. Environ.* **1998**, *32*, 2429–2441. [CrossRef]

34. Wenyou, L.; Lumei, Z. *Principle of Chemistry*; Chongqing University Press: Chongqing, China, 2015.

35. Chen, H.J.; Huang, Y.H.; Chen, C.C.; Maity, J.P.; Chen, C.Y. Microbial induced calcium carbonate precipitation (MICP) using pig urine as an alternative to industrial urea. *Waste Biomass Valorization* **2019**, *10*, 2887–2895. [CrossRef]

36. Choi, S.G.; Wu, S.; Chu, J. Biocementation for sand using an eggshell as calcium source. *J. Geotech. Geoenviron. Eng.* **2016**, *142*, t06016010. [CrossRef]

37. Cheng, L.; Shahin, M.A.; Cord-Ruwisch, R. Bio-cementation of sandy soil using microbially induced carbonate precipitation for marine environments. *Geotechnique* **2014**, *64*, 1010–1013. [CrossRef]

38. Zhao, Q.; Li, L.; Li, C.; Li, M.; Amini, F.; Zhang, H. Factors affecting improvement of engineering properties of MICP-treated soil catalyzed by bacteria and urease. *J. Mater. Civ. Eng.* **2014**, *26*, 04014094. [CrossRef]

39. Paassen, L.A.V.; Ghose, R.; Linden, T.J.M.V.D.; Star, W.R.L.V.D.; Loosdrecht, M.C.M.V. Quantifying biomediated ground improvement by ureolysis: Large-scale biogrout experiment. *J. Geotech. Geoenviron. Eng.* **2010**, *136*, 1721–1728. [CrossRef]

40. Ismail, M.A.; Joer, H.A.; Sim, W.H.; Randolph, M.F. Effect of cement type on shear behavior of cemented calcareous soil. *J. Geotech. Geoenviron. Eng.* **2002**, *128*, 520–529. [CrossRef]

41. Gomez, M.G.; Dejong, J.T. Engineering properties of bio-cementation improved sandy soils. *Grouting* **2017**, 23–33.

42. Cheng, X.; Ma, Q.; Yang, Z.; Zhang, Z.; Li, M. Dynamic response of liquefiable sand foundation improved by bio-grouting. *Chin. J. Geotech. Eng.* **2013**, *35*, 1486–1495.

43. Amarakoon, G.G.N.N.; Satoru, K. Factors affecting sand solidification using MICP with *pararhodobacter* sp. *Mater. Trans.* **2017**, *59*, 72–81. [CrossRef]

44. Choi, S.-G.; Wang, K.; Chu, J. Properties of biocemented, fiber reinforced sand. *Constr. Build. Mater.* **2016**, *120*, 623–629. [CrossRef]

45. Rong, H.; Qian, C. Characterization of microbe cementitious materials. *Chin. Sci. Bull.* **2012**, *57*, 1333–1338. [CrossRef]

46. Yang, D.; Fan, L.; Shi, F.; Liu, Q.; Wang, Y. Comparative study of cement manufacturing with different strength grades using the coupled LCA and partial LCC methods—A case study in China. *Resour. Conserv. Recycl.* **2016**, *119*, 60–68. [CrossRef]
47. Li, C.; Gong, X.Z.; Cui, S.P.; Wang, Z.H.; Zheng, Y.; Chi, B.C. CO_2 emissions due to cement manufacture. *Mater. Sci. Forum* **2011**, *1264*, 181–187. [CrossRef]
48. Dandautiya, R.; Singh, A.P. Utilization potential of fly ash and copper tailings in concrete as partial replacement of cement along with life cycle assessment. *Waste Manag.* **2019**, *99*, 90–101. [CrossRef] [PubMed]
49. Gong, Z.; Zhang, Z. A study on embodied environmental profile during the life cycle of cment. *China Civ. Eng. J.* **2004**, *37*, 86–91.
50. Tan, Q.; Guo, H.; Chen, X. Experimental study of strength and durability of microbial cement mortar. *Ind. Constr.* **2015**, *45*, 42–47.
51. Feng, S.T.; Lu, H.L.; Ye, F. Application of plaster powder in fired common brick strength testing. *Adv. Mater. Res.* **2014**, *1030-1032*, 705–708. [CrossRef]
52. Sun, B.; Liu, Y.; Nie, Z.; Zhang, Y.; Gao, F. Exergy-based model for quantifying land resource in China: A case study of sintered brick. *Int. J. Exergy* **2014**, *15*, 429–446. [CrossRef]
53. Kennedy, C.; Steinberger, J.; Gasson, B.; Hansen, Y.; Hillman, T.; Havránek, M.; Pataki, D.; Phdungsilp, A.; Ramaswami, A.; Mendez, G.V. Methodology for inventorying greenhouse gas emissions from global cities. *Energy Policy* **2009**, *38*, 4828–4837. [CrossRef]
54. Lan, M.; Nie, S.; Wang, J.; Zhang, Q.; Chen, Z. A state-of-the-art review on lime-based mortars for restoration of ancient buildings. *Mater. Rep.* **2019**, *33*, 1512–1516.
55. Kesimal, A.; Yilmaz, E.; Ercikdi, B. Evaluation of paste backfill mixtures consisting of sulphide-rich mill tailings and varying cement contents. *Cem. Concr. Res.* **2004**, *34*, 1817–1822. [CrossRef]
56. Zhang, Q.; Zhang, J.; Huang, Y.; Feng, J. Backfilling technology and strata behaviors in fully mechanized coal mining working face. *Int. J. Min. Sci. Technol.* **2012**, *22*, 151–157. [CrossRef]
57. Deng, X.; Zhang, J.; Klein, B.; Zhou, N.; de Wit, B. Experimental characterization of the influence of solid components on the rheological and mechanical properties of cemented paste backfill. *Int. J. Miner. Process.* **2017**, *168*, 116–125. [CrossRef]
58. Deng, X.; Klein, B.; Tong, L.; Wit, B.D. Experimental study on the rheological behavior of ultra-fine cemented backfill. *Constr. Build. Mater.* **2018**, *158*, 985–994. [CrossRef]

 *sustainability*

Article

Improved Mechanical Properties of Cement-Stabilized Soft Clay Using Garnet Residues and Tire-Derived Aggregates for Subgrade Applications

Patimapon Sukmak [1,*], Gampanart Sukmak [1], Suksun Horpibulsuk [2,3], Sippakarn Kassawat [4], Apichat Suddeepong [5] and Arul Arulrajah [6]

1   School of Engineering, Technology and Center of Excellence in Sustainable Disaster Management, Walailak University, Nakhonsithammarat 80161, Thailand; gampanart.su@wu.ac.th
2   School of Civil Engineering, and Center of Excellence in Innovation for Sustainable Infrastructure Development, Suranaree University of Technology, Nakhon Ratchasima 30000, Thailand; suksun@g.sut.ac.th
3   Academy of Science, Royal Society of Thailand, Bangkok 10300, Thailand
4   Faculty of Commerce and Management, Prince of Songkla University, Trang Campus, Trang 92000, Thailand; sippakarn.k@psu.ac.th
5   School of Civil and Infrastructure Engineering, and Center of Excellence in Innovation for Sustainable Infrastructure Development, Suranaree University of Technology, Nakhon Ratchasima 30000, Thailand; suddeepong@g.sut.ac.th
6   Department of Civil and Construction Engineering, Swinburne University of Technology, Melbourne, VIC 3000, Australia; aarulrajah@swin.edu.au
*   Correspondence: patimapon.su@wu.ac.th; Tel.: +66-7567-3000

Citation: Sukmak, P.; Sukmak, G.; Horpibulsuk, S.; Kassawat, S.; Suddeepong, A.; Arulrajah, A. Improved Mechanical Properties of Cement-Stabilized Soft Clay Using Garnet Residues and Tire-Derived Aggregates for Subgrade Applications. *Sustainability* **2021**, *13*, 11692. https://doi.org/10.3390/su132111692

Academic Editor: Castorina Silva Vieira

Received: 21 September 2021
Accepted: 20 October 2021
Published: 22 October 2021

Publisher's Note: MDPI stays neutral with regard to jurisdictional claims in published maps and institutional affiliations.

Abstract: The growth of the global economy in recent years has resulted in an increase in infrastructure projects worldwide and consequently, this has led to an increase in the quantity of waste generated. Two recycled materials, namely garnet residues (GR) and tire-derived aggregates (TDA), were used to improve mechanical properties of soft clay (SC) subgrade in this study. GR was evaluated as a replacement material in SC prior to Type I Portland cement stabilization. TDA was also studied as an elastic material in cement-stabilized SC–GR. The laboratory tests on the cement–TDA-stabilized SC–GR included unconfined compressive strength (UCS), indirect tensile stress (ITS) and indirect tensile fatigue (ITF). Microstructural analysis on the cement–TDA-stabilized SC–GR was also performed to illustrate the role of GR and TDA contents on the degree of hydration. The UCS of cement-stabilized SC–GR increased when cement content increased from 0% to 2%. Beyond 2% cement content, the UCS development was slightly slower, possibly due to the presence of insufficient water for hydration. The GR reduces the specific surface and particle contacts of the SC–GR blends to be bonded with cementitious products. The optimum SC:GR providing the highest UCS was found to be 90:10 for all cement contents. Increased amounts of GR led to a reduction in UCS values due to its high water absorption, resulting in the insufficient water for the cement hydration. Moreover, the excessive GR replacement ratio weakened the interparticle bond strength due to its smooth and round particles. The TDA addition can enhance the fatigue resistance of the cement-stabilized SC–GR. The maximum fatigue life was found at 2% TDA content. The excessive TDA caused large amounts of micro-cracks in cement–TDA-stabilized SC–GR due to the low adhesion property of TDA. The SC:GR = 90:10, cement content = 2% and TDA content = 2% were suggested as the optimum ingredients. The outcome of this research will promote the usage of GR and TDA to develop a green high-fatigue-resistant subgrade material.

Keywords: soil–cement; pavement geotechnics; ground improvement; recycled waste; fatigue life; subgrade; compressive strength

1. Introduction

The continuous growth of emerging and developed economies has led to an increase in infrastructure projects, such as roads. Nakhon Si Thammarat is one of the largest

115

economic cities in the southern region of Thailand and most of its population lives in coastal areas. These coastal areas are underlain by soft clay (SC) deposits with high organic matter contents and with poor geotechnical properties which are also sensitive to moisture change [1,2]. Therefore, ground improvement is normally required before the construction of highway and road projects.

A widely accepted soft ground improvement technique is chemical stabilization using Portland cement, calcium carbide, quicklime and geopolymer [1,3–7]. In the past century, cement has been extensively acceptable for pavement and road construction. However, cement production releases a large amount of carbon dioxide (CO_2), which is a critical cause of global warming issues. Therefore, the usage of low CO_2 emission cementing agents with an alternative method for ground improvement is an interesting issue in research and development in transportation geotechnics.

In the past few years, the coarse and fine waste aggregates from civil engineering projects and/or industries, e.g., recycled concrete aggregates, crushed masonry bricks, recycled glasses and melamine debris, have been successfully utilized for ground improvement projects [8–15]. These recycled materials are low in plastic and have potential for improving the stiffness and strength of soil, especially clayey soil.

Due to the rapid growth of the global economy, marine and land transportation and oil demand have been increasing. This causes the increased quantity of wastes from repair and maintenance industries, namely garnet residues (GR) and tire-derived aggregates (TDA) (Figure 1). Garnet refers to the most complex crystalline silicate structure group with various chemical compositions. GR is a waste generated from usage of garnet in restored applications such as pre-finishing surface preparation before paint or other coatings on ship structures [16]. GR causes a major environmental concern worldwide, including in Thailand. In 2019, the total estimated global production of raw garnets for industrial purposes was 1.2 million tonnes/year, and China, USA, India, South Africa and Australia were the major producers. The consumption of raw garnets in 2020 in the USA was a 32% increase from that of 2016 [17]. In Thailand, the quantity of raw garnets acquired from both local and foreign sources for the domestic industries is about 8000 tonnes annually, which is mainly imported by the Thai Beverage Distribution Co., Ltd. (TBD). The global consumption of raw garnets forecasted indicates that these numbers will continue to increase annually [18]. The contaminants in GR consist of old paint, oil and other residues from the surface during blasting. GR is mostly disposed of at landfills. These wastes could disrupt the balance of the natural environment system through the pollution of water sources caused by runoff or flooding in the landfills. Kunchariyakun and Sukmak [19] undertook research to reduce pollution and reported that mixing GR with cement reduced leaching of heavy metals. Therefore, the reuse of GR in civil infrastructure applications is an interesting issue.

Figure 1. Waste rubber tires.

Recently, several researchers [19–24] employed GR as a fine aggregate in an infrastructure construction. The replacement of GR in natural river sand of up to a maximum of 25% could produce geopolymer concretes that meet the required performance [21]. For road applications, GR can be used as a fine aggregate in asphalt concrete; the asphalt concrete with up to 25% GR replacement by weight of total aggregate had suitable Marshall properties comparable with conventional asphalt concrete using 100% granite aggregate [24]. Moreover, the GR replacement could improve California bearing ratio (CBR) of clayey sand for subgrade applications [23].

Automotive and truck tires and vulcanized rubbers have low elasticity and yield strain as well as high Young's modulus. Tires are made through the vulcanization process to form a crosslinked formation in the molecular structure of rubber to have high shear and temperature resistance for extreme environmental conditions. About one billion tonnes of TDA are generated annually around the world due to an increased number of vehicles [25]. TDA is a non-biodegradable material with a low degradation rate. Although the landfilling and combustion of TDA are a simple management technique, they cause recontamination of hazardous gases and dust in the atmosphere and underground water resources [25]. In past decades, the usage of TDA in road applications has become popular [26–28]. TDA as a fine aggregate in coarse recycled aggregates reduced the stiffness of concrete pavements; however, in turn, it could improve their performance, e.g., ductility and cracking and fatigue resistance. Moreover, the TDA could be used as an aggregate to improve geotechnical properties of highly expansive clay for subgrade applications [29]. The maximum unconfined compressive strength (UCS) and toughness were obtained at a 5% TDA replacement ratio. The higher TDA replacement ratio (>5%) caused a decrease in UCS. The swelling strain of expansive soil could also be minimized with the TDA replacement.

To the best of the authors' knowledge, there is no available research on the usage of combined GR and TDA in the mechanical strength improvement of soft clay to be a stabilized subgrade material. This research examined the feasibility of using GR as a replacement material in SC to improve its basic properties prior to cement stabilization to develop a green subgrade. TDA was used to improve the fatigue resistance of cement-stabilized SC–GR blends. The UCS, indirect tensile stress (ITS) and indirect tensile fatigue (ITF) of the cement–TDA-stabilized SC–GR were examined at various factors of SC:GR ratios and cement and TDA contents. Furthermore, the microstructural analysis was performed by using scanning electron microscopy and energy dispersive X-ray spectroscopy (SEM-EDX) to illustrate the role of GR in the interparticle bond strength improvement and TDA in the fatigue resistance improvement. Based on the authors' best knowledge, the investigation of cement–TDA-stabilized SC–GR blends under static and repeated tensile loading as well as their microstructural analysis has not been available, which is significant for road analysis and design. The outcome of this research will promote the usage of TDA and GR in road subgrade applications.

2. Materials and Methods

2.1. Materials

Tire derived-aggregates (TDA) was obtained from Union Commercial Development Co., Ltd., in Thailand, and air-dried before being used. The morphology and particle size distribution are shown in Figures 2a and 3, respectively. The TDA shape was irregular and prepared to have various single sizes of 2.830 mm, 2.000 mm, 0.841 mm, 0.595 mm, 0.400 mm, 0.297 mm and 0.250 mm. The TDA was then trial mixed to meet the gradation requirement for fine aggregates in accordance with ASTM C33/C33M-18 [30] (Figure 3). Table 1 presents the physical properties of TDA, indicating that the specific gravity and water absorption of TDA (ASTM C128-15 [31]) were 1.78% and 2.4%, respectively.

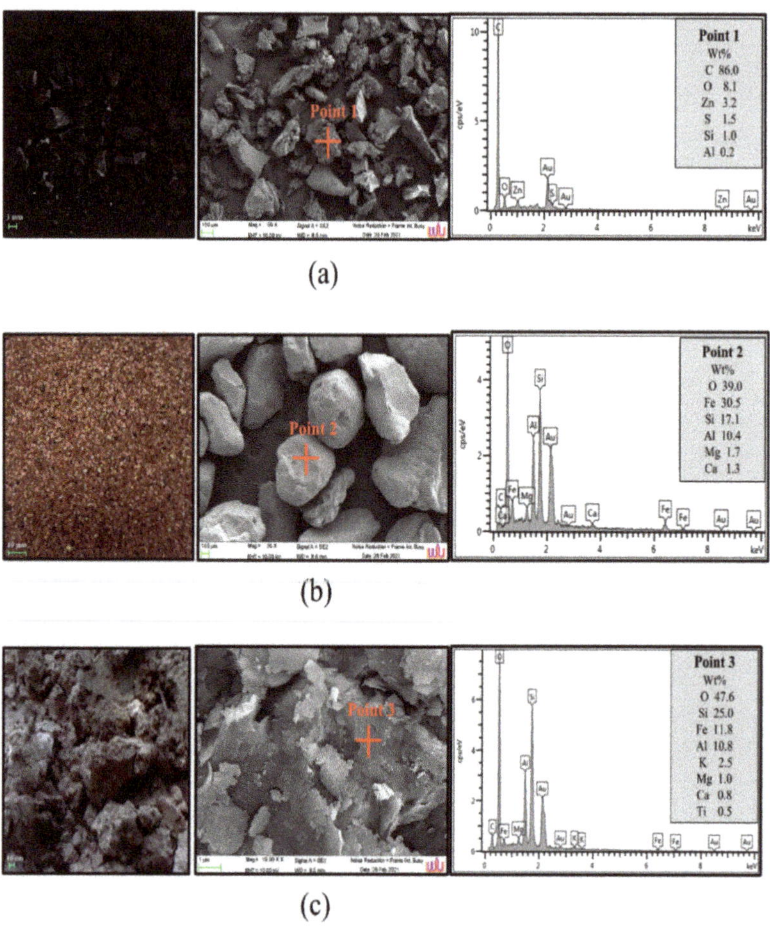

Figure 2. Appearance particles and SEM images of (**a**) tire-derived aggregates (TDA), (**b**) garnet residues (GR) and (**c**) soft clay (SC).

GR was sourced from Best Performance Engineering Co., Ltd. located in the south of Thailand; it came from the blasting and pre-finishing surface processes of ship and/or oil drilling tools. The GR was transferred to a laboratory and kept in sealed plastic bags for geotechnical tests. The physical properties, morphology and particle distribution curve are shown in Table 1 and Figures 2b and 3, respectively. The GR particles were relatively round in shape. The specific gravity and water absorption according to ASTM C128-15 [31] were 3.8% and 10.2%, respectively. The natural water content was approximately 0.2% based on ASTM D2216-19 [32]. GR has no liquid or plastic limits [33] due to its low plasticity. Based on ASTM C33/C33M-18 [30], the median diameter (D_{50}) of GR was 0.75 mm, similar to that of natural sand, as shown in Figure 3, whereas the specific gravity value of GR (=3.8) was greater than that of the natural sand (=2.7). The coefficient of uniformity (C_u) was 2.18 and the coefficient of curvature (C_c) was 1.35. The GR was therefore classified as poorly graded sand (SP) according to the Unified Soil Classification System (USCS) [34].

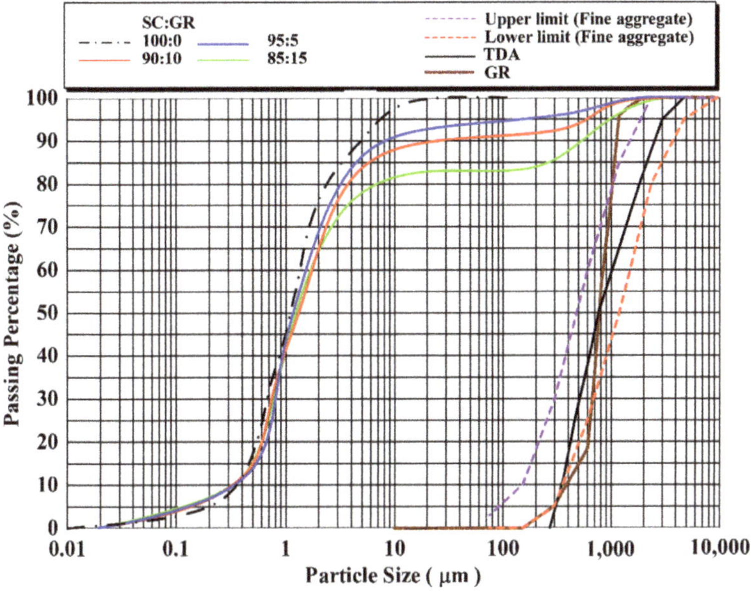

Figure 3. Gradation curves of GR and SC blends at SC:GR = 100:0, 95:5, 90:10 and 85:15.

Table 1. Physical properties of SC, GR and TDA.

Physical Properties	SC	GR	TDA
Specific gravity, SG	2.60	3.8	1.78
Water absorption (%)	-	10.2	2.4
Natural water content (%)	41.6	0.2	-
Liquid limit, LL (%)	65	N/A	-
Plastic limit, PL (%)	27.7	Non-plastic	-
Plastic index, PI (%)	37.3	N/A	-
Sand content (%)	-	100	-
Silt content (%)	25	-	-
Clay content (%)	78	-	-
D_{60} (mm)	-	0.95	1.01
D_{50} (mm)	-	0.75	0.75
D_{30} (mm)	-	0.52	0.52
D_{10} (mm)	-	0.29	0.35
C_u	-	2.18	2.88
C_c	-	1.35	0.76
Classification—USCS [34]	CH	SP	-
Classification—AASHTO [35]	A-7-6	A-1-b	
Maximum dry unit weight, $\gamma_{d,max}$ (kN/m^3) [36]	15.4	-	-
Optimum moisture content, OMC (%)	23.7	-	-

Soft clay (SC) samples studied were alluvial clay commonly found in the Pak-Phanang estuary, in Nakhon Si Thammarat, Thailand. It was taken from a depth of 3–4 m below ground level. The morphology of SC particles was found to be irregular in shape (Figure 2c). Table 1 presents physical properties of SC, indicating that the specific gravity (ASTM D854 [37]), natural water content (ASTM D2216-19 [32]) and liquid limit and plastic limit (ASTM C4318-10 [33]) were 2.60%, 41.6%, 65% and 27.7%, respectively. The particle size distribution of SC is also shown in Figure 3. The SC was classified as high plasticity (CH) according to USCS [34]. The maximum dry unit weight ($\gamma_{d,max}$) and optimum moisture content (OMC) of SC according to ASTM D 1557 [36] were 17.0 kN/m^3 and 19.8%, respectively.

The chemical compositions of TDA, GR and SC were examined under a scanning electron microscope with energy dispersive X-ray spectroscopy (SEM/EDX) and are shown in Figure 2. The major cation elements in GR and SC were Si, Al and Fe, whereas C was the domain cation element in TDA.

2.2. Mix Proportions and Preparation

GR was blended with SC at SC:GR ratios of 100:0 (only soft clay), 95:5, 90:10 and 85:15 to improve the basic properties and compactability. GR replacement is undertaken to reduce the specific surface of the SC to improve the UCS of cement-stabilized SC. However, with high water absorption, the excess GR replacement ratio might cause a negative contribution. As such, the SC:GR ratio was limited to 85:15 in this study. Type I Portland cement was employed to stabilize the SC–GR blends at five different cement contents of 0%, 1%, 2%, 3% 4% and 5% by the dry weight of the SC–GR blends. The mixture was compacted under modified Proctor energy [37] to determine the $\gamma_{d\,max}$ and OMC.

The cement–SC–GR blends of each ingredient were thoroughly mixed at the OMC until the homogenous mixture was achieved. The blends were compacted in a metal cylindrical mold with dimensions of 102 mm in diameter and 116.4 mm in height in five layers [37]. After 24 h, the cylindrical specimens were dismantled and sealed in plastic wraps to prevent evaporation. The cement-stabilized SC–GR specimens were kept at an ambient room temperature (27–30 °C) until the lapse of seven days of curing. The UCS tests were run on the cement-stabilized SC–GR specimens according to ASTM D1633 [38] to obtain the optimum of SC:GR ratio (highest UCS).

The TDA was blended with SC and GR at 1%, 2% and 3% by weight of the SC–GR mixtures at the optimum SC:GR ratio. The cement–SC–GR–TDA blends were then prepared at the OMC and compacted under modified Proctor energy to achieve the $\gamma_{d,max}$ state in a metal cylindrical mold with dimensions of 102 mm in diameter and 116.4 mm in height for UCS tests and in a metallic mold with dimensions of 101.60 mm diameter and 65.00 mm height for ITS and ITF tests. The specimens were dismantled, sealed in plastic wraps and kept at an ambient room temperature (27–30 °C) for seven days prior to the UCS, ITS, ITF and SEM-EDX testing. Figure 4 summarizes the steps of specimen preparation of cement–TDA-stabilized SC–GR specimens.

2.3. Testing Methods

2.3.1. Unconfined Compression Strength Test

The UCS tests were run according to ASTM D1633 [38] on the cement-stabilized SC–GR specimens with and without the TDA after seven days of curing, at a deformation rate of 1 mm/min. The UCS test was conducted on least five specimens to ensure testing consistency.

2.3.2. Indirect Tensile Strength Test

The indirect tensile strength (ITS) test in accordance with ASTM D6931 is performed to measure the tensile strength of pavement material for highway engineering design [39]. The ITS tests on cement–TDA-stabilized SC–GR specimens were conducted using a universal testing machine with a loading strip of 19 mm wide and 125 mm long at a deformation rate of 1 mm/min. According to the elastic theoretical approach, the ITS was calculated by using the following equation:

$$\text{ITS} = \frac{2P}{\pi DL} \tag{1}$$

where P is the is a maximum load (N), D is the specimen diameter (mm) and L is the specimen length (mm).

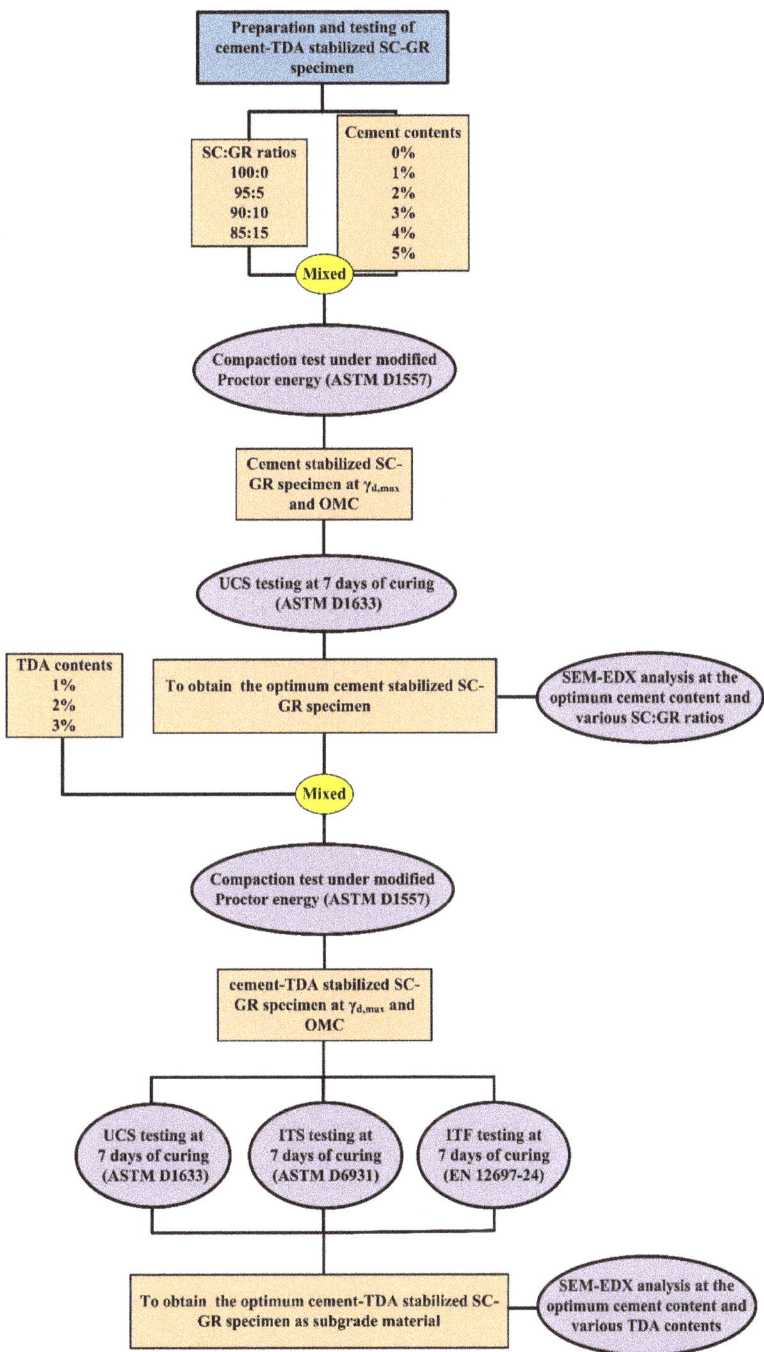

Figure 4. Summary of preparation and testing of cement–TDA-stabilized SC–GR specimens.

2.3.3. Indirect Tensile Fatigue Test

The indirect tensile fatigue (ITF) test according to EN 12697-24 is performed on road materials under controlled loading to examine the fatigue characterization. Kavussi and Modarres [40] recommended a loading frequency for the simulation of low traffic volume on rural roads of 0.66 Hz. Since rural roads are subject to the transportation of agricultural products such as livestock and agricultural products, which are relatively heavy, the applied stress level for the ITF specimens in this study was 80% of the corresponding ultimate ITS. The fatigue life of the ITF specimens is defined as the total number of loading cycles needed to damage the specimens. A linear variable differential transformer (LVDT) with an automatic recorder was used to measure horizontal deformations.

2.3.4. Scanning Electron Microscopy and Energy Dispersive X-ray Spectroscopy

The scanning electron microscopy and energy dispersive X-ray spectroscopy (SEM-EDX) analysis was achieved using a Merlin machine of Carl Zeiss, a company in Oberkochen, Germany, together with the Oxford Instruments Nano Analysis and the newest analytical system from Wycombe, U.K. The SEM-EDX specimen was a small fragment of the broken UCS specimen. It was frozen at $-195\ °C$ for five minutes in liquid nitrogen and evacuated at a pressure of 0.5 Pa at $-40\ °C$ to stop the hydration of cement. After drying, the SEM-EDX specimens were coated with gold to investigate the cementitious products and identify their chemical characterization by using the area mapping technique.

3. Results and Discussion

3.1. Cement-Stabilized SC–GR Blends

It is evident from Table 2 and Figure 3 that the basic properties of the SC–GR blends at SC:GR = 100:0, 95:5, 90:10 and 85:15, such as gradations and Atterberg limits, did not pass the requirements of the Thailand Department of Highways for stabilized base and subbase materials [41,42]. The SC–GR blends can only be used as stabilized subgrade material and its 7-day UCS must be greater than the minimum requirement of 294 kPa [43].

Table 2. Basic and mechanical properties of the SC–GR blends.

Properties	SC: GR Ratio				Standard for Stabilized Subbase (DH-S206/2532)	Standard for Stabilized Base (DH-S204/2556)
	100:0	95:05	90:10	85:15	Value	
Largest particle size (mm)	0.014	2.36	2.36	2.36	≤ 50	≤ 50
Passed at a 2.0 mm sieve (%)	100 *	100 *	100 *	100 *	NS	≤ 70
Passed at a 0.075 mm sieve (%)	100 *	94 *	91 *	83 *	≤ 40	≤ 25
Liquid limit, LL (%)	65.0	64.7	64.2	64.1	≤ 40	≤ 40
Plastic limit, PL (%)	27.7	30.7	31.8	31.7	NS	NS
Plasticity index, PI (%)	37.3	34	32.4	32.4	≤ 20	≤ 15
Maximum dry unit weight, $\gamma_{d,max}$ (kN/m^3) (ASTM D 15557)	15.4	15.6	15.8	16	NS	NS
Optimum moisture content, OMC (%) (ASTM D 15557)	23.7	23.1	21.8	21.5	NS	NS
Unconfined compression strength, UCS (kPa)	80 *	96 *	100 *	90 *	>689	>1724
Axial stress at 0.6% strain (kPa)	22	26	42	59	NS	NS
Secant modulus, E_{sec} (MPa)	3.7	4.3	7.0	9.8	NS	NS

Note: NS = not specified. * Did not meet requirement.

The change in Atterberg limits with cement content showed the impact of cement content on the specimens' plasticity characteristics (see Tables 2 and 3). The increase in cement content reduced the LL for all SC:GR ratios, for example, from 65% to 60.3% for cement contents from 0% to 5% for SC:GR = 100:0. This is because of the change in the SC's structure from dispersed to a flocculated structure. The increase in the plastic limit, PL, was caused by prominent flocculated structure and the development of cementation in the SC–GR structure [44,45].

Table 3. Basic and mechanical properties of the cement-stabilized SC–GR specimens.

Cement Content (%)	Properties	SC: GR Ratio			
		100:0	95:5	90:10	85:15
1	Liquid limit, LL (%)	64.1	63.6	63.4	63.1
	Plastic limit, PL (%)	28.6	31	31.5	31.9
	Plasticity index, PI (%)	35.5	32.6	31.9	31.2
	Maximum dry unit weight, $\gamma_{d,max}$ (kN/m^3)	16	16.2	16.4	16.6
	Optimum moisture content, OMC (%)	21.9	21.4	20.5	20
	Unconfined compression strength, UCS (kPa)	159	176	260	214
	Axial stress at 0.6% strain (kPa)	39	50	76	109
	Secant modulus, E_{sec} (MPa)	6.5	8.3	12.7	18.2
2	Liquid limit, LL (%)	62.3	61.8	61.5	61.2
	Plastic limit, PL (%)	29.6	31.2	31.5	31.8
	Plasticity index, PI (%)	32.7	30.6	30	29.4
	Maximum dry unit weight, $\gamma_{d,max}$ (kN/m^3)	16.2	16.6	17.1	17.4
	Optimum moisture content, OMC (%)	20.8	20.5	20.0	19.6
	Unconfined compression strength, UCS (kPa)	222	278	403	367
	Axial stress at 1% strain (kPa)	56	76	103	122
	Secant modulus, E_{sec} (MPa)	9.3	12.7	17.2	20.3
3	Liquid limit, LL (%)	61.1	60.7	60.4	60.2
	Plastic limit, PL (%)	30.1	32.1	32.6	33
	Plasticity index, PI (%)	31	28.6	27.8	27.2
	Maximum dry unit weight, $\gamma_{d,max}$ (kN/m^3)	16.8	17.2	17.3	17.5
	Optimum moisture content, OMC (%)	20.4	19.5	19.3	19
	Unconfined compression strength, UCS (kPa)	242	340	462	424
	Axial stress at 1% strain (kPa)	77	132	157	187
	Secant modulus, E_{sec} (MPa)	12.8	22.0	26.2	31.2
4	Liquid limit, LL (%)	60.3	59.7	59.5	59
	Plastic limit, PL (%)	31.6	33.8	34.1	34.8
	Plasticity index, PI (%)	28.7	25.9	25.4	24.2
	Maximum dry unit weight, $\gamma_{d,max}$ (kN/m^3)	17	17.4	17.9	18.1
	Optimum moisture content, OMC (%)	19.7	19.2	18.8	18.3
	Unconfined compression strength, UCS (kPa)	279	376	524	492
	Axial stress at 1% strain (kPa)	109	200	218	282
	Secant modulus, E_{sec} (MPa)	18.2	33.3	36.3	47.0
5	Liquid limit, LL (%)	60.3	59.7	59.5	59
	Plastic limit, PL (%)	33.2	33.8	35.7	34.8
	Plasticity index, PI (%)	27.1	25.9	23.8	24.2
	Maximum dry unit weight, $\gamma_{d,max}$ (kN/m^3)	17.2	17.8	18.2	18.4
	Optimum moisture content, OMC (%)	19.4	19	18.4	17.7
	Unconfined compression strength, UCS (kPa)	294	410	549	535
	Axial stress at 1% strain (kPa)	171	280	285	359
	Secant modulus, E_{sec} (MPa)	28.5	46.7	47.5	59.8

Table 2 and Figure 5 show the values of $\gamma_{d,max}$ and OMC and compaction curves, respectively, at SC:GR = 100:0, 95:5, 90:10 and 85:15 for various cement contents. For all SC:GR ratios, the addition of cement to the SC–GR blends increased the $\gamma_{d,max}$ but reduced the OMC, similar to the cement-stabilized coarse-grained soil [46] and the cement-stabilized fine-grained soil [44]. The cement had higher specific gravity than the SC; therefore, the density of the specimens increased when the cement content increased. The cement reaction mechanism consists of two stages: immediate and long-term reactions. In the immediate reaction, the Ca^{2+} ions from cement are adsorbed into negative charges of the SC surface and reduce the thickness of diffused double layers of the SC particles. The edge-to-face contacts of SC particles, on the other hand, are increased [44,45], thus resulting in an increase in PL with a decrease in LL. The decrease in LL reduces the OMC and increases $\gamma_{d,max}$ [44].

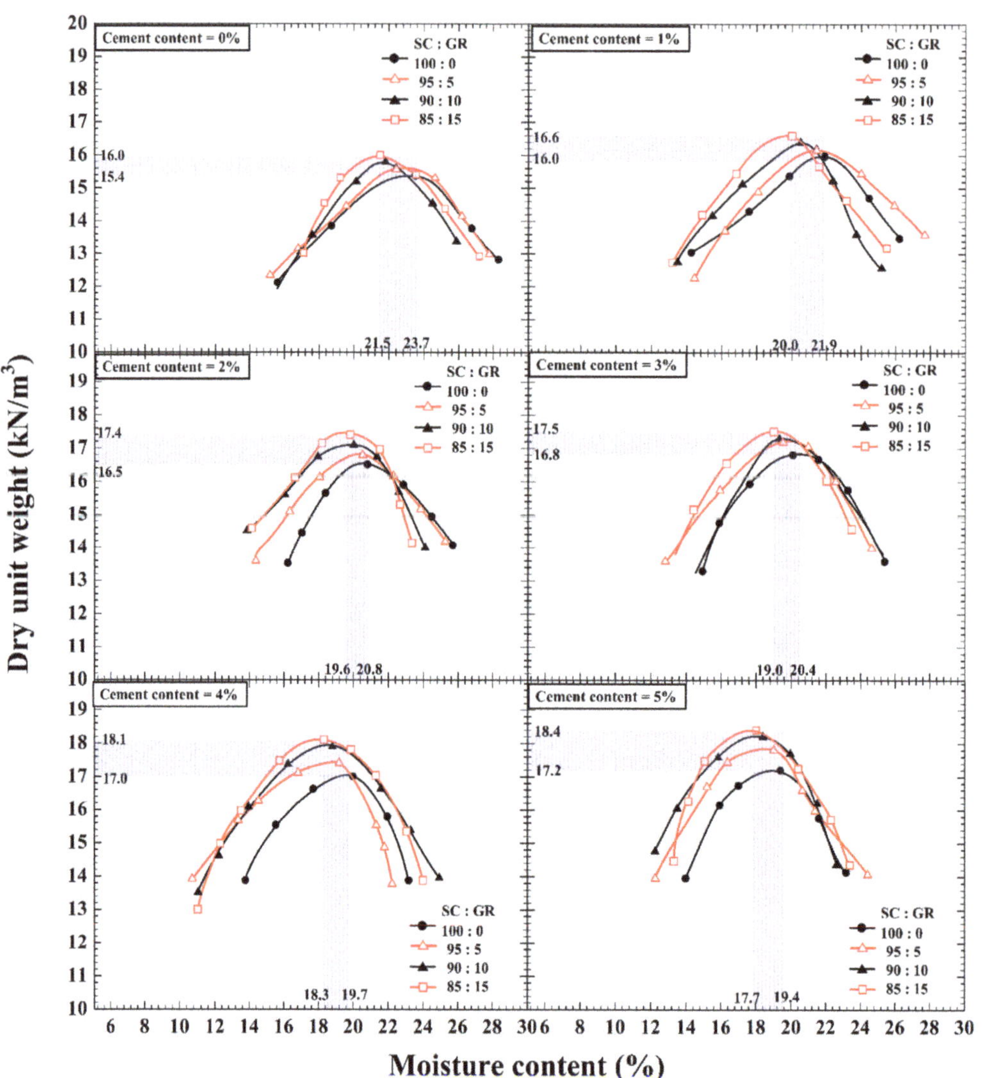

Figure 5. Dry unit weight versus moisture content curves of SC–GR blends with various cement contents.

For all cement contents, the $\gamma_{d,max}$ of the cement–SC–GR mixture increased with the increased GR because the specific gravity of GR was higher than that of SC and cement, as shown by the $\gamma_{d,max}$ of SC:GR = 100:0 being lower than the $\gamma_{d,max}$ of SC:GR = 85:15 for all cement contents. The increase in the $\gamma_{d,max}$ is associated with the decrease in the OMC, as presented in Figure 5.

Figure 6 shows stress–strain curves under the UCS tests for cement contents = 0 to 5% and SC:GR = 100:0 to 85:15. For all cement contents and SC:GR ratios tested, the cement-stabilized SC–GR specimens exhibited brittle behavior with a rapid drop in stress after peak. Since the strain levels developed in road subgrade material due to traffic load varies from 0.003% to 0.6% [47], in this research the secant modulus (E_{sec}) was calculated at 0.6% strain to describe the stiffness of a material. The equation for calculating E_{sec} is:

$$E_{SEC} = \frac{\sigma_{a@\,0.6\%\ strain} - \sigma_{a@\,0\%\ strain}}{\varepsilon_{a@\,0.6\%\ strain} - \varepsilon_{a@\,0\%\ strain}} \tag{2}$$

where $\sigma_{a@0.6\%\ strain}$ is the stress at 0.6% strain, $\sigma_{a@0\%\ strain}$ is the stress at 0% strain (equal to 0), $\varepsilon_{a@0.6\%\ strain}$ is the strain at 0.6% and $\varepsilon_{a\,@\,0.6\%\ strain}$ is the strain at 0% (equal to 0).

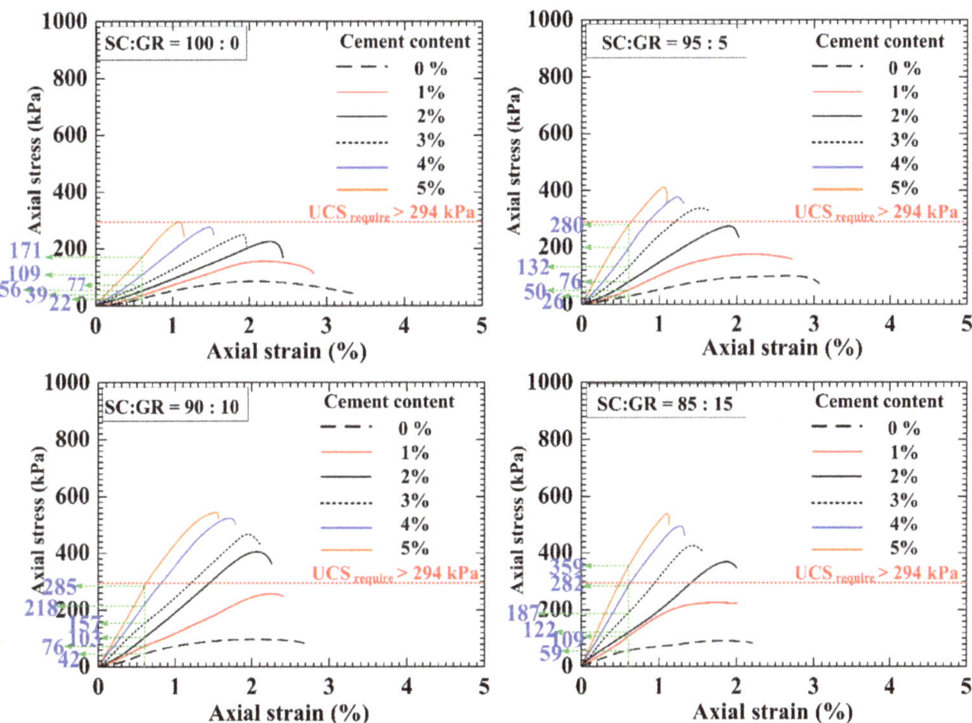

Figure 6. Stress–strain curves under UCS test for cement contents = 0–5% for various SC:GR ratios.

Table 3 shows the variation of E_{sec} of all cement-stabilized SC–GR specimens at 0.6% strain. The E_{sec} and UCS for all SC:GR ratios tended to increase with the increased cement content (Figure 6). Figure 7 shows the UCS development with the increased cement contents of the cement-stabilized SC–GR specimens. As the cement contents increased up to 2%, the cementation bonds at the contact points between the SC–GR particles were stronger due to predominant Calcium Silicate Hydrate (C-S-H, cementitious products). The amount of C-S-H products increased with an increase in the cement content. This range of cement contents could be termed as the active zone. When cement contents were between 2% and 5%, the UCS development was slightly slower, possibly because the water at OMC was not sufficient for hydration.

The role of GR is also clearly depicted in Figure 7. Without GR replacement, the UCS of cement-stabilized SC at cement contents = 1–5% could not meet the minimum requirement of 294 kPa. The UCS values at all cement contents were increased with the GR replacement ratio up to the optimum value of SC:GR = 90:10. This implies that the GR reduces the specific surface and particle contacts of the SC–GR blends to be bonded cementitious products, hence the stronger interparticle bond strength at the same input of cement.

However, when SC:GR > 90:10, the UCS decreased. The GR has high water absorption (refer to Table 1); the higher GR absorbed more water into its particles and therefore, the water is not sufficient cement hydration. Moreover, the excessive smooth and round GR particles caused the decrease in interparticle bond strength. The 2% cement content was found to be the most effective for the OMC when utilized with GR replacement. For all SC:GR ratios tested, the 2% cement-stabilized SC–GR blends met the strength requirement (UCS > 294.2 kPa) for stabilized subgrade specified by the DOH [43].

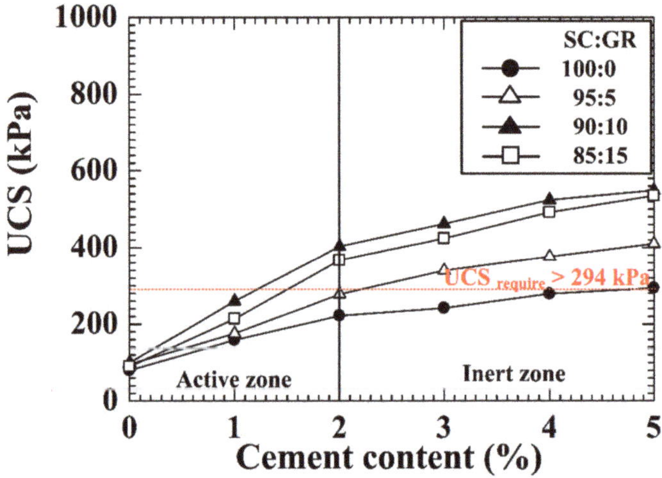

Figure 7. UCS development as a function of cement content.

Figure 8a,b show the SEM-EDX analyses of the specimen at cement content of 2% and SC:GR ratios of 90:10 and 85:15, respectively, to understand the role of GR replacement. The specimen at SC:GR = 90:10 had more C-S-H products (confirmed by EDX result in Area A) in pores than the specimens at SC:GR = 85:15. Moreover, the specimens at SC:GR = 85:15 had more micropores than the specimens at SC:GR = 85:15. The lower cementitious products in specimens at SC:GR = 85:15 were also confirmed by EDX results (refer to Area C (SC particles) and point B (GR particles)). These results confirmed the lower degree of cement hydration at the excessive GR replacement ratio (SC:GR = 85:15) due the high water absorption of GR particles, which resulted in lower strength and stiffness.

3.2. Cement–TDA-Stabilized SC–GR Blends

Figure 9 depicts dry unit weight versus moisture content relationship of the cement–TDA–SC–GR mixtures when SC:GR = 90:10 and cement content = 2% (optimum ingredient) with TDA contents. The $\gamma_{d,max}$ slightly reduced with an increase in the TDA content. Nonetheless, the OMC slightly increased with the increased TDA content (refer to Table 4). Figure 10 presents stress–strain curves under the UCS test when SC:GR = 90:10 and cement content = 2% for various TDA contents. The reduction in UCS and stiffness could be seen with an increase in the TDA content. Moreover, the TDA stabilization resulted in the increase in area under the curves and the decrease in E_{sec}, indicating the increased toughness and the energy absorption before rupture. This characteristic is associated with the higher fatigue resistance, which is required for durable roads. According to the UCS requirement, TDA > 2% cannot be accepted in practice.

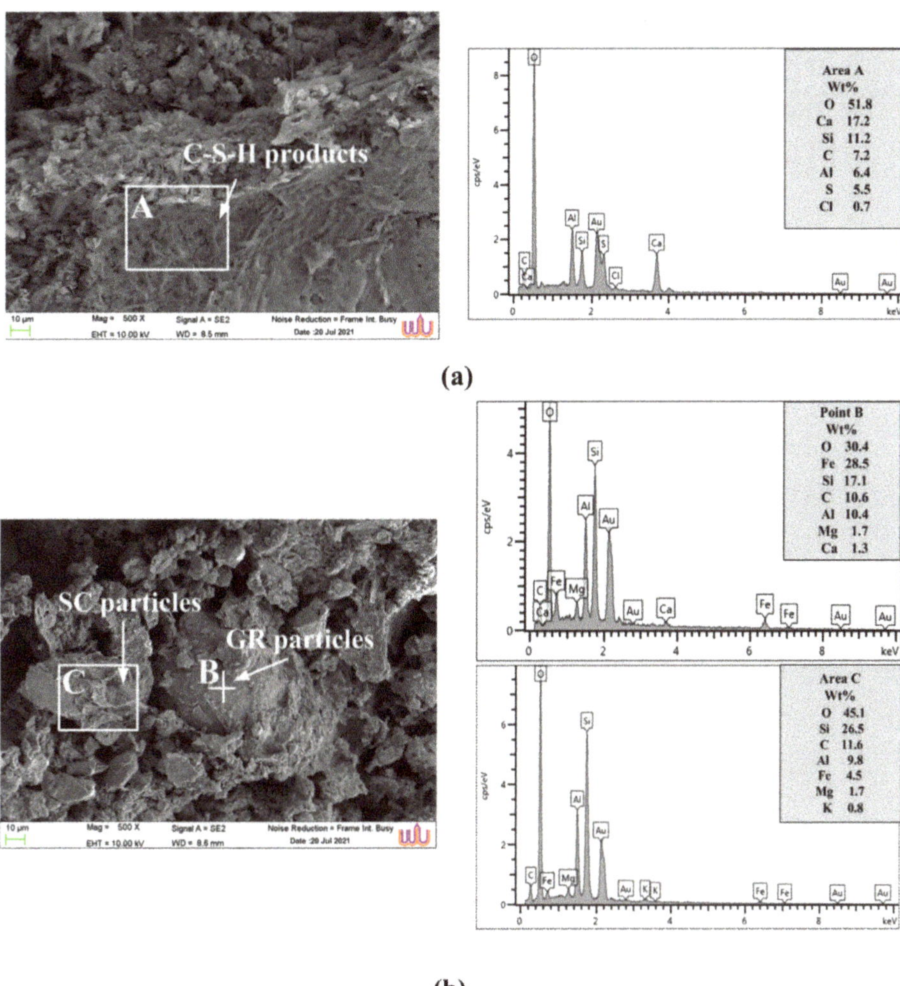

Figure 8. SEM-EDX analyses of specimens at cement content of 2% and when SC:GR ratio = (**a**) 90:10 and (**b**) 85:15.

Table 4. Basic and mechanical properties of the cement–TDA-stabilized SC–GR specimens when SC:GR ratio = 90:10 and cement content = 2% with various TDA contents.

Cement Content (%)	Properties	TDA Content (%)			
		0	**1**	**2**	**3**
2	Maximum dry unit weight, $\gamma_{d,max}$ (kN/m^3)	17.1	16.8	16.6	16.2
	Optimum moisture content, OMC (%)	20.0	21.3	22.0	22.5
	Unconfined compression strength, UCS (kPa)	403	379	339	231
	Axial stress at 0.6% strain (kPa)	103	79	42	21
	Secant modulus, E_{sec} (MPa)	17.2	13.2	6.7	5.2
	Indirect tensile stress, ITS (kPa)	113.6	132.3	137.2	119.3
	Indirect Tensile Fatigue, N_f (pulses)	22	95	115	72
	Initial deformation, Δp (mm)	0.19	0.79	1.18	0.94

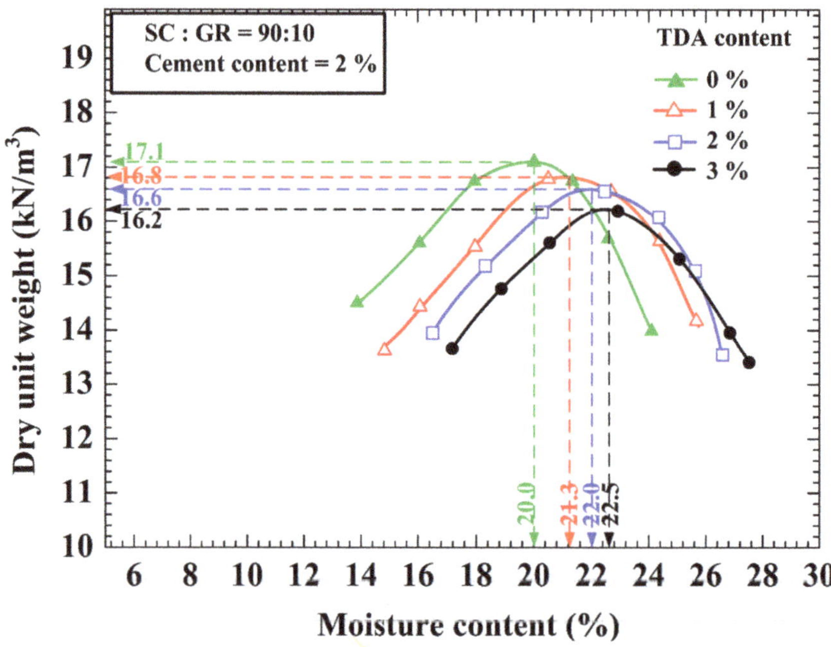

Figure 9. Dry unit weight versus moisture content relationship of cement–TDA-stabilized SC–GR mixtures when SC:GR = 90:10 and cement content = 2% with various TDA contents.

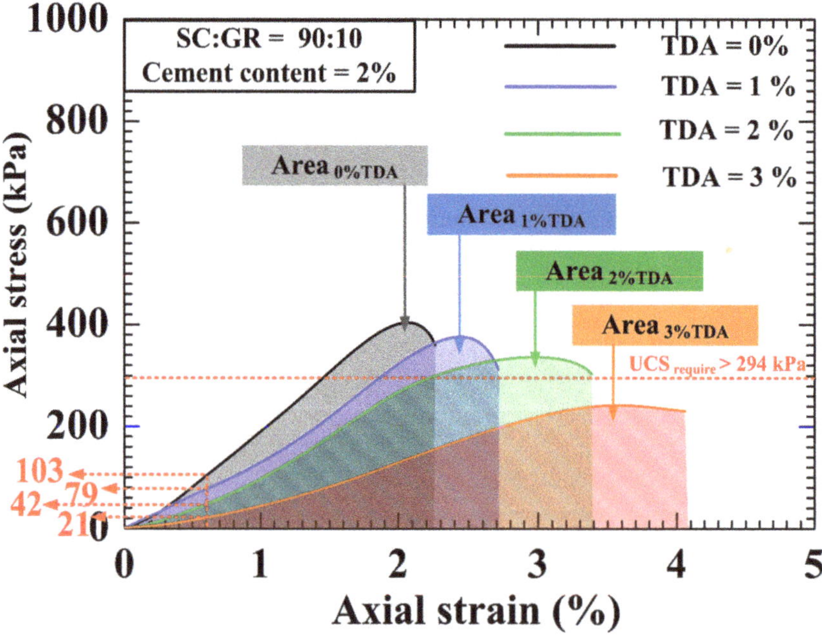

Figure 10. Stress–strain curves under the UCS test of cement–TDA-stabilized SC–GR mixtures when SC:GR = 90:10 and cement content = 2% for various TDA contents.

Figure 11a shows the relationship between the number of cycles versus horizontal deformation of the cement–TDA-stabilized SC–GR specimens when SC:GR = 90:10 and cement content = 2% for TDA contents of 0% to 3%. Figure 11b shows the typical relationship between the number of cycles versus horizontal deformation, which is divided into three zones. In the first zone, at a small number of cycles, high deformation occurred on the specimen because of the increase in plastic deformation. In the second zone, the increase in number of cycles is associated with the lower rate of deformation, whereby the micro-cracks are gradually formed and propagated. In the third zone, the complete splitting failure occurs because of the accumulated microcracks on specimen. Figure 11b also shows the method of determining fatigue life (N_f) and initial deformation in zone 2 (Δp). The initial deformation (Δp) is defined as the intersection of the straight lines extending from the linear portion in zone 1 and zone 2. The N_f is the number of cycles at the splitting failure of the specimen.

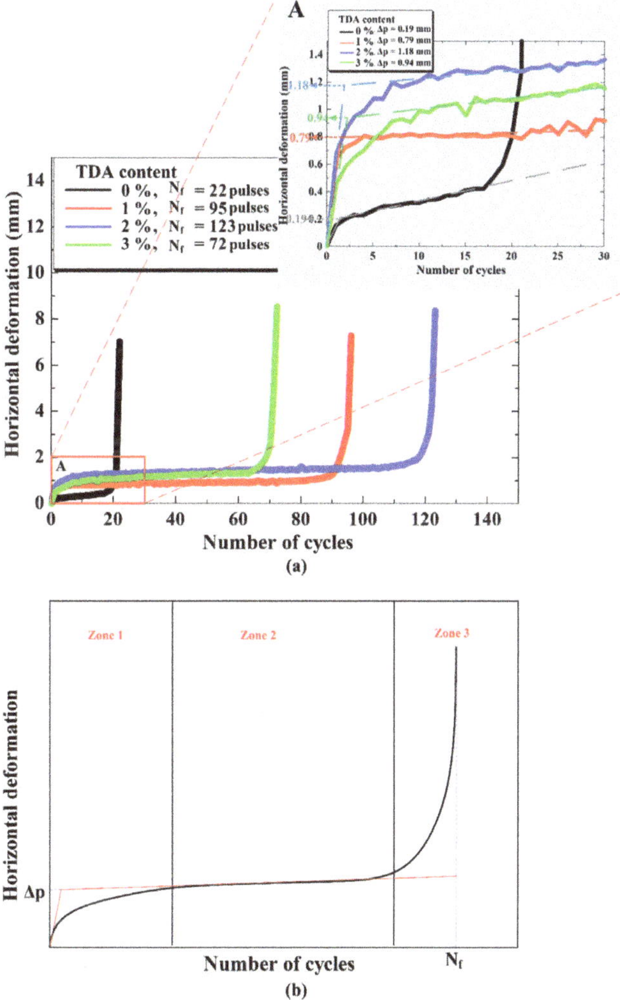

Figure 11. Relationship between the number of cycles versus horizontal deformation of (**a**) the cement–TDA-stabilized SC–GR specimen of SC:GR ratio = 90:10 and cement content = 2% and (**b**) the method of the determining of fatigue life (N_f) and initial deformation (Δp).

At zone 2, the cement–TDA-stabilized SC–GR specimens had longer N_f than the cement-stabilized SC–GR specimens (TDA = 0%). This indicated that the TDA improved the ductility behavior, whereas the specimen at TDA = 0% exhibited sudden failure. The Δp and N_f values increased with the TDA content up to the optimal TDA of 2%, after which they decreased. For example, the Δp values were increased from 0.19 to 1.18 mm for TDA contents from 0% to 2% and N_f values were increased from 22 to 123 pulses for TDA contents from 0% to 2%. The increase in both Δp and N_f is associated with the increase in ITS (Table 4). In other words, both Δp and N_f values are directly related to the ITS.

The role of TDA in the UCS and fatigue resistance can be explained by the SEM-EDX analyses shown in Figure 12. More C-S-H gels (Area E) bonding TDA (Area D) in SC–GR particles and in voids were observed at 1% TDA content (Figure 12a), when compared with 2% TDA content (Figure 12b) and 3% TDA content (Figure 12c). More micro-cracks within TDA–SC–GR clusters were, however, found (red dash line) for 2% and 3% TDA contents when compared with 1% TDA content. The cracks developed were attributed to the low adhesion property of TDA particles. As such, the UCS, which represents the static and short-term strength, decreased with increasing TDA content. Even with micro-cracks, the TDA particles at optimum content can absorb more cyclic load energy and result in larger N_f. However, the excessive TDA with more micro-cracks caused excessive plastic deformation and the reduction in energy absorption and hence, the reduction in N_f.

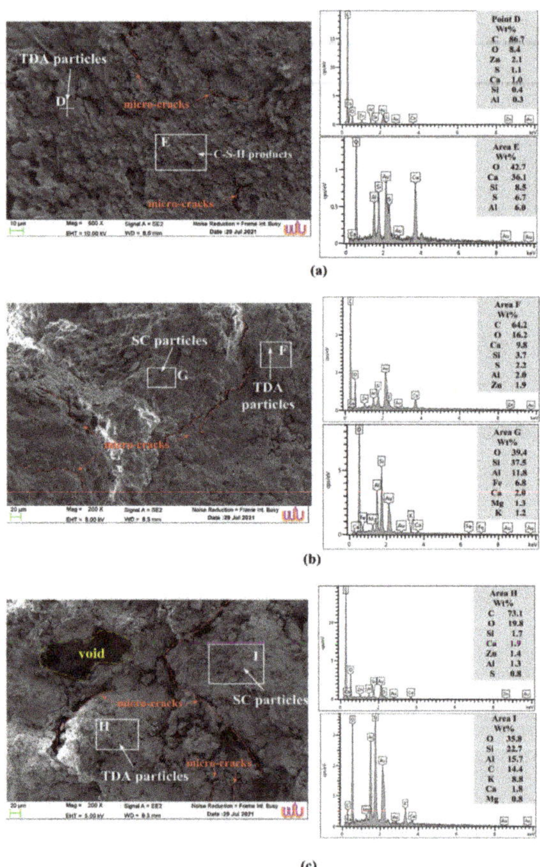

Figure 12. SEM-EDX analyses of the cement–TDA-stabilized SC–GR specimens when SC:GR ratio = 90:10 and cement content = 2%, and WRT content of (**a**) 1%, (**b**) 2% and (**c**) 3%.

3.3. Economic and Environmental Benefits

Table 5 shows the total construction costs of cement–TDA-stabilized SC–GR and lateritic soil as a pavement subgrade. The total construction cost of cement–TDA-stabilized SC–GR was 48.48% less than that of compacted lateral soil in 1 m^3 highway, indicating the cost savings. Moreover, the industry could reduce the GR and TDA disposal costs by approximately 58.03 USD/tonne (from GMA Garnet Group) and 10 USD/tonne [48], respectively, and also reduce environmental pollution from disposal in landfills.

Table 5. Material costs comparison between of using cement, GR and TDA in SC and compacted lateral soil for subgrade application in 1 cubic meter.

Section	Material	Volume(m^3) [a]	Weight (kg)	Price (USD/m^3) [b]	Total Cost (USD)
Cement–TDA-stabilized SC–GR at SC:GR = 90:10, cement content = 2% and TDA content = 2%.	cement	0.02	63	5.09 [49]	5.11
	GR	0.096	364.8	-	
	TDA	0.02	35.6	0.0178 [c]	
	SC	0.864	2246.4	-	
Lateral soil	lateral soil	1	-	10.54 [49]	10.54

[a] Based on the dry soil weight. [b] Not including shipping and labor costs. [c] The price from Union Commercial Development Co., Ltd., Samut Prakan, Thailand.

4. Conclusions

This research aims to examine the feasibility of using GR as a replacement material in soft clay (SC) prior to cement stabilization to be a subgrade material. TDA was used to improve the fatigue resistance of cement-stabilized SC–GR. The mechanical and microstructural investigation of the cement–TDA-stabilized SC–GR were performed to ascertain it as a sustainable subgrade material. The following conclusions can be drawn from this study.

1. The increase in $\gamma_{d,max}$ and the decrease in OMC were caused by changing the dispersed structure to a flocculated SC–GR structure with the addition of cement. Therefore, $\gamma_{d,max}$ increased with the GR replacement ratio. The GR replacement reduced the specific surface of SC, but at the same time, increased the water absorption. The optimum SC:GR ratio was found at 90:10. The 2% cement content for stabilized SC–GR at SC:GR of 90:10 was the optimum mixture.

2. The UCS and stiffness of cement-stabilized SC–GR were found to reduce with the increase in TDA content. This is due to the low adhesion property of TDA; the micro-cracks within SC–GR–TDA matrix were detected with the increased TDA content. However, the increased TDA content improved the ductile behavior and resulted in the increased energy absorption before rupture. The optimum TDA content was found to be 2%. When TDA content was greater than 2%, the excessive micro-cracks caused excessive plastic deformation and the reduction in energy absorption and, hence, the reduction in fatigue life.

3. The cement–TDA-stabilized SC–GR at SC:GR of 90:10, cement content of 2% and TDA content of 2% is suggested as a sustainable subgrade material. Its UCS met the strength requirements of the Department of Highways, Thailand (DH-S201/2532), and its fatigue life was found to be the highest when compared to other SC:GR ratios with the same cement content. The improved fatigue resistance of the cement-stabilized SC–GR is necessary for durable roads.

Author Contributions: Conceptualization, P.S., G.S., S.H., S.K., A.S. and A.A.; methodology, P.S.; formal analysis, P.S.; writing—original draft preparation, P.S.; writing—review and editing, S.H. and A.A.; supervision, S.H. and A.A.; funding acquisition, P.S. and S.H. All authors have read and agreed to the published version of the manuscript.

Funding: This research was supported by the new strategic research (P2P) project (grant no. CGS-P2P-2564-001), Walailak University, Thailand, and the National Science and Technology Development Agency under the Chair Professor Program (grant no. P-19-52303).

Institutional Review Board Statement: Not applicable.

Informed Consent Statement: Not applicable.

Data Availability Statement: The data presented in this study are available on request from the corresponding author. The data are not publicly available due to privacy restrictions.

Acknowledgments: The first author is grateful to a financial support from Walailak University. Facilities and equipment provided from the Walailak University are very much appreciated. The first and the third authors acknowledge the financial support from the new strategic research (P2P) project (grant no. CGS-P2P-2564-001), Walailak University, Thailand, and the National Science and Technology Development Agency under the Chair Professor Program (grant no. P-19-52303).

Conflicts of Interest: The authors declare no conflict of interest.

References

1. Sukmak, P.; Sukmak, G.; Horpibulsuk, S.; Setkit, M.; Kassawat, S.; Arulrajah, A. Palm oil fuel ash-soft soil geopolymer for subgrade applications: Strength and microstructural evaluation. *Road Mater. Pavement Des.* **2019**, *20*, 110–131. [CrossRef]
2. Du, Y.-J.; Jiang, N.-J.; Liu, S.-Y.; Horpibulsuk, S.; Arulrajah, A. Field evaluation of soft highway subgrade soil stabilized with calcium carbide residue. *Soils Found.* **2016**, *56*, 301–314. [CrossRef]
3. Arulrajah, A.; Yaghoubi, M.; Disfani, M.M.; Horpibulsuk, S.; Bo, M.W.; Leong, M. Evaluation of fly ash-and slag-based geopolymers for the improvement of a soft marine clay by deep soil mixing. *Soils Found.* **2018**, *58*, 1358–1370. [CrossRef]
4. Miura, N.; Horpibulsuk, S.; Nagaraj, T. Engineering behavior of cement stabilized clay at high water content. *Soils Found.* **2001**, *41*, 33–45. [CrossRef]
5. Horpibulsuk, S.; Phetchuay, C.; Chinkulkijniwat, A. Soil stabilization by calcium carbide residue and fly ash. *J. Mater. Civil Eng.* **2012**, *24*, 184–193. [CrossRef]
6. Dahal, B.K.; Zheng, J.-J.; Zhang, R.-J.; Song, D.-B. Enhancing the mechanical properties of marine clay using cement solidification. *Mar. Georesour. Geotechnol.* **2019**, *37*, 755–764. [CrossRef]
7. Sukmak, P.; Horpibulsuk, S.; Shen, S.-L. Strength development in clay–fly ash geopolymer. *Construct. Build. Mater.* **2013**, *40*, 566–574. [CrossRef]
8. Arulrajah, A.; Jegatheesan, P.; Bo, M.W.; Sivakugan, N. Geotechnical characteristics of recycled crushed brick blends for pavement sub-base applications. *Can. Geotech. J.* **2012**, *49*, 796–811. [CrossRef]
9. Arulrajah, A.; Ali, M.M.Y.; Disfani, M.M.; Piratheepan, J.; Bo, M.W. Geotechnical performance of recycled glass-waste rock blends in footpath bases. *J. Mater. Civ. Eng.* **2013**, *25*, 653–661. [CrossRef]
10. Arulrajah, A.; Piratheepan, J.; Disfani, M.M.; Bo, M.W. Geotechnical and geoenvironmental properties of recycled construction and demolition materials in pavement subbase applications. *J. Mater. Civ. Eng.* **2013**, *25*, 1077–1088. [CrossRef]
11. Donrak, J.; Arulrajah, A.; Rachan, R.; Horpibulsuk, S. Improvement of marginal lateritic soil using melamine debris replacement for sustainable engineering fill materials. *J. Clean. Prod.* **2016**, *134*, 515–522. [CrossRef]
12. Kianimehr, M.; Shourijeh, P.T.; Binesh, S.M.; Mohammadinia, A.; Arulrajah, A. Utilization of recycled concrete aggregates for light-stabilization of clay soils. *Construct. Build. Mater.* **2019**, *227*, 116792. [CrossRef]
13. Naeini, M.; Mohammadinia, A.; Arulrajah, A.; Horpibulsuk, S.; Leong, M. Stiffness and strength characteristics of demolition waste, glass and plastics in railway capping layers. *Soils Found.* **2019**, *59*, 2238–2253. [CrossRef]
14. Arulrajah, A.; Disfani, M.M.; Haghighi, H.; Mohammadinia, A.; Horpibulsuk, S. Modulus of rupture evaluation of cement stabilized recycled glass/recycled concrete aggregate blends. *Constr. Build. Mater.* **2015**, *84*, 146–155. [CrossRef]
15. Yaowarat, T.; Horpibulsuk, S.; Arulrajah, A.; Maghool, F.; Mirzababaei, M.; Safuan, A.A.; Chinkulkijniwat, A.R. Cement stabilisation of recycled concrete aggregate modified with polyvinyl alcohol. *Int. J. Pavement Eng.* **2020**, 1–9. [CrossRef]
16. Sidhardhan, S.; Sheela, S.J.; Meylin, J.S. Study on sea sand as a partial replacement for fine aggregate. *J. Adv. Chem.* **2017**, *13*, 6166–6171.
17. Olson, D.W. Garnet, industrial. In *Minerals Yearbook Metals and Minerals*; Government of India: Nagpur, India, 2021; p. 1.
18. Mines, I.B.O. *Indian Minerals Yearbook 2018 (Part-III: Mineral Reviews)*; Government of India, Ministry of Mines: Nagpur, India, 2018.
19. Kunchariyakun, K.; Sukmak, P. Utilization of garnet residue in radiation shielding cement mortar. *Constr. Build. Mater.* **2020**, *262*, 120122. [CrossRef]
20. Muttashar, H.L.; Bin Ali, N.; Ariffin, M.A.M.; Hussin, M.W. Microstructures and physical properties of waste garnets as a promising construction materials. *Case Stud. Constr. Mater.* **2018**, *8*, 87–96. [CrossRef]
21. Muttashar, H.L.; Ariffin, M.A.M.; Hussein, M.N.; Hussin, M.W.; Bin Ishaq, S. Self-compacting geopolymer concrete with spend garnet as sand replacement. *J. Build. Eng.* **2018**, *15*, 85–94. [CrossRef]

22. Huseien, G.F.; Sam, A.R.M.; Shah, K.W.; Budiea, A.; Mirza, J. Utilizing spend garnets as sand replacement in alkali-activated mortars containing fly ash and GBFS. *Constr. Build. Mater.* **2019**, *225*, 132–145. [CrossRef]
23. Sani, N.M.; Mohamed, A.; Khalid, N.H.A.; Nor, H.M.; Hainin, M.R.; Giwangkara, G.G.; Hassan, N.A.; Kamarudin, N.A.S.; Mashros, N. Improvement of CBR value in soil subgrade using garnet waste. *IOP Conf. Ser. Mater. Sci. Eng.* **2019**, *527*, 012060. [CrossRef]
24. Aletba, S.R.; Hassan, N.; Aminudin, E.; Hussein, A.A.; Jaya, R. Marshall properties of asphalt mixture containing garnet waste. *Adv. Res. Mater. Sci.* **2018**, *43*, 22–27.
25. Thomas, B.S.; Gupta, R.C. A comprehensive review on the applications of waste tire rubber in cement concrete. *Renew. Sustain. Energy Rev.* **2016**, *54*, 1323–1333. [CrossRef]
26. Hernández-Olivares, F.; Barluenga, G.; Parga-Landa, B.; Bollati, M.; Witoszek, B. Fatigue behaviour of recycled tyre rubber-filled concrete and its implications in the design of rigid pavements. *Constr. Build. Mater.* **2007**, *21*, 1918–1927. [CrossRef]
27. Meddah, A.; Beddar, M.; Bali, A. Use of shredded rubber tire aggregates for roller compacted concrete pavement. *J. Clean. Prod.* **2014**, *72*, 187–192. [CrossRef]
28. Arulrajah, A.; Mohammadinia, A.; Maghool, F.; Horpibulsuk, S. Tyre derived aggregates and waste rock blends: Resilient moduli characteristics. *Constr. Build. Mater.* **2019**, *201*, 207–217. [CrossRef]
29. Soltani, A.; Taheri, A.; Deng, A.; O'Kelly, B.C. Improved geotechnical behavior of an expansive soil amended with tire-derived aggregates having different gradations. *Minerals* **2020**, *10*, 923. [CrossRef]
30. ASTM. *Standard Specification for Concrete Aggregates*; ASTM International: West Conshohocken, PA, USA, 2013; C33/C33M;
31. ASTM. *Standard Test Method for Relative Density (Specific Gravity) and Absorption of Fine Aggregate*; ASTM International: West Conshohocken, PA, USA, 2015.
32. ASTM. *Standard Test Methods for Laboratory Determination of Water (Moisture) Content of Soil and Rock by Mass*; ASTM International: West Conshohocken, PA, USA, 2019.
33. ASTM. *Standard Test Methods for Liquid Limit, Plastic Limit, and Plasticity Index of Soils*; ASTM International: West Conshohocken, PA, USA, 2010.
34. ASTM. *Classification of Soils for Engineering Purposes (Unified Soil Classification System)*; American Society of Testing Materials: Philadelphia, PA, USA, 1992.
35. AASHTO. *145-91 Standard Specification for Classification of Soils and Soil-Aggregate Mixtures for Highway Construction Purposes*; American Association of State Highwway and Transporation Officials: Washington, DC, USA, 2008.
36. ASTM. *Standard Test Methods for Laboratory Compaction Characteristics of Soil Using Modified Effort (56,000 Ft-lbf/ft3 (2700 KN-m/m3))*; ASTM International: West Conshohocken, PA, USA, 2012; Volume D1557.
37. ASTM. *Test Methods for Specific Gravity of Soil Solids by Water Pycnometer*; ASTM International: West Conshohocken, PA, USA, 2010.
38. ASTM. *Standard Test Methods for Compressive Strength of Molded Soil-Cement Cylinders*; ASTM International: West Conshohocken, PA, USA, 2007.
39. Gnanendran, C.T.; Piratheepan, J. Characterisation of a lightly stabilised granular material by indirect diametrical tensile testing. *Int. J. Pavement Eng.* **2008**, *9*, 445–456. [CrossRef]
40. Kavussi, A.; Modarres, A. Laboratory fatigue models for recycled mixes with bitumen emulsion and cement. *Constr. Build. Mater.* **2010**, *24*, 1920–1927. [CrossRef]
41. Thailand Department of Highways. *Standard for Highway Construction (Soil Cement Subbase)*; Thailand Department of Highways: Bangkok, Thailand, 1989; DH-S206/2532;
42. Thailand Department of Highways. *Standard for Highway Construction (Soil Cement Base)*; Thailand Department of Highways: Bangkok, Thailand, 1996; DH-S204/2556;
43. Thailand Department of Highways. *Standard for Highway Construction (Subgrade)*; Thailand Department of Highways: Bangkok, Thailand, 1989; DH-S201/2532;
44. Horpibulsuk, S.; Rachan, R.; Chinkulkijiwat, A.; Raksachon, Y.; Suddeepong, A. Analysis of strength development in cement-stabilized silty clay from microstructural considerations. *Constr. Build. Mater.* **2010**, *24*, 2011–2021. [CrossRef]
45. Bhuvaneshwari, S.; Robinson, R.; Gandhi, S. Behaviour of lime treated cured expansive soil composites. *Indian Geotech. J.* **2014**, *44*, 278–293. [CrossRef]
46. Horpibulsuk, S.; Katkan, W.; Sirilerdwattana, W.; Rachan, R. Strength development in cement stabilized low plasticity and coarse grained soils: Laboratory and field study. *Soils Found.* **2006**, *46*, 351–366. [CrossRef]
47. Sawangsuriya, A.; Bosscher, P.; Edil, T. *Alternative Testing Techniques for Modulus of Pavement Bases and Subgrades*; INDOT Division of Research: West Lafayette, IN, USA, 2005; pp. 108–121.
48. Maley, S.M. *Tire Development for Effective Transportation and Utilization of Used Tires*; CRADA 01-N044 Final Report; National Energy Technology Laboratory: Pittsburgh, PA, USA, 2004.
49. The comptroller General's Department. *Cost of Construction Materials and Labor Costs 2021*; The Comptroller General's Department: Bangkok, Thailand, 2021.

 *sustainability*

Article

Effect of Polypropylene Fibers on the Shear Strength–Dilation Behavior of Compacted Lateritic Soils

Maitê Rocha Silveira [1], Sabrina Andrade Rocha [2], Natália de Souza Correia [2], Roger Augusto Rodrigues [1], Heraldo Luiz Giacheti [1] and Paulo César Lodi [1,*]

[1] Department of Civil and Environmental Engineering, São Paulo State University (UNESP), Av. Engenheiro Luiz Edmundo Carrijo Coube 14-01, Bauru 17033-360, SP, Brazil; maite81@hotmail.com (M.R.S.); roger.rodrigues@unesp.br (R.A.R.); h.giacheti@unesp.br (H.L.G.)

[2] Department of Civil Engineering, Federal University of Sao Carlos (UFSCar), Rodovia Washington Luiz, São Carlos 17033-360, SP, Brazil; sabrina-andrade@outlook.com (S.A.R.); ncorreia@ufscar.br (N.d.S.C.)

* Correspondence: paulo.lodi@unesp.br; Tel.: +1-646-755-2239

Abstract: The stress–dilatancy relationship for fiber-reinforced soils has been the focus of recent studies. This relationship can be used as a foundation for the development of constitutive models for fiber-reinforced soils. The present study aims to investigate the effect of recycled polypropylene fibers on the shear strength–dilation behavior of two lateritic soils using the stress–dilatancy relationship for direct shear tests. Results show that fibers improved the shear strength behavior of the composites, observed by increases in the friction angle. Fibers' orientation at the sheared interface could be observed. The volumetric change during shearing was altered by the presence of fibers in both soils. Overall, results indicate that the stress–dilatancy relationship is affected by inclusions in the soil mix. Results can be used to implement constitutive modeling for fiber-reinforced soils.

Keywords: polypropylene fibers; lateritic soil; shear strength; drained test; stress–dilatancy

Citation: Silveira, M.R.; Rocha, S.A.; Correia, N.d.S.; Rodrigues, R.A.; Giacheti, H.L.; Lodi, P.C. Effect of Polypropylene Fibers on the Shear Strength–Dilation Behavior of Compacted Lateritic Soils. *Sustainability* **2021**, *13*, 12603. https://doi.org/10.3390/su132212603

Academic Editor: Castorina Silva Vieira

Received: 4 October 2021
Accepted: 5 November 2021
Published: 15 November 2021

Publisher's Note: MDPI stays neutral with regard to jurisdictional claims in published maps and institutional affiliations.

1. Introduction

Aiming to reduce the production of waste generated worldwide in civil construction, the use of alternative materials has emerged as an urgent need in view of the current environmental challenges. Different alternatives appear in the geotechnical context for soil improvement, namely natural and synthetic fibers, rubber fibers, construction waste, and flakes [1–5]. The improvement of soil in many geotechnical applications (subgrades and subbases, the reinforcement of soft soils, the control of a soil's hydraulic conductivity, the improvement of erosion resistance, the prevention of piping, backfill in retaining structures, and shrinkage crack mitigation) has been done with the reinforcement of local soils with fibers [6–9]. The use of fibers to reinforce soils has been used as a sustainable reinforcement technique since it does not harm the environment and does not promote the removal of large volumes of soil for later compaction.

Extensive research has proven that reinforcing the soil with short, randomly distributed fibers (e.g., polypropylene and polyethylene terephthalate fibers) can improve the mechanical response of the soil, observed by an interception in the potential failure zone, fiber tensile strength mobilization, and an improvement in the soil ductility [10–13], as well as provide isotropic behavior and limit the development of weak planes [14]. Anagnostopoulos et al. [15] state that the bond's interfacial strength due to mechanical interlocking along the friction at the interface seems to be the dominant mechanism that controls the micromechanical benefits of the reinforcement of soil with fibers.

Recent studies available in the literature regarding clayey soils mixed with short, randomly distributed polymeric fibers provide evidence of the significant impact of the inclusion of fibers on the soil shear strength [13,16–20]. Most of these studies assessed the effectiveness of fiber reinforcement using uniaxial, direct shear, and triaxial

135

tests [10,13,21–29]. However, comparing the numerous published studies on sand–fiber mixtures, studies on clayey soils reinforced with fibers are limited despite the equal potential for geotechnical applications.

For Anagnostopoulos et al. [30], the existing studies on fiber-reinforced clays have not yet established the fundamental mechanisms or the conditions that may affect the behavior of these fiber mixtures. According to Freilich et al. [31], there is a need for advancing studies in this field of knowledge due to the greater complexity related to the fiber interaction mechanism in cohesive soils. In addition, most studies do not present a deeper analysis regarding the stress–dilatancy behavior of soils reinforced with fibers. As stated by Li and Zornberg [32], the main findings in fiber reinforcement research regard increases in the soil shear strength and the post peak strength and changes in the soil ductility.

The stress–dilatancy relationship for fiber-reinforced soils has been the focus of some recent studies. The concept of fiber space was introduced by Wood et al. [33] to describe significant changes in the dilatancy of fiber-reinforced sands [34]. For Kong [35], when a polypropylene fiber–soil assembly dilates in response to applied shear deformations, the work done by the driving stress will be dissipated by not only particle sliding but also the fiber deformation. Eldesouky et al. [14] conducted direct shear tests on polypropylene-fiber-reinforced sands and proved that as the specimen approaches failure, fiber-reinforced specimens have higher dilation angles than unreinforced ones, explained by an increase in the shear zone that leads to higher dilation angles. Kong et al. [34] conducted several triaxial compression tests to investigate the effect of uniformly distributed fiber reinforcements on the stress–dilatancy relationship of Nanjing sand. The authors propose a new stress–dilatancy relationship for fiber-reinforced sand based on Rowe's stress–dilatancy relationship [36] for granular materials and suggest that the results could be employed as a foundation for the development of a constitutive model for polypropylene-fiber-reinforced soils. According to Kong et al. [34], the extension of fibers due to rearrangement and microstructure disturbances during shearing provides an important contribution to the increase in strength; however, studies are not conclusive on the observed stress dilatancy of fiber-reinforced soils.

A recent study by Dołżyk-Szypcio [37] using the stress–dilatancy relationship for the direct shear tests developed by Szypcio [38] emphasizes that the stress–dilatancy relationship is a function of moisture, the degree of compaction, and normal stresses and can be affected by inclusions in the soil mix. Szypcio [38] suggests further experimental investigation, especially for cohesive soils on different stress–strain paths. Regarding the stress–dilatancy relationship of fine and cohesive soils, Yousefpour et al. [39] evaluated the shear strength–dilation characteristics of silty and clayey sands and demonstrate that the shear strength, dilation angle, and maximum friction angle decreased with an increase in the clay content and increased with an increase in the silt content. According to Yousefpour et al. [39], few studies are concerned with the relationship between the strength parameters and the dilation of silty and clayey sands. The authors suggest that the direct shear apparatus is a useful tool for the investigation of the shear strength and dilation characteristics of fine soils.

As exposed by the literature, there is a need for investigations regarding the shear stress–dilatancy behavior of fine soils and the influence of the inclusion of fibers on the strength and dilatancy behavior of fine soil mixtures. Studies related to fine-grained and clayey lateritic soils reinforced with polymeric fibers are scarce in the literature [4,5]. The present study combined an investigation of the shear strength–dilation behavior of two lateritic fine soils reinforced with short, randomly distributed polypropylene fibers. Analyses were conducted using the stress–dilatancy model for the drained direct shear tests developed by Szypcio [38].

2. Materials and Methods

Two lateritic soils (clayey sand and clay), taken from the state of Sao Paulo in Brazil, were chosen for this research since they represent typical soils that cover a large area in

Brazil. The lateritic soils, due to their formation process, underwent leaching processes. The clay fractions are essentially composed of clay minerals from the Kaolinite group and hydroxides and hydrated oxides of iron and/or aluminum [40,41]. The clayey soil was collected from the city of Santa Gertrudes, Sao Paulo and classified as CH soil according to the Unified Soil Classification System [42] although a significant percentage of sand was present. The clayey sand, classified as SC soil [42], was collected from the city of Bauru, Sao Paulo. According to X-ray diffraction analysis [43], the predominant clay minerals in both lateritic soils were Kaolinite, Illite, Gibbsite, and Hematite. In these soils, the formation of aggregates of finer soil particles is common due to the action of iron and aluminum oxides and hydroxides, a characteristic that explains the peculiar behavior of lateritic soils.

The soil samples were characterized by particle size analysis [44], specific gravity [45], Proctor tests [46], and consistency limits [47]. The physical properties of the soils are presented in Table 1. The particle distribution of the soils is shown in Figure 1.

Table 1. Characteristics of the natural lateritic soils used in this study.

Properties	Values		Specification
	CH	**SC**	
Clay fraction (%)	50	14	
Silt fraction (%)	14	5.8	ASTM D7928 [44]
Sand fraction (%)	36	80.2	
Specific gravity of solids	2.90	2.65	ASTM D854 [45]
Liquid limit (%)	51	16	
Plasticity limit (%)	29	2	ASTM D4318 [47]
Plasticity Index (%)	22	14	
Maximum dry unit weight (kN/m^3)	16.7	19.50	ASTM D698 [46]
Optimum water content (%)	24.0	10.6	

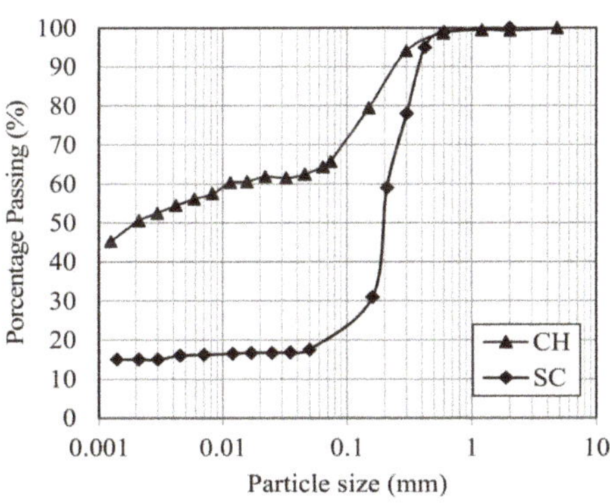

Figure 1. Particle size distribution curves of SC and CH soils.

The fibers used in this research were characterized as short, discrete recycled polypropylene (PP) fibers of light weight and high flexibility that were hydrophobic and inert. The PP fibers had an average diameter of 18 micrometers, a specific mass of 0.9 g/cm^3, an average length of 18 mm, zero water absorption, and a breaking tensile strength of 610 MPa.

Figure 2 presents the PP fibers used in this study and the samples' preparation. PP fibers were randomly distributed into the soil mass at a fiber content of 0.1% and 0.25% by soil dry weight. These fiber contents are representative of the contents used in soil mixtures in other studies [16,19,31,48–51]. The soil was homogenized using the optimum water content of the natural soil, which was obtained from the compaction test in normal Proctor energy by calculating the amount of water in relation to the total weight of the dry raw material (soil + fiber).

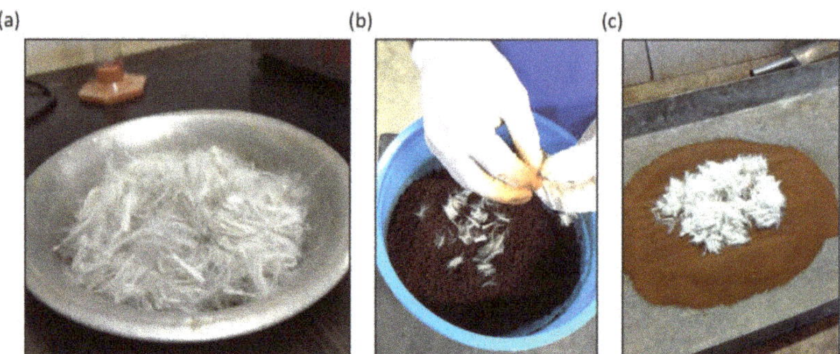

Figure 2. Sample preparation: (**a**) fibers; (**b**) mixture with CH soil; (**b**) mixture with SC soil.

Fibers were randomly distributed in the soil matrix and a mechanical mixer was used to reach a homogenous fiber distribution in the soil and to avoid any potential weakness plane. Fibers' mixture in the CH soil was done with humidified soil due to the difficulty of moistening and homogenizing the dry soil with fibers. In the field, special attention to the mixture's production is necessary. Some remarks on the limitations of quality control procedures can be found in Farloca et al. [52].

In order to obtain the target compaction parameters for each soil mixture, standard Proctor compaction tests [46] were also conducted on the soils with fiber reinforcements. Prior to compaction, prepared mixtures were preserved in air-proof bags for a minimum of 24 hours for moisture homogenization. The optimum compaction conditions were used as the target compaction parameters for the direct shear tests.

This study involved drained direct shear tests that were conducted according to ASTM D3080 [53] on the compacted soil specimens. The specimens (with and without fibers) were compacted using the optimum water content and a 95% compaction degree. The test was conducted on a shear box with dimensions of $100 \times 100 \times 25$ mm, where the lower part of the shear box is restrained, while the upper part is controlled by a motor to apply a horizontal shear load in displacement-controlled mode. Since the tests were carried out in different months, the research with clayey soil specimens was initially performed under vertical stresses of 100, 200, and 300 kPa prior to shearing. Then, the research continued, and the sandy soil specimens were consolidated under vertical stresses of 100, 200, and 400 kPa. Tests were conducted at a loading rate of 0.15 mm/min in all tests. We did not use duplicated soil samples to study the effect of structure on the stress–strain behavior of reconstituted soil samples. Only one test per normal stress level was conducted. During the tests, loads and displacements in the axial and horizontal directions were recorded automatically by a computer-controlled data collection system. Shear stress was recorded as a function of horizontal displacement up to a total displacement of 15 mm in order to observe the post-failure behavior.

3. Results and Discussion

3.1. Influence of PP Fibers on Compaction Properties

Figure 3 shows the compaction curves of both soil mixtures with 0.1% and 0.25% fiber content, compared to the respective natural soils. The behavior of the maximum dry unit weight did not change with the inclusion of fibers in the clayey soil, while the optimum gravimetric water content increased by 0.5% for the 0.25% fiber content. The behavior of the maximum dry unit weight did slightly reduce with the inclusion of fibers in the sandy soil, while the optimum gravimetric water content did not change. The results of other studies show similar compaction curves for soils with and without fiber reinforcement [54–56], where no significant alterations were evidenced.

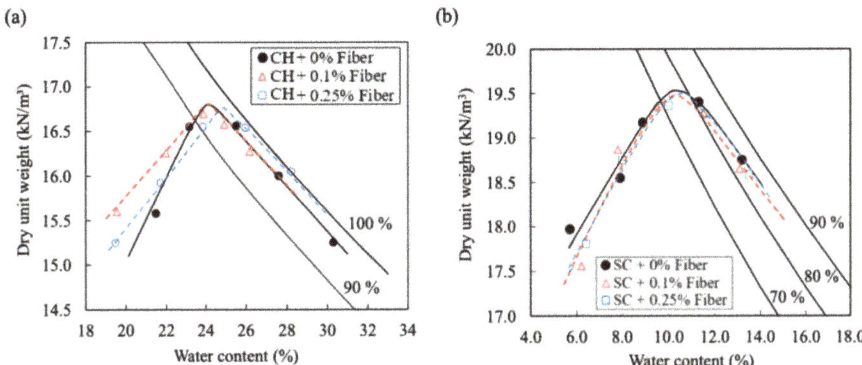

Figure 3. Standard Proctor compaction curves for soil with fiber contents of 0.0%, 0.1%, and 0.25%: (**a**) CH; (**b**) SC.

3.2. Influence of PP Fibers on Drained Shear Strength

The results of direct shear tests considering each combination of lateritic soil and PP fibers are presented in Figures 4 and 5, showing the variations in shear stress and volumetric change with shear displacements. In Figure 4, the inclusion of fibers in the CH soil improved the shear strength behavior of the composites, beyond which it remained constant for 0.1% fiber content. The contribution of the fibers to the increase in soil strength was superior for the highest fiber content (0.25%) and, after the mixtures underwent plastic deformation, the resistance of fibers was mobilized, and hardening was observed. This behavior is in accordance with the results presented in the research of Anagnostopoulos et al. [15] that studied the shear strength behavior of polypropylene-fiber-reinforced cohesive soils. The continuous increase in shear stress, mainly for the highest normal stresses, was justified in Khatri et al. [57] by the mobilization of the tensile strength of the fibers, which increases the deformation imposed on the material at failure. Regarding the volume change versus shear displacement plots for the CH soil (Figure 4), the results evidence, at all stress levels, a lower degree of contraction of fiber-reinforced specimens than that of unreinforced specimens, directly related to the fiber content. This behavior was also found in the research of Anagnostopoulos et al. [15]. Specimens exhibited a trend of dilation occurring under drained shearing for lower stresses and contraction occurring under higher stresses. The volumetric change was altered by the presence of fibers, mainly at higher stress levels, and indicates that the presence of fibers considerably limited the tendency for contraction. Similar results were obtained by Sadek et al. [58] and Ibraim et al. [59]. According to Anagnostopoulos et al. [15], these results suggest that the volumetric response, from contractive to dilative, could be a consequence of an apparent densification of the composite matrix resulting from the interaction mechanism between the fiber net and the soil particles.

Regarding SC soil (Figure 5), the inclusion of polymeric fibers increased the shear strength of the soil for higher normal stresses and the higher fiber content. The hardening behavior observed for CH soil was not evidenced in SC specimens. Regarding the volumetric change results, a trend of a decrease (contraction) in the volume variation was observed at all stress levels, although the addition of fibers did not produce a trend of a volume increase as in the CH soil. This behavior was also found in the results of Silveira et al. [5], who used polyethylene strips in SC lateritic soil. Maher and Gray [60] and Consoli et al. [61] state that the effect of the inclusion of fibers on the dilation and volume change is more pronounced at higher loads and strain levels. Regarding the initial tangent stiffness of the shear stress–displacement curves, the inclusion of polymeric fibers in both CH and SC soils practically did not affect this property.

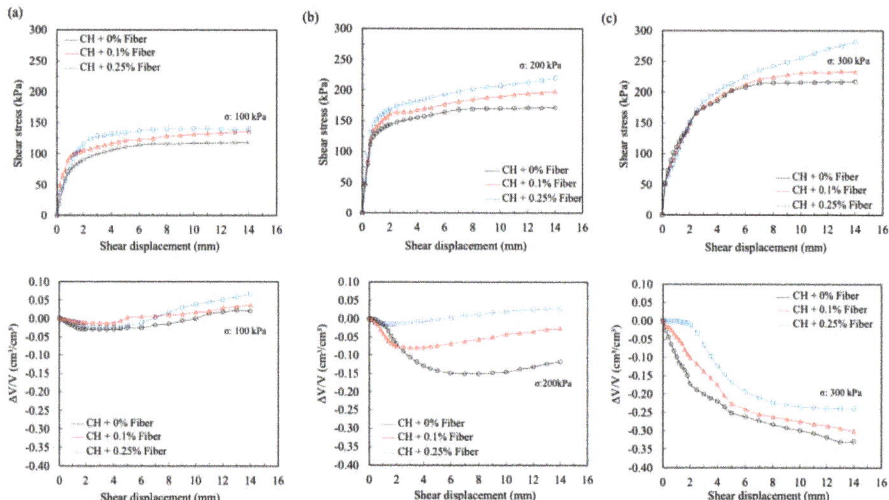

Figure 4. Shear stress and volume change vs. displacement for CH soil with fiber contents of 0.0%, 0.1%, and 0.25%: (**a**) 100 kPa; (**b**) 200 kPa; and (**c**) 300 kPa.

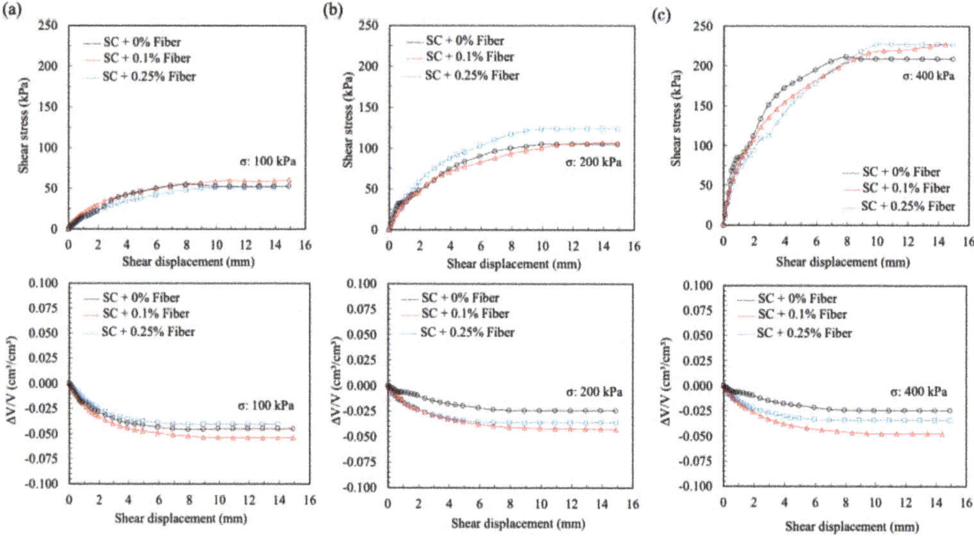

Figure 5. Shear stress and volume change vs. displacement for SC soil with fiber contents of 0.0%, 0.1%, and 0.25%: (**a**) 100 kPa; (**b**) 200 kPa; (**c**) 400 kPa.

Figure 6 presents the soil–fiber specimens after shear tests, showing the fibers' orientation at the sheared interface in the CH soil. No deformation or breakage of fibers was observed. Kumar and Singh [62] state that an improvement in the ductility of soil through the stretching of the fibers causes an increase in the soil's cohesion. According to Darvishi and Erken [51], because of the extensible nature of the fibers, they are stretched in the soil matrix during the shearing process, increasing the tension strength of fiber-reinforced soils. Kong et al. [34] state that the extension of fibers due to rearrangement and microstructure disturbances during shearing provides an important contribution to the increase in strength.

For Anagnostopoulos et al. [15], this is a confirmation that fibers withstand tension within the soil matrix without significantly deforming.

Figure 6. Fibers' orientation during shearing in lateritic CH soil.

The shear strength envelopes for mixtures with 0%, 0.1%, and 0.25% fiber contents are shown in Figure 7. The failure criterion adopted was the value of peak shear strength. The results shown in Figure 7 evidence the friction behavior of SC soil and the high cohesion (69.2 kPa) and friction angle (26.1°) of the lateritic clayey soil. Indeed, compacted tropical soils exhibit good shear strength behavior when unsaturated [63]. After the inclusion of fibers, an increase in the cohesion and the friction angle was observed with increasing fiber content in the CH soil. The results of Tang et al. [6] on clayey soil reinforced with 12-mm-long PP fibers showed that the values of cohesion and friction angles also increased with increasing fiber content.

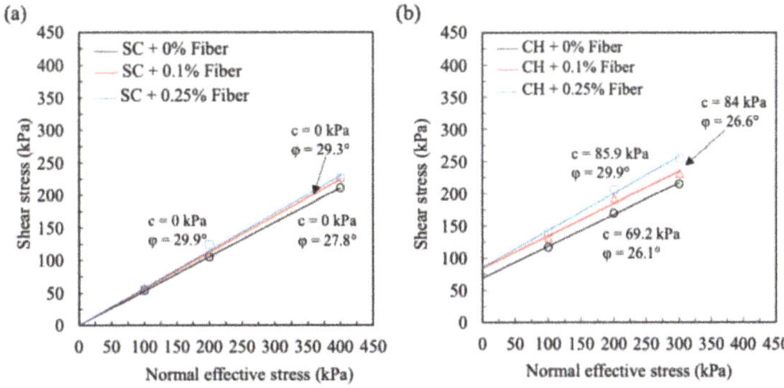

Figure 7. Shear strength envelopes of natural soil and soil with 0.1% and 0.25% fiber content: (a) CH; and (b) SC.

The effect of the addition of fibers on the shear strength parameters of SC soil was evidenced by an increase in the friction angle (Figure 7b), although it was not influenced by the increase in fiber content. The improvement in the friction angle is most probably associated with the mobilization of friction between the soil particles and the fibers [64,65] and due to the relative size of the fibers and soil grains [15].

In order to evaluate the variation in the shear strength response with fiber contents and normal stress levels, the following strength ratio parameter proposed by Darvishi and Erken [51] was used:

$$R = \frac{\tau}{\tau_{un}} \tag{1}$$

where τ and τ_{un} are the stresses of fiber-reinforced and unreinforced soil at the peak shear value, respectively.

Figure 8 presents results on the strength ratio in the evaluated lateritic soils with different PP fiber contents and with normal stress levels. For the CH soil (Figure 8a,c), the strength ratios were observed to be as high as 1.2 with increasing PP fiber content and the CH soil showed decreased strength ratios under higher normal stresses. The SC soil (Figure 8b,d) presented alterations in the strength ratio, such as a drop with higher normal stresses, which was also observed in the sand–fiber mixture evaluated by Darvishi and Erken [51]. In this case, reinforcement with fibers was more effective for specimens under low normal stress and 0.25% fiber content.

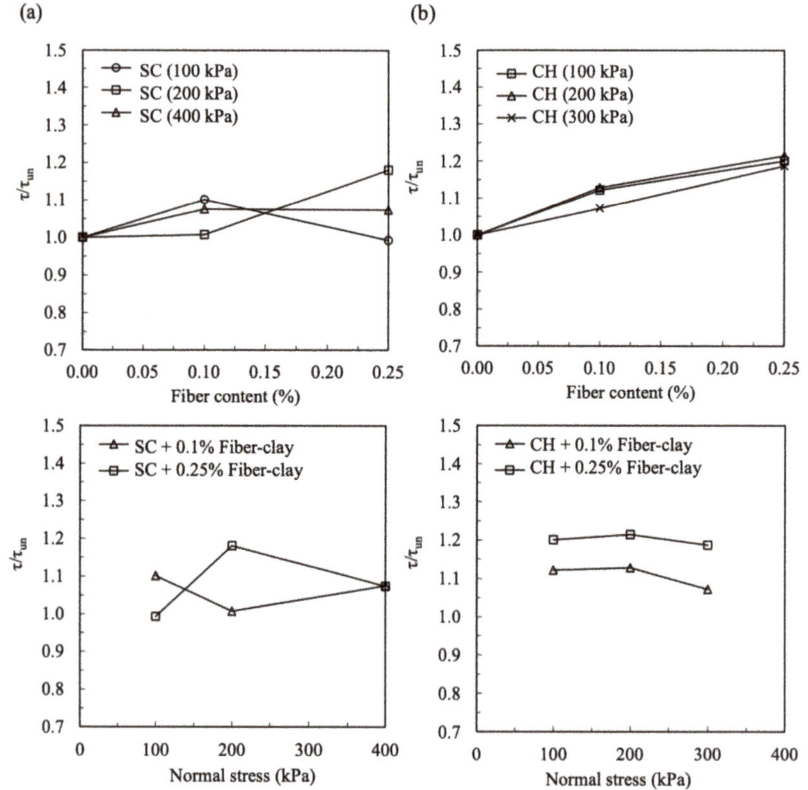

Figure 8. Variation in the shear strength ratio with fiber content and normal stresses: (**a**) SC; (**b**) CH.

3.3. Influence of PP Fibers on the Stress–Dilatancy of Soil Mixtures

Figures 9 and 10 show a plot of the stress ratio (τ/σ_n) against the dilatancy results of the natural soil and the fiber-reinforced mixtures for the different normal stress levels, respectively, for SC soil and CH soil. As shown in Figure 9, at all test stages, the specimens experienced high contraction rates (negative dilation angles) and the influence of the fibers is evidenced by the stress–dilatancy behavior of the SC soil mixtures. The results on the stress–dilatancy behavior of reinforced soil mixtures are in accordance with the results of Eldesouky et al. [14] in which the contraction rates increased with an increase in the fiber content. Figure 10 shows that adding 0.25% fiber content to the CH soil significantly alters the stress–dilatancy behavior of the soil mixture as compared with the addition of 0.1% fiber content. In this case, the clayey soil was more susceptible to alterations in the stress–dilatancy behavior of the soil than sandy soil.

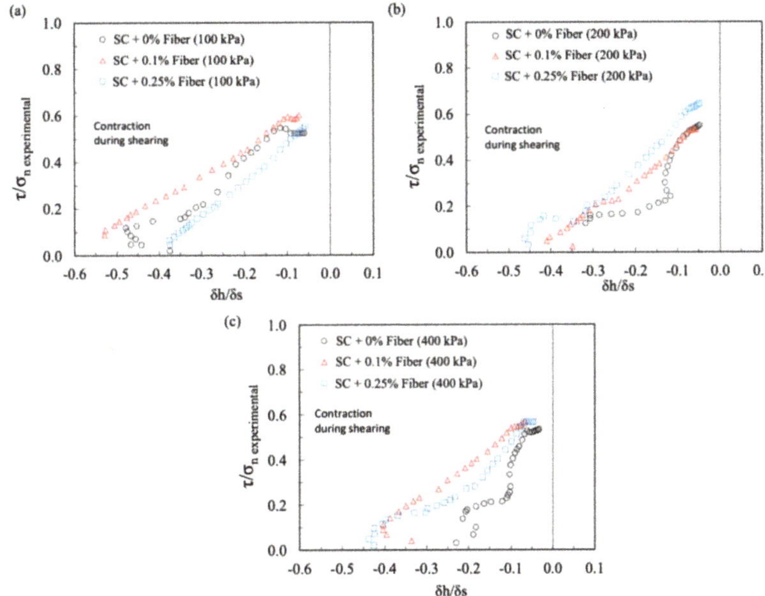

Figure 9. Influence of the inclusion of fibers on the stress–dilatancy behavior under direct shear conditions for SC soil: (**a**) 100 kPa; (**b**) 200 kPa; (**c**) 400 kPa.

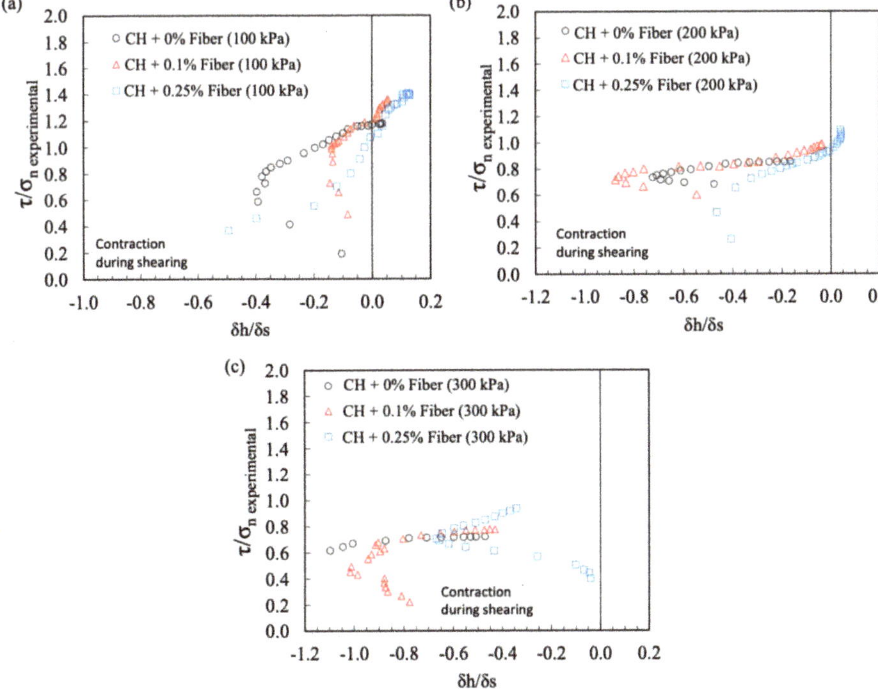

Figure 10. Influence of the inclusion of fibers on the stress–dilatancy behavior under direct shear conditions for CH soil: (**a**) 100 kPa; (**b**) 200 kPa; (**c**) 300 kPa.

Szypcio [38] developed a complex equation that represents the stress–dilatancy relationship for the simple shear condition, obtained from the frictional state theory. According to Szypcio [38], the stress–dilatancy Equation (4) for the simple shear test can be used for direct and ring shear tests, as follows:

$$\frac{\tau}{\sigma} = \frac{\sqrt{3}\eta\cos\Phi^{\circ}\cos\theta}{3 + \eta\left(\sin\theta - \sqrt{3}\sin\Phi^{\circ}\cos\theta\right)} \tag{2}$$

where:

$$\eta = Q - AD \tag{3}$$

$$Q = M^{\circ} - \alpha A^{\circ} \tag{4}$$

$$M^{\circ} = \frac{3\sin\Phi^{\circ}}{\sqrt{3}\cos\theta - \sin\Phi^{\circ}\sin\theta)} \tag{5}$$

for the drained condition,

$$A^{\circ} = \frac{1}{\cos(\theta - \theta_{\varepsilon})}\left\{1 - \frac{2}{3}M^{\circ}\sin\left(\theta + \frac{2}{3}\pi\right)\right\} \tag{6}$$

$$A = \beta A^{\circ} \tag{7}$$

$$\theta_{\varepsilon} = arc\tan\left\{\frac{1}{\sqrt{3}}\frac{\frac{\delta h}{\delta s}}{\sqrt{1 + \left(\frac{\delta h}{\delta s}\right)^2}}\right\} \tag{8}$$

$$D = -\sqrt{3}\frac{\frac{\delta h}{\delta s}}{\sqrt{1 + \frac{4}{3}\left(\frac{\delta h}{\delta s}\right)^2}} \tag{9}$$

where Φ° is the critical frictional state angle, α and β are frictional state theory parameters, h is the growth in the sample's height during shear, and s is the displacement of the shear box.

Based on the experimental data on non-cohesive soils, the stress–dilatancy relationship calculated using the frictional state theory with $\Phi^{\circ} = \Phi_{cv}$, $\theta = 15°$, $\alpha = 0$, and $\beta = 1.4$ is acceptable [38]. Figure 11 shows an example of the stress ratio–dilatancy relationship for the mentioned parameters.

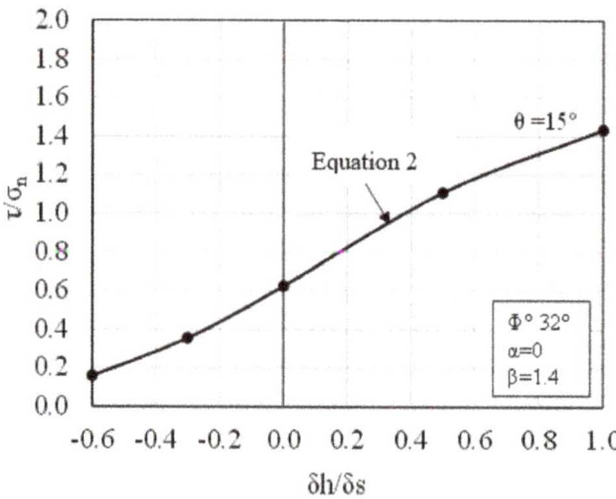

Figure 11. Stress–dilatancy relationship developed by Szypcio [38] (modified).

According to the proposed model, parameters α and β represent the mode of deformation in the shear band. Parameter α translates the reference curve obtained for $\alpha = 0$

upward for $\alpha < 0$ and downward for $\alpha > 0$. Parameter β significantly influences the stress ratio–dilatancy relationship for dilation and contraction during shearing. For almost all tests, in the initial phase of shearing, the relationship between $\tau/\sigma n$ and $\delta h/\delta s$ is not linear, while in the pre-peak and post-peak phases of shearing, a linear relationship $\tau/\sigma n$—$\delta h/\delta s$ is observed, and parameters α and β can be calculated by use of an approximation technique [37,38]. In the study of Dołżyk and Szypcio [37], the pre-peak and post-peak phases of shearing were obtained with parameter intervals of $0.1 < \alpha < 1.1$ and $2.1 < \beta < 6.0$ (large direct shear box) and $\alpha = 0$ and $\beta = 1.4$ (small direct shear box), showing a higher degree of non-homogeneous deformation in the shear band in the large box apparatus.

The results on stress–dilatancy relationships for the SC soil and the fiber-reinforced mixtures are shown in Figure 12. The best fit obtained using Equation (2) well approximates the experimental relationship at failure (pre-peak experimental data) for $\Phi° = 32°$, $-0.15 < \alpha < -0.1$, and $2.0 < \beta < 2.4$. According to Figure 12, the increase in parameter β after the inclusion of fibers in the SC soil demonstrates the influence on the stress ratio–dilatancy relationship during shear, which was superior for the higher fiber content. As also observed by Dołżyk and Szypcio [37], this indicates that the stress–dilatancy relationship is affected by inclusions in the soil mix.

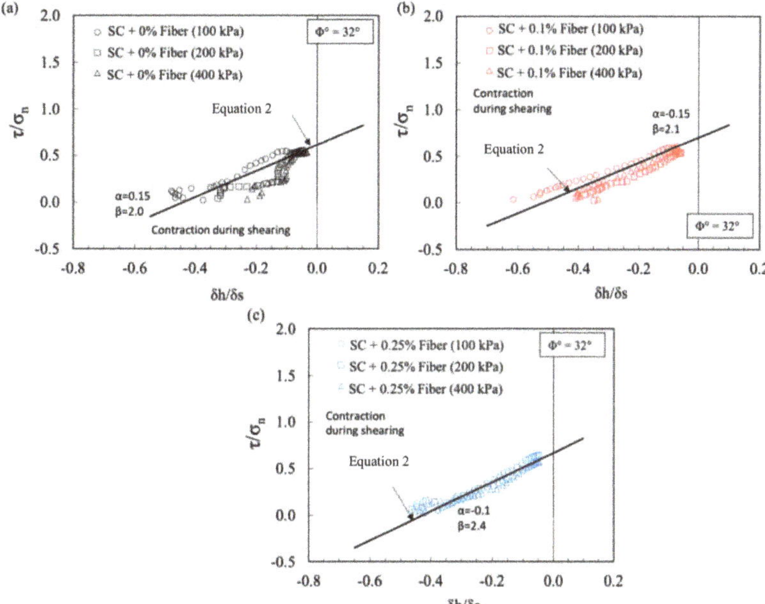

Figure 12. Influence of the inclusion of fibers on the stress–dilatancy relationship under direct shear conditions for SC soil: (**a**) 100 kPa; (**b**) 200 kPa; (**c**) 400 kPa.

An approach for cohesive soils was used to understand the stress–dilation behavior of the CH soil. According to BS8002 [66], for fine soils, the critical frictional state angle can be estimated based on Atterberg limits as follows:

$$\Phi° = 42 - 12.5\log(PI) \text{ for } 5\% < PI < 100\% \tag{10}$$

where PI is the soil plasticity index.

In the natural clayey soil (Figure 13), the best fit obtained using Equation (2) well approximates the experimental relationship at failure (pre-peak experimental data) for $\Phi° = 25.2°$, $-0.9 < \alpha < -0.8$, and $1.0 < \beta < 1.1$. In general, the stress–dilatancy behavior of the clayey samples was not affected by the inclusion of fibers, which was observed by the parameters with the same fit. However, the pre-peak and post-peak phases of shearing showed different behaviors when comparing the natural and fiber-reinforced samples. These results can be used to develop constitutive models for fiber-reinforced soils.

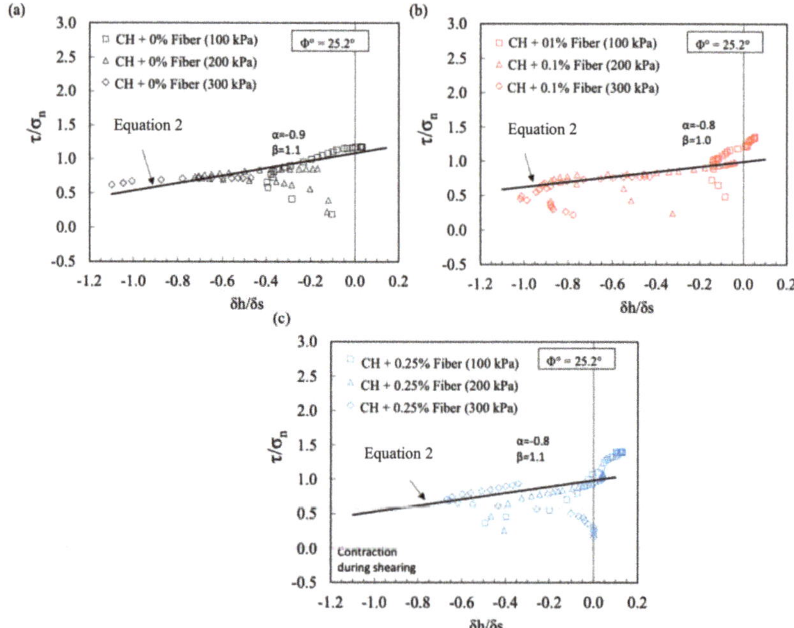

Figure 13. Influence of the inclusion of fibers on the stress–dilatancy relationship under direct shear conditions of CH soil: (**a**) 100 kPa; (**b**) 200 kPa; (**c**) 300 kPa.

4. Conclusions

This study investigated the effect of recycled polypropylene fibers on the shear strength–dilation behavior of two lateritic soils using the stress–dilatancy relationship for direct shear tests. The following conclusions can be drawn:

1. The inclusion of PP fibers improved the shear strength behavior of the composites in both soils, while the soils' initial stiffness was practically not affected;
2. The contribution of the fibers to the increase in the soil strength was superior in the clayey soil and for the highest content (0.25%), observed by a substantial increase in the friction angle. The resistance of fibers was mobilized, and soil hardening was observed. Fibers' orientation at the sheared interface could be observed;
3. The volumetric change in the clayey soil was altered by the presence of fibers under drained shear mainly at higher stress levels. The results indicate that the presence of fibers considerably limited the tendency for contraction. A trend of a decrease in the volume variation was observed for higher normal stresses in the sandy soil; and
4. The Szypcio [38] model demonstrated the influence on the stress ratio–dilatancy relationship during shear, which was superior for the higher fiber content in the sandy soil. In the clayey soil, the pre-peak and post-peak phases of shearing showed different behaviors when comparing natural and fiber-reinforced samples. Overall, the results indicate that the stress–dilatancy relationship is affected by inclusions in the soil mix.

Author Contributions: Conceptualization, M.R.S., S.A.R., P.C.L. and N.d.S.C.; methodology, M.R.S., S.A.R., P.C.L., N.d.S.C., R.A.R. and H.L.G.; formal analysis, M.R.S., S.A.R., P.C.L., N.d.S.C., R.A.R. and H.L.G.; investigation, M.R.S., S.A.R., N.d.S.C., P.C.L., H.L.G. and R.A.R.; resources, M.R.S., S.A.R., P.C.L., N.d.S.C., R.A.R. and H.L.G.; writing—original draft preparation, M.R.S., S.A.R.,

P.C.L., N.d.S.C., R.A.R. and H.L.G.; writing—review and editing, M.R.S., S.A.R., P.C.L. and N.d.S.C.; visualization, M.R.S., S.A.R., P.C.L., N.d.S.C., R.A.R. and H.L.G.; supervision, P.C.L. and N.d.S.C.; project administration, M.R.S., S.A.R., P.C.L., N.d.S.C., R.A.R. and H.L.G. All authors have read and agreed to the published version of the manuscript.

Funding: This research received no external funding.

Institutional Review Board Statement: Not applicable.

Informed Consent Statement: Not applicable.

Data Availability Statement: Part of the data presented in this study is available at the link: https://repositorio.ufscar.br/handle/ufscar/11891?show=full. Other data can be obtained by requesting the author for correspondence.

Acknowledgments: The authors are very thankful to the Capes/Print program and PROPG/UNESP.

Conflicts of Interest: The authors declare no conflict of interest.

References

1. Yoon, Y.W.; Heo, S.B.; Kim, K.S. Geotechnical performance of waste tires for soil reinforcement from chamber tests. *Geotext. Geomembr.* **2008**, *26*, 100–107. [CrossRef]
2. Luwalaga, J.G. Analysing the Behaviour of Soil Reinforced with Polyethylene. Ph.D. Thesis, Stellenbosch University, Stellenbosch, South Africa, 2016; p. 138.
3. Soltani, A.; Deng, A.; Taheri, A.; Mirzababaei, M.; Nikraz, H. Interfacial shear strength of rubber-reinforced clays: A dimensional analysis perspective. *Geosynth. Int.* **2019**, *26*, 164–183. [CrossRef]
4. Marçal, R.; Lodi, P.C.; de Souza Correia, N.; Giacheti, H.L.; Rodrigues, R.A.; McCartney, J.S. Reinforcing Effect of Polypropylene Waste Strips on Compacted Lateritic Soils. *Sustainability* **2020**, *12*, 9572. [CrossRef]
5. Silveira, M.R.; Lodi, P.C.; de Souza Correia, N.; Rodrigues, R.A.; Giacheti, H.L. Effect of recycled polyethylene terephthalate strips on the mechanical properties of cement-treated lateritic sandy soil. *Sustainability* **2020**, *12*, 9801. [CrossRef]
6. Tang, C.; Shi, B.; Gao, W.; Chen, F.; Cai, Y. Strength and mechanical behavior of short polypropylene fiber reinforced and cement stabilized clayey soil. *Geotext. Geomembr.* **2007**, *25*, 194–202. [CrossRef]
7. Shukla, S.K.; Sivakugan, N.; Das, B.M. Fundamental concepts of soil reinforcement—An overview. *Int. J. Geotech. Eng.* **2009**, *3*, 329–342. [CrossRef]
8. Shukla, S.K. *Fundamentals of Fibre-Reinforced Soil Engineering*; Springer Nature: Singapore, 2017; ISBN 978-981-10-3061-1.
9. Ehrlich, M.; Almeida, S.; Curcio, D. Hydro-mechanical behavior of a lateritic fiber-soil composite as a waste containment liner. *Geotext. Geomembr.* **2019**, *47*, 42–47. [CrossRef]
10. Consoli, N.C.; Montardo, J.P.; Marques Prietto, P.D.; Pasa, G.S. Engineering behavior of a sand reinforced with plastic waste. *J. Geotech. Geoenviron. Eng.* **2002**, *128*, 462–472. [CrossRef]
11. Heineck, K.S.; Coop, M.R.; Consoli, N.C. Effect of Microreinforcement of Soils from Very Small to Large Shear Strains. *J. Geotech. Geoenviron. Eng.* **2005**, *131*, 1024–1033. [CrossRef]
12. Sotomayor, J.M.G.; Casagrande, M.D.T. The performance of a sand reinforced with coconut fibers through plate load tests on a true scale physical model. *Soils Rocks* **2018**, *41*, 361–368. [CrossRef]
13. dos Santos Lopes Louzada, N.; Malko, J.A.C.; Casagrande, M.D.T. Behavior of Clayey Soil Reinforced with Polyethylene Terephthalate. *J. Mater. Civ. Eng.* **2019**, *31*, 04019218. [CrossRef]
14. Eldesouky, H.M.; Morsy, M.M.; Mansour, M.F. Fiber-reinforced sand strength and dilation characteristics. *Ain Shams Eng. J.* **2016**, *7*, 517–526. [CrossRef]
15. Anagnostopoulos, C.A.; Tzetzis, D.; Berketis, K. Shear strength behaviour of polypropylene fibre reinforced cohesive soils. *Geomech. Geoengin.* **2014**, *9*, 241–251. [CrossRef]
16. Mirzababaei, M.; Arulrajah, A.; Horpibulsuk, S.; Aldava, M. Shear strength of a fibre-reinforced clay at large shear displacement when subjected to different stress histories. *Geotext. Geomembr.* **2017**, *45*, 422–429. [CrossRef]
17. Mirzababaei, M.; Mohamed, M.; Arulrajah, A.; Horpibulsuk, S.; Anggraini, V. Practical approach to predict the shear strength of fibre-reinforced clay. *Geosynth. Int.* **2018**, *25*, 50–66. [CrossRef]
18. Murray, J.J.; Frost, J.D.; Wang, Y. Behavior of a sandy silt reinforced with discontinuous recycled fiber inclusions. *Transp. Res. Rec.* **2000**, *1714*, 9–17. [CrossRef]
19. Özkul, Z.H.; Baykal, G. Shear behavior of compacted rubber fiber-clay composite in drained and undrained loading. *J. Geotech. Geoenviron. Eng.* **2007**, *133*, 767–781. [CrossRef]
20. Suffri, N.; Jeludin, M.; Rahim, S. Behaviour of the Undrained Shear Strength of Soft Clay Reinforced with Natural Fibre. In *IOP Conference Series: Materials Science and Engineering*; IOP Publishing: Seoul, Korea, 2019; Volume 690, p. 012005. [CrossRef]
21. Botero, E.; Ossa, A.; Sherwell, G.; Ovando-Shelley, E. Stress–strain behavior of a silty soil reinforced with polyethylene terephthalate (PET). *Geotext. Geomembr.* **2015**, *43*, 363–369. [CrossRef]

22. Fathi, H.; Jamshidi Chenari, R.; Vafaeian, M. Shaking Table Study on PET Strips-Sand Mixtures Using Laminar Box Modelling. *Geotech. Geol. Eng.* **2019**, *38*, 683–694. [CrossRef]
23. Festugato, L.; Menger, E.; Benezra, F.; Kipper, E.A.; Consoli, N.C. Fibre-reinforced cemented soils compressive and tensile strength assessment as a function of filament length. *Geotext. Geomembr.* **2017**, *45*, 77–82. [CrossRef]
24. Onyelowe, K.C.; Bui Van, D.; Ubachukwu, O.; Ezugwu, C.; Salahudeen, B.; Nguyen Van, M.; Ikeagwuani, C.; Amhadi, T.; Sosa, F.; Wu, W.; et al. Recycling and reuse of solid wastes; a hub for ecofriendly, ecoefficient and sustainable soil, concrete, wastewater and pavement reengineering. *Int. J. Low-Carbon Technol.* **2019**, *14*, 440–451. [CrossRef]
25. Peddaiah, S.; Burman, A.; Sreedeep, S. Experimental Study on Effect of Waste Plastic Bottle Strips in Soil Improvement. *Geotech. Geol. Eng.* **2018**, *36*, 2907–2920. [CrossRef]
26. Santoni, B.R.L.; Tingle, J.S.; Members, A.; Webster, S.L. Engineering properties of sand-fiber mixtures for road construction. *J. Geotech. Geoenviron. Eng.* **2001**, *127*, 258–268. [CrossRef]
27. Dos Santos, A.P.S.; Consoli, N.C.; Baudet, B.A. The mechanics of fibre-reinforced sand. *Geotechnique* **2010**, *60*, 791–799. [CrossRef]
28. Sivakumar Babu, G.L.; Chouksey, S.K. Stress-strain response of plastic waste mixed soil. *Waste Manag.* **2011**, *31*, 481–488. [CrossRef] [PubMed]
29. Zhao, J.J.; Lee, M.L.; Lim, S.K.; Tanaka, Y. Unconfined compressive strength of PET waste-mixed residual soils. *Geomech. Eng.* **2015**, *8*, 53–66. [CrossRef]
30. Anagnostopoulos, C.A.; Papaliangas, T.T.; Konstantinidis, D.; Patronis, C. Shear Strength of Sands Reinforced with Polypropylene Fibers. *Geotech. Geol. Eng.* **2013**, *31*, 401–423. [CrossRef]
31. Freilich, B.J.; Li, C.; Zornberg, J.G. Effective shear strength of fiber-reinforced clays. In Proceedings of the 9th International Conference on Geosynthetics—Geosynthetics: Advanced Solutions for a Challenging World, ICG 2010, Guarujá, Brazil, 23 May 2010; pp. 1997–2000.
32. Li, C.; Zornberg, J.G. Shear Strength Behavior of Soils Reinforced with Weak Fibers. *J. Geotech. Geoenviron. Eng.* **2019**, *145*, 2–8. [CrossRef]
33. Muir Wood, D.; Diambra, A.; Ibraim, E. Fibres and soils: A route towards modelling of root-soil systems. *Soils Found.* **2016**, *56*, 765–778. [CrossRef]
34. Kong, Y.; Zhou, A.; Shen, F.; Yao, Y. Stress–dilatancy relationship for fiber-reinforced sand and its modeling. *Acta Geotech.* **2019**, *14*, 1871–1881. [CrossRef]
35. Kong, Y. Stress-Dilatancy Relationship for Fiber-Reinforced Soil. In Proceedings of the China-Europe Conference on Geotechnical Engineering, Vienna, Austria, 13–16 August 2018.
36. Rowe, P.W. The stress-dilatancy relation for static equilibrium of an assembly of particles in contact. *Proc. R. Soc. London. Ser. A Math. Phys. Sci.* **1962**, *269*, 500–527. [CrossRef]
37. Dołżyk-Szypcio, K. Direct Shear Test for Coarse Granular Soil. *Int. J. Civ. Eng.* **2019**, *17*, 1871–1878. [CrossRef]
38. Szypcio, Z. Stress-Dilatancy for Soils. Part IV: Experimental Validation for Simple Shear Conditions. *Stud. Geotech. Mech.* **2017**, *39*, 81–88. [CrossRef]
39. Yousefpour, V.; Hamidi, A.; Ghanbari, A. Shear Strength-Dilation Characteristics of Silty and Clayey Sands. *J. Eng. Geol.* **2020**, *13*, 177–205.
40. Townsend, F.C. Geotechnical Characteristics of Residual Soils. *J. Geotech. Eng.* **1985**, *111*, 77–94. [CrossRef]
41. Giacheti, H.L.; Bezerra, R.C.; Rocha, B.P.; Rodrigues, R.A. Seasonal influence on cone penetration test: An unsaturated soil site example. *J. Rock Mech. Geotech. Eng.* **2019**, *11*, 361–368. [CrossRef]
42. *ASTM D2487-17 Standard Practice for Classification of Soils for Engineering Purposes (Unified Soil Classification System)*; ASTM International: West Conshohocken, PA, USA, 2017; pp. 1–10.
43. *ASTM D4452-14 Standard Practice for X-ray Radiography of Soil Samples 2014*; ASTM International: West Conshohocken, PA, USA, 2014; pp. 1–14.
44. *ASTM D7928-17 Standard Test Method for Particle-Size Distribution (Gradation) of Fine-Grained Soils Using the Sedimentation (Hydrometer) Analysis*; ASTM International: West Conshohocken, PA, USA, 2017; pp. 1–25.
45. *ASTM D854-14 Standard Test Methods for Specific Gravity of Soil Solids by Water Pycnometer*; ASTM International: West Conshohocken, PA, USA, 2014; pp. 1–7.
46. *ASTM D698: Standard Test Methods for Laboratory Compaction Characteristics of Soil Using Standard Effort (12 400 ft-lbf/ft3 (600 kN-m/m3))*; ASTM International: West Conshohocken, PA, USA, 2012; Volume 3, pp. 1–13.
47. *ASTM D4318-17e1 Standard Test Methods for Liquid Limit, Plastic Limit, and Plasticity Index of Soils*; ASTM International: West Conshohocken, PA, USA, 2017; Volume 4, pp. 1–14.
48. Diambra, A.; Ibraim, E. Modelling of fibre–cohesive soil mixtures. *Acta Geotech.* **2014**, *9*, 1029–1043. [CrossRef]
49. Li, C.; Zornberg, J.G. Interface shear strength in fiber-reinforced soil. In Proceedings of the 16th International Conference on Soil Mechanics and Geotechnical Engineering, Osaka, Japan, 12–16 September 2005; Volume 3, pp. 1373–1376. [CrossRef]
50. Rowland Otoko, G. Stress–Strain Behaviour of an Oil Palm Fibre Reinforced Lateritic Soil. *Int. J. Eng. Trends Technol.* **2014**, *14*, 295–298. [CrossRef]
51. Darvishi, A.; Erken, A. Effect of Polypropylene Fiber on Shear Strength Parameters of Sand. In Proceedings of the 3rd World Congress on Civil, Structural, and Environmental Engineering (CSEE'18), Budapest, Hungary, 8–10 April 2018; p. 13.

52. Falorca, I.; Pinto, M.I.M. Effect of short, randomly distributed polypropylene microfibres on shear strength behaviour of soils. *Geosynth. Int.* **2011**, *18*, 2–11. [CrossRef]
53. *ASTM D3080/D3080M-11. Standard Test Method for Direct Shear Test of Soils under Consolidated Drained Conditions*; ASTM International: West Conshohocken, PA, USA, 2011; Volume 4, pp. 1–9. [CrossRef]
54. Gelder, C.; Fowmes, G.J. Mixing and compaction of fibre- and lime-modified cohesive soil. *Proc. Inst. Civ. Eng. Ground Improv.* **2016**, *169*, 98–108. [CrossRef]
55. Kumar, P.; Singh, S.P. Fiber-reinforced fly ash subbases in rural roads. *J. Transp. Eng.* **2008**, *134*, 171–180. [CrossRef]
56. Mirzababaei, M.; Miraftab, M.; Mohamed, M.; McMahon, P. Unconfined compression strength of reinforced clays with carpet waste fibers. *J. Geotech. Geoenviron. Eng.* **2013**, *139*, 483–493. [CrossRef]
57. Khatri, V.N.; Dutta, R.K.; Venkataraman, G.; Shrivastava, R. Shear strength behaviour of clay reinforced with treated coir fibres. *Period. Polytech. Civ. Eng.* **2016**, *60*, 135–143. [CrossRef]
58. Sadek, S.; Najjar, S.S.; Freiha, F. Shear strength of fiber-reinforced sands. *J. Geotech. Geoenviron. Eng.* **2010**, *136*, 490–499. [CrossRef]
59. Ibraim, E.; Diambra, A.; Muir Wood, D.; Russell, A.R. Static liquefaction of fibre reinforced sand under monotonic loading. *Geotext. Geomembr.* **2010**, *28*, 374–385. [CrossRef]
60. Mohamad, H.; Maher, M.H.; Gray, D.H. Static response of sands reinforced with randomly distributed fibers. *J. Geotech. Eng.* **1990**, *116*, 1661–1677.
61. Consoli, N.C.; Thomé, A.; Girardello, V.; Ruver, C.A. Uplift behavior of plates embedded in fiber-reinforced cement stabilized backfill. *Geotext. Geomembr.* **2012**, *35*, 107–111. [CrossRef]
62. Bera, A.K.; Chandra, S.N.; Ghosh, A.; Ghosh, A. Unconfined compressive strength of fly ash reinforced with jute geotextiles. *Geotext. Geomembr.* **2009**, *27*, 391–398. [CrossRef]
63. Futai, M.M.; Almeida, M.S.S.; Lacerda, W.A. Yield, Strength, and Critical State Behavior of a Tropical Saturated Soil. *J. Geotech. Geoenviron. Eng.* **2004**, *130*, 1169–1179. [CrossRef]
64. Yetimoglu, T.; Salbas, O. A study on shear strength of sands reinforced with randomly distributed discrete fibers. *Geotext. Geomembr.* **2003**, *21*, 103–110. [CrossRef]
65. Shao, W.; Cetin, B.; Li, Y.; Li, J.; Li, L. Experimental Investigation of Mechanical Properties of Sands Reinforced with Discrete Randomly Distributed Fiber. *Geotech. Geol. Eng.* **2014**. [CrossRef]
66. *BS 8002:1994. Code of Practice for Earth Retaining Structures*; British Standards Institution: London, UK, 1994; pp. 1–114.

 *sustainability*

Article

Modeling the Compaction Characteristics of Fine-Grained Soils Blended with Tire-Derived Aggregates

Amin Soltani [1,*], Mahdieh Azimi [2] and Brendan C. O'Kelly [3]

[1] School of Engineering, IT and Physical Sciences, Federation University, Churchill, VIC 3842, Australia
[2] School of Engineering and Technology, Central Queensland University, Melbourne, VIC 3000, Australia; m.azimiadehmortezapasha@cqu.edu.au
[3] Department of Civil, Structural and Environmental Engineering, Trinity College Dublin, D02 PN40 Dublin, Ireland; bokelly@tcd.ie
* Correspondence: a.soltani@federation.edu.au

Abstract: This study aims at modeling the compaction characteristics of fine-grained soils blended with sand-sized (0.075–4.75 mm) recycled tire-derived aggregates (TDAs). Model development and calibration were performed using a large and diverse database of 100 soil–TDA compaction tests (with the TDA-to-soil dry mass ratio $\leq$ 30%) assembled from the literature. Following a comprehensive statistical analysis, it is demonstrated that the optimum moisture content (OMC) and maximum dry unit weight (MDUW) for soil–TDA blends (across different soil types, TDA particle sizes and compaction energy levels) can be expressed as universal power functions of the OMC and MDUW of the unamended soil, along with the soil to soil–TDA specific gravity ratio. Employing the Bland–Altman analysis, the 95% upper and lower (water content) agreement limits between the predicted and measured OMC values were, respectively, obtained as +1.09% and −1.23%, both of which can be considered negligible for practical applications. For the MDUW predictions, these limits were calculated as +0.67 and −0.71 kN/m^3, which (like the OMC) can be deemed acceptable for prediction purposes. Having established the OMC and MDUW of the unamended fine-grained soil, the empirical models proposed in this study offer a practical procedure towards predicting the compaction characteristics of the soil–TDA blends without the hurdles of performing separate laboratory compaction tests, and thus can be employed in practice for preliminary design assessments and/or soil–TDA optimization studies.

Keywords: fine-grained soil; tire-derived aggregate; optimum moisture content; maximum dry unit weight; Bland–Altman analysis

Citation: Soltani, A.; Azimi, M.; O'Kelly, B.C. Modeling the Compaction Characteristics of Fine-Grained Soils Blended with Tire-Derived Aggregates. *Sustainability* **2021**, *13*, 7737. https://doi.org/10.3390/su13147737

Academic Editor: Castorina Silva Vieira

Received: 9 June 2021
Accepted: 8 July 2021
Published: 11 July 2021

Publisher's Note: MDPI stays neutral with regard to jurisdictional claims in published maps and institutional affiliations.

1. Introduction

Lately, many developed and developing countries have initiated the transition to 'sustainable infrastructure', a concept that (among other things) encourages the replacement of natural quarry-based aggregates with recycled solid waste materials. End-of-life tires (ELTs) from the automotive industry are among the largest and most problematic global waste streams, prompting recycled tire-derived aggregates (TDAs) to become one of the most targeted materials for civil engineering applications. Because of their physical and mechanical attributes, particularly in terms of their relatively low density, high energy absorption capacity, resilience and low water adsorption–retention potential, granulated TDA-based products (e.g., crumbs, buffings and fibers) have been well established as effective soil-blending agents for the development of high-performance (and sustainable) geomaterials for a variety of practical geotechnical applications, such as soil stabilization, highway embankment and pavement constructions, as well as for bridge abutment and retaining wall backfills [1–5]. Further, Shahrokhi-Shahraki et al. [6] investigated the use of pulverized waste tire, either on its own or mixed with soil (well-graded sand), to act as an adsorptive fill material, demonstrating adsorption of organic/inorganic contaminants,

151

namely benzene, toluene, ethylbenzene and xylene (BTEX) components, and two heavy metal ions (Pb^{2+} and Cu^{2+}).

Research on soil–TDA mixtures dates back to the early 1990s, where theoretical concepts governing the mechanical performance of this (then-emerging) geomaterial were first put into perspective. Earlier investigations were mainly focused on coarse-grained soils (mainly sands), demonstrating that the granular soil–TDA blend, resembling a rigid–soft matrix, can be optimized in terms of the TDA content and its particle geometry (i.e., its mean particle size and shape) to achieve any desired balance between the strength/stiffness and deformability parameters of the TDA-based blend [7–13]. These early investigations unanimously concluded that the addition of TDA to coarse-grained soils leads to notable reductions in the soils' mobilized strength and stiffness while enhancing their ductility characteristics, which was mainly ascribed to the lower stiffness (and higher deformability) of the soft TDA particles compared with that of the rigid soil grains. Furthermore, depending on the TDA content and its mean particle size (in relation to the rigid soil grains), the stress–strain response of a granular soil–TDA blend can fall into one of three behavioral categories [12,13]: (i) rigid-dominant; (ii) rigid–soft transitional; and (iii) soft-dominant. The transitional behavior (by definition) resembles a perfect balance between the blend's strength/stiffness and ductility/toughness—a review of the research literature indicates that the transitional behavior is often encountered at a volumetric TDA content (commonly defined as the 'TDA-to-granular soil + TDA' volume ratio) of 30–50% [14].

Later studies followed suit, confirming the suitability of TDA-based products, particularly when paired with chemical binders, as effective blending agents for compacted fine-grained soils (including expansive clays) capable of promoting improved shear strength performance, reduced swell–shrink (and hence desiccation-induced cracking) potential and improved damping [15–22]. In terms of shear strength, for instance, these studies concluded that the addition of TDA at low TDA-to-fine-grained soil mass ratios (mainly less than 10%) often produces relatively small improvements, attributed to 'arching' between the TDA particles within the soil–TDA agglomerations [19,22] and induced 'inter-particle friction' generated at the soil–TDA interfaces [16]. These studies also demonstrate that higher TDA contents tend to cause serious concerns for undrained strength and stiffness, largely due to the relatively lower stiffness (and higher deformability) of the soil–TDA agglomerations compared to individual soil agglomerations containing no TDA [17,19]. Accordingly, for projects where the strength and stiffness are of primary importance, the compacted fine-grained soil–TDA blend requires stabilization by means of conventional cementitious (e.g., Portland cement and quick lime [15,18]) or polymer (e.g., polyacrylamide and sodium alginate [22,23]) binders.

Like natural (unamended) fine-grained soils, an essential step towards the production and placement of suitable soil–TDA earth fills is compaction. The governing variables which control the compactability of TDA-blended fine-grained soils have been well documented in the research literature. It is generally accepted that the addition of TDA to fine-grained soils leads to notable reductions in the optimum moisture content (OMC) and the corresponding maximum dry unit weight (MDUW), mainly attributed to the TDA material's hydrophobic character (water adsorption being mainly less than 4%), relatively lower density and higher energy absorption capacity compared with that of the soil solids [24–26]. The OMC and MDUW parameters are commonly measured by standardized laboratory compaction tests which, though straightforward in terms of execution, are fairly labor-intensive and highly time-consuming. Accordingly, several attempts have been made to devise empirical-type correlations for indirect estimation of the OMC and MDUW of unamended fine-grained soils, all of which employ the soil consistency (Atterberg) limits as the primary predictors [27–33]. Common TDA-based products (e.g., crumbs, buffings and fibers) used in conjunction with fine-grained soils are mainly similar in size to predominantly medium–coarse sand (0.425–4.75 mm); as such, the soil consistency limit tests would not be applicable to most soil–TDA blends. This implies that the many well-established empirical correlations reported for indirect estimation of the OMC and

MDUW of unamended fine-grained soils cannot be extended to soil–TDA blends; this limitation highlighting the need to develop an entirely new modeling framework.

A review of the research literature indicates that no modeling framework exists for the compaction characteristics of fine-grained soil–TDA blends. Accordingly, this study aims at establishing practical empirical models for indirect estimation of the OMC and MDUW of TDA-blended fine-grained soils. Model development and calibration are carried out using a large and diverse database of 100 soil–TDA compaction tests assembled from the research literature. The empirical models proposed in this study offer a practical procedure towards predicting the compaction characteristics of soil–TDA blends without the hurdles of performing separate laboratory compaction tests, and thus can be used for preliminary design assessments and/or soil–TDA optimization studies.

2. Database of Soil–TDA Compaction Tests

Given that empirical models/correlations are purely data-driven, their predictive capability is highly dependent on the database from which they are developed. Accordingly, to establish practical empirical models for the OMC and MDUW of soil–TDA blends, a large and diverse database of 100 soil–TDA compaction tests was gathered from the authors' previous publications [19,22,23,34,35] as well as other recent literature sources [15–17,36–39]. Detailed descriptions of the assembled database are presented in Tables 1 and 2. The compiled database consisted of 21 datasets (designated as D1–D21), each defined as a collection of standard or modified (heavy) Proctor compaction tests for a given fine-grained soil mixed with a particular TDA material (constant particle size/shape) at varying TDA contents (denoted as f_T and defined as the TDA-to-soil dry mass ratio, here expressed as a percentage value).

Table 1. Soil properties for the compiled database of soil–TDA compaction tests.

Dataset	Source	Compaction Energy Level	N	Soil ID	G_s^S	LL (%)	PI (%)	f_{fines} (%)	f_{clay} (%)	A	Soil USCS
D1	[15]	Modified Proctor	4	S1	2.61	49.5	26.2	-	-	-	CI
D2	[36]	Standard Proctor	5	S2	2.65	74.0	38.0	97.0	70.0	0.54	MV
D3	[36]	Standard Proctor	5	S3	2.65	56.0	36.0	94.0	59.0	0.61	CH
D4	[37]	Standard Proctor	4	S4	2.72	53.0	14.0	63.6	52.8	0.27	MH
D5	[16]	Standard Proctor	7	S5	2.69	52.2	28.1	96.4	47.2	0.60	CH
D6	[16]	Modified Proctor	7	S5	2.69	52.2	28.1	96.4	47.2	0.60	CH
D7	[38]	Modified Proctor	5	S6	2.69	34.2	9.4	92.2	60.7	0.15	ML
D8	[17]	Modified Proctor	5	S6	2.69	34.2	9.4	92.2	60.7	0.15	ML
D9	[39]	Standard Proctor	5	S7	2.65	52.0	31.0	62.0	-	-	CH
D10	[39]	Standard Proctor	5	S8	2.65	60.0	38.0	54.0	-	-	CH
D11	[34]	Standard Proctor	5	S9	2.69	44.2	21.9	99.0	49.0	0.45	CI
D12	[34]	Standard Proctor	5	S10	2.67	47.2	29.2	69.0	37.0	0.79	CI
D13	[34]	Standard Proctor	5	S11	2.71	59.5	31.6	99.0	53.0	0.60	CH
D14	[34]	Standard Proctor	5	S12	2.72	77.6	57.0	80.0	44.0	1.30	CV
D15	[35]	Standard Proctor	5	S13	2.73	59.6	32.3	99.1	51.7	0.62	CH
D16	[35]	Standard Proctor	5	S13	2.73	59.6	32.3	99.1	51.7	0.62	CH
D17	[23]	Standard Proctor	5	S14	2.76	78.4	54.2	78.0	43.0	1.26	CV
D18	[19]	Standard Proctor	4	S15	2.77	43.6	21.5	80.0	43.0	0.50	CI
D19	[19]	Standard Proctor	4	S15	2.77	43.6	21.5	80.0	43.0	0.50	CI
D20	[19]	Standard Proctor	4	S15	2.77	43.6	21.5	80.0	43.0	0.50	CI
D21	[22]	Standard Proctor	5	S16	2.73	84.3	52.3	99.0	52.0	1.01	CV

Note: N = number of compaction tests; G_s^S = specific gravity of soil solids (in the absence of a reliable value, the typical representative specific gravity of 2.65 was considered); LL = liquid limit; PI = plasticity index; f_{fines} = fines content (<75 μm); f_{clay} = clay content (<2 μm); A = activity index (calculated as $A = PI/f_{clay}$); and USCS = Unified Soil Classification System, as per BS 5930 [40].

Table 2. TDA properties for the compiled database of soil–TDA compaction tests.

Dataset	Source	N	Soil ID	TDA Type	G_s^T	D_{50} (mm)	C_U	C_C	TDA USCS	TDA Content f_T (%)
D1	[15]	4	S1	Tire buffings	1.08	2.36	2.39	0.94	SP	0, 5.3, 11.1, 17.6
D2	[36]	5	S2	Crumb rubber (425–600 μm)	0.85	-	-	-	SP	0, 5, 10, 15, 20
D3	[36]	5	S3	Crumb rubber (425–600 μm)	0.85	-	-	-	SP	0, 5, 10, 15, 20
D4	[37]	4	S4	Granular rubber	1.12	1.04	2.55	1.08	SP	0, 5, 10, 20
D5	[16]	7	S5	Rubber powder (0.075–2 mm)	1.14	-	-	-	SP	0, 2.5, 5, 10, 15, 20, 25
D6	[16]	7	S5	Rubber powder (0.075–2 mm)	1.14	-	-	-	SP	0, 2.5, 5, 10, 15, 20, 25
D7	[38]	5	S6	Crumb rubber	1.13	1.54	2.20	1.31	SP	0, 2.5, 5, 7.5, 10
D8	[17]	5	S6	Rubber fiber	1.07	1.64	2.95	1.27	SP	0, 2.5, 5, 7.5, 10
D9	[39]	5	S7	Crumb rubber (75–425 μm)	0.85	-	-	-	SP	0, 5, 10, 15, 20
D10	[39]	5	S8	Crumb rubber (75–425 μm)	0.85	-	-	-	SP	0, 5, 10, 15, 20
D11	[34]	5	S9	Ground rubber	1.09	0.48	2.83	1.19	SP	0, 5, 10, 20, 30
D12	[34]	5	S10	Ground rubber	1.09	0.48	2.83	1.19	SP	0, 5, 10, 20, 30
D13	[34]	5	S11	Ground rubber	1.09	0.48	2.83	1.19	SP	0, 5, 10, 20, 30
D14	[34]	5	S12	Ground rubber	1.09	0.48	2.83	1.19	SP	0, 5, 10, 20, 30
D15	[35]	5	S13	Rubber crumbs	1.09	0.48	2.83	1.19	SP	0, 5, 10, 20, 30
D16	[35]	5	S13	Rubber buffings	1.09	1.58	1.56	1.03	SP	0, 5, 10, 20, 30
D17	[23]	5	S14	Ground rubber	1.09	0.48	2.83	1.19	SP	0, 5, 10, 20, 30
D18	[19]	4	S15	TDA–Fine	1.08	0.46	3.06	1.23	SP	0, 5, 10, 20
D19	[19]	4	S15	TDA–Medium	1.10	1.67	1.85	1.09	SP	0, 5, 10, 20
D20	[19]	4	S15	TDA–Coarse	1.11	3.34	1.89	1.10	SP	0, 5, 10, 20
D21	[22]	5	S16	Ground rubber	1.09	0.46	3.06	1.23	SP	0, 5, 10, 20, 30

Note: N = number of compaction tests; G_s^T = specific gravity of TDA particles; D_{50} = mean TDA particle size; C_U and C_C = coefficients of uniformity and curvature, respectively; f_T = TDA-to-soil dry mass ratio, here expressed as a percentage value; and USCS = Unified Soil Classification System, as per BS 5930 [40].

As demonstrated in Table 1, the 21 datasets included a total of sixteen fine-grained soils (designated as S1–S16), covering reasonably wide ranges of surface texture, plasticity and mineralogical properties—that is, f_{clay} = 37–70%, LL = 34.2−84.3% and $A = PI/f_{clay}$ = 0.15–1.30 (where f_{clay}, LL, PI and A denote clay content, liquid limit, plasticity index and activity index, respectively). In terms of classification, the database soils consisted of three silts and thirteen clays with the following USCS frequencies, as per BS 5930 [40]: ML = 1; MH = 1; MV = 1; CI = 4; CH = 6; and CV = 3. Referring to Table 2; the compiled database covers all major types of commercially available poorly-graded sand-sized (i.e., SP classification) TDA products (e.g., powder, crumbs and buffings), with the TDA mean particle size (or D_{50}) ranging between 0.46 and 3.34 mm. Furthermore, in terms of mix design, the TDA content varied between 2.5% and 30% (the latter considered to be the highest possible whilst still maintaining mixture homogeneity), and each of the 21 datasets, in addition to the unamended soil (f_T = 0), included a minimum of three compaction test data for three different TDA contents. It should be mentioned that the experimental OMC values ranged between 12.4% and 28.0% water content, with the complete results of the compaction tests presented in Figure A1 of the Appendix A.

3. Governing Mechanisms Controlling the Compactability of Soil–TDA Blends

It is generally accepted that the addition of (and content increase in) TDA, with constant particle size/shape, leads to a 'leftward–downward' translation of the soil compaction curve (for a given compactive effort), causing notable reductions in the OMC and MDUW parameters [24–26]. The TDA material's lower specific gravity (or density) compared with that of the soil solids has been reported to be the primary factor responsible for decreasing the MDUW [15]. For the compiled soil–TDA database used in this investigation (see Tables 1 and 2), the TDA-to-soil specific gravity ratio (i.e., G_s^T/G_s^S) was found to range between 0.31 and 0.44. Moreover, some researchers have postulated that, because of their high energy absorption capacity (attributed to their high elasticity), the compacted TDA particles may progressively recover their initial uncompacted shapes through a so-called 'elastic-rebound' recovery mechanism, thereby offsetting the efficiency of the imparted compactive effort, also contributing towards decreasing the MDUW [17,34,41]. The reductions reported for the OMC have been mainly ascribed to the TDA material's hydrophobic character and hence its lower water adsorption–retention capacity (being mainly less than 4%) compared with that of the fine soil particles, particularly clays [16]. It should be mentioned that, while the reported results for the MDUW are fairly consistent, a limited number of studies have reported either negligible or increasing trends for the OMC with respect to increasing the TDA content [42–44]. These unexpected trends may be attributed to TDA segregation (and hence TDA clustering) effects caused by inadequate soil–TDA mixing during sample preparation for the compaction test [19,39]. As such, in compiling our database for the present investigation, it was decided to include only those datasets which are consistent with the more unanimous 'OMC-decreasing' trend.

It is well accepted that the MDUW and OMC for unamended fine-grained soils (irrespective of compactive effort) are strongly correlated, following a unique 'path of optimums' somewhat parallel to the standard zero-air-voids (ZAV) saturation line (commonly obtained for a typical specific gravity of 2.65 for the soil solids) [27,29,31]. For instance, in their investigation, Gurtug and Sridharan [27] reported the following 'path of optimums' relationships based on a database of 181 compaction tests (with OMC water contents ranging w_{opt} = 7.4–49.0%) involving a variety of 'unamended' fine-grained soils tested at four different compaction energy levels (i.e., reduced, standard, reduced modified and modified Proctor): (i) γ_{dmax} = −0.28 w_{opt} + 22.26 (with R^2 = 0.941); and (ii) γ_{dmax} = 23.68 $exp[$ −0.018 $w_{opt}]$ (with R^2 = 0.960), where γ_{dmax} is the deduced MDUW value. Figure 1a illustrates the variations of MDUW against OMC for the compiled database of 100 soil–TDA compaction tests (data values presented in Figure A1). As is evident from this figure, the data points are significantly scattered, indicating that the MDUW and OMC are poorly correlated for the investigated soil–TDA blends. For unamended fine-grained soils, an increase in the coarse fraction (>75 μm) leads to an 'upward–leftward' translation of the compaction curve, with the optimum (peak) point translating along the previously described universal 'path of optimums'. For soil–TDA blends, however, an increase in TDA content (0.075–4.75 mm particle size range), which is essentially similar to increasing the soil coarse fraction, results in a 'downward–leftward' translation of the optimum point, implying that soil–TDA blends do not conform to the general 'path of optimums' correlation framework. This discrepancy can be attributed to the significant mismatch in density (and hence specific gravity) between the soil solids and TDA particles, allowing one to postulate that the lower MDUW values obtained for soil–TDA blends may not necessarily reflect their lower compactability potential. To achieve a more familiar visualization of the compaction characteristics of soil–TDA blends, consistent with the traditional soil compaction framework (for unamended soils), the conventional OMC and MDUW parameters, w_{opt} and γ_{dmax}, should be 'normalized' as follows [19]:

$$w_{opt}^* = w_{opt}\left(\frac{G_s^{ST}}{G_s^S}\right) \tag{1}$$

$$\gamma_{dmax}^* = \gamma_{dmax}\left(\frac{G_s^S}{G_s^{ST}}\right) \tag{2}$$

where w_{opt}^* and γ_{dmax}^* = normalized OMC and MDUW parameters, respectively; and G_s^S and G_s^{ST} = specific gravity of soil solids (values presented in Table 1) and soil–TDA mixture (values presented in Table A1 of the Appendix A), respectively.

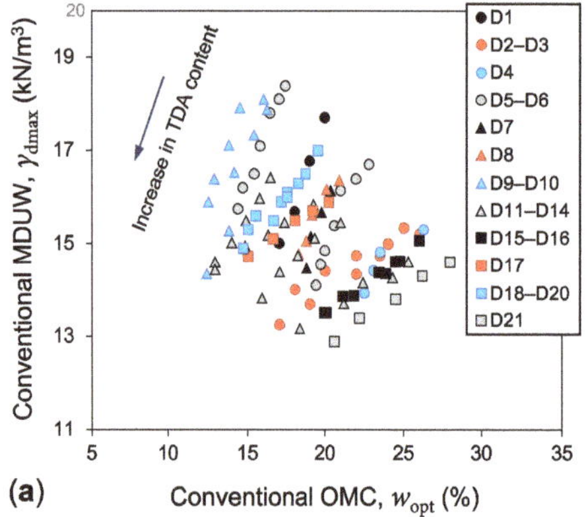
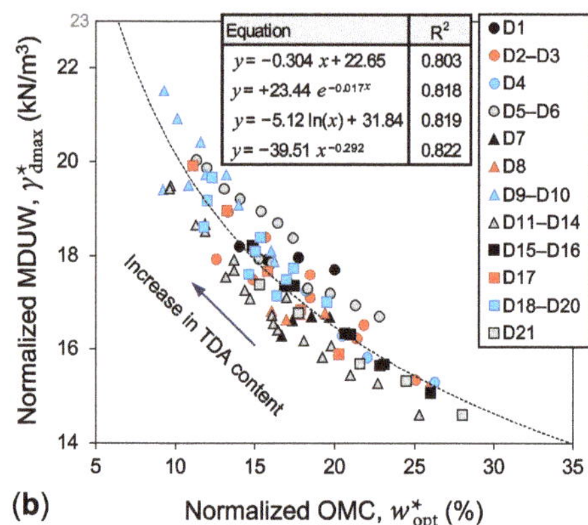

Figure 1. Variations of MDUW against OMC for the compiled database of 100 soil–TDA compaction tests: (**a**) Conventional definition; and (**b**) Normalized definition according to Equations (1) and (2).

It should be mentioned that the soil–TDA mixture specific gravity G_s^{ST}, which decreases with increasing the TDA content (i.e., $G_s^{ST} \sim f_T^{-1}$, as demonstrated in Table A1), was obtained as follows [34,45]:

$$G_s^{ST} = \frac{G_s^S G_s^T (M_S + M_T)}{G_s^S M_T + G_s^T M_S} = \frac{G_s^S G_s^T (1 + f_T)}{G_s^S f_T + G_s^T} \tag{3}$$

where G_s^T = specific gravity of TDA particles (values presented in Table 2); M_S and M_T = mass of oven-dried soil and TDA, respectively; and f_T = TDA content, defined as the TDA-to-soil dry mass ratio (or M_T/M_S).

Figure 1b illustrates the variations of γ_{dmax}^* against w_{opt}^* for the compiled database. As is evident from this figure, the normalized MDUW and OMC, in addition to showcasing a strong correlation, conform to the general 'path of optimums' framework described earlier for unamended fine-grained soils. As a typical example, exponential-fitting of the normalized compaction data resulted in $\gamma_{dmax}^* = 23.44\ exp[\ -0.017\ w_{opt}^*]$ (with $R^2 = 0.818$) for $f_T \leq 30\%$, which is essentially identical to γ_{dmax}–w_{opt} relationships previously reported for unamended fine-grained soils by Gurtug and Sridharan [27] and Sivrikaya et al. [29].

4. Results and Discussion

4.1. Modeling Premise

Following a comprehensive trial-and-error investigation employing the 21 soil–TDA compaction datasets (see Table A1 and Figure A1), it was observed that, for a given fine-grained soil mixed with a particular TDA material (constant particle size/shape), the conventional OMC and MDUW parameters (for the standard or modified Proctor energy level) can be expressed as follows:

$$w_{opt}^{ST} = w_{opt}^S \left(\frac{G_s^S}{G_s^{ST}} \right)^{\beta_M} \tag{4}$$

$$\gamma_{dmax}^{ST} = \gamma_{dmax}^S \left(\frac{G_s^S}{G_s^{ST}} \right)^{\beta_D} \tag{5}$$

where w_{opt}^{ST} and γ_{dmax}^{ST} = conventional OMC and MDUW for the soil–TDA mixture, respectively; w_{opt}^S and γ_{dmax}^S = intercept parameters for $f_T = 0$ (in % and kN/m^3, respectively); and β_M and β_D = reduction rate parameters (both < 0).

The dependent/input variable G_s^S/G_s^{ST} selected for model development captures the combined effects of TDA content and TDA specific gravity; the latter well-established to vary (i.e., 0.85–1.14 for the present investigation, as outlined in Table 2) depending on the source-tire composition, the adopted tire recycling process, and the TDA particle size/shape [24,26]. The intercept parameters w_{opt}^S and γ_{dmax}^S represent the OMC and MDUW of the unamended soil, since setting $f_T = 0$ in Equation (3) results in $G_s^{ST} = G_s^S$. Note that these intercept parameters can be either fixed based on measured values or set as independent fitting parameters. Further, since the soil–TDA mixture specific gravity decreases with increasing the TDA content (i.e., $G_s^{ST} \sim f_T^{-1}$ and hence $G_s^S/G_s^{ST} \sim f_T$; see Table A1), the parameters β_M and β_D (which are both negative) represent the rates of reduction in the OMC and MDUW, respectively, in relation to increasing the TDA content.

It should be mentioned that the trial-and-error investigation performed by the authors and leading to the proposal of Equations (4) and (5) involved applying various functional expressions (i.e., linear, logarithmic, polynomial, exponential and power) to the 21 soil–TDA compaction datasets and then cross-checking their predictive performances using routine fit-measure indices; namely, the mean absolute percentage error (MAPE) and the normalized root-mean-squared error (NRMSE), which were calculated as follows [46]:

$$\text{MAPE} = \frac{1}{N} \sum_{n=1}^{N} \left| \frac{y_n - \hat{y}_n}{y_n} \right| \times 100\% \tag{6}$$

$$\text{NRMSE} = \frac{\text{RMSE}}{\overline{y_n}} \times 100\% \tag{7}$$

$$\text{RMSE} = \sqrt{\frac{1}{N} \sum_{n=1}^{N} (y_n - \hat{y}_n)^2} \tag{8}$$

where RMSE = root-mean-squared error (in the same unit as OMC or MDUW); y_n = measured variable (OMC or MDUW); $\hat{y}_n$ = predicted variable (OMC or MDUW); $\overline{y_n}$ = arithmetic mean of y_n data; and N = number of observations (or compaction tests) in each dataset, as reported in Table 1.

The regression analysis outputs with respect to Equations (4) and (5) (with w_{opt}^S and γ_{dmax}^S set as independent fitting parameters) are presented in Tables 3 and 4, respectively. Judging by the high R^2 (with median values of 0.989 and 0.987 for the OMC and MDUW predictions, respectively) and the low MAPE or NRMSE (unanimously less than 4%) values, the functional expressions proposed in Equations (4) and (5) can be deemed acceptable. Quite clearly, to employ Equations (4) and (5) for routine prediction purposes, the fitting parameters β_M and β_D should be calibrated. In view of their definitions, the intercept parameters w_{opt}^S and γ_{dmax}^S can be simply fixed based on the measured OMC and MDUW of the unamended soil ($f_T = 0$). Provided that the reduction rate parameters β_M and β_D can be practically calibrated without the need for obtaining any specific soil–TDA compaction test data, it would follow that, having established the OMC and MDUW of an unamended fine-grained soil (along with the soil and TDA specific gravities; the latter often provided by the TDA manufacturer), one can predict the OMC and MDUW of the same soil mixed with any specified TDA content. The following sections describe practical calibration frameworks for obtaining β_M and β_D.

Table 3. Summary of the regression analysis outputs with respect to Equation (4) (OMC model).

Dataset	N	$w_{\text{opt}}^{\text{S}}(\%)$	β_{M}	R^2	MAPE (%)	NRMSE (%)
D1	4	20.06	-0.842	0.997	0.32	0.32
D2	5	25.99	-1.275	0.931	3.19	3.77
D3	5	26.56	-1.202	0.977	1.88	1.97
D4	4	25.45	-0.668	0.904	2.54	2.72
D5	7	22.37	-0.639	0.944	1.10	1.33
D6	7	17.37	-0.788	0.993	0.47	0.53
D7	5	20.93	-0.944	0.989	0.38	0.42
D8	5	20.83	-1.057	0.985	0.48	0.59
D9	5	15.87	-0.817	0.991	0.80	0.85
D10	5	16.44	-0.839	0.952	1.74	1.93
D11	5	20.74	-0.925	0.990	0.87	1.02
D12	5	16.53	-0.813	0.992	0.66	0.70
D13	5	25.74	-1.046	0.956	2.05	2.22
D14	5	19.17	-1.263	0.987	1.21	1.40
D15	5	26.28	-0.862	0.987	0.93	1.04
D16	5	26.08	-0.905	0.996	0.45	0.57
D17	5	20.41	-0.976	0.992	0.84	0.90
D18	4	19.80	-1.009	0.977	1.31	1.36
D19	4	19.65	-1.163	0.993	0.78	0.81
D20	4	19.34	-1.241	0.992	0.74	0.93
D21	5	28.04	-1.038	0.999	0.19	0.23

Note: N = number of compaction tests; R^2 = coefficient of determination; MAPE = mean absolute percentage error (Equation (6)); NRMSE = normalized root-mean-squared error (Equation (7)); $w_{\text{opt}}^{\text{S}}$ = OMC intercept parameter; and β_{M} = OMC reduction rate parameter for increasing TDA content.

Table 4. Summary of the regression analysis outputs with respect to Equation (5) (MDUW model).

Dataset	N	$\gamma_{\text{dmax}}^{\text{S}}$ (kN/m^3)	β_{D}	R^2	MAPE (%)	NRMSE (%)
D1	4	17.72	-0.874	0.996	0.35	0.39
D2	5	15.42	-0.481	0.984	0.51	0.65
D3	5	15.32	-0.265	0.937	0.69	0.73
D4	4	15.27	-0.440	0.995	0.23	0.24
D5	7	16.76	-0.714	0.996	0.34	0.40
D6	7	18.46	-0.655	0.996	0.29	0.35
D7	5	16.56	-1.030	0.944	1.01	1.05
D8	5	16.49	-0.846	0.966	0.62	0.74
D9	5	18.38	-0.454	0.942	1.08	1.24
D10	5	18.29	-0.734	0.938	1.79	2.05
D11	5	15.48	-0.368	0.987	0.36	0.42
D12	5	16.38	-0.404	0.996	0.21	0.27
D13	5	14.66	-0.336	0.968	0.61	0.64
D14	5	15.81	-0.290	0.976	0.42	0.48
D15	5	15.04	-0.360	0.998	0.13	0.16
D16	5	15.03	-0.359	0.997	0.18	0.22
D17	5	15.96	-0.254	0.984	0.33	0.35
D18	4	16.96	-0.371	0.995	0.21	0.22
D19	4	16.92	-0.460	0.989	0.39	0.41
D20	4	16.82	-0.580	0.964	0.91	0.93
D21	5	14.63	-0.415	0.990	0.41	0.44

Note: N = number of compaction tests; R^2 = coefficient of determination; MAPE = mean absolute percentage error (Equation (6)); NRMSE = normalized root-mean-squared error (Equation (7)); $\gamma_{\text{dmax}}^{\text{S}}$ = MDUW intercept parameter; and β_{D} = MDUW reduction rate parameter for increasing TDA content.

4.2. Predictive Models Employing Mean Reduction Rate Parameters

Figure 2a,b illustrates the variations of β_{M} and β_{D} for the 21 soil–TDA compaction datasets, respectively. In terms of absolute magnitude, β_{M} was found to be consistently greater than its β_{D} counterpart, indicating that the rate of OMC reduction with respect to increasing TDA content is greater than that of the MDUW. Judging by the low standard deviation (SD) for β_M and β_D (computed as 0.187 and 0.218, respectively), as well as the relatively small vertical distance between their upper and lower variation boundaries (see 'UB' and 'LB' in Figure 2), it may be possible to achieve reliable OMC and MDUW predictions (across different fine-grained soil types, TDA particle sizes/shapes and compaction energy levels) by adopting mean values for the β_M and β_D parameters. To examine this hypothesis, the arithmetic means for the 21 β_M and β_D values were calculated (i.e., $\overline{\beta_M} = -0.967$ and $\overline{\beta_D} = -0.509$, as outlined in Figure 2), and appointed to Equations (4) and (5), resulting in the following new relationships:

$$w_{opt}^{ST} = w_{opt}^{S} \left(\frac{G_s^S}{G_s^{ST}} \right)^{-0.967} \tag{9}$$

$$\gamma_{dmax}^{ST} = \gamma_{dmax}^{S} \left(\frac{G_s^S}{G_s^{ST}} \right)^{-0.509} \tag{10}$$

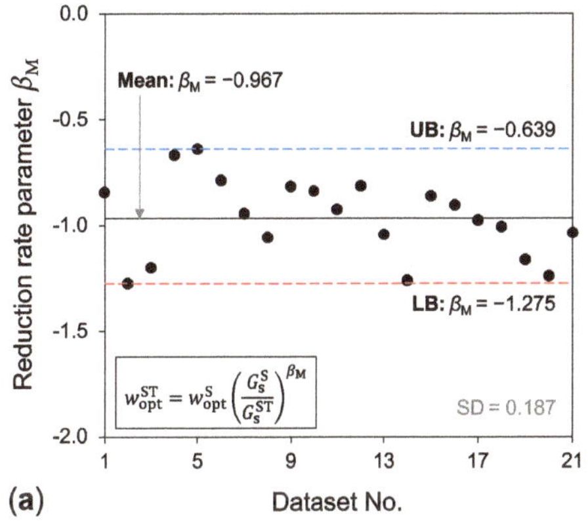

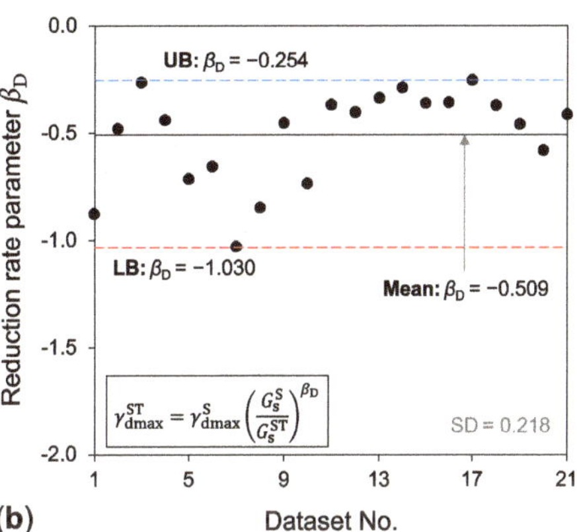

Figure 2. Variations of (**a**) β_M and (**b**) β_D for the 21 soil–TDA compaction datasets investigated. Note: UB and LB denote the upper and lower variation boundaries, respectively.

Scatter plots illustrating the variations of predicted (by Equations (9) and (10)) against measured OMC and MDUW values are presented in Figure 3a,b, respectively. As is evident from these figures, the predicted and measured values, particularly for the OMC, are strongly correlated with each other. The R^2, MAPE and NRMSE associated with these predictions were, respectively, calculated as 0.970, 2.7% and 3.2% for the OMC, and 0.908, 2.6% and 3.2% for the MDUW.

The excellent graphical correlation (high R^2), together with the low MAPE or NRMSE values, obtained for Equations (9) and (10) would normally lead to corroborating their predictive capability. However, a critical examination of the prediction residuals should also be performed to better perceive the true implications of these predictions for routine geotechnical engineering applications [47]. This can be achieved by quantifying and critically examining the statistical 'limits of agreement' between the predicted and measured values, which was conducted using the Bland–Altman (BA) analysis [48]. The BA analysis involves developing a scatter plot with the y-axis representing the difference between the two compared variables (e.g., $OMC_P - OMC_M$, where the subscripts 'P' and 'M' denote predicted and measured variables, respectively) and the x-axis showing the average of these variables (e.g., $[OMC_P + OMC_M]/2$). The 95% upper and lower agreement limits with respect to the BA plot can be, respectively, quantified as UAL = Mean + 1.96 × SD and LAL = Mean − 1.96 × SD (where 'Mean' and 'SD' are the arithmetic mean and standard deviation of the y-axis data, respectively).

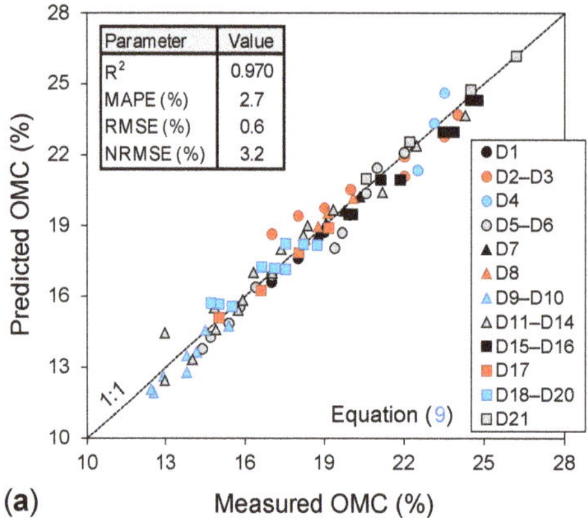 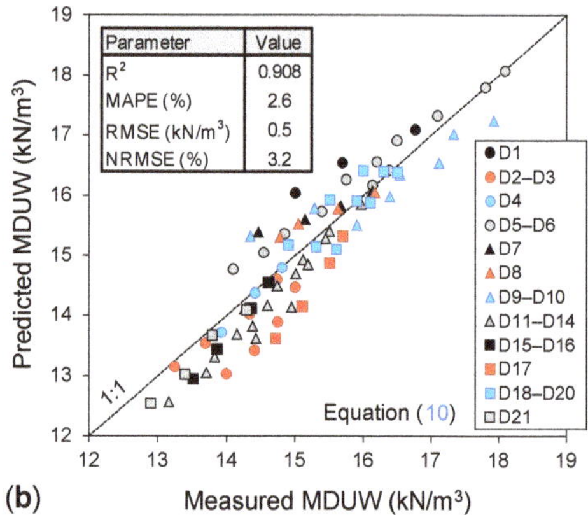

Figure 3. Variations of predicted against measured compaction parameters for the 21 soil–TDA compaction datasets: (**a**) OMC (Equation (9)); and (**b**) MDUW (Equation (10)).

BA plots for the OMC and MDUW predictions (Equations (9) and (10)) are presented in Figure 4a,b, respectively. The mean of differences between OMC_P and OMC_M was found to be -0.07%, indicating that the OMC predictions were on average 0.07% (water content) lower than their measured counterparts. The 95% agreement limits between OMC_P and OMC_M were calculated as UAL = +1.09% and LAL = −1.23%, implying that 95% of the predictions made by Equation (9) are associated with errors ranging between these two water content limits, both of which can be considered negligible for practical applications. As for the MDUW (see Figure 4b), the mean of differences, UAL and LAL were obtained as −0.12, +0.80 and −1.04 kN/m^3, respectively. Taking into account the nature of the MDUW parameter and its variations across different fine-grained soil types and also with standard and modified compaction energy levels (these variations being relatively smaller compared with that of the OMC [27,31,33]), the errors associated with Equation (10), though practically acceptable, may require further improvement. Alternatively, having predicted the OMC by Equation (9), the corresponding MDUW can be estimated with more accuracy through a practical single-point compaction test (performed at the predicted OMC).

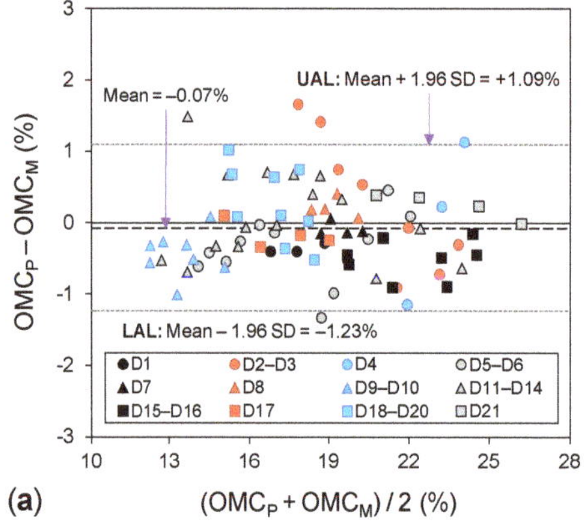 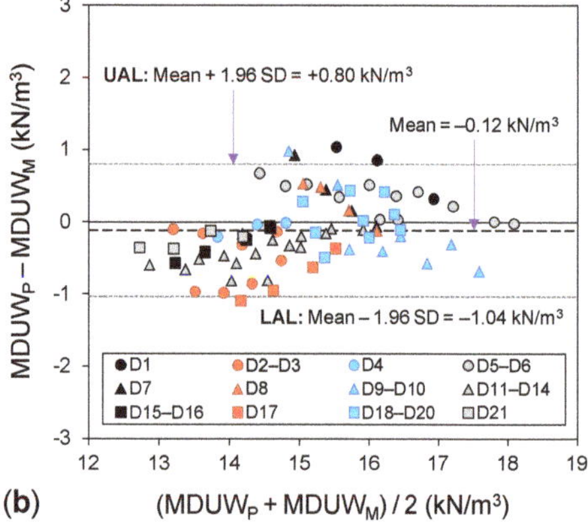

Figure 4. Bland–Altman plots for the (**a**) OMC and (**b**) MDUW predictions (made by Equations (9) and (10), respectively). Note: UAL and LAL denote the 95% upper and lower agreement limits, respectively.

4.3. Prediction Models Employing Empirical Reduction Rate Parameters

The authors postulate that the reduction rate parameters β_M and β_D may be systematically related to basic soil properties (namely those listed in Table 1). Accordingly, attempts were made to explore the existence of potential links/correlations between these fitting parameters and other parameters reflective of the soil gradation, plasticity and mineralogy. Following a comprehensive statistical analysis of the data, no meaningful correlation was found for β_M. However, it was observed that $|\beta_D|$ systematically decreases with increasing the soil activity index (i.e., $|\beta_D| \sim A^{-1}$). In other words, as the soil's principal clay mineral becomes more active (e.g., kaolinite to montmorillonite), the rate of reduction in the MDUW (with respect to increasing TDA content) decreases. Figure 5 illustrates the variations of $|\beta_D|$ against the activity index for the compiled database (excluding datasets D1, D9 and D10 for which the clay contents were not reported). As demonstrated in this figure, β_D can be expressed as follows (for $f_T \leq 30\%$):

$$\beta_D = 0.269 \ln\left(\frac{\text{PI}}{f_{\text{clay}}}\right) - 0.311 \tag{11}$$

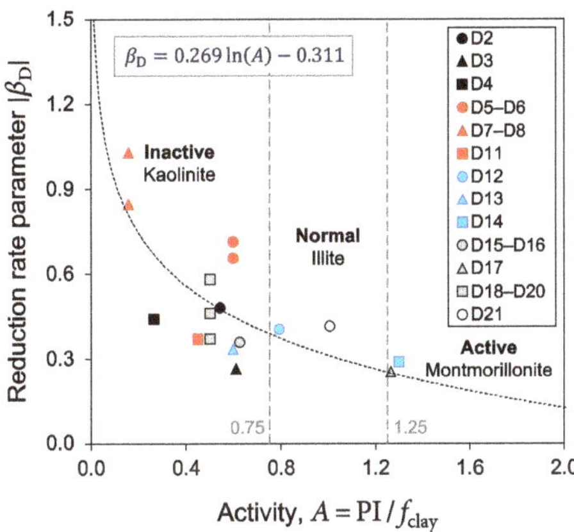

Figure 5. Variations of $|\beta_D|$ against the activity index for the compiled database of soil–TDA compaction tests (excluding datasets D1, D9 and D10 for which the soil activity index could not be calculated due to non-reporting of their clay contents).

Accordingly, substituting Equation (11) into Equation (5) leads to the following new relationship for the MDUW:

$$\gamma_{\text{dmax}}^{\text{ST}} = \gamma_{\text{dmax}}^{\text{S}} \left(\frac{G_s^{\text{S}}}{G_s^{\text{ST}}}\right)^{[0.269 \ln\left(\frac{\text{PI}}{f_{\text{clay}}}\right) - 0.311]} \tag{12}$$

Figure 6a illustrates the variations of predicted (by Equation (12)) against measured MDUW values for the compiled database. The R^2, MAPE and NRMSE for these new predictions were calculated as 0.936, 1.8% and 2.3%, respectively; corroborating the predictive capability of the newly proposed Equation (12). The 95% upper and lower agreement limits, as shown in Figure 6b, were obtained as UAL = +0.67 kN/m^3 and LAL = −0.71 kN/m^3, indicating that 95% of the MDUW predictions are associated with errors ranging between these two small unit weight limits. Note that Equations (10) and (12) were developed based on different dataset sizes (i.e., 21 and 18 datasets, respectively); as such, their predictive performances cannot be directly compared. However, a reliable comparison can be performed if the R^2, MAPE, NRMSE, UAL and LAL parameters for Equation (10) are recalculated based on the same 18 datasets (i.e., D2–D8 and D11–D21) used for the development of Equation (12). The outcome of this recalculation was R^2 = 0.920, MAPE = 2.5%,

NRMSE = 3.1%, UAL = +0.69 kN/m^3 and LAL = -1.0 kN/m^3, which appear to work (slightly) in favor of the more elaborate Equation (12). Even so, for prediction purposes, this performance improvement may not be sufficient to justify the use of Equation (12) over the more practical Equation (10); the latter making MDUW predictions without the need for PI and f_{clay} measurements.

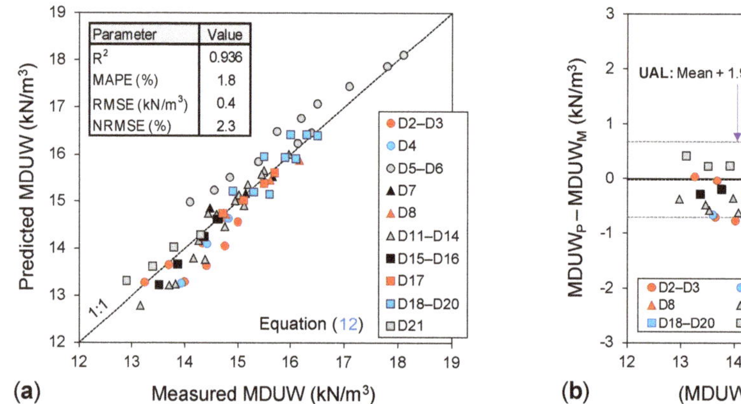

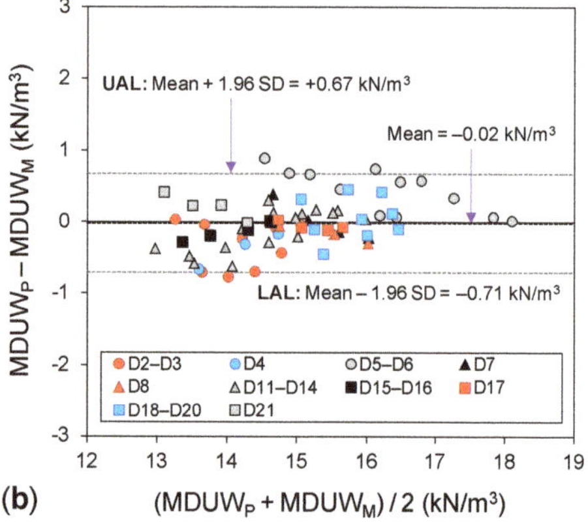

Figure 6. MDUW predictions made by Equation (12) (excluding datasets D1, D9 and D10): (**a**) Variations against measured MDUW; and (**b**) Bland–Altman plot. Note: UAL and LAL denote the 95% upper and lower agreement limits, respectively.

In addition to fundamental soil properties, the authors speculated that β_M and β_D may also be related to other variables, such as the TDA mean particle size (or D_{50}) and the imparted compaction energy level. Even though the compiled database did not permit a critical investigation of these variables to be performed (i.e., since only a small number of the database soils included compaction results for varying D_{50} and/or compaction energy levels), it is considered that both D_{50} and compaction energy would likely have minor effects on β_M and β_D. As mentioned in Section 4.1, for predominantly sand-sized TDA materials (0.075–4.75 mm), changes in the TDA mean particle size is normally reflected in the TDA specific gravity [24,26]. In other words, the dependent/input variable G_s^S/G_s^{ST} not only captures the combined effects of TDA content and TDA density, but it is also expected to account, at least in part, for changes in the TDA mean particle size. Moreover, a review of the admittedly limited literature (including those listed in Tables 1 and 2) indicates that, for TDA materials (i.e., powder, crumbs and buffings) having the same specific gravity but different D_{50} values, the variations in OMC and MDUW across the two TDA sizes are relatively small.

The elastic-rebound recovery exhibited by TDA particles in compacted soil–TDA mixtures has been reported to increase with increasing the compactive effort. In other words, the higher the imparted compaction energy level (from standard to modified Proctor), the lower the compaction efficiency of the soil–TDA matrix [25,41]. This may explain the limited soil–TDA compaction data reported for the modified (heavy) Proctor energy level (accounting for only four of the twenty-one cases listed in Tables 1 and 2). In view of this mechanism, it is speculated that the beneficial effects of compaction energy increase (from standard to modified Proctor) would likely be offset by the TDA material's increased energy dissipation potential, allowing one to postulate that the reduction rate parameters, particularly β_D, may not be significantly influenced by compactive effort. As such, the modeling framework proposed in this investigation allocates similar OMC and MDUW reduction rates for standard and modified compaction energy levels. Given that the bulk of the compiled database used for model development consisted of standard Proctor compaction data (17 datasets out of 21 examined), the predictions made for modified Proctor should be taken with some caution. Nevertheless, a systematically controlled test program involving a variety of TDA particle sizes and a range of compaction energy levels should be performed with the dual aims of checking the above postulations and potentially developing improved empirical correlations for the reduction rate parameters β_M and β_D.

5. Summary and Conclusions

This study aimed at modeling the compaction characteristics of fine-grained soils blended with sand-sized TDA products (e.g., powder, crumbs and buffings). Model development and calibration were carried out using a large and diverse database of 100 soil–TDA compaction tests (with $f_T \leq 30\%$) assembled from the literature. Following a comprehensive statistical analysis of the data, the following general and fundamental conclusions can be drawn from this study:

- Irrespective of the imparted compaction energy level (from standard to modified Proctor), the addition of (and content increase in) TDA leads to notable reductions in the OMC and MDUW parameters, both following exponentially decreasing trends with respect to increasing TDA content.
- The OMC and MDUW for soil–TDA blends (across different fine-grained soil types, TDA particle sizes and compaction energy levels) can be expressed as universal power functions of the OMC and MDUW of the unamended soil, together with the soil to soil–TDA specific gravity ratio; the latter capable of capturing the combined effects of TDA content and its lower density.
- Making use of the Bland–Altman analysis, the 95% upper and lower (water content) agreement limits between the predicted and measured OMC values were, respectively, obtained as +1.09% and −1.23%, both of which can be considered negligible for practical applications. For the MDUW predictions, these limits were calculated as +0.80 and −1.04 kN/m^3 and, employing a more elaborate correlation that also considers the soil activity, as +0.67 and −0.71 kN/m^3, which (like the OMC) can be deemed acceptable for prediction purposes. Accordingly, having established the OMC and MDUW of the unamended fine-grained soil, the various empirical models proposed in this study offer a practical procedure towards predicting the compaction characteristics of the soil–TDA blends without the hurdles of performing separate laboratory compaction tests, and thus can be used for preliminary design assessments and/or soil–TDA optimization studies.

Further investigations are warranted regarding the possible application of the new modeling framework developed for fine-grained soil–TDA blends in predicting the compaction characteristics of fine-grained soils when mixed with other recycled solid waste and/or virgin materials.

Author Contributions: Conceptualization, A.S. and M.A.; methodology, A.S. and M.A.; validation, A.S., M.A. and B.C.O.; formal analysis, A.S. and M.A.; investigation, A.S. and M.A.; writing—original draft preparation, A.S. and M.A.; writing—review and editing, A.S. and B.C.O.; visualization, A.S. and M.A.; supervision, A.S.; funding acquisition, B.C.O. All authors have read and agreed to the published version of the manuscript.

Funding: This research received no external funding.

Data Availability Statement: This study has not generated new experimental data.

Acknowledgments: Special thanks go to Mark B. Jaksa of the University of Adelaide for valuable suggestions to the authors.

Conflicts of Interest: The authors declare no conflict of interest.

Abbreviations

BA	Bland–Altman (analysis/plot)
BS	British Standard
BTEX	Benzene, Toluene, Ethylbenzene and Xylene
CH	Clay with high plasticity
CI	Clay with intermediate plasticity
CV	Clay with very high plasticity
ELT	End-of-life tire
LB	Lower (variation) boundary
MDUW	Maximum dry unit weight
MH	Silt with high plasticity
ML	Silt with low plasticity
MV	Silt with very high plasticity
OMC	Optimum moisture content
SP	Poorly-graded (sand)
TDA	Tire-derived aggregate
UB	Upper (variation) boundary
USCS	Unified Soil Classification System
ZAV	Zero-air-voids

Notations

A	Soil activity index
C_C	Coefficient of curvature
C_U	Coefficient of uniformity
D_{50}	TDA mean particle size (mm)
f_{clay}	Clay ($<2\ \mu m$) content (%)
f_{fines}	Fines ($<75\ \mu m$) content (%)
f_T	TDA content (i.e., TDA-to-soil dry mass ratio) (%)
G_s^S	Specific gravity of soil solids
G_s^T	Specific gravity of TDA particles
G_s^{ST}	Specific gravity of soil–TDA mixture
LAL	Lower (statistical) agreement limit (same unit as OMC or MDUW)
LL	Liquid limit (%)
MAPE	Mean absolute percentage error (%)
$MDUW_M$	Measured MDUW (kN/m^3)
$MDUW_P$	Predicted MDUW (kN/m^3)
M_S	Mass of oven-dried soil (g)
M_T	Mass of TDA (g)
n	Index of summation
N	Number of observations (or compaction tests)
NRMSE	Normalized root-mean-squared error (%)
OMC_M	Measured OMC (%)
OMC_P	Predicted OMC (%)
PI	Plasticity index (%)
R^2	Coefficient of determination
RMSE	Root-mean-squared error (same unit as OMC or MDUW)
SD	Standard deviation (same unit as OMC or MDUW)
UAL	Upper (statistical) agreement limit (same unit as OMC or MDUW)
w_{opt}	Conventional OMC (%)
w_{opt}^*	Normalized OMC (%)
w_{opt}^S	OMC of unamended soil (%)
w_{opt}^{ST}	OMC of soil–TDA mixture (%)
y_n	Measured variable (OMC or MDUW)
$\overline{y_n}$	Arithmetic mean of y_n data
$\hat{y}_n$	Predicted variable (OMC or MDUW)
β_D	Reduction rate parameter for increasing TDA content (MDUW model)

β_M	Reduction rate parameter for increasing TDA content (OMC model)
γ_{dmax}	Conventional MDUW (kN/m^3)
γ^*_{dmax}	Normalized MDUW (kN/m^3)
γ^S_{dmax}	MDUW of unamended soil (kN/m^3)
γ^{ST}_{dmax}	MDUW of soil–TDA mixture (kN/m^3)

Appendix A

The specific gravity values of the fine-grained soil–TDA mixtures—that is, G_s^{ST} calculated by Equation (3)—for the compiled database are presented in Table A1.

Table A1. Specific gravity values of the fine-grained soil–TDA mixtures for the compiled database.

Test ID	Dataset	Source	Soil ID	G_s^S	G_s^T	f_T (%)	G_s^{ST}	G_s^S/G_s^{ST}
T1						0	2.61	1
T2						5.3	2.44	1.070
T3	D1	Cabalar et al. [15]	S1	2.61	1.08	11.1	2.29	1.140
T4						17.6	2.15	1.214
T5						0	2.65	1
T6						5.0	2.41	1.100
T7	D2	Prasad et al. [36]	S2	2.65	0.85	10.0	2.22	1.194
T8						15.0	2.08	1.274
T9						20.0	1.96	1.352
T10						0	2.65	1
T11						5.0	2.41	1.100
T12	D3	Prasad et al. [36]	S3	2.65	0.85	10.0	2.22	1.194
T13						15.0	2.08	1.274
T14						20.0	1.96	1.352
T15						0	2.72	1
T16						5.0	2.55	1.067
T17	D4	Ramirez et al. [37]	S4	2.72	1.12	10.0	2.41	1.129
T18						20.0	2.20	1.236
T19						0	2.69	1
T20						2.5	2.60	1.035
T21						5.0	2.53	1.063
T22	D5	Signes et al. [16]	S5	2.69	1.14	10.0	2.39	1.126
T23						15.0	2.28	1.180
T24						20.0	2.19	1.228
T25						25.0	2.11	1.275
T26						0	2.69	1
T27						2.5	2.60	1.035
T28						5.0	2.53	1.063
T29	D6	Signes et al. [16]	S5	2.69	1.14	10.0	2.39	1.126
T30						15.0	2.28	1.180
T31						20.0	2.19	1.228
T32						25.0	2.11	1.275
T33						0	2.69	1
T34						2.5	2.60	1.035
T35	D7	Yadav and Tiwari [43]	S6	2.69	1.13	5.0	2.52	1.067
T36						7.5	2.45	1.098
T37						10.0	2.39	1.126
T33						0	2.69	1
T38						2.5	2.59	1.039
T39	D8	Yadav and Tiwari [38]	S6	2.69	1.07	5.0	2.51	1.072
T40						7.5	2.43	1.107
T41						10.0	2.36	1.140

Table A1. *Cont.*

Test ID	Dataset	Source	Soil ID	G_s^S	G_s^T	f_T (%)	G_s^{ST}	G_s^S/G_s^{ST}
T42						0	2.65	1
T43						5.0	2.41	1.100
T44	D9	Ravichandran et al. [39]	S7	2.65	0.85	10.0	2.22	1.194
T45						15.0	2.08	1.274
T46						20.0	1.96	1.352
T47						0	2.65	1
T48						5.0	2.41	1.100
T49	D10	Ravichandran et al. [39]	S8	2.65	0.85	10.0	2.22	1.194
T50						15.0	2.08	1.274
T51						20.0	1.96	1.352
T52						0	2.69	1
T53						5.0	2.51	1.072
T54	D11	Soltani et al. [34]	S9	2.69	1.09	10.0	2.37	1.135
T55						20.0	2.16	1.245
T56						30.0	2.01	1.338
T57						0	2.67	1
T58						5.0	2.50	1.068
T59	D12	Soltani et al. [34]	S10	2.67	1.09	10.0	2.36	1.131
T60						20.0	2.15	1.242
T61						30.0	2.00	1.335
T62						0	2.71	1
T63						5.0	2.53	1.071
T64	D13	Soltani et al. [34]	S11	2.71	1.09	10.0	2.39	1.134
T65						20.0	2.17	1.249
T66						30.0	2.02	1.342
T67						0	2.72	1
T68						5.0	2.54	1.071
T69	D14	Soltani et al. [34]	S12	2.72	1.09	10.0	2.39	1.138
T70						20.0	2.18	1.248
T71						30.0	2.02	1.347
T72						0	2.73	1
T73						5.0	2.55	1.071
T74	D15	Soltani et al. [35]	S13	2.73	1.09	10.0	2.40	1.138
T75						20.0	2.18	1.252
T76						30.0	2.03	1.345
T72						0	2.73	1
T77						5.0	2.55	1.071
T78	D16	Soltani et al. [35]	S13	2.73	1.09	10.0	2.40	1.138
T79						20.0	2.18	1.252
T80						30.0	2.03	1.345
T81						0	2.76	1
T82						5.0	2.57	1.074
T83	D17	Soltani et al. [23]	S14	2.76	1.09	10.0	2.42	1.140
T84						20.0	2.20	1.255
T85						30.0	2.04	1.353
T86						0	2.77	1
T87						5.0	2.58	1.074
T88	D18	Soltani et al. [19]	S15	2.77	1.08	10.0	2.43	1.140
T89						20.0	2.20	1.259
T86						0	2.77	1
T90						5.0	2.58	1.074
T91	D19	Soltani et al. [19]	S15	2.77	1.10	10.0	2.43	1.140
T92						20.0	2.21	1.253

Table A1. *Cont.*

Test ID	Dataset	Source	Soil ID	G_s^S	G_s^T	f_T (%)	G_s^{ST}	G_s^S/G_s^{ST}
T86						0	2.77	1
T93						5.0	2.59	1.069
T94	D20	Soltani et al. [19]	S15	2.77	1.11	10.0	2.44	1.135
T95						20.0	2.22	1.248
T96						0	2.73	1
T97						5.0	2.55	1.071
T98	D21	Soltani et al. [22]	S16	2.73	1.09	10.0	2.40	1.138
T99						20.0	2.18	1.252
T100						30.0	2.03	1.345

Note: f_T = TDA content (i.e., TDA-to-soil dry mass ratio, here expressed as a percentage value); G_s^S and G_s^T = specific gravity of soil solids and TDA particles, respectively; and G_s^{ST} = specific gravity of soil–TDA mixture (Equation (3)).

Figure A1 illustrates the variations of the conventional OMC and MDUW parameters, w_{opt} and γ_{dmax}, against TDA content f_T for the 21 fine-grained soil–TDA compaction datasets.

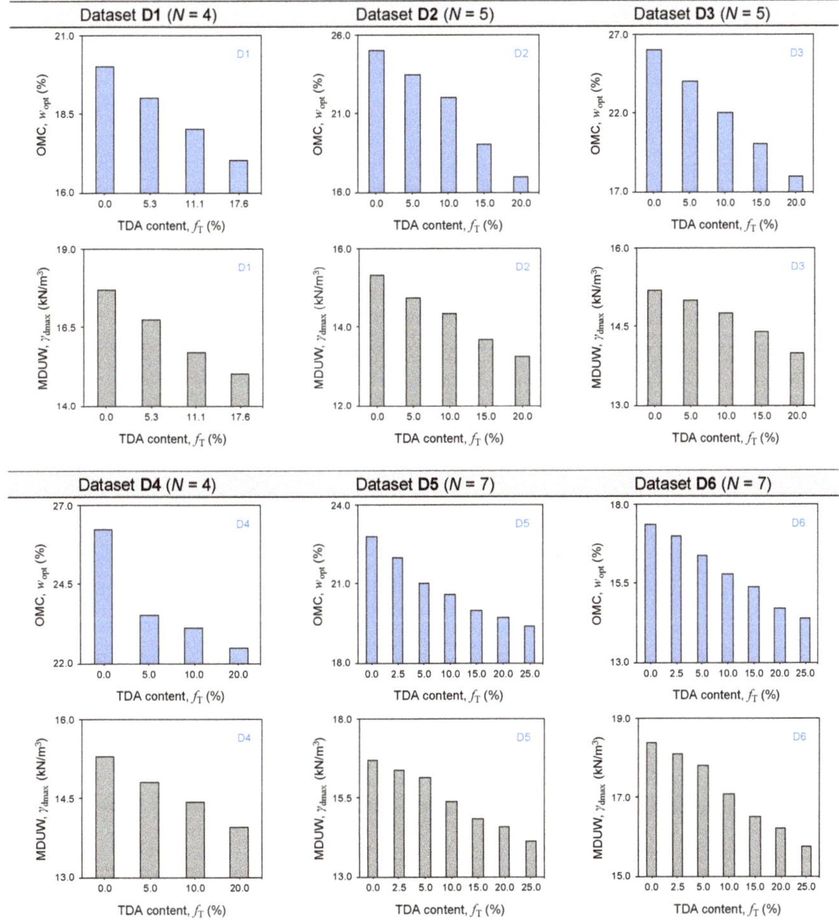

Figure A1. *Cont.*

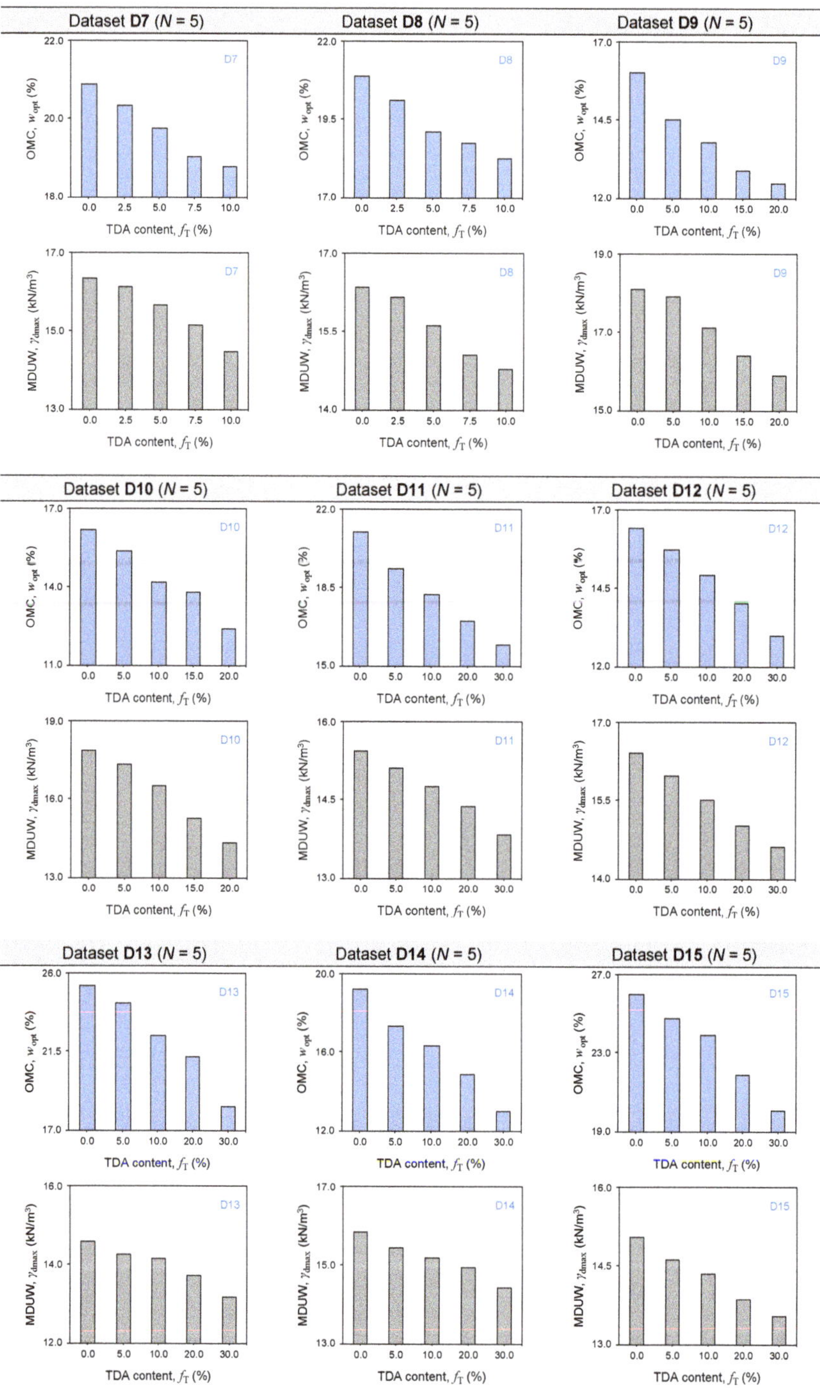

Figure A1. *Cont.*

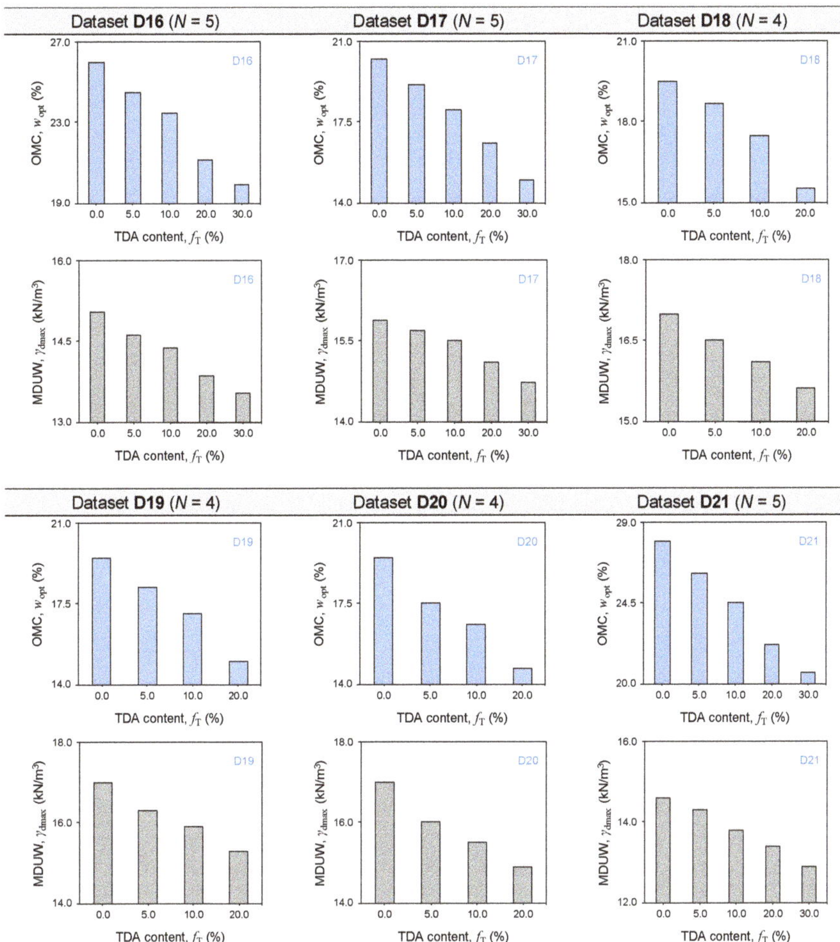

Figure A1. Variations of the conventional OMC and MDUW against TDA content for the 21 soil–TDA compaction datasets.

References

1. Humphrey, D.; Blumenthal, M. The use of tire-derived aggregate in road construction applications. In *Green Streets and Highways 2010: An Interactive Conference on the State of the Art and How to Achieve Sustainable Outcomes*; Weinstein, N., Ed.; ASCE: Reston, VA, USA, 2010; pp. 299–313. [CrossRef]
2. Reddy, S.B.; Krishna, A.M. Recycled tire chips mixed with sand as lightweight backfill material in retaining wall applications: An experimental investigation. *Int. J. Geosynth. Gr. Eng.* **2015**, *1*, 31. [CrossRef]
3. Li, L.; Xiao, H.; Ferreira, P.; Cui, X. Study of a small scale tyre-reinforced embankment. *Geotext. Geomembr.* **2016**, *44*, 201–208. [CrossRef]
4. Khatami, H.; Deng, A.; Jaksa, M. Passive arching in rubberized sand backfills. *Can. Geotech. J.* **2020**, *57*, 549–567. [CrossRef]
5. Raeesi, R.; Soltani, A.; King, R.; Disfani, M.M. Field performance monitoring of waste tire-based permeable pavements. *Transp. Geotech.* **2020**, *24*, 100384. [CrossRef]
6. Shahrokhi-Shahraki, R.; Kwon, P.S.; Park, J.; O'Kelly, B.C.; Rezania, S. BTEX and heavy metals removal using pulverized waste tires in engineered fill materials. *Chemosphere* **2020**, *242*, 125281. [CrossRef] [PubMed]
7. Edil, T.; Bosscher, P. Engineering properties of tire chips and soil mixtures. *Geotech. Test. J.* **1994**, *17*, 453–464. [CrossRef]
8. Foose, G.J.; Benson, C.H.; Bosscher, P.J. Sand reinforced with shredded waste tires. *J. Geotech. Eng.* **1996**, *122*, 760–767. [CrossRef]
9. Al-Tabbaa, A.; Blackwell, O.; Porter, S.A. An investigation into the geotechnical properties of soil–tyre mixtures. *Environ. Technol.* **1997**, *18*, 855–860. [CrossRef]

10. Lee, J.H.; Salgado, R.; Bernal, A.; Lovell, C.W. Shredded tires and rubber–sand as lightweight backfill. *J. Geotech. Geoenviron. Eng.* **1999**, *125*, 132–141. [CrossRef]

11. Zornberg, J.G.; Cabral, A.R.; Viratjandr, C. Behaviour of tire shred–sand mixtures. *Can. Geotech. J.* **2004**, *41*, 227–241. [CrossRef]

12. Lee, J.S.; Dodds, J.; Santamarina, J.C. Behavior of rigid–soft particle mixtures. *J. Mater. Civ. Eng.* **2007**, *19*, 179–184. [CrossRef]

13. Kim, H.K.; Santamarina, J.C. Sand–rubber mixtures (large rubber chips). *Can. Geotech. J.* **2008**, *45*, 1457–1466. [CrossRef]

14. Mohammadinia, A.; Disfani, M.M.; Narsilio, G.A.; Aye, L. Mechanical behaviour and load bearing mechanism of high porosity permeable pavements utilizing recycled tire aggregates. *Constr. Build. Mater.* **2018**, *168*, 794–804. [CrossRef]

15. Cabalar, A.F.; Karabash, Z.; Mustafa, W.S. Stabilising a clay using tyre buffings and lime. *Road Mater. Pavement Des.* **2014**, *15*, 872–891. [CrossRef]

16. Signes, C.H.; Garzón-Roca, J.; Fernández, P.M.; de la Torre, M.E.G.; Franco, R.I. Swelling potential reduction of Spanish argillaceous marlstone Facies Tap soil through the addition of crumb rubber particles from scrap tyres. *Appl. Clay Sci.* **2016**, *132–133*, 768–773. [CrossRef]

17. Yadav, J.S.; Tiwari, S.K. Effect of waste rubber fibres on the geotechnical properties of clay stabilized with cement. *Appl. Clay Sci.* **2017**, *149*, 97–110. [CrossRef]

18. Bekhiti, M.; Trouzine, H.; Rabehi, M. Influence of waste tire rubber fibers on swelling behavior, unconfined compressive strength and ductility of cement stabilized bentonite clay soil. *Constr. Build. Mater.* **2019**, *208*, 304–313. [CrossRef]

19. Soltani, A.; Taheri, A.; Deng, A.; O'Kelly, B.C. Improved geotechnical behavior of an expansive soil amended with tire-derived aggregates having different gradations. *Minerals* **2020**, *10*, 923. [CrossRef]

20. Akbarimehr, D.; Fakharian, K. Dynamic shear modulus and damping ratio of clay mixed with waste rubber using cyclic triaxial apparatus. *Soil Dyn. Earthq. Eng.* **2021**, *140*, 106435. [CrossRef]

21. Ghadr, S.; Samadzadeh, A.; Bahadori, H.; O'Kelly, B.C.; Assadi-Langroudi, A. Liquefaction resistance of silty sand with ground rubber additive. *Int. J. Geomech.* **2021**, *21*, 04021076. [CrossRef]

22. Soltani, A.; Raeesi, R.; Taheri, A.; Deng, A.; Mirzababaei, M. Improved shear strength performance of compacted rubberized clays treated with sodium alginate biopolymer. *Polymers* **2021**, *13*, 764. [CrossRef]

23. Soltani, A.; Deng, A.; Taheri, A.; O'Kelly, B.C. Engineering reactive clay systems by ground rubber replacement and polyacrylamide treatment. *Polymers* **2019**, *11*, 1675. [CrossRef]

24. Yadav, J.S.; Tiwari, S.K. The impact of end-of-life tires on the mechanical properties of fine-grained soil: A review. *Environ. Dev. Sustain.* **2019**, *21*, 485–568. [CrossRef]

25. Yang, Z.; Zhang, Q.; Shi, W.; Lv, J.; Lu, Z.; Ling, X. Advances in properties of rubber reinforced soil. *Adv. Civ. Eng.* **2020**, *2020*, 6629757. [CrossRef]

26. Mistry, M.K.; Shukla, S.J.; Solanki, C.H. Reuse of waste tyre products as a soil reinforcing material: A critical review. *Environ. Sci. Pollut. Res.* **2021**, *28*, 24940–24971. [CrossRef]

27. Gurtug, Y.; Sridharan, A. Compaction behaviour and prediction of its characteristics of fine grained soils with particular reference to compaction energy. *Soils Found.* **2004**, *44*, 27–36. [CrossRef]

28. Horpibulsuk, S.; Katkan, W.; Apichatvullop, A. An approach for assessment of compaction curves of fine grained soils at various energies using a one point test. *Soils Found.* **2008**, *48*, 115–125. [CrossRef]

29. Sivrikaya, O.; Togrol, E.; Kayadelen, C. Estimating compaction behavior of fine-grained soils based on compaction energy. *Can. Geotech. J.* **2008**, *45*, 877–887. [CrossRef]

30. Pillai, G.A.S.; Vinod, P.P. Re-examination of compaction parameters of fine-grained soils. *Proc. Inst. Civ. Eng. Gr. Improv.* **2016**, *169*, 157–166. [CrossRef]

31. Spagnoli, G.; Shimobe, S. An overview on the compaction characteristics of soils by laboratory tests. *Eng. Geol.* **2020**, *278*, 105830. [CrossRef]

32. Di Matteo, L.; Spagnoli, G. Predicting compaction properties of soils at different compaction efforts. *Proc. Inst. Civ. Eng. Geotech. Eng.* **2021**, in press. [CrossRef]

33. Shivaprakash, S.H.; Sridharan, A. Correlation of compaction characteristics of standard and reduced Proctor tests. *Proc. Inst. Civ. Eng. Geotech. Eng.* **2021**, *174*, 170–180. [CrossRef]

34. Soltani, A.; Deng, A.; Taheri, A.; Sridharan, A. Consistency limits and compaction characteristics of clay soils containing rubber waste. *Proc. Inst. Civ. Eng. Geotech. Eng.* **2019**, *172*, 174–188. [CrossRef]

35. Soltani, A.; Deng, A.; Taheri, A.; Sridharan, A. Swell–shrink–consolidation behavior of rubber-reinforced expansive soils. *Geotech. Test. J.* **2019**, *42*, 761–788. [CrossRef]

36. Prasad, A.S.; Ravichandran, P.T.; Annadurai, R.; Rajkumar, P.R.K. Study on effect of crumb rubber on behavior of soil. *Int. J. Geomatics Geosci.* **2014**, *4*, 579–584.

37. Ramirez, G.G.D.; Casagrande, M.D.T.; Folle, D.; Pereira, A.; Paulon, V.A. Behavior of granular rubber waste tire reinforced soil for application in geosynthetic reinforced soil wall. *Rev. IBRACON Estrut. Mater.* **2015**, *8*, 567–576. [CrossRef]

38. Yadav, J.S.; Tiwari, S.K. A study on the potential utilization of crumb rubber in cement treated soft clay. *J. Build. Eng.* **2017**, *9*, 177–191. [CrossRef]

39. Ravichandran, P.T.; Priyanga, G.; Krishnan, K.D.; Rajkumar, P.R.K. Compaction behavior of rubberized soil. In *Proceedings of the International Conference on Intelligent Computing and Applications—Advances in Intelligent Systems and Computing, Velammal Engineering College, Chennai, India, 2–3 February 2018*; Bhaskar, M., Dash, S., Das, S., Panigrahi, B., Eds.; Springer: Singapore, 2019; Volume 846, pp. 129–134. [CrossRef]
40. BS 5930. *Code of Practice for Ground Investigations*; British Standards Institution (BSI): London, UK, 2015; ISBN 9780539081350.
41. Özkul, Z.H.; Baykal, G. The influence of energy level on the effectiveness of compaction of clays with rubber fiber added. In *Proceedings of the Transportation Research Board 86th Annual Meeting, Washington, DC, USA, 21–25 January 2007*; Transportation Research Board (TRB): Washington, DC, USA, 2007; Paper No. 07-1685.
42. Al-Tabbaa, A.; Aravinthan, T. Natural clay–shredded tire mixtures as landfill barrier materials. *Waste Manag.* **1998**, *18*, 9–16. [CrossRef]
43. Priyadarshee, A.; Gupta, D.; Kumar, V.; Sharma, V. Comparative study on performance of tire crumbles with fly ash and kaolin clay. *Int. J. Geosynth. Gr. Eng.* **2015**, *1*, 38. [CrossRef]
44. Akbarimehr, D.; Eslami, A.; Aflaki, E. Geotechnical behaviour of clay soil mixed with rubber waste. *J. Clean. Prod.* **2020**, *271*, 122632. [CrossRef]
45. Trouzine, H.; Bekhiti, M.; Asroun, A. Effects of scrap tyre rubber fibre on swelling behaviour of two clayey soils in Algeria. *Geosynth. Int.* **2012**, *19*, 124–132. [CrossRef]
46. Soltani, A.; O'Kelly, B.C. Discussion of "The flow index of clays and its relationship with some basic geotechnical properties" by G. Spagnoli, M. Feinendegen, L. Di Matteo, and D. A. Rubinos, published in Geotechnical Testing Journal 42, no. 6 (2019): 1685–1700. *Geotech. Test. J.* **2021**, *44*, 216–219. [CrossRef]
47. Soltani, A.; O'Kelly, B.C. Reappraisal of the ASTM/AASHTO standard rolling device method for plastic limit determination of fine-grained soils. *Geosciences* **2021**, *11*, 247. [CrossRef]
48. Bland, J.M.; Altman, D.G. Measuring agreement in method comparison studies. *Stat. Methods Med. Res.* **1999**, *8*, 135–160. [CrossRef]

 *sustainability*

Article

Effect of Recycled Polyethylene Terephthalate Strips on the Mechanical Properties of Cement-Treated Lateritic Sandy Soil

Maitê Rocha Silveira [1], Paulo César Lodi [1,*] Natália de Souza Correia [2], Roger Augusto Rodrigues [1] and Heraldo Luiz Giacheti [1]

[1] Department of Civil and Environmental Engineering, São Paulo State University (UNESP), Av. Engenheiro Luiz Edmundo Carrijo Coube 14-01, Bauru, SP 17033-360, Brazil; maite81@hotmail.com (M.R.S.); roger.rodrigues@unesp.br (R.A.R.); h.giacheti@unesp.br (H.L.G.)

[2] Department of Civil Engineering, Federal University of Sao Carlos (UFSCar), Rodovia Washington Luiz, São Carlos, SP 17033-360, Brazil; ncorreia@ufscar.br

* Correspondence: paulo.lodi@unesp.br; Tel.: +1-646-755-2239

Received: 23 October 2020; Accepted: 13 November 2020; Published: 24 November 2020

Abstract: The civil engineering construction industry is nowadays one of the largest consumers of natural resources. Therefore, the proposal of using alternative materials that seek to reduce waste production or the use of previously generated waste is becoming increasingly necessary. This paper evaluated the effect of recycled polyethylene terephthalate (PET) strips on the mechanical properties of a cement-treated lateritic sandy soil. Unconfined compression strength (UCS) tests were conducted in natural and PET strips mixtures in different strips lengths and contents. In addition to UCS tests, compaction tests were also conducted in order to analyze the effect of these inclusions on the properties of a lateritic sandy soil. Lastly, direct shear tests were conducted on natural soil-strip, soil-cement, and soil-cement-strip composites using optimum UCS results. The addition of strips to the soil-cement composite showed an increase in the soil cohesion parameter. The inclusion of strips also provided a more ductile behavior to the soil, presenting greater deformations with fewer stress peaks. Results showed that the recycled strips' inclusion in soil-cement can provide a material with high strength, ductility, and a highly sustainable alternative.

Keywords: recycled pet strips; lateritic soil; cement; composite; uniaxial tests; shear strength

1. Introduction

Nowadays, the visible consequences of environmental degradation and the prediction of future environmentally catastrophic scenarios require drastic solutions for environmental conservation. Among them is the integrated management of solid wastes, in which a set of actions is proposed in order to promote sustainability and the preservation of natural resources through three main actions: reducing consumption, reusing consumed materials, and recycling generated waste.

The use of alternative materials in the civil engineering construction industry aiming to reduce the production of wastes or the use of previously generated wastes has become indispensable. In the context of the reduction of natural resource use in civil construction, such as soils, different materials appear as alternative options to compose soil-mixtures: synthetic or natural fibers, construction and demolition wastes, ashes or tire fibers. Another sustainable alternative is the addition of polymers, using previously generated wastes, such as plastic strips.

The study entitled "Fast facts about plastic pollution" by National Geographic written by Laura Parker [1] exposes the worrying situation of plastic in the world, showing the growing need for recycling and reusing this material. According to the study, 40% of plastic produced is packaging, used only once and then discarded. Worldwide, 448 million tons of plastic have been produced since the beginning of this materials manufacturing, of which 44% has been made since 2000. Currently, less than a fifth of all plastic is globally recycled. In Europe, plastic recycling rates are higher than 30 percent, while in China, the rate of plastic recycling is 25 percent. In the United States, plastic recycling is only 9 percent of total plastic trash.

After numerous studies indicating a high potential for the application of polymeric fiber reinforced-composites in improving the mechanical properties of soils, recent studies started to evaluate the influence of using polymeric strips as soil reinforcement materials, e.g., References [2–6].

The main difference between the two types of inclusions is in the shape of the materials. While polymeric fibers are very thin and elongated materials, such as filaments, polymeric strips are materials of greater width and thickness, usually cut from existing plastic structures. The use of Polyethylene Terephthalate (PET) strips as soil improvement has several advantages, such as the possibility of reusing plastic waste to increase soil strength without the need of a recycling process, as in the case of synthetic fibers.

Sivakumar Babu and Chouksey [2] evaluated the effect of including strips 12 mm long and 4 mm wide, in quantities of 0.50%, 0.75% and 1.0% in a sandy soil through unconfined strength tests and triaxial tests (consolidated and not drained). The authors noted significant increases in the soil shear strength parameters (cohesion and internal friction angle), which were greater for greater amounts of strips added. In addition, unconfined strength tests indicated an increase in ductility, proportional to the inclusion of strips. Soltani-Jigheh [3] studied the inclusion of plastic strips 4 mm wide and 8 mm long in quantities of 0.25; 0.50; 0.75; 1.0; 1.5 and 2% in relation to the mass of a clayey soil by performing triaxial tests (consolidated and not drained). The results showed a small increase in the shear strength of the soil. In general, changes obtained in the shear strength of soils were also small, resulting, in general, in an increase of cohesion and decrease of the friction angle.

Studies have also addressed the use of cement or lime to improve soil properties. The large number of studies that evaluated the use of soil-cement mixtures may be justified by the characteristics obtained with this composite, which presents a significant increase in natural soil strength and stiffness [6–17]. Although cement is considered a high-environmental-impact material, these studies aim to obtain a material with high strength and durability from the inclusion of small amounts of cement, reducing the use of other polluting materials. Specht et al. [9] suggest that a cement-treated soil usually shows an increase in soil strength and stiffness, turning this mixture into an ideal material for several geotechnical applications, such as the base of shallow foundations, slope protection and base/sub-base of flexible pavements. However, great brittleness and high cracking potential have discouraged the use of such material in pavement engineering. In this sense, the addition of fibers and strips in a soil-cement composite can provide a material with high strength and more ductile behavior.

Specht [9] studied the behavior of soil-cement-fiber mixtures submitted to static and dynamic loads aiming at paving. The author stated that the influence of the inclusion of polypropylene fibers on the properties of the composite essentially depends on the fiber and matrix. Still, according to Specht [9], more flexible fibers showed a more pronounced effect on post-peak behavior, increasing ductility, toughness, and resistance to fatigue, while fibers with greater stiffness pronounced the effect of increasing peak strength. In addition, longer fibers were more effective.

Guedes [14] analyzed the mechanical performance of a soil-cement micro reinforced with synthetic polypropylene fibers for use as a primary coating on unpaved roads. The study incorporated fibers of 6 and 24 mm lengths into the soil-cement in proportions of 0.25%, 0.50%, and 0.75%. The incorporation of fibers into the composite proved to be satisfactory, increasing the peak strength and the drop in strength after the peak. Other factors that were significantly affected by the inclusion are the increase in break deformations, decrease in stiffness, increase in elastic and plastic deformations, and reduction in deformability modules and strength to compaction. The 24 mm fiber incorporated into the soil at 0.75% content proved to be the most influential combination.

Girardello [15] studied the behavior of pullout tests of embedded plates in soil-cement-fiber layers using polymeric polypropylene fibers. The results indicated an increase in the strength required for the removal of the embedded sand-cement, sand-cement-fiber, and sand-fiber plates when compared to the removal of the embedded sand plates. Changes were observed in the form of soil rupture when reinforced with fibers and/or cement.

Cristelo et al. [12] analyzed microscopic images of the soil in its pure state, with the addition of fibers, cement, and with the addition of cement and fibers together. The authors observed the soil structure and concluded that there is an increase in the void index with the addition of fibers, which are responsible for a loss of mechanical strength, resulting from the friction between the particles. The addition of cement, unlike what occurs with the addition of fibers, is responsible for decreasing the void index of the composite, since it acts as a binder. In the case of the mixture of soil, cement and fibers, cement acts by reducing voids and increasing the bond strength at the soil-fiber interface, providing improvement in the mechanical strength of the mixture.

Regarding the inclusion of strips, Olutaiwo and Ezegbunem [16] evaluated the effect of cement and PET bottle strips in a lateritic soil using Modified Proctor and CBR tests. The amount of cement varied between 0%, 1%, 3%, 5%, and 7%, and the number of strips (5 mm wide and 10 mm long) in 0%, 5%, 10%, 15% and 20%. Results showed a decrease in the optimum humidity and an increase in the maximum dry density, which were shown to be greater in the inclusion of 10% of strips for all the cement quantities. As for the CBR tests, all cement-strip combinations evaluated increased the soil's bearing capacity. In all cement inclusions, the inclusion of 10% of strips generated the greatest increase in strength. The most effective inclusion was 7% cement with 10% strips, generating a 326.8% increase in soil support capacity. Overall, the addition of the strips to the soil-cement presented a beneficial alternative to the environment, in addition to being economical when compared to the single addition of cement.

Tang et al. [17] evaluated the inclusion of natural fiber reinforcements within cemented soils and found an increase in UCS and changes on the brittle behavior of cemented soil to a more ductile behavior. Olgun et al. [18] evaluated the effect of polypropylene (PP) fibers inclusion on the strength of cement-fly ash stabilized clay soil and found that the main advantage of fiber reinforcement was the improvement in material ductility, particularly above 0.5% fiber content, and with increased fibers length.

Therefore, the present study aims to expand the understanding of the shear strength and deformability behavior of a cement-treated lateritic soil mixture with recycled polymeric strips. Considering the high cost of the cement and the polluting potential of these materials, the recycled strips were added to a cement-treated soil seeking a low-cost material with high strength, ductility and a highly sustainable alternative. This practice seeks to assess how much strength can increase with the addition of the recycled strips in order to maximize their use and minimize the use of cement. Unconfined compression and direct shear tests were conducted using soil-strip, soil-cement, and soil-strip-cemented mixtures. A lateritic sandy soil was used in the present study for the evaluation of the effect of recycled strips and cement inclusion in soils found in tropical zones.

2. Materials and Methods

Sandy soil samples were collected from the experimental campus of the São Paulo State University (UNESP) at Bauru, in Sao Paulo State, Brazil. Soil samples were characterized according the following recommendations: particle size analysis—ABNT NBR 7181 [19], liquidity limit—ABNT NBR 6459 [20], plasticity limit—ABNT NBR 7180 [21], specific density of solids—ABNT NBR 6458 [22], unconfined compression strength—ABNT NBR 6457 [23] and ABNT 7182 [24], and direct shear—ASTM D3080 [25]. The soil was classified as a medium to fine, reddish brown, clayey sand, according to the classification adopted by ABNT NBR 6502 [26]. These materials are residual soils formed in humid tropical regions with a predominance of weathering. Lateritic soils are characterized in their formation by the intense migration of particles under the action of infiltrations and evaporations, giving rise to a porous surface horizon, remaining almost exclusively the most stable minerals (quartz, magnetite, illite and kaolinite). In these soils, the presence of aggregated clay and silt particles is common owing to the action of iron and aluminum oxides and hydroxides, which gives these soils characteristics of mechanical and hydraulic behavior not consistent with their texture. Lateritic fine-grained soils have superior properties when compacted; however, they can show unfavorable properties such as cracks, shrinkage, water sensitivity and irregular distribution [27–30]. Table 1 illustrates soil properties obtained in tests for characterization of this soil.

Table 1. Properties of the lateritic soil.

Parameter	Unit	Value
Sand	%	80.2
Silt	%	5.8
Clay	%	14.0
Liquid Limit	%	16.0
Plastic Limit	%	NP *
Maximum dry density, $\rho d_{máx}$	g/cm^3	1.950
Optimal moisture content, w_{opt}	%	10.6
Specific density of solids, ρ_s	-	2.649

* NP = Non-Plastic.

The Portland cement used in this research is type CP II-F-32, manufactured by CSN (Brazil). This cement was chosen, according to Brazilian Portland Cement Association (ABCP), due to its wide availability in Brazil. The cement was added to soil in percentages of 2, 4, 6, 8 and 10%.

Plastic wastes used in this research were composed by PET bottle. Strips were cut from Coca-Cola® plastic bottles and randomly distributed in the soil. Bottles were previously sanitized under running water. After this process, a portion of the bottle in which the label was placed was cut and separated from the rest of the packaging to form the strip sample. The sample was cut into strips with 1.5 mm width by 10, 15, 20, and 30 mm length. The percentual of strips (0.0, 0.25, 0.5, 0.75, 1.0, 1.5, and 2%) was added to soils in relation to the mass of dry soil (Figure 1). The PET strips have a specific mass of 1.30 g/cm^3, a tensile strength of 1000 MPa, and a tensile modulus of 15.0 GPa. The aspect ratio (AR) for the strips having a length of 10 mm, 15 mm, 20 mm, and 30 mm are 20, 30, 40, and 60, respectively. The cutting process of the PET strips, the final shape of the strips, and an example of soil mixed with strips are shown in Figure 1.

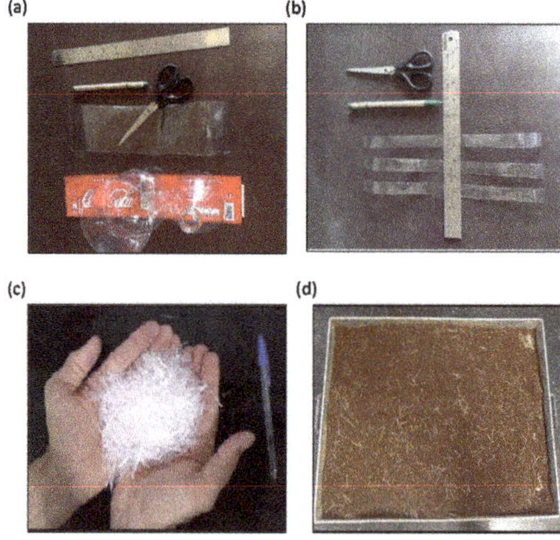

Figure 1. PET strips: (**a**) PET samples; (**b**) the cutting process; (**c**) PET strips after cutting; (**d**) soil mixed with PET strips.

Standard Proctor tests—ABNT NBR 7182 [24] were conducted with the addition of strips to the soil in order to evaluate the effect of including different sizes and percentages of strips to the soil compaction parameters (maximum dry density and optimum moisture content). Proctor tests with the addition of cement—ABNT NBR 12,023 [31] were also conducted in percentages of 2% and 10% to obtain the optimal parameters for the soil-cement mixture.

Unconfined compression tests—ABNT NBR 12,770 [32] were also conducted for natural soil and soil-strip mixtures. Due to the small variations obtained with the inclusion of strips in the compaction parameters (see item 3.1.1), a natural soil compaction curve was adopted for the molding of the specimens, which were compacted in the optimum moisture (±0.5) and with a 100% degree of compaction, with an acceptable variation of 3%. UCS tests were performed with 3 specimens for each length and respective strips content, totalizing 24 unconfined compression strength tests, allowing an analysis of the effect of the inclusion of strips in the soil in terms of strength improvement and variability.

The direct shear tests were conducted according to ASTM D3080 [25] with the addition of cement to soil in percentages of 2, 4, 6, 8, and 10%. In addition to the soil-cement tests, soil-cement-strips tests were also performed. The direct shear tests were performed with the molding of the specimens in the optimum soil content, using a previously calculated mass of soil or soil-cement according to the volume of each mold and the maximum density of the composite. The sought degree of compaction was 95% for the tests with the soil-cement and soil-cement-strip. For the tests with the soil-strip the degree of compaction of 100% was adopted for comparison with the unconfined compression test (UCS) results. The static compaction procedure was adopted, in which a metal plunger was attached to the same simple compression equipment and introduced into the metal mold of direct shear test at a constant speed, compacting the amount of soil, strip, and water previously calculated, weighed with a resolution of 0.01 gf, and homogenized to obtain maximum dry density and optimal moisture content according to compaction degree (Figure 2a,b). The specimens were then placed inside a capsule that was attached to the direct shear machine to initiate the shearing (Figure 2c,d). The specimens were densified with the use of loads of 1, 2 and 4 kg, which resulted in normal stresses of 30.56, 61.11, and 122.22 kPa, respectively. In the execution of tests with the use of soil-cement and soil-cement-strip composites, curing was carried out by placing the specimens in a humid chamber, with constant wetting for periods of 7, 14, and 28 days. The tests were carried out with the specimens of 7 days due to similarity with the results obtained considering the periods of 14 and 28 days.

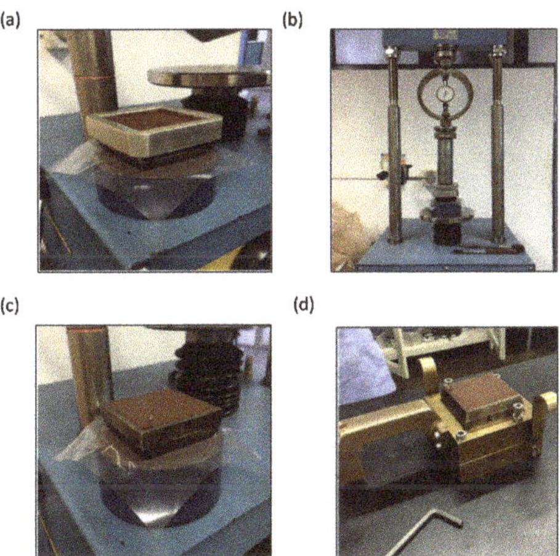

Figure 2. Molding of the specimens for the direct shear test. (**a**) Placing the soil on the mold; (**b**) compacting the soil using the machine; (**c**) the molded specimen; (**d**) placement of the specimen inside the capsule.

Considering the high cost of cement and the potential pollutant of the material, the strips were added to the sandy matrix composites with 2% cement, looking for a low-cost material, with high strength, ductility and high sustainable potential, seeking to assess how much the strength could increase with the addition of the strips so to maximize their use and minimize the use of cement. The analysis of the results was based on the analysis of the shear strength soil parameters (cohesion and internal friction angle) and the stress-displacement curves taking into account the soil, the soil-cement, and the soil-cement-strip.

3. Results and Discussion

3.1. Compaction Tests

3.1.1. Compaction Tests in Soil-Strip Mixtures

Table 2 shows the optimal compaction parameters obtained for soil-strip composites, with strip inclusions in different sizes and percentages.

Table 2. Proctor test results performed (soil-strip).

Strip Length (mm)	Inclusion (%)	Optimal Moisture Content (%)	Maximum Dry Density (g/cm^3)
	0.25	10.4	1.940
	0.50	9.4	1.945
10	0.75	10.1	1.915
	1.00	10.1	1.905
	1.50	10,4	1.915
	2.00	10.9	1.905
	0.25	10.4	1.925
	0.50	10.2	1.925
15	0.75	10.4	1.925
	1.00	10.4	1,930
	1.50	10.7	1.900
	2.00	10.9	1.905
	0.25	10.3	1.945
	0.50	10.8	1.900
20	0.75	9.6	1.925
	1.00	10.5	1.940
	1.50	10.8	1.895
	2.00	10.7	1.880
	0.25	10.2	1.940
	0.50	10.9	1.940
30	0.75	9.9	1.925
	1.00	10.6	1.930
	1.50	10.8	1.905
	2.00	10.6	1.880

In all cases analyzed it should be noted that there is a decrease in the maximum dry density obtained due to the inclusion of the strips regarding the results with the optimal parameters obtained from compaction test on natural lateritic sandy soil (Table 1). The decrease in the maximum dry density ranged from 3.58% to 0.26%. The highest specific mass was obtained for inclusions of strips in lower percentages (0.5% inclusions of 10 mm long strips and 0.25% 20 mm long strips). The lowest values were obtained for inclusions of longer length strips and in the highest percentages evaluated (20 and 30 mm long strips in additions of 2.0% in relation to the dry mass of the soil).

Given that the specific mass of PET (1.30 g/cm^3) is less than the maximum dry density of the analyzed soil (1.95 g/cm^3), it is not possible to conclude whether the addition of strips confers strength to compaction and increases the porosity of the mixture, as stated by Hoare [33] and Festugato et al. [34], or if it occurs due to the addition of a reduced specific mass material. Future evaluation of soil and soil-strip permeability is indicated to obtain more specific results.

3.1.2. Compaction Tests in Cement-Soil Composites

Table 3 shows the compaction parameters obtained for the natural lateritic sandy soil and with a 2% and 10% cement addition. The addition of cement to the soil showed no relevant variations in the soil compaction parameters, culminating in a very small variation in the maximum dry density and in a small reduction of de optimum moisture content.

Table 3. Proctor test results performed (soil-cement).

Cement Addition (%)	Optimal Moisture Content (%)	Maximum Dry Density (g/cm^3)
0.0	10.6	1.950
2.0	10.3	1.940
10.0	10.0	1.960

3.2. Unconfined Compressive Tests (UCS)

The medium values of unconfined strength (average from three tests for each combination of length and strip content) obtained by the inclusion of PET strips in the soil are shown in Table 4 and Figure 3.

Table 4. Values of unconfined compressive strength (kPa) (PET strip content—lateritic sandy soil).

	Length (mm)			
	10	**15**	**20**	**30**
(%)				
0	56.72	56.72	56.72	56.72
0.25	73.68	78.92	81.07	84.29
0.50	79.83	86.29	93.19	88.82
0.75	86.41	87.81	99.07	95.86
1.0	90.49	95.98	105.76	99.73
1.5	98.02	100.72	109.34 *	102.10
2.0	91.89	97.99	103.61	99.54

* Highest value obtained.

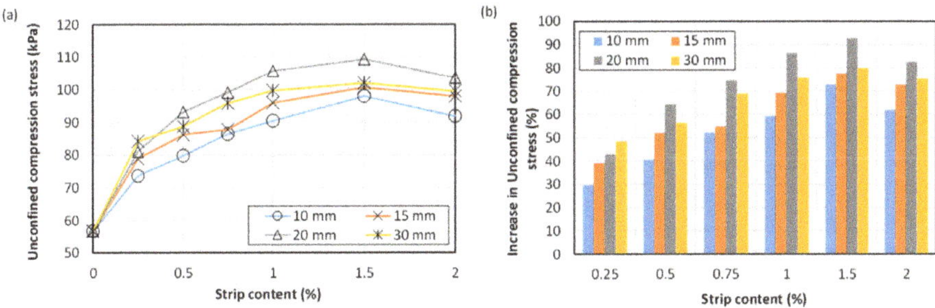

Figure 3. UCS results of lateritic sandy soil/PET strips. (**a**) UCS with fiber content increase; (**b**) the increase in UCS.

Results show that the unconfined strength of the samples varies for all sizes and percentages of added strips. In all analyzed cases there is an increase in strength due to the inclusion of strips, regardless of the length and percentages in which they were included. The optimum soil-strip parameter, that is, the inclusion of strips in which the size and percentage resulted in greater strength increase, was found for the inclusion of strips with 20 mm in length and 1.5% content in relation to the dry mass of the soil. The result is in an increase of 92.4% in relation to the strength of the natural soil.

Analyzing the behavior of the soil (Figure 3) for different types of strips (either in length or quantity), there is a tendency to increase soil strength the longer the length of the added strips.

Figure 4 shows the stress-strain curves obtained in the test considering the natural lateritic sandy soil and the lateritic sandy soil with the inclusion of the optimum soil-strip parameter (strips 20 mm long and 1.5% in relation to the dry mass of the soil). It is noted that the inclusion of this strip parameter not only increases the load capacity of the soil, but also the ductility, increasing the deformation of the soil before rupture by about 1.7%.

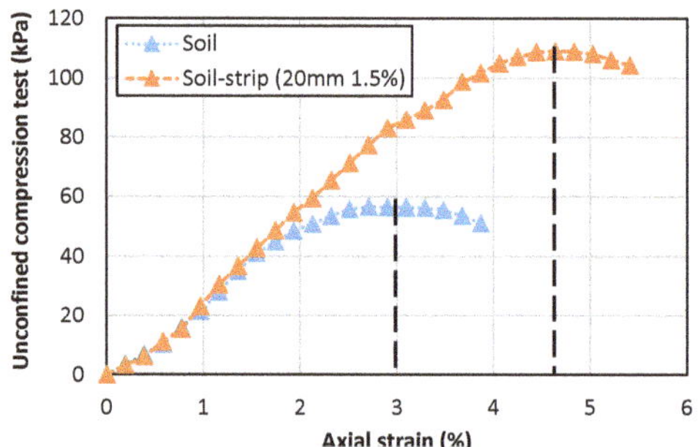

Figure 4. Stress-strain curves obtained: natural lateritic sandy soil and lateritic sandy soil with the inclusion of the optimum parameter soil-strips (strips 20 mm long and 1.5% content).

The strength of strip-reinforced soil increases with the increasing aspect ratio (A_R) of fibers. These results are in accordance with the literature, e.g., Shukla [5]. The shear strength of soil increased with increased PET content and the use of strips contributed to change soil brittle failure into ductile failure [2,35]. As discussed by Tang et al. [36], the increase in UCS might be related to the bridging effect of the fiber, which can efficiently avoid the later development of failure planes and strains in the soil.

3.3. Direct shear tests

3.3.1. Direct Shear Tests in Soil-Strip Composites

Direct shear tests were conducted on lateritic sandy soil (with a 100% degree of compaction) and on soil with the optimal soil-strip parameter inclusion (1.5% inclusions of 20 mm long strips obtained from UCS) (Table 5). The inclusion of strips in the evaluated parameter was effective in increasing the shear strength of the soil for normal stresses smaller than 300 kPa, presenting an increase of 66.7% in cohesion and a decrease of 3.5% in the friction angle in relation to the parameters obtained for the soil without the inclusion of strips.

Table 5. Shear strength parameters of soils obtained in direct shear tests for lateritic sandy soil with and without the inclusion of strips.

Sample	c (kPa)	ϕ (°)
Soil with 0% strips	11.7	31.4
Soil with 1.5% (L = 20 mm)	19.5	30.3

c = cohesion; ϕ = effective friction angle.

Figure 5 shows the shear stress-displacement curves for both soils (with and without strips). These curves are very similar considering the highest applied stresses and slightly more distant from each other considering the lowest stresses.

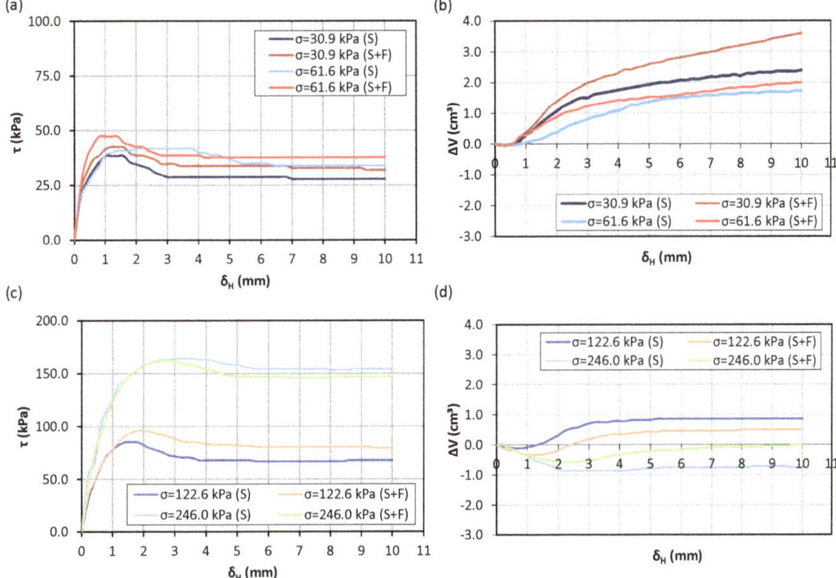

Figure 5. Direct shear test results: 1.5% content of 20 mm PET strips (S+F) and without strips (S). (**a**) shear stress-displacement (lower normal stress); (**b**) volume-displacement (lower normal stress); (**c**) shear stress-displacement (higher normal stress); (**d**) volume-displacement (higher normal stress);

Considering the volume-displacement curves, an increase in volume variation on the lateritic sandy soil at the shearing (dilatancy) was observed when lower normal stresses were applied. The soils with strips presented a slightly higher variation when compared to soil without strips. For higher normal stresses, there is a trend of decrease in volume variation. The volumetric variation showed that the soil presented a fragile rupture (with dilation) and the addition of the strips modified the behavior of the soil for the highest applied stress levels (the material starts to present a ductile behavior). Maher and Gray [37] and Consoli et al. [10] suggest that the effect of fiber inclusion on dilation and volume change is pronounced at higher load and strain levels, possibly due to the inhibiting action of the fibers. Many authors have reported that the addition of fibers tends to increase the ductility and strength of the soil-fiber composite as well, e.g., References [6,34,38–41].

3.3.2. Direct Shear Tests in Soil-Cement Composites

The tests using soil-cement and soil-cement-strip composites were performed with a 95% degree of compaction. Table 6 shows the shear strength parameters of the soil and the soil-cement composites found for each shear strength envelope, and Figure 6 presents the shear strength envelopes with and without inclusions of cement for comparison purposes.

Table 6. Shear strength parameters obtained for the soil with cement content.

Cement Addition (%)	c (kPa)	ϕ (^{o})
0	6.2	31.9
2	23.7	46.5
4	116.5	48.1
6	60.2	56.5
8	154.2	56.2
10	95.2	76.5

c = cohesion; ϕ = effective friction angle.

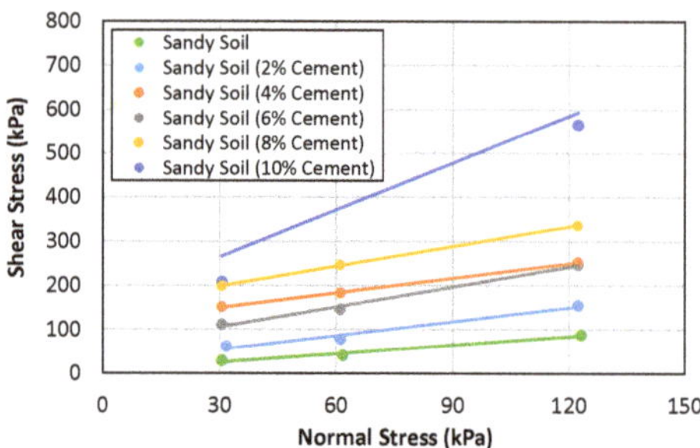

Figure 6. Failure envelopes with the cement addition.

In all the cases evaluated herein, the addition of cement was mostly effective in increasing soil shear strength via the increase in both the cohesion and friction angle parameters in large proportions, even in cases where small contents of cement were added. The reinforcement of the soil with the use of cement had its main effect in the creation of a cohesive intercept in the composite, making its application interesting in granular soils. Most of the shear strength of this type of matrix is due to friction between the particles, which also showed considerable increases. Regarding the increase in cohesion, the most pronounced effects were noted for additions of 8% and 4% of cement, while the friction angle shows larger changes for additions 10% and 6% of cement editions.

Figure 7 shows the stress-displacement curves obtained in direct shear tests of soil and soil-cement. Analyzing the curves, it is possible to notice that the rupture of the specimens with the addition of cement occurs after greater displacement than that of the soil without the cement addition. However, there is a trend for greater loss of strength after the peak. This behavior is clearly noticeable for inclusions of 4% and 6% of cement and accentuated for inclusions of 8% and 10% of cement. Thus, it can be concluded that the addition of cement alters the rupture of the material, making the rupture fragile so that the material strength decreases sharply as the deformation increases. In materials that present this type of rupture, the collapse process can be very fast, generating catastrophic situations.

These results obtained herein are in accordance with the literature [11–13,38]. Also, the results are in agreement of the conclusions of Festugato et al. [34] and Consoli [42], in which the use of cement percentages greater than 5%, in relation to the dry weight of the soil, gave a more significant stiffness to the composite. The addition of percentages of the order of 3% showed partial improvement in the matrix properties, mainly related to workability, with a certain increase in the carrying capacity.

In this sense, a good alternative to avoid this behavior is the addition of strips in soil-cement composite. In order to evaluate the influence of the strips on the strength, ductility, and mainly on the residual (post-peak) strength of the soil-cement composites, a new composite was tested: the soil-cement-strip. The tests on this composite were performed with the inclusion of PET strips in different lengths (10, 15, 20, and 30 mm) and percentages (0.75; 1.0; 1.5, and 2.0%—item 3.3.3). As previously mentioned, considering the high cost of the cement and the polluting potential of these materials, the strips were added only to a soil-cement with a 2% cement addition, seeking a low-cost material with high strength, ductility, and a highly sustainable alternative. This expedient seeks to assess how much strength can be increased with the addition of the strips in order to maximize their use and minimize the use of cement.

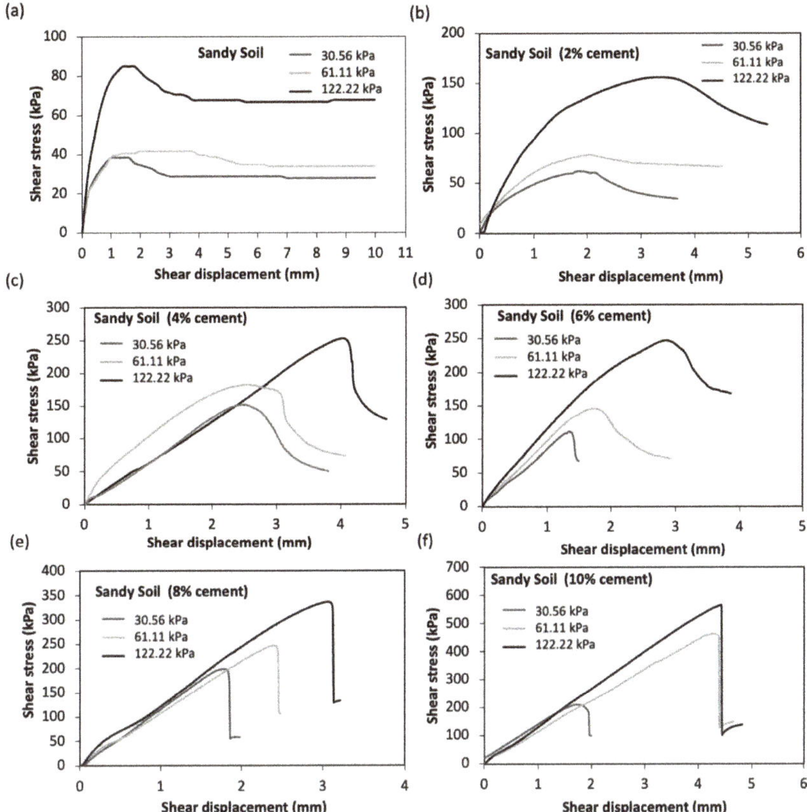

Figure 7. Stress-displacement curves obtained in soil and soil-cement.

3.3.3. Direct Shear Tests in Soil-Cement-Strip Composite

Figures 8–11 show the shear stress-displacement curves regarding the soil-cement-strip composite. Results from Figures 8–11 showed that the inclusion of strips provides, in general, a ductile behavior to the material, presenting greater deformations and lower stress peaks, in comparison with the data obtained in tests carried out with the addition of 2% cement to the lateritic sandy soil. This behavior is more pronounced for inclusions of larger strips, 30 mm long, and in larger quantities. The inclusion of 10 mm strips had a greater effect in decreasing the drop in the post-peak strength of the material, presenting deformations at peak slightly lower than those obtained for soil-cement with the addition of 2% cement. The inclusion of strips of 15 and 20 mm presented greater deformations of the material; however, the drops in strength after the peak were more pronounced in these types of inclusions. Finally, in relation to the stress-displacement behavior of the material, the inclusion of 30 mm strips in amounts of 1.5 and 2.0% proved to be more effective, presenting both greater deformations and lower post-peak strength drops compared to the data obtained in the tests with the addition of 2% cement to the soil.

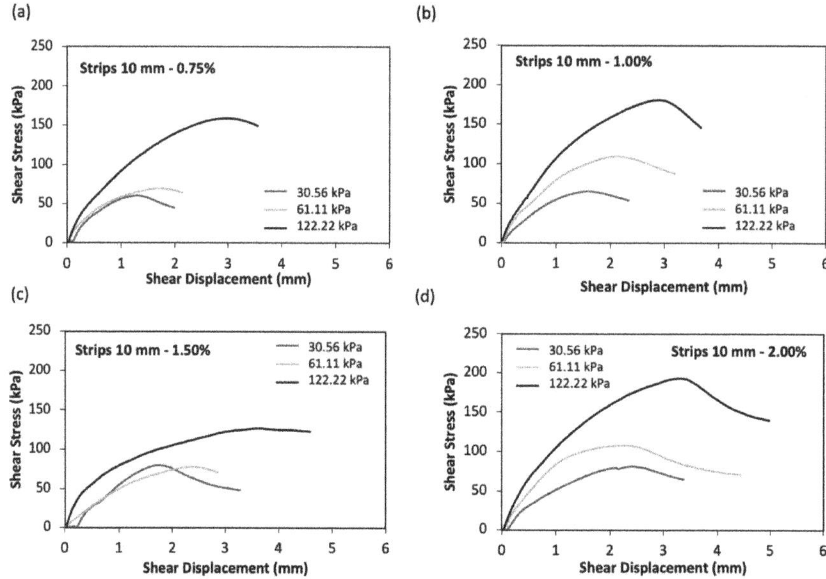

Figure 8. Direct shear results: the soil-cement-strip (2% cement—L = 10 mm).

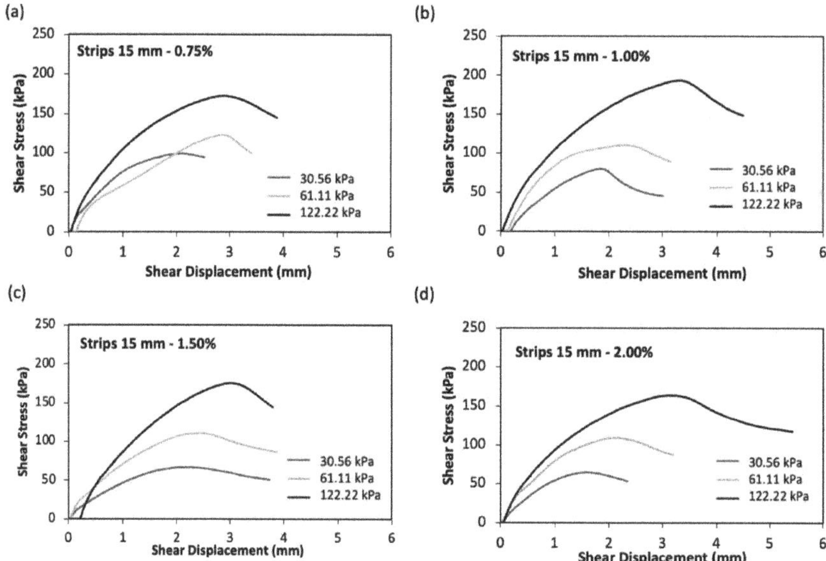

Figure 9. Direct shear results: the soil-cement-strip (2% cement—L = 15 mm).

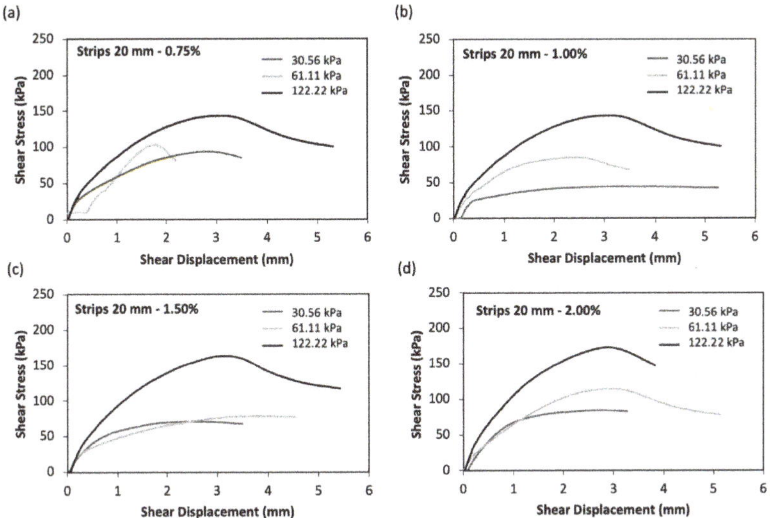

Figure 10. Direct shear results: the soil-cement-strip (2% cement—L = 20 mm).

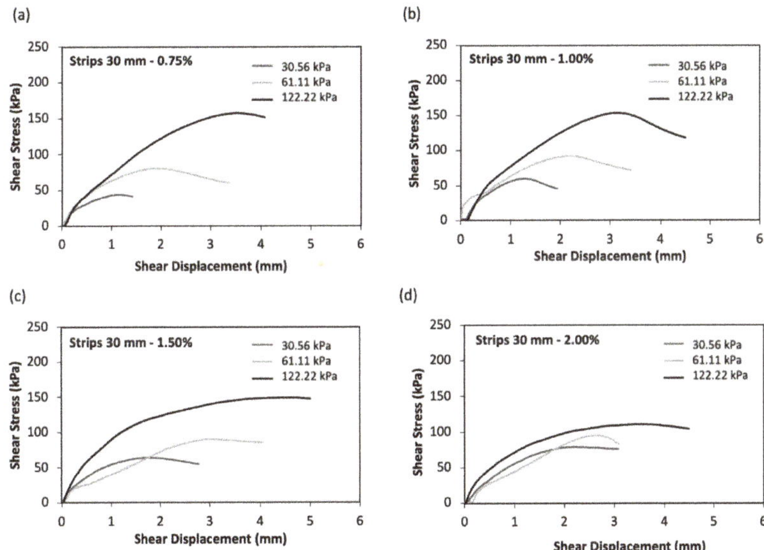

Figure 11. Direct shear results: the soil-cement-strip (2% cement—L = 30 mm).

The benefits of the addition of strips in increasing the ductility of the composite can be seen in terms of the variation of the secant elasticity modulus (E_{ps}). The secant modulus of elasticity was calculated from the strain at peak strength (E_{ps}).

The values presented in Table 7 and Figure 12 show the variation of E_{ps} with strip content and cement content. It can be noted that the secant modulus decreases with the increase of the strip content as well as with the increase of the cement content. This is because extensible fibers require an initial deformation to initiate strength mobilization, resulting in the reduction of fiber-reinforced cemented soil stiffness. Moreover, the addition of higher fiber contents into the cemented soil matrix may lead to a significant drop in stiffness [43].

Table 7. Secant modulus at peak strength for soil-cement-strip with 2% cement content.

Content (%)	L (mm)				
	0	**10**	**15**	**20**	**30**
0.0	510 *	-	-	-	-
0.75	-	500	470	430	400
1.00	-	480	454	410	390
1.50	-	430	420	380	365
2.00	-	400	390	360	352

* All values are given in kPa.

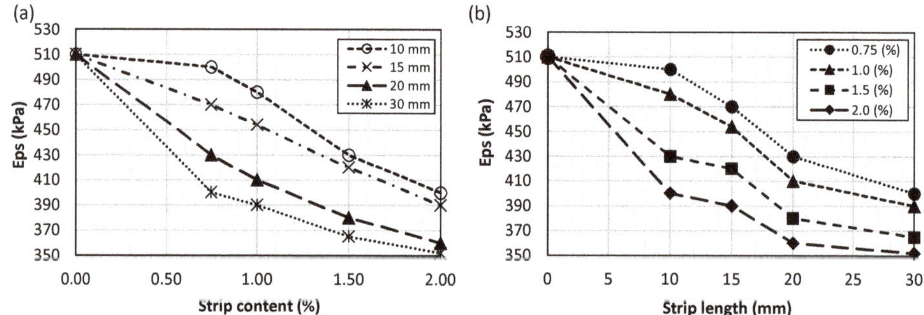

Figure 12. Variation of secant elastic modulus (Eps) of the strip-reinforced cemented specimen (2% cement and different strip contents). (**a**) E_{ps}—strip content (%); (**b**) E_{ps}—strip length (mm).

Another factor that can be used to check the effect of strips on the composite is the deformation strain index (DSI) [43]:

$$\text{DSI} = \frac{\varepsilon_f - \varepsilon_p}{\varepsilon_p} \tag{1}$$

where ε_f is the strain at the destruction stage (final test stage), ε_p is the strain at peak strength and DSI is the destruction strain index.

Table 8 presents the values of DSI and Figure 13 shows the DSI variation in terms of strip content and strip length.

Table 8. The DSI values for soil-cement-strip with 2% cement content.

Content (%)	L (mm)				
	0	**10**	**15**	**20**	**30**
0.0	0.35 *	-	-	-	-
0.75	-	0.27	0.33	0.63	0.75
1.00	-	0.39	0.41	0.65	0.77
1.50	-	0.42	0.45	0.70	0.80
2.00	-	0.45	0.6	0.73	1.0

* Value of the DSI for soil without fiber with 2% cement content.

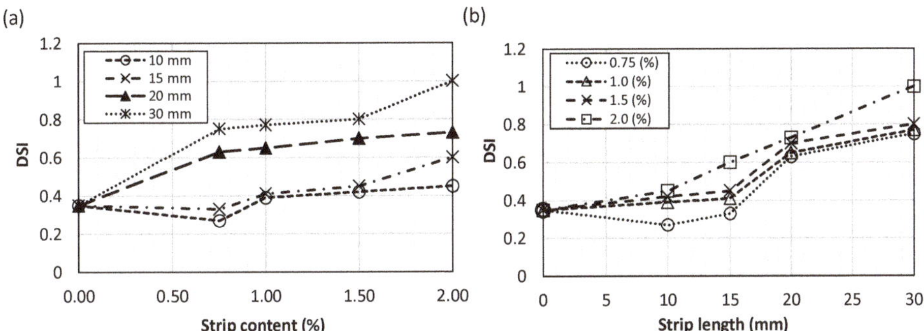

Figure 13. Variation of DSI (2% cement and different strip contents). (**a**) DSI-strip content (%); (**b**) DSI-strip length (mm).

DSI values showed an increase with the strip content as well as with the length when compared to the value of the soil with 2.0% cement without strips. A small initial decrease considering the strip content of 0.75% for the lengths of 10 and 15 mm was evidenced. However, in general, the value of the DSI increased for the different contents and lengths. For L = 20 mm (1.5%; 2.0%), the DSI value increased about two-fold. For L = 30 mm (2.0%), the DSI values increased about three-fold. The DSI shows the ductile response of the strips reinforced-cemented soil in comparison with unreinforced cemented soil mixture [17,43].

Tables 9 and 10 present the shear stresses at the peak obtained for each specimen and the shear strength parameters found for each soil-cement-strip envelope, respectively. Comparing the values obtained in Table 10 with the shear strength parameters obtained for the lateritic sandy soil with a 2% cement addition (23.7 kPa cohesion and 46.5° friction angle), it is possible to notice that all the analyzed inclusions showed improvement in at least one of the soil shear strength parameters, either in cohesion or in the friction angle. Only three of the analyzed inclusions presented a reduction in the cohesion parameters, whereas in relation to the friction angles obtained, eight of the sixteen analyzed inclusions presented decreases in this parameter.

Table 9. Normal and peak shear stresses: the soil-cement-strip (2% cement).

%/L (mm)	10		15		20		30	
	τ (kPa)	σ (kPa)	τ (kPa)	σ (kPa)	τ (kPa)	σ (kPa)	τ (kPa)	σ (kPa)
0.75	60.0	30.6	98.9	30.6	93.8	30.6	44.2	30.6
	69.7	61.1	123.1	61.1	103.5	61.1	80.9	61.1
	158.3	122.2	173.1	122.2	143.5	122.2	157.1	122.2
1.00	65.2	30.6	79.6	30.6	44.6	30.6	60.0	30.6
	109.2	61.1	110.2	61.1	84.6	61.1	92.3	61.1
	180.7	122.2	193.0	122.2	143.5	122.2	153.6	122.2
1.50	79.6	30.6	67.0	30.6	71.0	30.6	64.0	30.6
	78.0	61.1	110.2	61.1	78.6	61.1	90.0	61.1
	126.6	122.2	175.5	122.2	163.4	122.2	149.6	122.2
2.00	81.0	30.6	65.2	30.6	84.1	30.6	79.2	30.6
	107.9	61.1	109.2	61.1	104.0	61.1	95.7	61.1
	193.0	122.2	163.4	122.2	114.6	122.2	110.5	122.2

τ = shear stress; σ = normal stress; L = strip length.

Table 10. Shear strength parameters: the soil-cement-strip (2% cement).

Strip Length (mm)	Inclusion (%)	c (kPa)	φ (º)
10	0.75	15.7	48.4
	1.00	29.5	51.3
	1.50	55.3	28.9
	2.00	38.4	51.3
15	0.75	73.9	39.0
	1.00	38.2	51.4
	1.50	34.4	49.4
	2.00	38.1	46.2
20	0.75	73.8	29.2
	1.00	15.2	46.7
	1.50	28.6	46.7
	2.00	78.8	17.2
30	0.75	6.1	51.0
	1.00	29.4	45.5
	1.50	34.2	43.2
	2.00	71.8	38.6

c = cohesion; φ = effective friction angle.

In general, the inclusion of 2.0% of 10 mm long strips and 1.0% of 15 mm long strips proved to be more advantageous since it showed significant increases in both cohesion and friction angle. The cohesions values obtained were 38.4 kPa in the first case and 38.2 kPa in the second case, presenting increases of 62% and 61.2% in relation to the soil-cement composite and friction angles of 51.3° and 51.4° resulting in increases of 10.3% and 10.5%, respectively.

Results from Table 9 show an important characteristic of the soil-cement composite. This material may become heterogeneous. In some cases, adding strips to a cemented matrix made the behavior of the material unpredictable. The failure envelopes of some of the specimens molded with the lateritic sandy soil, added with cement and strips, were not consistent with the expected increase in shear strength. This may be related to a certain increase in the confining tension, as can be seen in the envelopes of the soil composites-strip-cement for inclusions of 10 mm strips in amounts of 0.75% and 1.5% and for strips of 20 mm in length in amounts of 1.5% and 2.0%.

Additionally, the material presented an altered rupture plain (Figure 14a). At the failure, the cemented matrix aggregates the strips in order to create a heterogeneous material, for which tensile strength and the rupture plane will depend on the distribution of the strips, especially on the number of strips that will be requested during the rupture and the way in which the efforts will be distributed. The analysis of the samples after the rupture indicated that only a small percentage of the included strips deformed definitively (showing folds), which occurred due to the efforts applied during loading or during compaction of the specimens (Figure 14b).

(a)   (b)

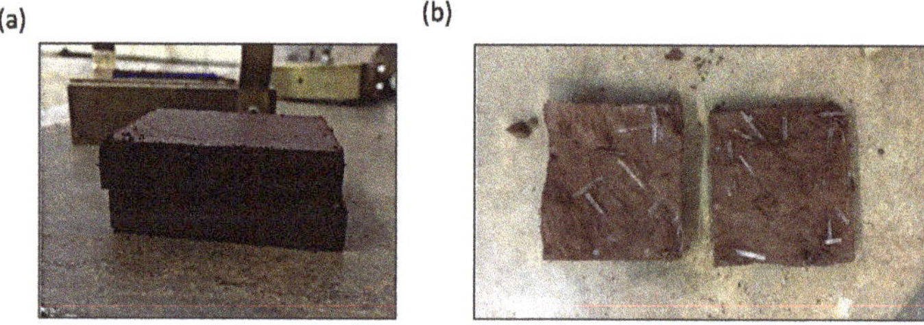

Figure 14. Direct shear tests. (**a**) The rupture plane; (**b**) random strip positions inside the specimen after failure.

In a more detailed analysis of the shear strength parameters, it can be seen that the friction angles and cohesions are not consistent with the expected values of a soil, so that the cement added to the analyzed composites leads to characteristics of cemented matrices, which is why the parameters obtained are so high. These factors are manifested in a more evident way for greater additions of cement or in cases in which the addition of strips occurred. Few studies have evaluated the effects of the composite formed by soil, cement, and strips. However, the results obtained are in accordance with the results of Tang et al. [17] and Olgun et al. [18].

4. Conclusions

This study evaluated the effect of PET strips on the mechanical properties of a cement-treated lateritic sandy soil. From the results obtained in this study, the main conclusions can be drawn:

- Regarding the uniaxial strength, all analyzed cases showed an increase in the soil strength due to the inclusion of fibers, regardless of the length and percentages in which they were included;
- Results from direct shear tests in soil-cement composites showed that in all analyzed cases the addition of cement was effective in increasing the shear strength of the soil, increasing both the cohesion parameter and the friction angle in large proportions, even when small amounts of cement have been added;
- As for the addition of fibers to the soil-cement composites, the pronounced effect occurred in increasing soil cohesion, often presenting a decrease in the friction angle. The inclusion of strips also provided a more ductile behavior to the material, presenting greater deformations with lower stress peaks;
- Results showed that the inclusion of recycled strips in soil-cement can provide a material with high strength, ductility, and a highly sustainable alternative and;
- In general, the fibers and cement addition the lateritic sandy soil mixture proved to be an excellent option for increasing the strength and deformability of the analyzed natural soil, showing high potential for applications of these materials in the civil construction industry.

Author Contributions: Conceptualization, M.R.S., P.C.L., and N.d.S.C.; methodology, M.R.S., P.C.L., N.d.S.C., R.A.R., and H.L.G.; formal analysis, M.R.S., P.C.L., N.d.S.C., R.A.R., and H.L.G.; investigation, M.R.S., P.C.L., H.L.G., R.A.R.; resources, M.R.S., P.C.L., N.d.S.C., R.A.R., and H.L.G; writing—original draft preparation, M.R.S., P.C.L., N.d.S.C., R.A.R., and H.L.G.; writing—review and editing, M.R.S., P.C.L., and N.d.S.C.; visualization, M.R.S., P.C.L., N.d.S.C., R.A.R., and H.L.G.; supervision, P.C.L. and N.d.S.C.; project administration, M.R.S., P.C.L., N.d.S.C., R.A.R., and H.L.G. All authors have read and agreed to the published version of the manuscript.

Funding: This research received no external funding.

Acknowledgments: The author are very thankful to Capes/Print program and PROPG/UNESP.

Conflicts of Interest: The authors declare no conflict of interest.

References

1. Parker, L. Fast facts about plastic pollution. *National Geographic*, 20 December 2018. Available online: https://www.nationalgeographic.com/news/2018/05/plastics-facts-infographics-ocean-pollution/(accessed on 11 August 2020).
2. Sivakumar Babu, G.L.; Chouksey, S.K. Stress-strain response of plastic waste mixed soil. *Waste Manag.* **2011**, *31*, 481–488. [CrossRef]
3. Soltani-Jigheh, H. Undrained Behavior of Clay—Plastic Waste Mixtures. In Proceedings of the 11th International Congress on Advances in Civil Engineering, Istanbul, Turkey, 21–25 October 2014.
4. Luwalaga, J.G. Analysing the Behaviour of Soil Reinforced with Polyethylene. Master's Thesis, Stellenbosch University, Stellenbosch, South Africa, 2016.
5. Shukla, S.K. *Fundamentals of Fibre-Reinforced Soil Engineering*; Springer: Berlin, Germany, 2017; ISBN 978-981-10-3061-1.

6. Peddaiah, S.; Burman, A.; Sreedeep, S. Experimental Study on Effect of Waste Plastic Bottle Strips in Soil Improvement. *Geotech. Geol. Eng.* **2018**, *36*, 2907–2920. [CrossRef]

7. Lima, D.C.; Bueno, B.S.; Thomasi, L. The mechanical response of soil-lime mixtures reinforced with short synthetic fiber. In Proceedings of the International Symposium on Environmental Geotechnology, San Diego, CA, USA, 1–3 March 1996; pp. 868–877.

8. Bueno, B.d.S.; Lima, D.C.; Teixeira, S.H.C.; Ribeiro, N.J. Soil fiber reinforcement: Basic understanding. In Proceedings of the International Symposium on Environmental Geotechnology, San Diego, CA, USA, 9–13 June 1996; Volume 3, pp. 878–884.

9. Specht, L.P. Comportamento de Misturas de Solo-Cimento-Fibra Submetidas a Carregamentos Estáticos e Dinâmicos Visando à Pavimentação. Master's Thesis, Universidade Federal do Rio Grande do Sul, Porto Alegre, Rio Grande do Sul, Brazil, 2000; p. 151. (In portuguese).

10. Consoli, N.C.; Thomé, A.; Girardello, V.; Ruver, C.A. Uplift behavior of plates embedded in fiber-reinforced cement stabilized backfill. *Geotext. Geomembr.* **2012**, *35*, 107–111. [CrossRef]

11. Fatahi, B.; Fatahi, B.; Le, T.M.; Khabbaz, H. Small-strain properties of soft clay treated with fibre and cement. *Geosynth. Int.* **2013**, *20*, 286–300. [CrossRef]

12. Cristelo, N.; Cunha, V.M.C.F.; Dias, M.; Gomes, A.T.; Miranda, T.; Araújo, N. Influence of Discrete Fibre Reinforcement on the Uniaxial Compression Response and Seismic Wave Velocity of a Cement-Stabilised Sandy-Clay. *Geotext. Geomembr.* **2015**, *43*, 1–13. [CrossRef]

13. Wei, L.; Chai, S.X.; Zhang, H.Y.; Shi, Q. Mechanical properties of soil reinforced with both lime and four kinds of fiber. *Constr. Build. Mater.* **2018**, *172*, 300–308. [CrossRef]

14. Guedes, S.B. Estudo do Desempenho Mecânico de um Solo-Cimento Microreforçado com Fibras Sintéticas Para Uso Como Revestimento Primário em Estradas Não Pavimentadas. Ph.D. Thesis, Universidade Federal de Pernambuco, Recife, Brazil, 2013; p. 515. (In portuguese).

15. Girardello, V. Comportamento de ensaios de Arrancamento de Placas Embutidas em Camadas de Solo-Cimento-Fibra. Ph.D. Thesis, Universidade Federal do Rio Grande do Sul, Porto Alegre, Rio Grande do Sul, Brazil, 2014; p. 195. (In portuguese).

16. Olutaiwo, A.O.; Ezegbunem, I.I. Effect of Waste PET Bottle Strips (WPBS) on the CBR of Cement-Modified Lateritic Soil. *Int. J. Sci. Res.* **2017**, *6*, 1098–1102.

17. Tang, Q.; Shi, P.; Zhang, Y.; Liu, W.; Chen, L. Strength and Deformation Properties of Fiber and Cement Reinforced Heavy Metal-Contaminated Synthetic Soils. *Adv. Mater. Sci. Eng.* **2019**, *2019*. [CrossRef]

18. Olgun, M. Effects of polypropylene fiber inclusion on the strength and volume change characteristics of cement-fly ash stabilized clay soil. *Geosynth. Int.* **2013**, *20*, 263–275. [CrossRef]

19. ABNT ABNT NBR 7181—Associação Brasileira de Normas Técnicas. *Solo—Análise Granulométrica*; Associação Brasileira de Normas Técnicas: Rio de Jainero, Brazil, 1984. (In Portuguese)

20. ABNT NBR 6459—Associação Brasileira De Norma Técnicas. *Solo—Determinação do Limite de Liquidez*; Associação Brasileira de Normas Técnicas: Rio de Jainero, Brazil, 2016. (In Portuguese)

21. ABNT NBR 7180—Associação Brasileira De Norma Técnicas. *Solo—Determinação do Limite de Plasticidade*; Associação Brasileira de Normas Técnicas: Rio de Jainero, Brazil, 2016. (In Portuguese)

22. ABNT NBR 6458—Associação Brasileira de Normas Técnicas. *Solo—Grãos de Pedregulho Retidos na Peneira de Abertura 4.8 mm—Determinação da Massa Específica, da Massa Específica Aparente e da Absorção de Água*; Associação Brasileira de Normas Técnicas: Rio de Jainero, Brazil, 2017. (In Portuguese)

23. ABNT NBR 6457—Associação Brasileira de Normas Técnicas. *Amostras de solo—Preparação Para Ensaios de Compactação e Ensaios de Caracterização*; Associação Brasileira de Normas Técnicas: Rio de Jainero, Brazil, 2016. (In Portuguese)

24. ABNT NBR 7182—Associação Brasileira de Normas Técnicas. *Solo—Ensaio de Compactação*; Associação Brasileira de Normas Técnicas: Rio de Jainero, Brazil, 2020. (In Portuguese)

25. ASTM. *Standard D3080/D3080M—11 Direct Shear Test of Soils Under Consolidated Drained Conditions*; ASTM International: West Conshohocken, PA, USA, 2012; pp. 1–9. [CrossRef]

26. ABNT NBR 6502—Associação Brasileira De Norma Técnicas. *Rochas e Solos*; Associação Brasileira de Normas Técnicas: Rio de Jainero, Brazil, 1995. (In Portuguese)

27. Miguel, M.G.; Bonder, B.H. Soil-Water Characteristic Curves Obtained for a Colluvial and Lateritic Soil Profile Considering the Macro and Micro Porosity. *Geotech. Geol. Eng.* **2012**, *30*, 1405–1420. [CrossRef]

28. Eberemu, A.O. Evaluation of bagasse ash treated lateritic soil as a potential barrier material in waste containment application. *Acta Geotech.* **2013**, *8*, 407–421. [CrossRef]
29. Bai, W.; Kong, L.; Guo, A. Effects of physical properties on electrical conductivity of compacted lateritic soil. *J. Rock Mech. Geotech. Eng.* **2013**, *5*, 406–411. [CrossRef]
30. Oyelami, C.A.; Van Rooy, J.L. A review of the use of lateritic soils in the construction/development of sustainable housing in Africa: A geological perspective. *J. Afr. Earth Sci.* **2016**, *119*, 226–237. [CrossRef]
31. ABNT NBR 12023—Associação Brasileira De Norma Técnicas. *Solo-cimento—Ensaio de compactação*; Associação Brasileira de Normas Técnicas: Rio de Jainero, Brazil, 2012. (In Portuguese)
32. ABNT NBR 12770—Associação Brasileira De Norma Técnicas. *Solo-coesivo—Determinação da Resistência à Compressão Não Confinada—Método de Ensaio*; Associação Brasileira de Normas Técnicas: Rio de Jainero, Brazil, 1992. (In Portuguese)
33. Hoare, D.J. Laboratory study of granular soils reinforced with randomly oriented discrete fibres. In Proceedings of the International Conference on Soil Reinforcement, Paris, France, 20–22 March 1979; pp. 47–52.
34. Festugato, L.; Menger, E.; Benezra, F.; Kipper, E.A.; Consoli, N.C. Fibre-reinforced cemented soils compressive and tensile strength assessment as a function of filament length. *Geotext. Geomembr.* **2017**, *45*, 77–82. [CrossRef]
35. Zhao, J.J.; Lee, M.L.; Lim, S.K.; Tanaka, Y. Unconfined compressive strength of PET waste-mixed residual soils. *Geomech. Eng.* **2015**, *8*, 53–66. [CrossRef]
36. Tang, C.; Shi, B.; Gao, W.; Chen, F.; Cai, Y. Strength and mechanical behavior of short polypropylene fiber reinforced and cement stabilized clayey soil. *Geotext. Geomembr.* **2007**, *25*, 194–202. [CrossRef]
37. Mohamad, H.; Maher, M.H.; Gray, D.H. Static response of sands reinforced with randomly distributed fibers. *J. Geotech. Eng.* **1990**, *116*, 1661–1677.
38. Velloso, R.Q.; Casagrande, M.D.T.; Vargas, E.A., Jr.; Consoli, N.C. 2012 Simulation of the mechanical behavior of fiber reinforced sand using the discrete element method. *Soils Rocks* **2010**, *33*, 81–93.
39. Botero, E.; Ossa, A.; Sherwell, G.; Ovando-Shelley, E. Stress-strain behavior of a silty soil reinforced with polyethylene terephthalate (PET). *Geotext. Geomembr.* **2015**, *43*, 363–369. [CrossRef]
40. Rasouli, H.; Fatahi, B. Geofoam blocks to protect buried pipelines subjected to strike-slip fault rupture. *Geotext. Geomembr.* **2019**, *48*, 257–274. [CrossRef]
41. Onyelowe, K.C.; Bui Van, D.; Ubachukwu, O.; Ezugwu, C.; Salahudeen, B.; Nguyen Van, M.; Ikeagwuani, C.; Amhadi, T.; Sosa, F.; Wu, W.; et al. Recycling and reuse of solid wastes; a hub for ecofriendly, ecoefficient and sustainable soil, concrete, wastewater and pavement reengineering. *Int. J. Low-Carbon Technol.* **2019**, *14*, 440–451. [CrossRef]
42. Consoli, N.C.; Arcari Bassani, M.A.; Festugato, L. Effect of fiber-reinforcement on the strength of cemented soils. *Geotext. Geomembr.* **2010**, *28*, 344–351. [CrossRef]
43. Janalizadeh Choobbasti, A.; Soleimani Kutanaei, S. Effect of fiber reinforcement on deformability properties of cemented sand. *J. Adhes. Sci. Technol.* **2017**, *31*, 1576–1590. [CrossRef]

Publisher's Note: MDPI stays neutral with regard to jurisdictional claims in published maps and institutional affiliations.

 *sustainability*

Article

Reinforcing Effect of Polypropylene Waste Strips on Compacted Lateritic Soils

Régis Marçal [1], Paulo César Lodi [1,*], Natália de Souza Correia [2], Heraldo Luiz Giacheti [1], Roger Augusto Rodrigues [1] and John S. McCartney [3]

[1] Department of Civil and Environmental Engineering, São Paulo State University (UNESP), Av. Engenheiro Luiz Edmundo Carrijo Coube 14-01, Bauru, SP 17033-360, Brazil; regis-ata@hotmail.com (R.M.); h.giacheti@unesp.br (H.L.G.); roger.rodrigues@unesp.br (R.A.R.)

[2] Department of Civil Engineering, Federal University of Sao Carlos (UFSCar), Rodovia Washington Luiz, São Carlos, SP 17033-360, Brazil; ncorreia@ufscar.br

[3] Department of Structural Engineering, University of California at San Diego (UCSD), 9500 Gilman Dr., SME 442J, La Jolla, CA 92093-0085, USA; mccartney@ucsd.edu

* Correspondence: paulo.lodi@unesp.br; Tel.: +55-646-755-2239

Received: 23 October 2020; Accepted: 16 November 2020; Published: 17 November 2020

Abstract: This study evaluated the strength properties of compacted lateritic soils reinforced with polypropylene (PP) waste strips cut from recycled plastic packing with the goal of promoting sustainability through using local materials for engineering work and reusing waste materials as low-cost reinforcements. Waste PP strips with widths of 15 mm and different lengths were uniformly mixed with clayey sand (SC) and clay (CL) soils with the goal of using these materials as low-cost fiber reinforcements. The impact of different PP strip contents (0.25% to 2.0%) and lengths (10, 15, 20, and 30 mm) on the unconfined compressive strength (UCS) of the soils revealed an optimum combination of PP strip content and length. Statistical analysis showed that PP strip content has a greater effect than the PP strip length on the UCS for both soils. Results led to the definition of an empirical equation to estimate the UCS of strip-reinforced soils. The results from direct shear tests indicate that the SC soil showed an increase in both apparent cohesion and friction angle after reinforcement, while the CL soil only showed an increase in friction angle after reinforcement. California bearing ratio (CBR) tests indicate that the SC soil experienced a 70% increase in CBR after reinforcement, while the CBR of the CL soil was not affected by strip inclusion.

Keywords: soil improvement; polypropylene strips; geotechnical properties; sustainable reuse of plastic waste

---

1. Introduction

Finding new ways to recycle plastic waste from water bottles, disposable cups, plates, or plastic packaging for foods has become a major challenge worldwide. According to the World Economic Forum (2016), a million plastic bottles are bought around the world every minute, and this number may jump 20% by 2021, potentially leading to an environmental disaster. As also pointed out in this report, plastic production has increased from 15 million tons in the 1960s to 311 million tons in 2014 and is expected to triple by 2050. Furthermore, the 2030 Agenda for Sustainable Development [1] sets out in its goals a substantial reduction in waste generation through recycling, reduction, and reuse and encourages the use of local materials in engineering works.

Environmental challenges have stimulated researchers to find techniques to improve the strength properties of geotechnical materials [2]. In the context of alternative or recycled waste materials in soil improvement, tire shreds and rubber fibers have been extensively studied [3–10]. Further, the use of fiber reinforcement, especially with local soils, has been recognized as a viable technique for soil improvement in numerous geotechnical engineering applications. Fiber reinforcement has been used in a range of applications, including as backfill in retaining structures, stabilization of subgrade and subbases, improvement in soil bearing capacity, reinforcement of soft soil embankments, control of soil hydraulic conductivity, improvement of erosion resistance, piping prevention, and shrinkage crack mitigation [11–15]. Fiber reinforcements can carry tensile stresses, which are mobilized by friction between the reinforcements and the soil. The mobilization of tensile stresses in the reinforcements generally leads to an increase in the shear strength of the soils, namely that generated by redistributing shear stresses in soils through their tensile strength. Randomly distributed polymeric additions, such as polypropylene (PP) and polyethylene terephthalate (PET), incorporated in soils improve their mechanical behavior.

Gathering the idea of plastic recycling and soil improvement, Consoli et al. [16] carried out one of the first experiments on the utilization of the polyethylene (PET) fibers derived from plastic wastes (stretched cylindrical shapes) in the reinforcement of natural and artificially cemented sand, showing plastic wastes improved soil mechanical response. Later, several studies reported the influence of PET fiber inclusion on the mechanical properties of soils [17–21]. The behavior of soils reinforced with PP fibers has also been extensively studied [12,22–28]. However, there is a lack of studies involving the inclusion of polymeric strips taken from recyclable materials as soil reinforcement.

The use of polymeric strips has several advantages, such as the possibility of reusing plastic waste to increase soil strength without the need to apply a recycling process, as in the case of synthetic fibers. However, the few available studies used PET strips and not PP strips, e.g., [2,17,29–32].

Sivakumar Babu and Choukey [17] evaluated the effect of including 12 mm long and 4 mm wide PET strips, in amounts of 0.50%, 0.75%, and 1.0%, in a sandy soil using unconfined compression strength (UCS) tests and triaxial tests (consolidated and undrained). The authors reported significant increases in soil shear strength parameters, which were greater for larger numbers of strips. In addition, UCS tests indicated an increase in ductility, proportional to the inclusion of strips. Soltani-Jigheh [31] studied the inclusion of PET strips (4 mm wide and 8 mm long) in quantities of 0.25%; 0.50%; 0.75%; 1.0%; 1.5%, and 2% (in relation to the clay soil mass) using consolidated undrained (CU) triaxial tests. Results showed an increase of around 11% in the shear strength of the soil, resulting from an increase in apparent cohesion and a decrease in friction angle.

Babu and Choukey [17] suggested a more economic and simple way of recycling plastic bottles as soil reinforcement using strips cut from PET water bottles. Plastic strips that were 12 mm long and 4 mm in width showed significant improvement in the strength of two soils due to an increase in friction and significant reduction in compression parameters. Chebet and Kalumba [30] evaluated soil improvement using HDPE plastic strips (0.1–0.3% by weight, 15 to 45 mm length, and 6 to 18 mm widths) obtained from shopping bags mixed with two sandy soils through direct shear tests. Findings showed that shear strength of sandy soils was sensitive and significantly affected with a small addition of strips. Luwalaga [2] evaluated sand reinforced with randomly mixed PET plastic waste flakes with varying percentages in terms of California Bearing Ratio CBR and direct shear box testing. Results concluded that the appropriate percentage of PET plastic waste to use while reinforcing sandy soil used is 22.5%. Peddaiah et al. [32] evaluated the addition of PET wastewater bottles cut into strips to locally available soils and showed enhanced soil engineering properties. Strips were cut with 15 mm width and lengths of 15, 25, and 35 mm in different contents of 0.2% to 0.8%. Strips randomly mixed with sandy soil improved the soil strength parameters. It was found that addition of PET strips to sand could reduce the soil brittleness under low overburden pressures.

According to Fathi et al. [33] recycling plastic waste as reinforcing material has become a cheap and viable alternative for soil improvement. Peddaiah et al. [32] concludes that the effect of plastic reinforcement in soil mass vitally depends on nature of the surface (i.e., plain/smooth or corrugated/undulated) and size of strips, plastic content, and type of soil. For Onyelowe et al. [34] the fundamental purpose of solving an engineering problem revolves around a sustainable, economic, efficient, and durable design, with optimal performance to meet certain desirable conditions.

Hence, the sustainable and economic alternative of plastic waste strips and local soils offers two advantages in geotechnical applications: reuse of plastic waste materials and reduction in the use of natural soils, producing materials with required engineering properties.

Although the use of strips from the reuse of waste bottles has high potential for improving soil characteristics, the field of study for these materials is relatively new, especially regarding lateritic soils. This fact generates a consensus among several authors regarding the need for a deeper assessment of different types of plastics and the characteristics of each type of inclusion in conjunction with different soils, in addition to full-scale studies [2,30,32].

Considering the experience from the literature, as well as the lack of research regarding polymeric strips as soil reinforcements, the strength properties of compacted lateritic soils reinforced with polypropylene waste strips cut from recycled plastic packing were evaluated in this study. A series of unconfined compressive strength (UCS), direct shear, and California bearing ratio (CBR) tests were conducted in order to evaluate an optimum combination of plastic waste strips in different soils. A statistical analysis of proposed equations to estimate the UCS of PP strip-stabilized soils is presented. Results were used to prepare samples for CBR and direct shear tests.

2. Materials and Methods

Lateritic soils (clayey sand and clay) were chosen in this research, since they represent typical soils that cover a large area in Brazil. These soils are residual sandstone soils, with low compressibility, unsaturated condition, and high porosity. The clayey sand was collected in Bauru, Sao Paulo, Brazil (22°21′6.03″ S; 49°01′57.68″ O), and the clay soil was collected in Pederneiras, also in Sao Paulo state (22°19′52.5″ S; 48°45′32.26″ O). The soil samples were characterized according to the following American Society for Testing and Materials (ASTM) recommendations: particle size analysis ASTM D7928 [35], soil classification (USCS) ASTM D2487 [36], Highway Research Board (HRB) classification ASTM D3282 [37], specific gravity (G_s) ASTM D854 [38], Proctor tests ASTM D698 [39], and consistency limits ASTM D4318 [40]. The physical properties of the soils including their classification from these tests are presented in Table 1. The particle distributions and the standard Proctor compaction test results for the sandy clay (SC) and clay low (CL) soils are shown in Figures 1 and 2, respectively.

Table 1. Physical properties of soils used in this research.

Property Value	Clayey Sand	Clay	Specification
Soil classification (USCS)	SC	CL	ASTM D2487 [36]
HRB classification	A-2-4	A-6	ASTM D3282 [37]
Percent sand (%)	80	8	ASTM D7928 [35]
Percent fines (<0.074 mm) (%)	20	92	
Specific gravity, Gs	2.65	2.69	ASTM D854 [38]
Maximum dry unit weight (kN/m^3)	19.50	18.4	ASTM D698 [39]
Optimum water content (%)	10.6	16.1	
Liquid limit	16	34	
Plasticity limit	NP	23	ASTM D4318 [40]
Plasticity index	NP	11	

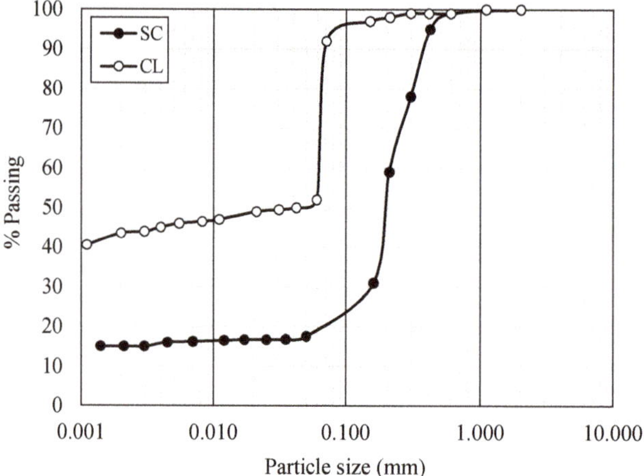

Figure 1. Particle size distribution of the two lateritic soils.

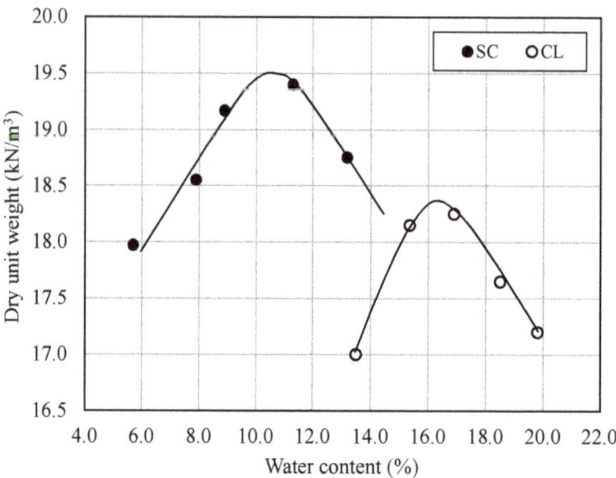

Figure 2. Compaction curves of the two lateritic soils under investigation.

The soil/water retention curves (SWRCs) of the two soils are presented in Figure 3, along with the fitted SWRC model of van Genuchten [41]. The SWRC data exhibit a bimodal behavior (two air entry suctions), while the van Genuchten [41] SWRC is unimodal, as follows:

$$w = w_r + (w_s - w_r) \times \left\{ \frac{1}{[1 + (\alpha.s)^n]^m} \right\} \tag{1}$$

where w_s and w_r are the saturation and residual water content (%), m and n are curvature parameters, and s is the matric suction (kPa).

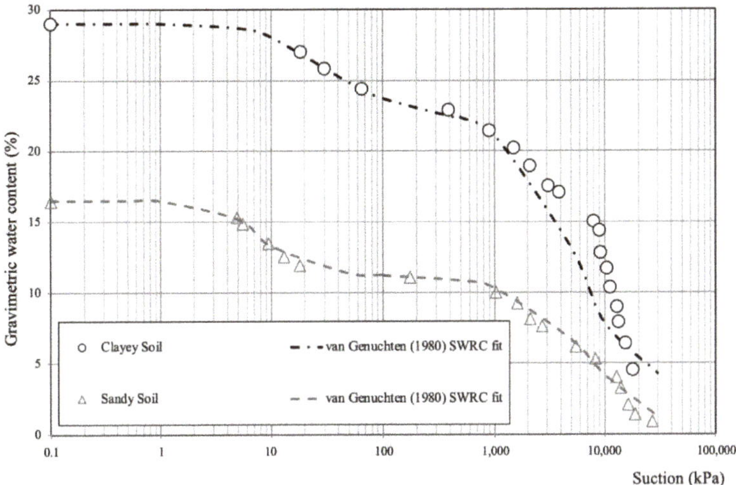

Figure 3. Soil water retention data for the two soils: sandy soil and clayey soil.

Accordingly, the van Genuchten [41] SWRC was fitted to both of the modes exhibited in the data. Specifically, the fits were performed in two parts for each curve. This behavior can be attributed to the presence of macro and micropores in the soil [42]. The fitting parameters of the SWRC of van Genuchten [41] are shown in Table 2. The curve for the clayey sand (SC) soil shows two air entry suctions, the first of approximately 3 kPa, and the second of approximately 2 MPa. The curves obtained for the clay (CL) soil, due to the greater retention capacity, show a great variation of suction pressures over a small range of gravimetric water content. Similar to the SC soil, two air entry suctions are observed for the CL soil, the first of approximately 11 kPa, and the second of approximately 6 MPa.

Table 2. Fitting parameters of the van Genuchten (1980) soil/water retention curve (SWRC).

Soil	Stretch	α (kPa^{-1})	m	n	w_r (%)	w_s (%)	R-Squared
Sandy	1	0.1520	0.6977	2.4762	11.2	16.5	0.996
	2	0.0001	1.4349	1.1890	0.0	11.3	0.976
Clayey	1	0.0669	0.3421	1.8113	21.4	29.0	0.985
	2	0.0003	0.4974	2.4974	3.00	22.6	0.976

Polypropylene (PP) strips were obtained from plastic packaging that would be discarded without any reuse. In order to avoid discrepancies in the results, only one specific brand of plastic packaging was used (without lids, labels, and other parts) in order to assure strip homogeneity. PP strips of 1.5 mm width and 0.5 mm thickness with lengths of 10, 15, 20, and 30 mm were added to the soil in different percentages by dry soil weight of 0.25%, 0.5%, 0.75%, 1.0%, 1.5%, and 2.0%. In order to achieve a uniform mixture, the soil and strips were homogenously distributed and mixed with the soil by hand mixing dry soil, water, and strips. The cutting process of the PP strips, the final shape of the strips, and an example of soil mixed with strips are shown in Figure 4. To prevent floating of the strips, the water was added before the strips. In addition, the specimens were destroyed after testing to verify segregation. In this sense, the visual inspection showed that the process of mixing the strips and soil provided an excellent integration of soil and strips. In field applications, mixing is performed according to the recommendations of Falorca et al. [43] and Shukla [14]. The aspect ratios (A_r) for the strips with a length of 10, 15, 20, and 30 mm are 20, 30, 40, and 60, respectively. The PP strips have a specific mass of 0.91 g/cm^3, a tensile strength of 150 MPa, and a tensile modulus of 3.5 GPa.

(a)   (b)   (c)

Figure 4. Polypropylene (PP) strips: (**a**) cutting process; (**b**) PP strips after cutting; (**c**) soil mixed with PP strips.

This study involved a combination of UCS, direct shear, and CBR tests to investigate the effect of strips on soil improvement. The UCS tests were conducted according to ASTM D2166 [44] with samples compacted at the optimum water content for each soil shown in Figure 2. Considering the importance of compaction parameters for each soil mixture in unconfined compression strength, standard Proctor compaction tests were conducted for each soil–strip mixture in order to compact soil specimens for UCS and shear strength tests. However, no significant alterations were observed in maximum dry unit weight and optimum water content (OWC) with PP strip addition, and the soil–strip samples were compacted at the OWC of natural soil conditions (Table 1). In order to examine the variability of the effect of waste strips in the UCS properties of both lateritic soils, triplicate specimens of 50 mm diameter and 100 mm height were tested. For each combination of optimum strip content obtained from the UCS results, drained direct shear tests were conducted according to ASTM D3080 [45] on the compacted unsaturated soils. Samples were consolidated under vertical stresses of 30, 60, and 125 kPa prior to shearing. Finally, CBR tests were conducted for each percentage of PP strips according to ASTM D1883 [46]. The specimens to be tested were also prepared with soil–strip samples compacted with optimum strip content properties in relation to UCS results.

3. Results and Discussion

3.1. Influence of PP Strips on Soil Unconfined Compression Strength (UCS)

The axial stress–strain curves from the UCS tests on the SC soil reinforced with PP strips are shown in Figure 5. Similar stress–strain curves were obtained for the CL soil. The curves in Figure 5 generally show that an increase in the peak value (the UCS) is observed after addition of PP strips. For both soils the results show that there is no big difference in the axial stress among the PP–soil mixtures as well as the pure soil before 2.5% of the axial strain. This behavior is in accordance with the literature [12,14,47]. As reported by Tang et al. [12], the addition of fibers does not affect the initial stiffness of unreinforced soil. Heineck et al. [47] concluded that the stiffness of soil–strip PP composite is not influenced at small strains. Shukla [14] states that only after a certain level of shear strain do the fibers begin to be more effective. The use of PP strips contributed to a change in the soil behavior from a brittle failure to a ductile failure, as shown in typical post-test photographs in Figure 6.

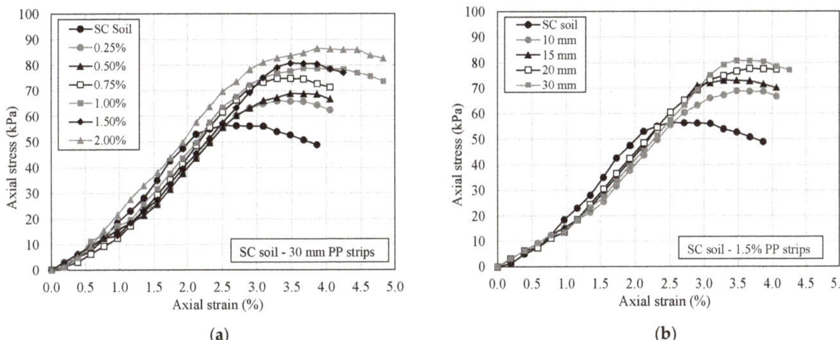

Figure 5. Axial stress–strain curves of clayey sand (SC) soil and PP strips: (**a**) increasing PP strip content; (**b**) increasing strip length.

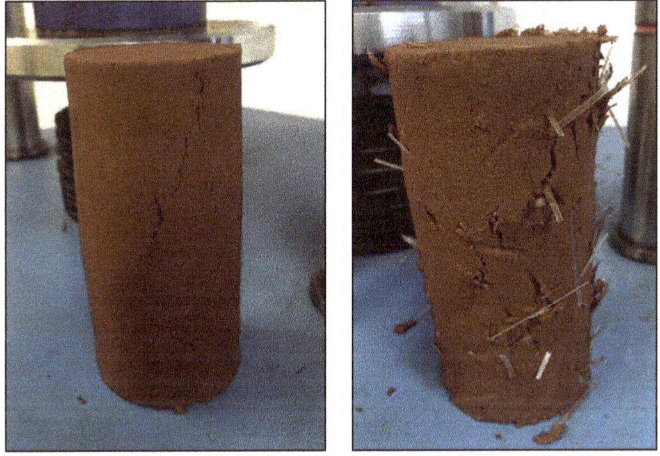

Figure 6. Specimens of natural and SC soil strips after failure.

The UCS values are shown in Figure 7 for the SC and CL soils as a function of PP strip contents for different strip lengths. For both soils, an increase in UCS was observed with increasing strip contents and lengths. No suction effects on strips results were noted. This can be explained by the fact that the strips are inert to the soil as well as by the gravimetric water content. An optimum combination of strip content and length was obtained for each soil from the UCS results. According to Figure 7a, the optimum combination for SC soil is 2% PP and 30 mm in length. In Figure 7b, the optimum combination for CL soil is 1.5% PP and 30 mm in length. These values were adopted, since previous UCS tests performed using contents of 2.5% and 3.0% of strips led the UCS values to a sharp drop for SC and CL soils considering all lengths and contents. Note that for CL soil (L = 20 and 30 mm) this occurred even before reaching 2.5%. This is probably due to the dimensions of the specimen, the length of the strips, and the excess of strips that accumulated in a concentrated manner in specific points in the specimen. Samples exhumed after the tests showed this agglomeration of the strips. This excess of strips complicates the process of compacting the specimens and can lead to lower density values, which decreases the UCS value. These results are in accordance with the literature; that is, the strength of fiber-reinforced soil increases with increasing aspect ratio of fibers [10].

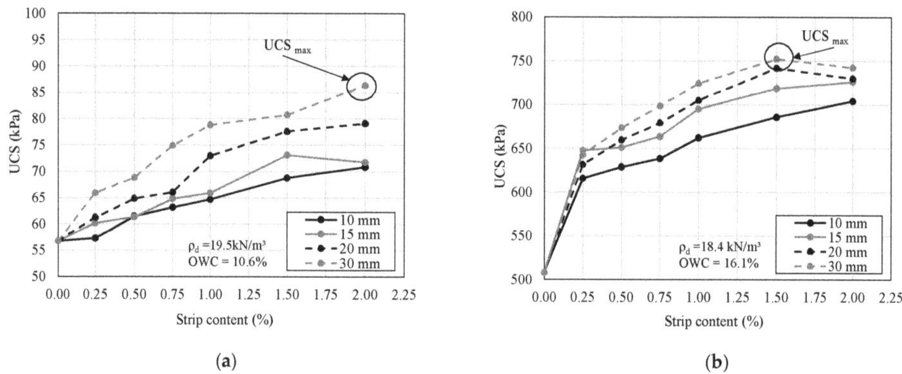

Figure 7. Unconfined compressive strength (UCS) results for different soils as a function of PP strip content for different PP strip lengths: (**a**) SC soil; (**b**) clay (CL) soil.

The UCS results for the two soils with different strip contents and strip lengths are shown in Figure 8. Both soils (with and without strips) were compacted at the corresponding optimum water content. It was observed that the soil highly influenced maximum UCS results. The SC soil presented higher increase in strength for increasing strip contents and length, showing that the soil friction is mobilized before mobilization of tension in the plastic strips. Higher strip lengths also indicated higher increase in SC shear strength, reaching the same strength increase of the clayey soil with 30 mm strip length. For the clayey soil, low contents of strips presented a significant strength increase, despite strip lengths. The increase in strip content also showed an increase in UCS.

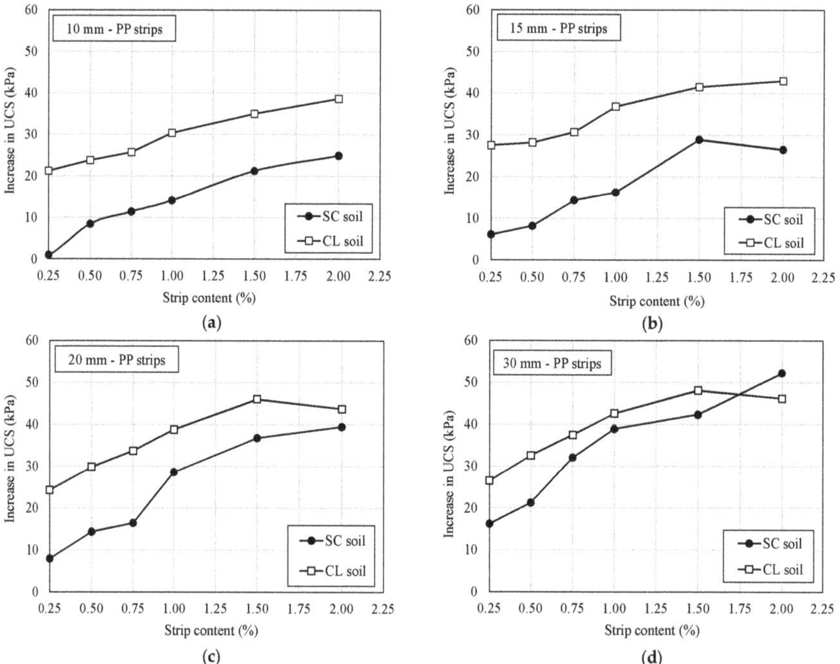

Figure 8. Influence of soil type on the UCS of soils with different strip lengths as a function of strip content: (**a**) 10 mm; (**b**) 15 mm; (**c**) 20 mm; (**d**) 30 mm.

As discussed, there are no results from the literature that discuss the use of PP strips in soil reinforcement. The literature only presents results of research using PP fibers. However, it can be seen that the results of this research are in accordance with previous results from the literature that evaluated PP fibers, e.g., [12,48–50]. Santoni et al. [48], for instance, concluded that the inclusion of randomly oriented discrete PP fibers significantly improves the UCS of sands. An optimum fiber length of 51 mm was identified for the reinforcement of sand specimens. A maximum performance is achieved at the fiber content between 0.6% and 1% by dry weight. The specimen performance is enhanced in both wet and dry optimum conditions. Tang et al. [12] evaluated the UCS of clayey soil cylindrical specimens (diameter = 39.1 mm, length = 80 mm) with inclusion of different contents of PP fibers (12 mm long). Fiber inclusion with 0.05% fiber content enhances the unconfined compressive/peak strength of soil. Kumar and Singh [49] used the random inclusion of PP fibers to evaluate the UCS of fly ash. At an aspect ratio (Ar) of 100, the unconfined compressive strength of fly ash increased from 128 to 259 kPa with an increment in fiber content from 0 to 0.5%. The results show that the variation of unconfined compressive strength with fiber content is linear, and the optimum fiber length and aspect ratio were found to be 30 mm and 100, respectively. Zaimoglu and Yetimoglu [50] investigated the UCS of fine-grained soil (MH, high plasticity soil) effects using randomly distributed PP fiber reinforcement (length = 12 mm; diameter = 0.05 mm). The main findings show that there is a tendency for UCS values to increase due to the increase in fiber content. The soil reinforced with a fiber content of 0.75% showed an expressive increase of 85% in the UCS value when compared to unreinforced soil. As Tang et al. [12] also discussed in their study, the increase in UCS might be due to the bridging effect of fiber, which can efficiently prevent the further development of failure planes and deformations of the soil.

The results from an analysis of variance (ANOVA) shown in Figure 9 indicate that the UCS is more affected by strip length or content. Results showed that strip content has a greater effect on results than strip length for both soils evaluated in this research. The equations were used to propose an analytical model to predict UCS of SC and CL soils reinforced with PP strips based on experimental results. The good agreement between the experimental data and the estimates indicates that the proposed model is adequate for estimating preliminary soil–strip UCS strength parameters. The limitations of the models include the type of soils used and PP strips with 15 mm width.

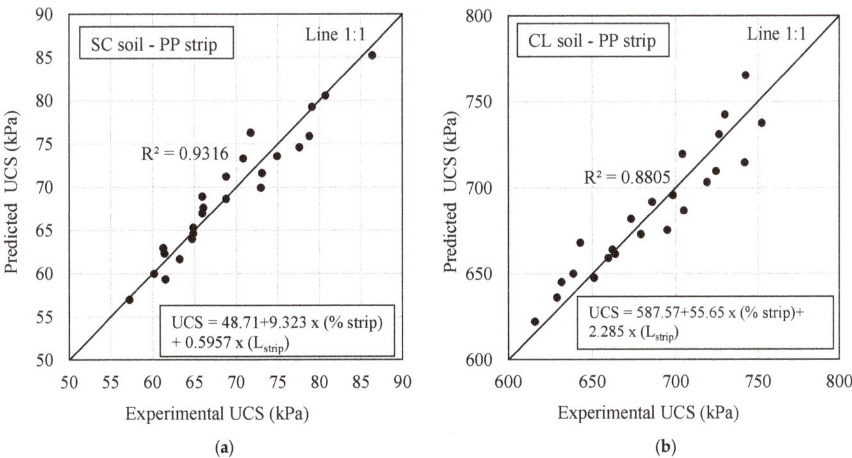

Figure 9. Prediction model for UCS of soil–strip mixtures: (**a**) SC; (**b**) CL.

An analysis showing the influence of compaction water content in UCS of soil–strip samples is shown in Figure 10. Samples at the optimum water content (OWC) had the best combination of strip length and content for each soil (Figure 7). UCS values were compared with the same mixtures compacted at OWC −2% and OWC +2% also using optimum strip combinations. The water content at compaction influenced the UCS of both soils. OWC −2% presented higher influence on UCS of both soils but with opposite results. Sandy soil showed superior UCS when compacted at OWC −2%, while clayey soil showed a lower increase in UCS. The best result for clayey soil in terms of UCS increase was seen for soil–strip samples compacted at OWC +2%. Results are more attributed to soil type than strip content.

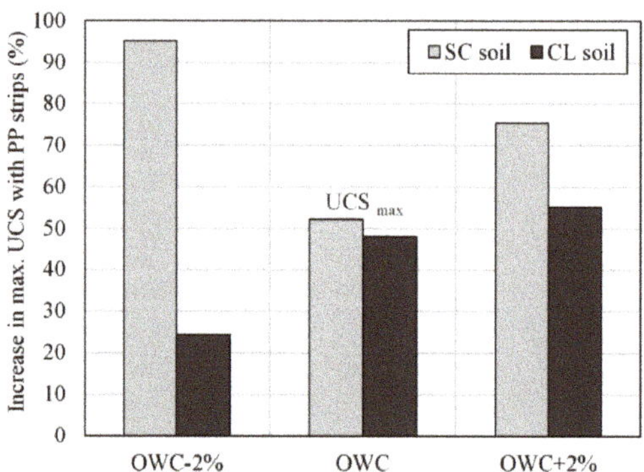

Figure 10. Influence of compaction water content on UCS results of soil mixtures at optimum strip combinations.

3.2. Influence of PP Strips on Drained Shear Strength

Results of direct shear tests considering each combination of soil and strips (15 × 30 mm) representing maximum UCS are presented in Figure 11. The specimens (with and without strips) were compacted at optimum water content. Figure 11a shows the shear strength envelopes of SC soil with and without PP strip reinforcement showing an increase in both apparent cohesion and friction angle. Figure 11b shows shear strength envelopes of the CL soil with and without PP strip reinforcement. In this case, results presented higher friction and no change in apparent cohesion. An improvement in shear strength parameters shown in Table 3 is observed with PP strip reinforcement, which can be attributed more attributed to friction than cohesion. Peddaiah et al. [32] showed results of increasing trend for apparent cohesion and friction angle with an increase in strip content, and the authors attribute this phenomenon to combined soil and plastic mass behavior during shearing. According to the author, increase in shear strength parameters is achieved because there is increase in frictional surface between soil particles and plastic strips.

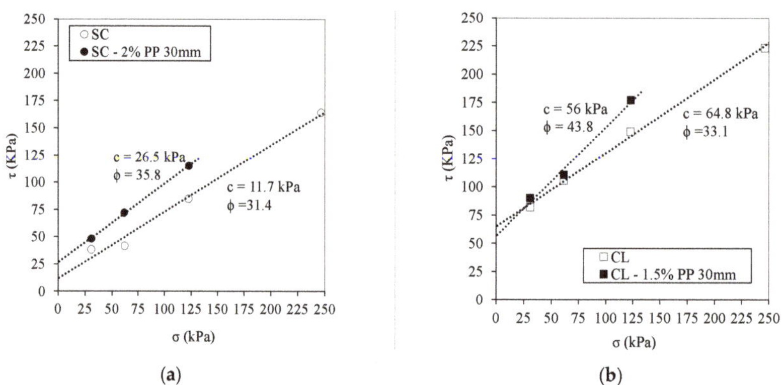

Figure 11. Shear strength envelopes of natural and PP strip-soils: (**a**) SC; (**b**) CL.

Table 3. Summary of shear strength parameters for polypropylene (PP) strips mixed with soils.

Soil Type	PP Strip Content (%)	PP Strip Length (mm)	Effective Friction Angle (Degrees)	Increase in Effective Friction (%)	Apparent Cohesion (kPa)	Increase in Apparent Cohesion (%)
SC	0.0	30	31.4	NA	11.7	NA
SC	2.0	30	35.8	1.18	26.5	2.26
CL	0.0	30	33.1	NA	56	NA
CL	1.5	30	43.8	1.47	64.8	0.86

It is important to note that, besides the fines contents, lateritic soils present good shear strength behavior when unsaturated. The natural clayey soil has a high friction angle (>30°), which is expected for lateritic soils. On the other hand, it is important to note that the soils are in an unsaturated condition, which could explain the high values of shear strength parameters, mainly the apparent cohesion (CL soil). The results presented in this research are in accordance with results of the literature, e.g., [12,14,43,51–54]. Falorca and Pinto [43] evaluated two soils very similar to the soils studied in this research. Authors carried out direct shear tests (60 mm square box) to evaluate the effect of short, randomly distributed PP microfibers on the shear strength behavior of two different types of soils: a poorly graded sandy (SP) soil and a clayey soil of low plasticity (CL). The main results show that the shear stress always increases up to the maximum deformation allowed, rather than reaching a peak or constant value typical for unreinforced soils. No significant difference was found when using straight or crimped fibers. The authors also concluded that the initial stiffness of the reinforced sand decreases with an increase in fiber content, whereas for reinforced clay there is no significant change. The reinforced sand is more compressive in the early stages of shear and more dilative subsequently compared with the unreinforced sand. There is much evidence that the influence of fiber content, fiber length, and normal stress level is due to the fibers' capacity to increase the number of contacts between soil particles and to mobilize a higher number of soil particles during shear. The number of fibers in the shear plane is a very important parameter.

Yetimoglu and Salbas [51] carried out a direct shear test (60 × 60 mm plane and 25 mm in depth) on sands reinforced with randomly distributed discrete PP fiber (length = 20 mm; diameter = 0.05 mm) reinforcements varying from 0.10% to 1%. The results of the tests indicated that the peak shear strength and initial stiffness of the clean, oven-dried, uniform river sand with particles of fine to medium size (0.075–2 mm) at a relative density of 70% were not affected significantly by the fiber reinforcement. Fiber reinforcements, however, could reduce soil brittleness providing smaller loss of post-peak strength and an increase in residual shear strength angle of the sand.

Tang et al. [12] conducted a series of direct shear tests on clayey soil cylindrical specimens (diameter = 61.8 mm, length = 20 mm) with inclusion of different percentages of PP fibers (12 mm long) at vertical normal stresses of 50, 100, 200, and 300 kPa. All the test specimens were compacted at their corresponding maximum dry unit weight and optimum water content. It was observed that the values of c and φ increased with increasing fiber content.

3.3. Influence of PP Strips on Soil CBR

Results of the CBR tests are shown in Figure 12. SC soil was highly influenced by plastic strips, with a 70% increase in CBR values. On the other hand, CL soil was not affected by strip inclusion, not altering CBR values.

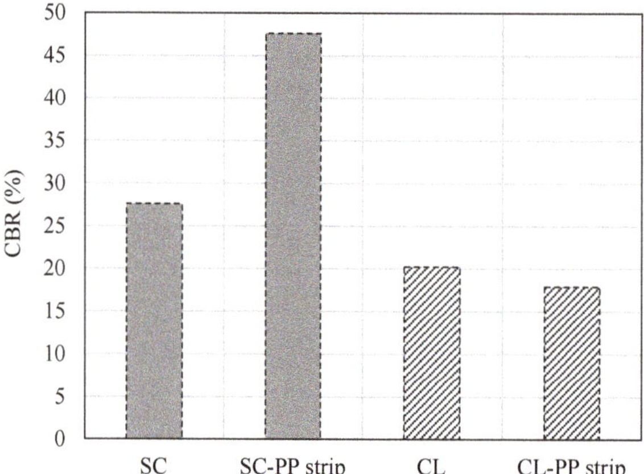

Figure 12. CBR values for SC and CL soils with and without PP strip reinforcement at their optimum combination identified from the UCS tests.

The results of the present research are in agreement with the results previously found in the literature for other soils and polymeric reinforcements, e.g., [49,50,55–58]. In this sense, as reported by Hoover et al. (1982), the CBR test values indicate that inclusion of fibers is most effective in sandy soils and less effective in fine-grained soils.

When evaluating the results obtained for the SC soil it was noted that they are in agreement with the results obtained by Fletcher and Humphries [55]. These authors showed that the CBR values of a silty soil increased significantly after the addition of PP fibers. According to the authors, PP fibers were used, and their content varied from 0%, 0.5%, 1%, and 1.5% in relation to the dry mass of soil, compacted with normal energy. The dimensions of the fibers used were 25 mm in length and 0.76 mm in diameter. According to the authors, there is an optimal fiber dosage that provides the highest CBR value. Higher than optimal dosages decrease the CBR value, since, with the increase in the number of fibers, there is a reduction in the amount of soil, which in turn affects the bonding forces at the soil–fiber interface. Finally, the authors concluded that the addition of fibers resulted in an increase in the CBR value of 133% when compared to the soil without the addition of fibers. Yetimoglu et al. [56] performed laboratory CBR tests to investigate the load-penetration behavior of a clean sand fill reinforced with randomly distributed discrete PP fibers (length = 20 mm; diameter = 0.50 mm) overlying a high plasticity inorganic clay with a nonwoven geotextile layer at the sand–clay interface as a separator. It was noticed that the peak load ratio (PLR) value increases with an increase in fiber content and becomes approximately five times as high as that of unreinforced sand.

Regarding the clayey soil, it was noted that the addition of fibers at the proposed optimum content generated an increase in expansion and a reduction in CBR due to the number of fibers present, impairing the contact (friction) between the particles. This behavior is in line with the results obtained by Pradhan et al. [58]. These authors evaluated the mechanical strength of a clayey soil reinforced with PP fibers by direct shear, unconfined compression, and CBR tests. The authors used PP fibers of 15, 20, and 25 mm in length with a 0.2 mm diameter, varying the fiber content from 0.1% to 1.0%, with an increase of 0.1%.

Chandra et al. [57] evaluated soils with PP fibers (length = 15, 25, 30 mm; diameter = 0.3 mm) and concluded that the CBR value of reinforced soils continues to increase with both fiber content and aspect ratio (Ar). However, they suggest that mixing soil and fibers is extremely difficult beyond the fiber content of 1.5%. The authors also suggest that 1.5% fiber content and an aspect ratio of 100 can be considered optimum values in the case of soils of low compressibility (classified as CL and ML), whereas 1.5% fiber content with an aspect ratio of 84 was found to be optimum for silty sand (classified as SM). Similarly, Kumar and Singh [49] studied fly ash (classified as silt of low compressibility, ML) with randomly distributed PP fibers. The soaked and unsoaked CBR values presented increases with an increase in fiber content at a particular aspect ratio (60, 80, 100, or 120). Zaimoglu and Yetimoglu [50] also investigated the effects of randomly distributed PP fiber reinforcement (length = 12 mm; diameter = 0.05 mm) on the soaked CBR behavior of a fine-grained soil (MH, high plasticity soil) by conducting a series of CBR tests. The main results show that the CBR value presented a significant increase with increasing fiber content up to around 0.75% and remained more or less constant thereafter.

According to the design of flexible pavements [59], based on CBR values of pavement layers, a subgrade thickness for the SC soil used in this area of research (CBR = 28%) is 16 cm for heavy traffic condition (55 kN wheel load), and it reduces to 10 cm for the same traffic condition for 2.0% plastic waste mixed with soil (CBR = 48%). The final reduction implies the reduction of natural resources (aggregate materials) and construction costs. The clayey soil–strip mixture does not meet the required 20% CBR for subbases but can be suitable for other applications.

4. Conclusions

An extensive experimental program was conducted in order to assess the effect of polypropylene waste strips (cut from recycled plastic packing) mixed with lateritic soils. The experimental program involved the evaluation of soil UCS properties and an optimum combination of soil–PP strips. Outcomes of these combinations were used in CBR and shear strength analysis. The following conclusions can be drawn from this research:

- The use of PP strips as reinforcements in both SC and CL lateritic soils led to an increase in UCS, as well as a clear influence of PP strip length on the soil stiffness. The use of PP strips contributed to a change in soil failure from a brittle to a ductile mode;
- The UCS results revealed an optimum combination of PP strip content and strip length: SC soil and 2% PP and 30 mm in length and CL soil with 1.5% PP and 30 mm in length. The SC soil had a higher increase in UCS for increasing strip content and strip length, indicating that the soil friction is mobilized before strip mobilization. For the CL soil, low strip contents led to a significant increase in UCS regardless of the strip length. Statistical analysis conducted showed that strip content has a greater effect on the UCS than the strip length for both soils evaluated;
- The compaction water content had an important effect on the UCS of both soils, although opposite effects were observed in the UCS for both soils when increasing and decreasing the compaction water content by +2% and −2% from the optimal value;
- Results from direct shear tests indicate that PP strip–SC soil showed an increase in both apparent cohesion and friction angle, while PP strip–CL soil presented a higher friction angle and no change in apparent cohesion;
- California bearing ratio (CBR) tests indicate that SC soil was highly influenced by plastic strips and experienced a 70% increase in CBR after reinforcement. On the other hand, the CBR of the CL soil was not affected by the addition of plastic strips.

Author Contributions: Conceptualization, R.M., P.C.L., N.d.S.C. and J.S.M.; methodology, R.M., P.C.L., N.D.S.C., H.L.G., R.A.R. and J.S.M.; formal analysis, R.M., P.C.L., N.d.S.C. and J.S.M.; investigation, R.M., P.C.L., H.L.G., R.A.R.; resources, R.M., P.C.L., N.d.S.C. and J.S.M; writing—original draft preparation, R.M., P.C.L., N.d.S.C., H.L.G., R.A.R. and J.S.M.; writing—review and editing, R.M., P.C.L., N.d.S.C. and J.S.M.; visualization, R.M., P.C.L., N.d.S.C. and J.S.M.; supervision, P.C.L., N.d.S.C. and J.S.M.; project administration, P.C.L., N.D.S.C. and J.S.M. All authors have read and agreed to the published version of the manuscript.

Funding: This research received no external funding.

Acknowledgments: The author are very thankful to the Capes/Print program and PROPG/UNESP.

Conflicts of Interest: The authors declare no conflict of interest.

References

1. United Nations General Assembly. *Transforming our World: The 2030 Agenda for Sustainable Development*; United Nations General Assembly: New York, NY, USA, 2015; pp. 1–35.
2. Luwalaga, J.G. Analysing the Behaviour of Soil Reinforced with Polyethylene. Ph.D. Thesis, Stellenbosch University, Stellenbosch, Stellenbosch, South Africa, 2016.
3. Hataf, N.; Rahimi, M.M. Experimental investigation of bearing capacity of sand reinforced with randomly distributed tire shreds. *Constr. Build. Mater.* **2006**, *20*, 910–916. [CrossRef]
4. Özkul, Z.H.; Baykal, G. Shear behavior of compacted rubber fiber-clay composite in drained and undrained loading. *J. Geotech. Geoenviron. Eng.* **2007**, *133*, 767–781. [CrossRef]
5. Yoon, Y.W.; Heo, S.B.; Kim, K.S. Geotechnical performance of waste tires for soil reinforcement from chamber tests. *Geotext. Geomembr.* **2008**, *26*, 100–107. [CrossRef]
6. Soltani, A.; Deng, A.; Taheri, A.; Mirzababaei, M.; Nikraz, H. Interfacial shear strength of rubber-reinforced clays: A dimensional analysis perspective. *Geosynth. Int.* **2019**, *26*, 164–183. [CrossRef]
7. Indraratna, B.; Qi, Y.; Heitor, A. Evaluating the properties of mixtures of steel furnace slag, coal wash, and rubber crumbs used as subballast. *J. Mater. Civ. Eng.* **2018**, *30*, 04017251. [CrossRef]
8. Qi, Y.; Indraratna, B.; Vinod, J.S. Behavior of steel furnace slag, coal wash, and rubber crumb mixtures with special relevance to stress–dilatancy relation. *J. Mater. Civ. Eng.* **2018**, *30*, 04018276. [CrossRef]
9. Qi, Y.; Indraratna, B.; Coop, M.R. Predicted behavior of saturated granular waste blended with rubber crumbs. *Int. J. Geomech.* **2019**, *19*, 04019079. [CrossRef]
10. Qi, Y.; Indraratna, B. Energy-based approach to assess the performance of a granular matrix consisting of recycled rubber, steel-furnace slag, and coal wash. *J. Mater. Civ. Eng.* **2020**, *32*, 04020169. [CrossRef]
11. Ziegler, S.; Leshchinsky, D.; Ling, H.I.; Perry, E.B. Effect of short polymeric fibers on crack development in clays. *Soils Found.* **1998**, *38*, 247–253. [CrossRef]
12. Tang, C.; Shi, B.; Gao, W.; Chen, F.; Cai, Y. Strength and mechanical behavior of short polypropylene fiber reinforced and cement stabilized clayey soil. *Geotext. Geomembr.* **2007**, *25*, 194–202. [CrossRef]
13. Shukla, S.K.; Sivakugan, N.; Das, B.M. Fundamental concepts of soil reinforcement—An overview. *Int. J. Geotech. Eng.* **2009**, *3*, 329–342. [CrossRef]
14. Shukla, S.K. *Fundamentals of Fibre—Reinforced Soil Engineering*; Springer: Singapore, 2017; ISBN 978-981-10-3061-1.
15. Ehrlich, M.; Almeida, S.; Curcio, D.; Almeida, M.S.S.; Curcio, D.; Almeida, S.; Curcio, D.; Almeida, M.S.S.; Curcio, D.; Almeida, S.; et al. Hydro-mechanical behavior of a lateritic fiber-soil composite as a waste containment liner. *Geotext. Geomembr.* **2019**, *47*, 42–47. [CrossRef]
16. Consoli, N.C.; Thomé, A.; Girardello, V.; Ruver, C.A. Uplift behavior of plates embedded in fiber-reinforced cement stabilized backfill. *Geotext. Geomembr.* **2012**, *35*, 107–111. [CrossRef]
17. Sivakumar Babu, G.L.; Chouksey, S.K. Stress-strain response of plastic waste mixed soil. *Waste Manag.* **2011**, *31*, 481–488. [CrossRef]
18. Botero, E.; Ossa, A.; Sherwell, G.; Ovando-Shelley, E. Stress-strain behavior of a silty soil reinforced with polyethylene terephthalate (PET). *Geotext. Geomembr.* **2015**, *43*, 363–369. [CrossRef]
19. Wu, H.; Huang, B.; Shu, X.; Zhao, S. Evaluation of geogrid reinforcement effects on unbound granular pavement base courses using loaded wheel tester. *Geotext. Geomembr.* **2015**, *43*, 462–469. [CrossRef]
20. Zhao, J.J.; Lee, M.L.; Lim, S.K.; Tanaka, Y. Unconfined compressive strength of PET waste-mixed residual soils. *Geomech. Eng.* **2015**, *8*, 53–66. [CrossRef]
21. Louzada, N.D.S.L.; Malko, J.A.C.; Casagrande, M.D.T. Behavior of clayey soil reinforced with polyethylene terephthalate. *J. Mater. Civ. Eng.* **2019**, *31*, 1–11. [CrossRef]
22. Cai, Y.; Shi, B.; Ng, C.W.W.; Tang, C. sheng Effect of polypropylene fibre and lime admixture on engineering properties of clayey soil. *Eng. Geol.* **2006**, *87*, 230–240. [CrossRef]
23. Plé, O.; Lê, T.N.H.H. Effect of polypropylene fiber-reinforcement on the mechanical behavior of silty clay. *Geotext. Geomembr.* **2012**, *32*, 111–116. [CrossRef]
24. Anagnostopoulos, C.A.; Papaliangas, T.T.; Konstantinidis, D.; Patronis, C. Shear strength of sands reinforced with polypropylene fibers. *Geotech. Geol. Eng.* **2013**, *31*, 401–423. [CrossRef]
25. Correia, A.A.S.; Venda Oliveira, P.J.; Custódio, D.G. Effect of polypropylene fibres on the compressive and tensile strength of a soft soil, artificially stabilised with binders. *Geotext. Geomembr.* **2015**, *43*, 97–106. [CrossRef]

26. Claria, J.J.; Vettorelo, P.V. Mechanical behavior of loose sand reinforced with synthetic fibers. *Soil Mech. Found. Eng.* **2016**, *53*, 12–18. [CrossRef]

27. Nguyen, L.; Fatahi, B.; Khabbaz, H. Predicting the behaviour of fibre reinforced cement treated clay. *Procedia Eng.* **2016**, *143*, 153–160. [CrossRef]

28. Mirzababaei, M.; Mohamed, M.; Miraftab, M. Analysis of strip footings on fiber-reinforced slopes with the aid of particle image velocimetry. *J. Mater. Civ. Eng.* **2017**. [CrossRef]

29. Sivakumar Babu, G.L.; Vasudevan, A.K.; Haldar, S. Numerical simulation of fiber-reinforced sand behavior. *Geotext. Geomembr.* **2008**, *26*, 181–188. [CrossRef]

30. Chebet, F.C.; Kalumba, D. Laboratory investigation on re-using polyethylene (plastic) bag waste material for soil reinforcement in geotechnical engineering. *Civ. Eng. Urban Plan.* **2014**, *1*, 1–16.

31. Soltani-Jigheh, H. Undrained behavior of clay—Plastic waste mixtures. In Proceedings of the 11th International Congress on Advances in Civil Engineering, Istanbul, Turkey, 21–25 October 2014.

32. Peddaiah, S.; Burman, A.; Sreedeep, S. Experimental study on effect of waste plastic bottle strips in soil improvement. *Geotech. Geol. Eng.* **2018**, *36*, 2907–2920. [CrossRef]

33. Fathi, H.; Jamshidi Chenari, R.; Vafaeian, M. Shaking table study on PET strips-sand mixtures using laminar box modelling. *Geotech. Geol. Eng.* **2020**, *38*, 683–694. [CrossRef]

34. Onyelowe, K.C.; Bui Van, D.; Ubachukwu, O.; Ezugwu, C.; Salahudeen, B.; Nguyen Van, M.; Ikeagwuani, C.; Amhadi, T.; Sosa, F.; Wu, W.; et al. Recycling and reuse of solid wastes; a hub for ecofriendly, ecoefficient and sustainable soil, concrete, wastewater and pavement reengineering. *Int. J. Low-Carbon Technol.* **2019**, *14*, 440–451. [CrossRef]

35. *ASTM D7928-17. Standard Test Method for Particle-Size Distribution (Gradation) of Fine-Grained Soils Using the Sedimentation (Hydrometer) Analysis*; ASTM: West Conshohocken, PA, USA, 2017; pp. 1–25.

36. *ASTM D2487-17. Standard Practice for Classification of Soils for Engineering Purposes (Unified Soil Classification System)*; ASTM: West Conshohocken, PA, USA, 2017; pp. 1–10.

37. *ASTM D3282-15. Standard Practice for Classification of Soils and Soil-Aggregate Mixtures for Highway Construction Purposes*; ASTM: West Conshohocken, PA, USA, 2015; pp. 1–7.

38. *ASTM D854-14. Standard Test Methods for Specific Gravity of Soil Solids by Water Pycnometer*; ASTM: West Conshohocken, PA, USA, 2014; pp. 1–7.

39. *ASTM D698-12e12. Standard Test Methods for Laboratory Compaction Characteristics of Soil Using Standard Effort (12 400 ft-lbf/ft3 (600 kN-m/m3))*; ASTM: West Conshohocken, PA, USA, 2012; pp. 1–13.

40. *ASTM D4318-17e1. Standard Test Methods for Liquid Limit, Plastic Limit, and Plasticity Index of Soils*; ASTM: West Conshohocken, PA, USA, 2017; Volume 04.08, pp. 1–14.

41. Van Genuchten, M.T. A Closed-form equation for predicting the hydraulic conductivity of unsaturated soils. *Soil Sci. Soc. Am. J.* **1980**, *44*, 892–898. [CrossRef]

42. Benatti, J.C.B.; Rodrigues, R.A.; Miguel, M.G. Aspects of mechanical behavior and modeling of a tropical unsaturated soil. *Geotech. Geol. Eng.* **2013**, *31*, 1569–1585. [CrossRef]

43. Falorca, I.M.C.F.G.; Pinto, M.I.M. Effect of short, randomly distributed polypropylene microfibres on shear strength behaviour of soils. *Geosynth. Int.* **2011**, *18*, 2–11. [CrossRef]

44. *ASTM D2166. Standard Test Method for Unconfined Compressive Strength of Cohesive Soil*; ASTM: West Conshohocken, PA, USA, 2016.

45. *ASTM Standard D3080/D3080M-11. Direct Shear Test of Soils Under Consolidated Drained Conditions*; ASTM: West Conshohocken, PA, USA, 2012; pp. 1–9.

46. *ASTM D1883. Standard Test Method for California Bearing Ratio (CBR) of Laboratory-Compacted*; ASTM: West Conshohocken, PA, USA, 1999.

47. Heineck, K.S.; Coop, M.R.; Consoli, N.C. Effect of microreinforcement of soils from very small to large shear strains. *J. Geotech. Geoenvironmental Eng.* **2005**, *131*, 1024–1033. [CrossRef]

48. Santoni, B.R.L.; Tingle, J.S.; Members, A.; Webster, S.L. Engineering properties of sand-fiber mixtures for road construction. *J. Geotech. Geoenviron. Eng.* **2001**, 258–268. [CrossRef]

49. Kumar, P.; Singh, S.P. Fiber-reinforced fly ash subbases in rural roads. *J. Transp. Eng.* **2008**, *134*, 171–180. [CrossRef]

50. Zaimoglu, A.S.; Yetimoglu, T. Strength behavior of fine grained soil reinforced with randomly distributed polypropylene fibers. *Geotech. Geol. Eng.* **2012**, *30*, 197–203. [CrossRef]

51. Yetimoglu, T.; Salbas, O. A study on shear strength of sands reinforced with randomly distributed discrete fibers. *Geotext. Geomembr.* **2003**, *21*, 103–110. [CrossRef]
52. Lovisa, J.; Shukla, S.K.; Sivakugan, N. Behaviour of prestressed geotextile-reinforced sand bed supporting a loaded circular footing. *Geotext. Geomembr.* **2010**, *28*, 23–32. [CrossRef]
53. Tang, C.-S.S.; Shi, B.; Zhao, L.-Z.Z. Interfacial shear strength of fiber reinforced soil. *Geotext. Geomembr.* **2010**, *28*, 54–62. [CrossRef]
54. Li, J.; Tang, C.; Wang, D.; Pei, X.; Shi, B. Effect of discrete fibre reinforcement on soil tensile strength. *J. Rock Mech. Geotech. Eng.* **2014**, *6*, 133–137. [CrossRef]
55. Fletcher, C.S.; Humphries, W.K. California bearing ratio improvement of remolded soils by the addition of polypropylene fiber reinforcement. *Transp. Res. Rec.* **1991**, 80–86.
56. Yetimoglu, T.; Inanir, M.; Inanir, O.E. A study on bearing capacity of randomly distributed fiber-reinforced sand fills overlying soft clay. *Geotext. Geomembr.* **2005**, *23*, 174–183. [CrossRef]
57. Chandra, S.; Viladkar, M.N.; Nagrale, P.P. Mechanistic approach for fiber-reinforced flexible pavements. *J. Transp. Eng.* **2008**, *134*, 15–23. [CrossRef]
58. Pradhan, P.K.; Kar, R.K.; Naik, A. Effect of random inclusion of polypropylene fibers on strength characteristics of cohesive soil. *Geotech. Geol. Eng.* **2012**, *30*, 15–25. [CrossRef]
59. Zuzulova, A.; Hodakova, D.; Capayova, S.; Schlosser, T. Design of pavement structures in tunnels. In Proceedings of the 19th International Multidisciplinary Scientific GeoConference SGEM 2019, Sofia, Bulgaria, 30 June–6 July 2019.

Publisher's Note: MDPI stays neutral with regard to jurisdictional claims in published maps and institutional affiliations.

 *sustainability*

Review

Sustainable Solutions with Geosynthetics and Alternative Construction Materials—A Review

Ennio M. Palmeira [1,*], Gregório L. S. Araújo [1] and Eder C. G. Santos [2]

[1] Department of Civil and Environmental Engineering, Faculty of Technology, University of Brasília, Brasilia 70910-900, Brazil; gregorio@unb.br

[2] School of Civil and Environmental Engineering, Federal University of Goiás, Goiânia 74605-220, Brazil; edersantos@ufg.br

* Correspondence: palmeira@unb.br; Tel.: +55-61-3107-0969

Abstract: Geosynthetics have proven to provide sustainable solutions for geotechnical and geoenvironmental problems when used with natural materials. Therefore, the expected benefits to the environment when geosynthetics are associated with unconventional or alternative construction materials will be even greater. This paper addresses the use of geosynthetics with wasted materials in different applications. The potential uses of alternative materials such as wasted tires, construction and demolition wastes, and plastic bottles are presented and discussed considering results from laboratory and field tests. Combinations of geosynthetics and alternative construction materials applied to reinforced soil structures, drainage systems for landfills, barriers, and stabilisation of embankments on soft grounds are discussed. The results show the feasibility of such combinations, and that they are beneficial to the environment and in line with the increasing trend towards a circular economy and sustainable development.

Keywords: geosynthetics; wastes; tires; CDW; PET bottles

Citation: Palmeira, E.M.; Araújo, G.L.S.; Santos, E.C.G. Sustainable Solutions with Geosynthetics and Alternative Construction Materials—A Review. *Sustainability* **2021**, *13*, 12756. https://doi.org/10.3390/su132212756

Academic Editors: Castorina Silva Vieira and Marc A. Rosen

Received: 30 September 2021
Accepted: 16 November 2021
Published: 18 November 2021

Publisher's Note: MDPI stays neutral with regard to jurisdictional claims in published maps and institutional affiliations.

1. Introduction

The preservation of the environment as a whole and specifically of its natural resources is of utmost importance for current and future generations. In this context, geosynthetics can provide sustainable engineering solutions for geotechnical and geoenvironmental problems, reducing the consumption of natural materials and causing less impact to the environment [1–4]. Geosynthetics can perform different functions in an engineering project, such as drainage, filtration, barrier, separation, reinforcement, and protection. Frischknecht et al. [1] showed reductions greater than 70% of some environmental impact parameters such as water consumption, renewable and non-renewable energy consumption, emissions of gases that contribute to global warming, etc. in some applications of geosynthetics in comparison with conventional geotechnical solutions. Significant reductions in energy consumption and CO_2 emissions were obtained by Damians et al. [2] when the environmental impacts caused by geosynthetic reinforced retaining structures were compared to those from conventional concrete retaining walls. Heerten [3] also presented two examples of construction infrastructures where the use of geosynthetics showed lower environmental impacts due to significant reductions of cumulated energy demand and CO_2 emissions. Touze-Foltz [4] presented several other examples of the environmental benefits of using geosynthetics in geotechnical and geoenvironmental works.

The benefits brought by the use of engineering solutions with geosynthetics discussed above were obtained using conventional soils as construction materials. When geosynthetics are combined with materials commonly considered wastes, solutions involving geosynthetics will be expected to be even more beneficial to the environment, since they will avoid or reduce the utilisation of good-quality natural materials (which are increasingly

209

scarce and expensive in several regions) and reduce their exploitation, with positive repercussions for the environment. These are the cases of combinations of different geosynthetic products with wasted tires, plastic objects, and construction and demolition wastes, for instance. However, one must bear in mind that some of these wastes can be harmful to the environment. Thus, due care must be exercised when using wastes in construction to avoid ground contamination due to the degradation of the waste with time or the presence of pollutants. In this context, the use of such wastes may still be feasible if appropriate geosynthetic barriers (geomembranes or GCLs) are employed.

Very little can be found in the literature on the combination of geosynthetic products and alternative/waste construction materials in geotechnical and geoenvironmental works. The same comment applies to the use of wastes to produce low-cost alternative geosynthetics. This paper presents the properties and relevant characteristics of some waste materials, treated hereafter as alternative construction materials, that can be combined with geosynthetics in geotechnical and geoenvironmental works. Advantages, limitations, and examples of such combinations are presented and discussed.

2. Some Examples of Alternative Construction Materials Commonly Used in Engineering Projects

2.1. Wasted Tires

Over the years, population growth has led to higher volumes of wastes, requiring large areas for disposal. One material that has been pointed out as an environmental hazard is discarded tire. Used tires have approximately 75% voids and, in some countries, there are several specific stockpile areas for their disposal. Therefore, these materials can generate an environmental problem due to the growing demand for space for their disposal. In geotechnical engineering applications, tires can be utilised as lightweight material after they are shredded to small pieces.

Tire stockpiles present large groundwater contamination potential once they are overall composed of rubbers, carbon black, metals, antioxidants, and polymers [5]. Besides the risk of ground contamination, there is also the possibility of combustion when the wasted tires are disposed in large areas (Figure 1a). Humphrey [6] reported self-heating reactions of tires in Washington and Colorado. Another occurrence of this type took place in Hagersville, Ontario, Canada (Figure 1b), where a large tire stockpile fire burned for days [7]. The huge amount of wasted tires is a global environmental concern and the related risk of either soil/groundwater contamination or stockpile fire poses the need to recycle and reuse these materials.

(a)

(b)

Figure 1. Environmental hazards produced by wasted tires [7]: (**a**) tire stockpile disposal; (**b**) tire stockpile fire in Canada in 1990 [7].

According to the United States Tires Association (USTA) [8], 303.5 million rubber tires were produced in 2019. The United States has made an effort to recycle these materials as much as possible and, in 2019, almost 76% of scrap tires were recycled to be used in rubber-modified asphalts, the manufacturing of automotive products, mulch for landscaping, and tire-derived fuel. The USTA also reported that tires are now one of the most recycled products in the U.S., but end-of-life markets are not keeping pace with their annual generation [8]. On the other hand, the Canadian Association of Tire Recycling Agencies reported that 82% of the 421,184 tires collected were recycled in 2019 [9]. In 2020, Brazil recycled 42 million scrap tires with a produced total amount of 59.5 million [10,11]. However, despite the efforts to recycle these materials, there are still tons of scrap tires discarded throughout the country.

Based on the above information, there is a need to recycle wasted tires as much as possible in order to reduce the environmental impacts caused by their disposal, and one way to tackle this problem is by mixing them with soil. Many applications of mixtures of wasted tires in geotechnical applications can be found in the literature. It is possible to find investigations involving laboratory and field tests with different types of soils and rubbers for different applications [6,12–26]. Despite the relevance of such mixtures, the present paper focuses on combinations of some wastes, including rubber, with geosynthetics in geotechnical and geoenvironmental applications.

2.2. Recycled Construction and Demolition Wastes (RCDW)

Important studies have been conducted for decades on raw source material conservation. The relevance of such studies is even more evident today, as human-made mass has exceeded all living biomass [27]. In 2020, the sub-groups of concrete and aggregates were estimated to represent approximately 80% of the total amount of anthropogenic mass—inanimate solid objects made by humans that have not been demolished or taken out of service. Nowadays, it is imperative to investigate solutions to improve the use of construction and demolition wastes (CDW).

The need for materials for future construction and renovation of buildings and/or infrastructure brings recycling as a fundamental strategy, once it can provide a material with a low-carbon footprint, hereafter called "recycled construction and demolition waste" (RCDW). Bearing in mind that over 10 billion tons of CDW are generated per year [28], recycling could significantly reduce the volume of this waste currently destined to final disposal in landfills. At the same time, it could also reintroduce RCDW into construction works at a very low cost and reduce the environmental impact since it could mitigate the demand for natural raw material supply.

Due to the introduction of various legislative measures and the adoption of several waste management strategies, CDW has been investigated in several disciplines, including phenomenology, environmental science and environmental engineering, material science and engineering, the industrial ecology perspective, management science, architecture, the construction and operation of buildings [28], pavement engineering, and geotechnical engineering. However, the applications of RCDW are mainly focused on the use of recycled aggregate in road pavements and concrete, but this is a consequence of the methods usually adopted by the recycling plants and some economic aspects.

With the potential for consuming a large amount of RCDW, investigations on the application in pavements have shown geotechnical properties equivalent or superior to those of typical quarry granular coarse base and subbase materials [29,30]. Laboratory tests have revealed that the use of recycled concrete aggregate in hot-mix asphalt for base courses is promising, given that the bitumen absorbed by the aggregate makes its whole surface be coated by the binder, reducing the porosity and, therefore, the sensitivity to moisture [31,32]. The adoption of new concepts when designing a cold asphalt mixture with recycled concrete aggregate may allow the application of such recycled products [33]. Regarding the potential environmental impacts of RCDW, field-site leaching from crushed concrete was measured after 10 years of exposure in a road subbase and the simplified risk assessment showed that the released trace elements did not exceed the pre-defined acceptance criteria for groundwater and fresh water [34].

Given that some countries classify soil as CDW, this material may correspond to a significant volume received by some recycling plants. Considering those regions where measures to promote an efficient on-site sorting process are not developed yet, the CDW will consist of a mixture of soils (from excavation), inert and non-inert materials, and materials from site clearance activities (organic matter and other debris). This fraction of the CDW, composed of soil, is not desirable for producing aggregates for concrete and pavements due to poor-quality properties. However, the recycling processes usually carried out by the recycling plants (sorting of non-inert material, crushing, and sieving) can enable the RCDW to be used as backfill material for geosynthetic reinforced soil structures (GRS). This use may allow the construction of such structures in places where natural materials are scarce, reducing the exploitation of new and far quarries and, consequently, their economic and environmental costs.

2.3. Plastic Bottles

Plastic wastes can also be combined with geosynthetics to fulfil different functions in geotechnical works. Despite its enormous advantages for society in different areas, when not properly disposed or confined, plastics can cause damage to the environment as well as serve as a habitat for organisms that are hazardous to public health. It is estimated that over 1,000,000 plastic bottles are sold every minute [35], with still a limited amount of recycling in most countries. In addition, it may take over 450 years for a plastic bottle to completely degrade in the environment. Plastic bottles may end up in rivers (Figure 2) [36], beaches, and even far out in the oceans, forming huge floating masses, or accumulating on remote islands and coral atolls [37], with evident harm to marine fauna. Thus, it is important to develop and encourage other uses for plastics in general, particularly for polyester (PET) bottles. As far as geosynthetics are concerned, PET bottles can be used as a drainage medium associated with a geotextile filter, for instance, as will be seen later in this paper, as well as being processed to produce recycled geosynthetics.

Figure 2. Plastic bottle accumulation in Tietê River, Brazil [36].

3. Some Combinations of Geosynthetics and Alternative Construction Materials

3.1. Recycled Construction and Demolition Wastes in Geosynthetic Reinforced Structures—Concerns and Relevant Properties

Some properties and characteristics of RCDW must be evaluated when considering the use of such material in geotechnical works. Below, some of these characteristics are described and discussed.

- Particle Crushing

Even when processing a homogenous CDW, the procedures adopted by the recycling plant may determine several of the properties of RCDW. Gomes et al. [38] investigated the influence of three different comminution and sizing processes (simple screening, crushing, and grinding) on the composition, shape, and porosity characteristic of a recycled concrete aggregate (obtained from concrete block wastes). The results revealed products with different chemical and mineralogical compositions, grain size distribution, particle shapes, and porosity. This case highlights the importance of considering the application of RCDW with commonly considered undesirable fractions (soil and powder) in works where material selection is technically more tolerant.

The compaction process, usually carried out during the construction of a GRS, improves the backfill material strength and promotes better interaction with the reinforcing element (e.g., geogrid or geotextile). However, this construction procedure may also cause additional crushing and breakage of RCDW particles, changing its grain size distribution. In a laboratory investigation, Leite et al. [29] found that the physical changes caused by compaction (changing grain size distribution and increasing the percentage of cubic grains) contributed to a better densification of the RCDW aggregate and consequently improved its bearing capacity, resilient modulus, and resistance to permanent deformation.

The degradation of aggregates and soils during shearing, when subjected to monotonic or cyclic stresses, has always been a concern for researchers and engineers even for natural materials such as decomposed granite soil [39], silica sand [40], and latite basalt [41]. When dealing with RCDW, this concern deserves special attention given that these materials present a very heterogeneous composition due to having different origins. According to Sivakumar et al. [42], cyclic direct shear tests on recycled aggregate revealed a reduction in friction angle due to particle crushing (from 43° to 38° for crushed concrete, and from 43° to 39° for crushed brickwork). Although special attention is needed for the construction and maintenance of some geotechnical works, the reported reduction in friction angle would not prevent the use of RCDW in most geosynthetic applications.

Domiciano et al. [43] conducted a laboratory investigation on RCDW with different grain size distributions—three products from a local recycling plant—subjected to static loading (ranging from 150 to 600 kPa). The RCDW more susceptible to particle breakage when subjected to the static loading process was the one presenting two main characteristics: (i) uniform grain size distribution and (ii) a composition marked by a significant presence of ceramic components. The RCDW products that were composed of concrete components showed a low occurrence of grain breakage. In general, the results revealed the occurrence of significant changes in the grain size distribution curves, but no abrupt particle breakage was noticed. It was observed that particles larger than 9.5 mm were the ones most affected by loading, showing a smooth surface (less rough) caused by the removal of fines particles around them.

- Interface Shear Strength

Soil–geosynthetic interaction is of utmost importance for the design and performance of GRS structures, and this interaction can be very complex depending on the nature and properties of the reinforcement and the soil [44]. The need to understand the interaction mechanisms in a GRS structure has encouraged researchers to develop new tests and to modify some classical ones, such as (i) a direct shear test with the geosynthetic specimen at the shear plane [45,46], (ii) a direct shear test with the geosynthetic specimen inclined to the shear plane [45,47,48], (iii) a confined tensile test [49,50], and (iv) a pull-out test [45,46]. Several studies have been performed using such testing apparatuses with different types of soils and geosynthetics, revealing all the main factors (e.g., boundary condition and scale factor) that may affect the test results. However, due to the few studies carried out with RCDW, such influencing factors will not be discussed in this paper.

A study conducted by Touahamia et al. [51] investigated the shear strengths of three waste materials (building debris, crushed concrete, and quarry waste). The tests were carried out using a medium-size shear box (305 mm × 305 mm) and considered different conditions of moisture content (dry or wet), the presence of reinforcement (reinforced or non-reinforced), and contamination (clean or smeared with clay slurry—a condition investigated only for concrete and quarry wastes). The waste materials were prepared in the laboratory with grain sizes between 20 and 40 mm. The contaminated condition was achieved with the addition of kaolin slurry (20% of kaolin powder by dry weight of the tested material; powder–water relation of 1:1.5). The results revealed that building debris presented friction angle values (dry, 37°; wet, 35°) and behaviour similar to those observed for crushed concrete. The presence of reinforcement (geogrid specimen at the box central plane) increased the dry friction angle by 12°. Among the wastes investigated, the increase in friction angle due to the presence of reinforcement was more pronounced for building debris.

Materials obtained from the demolition of single-family houses and the cleaning processes of land with illegal deposition of CDW were recycled and tested by Vieira et al. [52]. The RCDW was subjected to geotechnical characterisation, a leaching test, and direct shear tests with and without reinforcement. The grain size distribution revealed that the material was composed of fine particles (smaller than 20 mm) and the short-term contaminant release investigation revealed that the RCDW fulfils the acceptance criteria for inert landfills. The direct shear test results for the unreinforced RCDW showed values of peak friction angle and cohesion equals to 44.1° and 17.3 kPa, respectively. The reinforced condition presented values of peak friction angle and apparent adhesion equal to 35° and 12.6 kPa, respectively, for an extruded geogrid made from high-density polyethylene (ultimate tensile strength of 68 kN/m, and 16 × 219 mm aperture size). The tests with an extruded geogrid made of polyester (ultimate tensile strength of 80/20 kN/m, and 30 × 73 mm aperture size) revealed values of peak friction angle and apparent adhesion equal to 36.6° and 19.7 kPa, respectively.

The use of RCDW as backfill in a GRS structure was investigated by Santos and Vilar [53], who performed geotechnical characterisation and shear and pull-out tests. The RCDW was obtained from a local recycling plant and consisted of a mixture of a crushed material (consisting mainly of soil, bricks, and small particles of concrete). Two other materials were used as reference: (i) river sand (in accordance with the U.S. Federal Highway Administration—FHWA) and (ii) local soil (sandy clay soil). The RCDW presented low variability in its geotechnical properties (grain size distribution, specific gravity, unit dry weight, and moisture content) and an alkaline extract (mean pH = 9.1) that allowed its use with the PET geogrid tested (ultimate tensile strength, T_{ult}, of 61 kN/m × 30 kN/m, machine x cross-machine direction; 30 × 20 mm aperture size). The results of the pull-out tests showed that the RCDW presented a higher interface strength than that of the river sand and the values of the adherence factor—the ratio between the RCDW–geogrid interface strength and the RCDW shear strength—in a range (0.52 to 1.30) observed by other studies for conventional soil–geogrid interfaces.

- Geosynthetic Damage

The reduction in ultimate tensile strength (T_{ult}) caused during the installation process has been pointed out as the most critical mechanism affecting the short-term durability of geosynthetics [54]. Besides the intrinsic characteristics of a geosynthetic (e.g., geometry, shape, and polymer) that affect its durability, the backfill material composition and installation procedures influence the occurrence and severity of damages. Bearing in mind the proposal of using RCDW in GRS structures, the damage mechanisms may be influenced by the physical and chemical characteristics of such new backfill material.

To investigate the factors affecting the short-term damages of a polyester (PET) geogrid (T_{ult} of 20 kN/m at machine direction) and of a polypropylene (PP) non-woven geotextile (T_{ult} of 19 kN/m and mass per unit area of 300 g/m^2), Santos [55] simulated the construction procedures used to build two large-scale wrapped-face geosynthetic reinforced walls with RCDW as backfill material (classified as sand with gravel, pH equal to 8.84 at 25 °C). The criteria adopted to indicate the occurrence of damage was based on the mean value of the tensile strength for virgin specimens (T0) (not submitted to the installation procedure) and a level of confidence of 98% calculated using Student's t-distribution—given that only five specimens were tested for each scenario, characterising a small sample size. For scenarios where the mean tensile strength (Ti) presented values outside the confidence interval, the occurrence of damage was assumed and the reduction factor (RF) was calculated. The scenarios investigated were (i) compaction by a lightweight roller, (ii) a hand tamping plate, and (iii) compaction by a lightweight roller and burial in RCDW for 450 days. The results revealed that the PP geotextile tested was stronger than the PET geogrid and that more severe damages were observed for the specimens left in contact with RCDW. The results for the geogrid revealed the influences of compaction energy and contact with the RCDW. Table 1 presents the values of RF for all the scenarios investigated.

Table 1. Reduction factor (RF) for geosynthetics with RCDW backfill material [55].

Scheme	Nonwoven Geotextile	Geogrid
Lightweight roller	1.00	1.12
Hand tamping plate	1.00	1.28
Lightweight roller + 450 days burial	1.64	1.20

An extensive study on geogrid mechanical damage due to contact with RCDW was carried out by Fleury et al. [56] considering several factors of influence: (i) RCDW dropping height H (0.0, 1.0, and 2.0 m, and 2.0 m over a RCDW protection layer of 50 mm) and (ii) a compaction method (no compaction—to isolate the influence of dropping height, vibratory roller, and hand tamping plate) and geogrid type (polymer-T_{ult}: PVA-35 kN/m, PET-35 kN/m, and PET-55 kN/m). The RCDW (mainly composed of soil, concrete, mortar, and ceramic) was obtained from a local recycling plant and the final compacted layer was 200 mm thick. The combination of the influencing factors totalled 36 scenarios, with five specimens been tested for each scenario. The results of the tensile tests (wide strip specimens) were obtained according to ASTM D-6637 [57], and the method for determining damage occurrence followed the one presented by Santos [55]. The results showed that (i) the RCDW presented variability in its geotechnical properties, (ii) the dropping process caused slight damages and the increase in the dropping height showed limited influence in the damage intensity (RF = 0.94 to 1.21), (iii) the adoption of fine-grained RCDW can be seen as an attractive alternative as a protective layer, (iv) the compaction method was the most important factor for geogrid installation damage (RF = 0.98 to 1.22) once the severity of the damage seemed to be directly associated with the compaction degree reached during the tests, and (v) the multiplication of individual RF values for the investigated factors (dropping height and compaction method) was conservative. The study conclusions highlight the importance of obtaining RF for specific situations when RCDW is used, the complexity of damage mechanisms, and the positive technical, economic, and environmental aspects of using such non-conventional backfill material in GRS structures. Testing the same geogrids, Domiciano et al. [43] reported no influence of damage (RF = 1.0) on T_{ult}. The tests were carried out by subjecting the specimen to different values of static loading (ranging from 150 to 600 kPa) and using RCDW with different compositions and grain size distributions. However, Domiciano et al. [43] showed different RF values for other parameters of interest (strain at failure, ε_{rup}; stiffness at 2%, $J_{2\%}$; and stiffness at 5%, $J_{5\%}$).

3.2. Recycled Construction and Demolition Wastes in Unpaved Roads

Recycled construction and demolition wastes (RCDW) can be effectively used as fill materials in environmentally friendly solutions for unpaved roads. Góngora [58] carried out a series of large-scale tests on unreinforced and geosynthetic reinforced unpaved roads on a weak subgrade. Figure 3 shows the characteristics of the equipment used in these tests, which consisted of a rigid steel tank (750 mm diameter, 550 mm high). A rigid circular platen (200 mm diameter) applied the repeated loading (frequency of 1 Hz) on the fill surface. Geogrids were used as reinforcement, whose main properties are listed in Table 2. The secant tensile stiffness at a 5% strain of the geogrids tested varied between 130 kN/m and 1500 kN/m, with varying aperture sizes. The subgrade soil consisted of a fine-grained soil (Table 3) with a California Bearing Ratio (CBR) value of 4.2%. Two materials were investigated as fill for the roads. The first one was a natural gravel with an average particle diameter (D_{50}) equal to 10.5 mm, which served as a reference fill material. The other fill material was a recycled construction and demolition waste (RCDW) with an average particle diameter of 34 mm. The main geotechnical properties of the soils tested are presented in Table 3. Additional information can be found in Góngora [58] and Góngora and Palmeira [59].

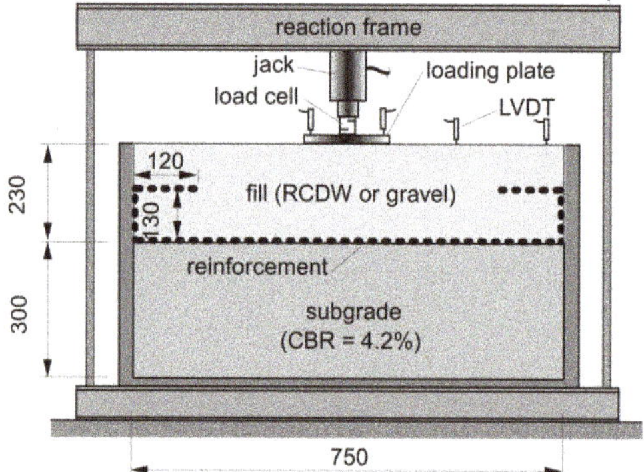

Figure 3. Equipment used in the tests on unpaved roads [58].

Table 2. Reinforcement properties.

Property [1]	G1	G2	G3
Aperture dimensions (mm)	30 × 30	20 × 20	40 × 40
Percentage of grid open area (%)	33	73	84
Tensile stiffness at 5% strain	1500	260	130
Tensile strength (kN/m)	200	35	17.5
Aperture stability modulus (N-m/deg.) [2]	1.54	0.029	0.019

Note: [1] Data from manufacturers' catalogues. Geogrids manufactured with polyester fibres protected by a PVC cover; [2] also known as in-plane torsional rigidity modulus [60,61], obtained at 2 N-m torque, minimum value.

Table 3. Properties of the soils.

Property	Subgrade	Gravel	RCDW [1]
D_{85} (mm) [2]	0.19	16.7	47.4
D_{50} (mm)	0.025	10.5	34
D_{10} (mm)	—	1.6	5.9
Percent of fines (<0.075 mm) (%) [3]	63	1.0	1.0
Soil coefficient of uniformity	—	7.1	6.8
Liquid limit (%)	39	—	—
Plastic limit (%)	29	—	—
Moisture content (%)	27.1	—	—
Dry unit weight (kN/m^3)	14.0	17.6	16.9
Soil particles density	2.68	2.65	2.74
California Bearing Ratio (%)	4.2	—	—
Los Angeles Abrasion Test (%)	—	36.0	56

Notes: [1] RCDW = recycled construction and demolition waste; [2] D_n = diameter for which *n* percent in mass of the remaining particle diameters is smaller than that diameter, coefficient of uniformity of the soil = D_{60}/D_{10}; [3] tests using a dispersing agent for the subgrade soil.

Figure 4 shows the results of tests on unreinforced roads, where the target surface rut depth of 25 mm at the fill surface was reached for close values of a number of load repetitions (N) for both fill materials (N equal to 1630 and 1710 for the gravel and RCDW roads, respectively). However, the presence of reinforcement (geogrid G1) made a significant difference on the road performance, as can be seen in Figure 4. In this case, the 25 mm-deep rut was reached in the reference (gravel) road for a value of N of 24,064, whereas in the case of the reinforced RCDW road it was reached for a value of N of 57,235. It should be noted that up to a rut depth value of 22 mm (N $\cong$ 11,000) the behaviour of the two fill materials was very similar. As that rut value came closer to the target maximum rut depth, it may be considered that both materials behaved similarly under reinforced conditions, with a TBR (traffic benefit ratio = N_r/N_{unr}, where N_r and N_{unr} are the values of N under reinforced and unreinforced conditions, respectively, for a given rut depth) of the order of 15 at the end of the tests. Similar tests after road surface repair showed the good performance of the reinforced roads (gravel and RCDW fills) in terms of TBR values, particularly for tests with reinforcements less prone to suffering mechanical damage [59].

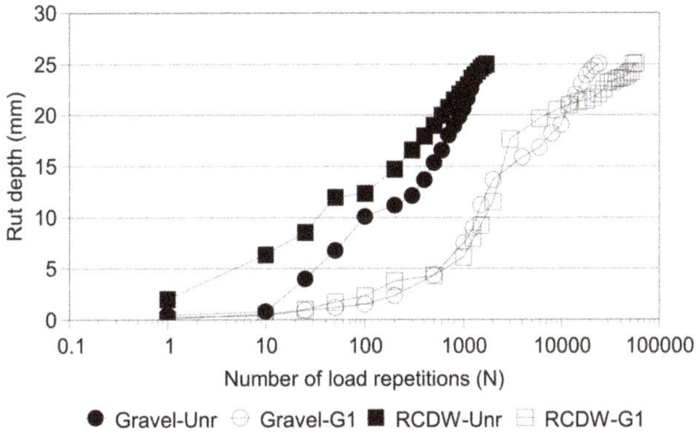

Figure 4. Rut displacement versus number of load repetitions.

Mehrjardi et al. [62] reported the use of construction and demolition wastes as fill materials combined with geocell reinforcement in unpaved roads on compressible subgrade. Figure 5 shows a schematic of the test equipment used in the investigation. A natural aggregate (gravel, USCS classification = GP, D_{50} = 4.6 mm, D_{10} = 0.2 mm and coefficient of uniformity (CU) of 26.3) layer was also used as fill material for comparison purposes. The alternative base materials consisted of a waste soil (fine-grained fraction of the CDW, USCS classification = GW, D_{50} = 1.3 mm, D_{10} = 0.12 mm and coefficient of uniformity (CU) of 19) and a recycled concrete aggregate from CDW (USCS classification = GP, D_{50} = 3.0 mm, D_{10} = 0.21 mm and coefficient of uniformity (CU) of 18). The subgrade soil consisted of the wasted soil with a moisture content of 1.5% and a relative density of 60%. The geocells utilised were manufactured from a heat-bonded nonwoven geotextile made of polypropylene, with cells with equivalent diameter and height of 55 mm and 50 mm, respectively; a mass per unit area of 690 g/m^2; and a secant tensile stiffness at 5% strain of 5.7 kN/m.

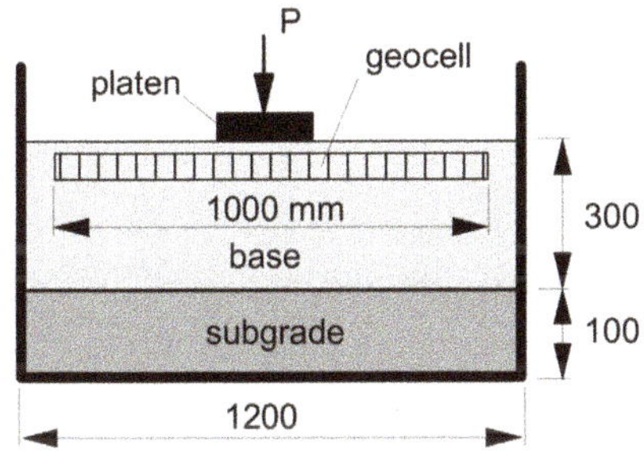

Figure 5. Tests on unpaved roads reinforced with geocells [62], modified.

Figure 6 shows some test results obtained by Mehrjardi et al. [62] in terms of permanent settlement at the fill surface versus number of load repetitions for the tests with the waste soil and natural aggregate. This figure shows that the geocell-reinforced road built with the waste soil performed significantly better than its unreinforced counterpart and presented similar performance to that of the natural aggregate at the end of the test.

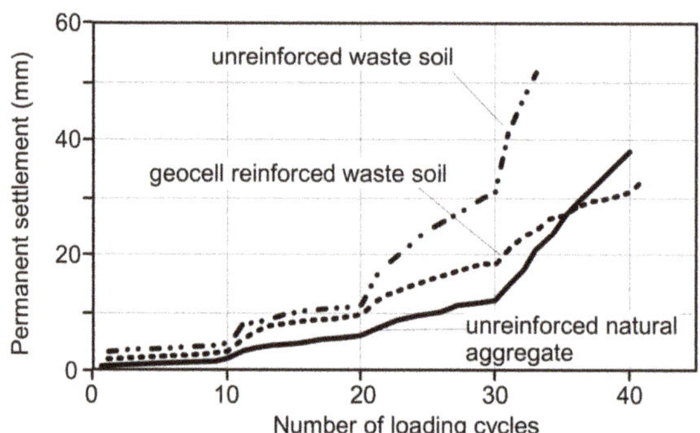

Figure 6. Settlement versus number of loading cycles [62], modified.

The results presented in Figures 4 and 6 show the potential of the use of recycled construction and demolition wastes reinforced with geosynthetics in unpaved roads on poor subgrades. The utilisation of RCDW will avoid or reduce the use and exploitation of more expensive natural materials, with favourable repercussions to the environment.

3.3. Recycled Construction and Demolition Wastes in Geosynthetic-Encased Granular Columns

Alkhorshid [63] investigated the use of geosynthetic-encased columns (GEC) to reduce excessive settlements and failure of embankments built on soft soils (Figure 7a,b). In this type of subgrade, there is low confinement at the upper part of the granular column, reducing its load capacity. A technique that has been applied to increase column lateral confinement is the use of geotextile encasement (Figure 7b). Large scale tests were carried out by Alkhorshid [63] on conventional and geotextile-encased granular columns using a large box (1.6 m × 1.6 m × 1.2 m) to investigate the use of three types of woven geotextile encasements and three types of infill materials, including RCDW. The GEC models were 1000 mm in height and 150 mm in diameter. Sand, calcareous gravel, and recycled construction and RCDW were used as infill materials for the columns. Bentonite (4% in mass) was added to the subgrade soil to increase its plasticity and workability, resulting in a soft subgrade classified as CH by the Unified Soil Classification System (USCS). The properties of the soft soil and filling materials and the geotextile characteristics are presented in Tables 4 and 5. If a scale factor of 4 is considered, the tests would simulate a prototype problem of a 0.6 m-diameter encased column in a soft clay with an undrained strength of 20 kPa. During the tests, monotonically increasing vertical loads were applied to the column top by a rigid steel plate.

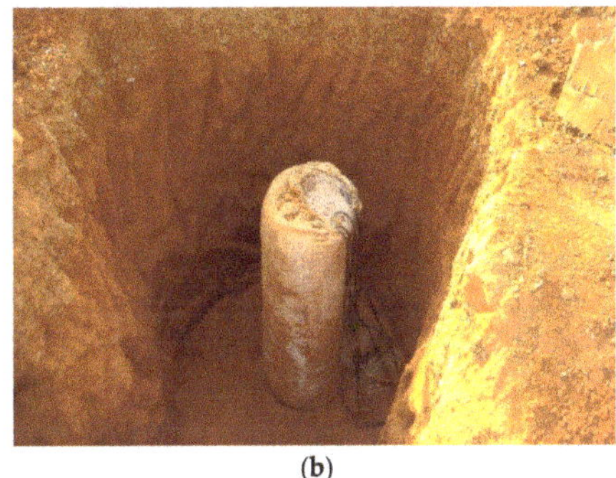

(**a**)   (**b**)

Figure 7. Geosynthetic-encased granular columns to stabilise embankments on compressible ground: (**a**) typical differential settlement between an abutment on soft ground and a bridge; (**b**) geosynthetic-encased granular column.

Table 4. Properties of the soft subgrade.

Property	Soil			
	Clay	Sand	Gravel	RCDW
Liquid limit (%)	60	-	-	-
Plasticity index (%)	21	-	-	-
Soil particle density (-)	2.7	2.65	2.66	2.65
Compression index	0.47	-	-	-
Expansion index	0.03	-	-	-
Undrained strength (kPa)	5	-	-	-
Coefficient of uniformity	5.5	3.51	1.6	1.5
Coefficient of curvature	1.05	0.825	0.98	0.92
D_{50} (mm)	0.304	0.50	6.55	6.64
D_{10} (mm)	0.073	0.179	4.44	4.78
D_{30} (mm)	0.175	0.305	5.56	5.64
D_{60} (mm)	0.401	0.63	7.11	7.21
Maximum void ratio	-	0.87	0.74	0.76
Minimum void ratio	-	0.6	0.41	0.45
Friction angle (degrees)	-	41	43	42
Dilation angle (degrees)	-	11	12	12

Table 5. Properties of the geosynthetic encasements.

Property	Geotextile		
	G-1	G-2	G-3
Tensile strength (kN/m)	30	16	8
Maximum tensile strain (%)	22	16	15
Tensile stiffness (k/m)	120	107	53.4

The G-1-encased column (the column using geotextile G-1, Table 5, with the highest tensile strength and stiffness) presented higher bearing capacity and lower settlements than those encased with G-2 and G-3. Moreover, the maximum lateral bulging of the encased columns was observed for the column encased with the most extensible geotextile (G-3), at a depth ranging from 1 to 1.5 times the column diameter. Figure 8 shows values of load capacities of the columns encased by geotextile G1 and with the three different infill materials tested (sand, gravel, and RCDW). This figure also shows the results for the conventional materials (no encasement). A significant increase in column load capacity was observed with the use of geotextile encasement. It is also noticeable that the values of load capacity of the RCDW-encased columns were very similar to those of the traditional natural infill materials, which highlights the potential use of RCDW as infill material in encased granular columns for the stabilisation of embankments on soft soils. Additional information on this type of application of RCDW can be found in Alkshorshid [63] and Alkhorshid et al. [63–65].

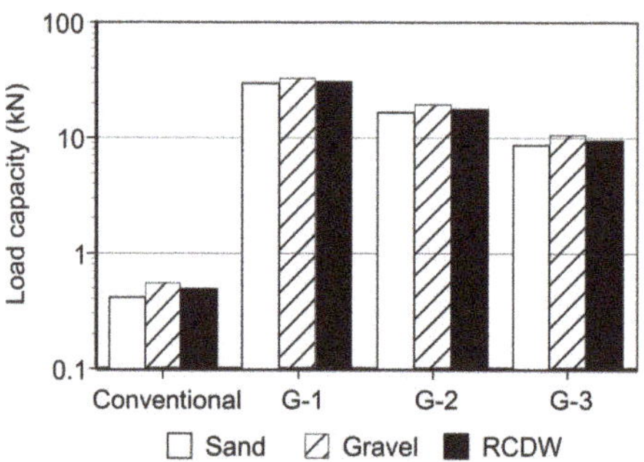

Figure 8. Comparison between load capacities of granular columns [63], modified.

3.4. Reinforced Soil Structures Constructed with RCDW: Performance

As part of a research programme on the combined use of CDW and geosynthetics in geotechnical and geoenvironmental works at the University of Brasília, Brasilia, Brazil, a research project was responsible for the conception, construction, instrumentation, and monitoring of two full-scale GRS structures using RCDW as backfill material. The 3.6 m-high wrapped face walls were constructed over a porous collapsible foundation soil located at the Foundation and Field Investigation Site of the Graduate Programme of Geotechnics of the University of Brasilia. Because the walls were monitored through dry and rainy seasons, it was possible to observe the influence of the properties and performance of the foundation soil on walls deformations, settlements, horizontal earth pressures, and reinforcement strains [55,66,67]. The walls were constructed with different geosynthetic reinforcements (Wall 1: PET geogrid and Wall 2: PP non-woven geotextile) and back-to-back in a reinforced masonry block facility (Figure 9). Three layers of lubricated polyethylene sheets were placed on the internal faces of the facility to minimise the influence of soil-side wall friction on the test results. Tables 6 and 7 present the properties of RCDW and geosynthetics used in these experiments, respectively.

Figure 9. Complete view of the test facili showing the water reservoirs and main reservoir [55].

Table 6. Properties of the RCDW backfill material (Santos [55]).

Property	Value
D_{85} (mm)	15.0
D_{50} (mm)	2.1
D_{10} (mm)	0.032
CU	106
pH of backfill	8.9
Unit weight (kN/m^3)	17.8
Moisture content (%)	6.6
Friction angle (degrees)	41
Cohesion (kPa)	6

Notes: D_n, diameter of particles for which $n\%$ in mass of the remaining particles is smaller than that diameter; CU, soil coefficient of uniformity (CU = D_{60}/D_{10}).

Table 7. Geosynthetic properties [55].

Property	Geogrid (Wall 1)	Nonwoven Geotextile (Wall 2)
Polymer	Polyester	Polypropylene
Aperture size (mm × mm)	20 × 20	NA [2]
Tensile strength (kN/m)	20 [1]	24 [3]
Secant stiffness (kN/m)	300 at 5% strain [1]	15 at 5% strain [3]
Strain at break (%)	12	70

Notes: [1] In direction of loading (transverse members); [2] NA = not applicable; data from manufacturer's literature unless stated otherwise; tensile properties as per ASTM D6637 [57]; [3] Santos [55].

Besides the expected collapse of the foundation soil due to the natural infiltration of water through the wall face and backfill surface during the rainy season, artificial inundation was also carried out to enhance foundation soil collapse in a controlled manner and thus further investigate its effect on the performance of the walls. To assure the complete infiltration of water into the foundation soil, a granular drainage layer was installed at the base of the walls and water reservoirs were constructed adjacent to the test facility (Figure 9).

Because the walls were built during the dry season, the following rainy season presented a relevant effect on the performance of the walls, mainly during the first 100 days, which resulted in approximately 250 mm of cumulative precipitation. Concerning the horizontal displacements of the wall face, it was noted that after being subjected to the first rainy season or to artificial inundation (the latter for Wall 1) the values measured tended toward stabilisation. The artificial inundation of the foundation soil presented a greater influence on Wall 1 (geogrid) and were shown to be negligible for Wall 2 (nonwoven geotextile wall). Figure 10 presents the normalised horizontal displacements measured (at a normalised elevation of 0.83) at the wall faces from the end of construction up to 587 days after construction. Measures of inward displacement at the crest of both walls were also observed, which may have been a consequence of the compressibility of the foundation soil at the wall toe as reported in other studies not related to the use of RCDW as backfill material [68–71]. Although both walls showed the largest outward displacement at an elevation equal to 3.00 m (h/H = 0.83, where h is the elevation and H is the height of the wall), the final values for Wall 2 ($\Delta x/H$ = 6%, where Δx is the face horizontal displacement) were about twice those of Wall 1 ($\Delta x/H$ = 3%). This performance of Wall 2 can be explained by the fact that nonwoven geotextile presents much lower stiffness under the unconfined conditions at and closer to the wall face, where the occurrence of greater reinforcement bulging was marked. The maximum outward post construction movement observed was greater than the values recorded for conventional wrapped-face walls in a database of wall case studies collected by Bathurst et al. [72]. Clearly, the performance of the walls can be attributed to the compressibility of the foundation soil. However, the magnitudes of face displacements recorded by Santos et al. [67] could be acceptable for temporary structures in some jurisdictions (e.g., WSDOT [73]) or even prevented by means of some foundation soil treatment, which could be carried out before or simultaneously to the wall construction.

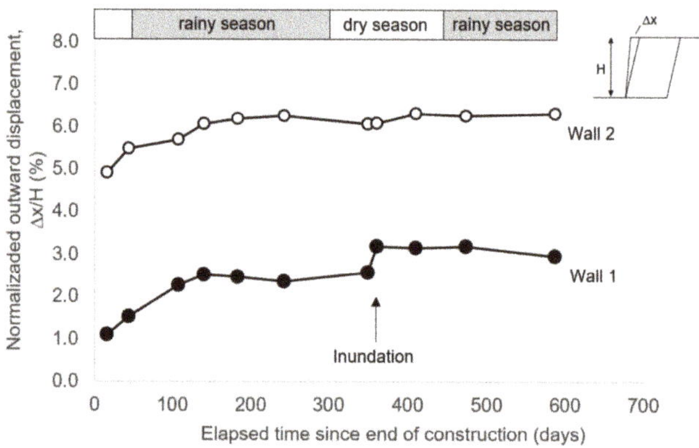

Figure 10. Horizontal outward displacement of the wall face at an elevation of 3.0 m.

The displacement of the wall top was influenced by the first rainy season (following the construction of the walls) in a very significant manner, associated with a value of 250 mm of cumulative precipitation as a trigger to the process. Differences in settlement profiles were observed, which can be attributed to differences in the compressibility of the foundation soil under each wall. The wall crest presented the most significant vertical displacements for both walls, with the highest values observed for Wall 1 (Figure 11). Considering the wall top, for $x/B = 0.24$, where x is the distance from the wall face and B is the wall base width, the stabilisation of displacements was observed at 261 days after construction, which was associated with approximately 1250 mm of cumulative precipitation. However, the artificial inundation showed that this stabilisation was due to the reduction in precipitation and infiltration of water, because of the dry season's proximity. In a period of just 48 h after the artificial inundation process, the vertical displacement at $x/B = 0.24$ increased by 45% and 39% for Walls 1 (geogrid) and 2 (nonwoven geotextile), respectively. The effects caused by the artificial inundation continued for approximately 48 days (during the dry season), with increments of 8% and 15% for Walls 1 and 2, respectively. The second rainy season showed no significant influence on the vertical displacements at the wall top up to the end of monitoring of the walls (587 days after the end of construction).

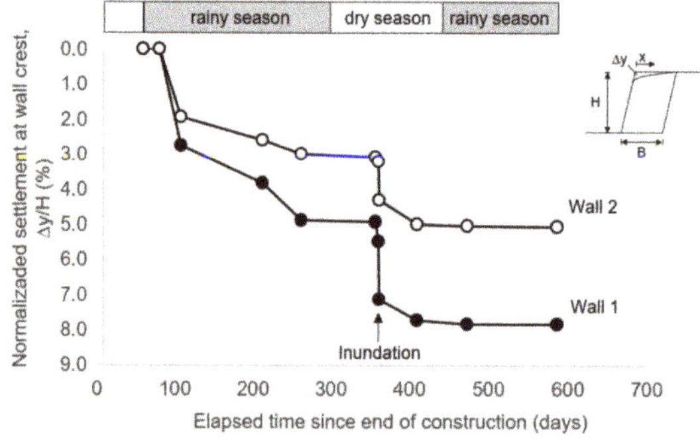

Figure 11. Vertical displacements at the wall top ($x/B = 0.24$).

Concerning the reinforcement strain, much higher values were observed close to the face of Wall 2 (nonwoven geotextile) compared to Wall 1, particularly for reinforcement layer 5 (elevation of 2.4 m). Maximum strains in this layer in Wall 1 reached 0.23% just before the artificial inundation of the foundation, whereas for the same reinforcement layer in Wall 2 the maximum strain was equal to 12.4%. The maximum reinforcement strains in Wall 1 reveal the viability of using RCDW as backfill material once such values are consistent with the low strain levels reported in databases of monitored geosynthetic soil walls under operational conditions by Miyata and Bathurst [74] and Bathurst et al. [75]. The larger values of strains close to the face of Wall 2 (nonwoven geotextile) are a consequence of the low confinement in that region. Beyond half of the base length, the RCDW provided enough confinement to the geotextile reinforcement and the strains were low (below 1%) in both walls. The effects of the rainy season on reinforcement strains were noted more significantly in Wall 1, with the cumulative precipitation of 250 mm working again as a trigger. On the other hand, the results indicate that reinforcement strain mobilisation in Wall 2 occurred during the construction process, with the rainy season and artificial inundation not causing relevant changes.

4. Combinations between Geosynthetics and Alternative Materials in Waste Disposal

Landfills have plenty of materials that can be reused in civil engineering works and to fulfil some needs of the landfills themselves. This section addresses some combinations of geosynthetics and waste materials for applications in landfills.

Palmeira and Silva [76] carried out tests to investigate the performance of the combination of geotextile filters and alternative drainage materials using large-scale waste containers. The containers were made of steel plates and were 3.5 m long, 2.5 m wide, and 1.0 m high, as shown in Figure 12. A lateral drainage trench allowed the flow of leachate to external tanks for leachate volume and property measurements. The drainage trench in each container was 300 mm wide and 300 mm deep. Perforated pipes at the top of the containers allowed the pluviation of water on the waste under controlled conditions. In cell C1 (Figure 12a) the drainage trench was filled with recycled construction and demolition wastes (RCDW). A nonwoven geotextile layer covered the entire plan area of the container. The main properties of the RCDW are listed in Table 8. In cell C2 the drainage system consisted of a 100 mm-thick layer of shredded wasted tires with a geotextile filter layer on top (Figure 12b). In both containers the geotextile filter was a nonwoven, needle-punched geotextile made of polyester, with a thickness of 2.3 mm and a mass per unit area equal to 200 g/m^2. The mass of domestic waste in each cell was approximately equal to 2.5 tons, with the following composition: 44% organic matter, 26% paper and cardboard, 16% plastics, 4% metal, and 10% other materials. The initial height of the waste was equal to 900 mm and the waste was disposed in a loose state, with a unit weight of 4 kN/m^3 and an initial moisture content of 58%. Figure 13 shows views of the drainage trenches in containers C1 and C2 before the installation of the geotextile filter.

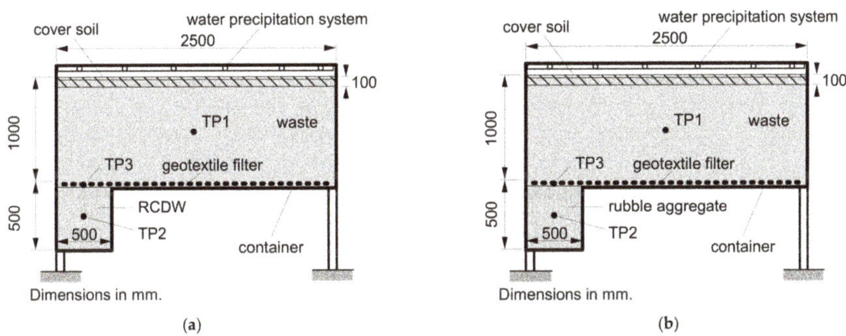

Figure 12. Waste containers: (**a**) container C1; (**b**) container C2 [77].

Table 8. Properties of the drainage materials.

Property	RCDW	Tire Shreds
D_{90} (mm) [1]	33	45
D_{85} (mm)	31	—
D_{50} (mm)	21.4	—
D_{10} (mm)	2.0	—
Coefficient of uniformity	12.5	1.5
Dry unit weight (kN/m^3)	11.13	3.67
Void ratio	0.95	2.19
Permeability (cm/s) [2,3]	10	15
Compression index [2]	1.18	0.79

Notes: [1] D_n = diameter for which n percent in mass of the remaining particle diameters is smaller than that diameter, coefficient of uniformity of the soil = D_{60}/D_{10}; [2] from large-scale laboratory tests [77]; [3] from large-scale constant head permeability tests [77].

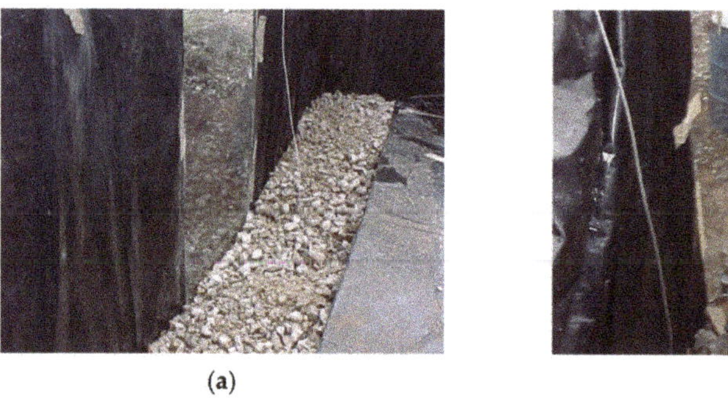

(a)

(b)

Figure 13. Drainage trenches in containers C1 and C2: (**a**) C1; (**b**) C2.

The water pluviation on the waste started 75 days after the waste disposal in the containers, under pluviation rates of 10 L/week to simulate wet seasons and 2.5 L/week to simulate dry seasons. The instrumentation used allowed the measurement of waste settlements and temperature in the waste mass below the geotextile filter and inside the drainage trench. Effluent volumes were collected for measurements of effluent flow rates as well as chemical analyses such as the determination of pH, chemical oxygen demand, sulphate content, ammonium content, nitrate content, and solids content. Additional information on materials and testing methodology can be found in Paranhos and Palmeira [77], Silva [78], and Palmeira et al. [79].

Figure 14 depicts the variation of cumulative effluent volume from the containers with the cumulative volume of precipitation. The results show a greater discharge capacity of the drainage trench with tire shreds in comparison with that of the RCDW aggregate. Some level of waste heterogeneity may have influenced the different responses of the drainage systems. However, even before the start of the water pluviation on the waste, effluent flow was already noticed in container C2, whereas effluent in container C1 only started after water pluviation. This late response from C1 may also be a consequence of the presence of the horizontal drainage layer in container C2, which was not present in C1, besides the large coefficient of permeability of the tire shreds (Table 8). It can also be noted that after 130 L of water pluviation (approximately 138 days since the beginning of pluviation) the rate of increase in the effluent volume with pluviation volume was very similar for both drainage systems.

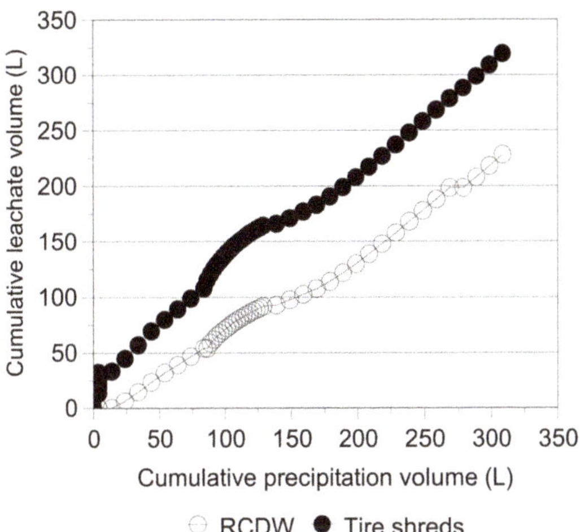

Figure 14. Variation of cumulative effluent volume with cumulative precipitation volume in containers C1 and C2.

Figures 15 and 16 show the variation of chemical oxygen demand (COD) and pH with cumulative precipitation for the two drainage systems investigated. These figures also present the variation in cumulative precipitation volume with time. Significant differences between COD values can be noted, and this may have been due to the influence of the delay in the liberation of leachate from container C1, as mentioned before; the interaction between the particles and leachate, and the better filtering action of the RCDW layer. A significant amount of COD reduction in the effluent from container C1 was noted during the dry season, when the rate of precipitation was smaller, probably due to more favourable conditions for bacterial activity, as well as when water pluviation was stopped on day 600 (Figure 15). At the end of the tests, some level of degradation of the RCDW grains was visually observed, which must be considered when using this type of material in drainage systems under long-term conditions. The initially larger pH values of the effluent from container C1 (Figure 16) was likely due to the interaction between the concrete particles and the leachate. As time passed, the pH values became similar for both containers. The reduction in the pH values with time for container C1 may have been a consequence of the particles of the drainage system having been covered with a layer of leachate, which avoided or minimised continuous direct contact between the leachate and concrete particles. The effluents of both systems presented pH values close to the neutral value at the end of the experiments. Additional information on the behaviour of the drainage systems in the containers can be found in Silva [76] and in Palmeira and Silva [80].

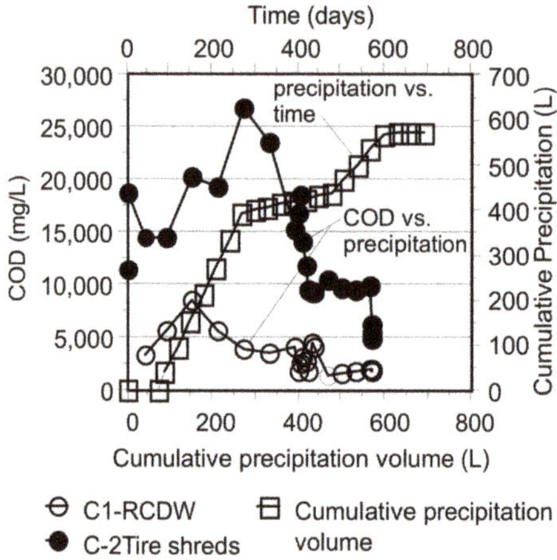

Figure 15. COD variation with cumulative precipitation volume.

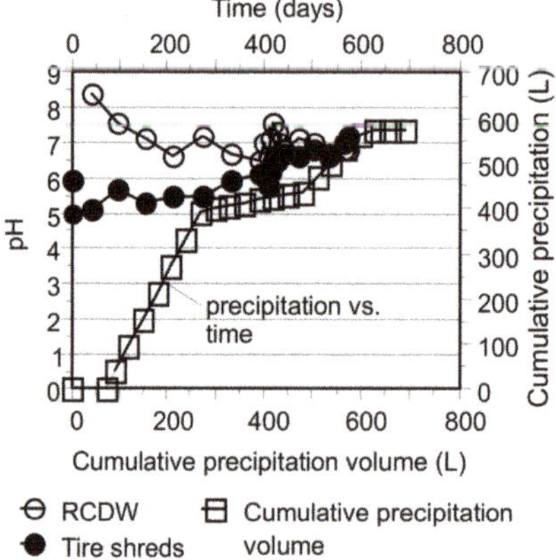

Figure 16. Variation of COD and pH of the effluent with cumulative precipitation volume [78], modified.

Junqueira et al. [16] and Silva [78] described the use of wasted tires as drainage layers in landfills. The authors compared the performance of a conventional gravel layer at the base of an experimental domestic waste cell to an alternative drainage system consisting of a layer of whole wasted tires and a nonwoven geotextile filter. The use of the geotextile filter aimed also to investigate the possibility of its clogging, since clogging of either geotextile or sand filters and even of gravel drainage layers have been observed in landfills. Figure 17 shows the geometrical characteristics of these waste cells, where each cell contained approximately 50 tons of municipal solid waste collected in the Federal District, Brazil. The main properties of the soils and alternative drainage material involved in this study are presented in Table 9. The municipal solid waste in the cells had a composition consisting of 49% (in mass) organic matter, 18% plastics, 22% paper and cardboard, 2% metals, and 9% other materials. The waste had a unit weight of 5.4 kN/m^3. Cell Gr had a conventional drainage layer at its bottom consisting of a gravel layer 200 mm thick (Figure 17). Cell T had an alternative drainage system formed by a layer of wasted tires as the system drainage core with a geotextile filter (nonwoven, needle-punched, 150 g/m^2 mass per unit area, 1.5 mm thick, 2.5 s^{-1} permittivity, and filtration opening size of 0.15 mm). Figure 18 shows images of cell T during construction and filling with waste. HDPE geomembranes were used as barriers at the bottom of both cells to avoid the contamination of the subgrade.

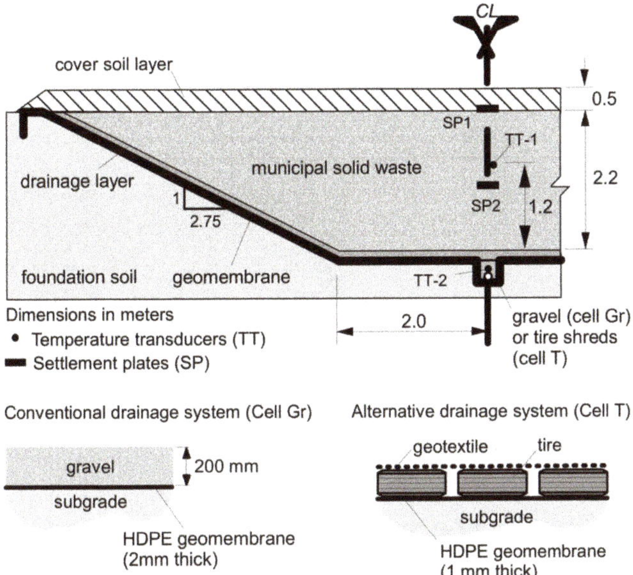

Figure 17. Geometrical characteristics of the experimental waste cells [78], modified.

Table 9. Properties of the materials tested.

	Gravel	Tire Shreds	Cover Soil [3]
D_{10} (mm) [1]	53	NA [6]	NA
D_{50} (mm)	60	NA	NA
D_{85} (mm)	65	41	0.20
CU [2]	1.2	1.5	NA
Dry unit weight (kN/m^3)	19.7	3.67	13.4
Permeability (cm/s) [4]	24 [5]	15 [5]	0.003 to 0.009

Notes: [1] D_n is the diameter for which $n\%$ in weight of the soil has particles with diameters smaller than that value; [2] CU = soil coefficient of uniformity = D_{60}/D_{10}; [3] 79% in mass with particles smaller than 0.074 mm; [4] from field infiltration [78,81]; [5] from large-scale constant head permeability tests [77]; [6] NA = not applicable.

Figure 18. Construction and filling of one of the experimental waste cells [78]: (**a**) drainage layer; (**b**) installation of the geotextile filter; (**c**) filling.

The experimental cells were instrumented with settlement plates and temperature transducers for waste mass settlement and temperature measurements, respectively. The variation of the effluent flow rate with time and the pluviometry in the region was also assessed. In should be pointed out that the region has very well-defined wet and dry seasons. The dry season runs from May to September each year, whereas the rainy season runs from October to April. Data acquisition from the instrumentation started in the month of August, during the dry season. Additional information on the experiments can be found in Silva [16] and Junqueira et al. [78].

The variation of cumulative effluent volume with cumulative precipitation volume is presented in Figure 19. It can be observed that the results obtained are very similar, showing the good performance of the alternative drainage system in comparison with that of the conventional one. The rather constant value of effluent volume between months 7 and 14 corresponds to the dry season in the region. Hence, the response of the drainage systems in terms of effluent volumes was directly related to the pluviation in the region. Figure 20 shows the ratio between the effluent volume and the volume of water precipitated on the cells during rainy periods, showing again the similar behaviour of both drainage systems investigated. The rather constant rate between the effluent and precipitated volumes in the first three months of monitoring was due to fact that the initiation of flow took place during a dry season, thus with the effluent from both cells being mainly a consequence of the decomposition of organic matter. Junqueira et al. [16] showed that similar variations of settlement and chemical parameters of the effluent with time were also found for both cells. However, the amount of suspended solids in the effluent from cell T was smaller than that from cell Gr due to the filtering action of the geotextile layer in the former. For the duration of the experiment, there was no indication of clogging of the geotextile filter.

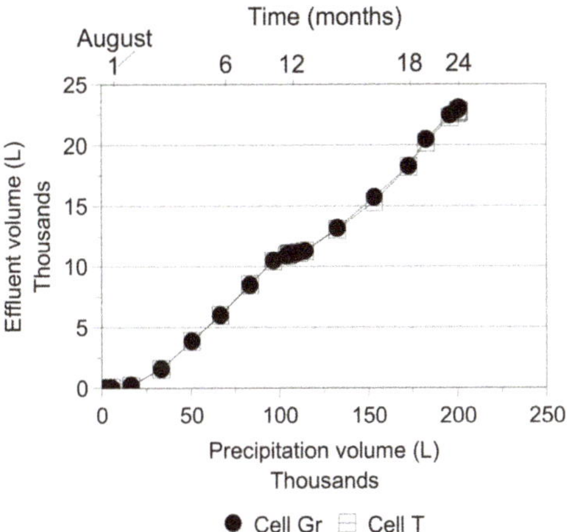

Figure 19. Cumulative effluent volume versus cumulative precipitated volume.

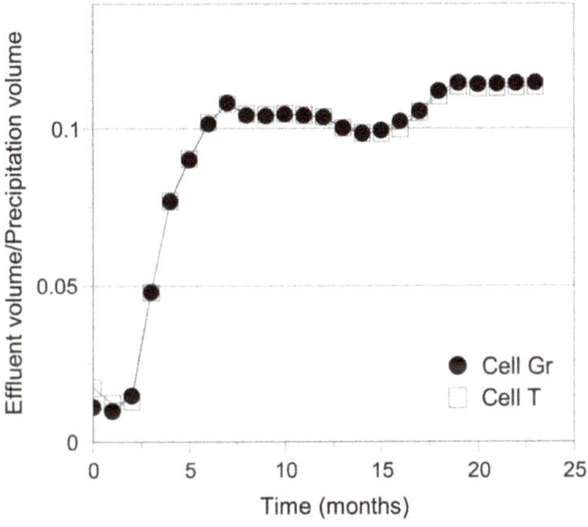

Figure 20. Ratio between effluent and precipitated volumes versus time.

5. Other Sustainable Combinations of Geosynthetics and Alternative Materials

Other waste materials can be combined with geosynthetics to provide sustainable engineering solutions. This is the case for PET bottles, which can be used to form a drainage layer. Figure 21 shows a drainage system consisting of compressed PET bottles enveloped by a geotextile filter. The compressed bottle layer presented high permeability coefficients under low stress levels, even greater than those of conventional granular materials commonly used in drainage systems. Table 10 lists typical values of relevant geotechnical properties of some alternative drainage materials. These results were obtained in large-scale permeability (constant head) and compression tests [77,79]. The coefficient of permeability (50 cm/s) of the compressed PET bottles was of the order of 3.5 to 5 times those of the tire shreds and rubble aggregate tested. However, PET bottles were significantly more compressible (Table 10), and this may be an issue depending on the stress level on the drainage layer.

(a)   (b)

Figure 21. Alternative drainage system with a core of compressed PET bottles (images courtesy of H. Paranhos): (**a**) compressed PET bottles; (**b**) geocomposite for drainage formed by compressed PET bottles enveloped by a geotextile filter in a drainage trench.

Table 10. Properties of some alternative drainage materials [77].

Property	Compressed PET Bottles	Shredded Tires	Rubble Aggregate
Shape	Round	Lamellar	Irregular
D_{90} (mm)	95	45	35
CU	1.0	1.5	2.0
Unit weight (kN/m^3)	92.0	367.0	1113.0
k (cm/s)	50.0	15.0	10.0
e	2.11	2.19	0.95
C_c	4.7	0.79	1.18

Notes: D_{90} diameter for which 90% of the remaining particles have smaller diameters than that value; CU = coefficient of uniformity; k = coefficient of permeability (from large-scale constant head tests); e = void ratio; C_c = compression index (from large-scale compression tests).

Another interesting application of PET bottles is in infiltration trenches for runoff water to avoid or minimise the consequences of floods in urban areas. Figure 22 shows this type of application, where a geotextile filter is associated with whole PET bottles to favour the infiltration of surface runoff water into the ground [82].

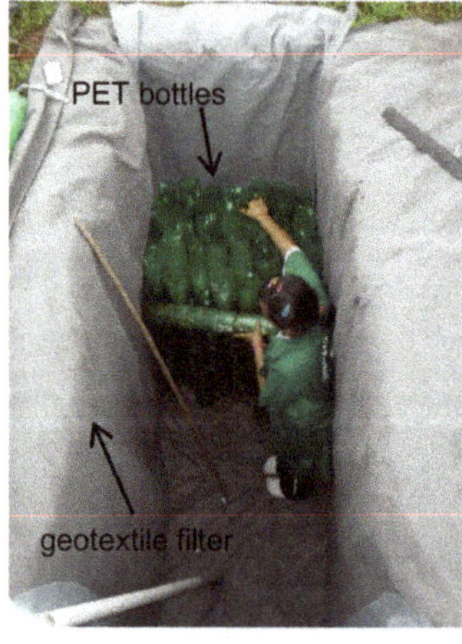

Figure 22. Drainage trench with PET bottles for flood control [82].

Geotextiles can also be combined with alternative drainage materials to produce low-cost biplanar drainage geocomposites. Figure 23 shows alternative geocomposites for drainage consisting of a nonwoven filter layer and drainage cores made of PET bottle caps or rubber strips from wasted tires [83,84]. Transmissivity tests performed on these alternative geocomposites showed values of transmissivity under pressure similar to those of some conventional commercially available products [84].

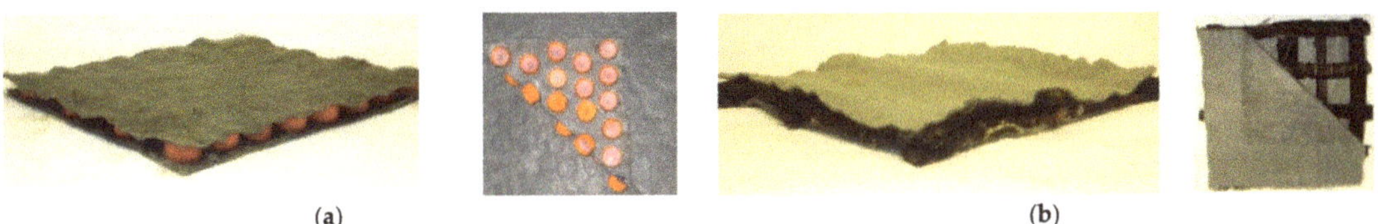

(a)   (b)

Figure 23. Alternative geocomposites for drainage [83]: (**a**) drainage core consisting of PET bottle caps; (**b**) drainage core consisting of rubber strips from wasted tires.

Rubber grains from wasted tires (Figure 24a) can also be utilised to manufacture low-cost alternative geocomposite clay liners (GCL, Figure 24b). In this case, the rubber grains can be mixed with the bentonite to save bentonite. However, this mixture will swell less than pure bentonite when subjected to moisture content increase. Viana et al. [85] investigated the influence of the amount of rubber grains (D_{85} = 0.6 mm, D_{10} = 0.12 mm, CU = 4.0) from wasted tires on the permittivity and expansibility of an alternative GCL. Figure 25a shows that significant permittivity increased with increasing rubber content. On the other hand, the permittivity decreased with increasing normal stress. The variation of expansion of rubber-bentonite mixtures with vertical stress is depicted in Figure 25b. The results show less expansion of the alternative barrier due to the addition of rubber grains. Despite the increase in permittivity and decrease in expansibility, the use of a low-cost GCL incorporating rubber grains may be interesting as additional barrier layers or as bedding or protective layers underneath geomembranes in landfills, particularly in less critical situations.

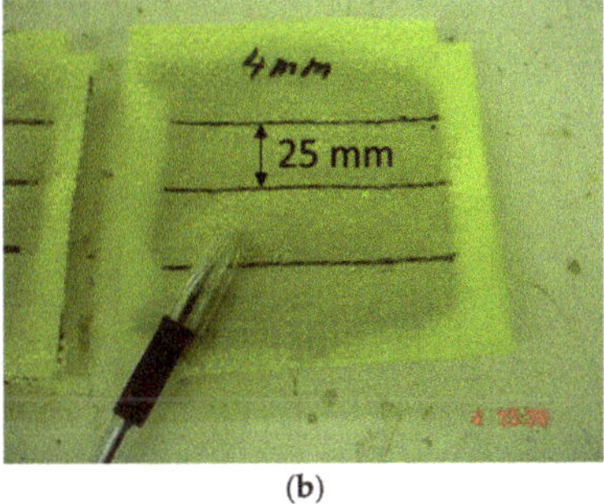

(a)   (b)

Figure 24. Alternative GCL [85], modified: (**a**) rubber crumbs; (**b**) alternative GCL.

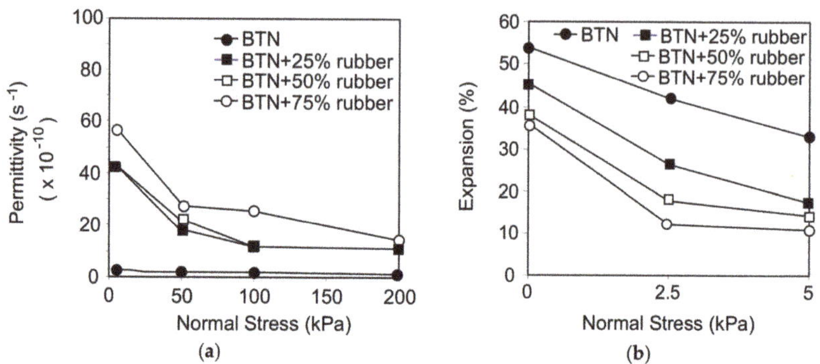

Figure 25. Results of tests on an alternative GCL [85], modified: (a) permittivity versus normal stress; (b) expansion versus normal stress.

Inert mining wastes can also be used as alternative construction materials if the requirements regarding geotechnical and geoenvironmental properties are fulfilled. Fernandes et al. [86] reported on the use of geogrid-reinforced fine mining waste mixed with local soils in an alternative sub-ballast layer of a railway. Instrumented experimental sections showed the good performance of the alternative geogrid reinforced sub-ballast layer in comparison with the conventional and more expensive material traditionally used in that railway. Some of the benefits brought by the combination of mining waste and geosynthetic reinforcement were less vertical and horizontal strains in the sub-ballast layer, an increase in the track stiffness, and less breakage of ballast particles, with positive repercussions in reducing the maintenance costs of the railway track.

Martins [87] and Gomes and Martins [88] described the use of mining waste as fill material in a 28 m-high retaining structure (Figure 26). The reinforced mass was 18 m high and was constructed using ore mining waste as backfill material. This mining waste had up to 40% fine fraction, with a unit weight of 20.3 kN/m³, a cohesion of 19.7 kPa, and a friction angle of 39°. Nonwoven and woven geotextiles were used as reinforcement with tensile strengths of 40 kN/m and 70 kN/m, respectively. The reinforced structure is already 24 years old and has behaved very well.

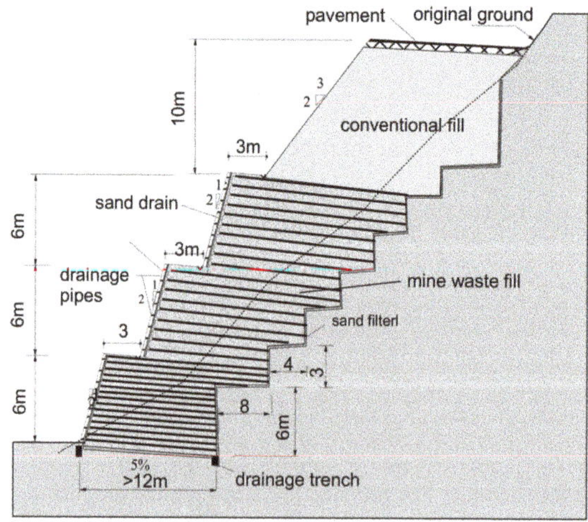

Figure 26. Geosynthetic reinforced structure built with mining waste backfill [87].

6. Conclusions

The preservation of the environment and natural resources is of major importance for current and future generations. Therefore, the concepts of sustainability and circular economy have come to stay, and engineers must be encouraged to propose engineering solutions in line with this new reality. In this context, geosynthetics can already provide sustainable solutions when used in combination with conventional natural materials. Hence, if combined with properly selected wastes and low-grade construction materials, the benefit to the environment will be even more significant. This paper presented a review on the use of wastes as alternative construction materials in geotechnical and geoenvironmental works, with particular emphasis on their combination with geosynthetics. The main conclusions obtained are summarised below.

- Although there are environmental concerns regarding the use of rubber tire waste mixed with soils, this type of mixture has been evaluated over the last decades in different countries and the investigations have shown that the risk of contamination is low for constructions involving low rates of waterflow.
- Rubber tires mixed with soils are considered suitable for different types of granular materials, although the mechanical properties of the mixtures vary depending on the type of soil and the size and shape of the rubber particles.
- The performance of unpaved roads on soft ground where geosynthetic reinforced construction and demolition wastes (CDW) were used as fill materials compared well to the performance of unpaved roads constructed with conventional natural granular fills.
- Recycled construction and demolition wastes, wasted tires, and plastic bottles may also provide low-cost and environmentally friendly solutions for drainage systems in geotechnical and geoenvironmental works, particularly when used in landfills, where these wastes are abundant.
- Wasted tires and plastic elements such as plastic bottle caps can be combined with geotextiles to produce low-cost geocomposites for drainage. The use of recycled plastics to produce geosynthetics should also be encouraged, particularly for less critical and less severe applications.
- Properly selected recycled construction and demolition waste can be a feasible infill material for geosynthetic-encased granular columns.
- The reported studies show the influence of waste origin and recycling procedures on the variability of RCDW geotechnical properties and chemical behaviour. The need for specific characterisation when RCDW is subjected to surcharge (static or cyclic loading) and the influence of its component properties (e.g., concrete, ceramic, mortar, etc.) was observed. However, the results reported in the literature present physical, mechanical, and chemical properties that allow the effective application of RCDW in several geotechnical works involving the use of geosynthetics.
- The compaction method was revealed to be of utmost importance for short-term mechanical damage of geosynthetic reinforcement when RCDW is used as backfill material. Due to the intrinsic complex mechanisms that are involved with geosynthetic damage, it is important to use appropriate values of reduction factors based on the properties of the RCDW, construction site conditions, and reinforced soil structure characteristics to obtain low-cost and safe designs.
- The performance of two full-scale wrapped-face geosynthetic-reinforced soil walls constructed over a collapsible foundation soil revealed the influences of the rainfall regime (dry or wet season), foundation inundation, and geosynthetic type on the behaviour of the walls as a whole (face and top displacements and reinforcement strains). Even under extreme conditions, the walls built with RCDW as backfill material showed satisfactory performance. The proposal of using such a non-conventional backfill material further enhances the recognised advantages of geosynthetic-reinforced walls (cost effectiveness and reduced environmental impacts, for instance) and encourages

the growth of geotechnical engineering practices in accordance with the concept of sustainable development. The same comments apply to the use of mining wastes.

- Mining wastes can also be successfully combined with geosynthetics in reinforced soil structures.

Despite the relevance and benefits to the environment in reusing wastes in geotechnical engineering, one must bear in mind that some of these wastes will degrade with time or can contain substances that may cause ground contamination. Thus, a careful evaluation of such aspects must be carried out before using wastes as construction materials in geotechnical and geoenvironmental works. In this context, geosynthetics such as geomembranes and geocomposite clay liners can be properly specified to provide efficient barriers for such contaminants.

Author Contributions: Conceptualization, E.M.P., G.L.S.A. and E.C.G.S.; Methodology, E.M.P., G.L.S.A. and E.C.G.S.; Validation, E.M.P., G.L.S.A. and E.C.G.S.; Formal analysis, E.M.P., G.L.S.A. and E.C.G.S.; Investigation, E.M.P., G.L.S.A. and E.C.G.S., Resources: E.M.P., G.L.S.A. and E.C.G.S.; Writing-Review and Editing, E.M.P., G.L.S.A. and E.C.G.S.; Funding acquisition, E.M.P. and G.L.S.A. All authors have read and agreed to the published version of the manuscript.

Funding: This research was funded by the National Council for Scientific and Technological Development (CNPq, grant Nos. 301436/2009-6, 474551/2010-5, 472564/06-4, and 461018/2014-4) and the Research Support Foundation of the Federal District (FAP-DF, grant Nos. 193.000.263/2007 and 193.000.540/2009).

Institutional Review Board Statement: Not applicable.

Informed Consent Statement: Not applicable.

Data Availability Statement: Not applicable.

Acknowledgments: The authors are indebted to the following institutions for the support given to this research: University of Brasília, Federal University of Goiás, the National Council for Scientific and Technological Development (CNPq), the Capes/Brazilian Ministry of Education and Research Support Foundation of the Federal District (FAP-DF), geosynthetic manufacturers, CDW recycling plants, and engineering companies.

Conflicts of Interest: The authors declare no conflict of interest.

References

1. Frischknecht, R.; Stucki, M.; Büsser, S.; Itten, R. Comparative life cycle assessment of geosynthetics versus conventional construction materials. *Ground Engng.* **2012**, *45*, 24–28.
2. Damians, I.P.; Bathurst, R.J.; Adroguer, E.G.; Josa, A.; Lloret, A. Environmental assessment of earth retaining wall structures. *Env. Geotech.* **2017**, *4*, 415–431. [CrossRef]
3. Heerten, G. Reduction of climate-damaging gases in geotechnical engineering practice using geosynthetics. *Geotex. Geomemb.* **2012**, *30*, 43–49. [CrossRef]
4. Touze-Foltz, N. Healing the world: A geosynthetic solution. In Proceedings of the 11th International Conference on Geosynthetics, Seoul, Korea, 16–21 September 2018. 59p.
5. Khaloo, A.R.; Dehestani, M.; Rahmatabadi, P. Mechanical properties of concrete containing a high volume of tire–rubber particles. *Waste Manag.* **2008**, *28*, 2472–2482. [CrossRef] [PubMed]
6. Humphrey, D.N. Effectiveness of design guidelines for use of tire derived aggregate as lightweight embankment fill. Recycled Materials in Geotechnics (GSP 127). In Proceedings of the ASCE Civil Engineering Conference and Exposition 2004, ASCE, Reston, VA, USA, 19–21 October 2004.
7. CBC. End in Sight for Used Tire Stockpile. 2013. Available online: https://www.cbc.ca/news/canada/newfoundland-labrador/end-in-sight-for-used-tire-stockpile-1.1402610 (accessed on 15 June 2021).
8. USTA. *Scrap Tire Management Summary*; United States Tires Association: Washington, DC, USA, 2019; 20p.
9. CATRA—Canadian Association of Tire Recycling Agencies. 2021. Available online: https://www.catraonline.ca/ (accessed on 25 May 2021).
10. Reciclalimp. Brazilian National Association of Pneumatic Industry. 2020. Available online: https://www.reciclanip.org.br/ (accessed on 25 May 2021).
11. ANIP. *Technical Report*; Brazilian National Society of Pneumatics: São Paulo, Brazil, 2018. (In Portuguese)

12. Bosscher, P.J.; Edil, T.B.; Eldin, N.N. Construction and performance of a shredded waste tire test embankment. *Transp. Res. Rec.* **1992**, *1345*, 44–52.

13. Yang, S.; Lohnes, R.A.; Kjartanson, B.H. Mechanical properties of shredded tires. *Geotech. Test. J.* **2002**, *25*, 44–52. [CrossRef]

14. Cokca, E.; Yilmaz, Z. Use of rubber and bentonite added fly ash as a liner material. *Waste Manag.* **2004**, *24*, 153–164. [CrossRef] [PubMed]

15. Attom, M.F. The use of shredded waste tires to improve the geotechnical engineering properties of sands. *Env. Geol.* **2006**, *49*, 497–503. [CrossRef]

16. Junqueira, F.F.; Silva, A.R.L.; Palmeira, E.M. Performance of drainage systems incorporating geosynthetics and their effect on leachate properties. *Geotex. Geomemb.* **2006**, *24*, 311–324. [CrossRef]

17. Chrusciak, M.R. Soil Tire Shreds Mixtures for Ground Improvement. Master's Thesis, Graduate Program on Geotechnical Engineering, University of Brasilia, Brasília, Brazil, 2012; 109p. (In Portuguese).

18. Xiao, M.; Ledezma, M.; Hartman, C. Shear resistance of tire-derived aggregate using large-scale direct shear tests. *J. Mater. Civ. Eng.* **2015**, *27*, 1. [CrossRef]

19. Nimbalkar, S.S.; Indraratna, B. Improved performance of ballasted rail track using geosynthetics and rubber. *J. Geotech. Geoenv. Eng.* **2016**, *142*, 04016031. [CrossRef]

20. Suárez, M.J. Resistance Evaluation of the Lightweight Composite Using Soil of the Brazilian Federal District and Tire Rubbers. Master's Thesis, Graduate Program on Geotechnical Engineering, University of Brasilia, Brasília, Brazil, 2016. Publication G.DM-904 268/15. 116p. (In Portuguese).

21. Yadav, J.S.; Tiwari, S.K. Effect of waste rubber fibres on the geotechnical properties of clay stabilized with cement. *Appl. Clay Sci.* **2017**, *149*, 97–110. [CrossRef]

22. Mittal, R.K.; Gill, G. Sustainable application of waste tire chips and geogrid for improving load carrying capacity of granular soils. *J. Clean. Prod.* **2018**, *200*, 542–551. [CrossRef]

23. Pasha, S.M.K.; Hazarika, H.; Yoshimoto, N. Physical and mechanical properties of gravel-tire chips mixture (GTCM). *Geosynth. Int.* **2019**, *26*, 92–110. [CrossRef]

24. Zhang, H.; Yuan, X.; Liu, Y.; Wu, J.; Song, X.; He, F. Experimental study on the pullout behavior of scrap tire strips and their application as soil reinforcement. *Constr. Build. Mater.* **2020**, *254*, 119288. [CrossRef]

25. Araujo, G.L.S.; Moreno, J.A.S.; Zornberg, J.G. Shear behavior of mixtures involving tropical soils and tire shreds. *Constr. Build. Mater.* **2021**, *276*, 122061. [CrossRef]

26. El Naggar, H.; Iranikhah, A. Evaluation of the Shear Strength Behavior of TDA Mixed with Fine and Coarse Aggregates for Backfilling around Buried Structures. *Sustainability* **2021**, *13*, 5087. [CrossRef]

27. Elhacham, E.; Ben-Uri, L.; Grozovski, J.; Bar-On, Y.M.; Milo, R. Global human-made mass exceeds all living biomass. *Nature* **2020**, *558*, 442–444. [CrossRef]

28. Wu, H.; Zuo, J.; Zillante, G.; Wang, J.; Yuan, H. Status quo and future directions of construction and demolition waste research: A critical review. *J. Clean. Prod.* **2019**, *240*, 118163. [CrossRef]

29. Leite, F.C.; Motta, R.S.; Vasconcelos, K.L.; Bernucci, L. Laboratory evaluation of recycled construction and demolition waste for pavements. *Constr. Build. Mater.* **2011**, *25*, 2972–2979. [CrossRef]

30. Arulrajah, A.; Piratheepan, J.; Disfani, M.; Bo, M. Geotechnical and geoenvironmental properties of recycled construction and demolition materials in pavement subbase applications. *J. Mater. Civ. Eng.* **2013**, *25*, 1077–1088. [CrossRef]

31. Pasandín, A.R.; Pérez, I. Laboratory evaluation of hot-mix asphalt containing construction and demolition waste. *Constr. Build. Mater.* **2013**, *43*, 497–505. [CrossRef]

32. Pasandín, A.R.; Pérez, I. Mechanical properties of hot-mix asphalt made with recycled concrete aggregates coated with bitumen emulsion. *Constr. Build. Mater.* **2014**, *55*, 350–358. [CrossRef]

33. Gómez-Meijide, B.; Pérez, I.; Passandín, A.R. Recycled construction and demolition waste in cold asphalt mixtures: Evolutionary properties. *J. Clean. Prod.* **2016**, *112*, 588–598. [CrossRef]

34. Engelsen, C.J.; Van Der Sloot, H.A.; Petkovic, G. Long-term leaching from recycled concrete aggregate applied as sub-base material in road construction. *Sci. Total Env.* **2017**, *587–588*, 94–101. [CrossRef]

35. National Geographics. How the Plastic Bottle Went from Miracle Container to Hated Garbage. 2019. Available online: https://www.nationalgeographic.com/environment/article/plastic-bottles (accessed on 18 September 2021).

36. G1. River Tietê Floods a City in the Interior of São Paulo and Covers Streets with a 'Sea of Garbage'. Photo by Prefeitura de Salto, S.P., Brazil. 2020. Available online: https://g1.globo.com/sp/sorocaba-jundiai/noticia/2020/02/11/rio-tiete-inunda-cidade-no-interior-de-sp-e-cobre-ruas-com-mar-de-lixo.ghtml (accessed on 18 September 2021). (In Portuguese).

37. The Guardian. A Million Bottles a Minute: World's Plastic Binge 'as Dangerous as Climate Change'. 2017. Available online: https://www.theguardian.com/environment/2017/jun/28/a-million-a-minute-worlds-plastic-bottle-binge-as-dangerous-as-climate-change (accessed on 18 September 2021).

38. Gomes, P.C.C.; Ulsen, C.; Pereira, F.A.; Quattrone, M.; Angulo, S.C. Comminution and sizing processes of concrete block waste as recycled aggregates. *Waste Manag.* **2015**, *45*, 171–179. [CrossRef]

39. Miura, N.; O-Hara, S. Particle-crushing of decomposed granite soil under shear stresses. *Soils Found.* **1979**, *19*, 1–14. [CrossRef]

40. Hyodo, M.; Hyde, A.F.L.; Arakami, N.; Nakata, Y. Undrained monotonic and cyclic shear behaviour of sand under low and high confining stresses. *Soils Found.* **2002**, *42*, 63–76. [CrossRef]

41. Indraratna, B.; Salim, W. Modelling of particle breakage of coarse aggregates incorporating strength and dilatancy. *Proc. Inst. Civ. Eng.* **2002**, *155*, 243–252. [CrossRef]
42. Sivakumar, V.; McKinley, J.D.; Ferguson, D. Reuse of construction waste: Performance under repeated loading. *Proc. Inst. Civ. Eng.* **2004**, *157*, 91–96. [CrossRef]
43. Domiciano, M.L.; Santos, E.C.G.; Lins da Silva, J. Geogrid mechanical damage caused by recycled construction and demolition waste (RCDW): Influence of grain size distribution. *Soils Rocks* **2020**, *43*, 231–246. [CrossRef]
44. Palmeira, E.M. Soil-geosynthetic interaction: Modelling and analysis. *Geotex. Geomembr.* **2009**, *27*, 368–390. [CrossRef]
45. Palmeira, E.M. A Study on Soil-Interaction by Means of Large Scale Laboratory Tests. Ph.D. Thesis, University of Oxford, Oxford, UK, 1987.
46. Palmeira, E.M.; Milligan, G.W.E. Scale and other factors affecting the results of pull-out tests of grids buried in sand. *Geotéchnique* **1989**, *39*, 511–524. [CrossRef]
47. Jewell, R.A. Some Effects of Reinforcement on Soils. Ph.D. Thesis, University of Cambridge, Cambridge, UK, 1980.
48. Dyer, M. Observation of the Stress Distribution in Crushed Glass with Applications to Soil Reinforcement. Ph.D. Thesis, University of Oxford, Oxford, UK, 1985.
49. McGown, A.; Andrawes, K.Z.; Kabir, M.H. Load-extension testing of geotextiles confined in soil. In Proceedings of the 2nd International Conference on Geotextiles, Las Vegas, NV, USA, 1–6 August 1982; Volume 2, pp. 793–798.
50. Mendes, M.J.A.; Palmeira, E.M.; Matheus, E. Some factors affecting the in-soil load–strain behaviour of virgin and damaged nonwoven geotextiles. *Geosynth. Int.* **2007**, *14*, 39–50. [CrossRef]
51. Touahamia, M.; Sivakumar, V.; McKelvey, D. Shear strength of reinforced recycled material. *Constr. Build. Mater.* **2002**, *16*, 331–339. [CrossRef]
52. Vieira, C.S.; Pereira, P.P.; Lopes, M.L. Behaviour of geogrid-recycled construction and demolition waste interfaces in direct shear mode. In Proceedings of the 10th International Conference on Geosynthetics, Berlin, Germany, 21–25 September 2014.
53. Santos, E.C.G.; Vilar, O.M. Use of recycled construction and demolition wastes (RCDW) as backfill of reinforced soil structures. In Proceedings of the 4th European Conference on Geosynthetics—EUROGEO 4, Edinburgh, UK, 7–10 September 2008; p. 199.
54. Hufenus, R.; Rüegger, R.; Flum, D.; Sterba, I. Strength reduction factors due to installation damage of reinforcing geosynthetics. *Geotex. Geomemb.* **2005**, *23*, 401–424. [CrossRef]
55. Santos, E.C.G. Experimental Evaluation of Reinforced Walls Built with Recycled Construction and Demolition Wastes (RCDW) and Fine Soil. PhD. Thesis, Graduate Programme of Geotechnics, University of Brasilia, Brasilia, Brazil, 2011. (In Portuguese).
56. Fleury, M.P.; Santos, E.C.G.; Lins da Silva, J.; Palmeira, E.M. Geogrid installation damage caused by recycled construction and demolition waste. *Geosynth. Int.* **2019**, *26*, 641–656. [CrossRef]
57. ASTM. D6637. In *Standard Test Method for Determining Tensile Properties of Geogrids by the Single or Multi-rib Tensile Method*; American Society for Testing and Materials: West Conshohocken, PA, USA, 2015.
58. Góngora, I.A.G. Utilization of Geosynthetics as Reinforcement in Unpaved Roads: Influence of the Type of Reinforcement and Fill Material. Master's Thesis, University of Brasilia, Brasília, Brazil, 2011. (In Portuguese).
59. Góngora, I.A.G.; Palmeira, E.M. Influence of fill and geogrid characteristics on the performance of unpaved roads on weak subgrades. *Geosynth. Int.* **2012**, *19*, 1–9. [CrossRef]
60. Kinney, T.C.; Xiaolin, Y. Geogrid Aperture Rigidity by in-Plane Rotation. In Proceedings of the Geosynthetics'95, Nashville, TN, USA, 21–23 February 1995; Volume 2, pp. 525–537.
61. Kinney, T.C. *Standard Test Method for Determining the "Aperture Stability Modulus" of a Geogrid*; Shannon & Wilson, Inc.: Seattle, WA, USA, 2000.
62. Mehrjardi, G.T.; Azizi, A.; Haji-Azizi, A.; Asdollafardi, G. Evaluating and improving the construction and demolition waste technical properties to use in road construction. *Transp. Geotech.* **2020**, *23*, 1–13. [CrossRef]
63. Alkhorshid, N.R. Analysis of Geosynthetic Encased Columns in Very Soft Soil. Ph.D. Thesis, Graduate Programme of Geotechnics, University of Brasilia, Brasilia, Brazil, 2017.
64. Alkhorshid, N.R.; Araújo, G.L.; Palmeira, E.M. Behavior of geosynthetic-encased stone columns in soft clay: Numerical and analytical evaluations. *Soils Rocks* **2018**, *41*, 333–343. [CrossRef]
65. Alkhorshid, N.R.; Araujo, G.L.S.; Palmeira, E.M.; Zornberg, J.G. Large-scale load capacity tests on a geosynthetic encased column. *Geotex. Geomembr.* **2019**, *47*, 632–641. [CrossRef]
66. Santos, E.C.G.; Palmeira, E.M.; Bathurst, R.J. Behaviour of a geogrid reinforced wall built with recycled construction and demolition waste backfill on a collapsible foundation. *Geotex. Geomembr.* **2013**, *39*, 9–19. [CrossRef]
67. Santos, E.C.G.; Palmeira, E.M.; Bathurst, R.J. Performance of two geosynthetic reinforced walls with recycled construction waste backfill and constructed on collapsible ground. *Geosynth. Int.* **2014**, *21*, 256–269. [CrossRef]
68. Jones, C.J.F.P.; Edwards, L.W. Reinforced earth structures situated on soft foundations. *Geotéchnique* **1980**, *30*, 207–213. [CrossRef]
69. Dellabianca, L.M.; Palmeira, E.M. Parametric study of geosynthetic reinforced walls. In Proceedings of the 1st South American Symposium on Geosynthetics/3rd Brazilian Symposium on Geosynthetics, Rio de Janeiro, Brazil, 19–22 October 1999; Volume 1, pp. 99–106.
70. Jones, C.J.F.P. Case history—Deflection of structures reinforced with materials of different stiffness. In Proceedings of the 9th International Conference on Geosynthetics, Guarujá, Brazil, 23–27 May 2010; Volume 3, pp. 1315–1318.

71. Damians, I.P.; Bathurst, R.J.; Josa, A.; Lloret, A. Numerical study of the influence of foundation compressibility and reinforcement stiffness on the behavior of reinforced soil walls. *Int. J. Geotech. Engng.* **2014**, *3*, 247–259. [CrossRef]
72. Bathurst, R.J.; Miyata, Y.; Allen, T.M. Facing displacements in geosynthetic reinforced soil walls. In Proceedings of the Earth Retention Conference 3 (ER2010), Bellevue, WA, USA, 1–4 August 2010. 18p.
73. WSDOT. *Geotechnical Design Manual M 46e03, Chapter 15 Abutments, Retaining Walls, and Reinforced Slopes*; Washington State Department of Transportation: Olympia, WA, USA, 2005.
74. Miyata, Y.; Bathurst, R.J. Evaluation of K-stiffness method for vertical geosynthetic reinforced granular soil walls in Japan. *Soils Found.* **2007**, *47*, 319–335. [CrossRef]
75. Bathurst, R.J.; Miyata, Y.; Nernheim, A.; Allen, A.M. Refinement of K-stiffness method for geosynthetic-reinforced soil walls. *Geosynth. Int.* **2008**, *15*, 269–295. [CrossRef]
76. Palmeira, E.M.; Silva, A.R.L. A study on the behaviour of alternative drainage systems in landfills. In Proceedings of the 11th International Waste Management and Landfill Symposium-Sardinia 2007, Cagliari, Italy, 1–5 October 2007. 10p.
77. Paranhos, H.; Palmeira, E.M. *Alternative Low-Cost Drainage Systems for Geotechnical and Geoenvironmental Works*; Research Report-RHAE/CNPq/UnB; Graduate Programme of Geotechnics, University of Brasilia: Brasilia, Brazil, 2002; 79p. (In Portuguese)
78. Silva, A.R.L. A study on the Behaviour of Drainage Systems of Waste Disposal Areas under Different Scales. Ph.D. Thesis, Graduate Programme on Geotechnical Engineering, University of Brasilia, Brasília, Brazil, 2004; 329p. (In Portuguese).
79. Palmeira, E.M.; Silva, A.R.L.; Paranhos, H. Recycled and alternative materials in drainage systems of waste disposal areas. In Proceedings of the 5th International Congress on Environmental Geotechnics, Cardiff, UK, 26–30 June 2006; Volume 2, pp. 1511–1518.
80. Palmeira, E.M. Sustainability and innovation in geotechnics: Contributions from geosynthetics. *Soils Rocks* **2016**, *39*, 113–135.
81. Junqueira, F.F. Analyses of Domestic Waste Behaviour and Draining Systems with Reference to the Jóquei Club Dump. Ph.D. Thesis, Graduate Programme on Geotechnical Engineering, University of Brasilia, Brasília, Brazil, 2000; 288p. (In Portuguese).
82. Silva, J.P. Infiltration Structures Using Alternative Materials to Control Flooding and for the Prevention of Erosive Processes. DSc. Thesis, Graduate Programme of Geotechnics, University of Brasília, Brasília, Brazil, 2012. (In Portuguese).
83. da Silva, C.A. Transmissivity Tests on Geocomposites for Drainage. MSc. Dissertation, Graduate Programme of Geotechnics, University of Brasília, Brasília, Brazil, 2007. (In Portuguese).
84. da Silva, C.A.; Palmeira, E.M. Performance comparison of conventional biplanar and low-cost alternative geocomposites for drainage. *Geosynth. Int.* **2013**, *20*, 226–237. [CrossRef]
85. Viana, P.M.F.; Palmeira, E.M.; Viana, H.N.L. Evaluation on the use of alternative materials in geosynthetic clay liners. *Soils Rocks* **2011**, *34*, 65–77.
86. Fernandes, G.; Palmeira, E.M.; Gomes, R.C. Performance of geosynthetic-reinforced alternative sub-ballast material in a railway track. *Geosynth. Int.* **2008**, *15*, 311–321. [CrossRef]
87. Martins, C.C. Analysis and Revaluation of Geotextile Reinforced Structures. MSc. Dissertation, Federal University of Ouro Preto, Ouro Preto, Brazil, 2000. (In Portuguese).
88. Gomes, R.C.; Martins, C.C. Design of geotextile-reinforced structures with residual soils and mining residues in highway applications in Brazil. In Proceedings of the 12th Panamerican Conference on Soil Mechanics and Geotechnical Engineering, Cambridge, MA, USA, 22–26 June 2003; Volume 2, pp. 1773–1778.

sustainability

Article

Substitute Building Materials in Geogrid-Reinforced Soil Structures

Sven Schwerdt *, Dominik Mirschel, Tobias Hildebrandt, Max Wilke and Petra Schneider

Department for Water, Environment, Civil Engineering and Safety, University of Applied Sciences Magdeburg-Stendal, Breitscheidstraße 2, D-39114 Magdeburg, Germany; dms14fh@hotmail.com (D.M.); Tobs10@gmx.de (T.H.); wilke.max@freenet.de (M.W.); petra.schneider@h2.de (P.S.)

* Correspondence: sven.schwerdt@h2.de

Abstract: The feasibility of substitute building materials (SBMs) in engineering applications was investigated within the project. A geogrid-reinforced soil structure (GRSS) was built using SBM as the fill material as well as vegetated soil for facing and on top of the construction. Four different SBMs were used as fill material, namely blast furnace slag (BFS), electric furnace slag (EFS), track ballast (TB), and recycled concrete (RC). For the vegetated soil facing, a mixture of either recycled brick (RB) material or crushed lightweight concrete (LC) mixed with organic soil was used. The soil mechanical and chemical parameters for all materials were determined and assessed. In the next step, a GRSS was built as a pilot application consisting of three geogrid layers with a total height of 1.5 m and a slope angle of 60°. The results of the soil mechanical tests indicate that the used fill materials are similar or even better than primary materials, such as gravel. The results of the chemical tests show that some materials are qualified to be used in engineering constructions without or with minor restrictions. Other materials need a special sealing layer to prevent the material from leakage. The vegetation on the mixed SBM material grew successfully. Several ruderal and pioneer plants could be found even in the first year of the construction. The porous material (RB and LC) provide additional water storage capacity for plants especially during summer and/or heat periods. With regard to the results of the chemical analyses of the greening layers, they are usable under restricted conditions. Here special treatment is necessary. Finally, it can be stated that SBMs are feasible in GRSS, particularly as fill material but also as a mixture for the greenable soil.

Keywords: geogrid-reinforced soil structure; geogrid; substitute building material; recycled material; green infrastructure

Citation: Schwerdt, S.; Mirschel, D.; Hildebrandt, T.; Wilke, M.; Schneider, P. Substitute Building Materials in Geogrid-Reinforced Soil Structures. *Sustainability* **2021**, *13*, 12519. https://doi.org/10.3390/su132212519

Academic Editor: Castorina Silva Vieira

Received: 9 September 2021
Accepted: 5 November 2021
Published: 12 November 2021

Publisher's Note: MDPI stays neutral with regard to jurisdictional claims in published maps and institutional affiliations.

1. Introduction

Mineral waste, and especially construction and demolition waste (CDW) as well as soil material, will be the largest waste stream in terms of volume once a certain level of urbanization has been reached. At the EU level, this state has already occurred, while at the global level, developing countries are still in the process of urbanization and a large material flow of CDW and soil materials will occur with a time lag. Nevertheless, this material flow represents both a global challenge and a significant resource potential for replacing mineral primary raw materials. In Germany, the material flow falls under the class of substitute building materials, i.e., "building materials from industrial manufacturing processes or from processing/treatment plants (waste, products) used instead of primary raw materials, such as recycled building materials (rubble), soil material, slags, ashes, railway ballast" [1]. In order to conserve natural raw material resources, substitute building materials should be preferentially used in construction projects in the future.

In Germany, around 350 million tons of waste are generated every year. Among other materials, the waste stream consists of about 250 million tons of mineral waste, such as 100 million tons of soil and stones, 73 million tons of CDW, 15 million tons of ashes and

239

slags from energy plants, as well as 7 million tons of blast furnace slag, and 6 million tons of steel slag.

While the majority of CDW is reused [2], other materials are often used for low-value purposes, such as landfill cover materials or for backfilling of open-cast mines [3]. This kind of use is considered a downcycling process. Materials like slags, ashes, and CDW which are reused in the construction process are considered so-called substitute building materials (SBMs) in Germany. Therefore, the presented study will focus on upcycling applications, especially on the use of SBMs in green applications. Green applications refer to structures that will become vegetated. These applications are chosen because green elements will have a significant positive impact on lowering the local temperature and improving the living climate in cities. These green applications are called "Green Infrastructure" (GI), according to the European Commission (2013) [4]. Green infrastructure is "a strategically planned network of high quality natural and semi-natural areas with other environmental features, which is designed and managed to deliver a wide range of ecosystem services and protect biodiversity in both rural and urban settings".

Widely used examples for GI in civil engineering are green roofs, vegetated gabions, or vegetated reinforced soil constructions. While these constructions are well established, the simultaneous use of SBMs in these constructions has rarely been carried out. Examples of used SBMs (waste silica) are described in Krawczyk et al. [5]. Molineux et al. [6] investigated the use of different recycled aggregates and combinations of them in green roof growing substrate. Carson et al. [7] used recycled and waste materials as substrates for roof greening. They described how to process aggregates from waste drywall, concrete, roof shingles, glass, and lumber cuttings which can be further used for roof substrates. Gagari et al. [8] provided a life-cycle assessment for the use of brick material with additional clay and compost ingredients. Several articles have described the experiences with recycled materials in geogrid-reinforced earth structures (GRSS), such as Santos et al. [9]. In this article, the authors reported the results of a 3.6 m high GRSS construction built with CDW material. Despite the construction being carried out on collapsible soil, the behavior of the wall was found to be satisfactory. Vieira et al. [10] described the performance of CDW as backfill in reinforced soil structures (RSS) and concluded that the resistance of the geogrid increases with the specimen size. Fleury et al. [11] investigated the influence of the drop and compaction of CDW material on the factor of built-in damage (RF_{ID}). They found out that grain size, drop height, and applied compaction method must be considered when calculating the factor of safety. Sachidanand and Divya [12] described the use of CDW material in RSS. The authors found that the geotechnical properties of the CDW met the requirements of an ideal backfill material for RSS. Ferreira et al. [13] reported on the results of time-dependent shear tests on CDW materials in connection with an underlying geotextile material. A stress relaxation was discovered in the tests. However, it was found that the peak and residual shear strength parameters in the long-time tests were similar to those in the benchmark tests. Viera [14] described the shear and pull-out behavior of different geosynthetic materials in connection with CDW material.

Most papers dealing with recycled materials in GRSS described the use of CDW. No applications of materials such as slag or track ballast to be used in RSS are published yet. This paper fills this gap as not only soil mechanics tests were carried out. In addition, the chemical constituents were analyzed and assessed with regard to potential hazards. Possible uses or additional necessary measures were derived. The present study describes the results of a pilot application in which a GRSS was constructed. For its construction, only SBM was used as fill materials and soil material for facing. Cost–benefit analyses as a comparison between conventional materials and SBM as well as life-cycle assessment were not within the scope of this work and will be subjects of future studies.

2. Materials and Methods

2.1. General Approach

This study consists of three parts. Firstly, additional soil mechanical tests were carried out. These tests complement the test results from preliminary tests. The results of these tests were reported in [15]. Additionally to the soil mechanical tests, greening tests were carried out to determine the best seed mixture for the facing material. Based on the results of these tests, a pilot application was erected.

The SBM materials used in the study are recycled brick material (RB), crushed lightweight concrete material (LC), blast furnace slag (BFS) electric furnace slag (EFS), track ballast (TB), and recycled concrete material (RC). The crushed brick material and lightweight concrete materials were used as greenable material for the GRSS facing. The remaining materials were tested for use as fill material in GRSS construction. Several geosynthetic materials, as well as steel grids, were also used.

2.2. Materials for Pilot Application

2.2.1. Fill Materials for Pilot Application

Four types of SBM were used as fill materials. The materials are shown in Figure 1.

Figure 1. Tested SBM as fill material from left to right: recycled concrete (RC), track ballast (TB), blast furnace slag (BFS), electric furnace slag (EFS). Source: Schneider.

The following properties and parameters were determined for the materials:

- Soil type acc. to EN ISO 14688-1 [16] and EN ISO 14688-2 [17];
- Density of soil particles (g/cm^3);
- Ignition loss (%);
- Proctor density (g/cm^3);
- Shear resistance acc. to EN ISO 17892-10 [18];
- Friction characteristics acc to EN ISO 12957-1 [19];
- Field capacity (%);
- pH value;
- Chemical analyses.

2.2.2. Materials for Facing Elements of Pilot Application

The materials used in the pilot application for facing and greening purposes consist either of crushed brick material (RB) or crushed lightweight concrete (LC) which was mixed with organic soil with a mixture rate of two parts SBM and one part organic soil (Figure 2). The mixed materials were tested to determine soil mechanical and chemical properties and parameters. The following tests were carried out:

- Soil type acc. to EN ISO 14688-1 [16] and EN ISO 14688-2 [17];
- Density of soil particles (g/cm^3);
- Ignition loss (%);
- Proctor density (g/cm^3);
- Field capacity (%);
- pH value;
- Chemical analyses.

Figure 2. SBM mixed with organic soil, **left**: before installation, **right**: after installation in a pilot application. Pictures: Schneider.

For the initial greening tests, RB and LC were mixed with different materials. There was a total of 32 test samples. In each of the test pits, either RB material or LC was mixed with one or several components of compost, bark mulch, expanded clay, peat, and/or lava mulch. The used seed was a lawn mixture. Additional tests were carried out using a mixture of two parts RB or LC material and one part organic soil. In these tests, flower seed was used for greening. In total, 15 test samples were used in this pre-test. The organic soil was topsoil material from the site of the pilot application.

2.2.3. Additional Materials for Pilot Application and for Greening Tests

Several geosynthetics were used for the pilot application. The reinforcement of the GRSS was carried out using a laid uniaxial PET geogrid made of extruded monolithic flat bars which are factory-welded at the crossing points. The maximum tensile force of this material is 80 kN/m at a strain of less than 7%.

At the surface, a geosynthetic erosion mat was placed. At the bottom of the pilot construction and between the different chambers a geosynthetic seal layer was placed. The facing was stabilized with galvanized steel grid with a mesh wide of 5 cm in each direction. For the prior greening tests, additional materials like compost, bark mulch, expanded clay, peat, and lava mulch were used. The used seed was a lawn mixture in both tests.

2.3. Pilot Application

The layout of the pilot application can be seen in Figure 3.

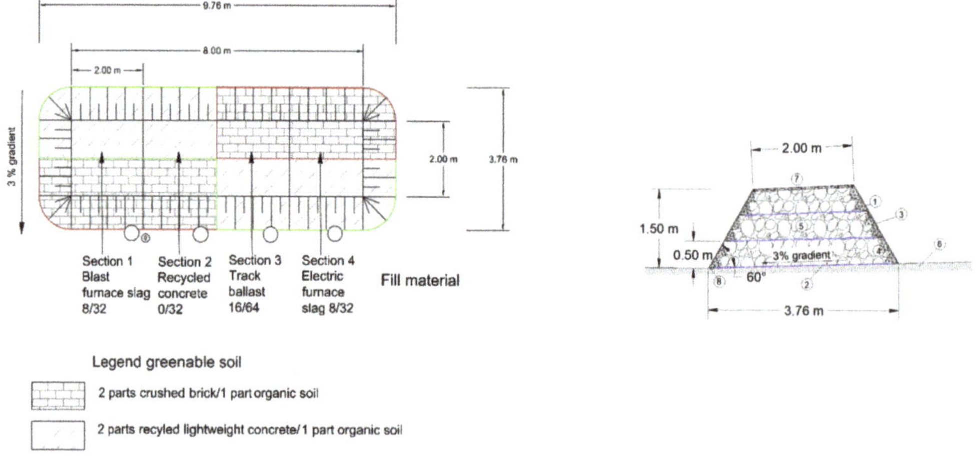

Figure 3. Top view and crosssection of the GRSS with recycled materials. Legend: 1—geogrid; 2—horizontal sealing; 3—facing (steel grid and erosion protection mat); 4—facing material/greenable soil; 5—fill material; 6—underground; 7—greenable soil; 8—drainage pipe; 9—collection shaft for leachate. Facing material 1:2 parts RB+1 part soil; facing material 2:2 part LC+1 part soil; Drawing: Schwerdt, Mirschel.

The construction is divided into four parts. On each part, the fill material was foreseen according to the findings in Section 2.2. The construction height was 1.5 m, the slope angle was 60° (Figure 4).

Figure 4. Left: View at the base layer of the pilot application before installation; **Right**: Northwest view at the greened construction in October 2020 (6 months after erection). Pictures: Schneider.

At the ground of the steep slope construction, a horizontal sealing layer with a slight slope to the north side of the construction was placed. The leachate from the rainfall was sampled and transported in gutters and stored in shafts. The four sections are divided by vertical panels to make sure the leachate cannot be mixed between the sections. The leachate was analyzed several times during the lifespan of the construction. The aim was to identify potential differences in the water quality between the leachate and the results of the chemical analyses carried out on recycled materials.

The geogrid layers were placed with a 50 cm vertical difference. A PET material was used as geogrid reinforcement. This material is usable in alkaline environmental conditions. Additional tests were carried out to determine the reduction factor for installation damage (RF_{ID}/A2 acc. to [20]). Beneath the grid elements, an erosion protection mat and greenable soil were placed. The greenable soil was only slightly compacted. The facing was carried out using galvanized steel grid elements.

The RF_{ID} tests were carried out according to EBGEO [20] beside the pilot application. The geogrid was placed on a 15 cm base layer made from the 4 used SBM fill materials. The layout can be seen in Figure 5. Above the geogrid layer additional 15 cm of SBM was placed. The SBM materials were compacted using a vibrating plate with a mass of 135 kg and a performance of 3.1 kW (4.1 HP). The vibration plate was the same used for compaction in the pilot application. After compacting the top layer, the material was removed and geogrid samples were taken for tensile tests according to EN ISO 10319 [21].

 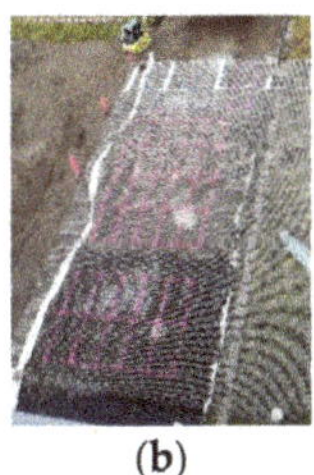  

| (a) | (b) | (c) | (d) |

Figure 5. RF_{ID} tests: (**a**) base layer; (**b**) geogrid layer with test samples; (**c**) top layer; (**d**) vibrating plate.

After the installation, the construction was covered with greenery. A seed mixture consisting of grass seeds and flowers seeds was used for this purpose.

3. Results

3.1. Soil Mechanical and Chemical Test Results

The soil mechanical and chemical test results for the GRSS materials are summarized in Tables 1 and 2 as well as in Figures 6 and 7.

Table 1. Results of soil mechanical and chemical tests of the fill materials [22,23].

	BFS	EFS	RC	TB
Soil classification (EN ISO 14688-1)	mgrCGr	cgrMGr	Gr	mgrCGr
Soil classification (EN ISO 14688-2)	Uniformly graded gravel	Uniformly graded gravel	Medium to well-graded gravel	Uniformly graded gravel
Density of soil particles (g/cm^3)	2.41–2.83	3.84–3.96	2.55–2.57	2.66
Ignition loss (%)	0	0	0	0
Water absorption (%)	24.5	25.8	n.d.	n.d.
Proctor density (g/cm^3)	1.51–1.58	2.10–2.16	1.78	1.61
pH value	10.2	10.7	9.3	9.3
Field capacity (%)	2.29	1.81	9.97	n.d.
Air capacity (%)	7.90	6.20	9.53	n.d.
Shear parameter (soil) (φ'/c') ($°$/kN/m^2)	54.3/0	53.6/0	53.2/0	59.6/0
Friction ratio (–)	0.91	0.94	0.81	0.75
Chemical classification according to LAGA M20 * [24]	Z2 (sulfate)	Z0	Z1.2 (sulfate)	Z0

* Legend Chemical classification LAGA M 20: Z0—usable without restrictions; Z1.2—usable with minor restrictions above groundwater level; Z2—usable with restrictions (sealing layer).

Table 2. Results of soil mechanical and chemical tests of facing soil materials [22,23].

	LC + Soil	RB + Soil
Soil classification (EN ISO 14688-1)	grcsiSa	csisaGr
Soil classification (EN ISO 14688-2)	Well-graded sand	Well-graded gravel
Density of soil particles (g/cm^3)	1.89	2.47–2.64
Proctor density (g/cm^3)	1.24	1.96
Ignition loss (%)	7.04	2.5
Field capacity (%)	14.96	13.96
Air capacity (%)	2.13	5.55
pH value	8.8	8.0
Chemical classification according to LAGA M20 * [24]	Z2 (sulfate)	Z1.2 (sulfate)

* Legend Chemical classification LAGA M 20: Z1.2—usable with minor restrictions above groundwater level; Z2—usable with restrictions (sealing layer).

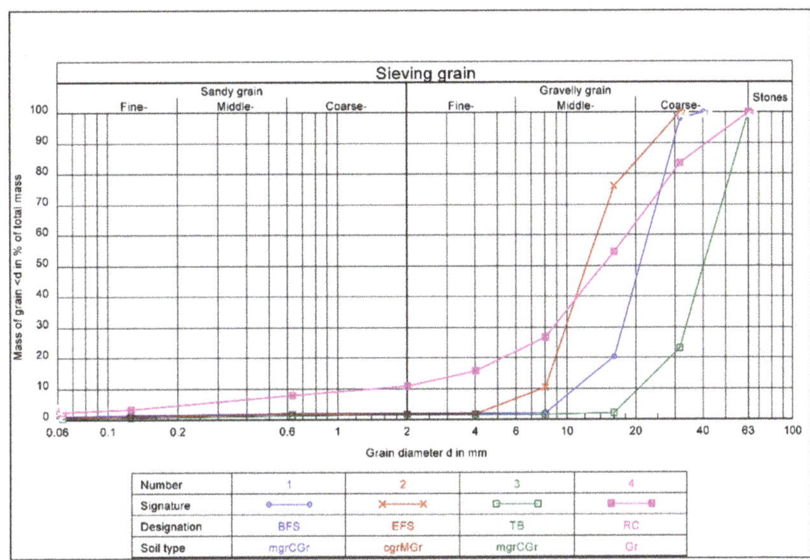

Figure 6. Grain size distribution of fill material.

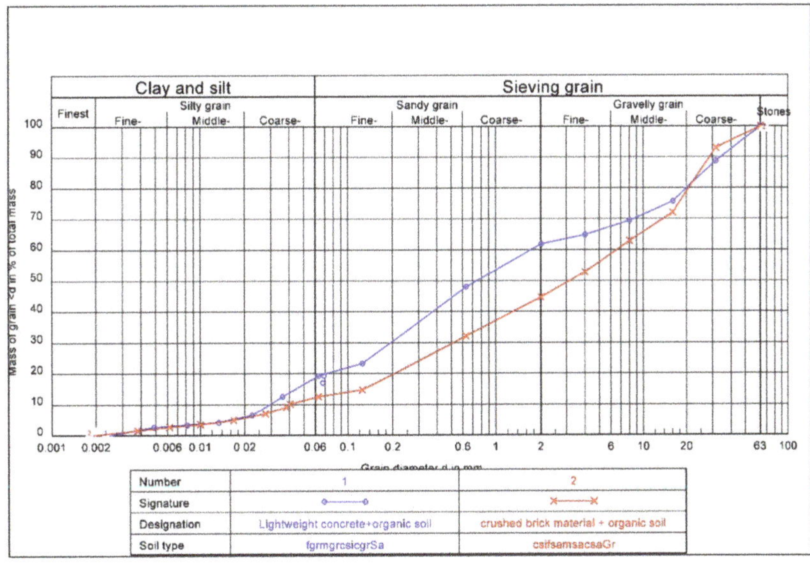

Figure 7. Grain size distribution of greening soil material.

The results show that the soil mechanical parameters of the tested SBM are without exception within the range of comparable gravel. Differences are only noticeable for some Proctor densities or grain densities. The maximum dry density that a material can reach is called Proctor density. The achieved Proctor density values are shown in Figure 8. This is due to the source materials.

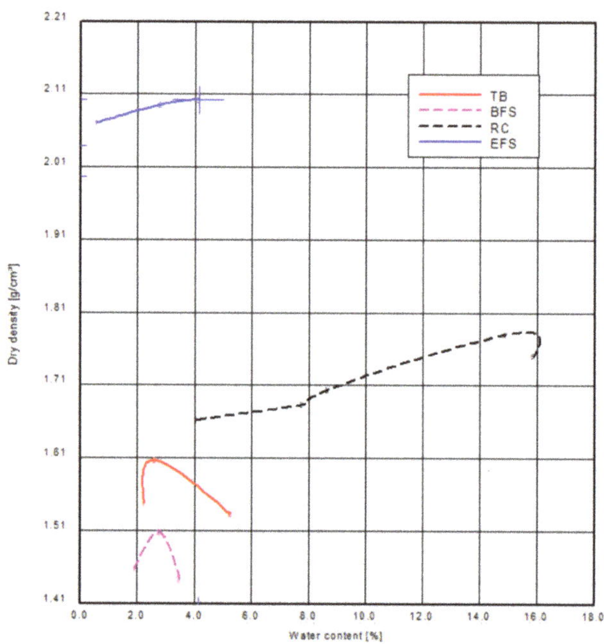

Figure 8. Proctor density of the soil materials.

For the shear tests, a large shear box with dimensions of length × width × height of 50 × 50 × 20 cm was used. Each test consisted of three sub-tests with normal stresses of 25, 50, and 100 kPa. The tests were carried out both exclusively on the soil material used in the construction and additionally with a geogrid layer in the joint between the upper and lower shear box. The results of the shear tests can be seen in Figure 9. The shear tests show that the materials can be used without restriction for KBE constructions in terms of soil mechanics.

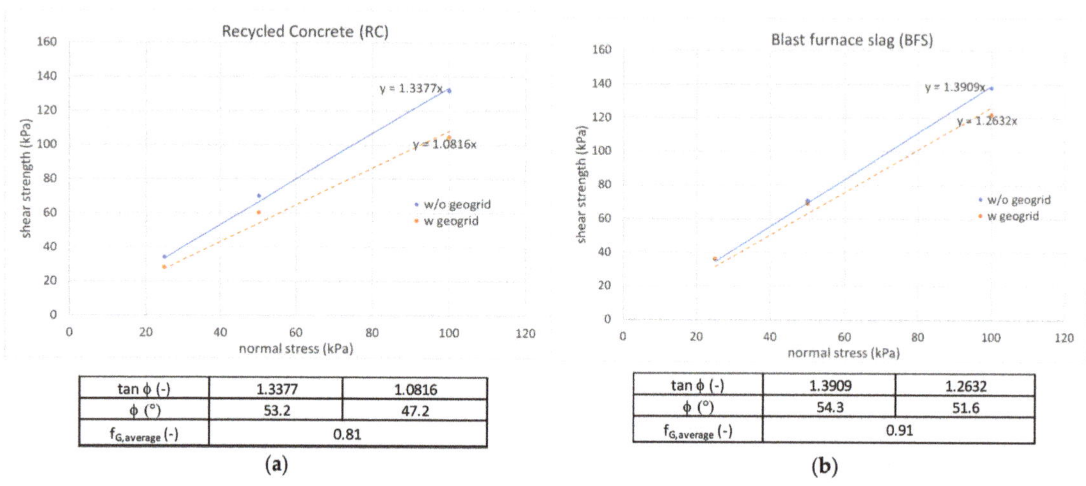

Figure 9. *Cont.*

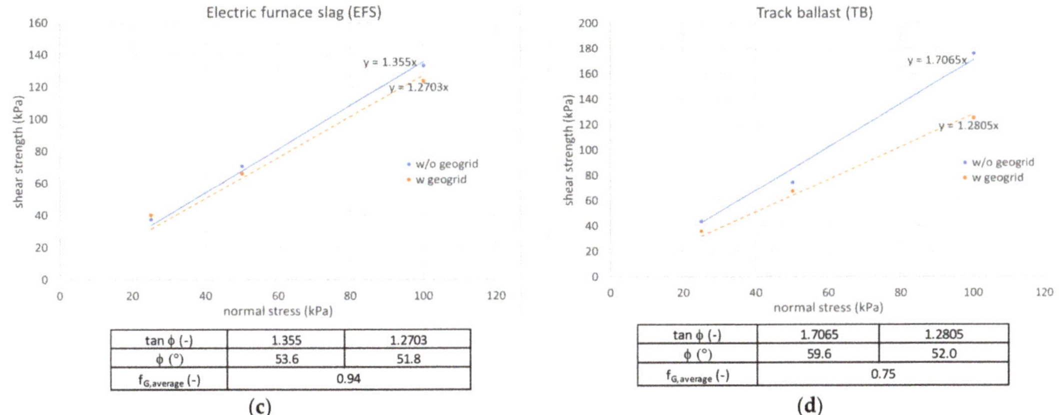

Figure 9. Results of shear tests: (a) recycled concrete (RC); (b) blast furnace slag (BFS); (c) electric furnace slag (EFS); (d) track ballast (TB).

The field capacity refers to the field the maximum water content that an unsaturated soil can retain against gravity under undisturbed soil conditions, according to ISO 11074 [25].

With regard to the chemical investigations, it can be stated that the sulfate contents were high in nearly all materials (with exception of track ballast and EOS) and mostly resulted in a classification in recovery class Z2 according to LAGA M20 [24]. This also resulted in comparatively high specific electrical conductivities (leachate mineralization). Since sulfate contents are generally not critical, the following evaluation was carried out neglecting the sulfate values. The RC material has an allocation class of Z1.2 according to LAGA M 20. Exceeded benchmark values concern polycyclic aromatic hydrocarbons (PAH) according to EPA, the specific electrical conductivity, and sulfate. RB has an allocation class of Z1.2 because the values exceeded the benchmarks for PAH according to EPA, electrical conductivity, chloride, and sulfate. Blast furnace slag (BFS) has an allocation class of Z0 (neglecting the sulfate value, being Z2). Electric furnace slag (EFS) has an allocation class of Z0. Track ballast (TB) is to be assigned to allocation class Z0. Lightweight concrete (LC) is to be assigned to allocation class Z2, having exceeded benchmark values for lead and sulfate.

With regard to general feasibility, it can be concluded that the materials with a classification value Z2 can only be installed using a water-impermeable top layer or using additional technical safety measures. In such a case, Code of Practice M TS E [26] is authoritative. If sulfate is not considered critical, the materials are to be allocated to allocation classes Z 0 or Z 1.1. This means that the use of the materials is feasible in the following kinds of applications [24]:

- Roads, paths, traffic areas (superstructure and substructure);
- Industrial, commercial, and storage areas (superstructure and substructure);
- Substructures of buildings;
- Below the rootable soil layer of earthworks (noise and protection walls);
- Substructures of sports facilities.

The SBM investigation results for the usable field capacity and air capacity show that water can be stored in RB/topsoil and LC/topsoil mixtures, but not in other materials. The air capacity values can be considered average due to the large pore space. The use of blast furnace slag, electric furnace slag, and recycled concrete as a greening layer is not recommended. Field and air capacity are affected by the capillarity of the material in relation to the fine particle content. Coarse structures are suspicious of washing and producing leachate. The components RB/topsoil and LC/topsoil mixtures are suitable for this purpose. In summary, it can be stated that the tested SBMs, if necessary with the use of technical safety measures, are also suitable in chemical terms for the use in plastic-reinforced earth constructions. The leachate pH value was measured to be alkaline, which means not usable for use in all geosynthetic GRSS constructions.

The greening material was only slightly compacted in the pilot application. For this reason, the pore content of the material could not be determined in a meaningful way. The organic content provides nutrient input for plants. The porosity of the RB and LC material provides additional water storage capacity. The pH value is also alkaline, which has some implications for the usability of the materials in GRSS (geogrid, erosion mat) as well as for the selection of revegetation materials. The chemical results show an expected outcome at high sulfate levels. The other ingredients except sulfate are within the benchmark ranges.

3.2. Preliminary Greening Test Results

In the greening tests carried out before the erection of the pilot application, the best results could be found with material mixtures supplemented with expanded clay and lava mulch. Other mixtures were also successful to different degrees. Even on pure brick or lightweight concrete material, some plants could be found. In the second part with either RB or LC mixed with organic soil the most of the greening tests were successful. Finally, the decision was made to use the ladder mixture for the pilot application. With a focus on further applications, this mixture will provide the most cost-effective solution.

3.3. Pilot Application

During the erection of the pilot application, the compaction grade of the different fill materials was determined. The values are within a range of 97 to 98%.

In the RF$_{ID}$ tests (A2 value acc. to EBGEO) the results in Table 3 could be achieved:

Table 3. A2 values.

SBM	A2–Value (md) (–)	A2–Value (cmd) (–)
EFS	1.01	1.03
BFS	1.00	1.00
RC	1.01	1.00
TB	1.08	1.06

It can be concluded that the damages on the geogrid are only minor. This finding can be confirmed by the results of the visual inspection (see next figures). No damage to the material was found in the tests with BFS, EFS, and RC. Only in the tests with TB were minor damages found as can be seen in Figure 10.

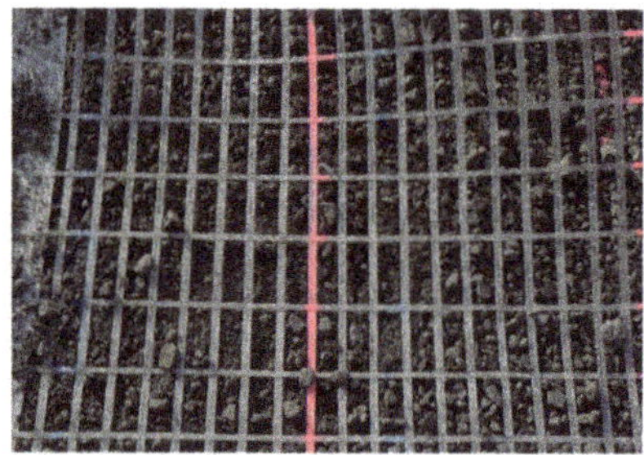

Figure 10. Visual inspection of geogrid after RFID test: **left**: test with TB; some smaller damages can be seen; **right**: no damages in tests with BFS, EFS, and RC.

In terms of vegetation analysis, each GRSS side was investigated separately. The sowing was carried out in April 2020. On the east side, some germination could be seen already at the end of April, but mainly on the RB side. From May to June was observed that the RB/soil side was more densely vegetated than the LC/soil side. On the other hand, more vegetation appeared on the RB/soil side at the end of September and in mid-October. The trend shows that the RB/soil is suitable for rapid vegetation and LC/soil has more of a long-term positive effect.

At the south side of the GRSS, the findings from the east side were confirmed. In addition to the lush greening, however, it can also be seen that most of the plants have been dried out by the end of September. The reason is the higher sunlight exposure on the south side. In autumn (October to December), further greening was observed on the south side, with particular grass seed sprouting. No distinction can be made between the greening results of the RB/soil and LC/soil material mixtures. On the west side, the same setting like on the east and south sides is evident. The rapid vegetation of May and June is clearly visible on the RB/soil mixture (right) and only partially on the LC/soil mixtures (left). On the other hand, the vegetation withers faster on the RB/soil mixture than on the LC/soil mixtures. The north side is by far the best-vegetated slope. Initially, it is again recognizable that greening and blossoms can be seen earlier on the RB/soil mixtures (left) than on the LC/soil mixtures. In contrast to the other slopes, the shaded locations did not cause dry out or wither in summer, so that the slope has the highest degree of greening at the end.

The differences between the building materials can be assessed as follows,

- RB material is coarser-grained compared to LC. As a result, it has a higher pore volume, which supports water flow and air capacity. The large pore spaces also give roots better opportunities to grow. The brick acts as a drainage layer. The RB's disadvantage is that it has no real storage capacity due to high water permeability. A long-lasting, stable moisture level can therefore not be expected;
- LC, on the other hand, is very fine-pored, and this property gives it a high storage capacity. With many small cavities, the water has plenty of room to spread and to be stored. Due to its good heat capacity, this SBM can store heat well when the outside temperature is too low and can insulate against heat when it is very hot. This offers the advantage that plants can thrive even on colder days/months and survive better in hot conditions. The disadvantage of the fine pores is that there is no good water permeability. When water enters LC, it runs off on the surface because the pores are too fine to allow the water to penetrate quickly. This characteristic does not guarantee fast plant growth. Furthermore, the pores block possible space for rapid root development. They, therefore, need more time to break through.

Looking at the GRSS as an overall structure, it can be seen that vegetation progressed in May, peaked in June (vegetation period 1), flowered from July to September, and re-flowered again in October (vegetation period 2). Figure 11 shows pictures of the pilot application at different points in time. Not only did the different SBMs have an influence on plant growth, but weather conditions did as well. The temperature rose since March. The peak was observed in August with an average of 21.9 °C. Since flowering took place in the period from June to September, the temperatures were possibly too high for the first vegetation. Comparing the precipitation total with the previous year, the year 2020 was drier. Only in June, August, and September was a higher precipitation level observed. However, it must be taken into account that the construction was irrigated. This took place especially in the very dry spring months and, to a lesser extent, in summer. A closer look at the precipitation patterns shows that there were no continuous rain phases. The final amount came mainly from short heavy rainfall events. In this situation, the soil cannot absorb the water fast enough. The rainwater runs off aboveground or into the sewage system. With normal precipitation events, the greening of the reinforced earth might have been maintained for a longer period (midsummer).

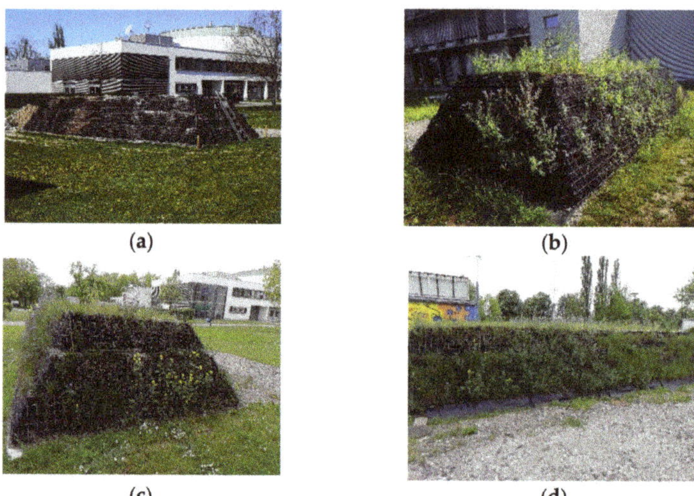

Figure 11. Pictures from the pilot application: (**a**) 1 month after installation view from the southeast; (**b**) 3 months after installation view from the southwest; (**c**) 1 year after installation view from the east; (**d**) 3 months after installation view from north.

When looking at the flowering plants, all have low soil quality requirements and are very undemanding. Among others, the followings plants were found at the GRSS: White Mustard, Common Sheep weed, Field Bindweed, White Goosefoot, Field Thistle, Jacob's Grasswort, and Mallow. These plants are considered pioneer or ruderal plants. Pioneer plants have a special adaptability for colonizing new, still vegetation-free habitats. These species occur more frequently in newly created habitats than in existing ecosystems. The two main characteristics for pioneer plants are,

- Effective long-distance dispersal mechanism: Pioneer habitats emerge unpredictably and in isolation. Therefore, species with high seed numbers and with dispersal by wind (anemochory) and animal (ornithochory) are typical pioneers;
- High hardiness: Tolerate extreme environmental conditions, established vegetation stands reduces occurring maxima e.g., in terms of temperature and soil water; furthermore, the soils usually show nutrient deficiencies or imbalances.

In addition to pioneer plants, there are also ruderal plants. These species prefer to settle on rubble and debris sites, stony slopes, disturbed roadsides, and similar territories. All the above criteria apply to the GRSS.

4. Conclusions

From a soil mechanics point of view, the proof was provided that SBM can be used in GRSS. This can be considered proof of the feasibility of SBM in GRSS and other engineering structures, such as gabion walls, green roofs, bridge abutments, as well as the construction of noise barrier dams or for backfilling structures. Further investigations are required in this regard. The soil mechanical parameters Proctor density, grain distribution, grain bulk density, and shear strength (friction angle) are comparable to those of primary building materials.

When considering the chemical properties, differentiation is necessary as well as an adaptation to the detailed conditions. The investigated materials are all suitable for recycling or reuse respectively, although the recycling feasibility differs depending on material type. The current legal situation also permits the recovery of Z2 materials in some areas. The technical prerequisites to prevent precipitation from penetrating the construction are given. For example, the use of plastic sealing membranes is a suitable measure. For the greening layer, other solutions must be sought here. While the RB material only has to fulfill minor additional requirements, in terms of LC, an additional treatment step must be interposed in order to immobilize potentially environmentally harmful constituents that cause classification in recycling class Z2. For obvious reasons, a cover as in the case of fill soils is ruled out. However, the effort to undertake such developments is justified, as crushed LC has proven to be an ideal material for greening layers, at least in combination with an organic admixture/soil material, respectively.

The results of the greening tests have shown that RB/soil and LC/soil mixtures are well suited for revegetation. First colonizers and ruderal plants can grow and spread very well. The different SBM properties are reflected in the vegetation. RB is well suited for quick vegetation success and LC, on the other hand, offers good conditions for long-term revegetation. The results support the conclusion of the feasibility of SBM-based GRSS as green infrastructure. In this way, the project results contribute to the closure of SBM loops for climate adaptation purposes.

Further tests must prove the durability and resistance to deterioration e.g., by freezing over a longer period of time. For this purpose, the pilot application will continue to be monitored regularly. The results will be reported in due course.

Author Contributions: Conceptualization, S.S. and P.S.; methodology, S.S., D.M. and P.S.; validation, S.S. and P.S.; formal analysis, D.M.; investigation, D.M., M.W. and T.H.; resources, D.M.; data curation, D.M.; writing—original draft preparation, S.S.; writing—review and editing, S.S. and P.S.; visualization, S.S. and D.M.; supervision, S.S. and P.S.; project administration, S.S.; funding acquisition, S.S. and P.S. All authors have read and agreed to the published version of the manuscript.

Funding: This research was funded by the Ministry of Environment, Agriculture and Energy Saxony-Anhalt, Germany, grant number FKZ U02/2019. Parts of the APC were funded by Naue GmbH & Co. KG, Espelkamp.

Institutional Review Board Statement: Not applicable.

Informed Consent Statement: Not applicable.

Data Availability Statement: The data presented in this study are available upon request from the corresponding author. The data are not publicly available due to privacy reasons.

Acknowledgments: The authors wish to thank the Ministry of Environment, Agriculture and Energy Saxony-Anhalt, Germany for funding the project, Naue GmbH & Co. KG, Espelkamp Germany for providing the used geogrid and funding parts of APC, and Salzgitter Flachstahl GmbH, Salzgitter, E. Friedrich GmbH, Salzgitter and Storck Umweltdienste GmbH, Magdeburg for providing the used SBM, respectively.

Conflicts of Interest: The authors declare no conflict of interest. The funders had no role in the design of the study; in the collection, analyses, or interpretation of data; in the writing of the manuscript, or in the decision to publish the results.

Abbreviations

BFS	Blast furnace slag
CDW	Construction and demolition waste
cmd	Cross machine direction (secondary direction of geosynthetics reinforcement)
EFS	Electric furnace slag
EPA	Environmental Protection Agency
GRSS	Geogrid-reinforced soil structure
LC	(crushed) Lightweight concrete material
md	Machine direction (main direction of geosynthetics reinforcement)
PAH	Polycyclic aromatic hydrocarbons
PET	Polyethylenterephthalat
RB	Recycled brick material

RC	Recycled concrete material
RF_{ID}	Reduction factor of installation damage
RSS	Reinforced soil structure
SBM	Substitute building material
TB	Track ballast

References

1. Bundesministerium für Umwelt, Entwurf der Mantelverordnung. Available online: https://www.bmu.de/fileadmin/Daten_BMU/Download_PDF/Gesetze/mantelv_text.pdf (accessed on 3 October 2019).
2. Kreislaufwirtschaft Bau. Kreislaufwirtschaft-Bau.de. Available online: http://kreislaufwirtschaft-bau.de/#aktuelleDaten (accessed on 30 October 2019).
3. Ministerium für Umwelt, Landwirtschaft und Energie, Sachsen-Anhalts. *Leitfaden zur Wiederverwendung und Verwertung von Mineralischen Abfällen in Sachsen-Anhalt*; Ministerium für Umwelt, Landwirtschaft und Energie, Sachsen-Anhalts: Magdeburg, Germany, 2018.
4. European Comission. Communication from the Commission to the European Parliament, the Council, the European Economic and Social Committee and the Committee of the Regions-Green Infrastructure (GI)—Enhancing Europe's Natural Capital. 2013. Available online: https://eur-lex.europa.eu/resource.html?uri=cellar:d41348f2-01d5-4abe-b817-4c73e6f1b2df.0014.03/DOC_1&format=PDF (accessed on 3 October 2019).
5. Krawczyk, A.; Domagała-Świątkiewicz, I.; Lis-Krzyścin, A.; Daraż, M. Waste Silica as a Valuable Component of Extensive Green-Roof Substrates. *Pol. J. Environ. Stud.* **2017**, *26*, 643–653. [CrossRef]
6. Molineux, C.; Gange, A.C.; Connop, S.P.; Newport, D. Using recycled aggregates in green roof substrates for plant diversity. *Ecol. Eng.* **2015**, *82*, 596–604. [CrossRef]
7. Carson, T.; Hakimdavar, R.; Sjoblom, K.; Culligan, P. Viability of Recycled and Waste Materials as Green Roof Substrates. In Proceedings of the GeoCongress 2012, Oakland, CA, USA, 25–29 March 2012. [CrossRef]
8. Gargari, C.; Bibbiani, C.; Fantozzi, F.; Campiotti, C.A. Environmental Impact of Green Roofing: The Contribute of a Green Roof to the Sustainable use of Natural Resources in a Life Cycle Approach. *Agric. Agric. Sci. Procedia* **2016**, *8*, 646–656. [CrossRef]
9. Santos, E.; Palmeira, E.; Bathurst, R. Performance of two geosynthetic reinforced walls with recycled construction waste backfill and constructed on collapsible ground. *Geosynth. Int.* **2014**, *21*, 256–269. [CrossRef]
10. Vieira, C.S.; Pereira, P.; Ferreira, F.; Lopes, M.D.L. Pullout behaviour of Geogrids embedded in a recycled construction and demolition material. Effects of specimen size and displacement rate. *Sustanability* **2020**, *12*, 3825. [CrossRef]
11. Fleury, M.P.; Santos, E.C.G.; Da Silva, J.L.; Palmeira, E.M. Geogrid installation damage caused by recycled construction and demolition waste. *Geosynth. Int.* **2019**, *26*, 641–656. [CrossRef]
12. Vibha, S.; Divya, P.V. Geosynthetic-Reinforced Soil Wall with Sustainable Backfills. *Indian Geotech. J.* **2020**, *51*, 1–10. [CrossRef]
13. Ferreira, F.; Pereira, P.; Vieira, C.; Lopes, M. Time-Dependent Response of a Recycled C&D Material-Geotextile Interface under Direct Shear Mode. *Materials* **2021**, *14*, 3070. [CrossRef] [PubMed]
14. Vieira, C.S. Valorization of Fine-Grain Construction and Demolition (C&D) Waste in Geosynthetic Reinforced Structures. *Waste Biomass-Valoriz.* **2018**, *11*, 1615–1626. [CrossRef]
15. Schneider, P.; Schwerdt, S.; Schulz, K.; Fiebig, S.; Mirschel, D. Feasibility of substitute building materials for circular use in urban green infrastructure. *Civ. Eng. Des.* **2020**, *2*, 159–168. [CrossRef]
16. *EN ISO 14688-1: Geotechnical Investigation and Testing-Identification and Classification of Soil-Part 1: Identification and Description (ISO 14688-1:2017)*; German version EN ISO 14688-1:2018; Beuth-Verlag: Berlin, Germany, 2018.
17. *EN ISO 14688-2: Geotechnical Investigation and Testing-Identification and Classification of Soil-Part 2: Principles for a Classification (ISO 14688-2:2017)*; German version EN ISO 14688-2:2018; Beuth-Verlag: Berlin, Germany, 2018.
18. *EN ISO 17892-10: Geotechnical Investigation and Testing-Laboratory Testing of Soil-Part 10: Direct Shear Test (ISO 17892-10:2018)*; German Version EN ISO 17892-10:2018; Beuth-Verlag: Berlin, Germany, 2019.
19. *EN ISO 12957-1: Geosynthetics-Determination of Friction Characteristics-Part1: Direct Shear Test (ISO/DIS 12957-1: 2018)*; German and English Version prEN ISO 12957-1:2018; Beuth-Verlag: Berlin, Germany, 2018.
20. DGGT. *EBGEO Empfehlungen für Planung und Bemessung von Bewehrungen mit Geokunststoffen*; Deutsche Gesellschaft für Geotechnik, Ed.; Verlag W. Ernst und Sohn: Berlin, Germany, 2010; ISBN 978-3-433-02950-3.
21. *EN ISO 10319-Geotextiles-Wide-Width Tensile Test (ISO 10319: 1993)*; German Version EN ISO 10319:1996; Beuth-Verlag: Berlin, Germany, 1996.
22. Schwerdt, S.; Schneider, P.; Mirschel, D. *Abschlussbericht Forschungsprojekt Verbesserung und Stärkung der Urbanen Grünen Infrastruktur durch Einsatz von Ersatzbaustoffen in Kunststoff-Bewehrte-Erde-Konstruktionen (Recycle-KBE)*; Final Report; Magdeburg-Stendal University of Applied Sciences: Magdeburg, Germany, 2020.
23. Hildebrandt, T.; Wilke, M. Verwendbarkeit von Recyclingbaustoffen als Ersatz zu Primärbaustoffen in Kunststioff-Bewehrte-Erde-Konstruktionen. Master's Thesis, Magdeburg-Stendal University of Applied Sciences, Magdeburg, Germany, 2020, unpublished.
24. Länderarbeitsgemeinschaft Abfall-LAGA. Mitteilungen der Länderarbeitsgemeinschaft (LAGA) 20 Anforderungen an die stoffliche Verwertung von mineralischen Reststoffen/Abfällen-Technische Regeln. Available online: https://www.laga-online.de/documents/m20_nov2003u1997_2_1517834540.pdf (accessed on 3 October 2019).

25. Schulz, K. Untersuchung zu Einsatzmöglichkeiten von Ersatzbaustoffen in Bewehrter Erde als Grüne Infrastruktur. Bachelor's Thesis, Magdeburg-Stendal University of Applied Sciences, Magdeburg, Germany, 2019, unpublished.
26. M TS E. *Merkblatt über Bauweisen für Technische Sicherungsmaßnahmen beim Einsatz von Böden und Baustoffen mit umweltrelevanten Inhaltsstoffen im Erdbau*; FGSV Verlag: Köln, Germany, 2017; ISBN 978-3-86446-203-0. Available online: https://www.fgsv-verlag.de/pub/media/pdf/559.i.pdf (accessed on 30 October 2019).

Article

Engineering Characteristics and Environmental Risks of Utilizing Recycled Aluminum Salt Slag and Recycled Concrete as a Sustainable Geomaterial

Youli Lin [1], Farshid Maghool [1], Arul Arulrajah [1] and Suksun Horpibulsuk [1,2,3,*]

[1] Department of Civil and Construction Engineering, Swinburne University of Technology, Melbourne 3122, Australia; youlilin@swin.edu.au (Y.L.); fmaghool@swin.edu.au (F.M.); aarulrajah@swin.edu.au (A.A.)

[2] Center of Excellence in Innovation for Sustainable Infrastructure Development, School of Civil Engineering, Suranaree University of Technology, Nakhon Ratchasima 30000, Thailand

[3] Academy of Science, Royal Society of Thailand, Bangkok 10300, Thailand

[*] Correspondence: suksun@g.sut.ac.th; Tel.: +66-4422-4322

Citation: Lin, Y.; Maghool, F.; Arulrajah, A.; Horpibulsuk, S. Engineering Characteristics and Environmental Risks of Utilizing Recycled Aluminum Salt Slag and Recycled Concrete as a Sustainable Geomaterial. *Sustainability* **2021**, *13*, 10633. https://doi.org/10.3390/su131910633

Academic Editor: Castorina Silva Vieira

Received: 1 September 2021
Accepted: 21 September 2021
Published: 24 September 2021

Abstract: Recycled aluminum salt slag (RASS) is an industrial by-product generated from the melting of white dross and aluminum scraps during the secondary smelter process. Insufficient knowledge in the aspects of engineering characteristics, and the environmental risks associated with RASS, is the primary barrier to the utilization of RASS as a substitute material for natural quarry materials in the field of geotechnical construction. In this research, comprehensive geotechnical and environmental engineering tests were conducted to evaluate the feasibility of utilizing RASS as a sustainable geomaterial. This was undertaken by comparing the laboratory testing results for RASS with a well-known recycled material, namely recycled concrete aggregate (RCA), and the relevant specifications set forth by the local road authority. The geotechnical engineering assessment included particle size distribution, flakiness index, organic content, pH, particle density, water absorption, modified Proctor compaction, aggregate impact value, Los Angeles (LA) abrasion, hydraulic conductivity, and California bearing ratio (CBR). The CBR results of the RASS samples satisfied the minimum CBR value (>80%) for usage as pavement subbase material in road construction. In addition, the repeated load triaxial (RLT) tests were carried out on the RASS samples to assess the response of the RASS under cyclic loading conditions. Furthermore, a range of chemical tests, consisting of leaching and polycyclic aromatic hydrocarbon tests, were also performed on the RASS to address the environmental concerns. Comparing the chemical test results with the environmental protection authorities' guidelines provided satisfactory evidence that RASS will not pose any environmental and health issues throughout its service life as a geotechnical construction material.

Keywords: recycled aluminum salt slag; recycled materials; resilient modulus; leachate analysis; pavement geotechnics

1. Introduction

The sophisticated lifestyles and urbanization have dramatically increased the generation of waste materials from both the construction and industrial sectors. Recycling of waste materials, produced by construction and industrial activities, is becoming one of the key strategies globally in waste reduction, landfill avoidance, and to facilitate the movement towards sustainable development [1]. Many countries, such as Australia and New Zealand, are encouraging the recycling and reutilizing of waste materials by implementing a landfill levy on the disposal of waste to alleviate the pressure on natural resources and the shortage of landfill sites [2]. Over the last few years, many attempts have been made to utilize construction and demolition (C&D) waste materials, such as recycled concrete aggregate, crushed brick, reclaimed asphalt pavement, and waste rock, as substitution materials of virgin quarry materials in various geotechnical applications:

255

for instance, road, embankment, pipe bedding, and ground improvements in the aspects of soil stabilization and the construction of stone columns [3–7]. Research on stockpiled industrial by-products, in particular, granulated blast furnace slag, electric arc furnace slag, ladle furnace slag, copper slag, and gypsum-based waste, have also garnered significant interest in the past years, used as supplementary or substitution materials in construction, concrete production, and pavement applications [8–11]. Yi et al. [8] compared carbide slag-activated ground granulated blast furnace slag (CS-GGBS) with Portland cement (PC), for use in soft clay stabilization, and reported that the unconfined compressive strength of the optimum CS-GGBS blend was twice as high as the corresponding PC stabilized clay, demonstrating the feasibility of utilizing CS-GGBS for the replacement of PC in soft clay stabilization. Imteaz et al. [9] conducted a series of geotechnical and environmental tests to evaluate the viability of utilizing gypsum-based waste in geotechnical applications. The results of the study found that gypsum-based plasterboard waste satisfied the engineering requirements and caused no environmental impacts for usage as road subgrade and pipe bedding material. Maghool et al. [10] carried out a study to investigate the shear strength and stiffness of electric arc furnace slag (EAFS), ladle furnace slag (LFS), and their mixture (50% LFS + 50% EAFS). The study results demonstrated that both LFS and the mixture of (50% LFS + 50% EAFS) exhibit high friction angles similar to that of natural quarry materials. The test results of California bearing ratio and resilient modulus have also met the requirements for usage as a geomaterial in road applications, such as pavement base and subbase. Prem et al. [11] performed research to evaluate the influence of substituting 100% natural sand with high volume copper slag in concrete and reported that concrete samples on full replacement of natural sand with copper slag improved the compressive/flexural strength, toughness, and energy-absorbing capacity compared to the concrete made with river sand. In addition, the durability performance, in terms of chloride penetration and water absorption, remains unaffected after substituting 100% of natural sand with copper slag in concrete.

Aluminum is a non-ferrous metal, an indispensable raw material used for making a wide range of metallic products. According to the data provided by the International Aluminium Institute Statistics, about 65,000 thousand tons of aluminum yielded from primary aluminum production in 2020 [12]. During the aluminum making process, three types of by-products are generated at various stages, including white dross, very rich in aluminum content ranging from 40–80% produced during the primary aluminum process and downstream ingot smelting operations. Black dross, also known as aluminum salt slag, contains a lower aluminum content, typically 5–20% generated from the melting of white dross and aluminum scraps during the secondary smelters process with the presence of a salt flux composed of potassium chloride (KCl), sodium chloride (NaCl), and a small amount of fluorides. Non-metallic products are created from the tertiary smelters during the water leaching process of black dross/aluminum salt slag [13]. Depending on the type of furnace used and the manufacturing technique, between 300 to 600 kg of aluminum slat slag is generated for the production of one ton of aluminum [14]. However, about 95% of aluminum salt slag was stockpiled in landfill, leading to extensive pressure on landfill sites and severe environmental concerns [15]. In the past, the use of aluminum salt slag as a substitute material to natural aggregates, in the field of civil construction, was not so favorable. It is because the aluminum slat slag is prone to leaching chlorides and releases unpleasant odorous gases such as methane, ammonia, phosphide, and hydrogen sulphide when in contact with water [16]. However, with the improvement in aluminum production technology, black dross/aluminum salt slag can be treated by a wet/dry separation process, followed by water leaching and filtering to wash away and separate the soluble salt flux from insoluble oxides. Harmful gases are eliminated through conveyed and burnt odorous gases (typically hydrogen, ammonia, phosphine, and methane), generated during the leaching and dissolution process in a combustor, consequently transforming the harmful gases into water and inert gases. Finally, the separated salt flux is converted into a useable form through evaporation and crystallization processes. Karvelas et al. [17] conducted an economic and technical assessment of recycling systems for black dross/aluminum salt slag in the secondary aluminum industry. The study found that, if the presence of aluminum content in the black dross/aluminum salt slag is higher than 10%, the net revenues from the recovered aluminum and salts are sufficient to cover the operating and capital cost of the wet/dry separation process [17]. By considering the scarcity of landfill sites, rapid increase in landfill levy, along with all the environmental concerns, the treatment for such aluminum industrial by-products that convert aluminum waste into a useable form is rewarding and worth performing to improve sustainability in the aluminum industry. As a result, the aluminum salt slag treated by wet/dry separation technique exhibits very low chloride content, harmful gases are eliminated, and thus, it is safe to be utilized as a civil construction material and by the cement industry [14].

To date, some studies have initiated the utilization of recycled aluminum salt slag (RASS) as a supplementary material for making cement composites. A number of recent studies have demonstrated that incorporating RASS as an ingredient in mortar and concrete, up to 15%, caused no negative impact on the mechanical performance of the cement composites; the addition of RASS in mortar and concrete can also enhance the durability of the cement composites in terms of controlling microstructural cracks, improving abrasion, and corrosion resistivity [15,16,18–22]. Dunster et al. [23] stated that black dross/RASS with smaller grain size (<700 µm) had the potential to be used as filler aggregate in asphalt to enhance the hardness, abrasion, and skid resistance of the pavement. Similarly, López-Alonso et al. [24] demonstrated the feasibility of using aluminum waste, blended with recycled aggregates, as a road construction material to enhance the resistance properties and long-term mechanical behavior of the unbound road layer. Furthermore, several studies have utilized aluminum waste in stabilizing expansive soil and tropical lateritic soil. The results of the studies indicated that infusing aluminum waste in the soil specimen could reduce the swelling and shrinkage behavior of the expansive soil and enhance the bearing capacity of the tropical lateritic soil notably [25,26].

Over the years, several million tons of RASS were produced annually, and about 95% of them were sent to landfill instead of being recycled and reused [16]. This is believed to be the consequence of insufficient knowledge and understanding of RASS in terms of its engineering and environmental properties. Research on the utilization of RASS as a geomaterial is still very limited. There have been no known studies thoroughly investigating the geotechnical and geo-environmental characteristics of the RASS and its potential to be utilized as an alternative source of aggregate in geotechnical applications. To fulfil the current research gap, comprehensive geotechnical and environmental engineering tests were conducted on RASS to assess the feasibility of utilizing RASS as a sustainable geomaterial. The usage of RASS, as a geomaterial in civil construction, can add value to the aluminum industrial waste by-products, reduce the cost for landfilling, and lower the usage of raw quarry materials and the carbon footprint of future civil infrastructure projects.

2. Materials and Methods

2.1. Geotechnical Engineering Tests

The materials used in this research were comprised of recycled aluminum salt slag (RASS) and recycled concrete aggregate (RCA), one of the well-known recycled materials available in the market. RASS was derived from a major aluminum manufacturer located in Melbourne, Australia. RCA is a by-product of (C&D) activities of concrete structures. It is produced by crushing large concrete chunks into aggregates of various sizes depending on the field application. The RCA used in this research was sourced from a recycling plant in Melbourne, Australia and had a maximum particle size of 20 mm. The RCA was comprises of 1–2 wt% of crushed brick, reclaimed crushed asphalt pavement aggregates, and a very small fraction (<1 wt%) of other foreign materials such as glass, wood, plastic, and gypsum.

An extensive suite of laboratory tests was conducted to assess the geotechnical and geo-environmental characteristics of RASS and RCA as per relevant international standards, including the American Society for Testing Materials (ASTM), Australian Standards (AS), and British Standards (BS). The experimental program included particle size distribution, unified soil classification, flakiness index, organic content, pH, particle density, water absorption, modified Proctor compaction, aggregate impact value, Los Angeles (LA) abrasion, hydraulic conductivity, California bearing ratio (CBR), resilient modulus (M_R), as well as a range of chemical and environmental tests. The morphology studies of RASS and RCA were also performed by employing scanning electron microscopy (SEM). The testing results of RASS and RCA were compared to provide a synthetic examination on the viability of utilizing RASS as a sustainable geomaterial.

Necessary precautions were taken to obtain RASS and RCA samples from different stockpiles, as per the procedures mentioned in ASTM D75 [27]. The collected materials were dried in an oven at the standard drying temperature ($100 \pm 5\ °C$) for 24 h to remove the natural moisture content. The oven-dried materials were then thoroughly mixed, split, sieved through a 20 mm sieve, and riffled by different sizes of riffle splitter to prepare the representative samples for further laboratory testing.

Particle size distribution (PSD) test was carried out on RASS and RCA as per AS 1141.11 [28]. Since both RASS and RCA samples contained fine fraction of less than 5%, the hydrometer test was not performed further. The particle breakage of the material was also assessed through conducting additional PSD tests on the RASS and RCA samples, which underwent the modified Proctor compaction effort, and the PSD curves before and after the compaction were plotted for comparison and analysis.

Unified Soil Classification System (USCS) was employed for the classification of the recycled materials according to the sieve analysis results [29]. A flakiness index test was implemented to examine the particle shape of the recycled materials, in line with the procedures outlined in BS 812–105 [30]. The percentage of organic content present in RASS and RCA were determined using the loss of ignition methods described in ASTM D2974 [31], and an electrometric method was utilized to measure the pH value of both materials as per AS 1289.4.3.1 [32]. Triplicate samples of both recycled materials were prepared and tested for their flakiness index, organic content, and pH value to minimize error margins.

The particle density and water absorption of RASS and RCA samples for both fine and coarse fraction were established in line with the steps described in AS 1141.5 [33] and AS 1141.6.1 [34]. Three samples for each recycled material were prepared for the particle density and water absorption tests, and the average values of three tests were adopted as final results. Modified Proctor compaction tests were performed to investigate the relationship between the moisture content and dry density of the samples, according to the AS 1289.5.2.1 [35]. By following the procedures outlined in AS 1289.5.2.1 [35], the samples were compacted in five layers in a steel mold with an internal diameter of 105 mm and 115 mm in height. An automatic Proctor compactor was employed to apply constant compaction energy to ensure each layer is evenly compacted when making the samples. Each layer was compacted with a 4.9 kg rammer falling freely from a height of 450 mm by 25 blows. The optimum moisture content (OMC) and the maximum dry density (MDD) were determined by plotting the compaction curves accordingly.

The aggregate impact value (AIV) of the samples was measured by an impact testing machine to identify the resistance of the RASS and RCA aggregates subject to sudden shock as per BS 812-112 [36]. Los Angeles (LA) abrasion tests were conducted to assess the resistance to degradation performance of RASS and RCA aggregates by employing the Los Angeles testing machine in accordance with ASTM C131 [37]. Hydraulic conductivity tests were performed to confirm the coefficient of permeability for the flow of water through a remolded specimen, according to AS 1289.6.7.1 [38] and AS 1289.6.7.2 [39], in which a constant head method was selected for a more permeable material (RASS), and a falling head method was adopted for RCA sample with an expected hydraulic conductivity between 10^{-7} to 10^{-9} m/s. Triplicate samples of both recycled materials were prepared for

AIV, LA abrasion, and hydraulic conductivity tests to ensure the repeatability and accuracy of the results.

Three samples for each recycled material were prepared in line with the procedures detailed in the ASTM D1883 [40] for conducting the California bearing ratio (CBR) test. The samples were compacted by the modified Proctor compaction effort at the OMC to achieve the targeted MDD of 98%. The compacted samples were immersed in a water tank for 96 h to simulate the worst-case scenario of the field conditions. A mechanical dial gauge was installed on top of the CBR mold to measure the swelling of each material during the soaking period, and the penetration tests were carried out on the samples after being soaked for 4 days at a rate of 1 mm/min.

The repeated load triaxial (RLT) test was undertaken to measure the resilient modulus (M_R) of RASS and RCA, as per the procedures outlined in AASHTO T 307 [41] for unbound granular base/subbase materials. Triplicate samples were prepared using a split mold with an internal diameter of 100 ± 1 mm and 200 ± 1 mm in height. An automatic Proctor compactor was used to apply constant Proctor compaction energy of 25 blows per layer in a total of 8 layers at the OMC to attain the targeted MDD of 98%. Since the RASS sample contains insufficient fine fraction and cohesion among its particles, the compacted sample was not able to retain its shape and slumped on the removal of the split mold. A vacuum pump was employed to produce vacuum pressure inside the membrane to hold the sample together during the assembling of the triaxial chamber. The vacuum pressure inside the sample was then released after initiating the RLT test, as the applied air confining pressure inside the chamber will take place and hold the sample together during the testing period. According to the RLT testing protocol for unbound granular base/subbase materials, outlined in AASHTO T 307 [41], the testing sample was subject to five different confining pressures and 15 load sequences. For each load sequence, 100 repetitions of cyclic load were applied on the sample with repeated cycles of a Haversine-shaped loading pulse of 0.1 s and a resting period of 0.9 s to simulate the dynamic traffic loads acting on the unbound granular base/subbase materials.

2.2. Geo-Environmental Tests

Apart from the above-mentioned tests evaluating the geotechnical properties of the RASS sample, the chemical characteristics of the RASS, with regard to the environmental risks and health hazards, are crucial in consideration of utilizing RASS as a geomaterial. A total contaminant and leachate concentration test, to determine the possible existence of contaminant constituents such as heavy metals and polycyclic aromatic hydrocarbons, were conducted in accordance with the AS 4439.3-1997 [42]. For the preparation of leachate, the procedures detailed in the AS 4439.3-1997 [42] were followed. A marginally acidic and alkaline leaching fluid, with pH values of 5.0 and 9.2, were employed as leachate for the test. The results from the total contaminant and leachate concentration tests of the RASS sample were then compared with the requirements specified by the Environmental Protection Authority of Victoria (EPA Victoria) [43,44] and the Environmental Protect Agency of Washington (EPA USA) [45] to ascertain the environmental impacts of using RASS as a sustainable geomaterial.

3. Results and Discussion

3.1. Morphology of RASS and RCA

The physical appearance of as-received RASS and RCA, with maximum grain size smaller than 20 mm, is shown in Figure 1a,d, respectively. The RASS is a greyish and silvery color as illustrated in Figure 1a, whereas the RCA is of a brownish color depicted in Figure 1d. Different types of impurities were present in both recycled materials, of which wooden particles and some unmelted aluminum products, such as nails, screws and residual parts of the beverage cans were found in the RASS sample. Natural aggregate, crushed brick, timber, and plastic were discovered in the RCA sample. For a better understanding of the particle shapes, and microstructure of RASS and RCA, the SEM images were also taken at a magnification of 100 μm and 20 μm to further investigate the morphology of both recycled materials. SEM image in Figure 1b shows that the RASS particles appear to be angular and irregular in shape. Some flaky particles are also observed in the RASS sample. Upon closer inspection of Figure 1c, the surface of RASS particles is reasonably rough. On the other hand, Figure 1e demonstrates that the RCA particles are more rounded and oval in shape, and the surface is found to be cratered. Additionally, noticeable microcracks are observed on the RCA particles at higher magnification, as depicted in Figure 1f, and the formation of the microcracks could be mainly attributed to the crushing process of the concrete blocks.

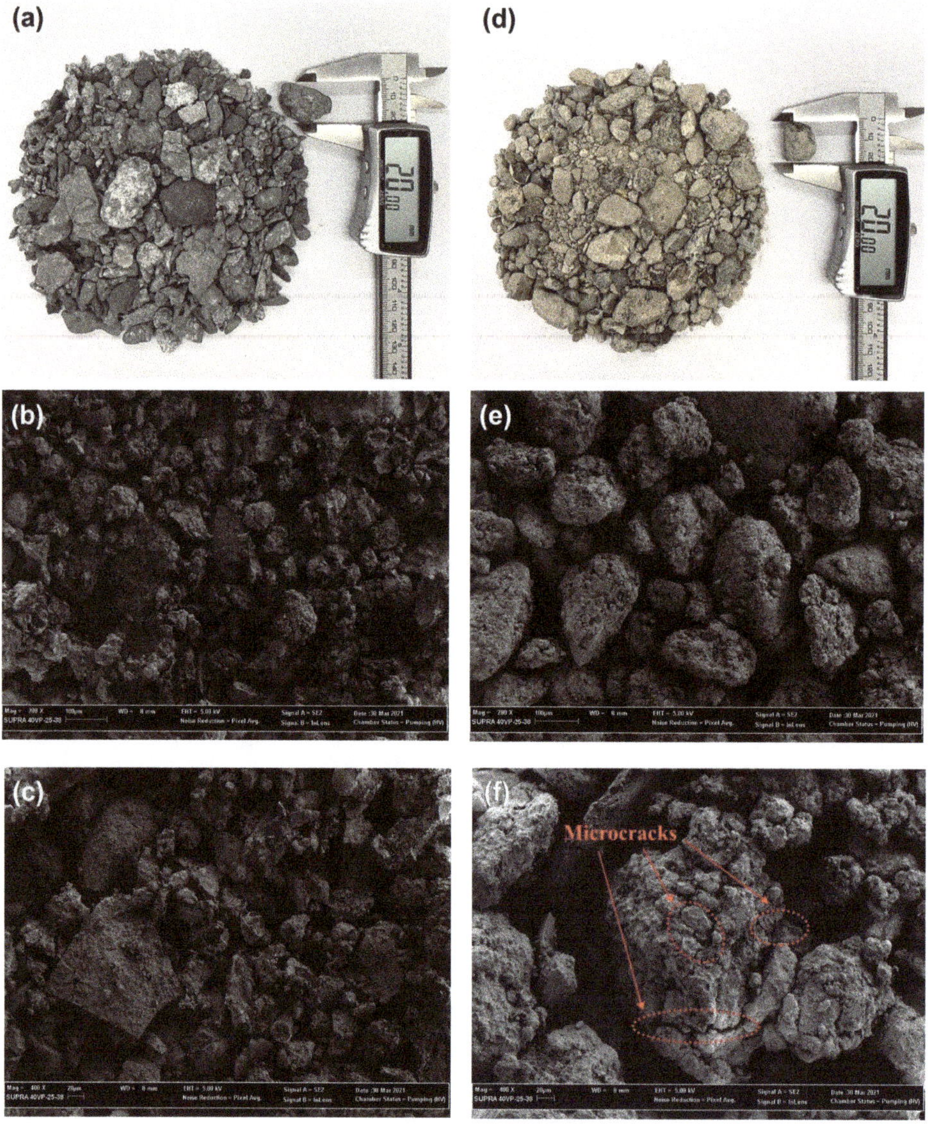

Figure 1. Physical appearance and SEM images of RASS and RCA at 100 μm and 20 μm magnification: (**a**) RASS actual sample; (**b**) RASS 100 μm; (**c**) RASS 20 μm; (**d**) RCA actual sample; (**e**) RCA 100 μm; (**f**) RCA 20 μm.

3.2. Geotechnical Characteristics

3.2.1. Particle Size Distribution

The geotechnical characteristics of RASS and RCA were evaluated through undertaking a series of laboratory assessments in accordance with relevant international standards (ASTM, Australian, and British). The testing results of RASS and RCA are tabulated in Table 1. Figure 2 illustrates the PSD curves of RASS and RCA samples before and after applying modified Proctor compaction effort. The gradation curves of RCA lay well within the grading specification recommended by the local road authority for unbound subbase materials [46], whereas the PSD curves of RASS are off the subbase lower limit due to the lack of fine fraction in the sample. According to the USCS, the RASS and RCA sample were classified as GP and GW, the percentage of fine, sand, and coarse fraction were also computed and shown in Table 1. By comparing the gradation curves before and after the modified compaction effort, the (after compaction) gradation curve of RCA shifts upward, indicating the degradation and particle breakage of the granular material subject to impact loading. The reason can be ascribed to the formation of microcracks on the RCA granular material during the crushing process of the concrete blocks (illustrated in Figure 1f). The presence of microcracks weaken the stiffness of RCA particles and make RCA more susceptible to degradation under shock loading. Interestingly, no significant alteration was observed for the gradation curve of the RASS sample after the compaction, which indicates superior stiffness and excellent performance on resistance to degradation of RASS material.

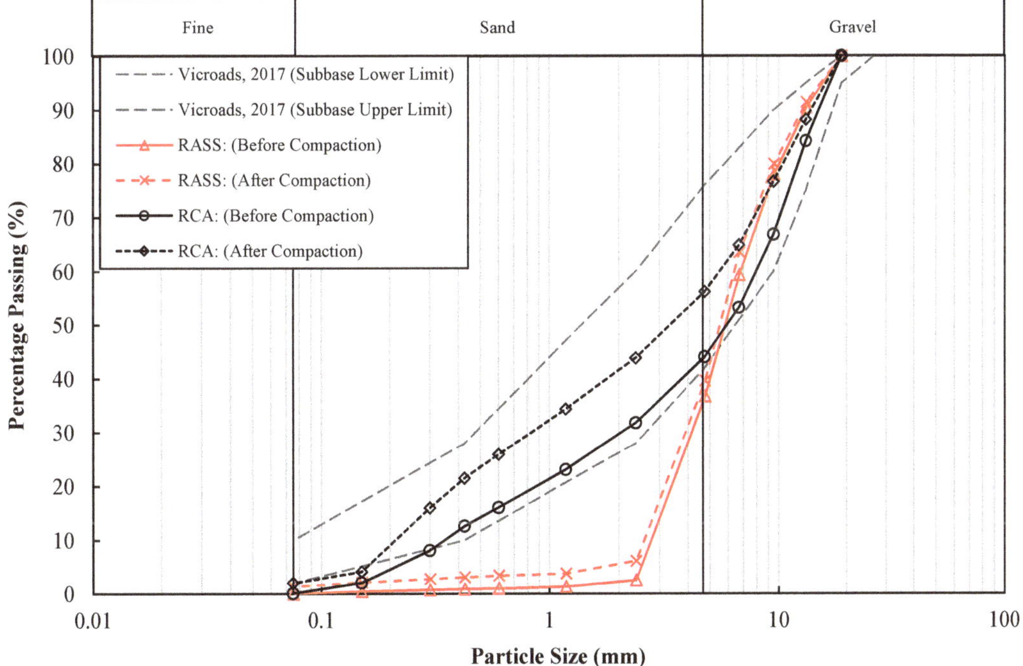

Figure 2. Particle size distribution curves of RASS and RCA.

Table 1. Geotechnical Characteristics of RASS and RCA.

Geotechnical Parameters	RASS	RCA
Fine content: <0.075 mm (%)	1.61	3.01
Sand content: 0.075–4.75 mm (%)	36.18	42.67
Gravel content: >4.75 mm (%)	62.20	54.31
Coefficient of uniformity (C_u)	2.59	25.71
Coefficient of curvature (C_c)	0.93	1.54
Soil classification (USCS)	GP	GW
OMC (%)—modified Proctor compaction	6.6	10.6
MDD (Mg/m^3)—modified Proctor compaction	1.66	1.94
Particle density—coarse fraction (Mg/m^3)	2.83	2.72
Particle density—fine fraction (Mg/m^3)	2.95	2.85
Water absorption—coarse fraction (%)	3.29	6.49
Water absorption—fine fraction (%)	4.75	9.2
Organic content (%)	1.91	3.51
pH	8.01	12.39
Flakiness index	32.63	8.81
LA abrasion loss (%)	6.23	31.77
Hydraulic conductivity (m/s)	4.16×10^{-6}	4.16×10^{-8}
Aggregate impact value (%)	2.71	32.36
CBR range (%)	91–106	189–220
CBR swell (%)	≈ 0	≈ 0
Resilient modulus, M_R range (MPa)	88.5–288.1	201.5–402.0

3.2.2. Particle Density, Water Absorption, pH and Organic Content

The particle density and water absorption test results for both fine and coarse fraction, of RASS and RCA, are shown in Table 1. The apparent density of RASS for both coarse (2.83 Mg/m^3) and fine (2.95 Mg/m^3) fraction is slightly higher than that of RCA (2.72 Mg/m^3 for coarse and 2.85 Mg/m^3 for fine fraction). The slightly lower particle density of RCA can mainly be attributed to the presence of a small fraction of low-density materials such as timber, plastic, and gypsum. As RCA is derived from construction and demolition site, such low-density materials including timber, plastic, and gypsum have a higher chance to be mixed with RCA during the demolishing, recycling, and crushing process. Interestingly, the percentage of water absorption of RCA is almost twice as high as that of the RASS for both fine and coarse fractions, indicating a relatively lower water absorption characteristic of RASS material. In addition, the water absorption of the fine fraction of both materials was found to be higher than that of the coarse fraction. The reason can be ascribed to the larger specific surface area of fine particles that absorbs more water to wet their surface compared to the coarse ones. The average pH value of the RCA sample was found to be 12.39, which is higher than the pH value of RASS (pH $\approx$ 8), as the concrete is alkaline by nature. The percentage of organic content of both materials are manifested in Table 1. Wooden particles and other impurities were observed in both materials during the preparation of samples for the organic content test. Since the RCA is derived from the demolition wastes, there is a higher chance for the impurities, such as wood and plastic, to be mixed with the RCA during the demolishing, recycling, and crushing process, which explains the reason for the greater percentage of organic content in comparison with RASS.

3.2.3. Aggregate Impact Value and LA Abrasion Loss

The results of LA abrasion loss and aggregate impact value (AIV) of RASS and RCA are depicted in Table 1. A lower percentage of LA abrasion loss and AIV is obtained for the RASS (6.23% and 2.71%), whereas higher values were acquired for the RCA (31.77% and 32.36%). The lower percentage of LA abrasion loss and AIV indicates an excellent stiffness and better resistance to degradation properties of the RASS. However, the hardness of RCA particles is lower than that of the RASS, based on the experimental outcomes of

LA abrasion loss and AIV. The RCA particles are more susceptible to degradation under the action of attrition and impact, attributed to the presence of microcracks developed during the crushing process (shown in Figure 1f), and consequently weaken the structural integrity of RCA particles. The results of LA abrasion loss and AIV are also consistent with the findings from the sieving analysis discussed previously, revealing the same intrinsic properties of both materials. Nevertheless, both recycled materials satisfy the limits (LA abrasion loss <40%) recommended in VicRoads [46] for pavement subbase applications.

3.2.4. Modified Proctor Compaction

Figure 3 illustrates the modified Proctor compaction curves of RASS and RCA. Typical bell shape compaction curves are attained for both materials. Since the OMC of a material is typically influenced by its water absorption characteristic, which explains the reason for a lower OMC value obtained for the RASS compared to the RCA. Additionally, well-graded RCA also contains a greater amount of fine particles, with larger specific surface areas to absorb more water than a coarser material (RASS), and subsequently leading to a higher OMC value as a result. Despite the fact that RASS has a slightly higher apparent density than that of RCA, the modified Proctor compaction test results indicated that RASS has a lower MDD value compared to the RCA. The particle breakage of aggregates in RCA is more significant compared to RASS. As discussed in Section 3.2.3, the RCA has a much higher LA abrasion loss and AIV value compared to RASS, which is also consistent with the occurrence of particle breakage observed from the PSD tests (illustrated in Figure 2) for the RCA sample that underwent the modified Proctor compaction effort. A previous study reported that particle breakage, during the modified Proctor compaction process, can aid the compaction of a material into a denser state [47]. A higher degree of particle breakage of RCA will lead to a denser arrangement in the fabric of RCA compared to RASS, and enhance, contributing a higher MDD value of RCA. Apart from that, the substantial amount of voids retained among the coarse aggregates during the compaction process, due to the poor gradation of RASS, was considered to be another factor that significantly limits the compressibility of the material.

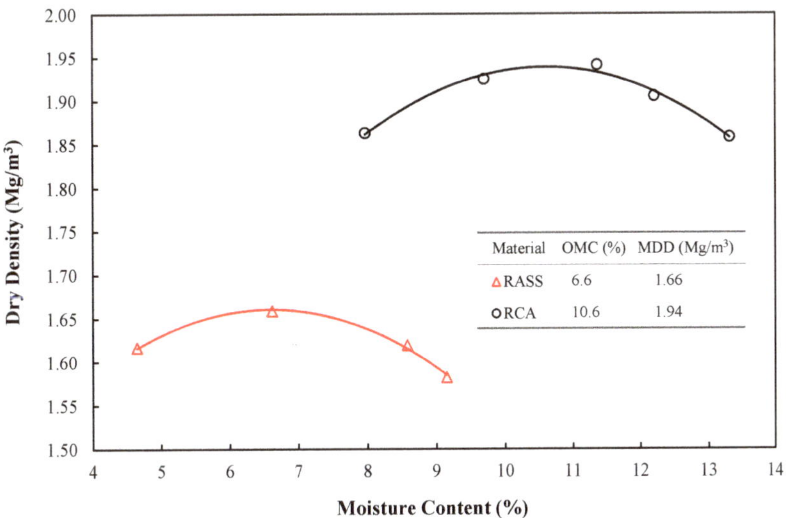

Figure 3. Modified Proctor compaction curves of RASS and RCA.

3.2.5. Flakiness Index and Hydraulic Conductivity

The flakiness index value of RASS was established to be considerably higher than that of the RCA, indicating a more flaky shape of particles present in the RASS sample. The flakiness index values of both recycled materials are still within the maximum limit (<35) specified by the local road authority for usage as a road construction base material [46]. The hydraulic conductivity of both materials was computed and shown in Table 1. The hydraulic conductivity value of RASS was determined to be higher than that of RCA. The reason can be ascribed to the insufficient fine particles in the RASS sample to fill up the voids among coarser aggregates, eventually leading to high voids ratio, which allows the water to seep through more easily. In contrast, the better compressibility of RCA material, owing to its excellent gradation accompanied by a low void ratio, reduces the water seepage rate notably, hence yielding a lower hydraulic conductivity as a result.

3.2.6. California Bearing Ratio (CBR)

Three samples for each material were prepared for the CBR test, and the range of CBR test results for both RASS and RCA are manifested in Table 1. In light of the specifications recommended by the local road authority, a minimum CBR value of 80% is typically required for a subbase material [46]. The results of the soaked CBR tests suggest that both recycled materials satisfy the minimum CBR value of 80% for usage as subbase material in road construction. In addition, the swelling behavior was not observed for the RASS and RCA samples during the soaking period. The CBR values of RASS samples are confirmed to be noticeably lower than those of RCA samples. This can be related to the lower MDD value of RASS achieved during the modified Proctor compaction effort, due to the poor gradation of the RASS. As a consequence, the formation of abundant voids inside the RASS samples weaken the particle interlocking performance, reduce the friction between the particles, leading to a poor stress distribution behavior of the RASS samples, and tend to lower the CBR value. The other reason that could contribute to a lower CBR value on the RASS samples can be ascribed to the cohesionless nature of the RASS material to bond the particles together, which also explains the greater CBR value observed on the RCA material with higher cohesion. It is suggested that RASS can be blended with other well-graded high-quality recycled material, such as RCA, to improve its CBR value and widen its applications in road construction.

3.2.7. Repeated Load Triaxial (RLT)

The RLT tests were performed on the RASS and RCA samples to simulate the dynamic traffic loads acting on the unbound granular base/subbase materials, under a combination of different confining pressures and deviator stresses. Figure 4 illustrates the M_R values of RASS and RCA at each load sequence, accompanied with different confining pressures. Figure 5 demonstrates the influence of various axial stresses on the resilient modulus response of the RASS and RCA samples. According to the RLT results, the M_R values of the RASS were found to be lower than those of the RCA. The reason is mainly attributed to the poor gradation of RASS with insufficient fine particles to fill up the voids among the coarser aggregates during the compaction process. This weakens the particle interlocking performance, and the sensitivity of resilient modulus response of the tested granular samples due to the reduction in the contact area between the particles, eventually contributing to lower M_R values of RASS samples. Additionally, a higher content of flaky-shape particles present in the RASS was considered to be another possible factor leading to lower M_R values of the RASS, in comparison with the RCA, that contains a higher portion of rounded particles [48]. However, further investigation is needed to verify this hypothesis. Moreover, the M_R values of both materials are increased accompanied by an increase in confinement, which is similar to the trend observed in previous studies [49,50]. The reason is due to the fact that the tested samples inside the triaxial chamber became denser and stronger with an increase in confining pressure, thus contributing to greater M_R values. Apart from that, both materials also exhibited higher M_R values when the deviator stress was increased under constant confinement. A similar trend was also revealed in previous studies, and the reason for this phenomenon can be owed to the stress hardening behavior on the granular samples, in which the material tends to get stiffer under higher axial stresses. As a consequence, the samples yield lower axial strain, with increments in deviator stress, and achieved higher M_R values [51,52].

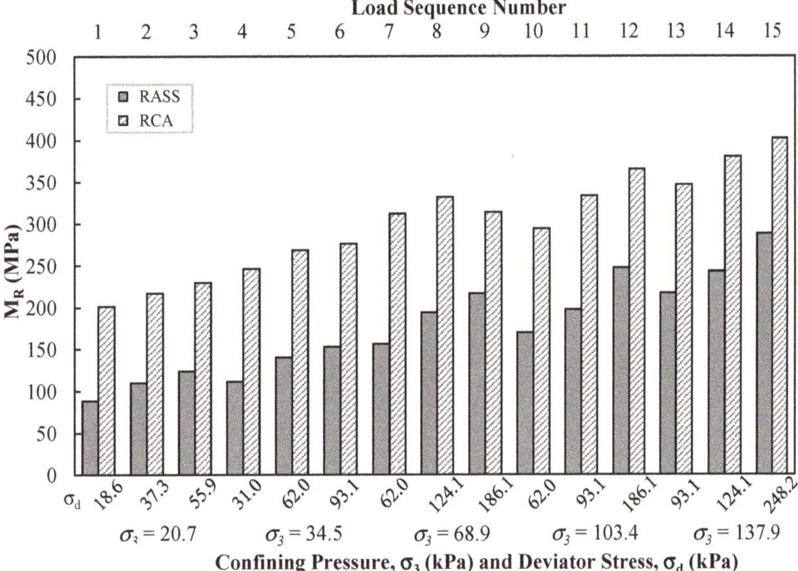

Figure 4. M_R of RASS and RCA at different confining pressures and deviator stresses.

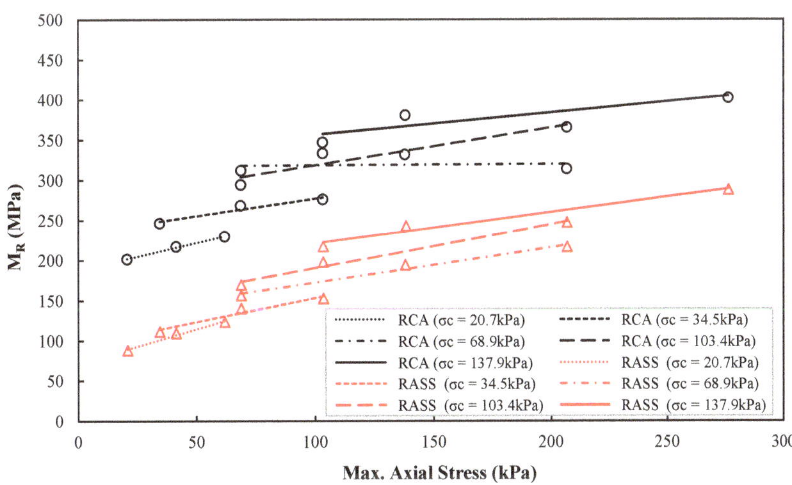

Figure 5. Influence of different axial stresses on the M_R of RASS and RCA.

Apart from the above-mentioned findings, a slightly different trend was observed for the RCA sample at the ninth load sequence. A sudden drop in M_R values of the RCA was observed at the ninth load sequence with applied deviator stress higher than the eighth load sequence under the same confinement, suggesting that granular sample does not always follow the stress hardening behavior under the aforementioned testing condition.

A softening behavior can also occur on the granular sample, with increments in the deviator stress, whilst maintaining the same confining pressure. A similar phenomenon was also observed by Attia and Abdelrahman [53] for different C&D recycled materials experiencing softening behavior with an increase in deviator stress under constant confinement. The range of the M_R values of RASS and RCA are tabulated in Table 1. The M_R values of the RCA were above the minimum requirements (125 MPa), specified by the local road authority for an unbound base and subbase material at all load sequences, whereas the M_R value of the RASS was slightly below the limit at load sequences 1 to 4 [54]. Nevertheless, due to the superior hardness of the RASS, it is suggested that RASS can be blended with other well-graded recycled materials to overcome the limitation of RASS material, in terms of poor gradation, and further improve its resilient modulus response to a great extent.

According to the AASHTO test procedures, a two-parameter theta (bulk stress) regression model is recommended for the analysis of M_R test results. The bulk stress model is expressed by the following equation [55]:

$$M_R = k_1 \times \theta^{k_2} \tag{1}$$

The Equation (1) can be rearranged into a logarithmic form listed as:

$$\log M_R = \log k_1 + k_2 \times \log \theta \tag{2}$$

in which M_R is the resilient modulus; θ = bulk stress = $(\sigma_1 + \sigma_2 + \sigma_3)$ = $3\sigma_3 + \sigma_d$, representing the triaxial test conditions; k_1 and k_2 are the regression model parameter. The regression results of present tests for both materials are depicted in Figure 6. Table 2 demonstrates the regression model parameters, $\log k_1$ and k_2, as well as the coefficient of determination R^2. The R^2 values of the present resilient modulus tests were computed to be very close to 0.96 for both recycled materials, indicating that a good fit was achieved by employing the bulk stress model.

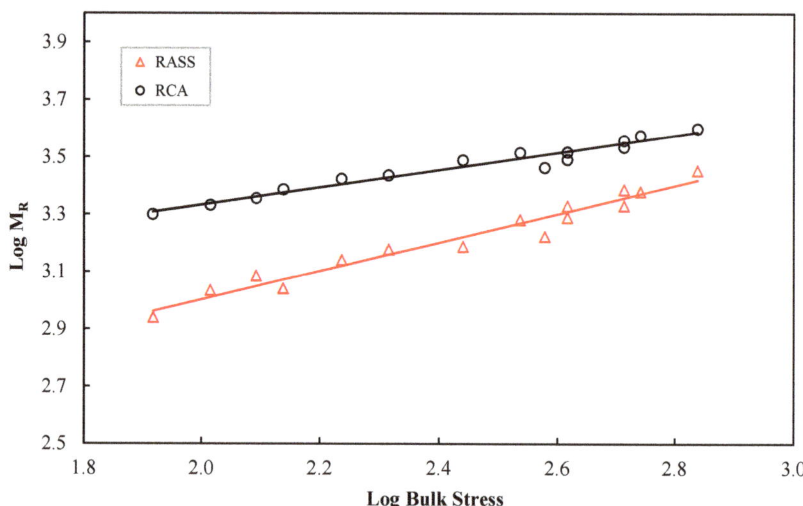

Figure 6. Bulk stress regression model of RASS and RCA.

Table 2. Regression parameters of bulk stress model.

Material	Log k_1	k_2	Coefficient of Determination, R^2
RASS	2.01	0.50	0.96
RCA	2.73	0.30	0.96

Although the concise bulk stress model yields a reliable result, the major drawback of the bulk stress model is owing to the incapability of separating the influence of different confining pressures and deviator stresses on the corresponding M_R. A three-parameter theta model that takes into consideration both confining pressures and deviator stresses, proposed by Puppala et al. [52], was also used for the analysis of the present M_R test results. The three-parameter theta model is given by:

$$\frac{M_R}{\sigma_{atm}} = k_1 \left(\frac{\sigma_3}{\sigma_{atm}} \right)^{k_4} \times \left(\frac{\sigma_d}{\sigma_{atm}} \right)^{k_5} \tag{3}$$

The logarithmic format of Equation (3) is described as:

$$\log \left(\frac{M_R}{\sigma_{atm}} \right) = log k_1 + k_2 log \left(\frac{\sigma_3}{\sigma_{atm}} \right) + k_3 log \left(\frac{\sigma_d}{\sigma_{atm}} \right) \tag{4}$$

where σ_{atm} = atmospheric pressure; σ_3 = confining pressure; σ_d = deviator stress; $\log k_1$, k_2 and k_3 are the regression constants of the three-parameter theta model. Table 3 lists the regression model constant and the coefficient of determination R^2. By comparing the results obtained from both regression models, the coefficient of determination, yielded by the three-parameter model, was higher than that of the bulk stress model, which is similar to the findings reported by Puppala et al. [52] and Mohammadinia et al. [56]. The consideration of the effects, induced by different confinement and deviator stresses on the M_R, produces a better fit and further enhances the accuracy of the results compared to the bulk stress model.

Table 3. Regression parameters of three-parameter model.

Material	Logk_1	k_2	k_3	Coefficient of Determination, R^2
RASS	3.29	0.22	0.30	0.99
RCA	3.52	0.19	0.12	0.97

3.3. Environment Risks and Health Hazards

In considering the usage of RASS as a geomaterial in civil construction, any possible environmental risks and health hazards, in terms of leaching hazards and exposure of contaminant constituents from such material into the soil, surrounding surface areas, and underground aquifers during its lifecycle in a project must be ascertained [57,58]. As per the EPA Victoria regulations, wastes need to be categorized into one of the following types: (a) fill material, typically made up of natural soil, sand, gravel, and rock, (b) solid inert waste, comprised of industrial and municipal wastes, such as industrial by-products and construction wastes (the RASS used in this research seems to fall into this category based on its source and production method), (c) putrescible waste that can be decomposed by bacteria activities, and (d) prescribed industrial waste that could have adverse effects on the environment and human health, sourced from manufacturing sectors or contaminated soils [43]. The assessments to determine the existence of contaminant constituents of the wastes are required before the wastes can be recycled, reused, and disposed in a landfill. The hazard categorization of the RASS sample was evaluated as per the guidelines recommended by the EPA Victoria and the Australian Standard Leaching Procedure (ASLP) [43]. The total contaminant (TC) and leachate concentration tests were carried out on the RASS to identify the presence of any possible contaminant constituents from such material. According to AS 4439.3-1997 [42], an appropriate type of leaching fluid, used for the leaching analysis on the material, should be determined based on the field application of the tested material. The kinds of leaching fluids recommended for conducting the leaching analysis on material, as per AS 4439.3-1997 [42], include reagent water and three other leaching fluids with a different pH value of 2.9, 5.0, and 9.2. Since the aim of this research is to evaluate the suitability of utilizing RASS as a geomaterial in applications such as pavement base/subbase material and embankment fills, reagent water that resembles rainwater seems to be the most suitable leaching fluid for such material. However, it should be noted that there is a possibility of acid rain (pH 3.6–4.9) resulting from the presence of organic acids in the atmosphere [59]. Graham [59] states that such organic acids are formed in the atmosphere by the photochemistry of organic compounds volatilized from terrestrial vegetation. In addition, the possible occurrence of acidic freshwater (pH 4.0–5.5), in the first flush at the start of the wet season, should also be considered for determining the suitable leaching fluid [59]. According to the National Land and Water Resources Audit [60], about 50 million hectares of Australian agricultural land has a surface pH of 4.3–5.5, which implies the possible occurrence of acidic freshwater in the first flush of the wet season.

With these considerations in mind, it is safer to use a slightly acidic leaching fluid (pH 5.0), based on the field application of RASS, rather than a reagent water. Furthermore, take into consideration that RASS could be utilized in blend with other types of recycled materials such as crushed brick, crushed concrete aggregate, and reclaimed asphalt pavement aggregate to meet specific engineering requirements for various applications. There is a possibility for RASS to be exposed to a slightly alkaline environment throughout its service life as a geomaterial. By considering all the aspects, a slightly acidic and alkaline leaching fluid with a pH value of 5.0 and 9.2 were selected to be used for the leachate analysis of the RASS. The leachate was prepared in line with the procedures described in the AS 4439.3-1997 [42]. The results of TC and ASLP values of RASS were compared with the limits set by EPA Victoria and EPA USA to ascertain the environmental risks and health hazards of utilizing RASS as a sustainable geomaterial [43–45]. Since RCA's leachate analysis was reported in the previous study, the leachability test of RCA was not repeated in this research [54].

The total contaminant and leachate concentration test results of RASS, illustrated in Table 4, were compared to the threshold values of TC and ASLP of fill material, solid inert waste, and drinking water standards specified by the EPA Victoria and EPA USA, manifested in Table 5. The TC values of the RASS were found to be far below the threshold values of fill material demanded by the EPA Victoria, suggesting that RASS falls into the category of fill material, as per the requirements recommended by the EPA Victoria. Such material can be safely utilized as fill material in geotechnical applications such as the construction of stone column for ground improvement, pavement layers, and embankment fill. The leachate concentration test results of RASS, expressed as TC and ASLP values, by use of both acidic and alkaline leaching fluid listed in Table 4 were, again, confirmed to be much lower than the acceptable values of solid inert waste according to the EPA Victoria specifications. The comparison implies that the environmental risks and health hazards concerning the exposure of contaminant constituents from such material into the soil, surrounding surface areas, and underground aquifers throughout its service life as a geotechnical construction material are negligible. Furthermore, based on the EPA USA specifications, wastes are classified as hazardous material if one's heavy metal content was detected at a concentration 100 times higher than the acceptable values of drinking water standards (illustrated in Table 5) specified by the EPA USA. By following this criterion, the ASLP values (both acidic and alkaline) of the RASS were also confirmed to satisfy the limit recommended by the EPA USA. Thus, the comparison indicates that RASS will not pose any environmental risks to the surface water stream or underground aquifer and is safe to use as a construction material in geotechnical applications.

Table 4. TC and ASLP test results of RASS and RCA.

Contaminant	RASS			RCA [c]		
	TC [a]	ASLP [b] (Acetate)	ASLP [b] (Tetraborate)	TC [a]	ASLP [b] (Acetate)	ASLP [b] (Tetraborate)
Arsenic	<0.01	<0.01	0.05	<5	<0.01	<0.1
Barium	N/A	N/A	N/A	88	0.34	<0.1
Cadmium	<0.01	<0.01	<0.01	<0.2	<0.002	<0.02
Chromium	0.06	0.01	0.03	15	0.05	<0.1
Copper	0.02	0.03	0.05	N/A	N/A	N/A
Lead	0.11	0.10	<0.01	11	<0.01	<0.1
Mercury	<0.005	<0.001	<0.005	<0.05	<0.001	<0.01
Selenium	N/A	N/A	N/A	<3	<0.01	<0.1
Silver	N/A	N/A	N/A	<5	<0.01	<0.1
Nickel	0.06	0.02	<0.01	N/A	N/A	N/A
Zinc	0.46	0.18	0.02	N/A	N/A	N/A
Cyanide	<0.25	<0.25	<0.25	N/A	N/A	N/A
PAH	<0.001	<0.001	<0.001	N/A	N/A	N/A

[a] mg/kg of dry weight. [b] mg/L. [c] Data from Arulrajah et al. [54]. N/A: Not available. PAH: Polycyclic aromatic hydrocarbons.

Table 5. EPA Victoria and EPA USA requirements.

Contaminant	Maximum TC Allowed for Fill Material	Allowable TC and ASLP Value for Solid Inert Waste		US EPA Drinking Water Standard [b,e]
	TC [a,c]	TC [a,d]	ASLP [b,d]	
Arsenic	20	500	0.35	0.05
Barium	N/A	6250	35	2.0
Cadmium	3	100	0.1	0.005
Chromium	1 (Chromium VI)	500	2.5	0.1
Copper	100	5000	100	1.3
Lead	300	1500	0.5	0.015
Mercury	1	75	0.05	0.002
Selenium	10	50	0.5	0.05
Silver	10	180	5	0.05
Nickel	60	3000	1	N/A
Zinc	200	35,000	150	N/A
Cyanide	50	2500	4	N/A
PAH	20	50	N/A	N/A

[a] mg/kg of dry weight. [b] mg/L. [c] EPA Victoria [43]. [d] EPA Victoria [44]. N/A: Not available. PAH: Polycyclic aromatic hydrocarbons. [e] EPA USA [45].

4. Conclusions

The geotechnical and environmental characteristics of RASS were investigated, in this research, to fill the knowledge gap on the feasibility of utilizing such aluminum industrial waste by-products as sustainable geomaterials. The following conclusion can be drawn based on the laboratory testing results:

1   The RASS was classified as a poorly-graded material, due to the insufficient fine fraction, and the pH value of RASS was found to be slightly alkaline. The particle density of RASS was slightly higher that of RCA, whereas the water absorption of RASS was lower than that of the RCA. In addition, the RASS was also confirmed to be free-draining, cohesionless, and exhibited a very low organic content.

2   The RASS exhibited much higher stiffness compared to the RCA, based on the experimental outcomes of AIV and LA abrasion loss tests. The test results of flakiness index, LA abrasion loss, and CBR value of the RASS samples satisfy the requirements specified by the local road authority for usage as a pavement base and subbase material in road construction.

3   The M_R values of the RASS were found to be lower than those of the RCA and are slightly below the minimum M_R requirement (125 MPa) recommended by the local road authority, Victoria, for an unbound base/subbase material, at load sequences 1 to 4, under triaxial test conditions. The reasons contributing to the lower M_R values of the RASS is mainly owing to the poor gradation of RASS material, which weakens the particle interlocking performance and the sensitivity of resilient modulus response of the tested granular samples, due to the reduction in the contact area between the particles. Considering the superior stiffness of RASS, it is suggested that RASS can be blended with other well-graded recycled materials, RCA for instance, to overcome the limitation in terms of its poor gradation and further enhance the M_R value of the RASS significantly.

4   The results of the total contaminant and leachate concentration tests have shown that TC and ASLP values of RASS are far below the threshold values of fill material and solid inert waste specified by the EPA Victoria. This suggests that RASS can be safely employed as a geotechnical construction material and cause no harm to the surrounding environment. Furthermore, the ASLP values of the RASS are also within the acceptable limits of drinking water standards, according to the EPA USA specifications, indicating that RASS will not pose any health hazards by contaminating the surface water stream or underground aquifer.

In this research, the comparison between RASS and RCA provides satisfactory evidence that RASS possesses the capability to be utilized as a sustainable geomaterial in civil construction. However, it should be noted that the quality of the aluminum industrial by-product, RASS, is dependent on the machinery used and manufacturing technique. In developing countries, where manufacturing technique and equipment are outdated, the percentage of unmelted aluminum products (such as nails, screws, and residual parts of beverage cans) in the RASS stockpile could be various, which will result in fluctuating in the overall geotechnical properties of such recycled material. Geotechnical and engineering assessment should be performed on the RASS sample from a stockpile before utilizing the RASS in geotechnical applications. The research outcomes of this study can be used as guidance, assisting relevant bodies in comparing the required geotechnical and engineering parameters, and determining the suitability of a RASS stockpile to be utilized in geotechnical applications. It is also recommended that future research can focus on the stabilization of RASS by introducing additive binders, such as cement or alkali-activated cementitious material, to further strengthen the mechanical performance of RASS and widen its applications in geotechnical construction. The usage of RASS as a sustainable geomaterial in civil construction provides a feasible end-of-life option to convert the aluminum industrial waste from landfill into a usable material, mitigate the pressure on land and natural resources, as well as improve sustainability in the civil construction industry.

Author Contributions: Conceptualization: Y.L., F.M., A.A. and S.H.; methodology: Y.L.; formal analysis: Y.L.; writing—original draft preparation: Y.L.; writing—review and editing: F.M., A.A. and S.H.; supervision: F.M. and A.A.; funding acquisition: F.M., A.A. and S.H. All authors have read and agreed to the published version of the manuscript.

Funding: This research was funded by the SmartCrete CRC Ltd. (Macquarie Park, Australia).

Institutional Review Board Statement: Not applicable.

Informed Consent Statement: Not applicable.

Data Availability Statement: The data presented in this study are available on request from the corresponding author. The data are not publicly available due to privacy restrictions.

Acknowledgments: The authors are grateful for the financial support provided by the SmartCrete CRC Ltd. and the industry partners (Hawks Excavation Pty Ltd. (Doncaster East, Australia), Stretford Civil Constructions Pty Ltd. (Balwyn North, Australia)) for conducting this research project.

Conflicts of Interest: The authors declare no conflict of interest.

References

1. Li, C.Z.; Zhao, Y.; Xiao, B.; Yu, B.; Tam, V.W.; Chen, Z.; Ya, Y. Research trend of the application of information technologies in construction and demolition waste management. *J. Clean. Prod.* **2020**, *263*, 121458. [CrossRef]
2. Akhtar, A.; Sarmah, A.K. Construction and demolition waste generation and properties of recycled aggregate concrete: A global perspective. *J. Clean. Prod.* **2018**, *186*, 262–281. [CrossRef]
3. Silva, R.; De Brito, J.; Dhir, R. Use of recycled aggregates arising from construction and demolition waste in new construction applications. *J. Clean. Prod.* **2019**, *236*, 117629. [CrossRef]
4. McKelvey, D.; Sivakumar, V.; Bell, A.; McLaverty, G. Shear strength of recycled construction materials intended for use in vibro ground improvement. *Proc. Inst. Civ. Eng.-Ground Improv.* **2002**, *6*, 59–68. [CrossRef]
5. Vieira, C.S.; Pereira, P.M. Use of recycled construction and demolition materials in geotechnical applications: A review. *Resour. Conserv. Recycl.* **2015**, *103*, 192–204. [CrossRef]
6. Henzinger, C.; Heyer, D. Soil improvement using recycled aggregates from demolition waste. *Proc. Inst. Civ. Eng.-Ground Improv.* **2018**, *171*, 74–81. [CrossRef]
7. Brooks, R.; Cetin, M. Application of construction demolition waste for improving performance of subgrade and subbase layers. *Int. J. Res. Rev. Appl. Sci.* **2012**, *12*, 375.
8. Yi, Y.; Gu, L.; Liu, S.; Puppala, A.J. Carbide slag–Activated ground granulated blastfurnace slag for soft clay stabilization. *Can. Geotech. J.* **2015**, *52*, 656–663. [CrossRef]
9. Imteaz, M.A.; Arulrajah, A.; Maghool, F. Environmental and geotechnical suitability of recycling waste materials from plasterboard manufacturing. *Waste Manag. Res.* **2020**, *38*, 383–391. [CrossRef] [PubMed]
10. Maghool, F.; Arulrajah, A.; Suksiripattanapong, C.; Horpibulsuk, S.; Mohajerani, A. Geotechnical properties of steel slag aggregates: Shear strength and stiffness. *Soils Found.* **2019**, *59*, 1591–1601. [CrossRef]
11. Prem, P.R.; Verma, M.; Ambily, P. Sustainable cleaner production of concrete with high volume copper slag. *J. Clean. Prod.* **2018**, *193*, 43–58. [CrossRef]
12. International Aluminium Institute. Primary Aluminium Production Statistics Report. 2021. Available online: https://www.world-aluminium.org/statistics/ (accessed on 10 March 2021).
13. Shinzato, M.; Hypolito, R. Solid waste from aluminum recycling process: Characterization and reuse of its economically valuable constituents. *Waste Manag.* **2005**, *25*, 37–46. [CrossRef]
14. Gil, A.; Korili, S. Management and valorization of aluminum saline slags: Current status and future trends. *Chem. Eng. J.* **2016**, *289*, 74–84. [CrossRef]
15. Adeosun, S.O.; Sekunowo, O.I.; Taiwo, O.O.; Ayoola, W.A.; Machado, A. Physical and Mechanical Properties of Aluminum Dross. *Adv. Mater.* **2014**, *3*, 6–10. [CrossRef]
16. Mahinroosta, M.; Allahverdi, A. Hazardous aluminum dross characterization and recycling strategies: A critical review. *J. Environ. Manag.* **2018**, *223*, 452–468. [CrossRef]
17. Karvelas, D.; Daniels, E.; Jody, B.; Bonsignore, P. *An Economic and Technical Assessment of Black-Dross and Salt-Cake-Recycling Systems for Application in the Secondary Aluminum Industry*; Argonne National Lab., Energy Systems Div.: Chicago, IL, USA, 1991.
18. Dai, C. Development of Aluminum Dross-Based Material for Engineering Application. Master's Thesis, Worcester Polytechnic Institute, Worcester, MA, USA, January 2012.
19. Ozerkan, N.; Maki, O.; Anayeh, M.; Tangen, S.; M Abdullah, A. The effect of aluminium dross on mechanical and corrosion properties of concrete. *Int. J. Innov. Res. Sci.* **2014**, *3*, 9912–9922.
20. Pereira, D.; de Aguiar, B.; Castro, F.; Almeida, M.; Labrincha, J. Mechanical behaviour of Portland cement mortars with incorporation of Al-containing salt slags. *Cem. Concr. Res.* **2000**, *30*, 1131–1138. [CrossRef]
21. Reddy, M.S.; Neeraja, D. Mechanical and durability aspects of concrete incorporating secondary aluminium slag. *Resour.-Effic. Technol.* **2016**, *2*, 225–232. [CrossRef]
22. Reddy, M.S.; Neeraja, D. Aluminum residue waste for possible utilisation as a material: A review. *Sādhanā* **2018**, *43*, 124. [CrossRef]
23. Dunster, A.; Moulinier, F.; Abbott, B.; Conroy, A.; Adams, K.; Widyatmoko, D. *Added Value of Using New Industrial Waste Streams as Secondary Aggregates in both Concrete and Asphalt*; The Waste & Resources Action Programme: Banbury, UK, 2005.
24. López-Alonso, M.; Martinez-Echevarria, M.; Garach, L.; Galán, A.; Ordoñez, J.; Agrela, F. Feasible use of recycled alumina combined with recycled aggregates in road construction. *Constr. Build. Mater.* **2019**, *195*, 249–257. [CrossRef]
25. Busari, A.A.; Akinwumi, I.; Awoyera, P.O.; Olofinnade, O.; Tenebe, T.; Nwanchukwu, J. In Stabilization effect of aluminum dross on tropical lateritic soil, International Journal of Engineering Research in Africa. *Trans. Tech. Publ.* **2018**, *39*, 86–96.
26. Gayatri, A.; Verma, A.K. Geotechnical Characterization of Expansive Soil and Utilization of Waste to Control Its Swelling and Shrinkage Behaviour. In *Recent Developments in Waste Management*; Springer: Singapore, 2020; pp. 11–21. [CrossRef]
27. ASTM D75/D75M–19. *Standard Practice for Sampling Aggregates*; ASTM International: West Conshohocken, PA, USA, 2014.
28. AS 1141.11. *Method for Sampling and Testing Aggregates Method 11: Particle Size Distribution by Sieving*; Standards Australia: Sydney, Australia, 2020.
29. ASTM D2487-17. *Standard Practice for Classification of Soils for Engineering Purposes (Unified Soil Classification System)*; ASTM International: West Conshohocken, PA, USA, 2020.
30. BS 812-105.1. *Method for Determination of Particle Shape; Flakiness Index*; British Standards Institution: London, UK, 2000.

31. ASTM D2974. *Standard Test Methods for Moisture, Ash, and Organic Matter of Peat and other Organic Soils*; ASTM International: West Conshohocken, PA, USA, 2007.

32. AS 1289.4.3.1. *Soil Chemical Tests-Determination of the pH Value of a Soil-Electrometric Method*; Standards Australia: Sydney, Australia, 1997.

33. AS 1141.5. *Particle Density and Water Absorption of fine Aggregate*; Standards Australia: Sydney, Australia, 2000.

34. AS 1141.6.1. *Particle Density and Water Absorption of Coarse Aggregate-Weighing-in-Water Method*; Standards Australia: Sydney, Australia, 2000.

35. AS 1289.5.2.1. *Soil Compaction and Density Tests-Determination of the Dry Density/Moisture Content Relation of a Soil Using Modified Compactive Effort*; Standards Australia: Sydney, Australia, 2003.

36. BS 812-112. *Testing Aggregates-Part 112: Methods for Determination of Aggregate Impact Value (AIV)*; British Standards Institution: London, UK, 1990.

37. ASTM C131. *Standard Test Method for Resistance to Degradation of Small-Size Coarse Aggregate by Abrasion and Impact in the Los Angeles Machine*; ASTM International: West Conshohocken, PA, USA, 2006.

38. AS 1289.6.7.1. *Methods of Testing Soils for Engineering Purposes-Soil Strength and Consolidation Tests–Determination of Permeability of a Soil-Constant Head Method for a Remoulded Specimen*; Standards Australia: Sydney, Australia, 2001.

39. AS 1289.6.7.2. *Methods of Testing Soils for Engineering Purposes-Soil Strength and Consolidation Tests–Determination of Permeability of a Soil-Falling Head Method for a Remoulded Specimen*; Standards Australia: Sydney, Australia, 2001.

40. ASTM D1883. *Standard Test Method for CBR (California Bearing Ratio) of Laboratory-Compacted Soils*; ASTM International: West Conshohocken, PA, USA, 2007.

41. AASHTO T 307-99. *Standard Method of test for Determining the Resilient Modulus of Soils and Aggregate Materials*; American Association of State Highway and Transportation Officials: Washington, DC, USA, 2007.

42. AS 4439.3. *Wastes, Sediments and Contaminated Soils, Part 3: Preparation of Leachates-Bottle Leaching Procedure*; Standards Australia: Sydney, Australia, 1997.

43. EPA Victoria. *Waste Categorization, Industrial Waste Resource Guidelines*; Environmental Protection Agency of Victoria: Melbourne, Australia, 2010.

44. EPA Victoria. *Solid Industrial Waste Hazard Categorization and Management, Industrial Waste Resource Guidelines*; Environmental Protection Agency of Victoria: Melbourne, Australia, 2009.

45. U.S. EPA. *National Primary Drinking Water Standards*; Environmental Protection Agency: Washington, DC, USA, 1999.

46. VicRoads. *Registration of Crushed Rock Mixes, Code of Practice RC 500.02*; VicRoads: Melbourne, Australia, 2017.

47. Arulrajah, A.; Piratheepan, J.; Ali, M.; Bo, M. Geotechnical properties of recycled concrete aggregate in pavement sub-base applications. *Geotech. Test. J.* **2012**, *35*, 743–751. [CrossRef]

48. Lekarp, F.; Isacsson, U.; Dawson, A. State of the art. I: Resilient response of unbound aggregates. *J. Transp. Eng.* **2000**, *126*, 66–75.

49. Maghool, F.; Arulrajah, A.; Horpibulsuk, S.; Du, Y.-J. Laboratory evaluation of ladle furnace slag in unbound pavement-base/subbase applications. *J. Mater. Civ. Eng.* **2017**, *29*, 04016197. [CrossRef]

50. Yaghoubi, E.; Sudarsanan, N.; Arulrajah, A. Stress-strain response analysis of demolition wastes as aggregate base course of pavements. *Transp. Geotech.* **2021**, *30*, 100599. [CrossRef]

51. Zhang, J.; Gu, F.; Zhang, Y. Use of building-related construction and demolition wastes in highway embankment: Laboratory and field evaluations. *J. Clean. Prod.* **2019**, *230*, 1051–1060. [CrossRef]

52. Puppala, A.J.; Hoyos, L.R.; Potturi, A.K. Resilient moduli response of moderately cement-treated reclaimed asphalt pavement aggregates. *J. Mater. Civ. Eng.* **2011**, *23*, 990–998. [CrossRef]

53. Attia, M.; Abdelrahman, M. Effect of state of stress on the resilient modulus of base layer containing reclaimed asphalt pavement. *Road Mater. Pavement Des.* **2011**, *12*, 79–97. [CrossRef]

54. Arulrajah, A.; Piratheepan, J.; Disfani, M.M.; Bo, M.W. Geotechnical and geoenvironmental properties of recycled construction and demolition materials in pavement subbase applications. *J. Mater. Civ. Eng.* **2013**, *25*, 1077–1088. [CrossRef]

55. AASHTO. *Guide for Design of Pavement Structures*; American Association of State Highway and Transportation Officials: Washington, DC, USA, 1993.

56. Mohammadinia, A.; Arulrajah, A.; Sanjayan, J.; Disfani, M.M.; Bo, M.W.; Darmawan, S. Laboratory evaluation of the use of cement-treated construction and demolition materials in pavement base and subbase applications. *J. Mater. Civ. Eng.* **2015**, *27*, 04014186. [CrossRef]

57. Disfani, M.; Arulrajah, A.; Bo, M.; Sivakugan, N. Environmental risks of using recycled crushed glass in road applications. *J. Clean. Prod.* **2012**, *20*, 170–179. [CrossRef]

58. Kua, T.-A.; Imteaz, M.A.; Arulrajah, A.; Horpibulsuk, S. Environmental and economic viability of Alkali Activated Material (AAM) comprising slag, fly ash and spent coffee ground. *Int. J. Sustain. Eng.* **2019**, *12*, 223–232. [CrossRef]

59. Graham, P. Current waste soil disposal practices in NSW. In Proceedings of the SuperSoil 2004, 3rd Australian New Zealand Soils Conference, Sydney, Australia, 5–9 December 2004.

60. National Land and Water Resources Audit. *Australian Agriculture Assessment 2001*; Australian Government, Land & Water Australia: Canberra, Australia, 2001.

Article

Generalized Interface Shear Strength Equation for Recycled Materials Reinforced with Geogrids

Artit Udomchai [1], Menglim Hoy [1,2,3,*] Apichat Suddeepong [2,3], Amornrit Phuangsombat [4], Suksun Horpibulsuk [1,2,3,5,*], Arul Arulrajah [6] and Nguyen Chi Thanh [7]

[1] School of Civil Engineering, Suranaree University of Technology, Nakhon Ratchasima 30000, Thailand; artit.u@g.sut.ac.th
[2] School of Civil and Infrastructure Engineering, Suranaree University of Technology, Nakhon Ratchasima 30000, Thailand; suddeepong@g.sut.ac.th
[3] Center of Excellence in Innovation for Sustainable Infrastructure Development, Suranaree University of Technology, Nakhon Ratchasima 30000, Thailand
[4] Graduate Program in Construction and Infrastructure Management, Suranaree University of Technology, Nakhon Ratchasima 30000, Thailand; chaninnun@outlook.co.th
[5] Academy of Science, and Royal Society of Thailand, Bangkok 10300, Thailand
[6] Department of Civil and Construction Engineering, Swinburne University of Technology, Melbourne, VIC 3122, Australia; aarulrajah@swin.edu.au
[7] Department of Materials Technology, Faculty of Applied Sciences, Ho Chi Minh City University of Technology and Education, Ho Chi Minh City 70000, Vietnam; thanhnc@hcmute.edu.vn
* Correspondence: menglim@g.sut.ac.th (M.H.); suksun@g.sut.ac.th (S.H.)

Citation: Udomchai, A.; Hoy, M.; Suddeepong, A.; Phuangsombat, A.; Horpibulsuk, S.; Arulrajah, A.; Thanh, N.C. Generalized Interface Shear Strength Equation for Recycled Materials Reinforced with Geogrids. *Sustainability* 2021, 13, 9446. https://doi.org/10.3390/su13169446

Academic Editor: Castorina Silva Vieira

Received: 14 July 2021
Accepted: 18 August 2021
Published: 23 August 2021

Abstract: In this research, large direct shear tests were conducted to evaluate the interface shear strength between reclaimed asphalt pavement (RAP) and kenaf geogrid (RAP–geogrid) and to also assess their viability as an environmentally friendly base course material. The influence of factors such as the gradation of RAP particles and aperture sizes of geogrid (D) on interface shear strength of the RAP–geogrid interface was evaluated under different normal stresses. A critical analysis was conducted on the present and previous test data on geogrids reinforced recycled materials. The D/F_D, in which F_D is the recycled materials' particle content finer than the aperture of geogrid, was proposed as a prime parameter governing the interface shear strength. A generalized equation was proposed for predicting the interface shear strength of the form: $\alpha = a(D/F_D) + b$, where α is the interface shear strength coefficient, which is the ratio of the interface shear strength to the shear strength of recycled material, and a and b are constants. The constant values of a and b were found to be dependent upon types of recycled material, irrespective of types of geogrids. A stepwise procedure to determine variable a, which is required for analysis and design of geogrids reinforced recycled materials in roads with various gradations was also suggested.

Keywords: ground improvement; geogrid; recycled materials; interface shear strength; large-direct shear test; base course reinforcement; pavement geotechnics

1. Introduction

Roadways and highways are commonly categorized based on the traffic volumes and service life into two main categories—namely, permanent roads and temporary roads. Permanent roads are subjected to heavy traffic volumes of more than a million traffic loads during their service life. Temporary roads, on the other hand, are subject to lower traffic volumes of less than 10,000 load applications during their service life. Temporary roads include access roads, haul, detours, and construction platforms, which are used to construct permanent roads on weak soil layers [1].

Due to the scarcity of high-quality natural materials, marginal soils have been used for road construction with some form of mechanical or chemical treatment. Chemical stabilization such as with cement, natural rubber latex stabilization [2–4], and geopolymer stabilization [5–7] are often used to enhance the mechanical properties of marginal materials.

Geogrid applications have also been found to improve the mechanical properties and performance of marginal materials for base/subbase courses [8–11]. Research on geosynthetics in pavement reinforcement application has reported that the use of a geogrid within an unbound layer of a pavement structure can improve the stiffness of pavement layers, especially below and above the location of the geogrid [1,12,13]. Geogrids stabilize the aggregate layer by increasing aggregate interlocking, enhancing confinement, and reducing the lateral movement of the pavement structure, leading to deformation reduction.

The advantages of utilizing geogrids in road construction include a decrease in the thickness of the pavement structure layers and prolonging the durability of the road structure. Geosynthetics reinforcement in asphalt layers has also been reported to reduce rutting, pavement material fatigue, as well as thermal and reflective cracking. Geocomposite materials, such as geotextiles sandwiched within geogrids are also used as a separation layer to prevent the movement of small particles into open-graded base layers, resulting in improving the drainage systems and enhancing the road performance [14].

In temporary roads, geogrids are used within the weak foundation to support the initial construction work. The geogrid-reinforced aggregates are used as a working platform to mobilize the heavy machinery into the construction sites. For a particular subgrade stabilization application, geogrids are used to reinforce the soft subgrade and to decrease the excessive deformation of pavement structures due to the traffic loads [15–18]. For basal reinforcement applications, geogrids are installed under or within unbound layers of a flexible road to enhance the bearing capacity of the pavement against cyclic loads [19,20]. For pavement surface reinforcement, geogrids are used within the asphalt layer to decrease fatigue and rutting of the pavement surface using the marginal quantity aggregate [21,22].

Annually across the globe, the construction industry generates large quantities of construction and demolition (C&D) wastes, including recycled concrete aggregate (RCA), as well as recycled glass and brick. Similarly, reclaimed asphalt pavement (RAP) is generated when asphalt pavements are removed for reconstruction and/or resurfacing. Road authorities in many countries have been seeking to develop innovative methods of recycling and reusing recycled aggregates for partial or total replacement of natural aggregates in roadwork applications. The use of recycled aggregates including RAP [23–25], RCA [26–28], and recycled glass [29–31] as an alternative aggregate is widely accepted for road construction, especially as pavement base/subbase materials. Reusing recycled materials can decrease waste and energy consumption and therefore significantly contributes toward the sustainable road construction industry [32,33]. However, these materials sometimes require mechanical improvement to meet the local and international standards for both design and construction.

Several researchers have reported on the successful application of commercial synthetic geogrids reinforced natural materials in road construction. However, the applications of geogrid-reinforced recycled aggregates remain limited due to the lack of research studies, accepted design methodology, and construction guidelines. Pioneering research on the commercial geogrid-reinforced recycled aggregates was recently undertaken by several researchers [29,34–39]. Suddeepong et al. (2021) [40] investigated the interface shear behavior of natural kenaf geotextiles and RCA to promote the use of natural geotextiles with recycled aggregates for sustainable development of road construction and environmentally sound technologies. The performance of geogrid-reinforced recycled aggregates relies on various factors such as geometric forms and stiffness of geogrid, location and depth of geogrid installation, and particle sizes of aggregates [34,35,40].

This research aims to further contribute to the increased utilization of recycled aggregates in the pavement structure and to also facilitate the analysis and design by developing a generalized predictive equation of interface shear strength between geogrid and recycled aggregates. A large direct shear test (LDST) was first conducted to determine the interface shear strength behavior of RAP reinforced with natural kenaf geogrid in this research.

The influence of gradation of RAP and aperture size of geogrid on the interface shear responses of RAP—geogrid under different normal stresses was investigated. The results were then compared with the previous results to introduce a prime factor for developing a generalized predictive equation that can be used for rapid estimation of the interface shear strength of recycled aggregates reinforced with both commercial and natural geogrids.

2. Materials and Methods

2.1. Materials

Reclaimed asphalt pavement (RAP) samples were obtained from the Bureau of Highways, Nakhon Ratchasima, Thailand. A cold milling machine was used to remove the asphalt pavement for resurfacing in the cold in-place recycling process. The asphalt content in RAP aggregate is approximately 3–5% by weight. Figure 1 indicates the gradations of large-sized RAP and small-sized RAP samples. The large-sized RAP is on the lower boundary and the small-sized RAP is on the upper boundary, designated by the Department of Highways, Thailand (DOH, 2001) [41].

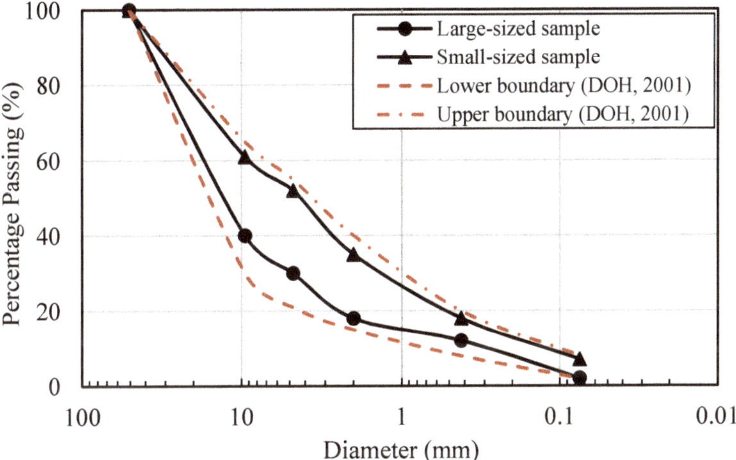

Figure 1. Grain size distribution of RAP.

Table 1 summarizes the basic and engineering properties of the RAP samples. The large-sized RAP and small-sized RAP samples were classified as poorly graded gravel (GP) and well-graded gravel (GW), respectively, according to the Unified Soil Classification System. Although the average particle size of the large-sized RAP sample and small-sized RAP sample was different, the specific gravity, maximum dry density (MDD) at optimum water content, California bearing ratio, internal friction angles, and cohesion were almost the same.

Table 1. Basic engineering properties of RAP samples.

Parameter	Recycled Asphalt Pavement (RAP)	
	Large Sized	**Small Sized**
Bulk specific gravity	2.6	2.6
Soil classification (USCS)	GP	GW
Average particle size (mm)	17	3.7
Optimum water content (%)	13.70	13.80
Maximum dry unit weight (kN/m^3)	19.56	19.48
California bearing ratio (%)	20	20
Internal friction angle (degree)	56.99	54.81
Cohesion (kPa)	53.68	56.98

The natural kenaf fibers were obtained from Tai Song Huad Co., Ltd., Sai Mai, Bangkok, Thailand, and were used to fabricate kenaf geogrid in this research. The handmade biaxial kenaf geogrid was a planar grid, which possesses the same strength in both ortho-directions (longitudinal and transversal) (Figure 2). The single rib tensile strength of kenaf geogrid was 43 MPa, which was obtained from the tensile test using a universal testing machine with a capacity of 2.5 kN based on ASTM-D6637 (2015) [42]. Two different aperture sizes of kenaf geogrids: 7 × 7 mm and 21 × 21 mm with a 3 mm rib thickness were prepared.

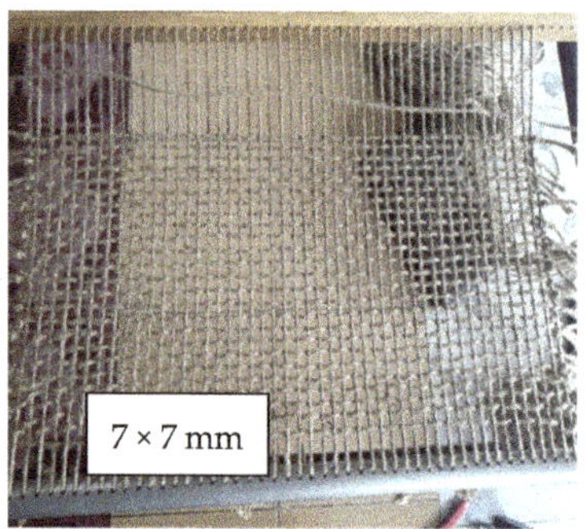

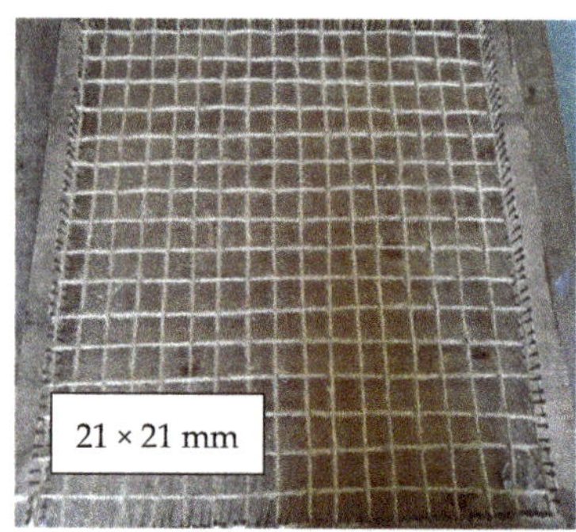

Figure 2. Photos of the planar grid of handmade biaxial kenaf geogrids.

2.2. Experimental Program

The LDST was undertaken in accordance with ASTM-D5321 (2020) [43] to investigate the interface shear strength of RAP–geogrid samples ($\tau_{reinforced}$) and the shear strength of unreinforced RAP samples ($\tau_{unreinforced}$). The LDST shear box apparatus with a dimension of 305 × 305 mm^2 and 204 mm high was divided into two parts, whereby the stationary upper half provides a confined vertical load to the sample, while the lower half of the box allows the application of horizontal shearing stress. To conduct the shear test, hand compaction was first carried out on RAP samples at optimum water content in three layers in the shear box under the modified Proctor effort to attain the MDD. For the consolidation process, the lower shear box and half of the upper one were filled with de-aired water to saturate the compacted RAP samples under different normal stresses (σ_n = 50, 100, and 200 kPa) for 12 h. LDSTs were conducted at the same σ_n levels with a constant shear rate of 0.025 mm/min at a controlled temperature of 20 ± 1 °C. The tests were completed when the horizontal shear displacement (HSD) attained 40 mm. Three samples were carried out for each direct shear test, and the mean value was reported in this study. The results under the same testing condition were reproducible with a low mean standard deviation, SD (SD/$\bar{x}$ < 10%, where $\bar{x}$ is the mean value). Table 2 illustrates the names of the prepared sample for LDST. Figure 3 illustrates the LDST apparatus and a photo of the tested kenaf geogrid.

Table 2. Summary of LDST testing program.

RAP Sample	Reinforcement	Normal Stress (kPa)
Large size	No reinforcement	50, 100, 200
	RAP + 7 × 7 mm geogrid	50, 100, 200
	RAP + 21 × 21 mm geogrid	50, 100, 200
Small size	No reinforcement	50, 100, 200
	RAP + 7 × 7 mm geogrid	50, 100, 200
	RAP + 21 × 21 mm geogrid	50, 100, 200

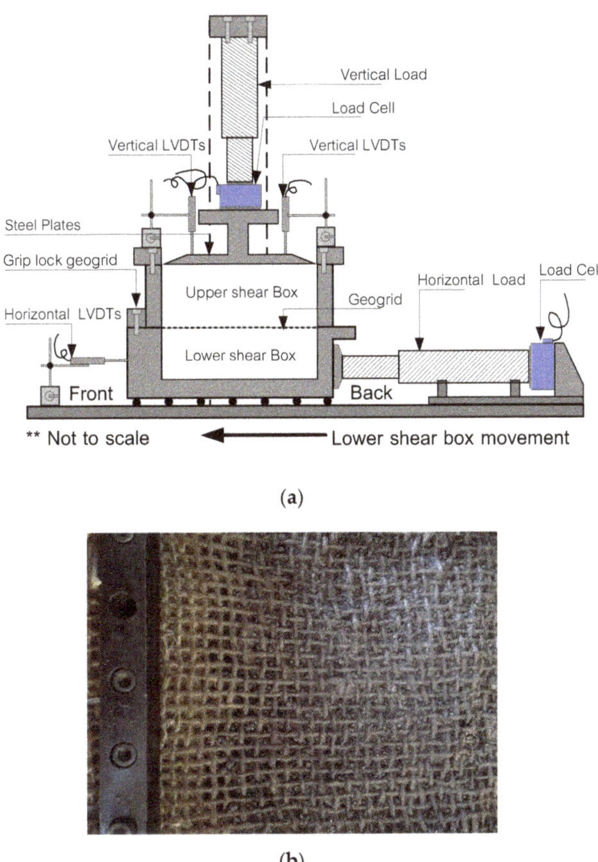

(**a**)

(**b**)

Figure 3. (**a**) LDST apparatus and (**b**) a photo of the tested kenaf geogrid.

3. Results and Discussion

The shear stresses and dilatation characteristics of unreinforced RAP obtained from the LDST were demonstrated in Figure 4. The shear stress behaviors of large-sized RAP and small-sized RAP were similar. The shear strength of unreinforced material ($\tau_{unreinforced}$) versus horizontal shear displacement (HSD) relationship exhibited strain-hardening behavior, whereby the shear stress increased with horizontal displacement and then became almost constant after HSD = 20-mm. The maximum $\tau_{unreinforced}$ increased with the increased σ_n.

The vertical shear displacement (VSD) versus HSD relationship of both large-sized RAP and small-sized RAP exhibited dilative behavior, which behaved similar to dense recycled glass [44] and RCA [35], at high σ_n of 100–200 kPa. The VSD of large- and small-sized RAP samples were similar when the HSD < 20 mm. However, the VSD of the small-sized RAP was higher than that of the large-sized RAP when the HSD was > 20 mm, especially at a high σ_n = 200 kPa.

In accordance with the Mohr–Coulomb failure criterion, the friction angle (ϕ) and cohesion (c) at the peak for both large-sized RAP and small-sized RAP samples were determined and are illustrated in Figure 5. The ϕ and c of small-sized RAP samples ($\phi = 54.81°$ and $c = 56.99$ kPa) and large-sized RAP samples ($\phi = 56.99°$ and $c = 53.68$ kPa) were similar. The high shear strength properties of RAP samples demonstrate that the material is stiff to withstand the traffic load and can be used as a base/subbase material based on the Department of Highways (DOH) specification [41]. The results also indicated that the RAP samples with gradation within the boundary specified by DOH can be used as base/subbase materials.

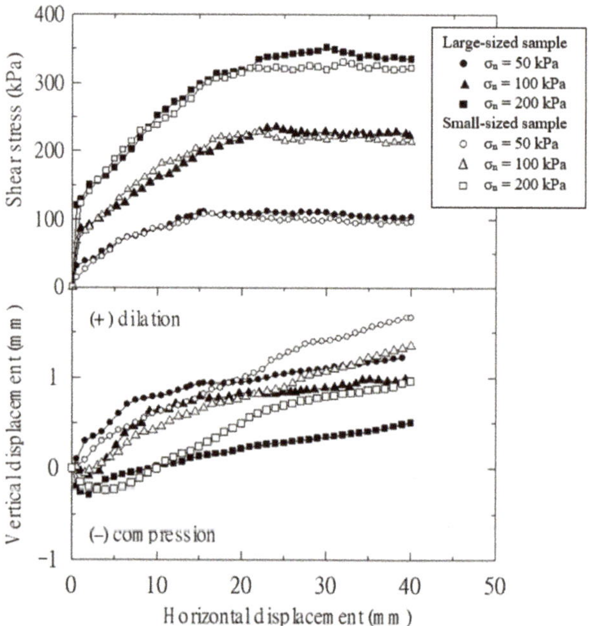

Figure 4. LDST test results of unreinforced RAP.

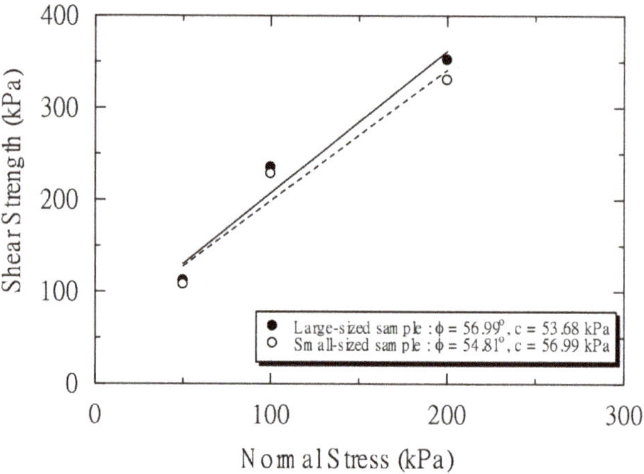

Figure 5. Shear strength failure envelope for unreinforced RAP.

Figures 6 and 7, respectively, indicate the influence of aperture sizes of kenaf geogrid (D) on the $\tau_{reinforced}$ behaviors of large-sized RAP–geogrid and small-sized RAP–geogrid. The $\tau_{reinforced}$ behavior of large-sized RAP samples was similar to that of the small-sized ones. $\tau_{reinforced}$, stiffness, and its peak values were found to increase with the increase in σ_n from 50 to 200 kPa. The relationship between $\tau_{reinforced}$ versus HSD of RAP–geogrid samples for both RAP gradations indicated strain-hardening behavior, similar to unreinforced RAP samples.

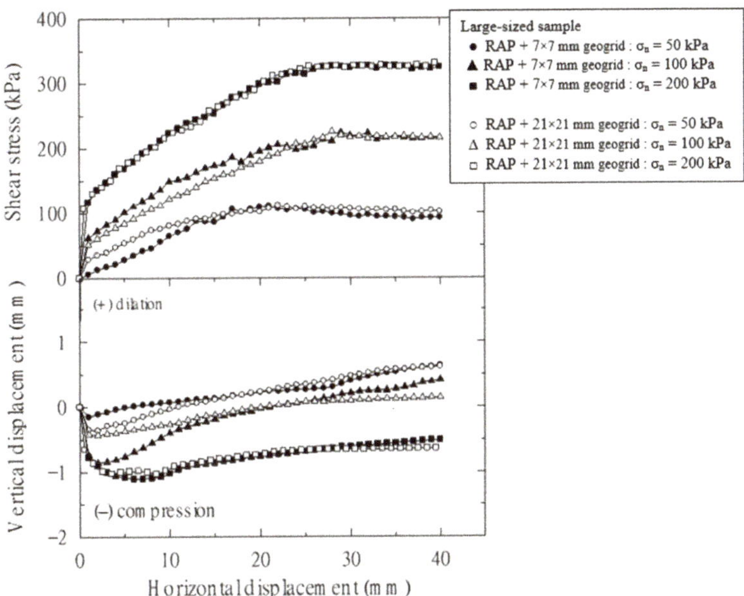

Figure 6. Effect of aperture size of geogrid on shear interface between geogrid and large-sized RAP sample.

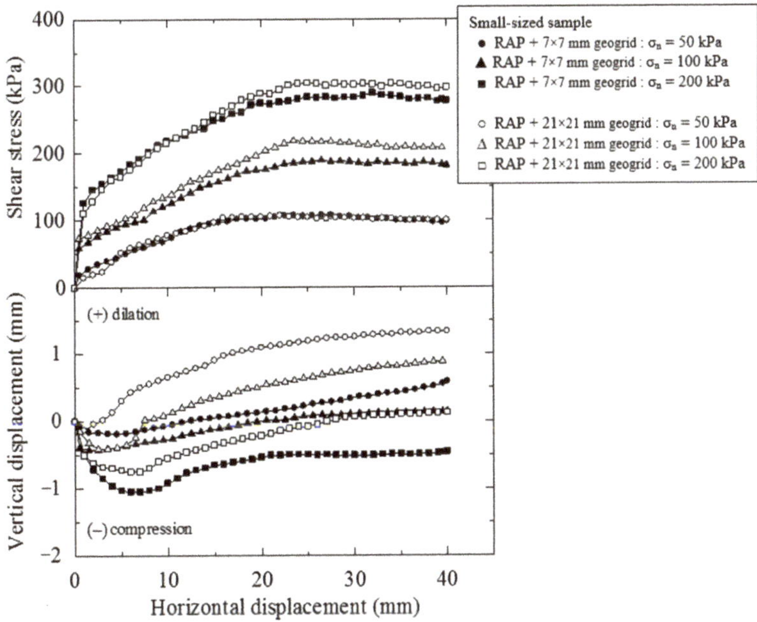

Figure 7. Effect of aperture size of geogrid on shear interface between geogrid and small-sized RAP sample.

For the large-sized RAP samples (Figure 6), the peak $\tau_{\text{reinforced}}$ of the samples at $\sigma_n = 50$ kPa was found at an HSD of approximately 20 to 25 mm. The peak $\tau_{\text{reinforced}}$ of the samples at σ_n of 100–200 kPa was, however, found at a large HSD of approximately

30 to 40 mm. For the small-sized RAP samples, the peak $\tau_{\text{reinforced}}$ of the samples was at an HSD of approximately 25 to 35 mm for all σ_n (Figure 7). The peak $\tau_{\text{reinforced}}$ of kenaf geogrid-reinforced small-sized RAP samples was slightly lower than those of kenaf geogrid-reinforced large-sized RAP samples at the same σ_n.

The relationship between VSD and HSD of RAP–geogrid with large- and small-sized RAP samples is also shown in Figures 6 and 7, respectively. A contraction behavior is noticed at an early stage, followed by continuous dilative behavior at the final stage. The dilative vertical displacements of kenaf geogrid-reinforced both large- and small-sized RAP samples were higher than those of unreinforced RAP samples at the same σ_n. This implies that the interaction between geogrid and RAP particles was improved. The influence of kenaf geogrid aperture size on VSD versus HSD relation was clearly apparent for the small-sized RAP samples. At a particular σ_n, the VSD of kenaf geogrid-reinforced small-sized RAP samples with 7×7 mm geogrid was higher than that of samples with 21×21 mm geogrid at the same HSD. The $\tau_{\text{reinforced}}$ value of RAP–geogrid samples was dependent upon both RAP aggregate interlocking and the contact surface area between RAP particles and geogrid.

Figures 8 and 9 show the Mohr–Coulomb failure envelopes of RAP–geogrids with different aperture sizes of geogrid and different gradations of RAP samples, compared with the failure envelopes of unreinforced RAP samples. For the large-sized RAP sample (Figure 8), the friction angle of unreinforced RAP samples ($\phi = 56.99°$) was slightly higher than the interface friction angles of kenaf geogrid-reinforced RAP samples with 21×21 mm geogrid ($\delta = 55.06°$) and 7×7 mm ($\delta = 54.18°$). In contrast, the adhesion values of kenaf geogrid-reinforced RAP samples with 21×21 mm geogrid ($c_a = 56.92$ kPa), 7×7 mm geogrid ($c_a = 59.29$ kPa) were higher than the cohesion of unreinforced RAP samples ($c = 53.68$ kPa). For small-sized RAP samples (Figure 9), the interface friction angles of kenaf geogrid-reinforced RAP samples were similar for both aperture sizes of 21×21 mm geogrid ($\delta = 51.26°$) and 7×7 mm geogrid ($\delta = 50.34°$). These values were lower than the friction angle of the unreinforced RAP samples ($\phi = 54.81°$). The adhesion values of kenaf geogrid-reinforced RAP samples with 21×21 mm geogrid and 7×7 mm geogrid were 64.41 kPa and 53.12 kPa, respectively, while the cohesion value of the unreinforced RAP sample was 56.99 kPa. This reveals that the aperture size of geogrid and gradation of RAP particles had a significant influence on the $\tau_{\text{reinforced}}$ value of kenaf geogrid-reinforced RAP samples. The interface shear strength ($\tau_{\text{reinforced}}$) of RAP–geogrid samples was found to be lower than the $\tau_{\text{unreinforced}}$ of unreinforced RAP samples, which are consistent with the previous findings [29,35,40,45]. For all sizes of RAP samples, the higher aperture size of kenaf geogrid resulted in the higher adhesion but insignificantly affected the interface friction angles.

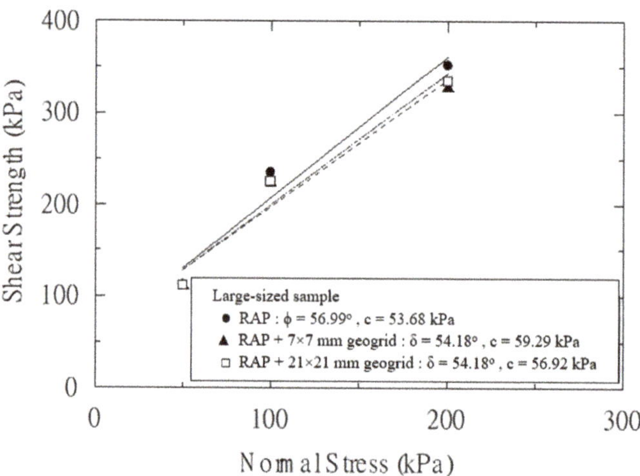

Figure 8. Interface stress failure envelopes for large-sized RAP samples.

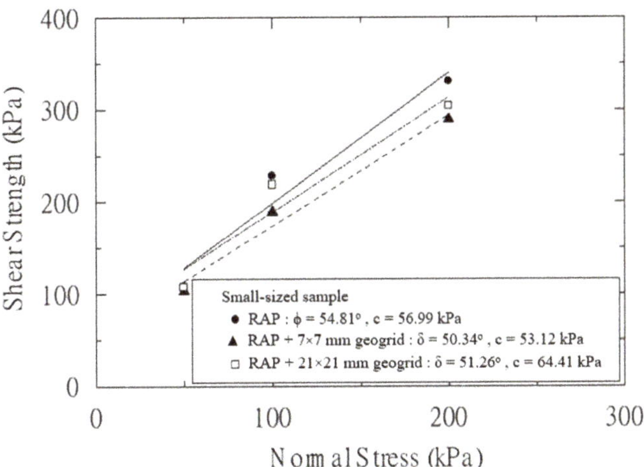

Figure 9. Interface stress failure envelopes for small-sized RAP samples.

To facilitate the analysis and design for pavement geotechnics applications, particularly by the finite element method, it is useful to interpret the $\tau_{reinforced}$ using the interface shear strength coefficient (α) in the following expression [46,47]:

$$\alpha = \frac{\tau_{reinforced}}{\tau_{unreinforced}} \tag{1}$$

The correlation between α and σ_n of RAP–geogrid samples for large- and small-sized RAP samples is presented in Figure 10. Though the $\tau_{unreinforced}$ value of unreinforced RAP directly influenced the $\tau_{unreinforced}$ value of RAP–geogrid samples, it was found that the α was irrespective of σ_n. The aperture size of geogrid (D) and gradation of RAP samples were found to strongly affect the interlocking mechanism of geogrid reinforcement and aggregates. The RAP particle content finer than the geogrid aperture size (F_D), which is related to the influence of gradation of RAP samples on the interface shear strength is investigated. In other words, F_D is the percentage passing obtained from the grain size distribution of RAP that is smaller than the aperture size of geogrid ($D = 7 \times 7$ mm and 21×21 mm). The relationship between α versus D and between α versus F_D is depicted in Figures 11 and 12, respectively.

The effect of F_D on the α values of RAP–geogrid samples with different D and gradations of RAP samples is depicted in Figure 12. The effect of F_D on the α values was found to be similar to the effect of D on α values (Figure 11). For the small-sized RAP samples, the large aperture size (21×21 mm) of geogrid exhibited higher α values than the small aperture size (7×7 mm), while the α values were found to be practically the same for both aperture sizes (21×21 mm and 7×7 mm) of geogrid-reinforced RAP samples, although F_D was varied from 0.28 to 0.6. However, for the same F_D of 0.6, the large-sized RAP + 21×21 mm geogrid had higher α than the small-sized RAP + 7×7 mm geogrid. It seems that $F_D = 0.28$ for large-sized RAP and $F_D = 0.8$ for small-sized RAP yielded the same α value of 0.96. In other words, both F_D and D control the α value.

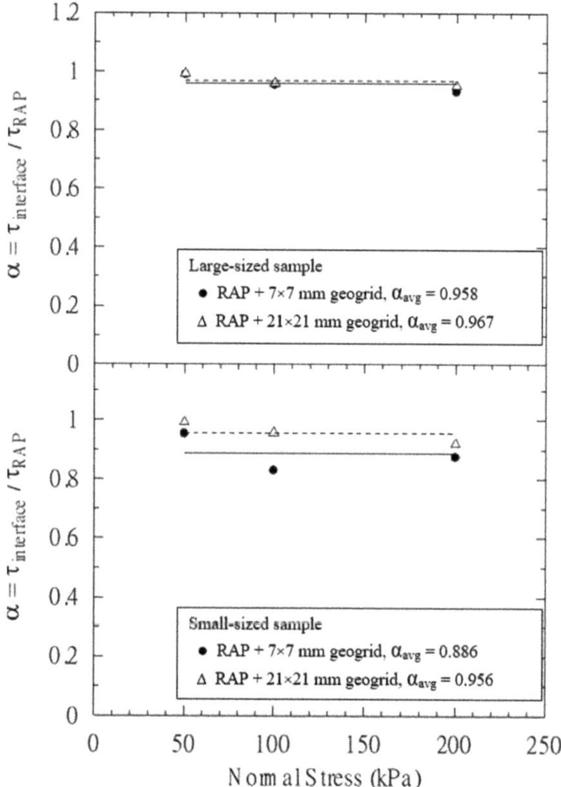

Figure 10. Relationship between α and normal stress.

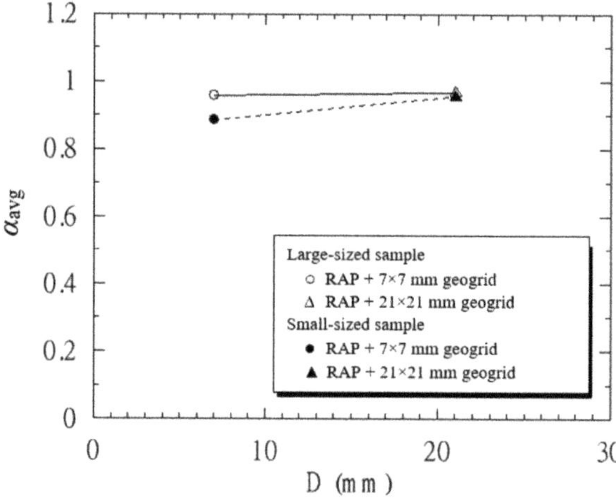

Figure 11. Effect of the aperture width of geogrid on the interface shear strength coefficient.

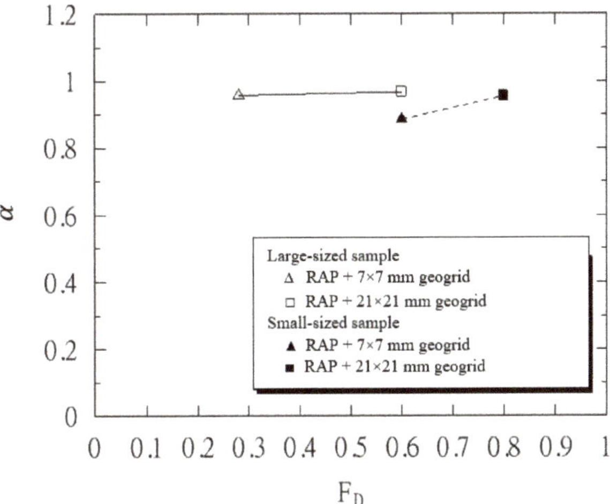

Figure 12. Effect of particle content finer than the aperture width of geogrid on the interface shear strength coefficient.

The relationship between α and D for the small-sized RAP samples (Figure 11) showed that the large aperture size (21 × 21 mm) of geogrid resulted in higher α values than the small aperture size (7 × 7 mm) of geogrid. In contrast, the α values of the large-sized RAP samples were found to be essentially the same for both large and small aperture sizes of geogrid. In addition, the α values of the large-sized RAP and small-sized RAP samples were similar for the 21 × 21 mm aperture size of geogrid, while the α value of small-sized RAP samples was lower than that of large-sized RAP samples with 7 × 7 mm geogrid. This implies that the aperture size of geogrid influences the α values of kenaf geogrid-reinforced RAP samples and ideally, a very large D results in the same α for different RAP gradations.

Several researchers have investigated the effect of a ratio of D to average aggregate particle size (D_{50}) on the $\tau_{reinforced}$ value of geogrid-reinforced aggregates [8,48,49]. However, the use of D_{50} to interpret the influence of the $\tau_{reinforced}$ behavior of geogrid-reinforced aggregate remains elusive. The proportion of aggregates indicated by D_{50} might have a large variation in large- and small-sized particles, which can significantly influence the gradation of the recycled materials. Consequently, excessively small or large particles of aggregates impact the effectiveness of the interlock mechanism or the $\tau_{reinforced}$ value of geogrid-reinforced recycled materials [50]. Some researchers studied the effect of a ratio of D to a single-sized gradation on the $\tau_{reinforced}$ behavior of geogrid-reinforced aggregates [51]. On the other hand, the use of a single-size particle or a poorly gradation of aggregate might not be suitable for pavement material in some road projects. Therefore, the use of the correlation between α versus D/F_D compliance for interpreting $\tau_{reinforced}$ of geogrid-reinforced recycled materials is a sound principle in this study.

Using the D/F_D as a prime parameter and integrating the contribution from D and F_D, the correlation between α and D/F_D is presented in Figure 13 and Equation (2) in the following expression:

$$\alpha = 0.0037 \left(\frac{D}{F_D} \right) + 0.85; \ 10 < \frac{D}{F_D} < 35 \, (\text{mm}) \tag{2}$$

where D is expressed in mm, and F_D is expressed in decimal with a high degree of coefficient, determined as 0.94.

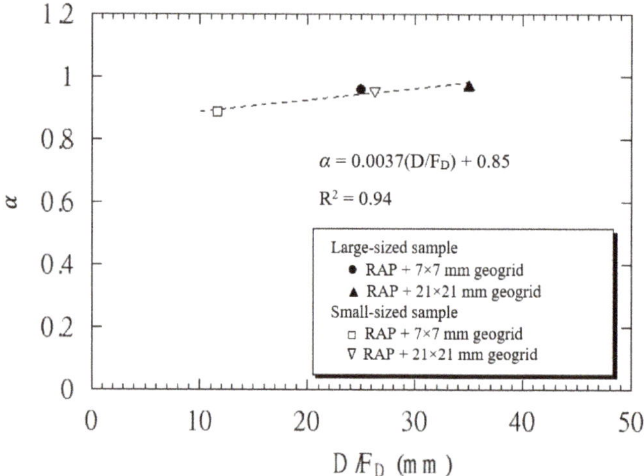

Figure 13. Relationship between α and D/F_D of kenaf geogrid-reinforced RAP samples.

Equation (2) is a useful practical tool for geotechnical and pavement engineers and the rational development of the equation can be extended to develop a generalized equation for different types of recycled materials and geogrids. Therefore, the separate set of data from the previous studies on the $\tau_{reinforced}$ behavior of the commercial polymer and natural kenaf geogrid-reinforced recycled aggregates such as RAP and RCA were taken and reanalyzed. Figure 14 shows the relationship between α and D/F_D of geogrid-reinforced RAP and RCA samples. The general form of the relationship can be expressed as the following equation:

$$\alpha = a\left(\frac{D}{F_D}\right) + b;\ 10 < \frac{D}{F_D} < 40\,(\text{mm}) \tag{3}$$

where a and b are constant. It is worthwhile mentioning that values of a and b are irrespective of geogrid types (natural kenaf or commercial polymer), while they were mainly dependent upon the recycled materials. From the regression analysis, the values of a = 0.0046 and b = 0.8336 were obtained for geogrid-reinforced RAP samples, while values of a = 0.0057 and b = 0.7185 were for geogrid-reinforced RCA samples. This implies that the geogrid-reinforced RAP has a higher α value than the geogrid-reinforced RCA at the same D/F_D for both natural kenaf and commercial polymer type. This might be due to the difference in shear strength, stiffness, and impurity of the recycled materials. The shear strength and stiffness of unreinforced RCA were higher than that of the unreinforced RAP material [34,51,52]. In other words, the geogrid-reinforced RCA samples exhibited a better interlocking mechanism than that geogrid-reinforced RAP samples. The α of RAP–geogrid is found to be more sensitive to the D/F_D than that of RAP–geogrid sample. Logically, there is no interaction between kenaf geogrid and RAP particles when $\tau_{reinforced}$ and $\tau_{unreinforced}$ are equal (α = 1.0). Based on equation (3), the geogrid-reinforced RAP and geogrid-reinforced RCA have no interaction when $D = 36.2F_D$ and $D = 49.4F_D$, respectively.

The proposed equation was developed based on sound principles and can therefore be used to predict the α of the recycled materials reinforced with various geogrids once the values of the constants a and b are known. In practice, a stepwise procedure to determine α values for the design of geogrids stabilized base/subbase with recycled aggregates is proposed as follows:

(1) From a selected recycled aggregate, adjust its gradation to meet the requirement for base/subbase courses specified by local or international standards;
(2) From the gradation, which might be varied along the constructing road, select at least two gradations to determine D/F_D values for a selected geogrid;

(3)   Perform the direct shear test on the selected recycled aggregate at various normal stresses in the range of field working stress;

(4)   Perform the direct interface shear test on the recycled aggregate reinforced with geogrid at various normal stress and D/F_D values;

(5)   From Equation (3), determine values of a and b. With these two values, the α values of the selected geogrid and recycled aggregate can be approximated.

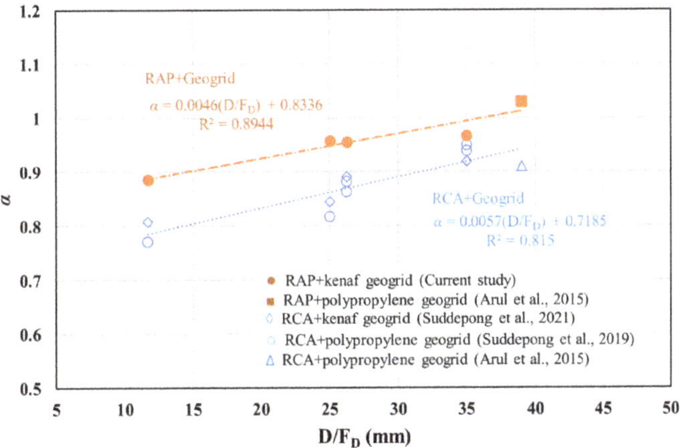

Figure 14. Relationship between α and D/F_D of geogrid-reinforced C&D samples.

This proposed general equation can be used to estimate the α for analysis and design of related geotechnical projects including pavement projects, mechanically stabilized earth (MSE) wall design, embankment reinforcement construction, and foundation design, which deal with various types of geogrid-reinforced recycled aggregates. Furthermore, to fully understand the behavior of natural geogrid-reinforced recycled aggregates, the relevant experimental program including dynamic flexural strength and fatigue tests are suggested for further research [53]. The outcome of this research will lead to the promotion of recycled aggregates as green aggregate for sustainable geotechnical and pavement applications.

4. Conclusions

In this research, a large direct shear test (LDST) was conducted to investigate the interface shear strength ($\tau_{reinforced}$) between reclaimed asphalt pavement (RAP) and kenaf geogrid (RAP–geogrid) as a sustainable base course material. The influence of gradation of RAP particles and aperture sizes of geogrid (D) on $\tau_{reinforced}$ of RAP–geogrid was evaluated under different normal stresses. Based on the critical analysis of the present and previous test data on both natural and commercial geogrid-reinforced C&D materials including RAP and RCA, it is found that the $\tau_{reinforced}$ value of geogrid-reinforced recycled materials was controlled by the D/F_D, where F_D is the recycled materials' particle content finer than the aperture of geogrid. The generalized equation for predicting $\tau_{reinforced}$ is proposed in the form: $\alpha = a(D/F_D) + b$, where α is interface shear strength coefficient, which is the ratio of $\tau_{reinforced}$ to $\tau_{unreinforced}$ of recycled material, and a and b are constant. The values of a and b were found to be dependent upon types of recycled material, irrespective of types of geogrids. This proposed generalized equation is useful to determine α, a required parameter for analysis and design of geotechnical and pavement work dealt with the geogrid-reinforced recycled aggregates. It is advantageous to the designer to select the various aperture sizes of geogrids and gradations of recycled aggregates for geotechnical and pavement projects.

Author Contributions: Conceptualization, A.U., A.S., M.H. and S.H.; methodology, A.U., M.H., A.P. and S.H.; software, A.U., A.P. and M.H.; formal analysis, A.U., A.S. and M.H.; writing—review and editing, A.U., A.S., S.H., M.H., N.C.T. and A.A.; visualization, M.H., A.A. and S.H.; super-vision, M.H., A.A. and S.H. All authors have read and agreed to the published version of the manuscript.

Funding: Thailand Research Fund under Research and Researcher for Industries-RRI Program (Grant Number PhD62I0026).

Institutional Review Board Statement: Not applicable.

Informed Consent Statement: Not applicable.

Data Availability Statement: The data presented in this study are available on request from the corresponding author.

Acknowledgments: This work was financially supported by the Thailand Research Fund under Research and Researcher for Industries-RRI Program (Grant Number PhD62I0026).

Conflicts of Interest: The authors declare no conflict of interest.

References

1. Holz, R.; Christopher, B.R.; Berg, R.R. *Geosynthetic Design and Construction Guidelines*; FHWA Publication No. FHWA HI-07-092; Federal Highway Administration: Washington, DC, USA, 1998.
2. Buritatum, A.; Horpibulsuk, S.; Udomchai, A.; Suddeepong, A.; Takaikaew, T.; Vichitcholchai, N.; Arulrajah, A. Durability improvement of cement stabilized pavement base using natural rubber latex. *Transp. Geotech.* **2021**, 100518. [CrossRef]
3. Buritatum, A.; Takaikaew, T.; Horpibulsuk, S.; Udomchai, A.; Hoy, M.; Vichitcholchai, N.; Arulrajah, A. Mechanical Strength Improvement of Cement-Stabilized Soil Using Natural Rubber Latex for Pavement Base Applications. *J. Mater. Civ. Eng.* **2020**, *32*, 04020372. [CrossRef]
4. Jose, A.; Kasthurba, A. Laterite soil-cement blocks modified using natural rubber latex: Assessment of its properties and performance. *Constr. Build. Mater.* **2021**, *273*, 121991. [CrossRef]
5. Avirneni, D.; Peddinti, P.R.; Saride, S. Durability and long term performance of geopolymer stabilized reclaimed asphalt pavement base courses. *Constr. Build. Mater.* **2016**, *121*, 198–209. [CrossRef]
6. Phummiphan, I.; Horpibulsuk, S.; Rachan, R.; Arulrajah, A.; Shen, S.-L.; Chindaprasirt, P. High calcium fly ash geopolymer stabilized lateritic soil and granulated blast furnace slag blends as a pavement base material. *J. Hazard. Mater.* **2018**, *341*, 257–267. [CrossRef]
7. Xiao, R.; Polaczyk, P.; Zhang, M.; Jiang, X.; Zhang, Y.; Huang, B.; Hu, W. Evaluation of glass powder-based geopolymer stabilized road bases containing recycled waste glass aggregate. *Transp. Res. Rec.* **2020**, *2674*, 22–32. [CrossRef]
8. Giroud, J.P.; Han, J. Design Method for Geogrid-Reinforced Unpaved Roads. I. Development of Design Method. *J. Geotech. Geoenviron. Eng.* **2004**, *130*, 775–786. [CrossRef]
9. Maghool, F.; Arulrajah, A.; Mirzababaei, M.; Suksiripattanapong, C.; Horpibulsuk, S. Interface shear strength properties of geogrid-reinforced steel slags using a large-scale direct shear testing apparatus. *Geotext. Geomembr.* **2020**, *48*, 625–633. [CrossRef]
10. Saberian, M.; Li, J.; Perera, S.T.A.M.; Zhou, A.; Roychand, R.; Ren, G. Large-scale direct shear testing of waste crushed rock reinforced with waste rubber as pavement base/subbase materials. *Transp. Geotech.* **2021**, *28*, 100546. [CrossRef]
11. Vieira, C.S. Valorization of Fine-Grain Construction and Demolition (C&D) Waste in Geosynthetic Reinforced Structures. *Waste Biomass Valorization* **2020**, *11*, 1615–1626.
12. Christopher, B.; Perkins, S. Full scale testing of geogrids to evaluate junction strength requirements for reinforced roadway base design. In Proceedings of the Fourth European Geosynthetics Conference, Edinburgh, UK, 7–10 September 2008.
13. Perkins, S.; Christopher, B.; Thom, N.; Montestruque, G.; Korkiala-Tanttu, L.; Want, A. Geosynthetics in pavement reinforcement applications. In Proceedings of the 9th International Conference on Geosynthetics, Guarujá, Brazil, 23–27 May 2010.
14. Rajagopal, K.; Krishnaswamy, N.R.; Madhavi Latha, G. Behaviour of sand confined with single and multiple geocells. *Geotext. Geomembr.* **1999**, *17*, 171–184. [CrossRef]
15. Al-Qadi, I.L.; Brandon, T.L.; Valentine, R.J.; Lacina, B.A.; Smith, T.E. Laboratory evaluation of geosynthetic-reinforced pavement sections. *Transp. Res. Rec.* **1994**, *1439*, 25–31.
16. Cuelho, E.V.; Perkins, S.W. Geosynthetic subgrade stabilization—Field testing and design method calibration. *Transp. Geotech.* **2017**, *10*, 22–34. [CrossRef]
17. Perkins, S.W. Numerical Modeling of Geosynthetic Reinforced Flexible Pavements. *Constr. Build. Mater.* **2016**, *122*, 214–230.
18. Shirazi, M.G.; Rashid, A.S.A.; Nazir, R.; Abdul Rashid, A.H.; Horpibulsuk, S. Enhancing the Bearing Capacity of Rigid Footing Using Limited Life Kenaf Geotextile Reinforcement. *J. Nat. Fibers* **2020**, 1–17. [CrossRef]
19. Abu-Farsakh, M.Y.; Chen, Q. Evaluation of geogrid base reinforcement in flexible pavement using cyclic plate load testing. *Int. J. Pavement Eng.* **2011**, *12*, 275–288. [CrossRef]
20. Alimohammadi, H.; Zheng, J.; Schaefer, V.R.; Siekmeier, J.; Velasquez, R. Evaluation of geogrid reinforcement of flexible pavement performance: A review of large-scale laboratory studies. *Transp. Geotech.* **2021**, *27*, 100471. [CrossRef]

21. Ling, H.I.; Liu, Z. Performance of Geosynthetic-Reinforced Asphalt Pavements. *J. Geotech. Geoenviron. Eng.* **2001**, *127*, 177–184. [CrossRef]

22. Mirzapour Mounes, S.; Karim, M.R.; Khodaii, A.; Almasi, M.H. Improving Rutting Resistance of Pavement Structures Using Geosynthetics: An Overview. *Sci. World J.* **2014**, *2014*, 764218. [CrossRef] [PubMed]

23. Horpibulsuk, S.; Hoy, M.; Witchayaphong, P.; Rachan, R.; Arulrajah, A. Recycled asphalt pavement—Fly ash geopolymer as a sustainable stabilized pavement material. In *The IOP Conference Series: Materials Science and Engineering*; IOP Publishing: Bristol, UK, 2017.

24. Hoy, M.; Horpibulsuk, S.; Arulrajah, A.; Mohajerani, A. Strength and microstructural study of recycled asphalt pavement: Slag geopolymer as a pavement base material. *J. Mater. Civ. Eng.* **2018**, *30*, 04018177. [CrossRef]

25. Hoy, M.; Horpibulsuk, S.; Arulrajah, A. Strength development of Recycled Asphalt Pavement—Fly ash geopolymer as a road construction material. *Constr. Build. Mater.* **2016**, *117*, 209–219. [CrossRef]

26. Maghool, F.; Senanayake, M.; Arulrajah, A.; Horpibulsuk, S. Permanent Deformation and Rutting Resistance of Demolition Waste Triple Blends in Unbound Pavement Applications. *Materials* **2021**, *14*, 798. [CrossRef] [PubMed]

27. Yaowarat, T.; Horpibulsuk, S.; Arulrajah, A.; Maghool, F.; Mirzababaei, M.; Rashid, A.S.A.; Chinkulkijniwat, A. Cement stabilisation of recycled concrete aggregate modified with polyvinyl alcohol. *Int. J. Pavement Eng.* **2020**, 1–9. [CrossRef]

28. Yaowarat, T.; Horpibulsuk, S.; Arulrajah, A.; Mohammadinia, A.; Chinkulkijniwat, A. Recycled concrete aggregate modified with polyvinyl alcohol and fly ash for concrete pavement applications. *J. Mater. Civ. Eng.* **2019**, *31*, 04019103. [CrossRef]

29. Arulrajah, A.; Horpibulsuk, S.; Maghoolpilehrood, F.; Samingthong, W.; Du, Y.-J.; Shen, S.-L. Evaluation of interface shear strength properties of geogrid reinforced foamed recycled glass using a large-scale direct shear testing apparatus. *Adv. Mater. Sci. Eng.* **2015**, *2015*, 235424. [CrossRef]

30. Arulrajah, A.; Kua, T.-A.; Horpibulsuk, S.; Mirzababaei, M.; Chinkulkijniwat, A. Recycled glass as a supplementary filler material in spent coffee grounds geopolymers. *Constr. Build. Mater.* **2017**, *151*, 18–27. [CrossRef]

31. Naeini, M.; Mohammadinia, A.; Arulrajah, A.; Horpibulsuk, S. Recycled Glass Blends with Recycled Concrete Aggregates in Sustainable Railway Geotechnics. *Sustainability* **2021**, *13*, 2463. [CrossRef]

32. Hoy, M.; Horpibulsuk, S.; Rachan, R.; Chinkulkijniwat, A.; Arulrajah, A. Recycled asphalt pavement—Fly ash geopolymers as a sustainable pavement base material: Strength and toxic leaching investigations. *Sci. Total Environ.* **2016**, *573*, 19–26. [CrossRef]

33. Yeheyis, M.; Hewage, K.; Alam, M.S.; Eskicioglu, C.; Sadiq, R. An overview of construction and demolition waste management in Canada: A lifecycle analysis approach to sustainability. *Clean Technol. Environ. Policy* **2013**, *15*, 81–91. [CrossRef]

34. Arulrajah, A.; Rahman, M.; Piratheepan, J.; Bo, M.; Imteaz, M. Interface shear strength testing of geogrid-reinforced construction and demolition materials. *Adv. Civ. Eng. Mater.* **2013**, *2*, 189–200. [CrossRef]

35. Suddeepong, A.; Sari, N.; Horpibulsuk, S.; Chinkulkijniwat, A.; Arulrajah, A. Interface shear behaviours between recycled concrete aggregate and geogrids for pavement applications. *Int. J. Pavement Eng.* **2020**, *21*, 228–235. [CrossRef]

36. Han, J.; Thakur, J.K. Use of geosynthetics to stabilize recycled aggregates in roadway construction. In Proceedings of the ICSDEC 2012: Developing the Frontier of Sustainable Design Engineering, and Construction, Fort Worth, TX, USA, 7–9 November 2012; pp. 473–480.

37. Han, J.; Thakur, J.K. Sustainable roadway construction using recycled aggregates with geosynthetics. *Sustain. Cities Soc.* **2015**, *14*, 342–350. [CrossRef]

38. Vieira, C.; Pereira, P. Interface shear properties of geosynthetics and construction and demolition waste from large-scale direct shear tests. *Geosynth. Int.* **2016**, *23*, 62–70. [CrossRef]

39. Vieira, C.S.; Pereira, P.M. Short-term tensile behaviour of three geosynthetics after exposure to Recycled Construction and Demolition materials. *Constr. Build. Mater.* **2021**, *273*, 122031. [CrossRef]

40. Suddeepong, A.; Hoy, M.; Nuntasena, C.; Horpibulsuk, S.; Kantatham, K.; Arulrajah, A. Evaluation of interface shear strength of satural kenaf geogrid and recycled concrete aggregate for sustainable pavement applications. *J. Nat. Fibers* **2021**. [CrossRef]

41. DOH. *Standard No. DH-S 201/2544, Standard of Crusher Rock Base*; Department of Highways: Bangkok, Thailand, 2001.

42. ASTM-D6637. *Standard Test Method for Determining Tensile Properties of Geogrids by the Single or Multi-Rib Tensile Method*; ASTM International: West Conshohocken, PA, USA, 2015.

43. ASTM-D5321. *D5321M-20, Standard Test Method for Determining the Shear Strength of Soil-Geosynthetic and Geosynthetic-Geosynthetic Interfaces by Direct Shear*; ASTM International: West Conshohocken, PA, USA, 2020.

44. Disfani, M.M.; Arulrajah, A.; Bo, M.W.; Hankour, R. Recycled crushed glass in road work applications. *Waste Manag.* **2011**, *31*, 2341–2351. [CrossRef]

45. Liu, X.; Scarpas, A.; Blaauwendraad, J.; Genske, D.D. Geogrid Reinforcing of Recycled Aggregate Materials for Road Construction: Finite Element Investigation. *Transp. Res. Rec.* **1998**, *1611*, 78–85. [CrossRef]

46. Horpibulsuk, S.; Niramitkornburee, A. Pullout Resistance of Bearing Reinforcement Embedded in Sand. *Soils Found.* **2010**, *50*, 215–226. [CrossRef]

47. Sukmak, G.; Sukmak, P.; Joongklang, A.; Udomchai, A.; Horpibulsuk, S.; Arulrajah, A.; Yeanyong, C. Predicting Pullout Resistance of Bearing Reinforcement Embedded in Cohesive-Frictional Soils. *J. Mater. Civ. Eng.* **2020**, *32*, 04019379. [CrossRef]

48. Indraratna, B.; Karimullah Hussaini, S.; Vinod, J. On The Shear Behavior of Ballast-Geosynthetic Interfaces. *Geotech. Test. J.* **2012**, *35*, 305–312. [CrossRef]

49. Liu, C.-N.; Ho, Y.-H.; Huang, J.-W. Large scale direct shear tests of soil/PET-yarn geogrid interfaces. *Geotext. Geomembr.* **2009**, *27*, 19–30. [CrossRef]

50. Han, B.; Ling, J.; Shu, X.; Gong, H.; Huang, B. Laboratory investigation of particle size effects on the shear behavior of aggregate-geogrid interface. *Constr. Build. Mater.* **2018**, *158*, 1015–1025. [CrossRef]
51. Arulrajah, A.; Rahman, M.A.; Piratheepan, J.; Bo, M.W.; Imteaz, M.A. Evaluation of interface shear strength properties of geogrid-reinforced construction and demolition materials using a modified large-scale direct shear testing apparatus. *J. Mater. Civ. Eng.* **2014**, *26*, 974–982. [CrossRef]
52. Soleimanbeigi, A.; Likos, W. Mechanical Properties of Recycled Concrete Aggregate and Recycled Asphalt Pavement Reinforced with Geosynthetics. In *Geo-Congress 2019: Earth Retaining Structures and Geosynthetics*; American Society of Civil Engineers: Reston, VA, USA, 2019; pp. 284–292.
53. Pasetto, M.; Pasquini, E.; Giacomello, G.; Baliello, A. Innovative composite materials as reinforcing interlayer systems for asphalt pavements: An experimental study. *Road Mater. Pavement Des.* **2019**, *20* (Suppl. 2), S617–S631. [CrossRef]

 *sustainability*

Article

Pullout Behaviour of Geogrids Embedded in a Recycled Construction and Demolition Material. Effects of Specimen Size and Displacement Rate

Castorina Silva Vieira *, Paulo Pereira, Fernanda Ferreira and Maria de Lurdes Lopes

CONSTRUCT, Faculty of Engineering, University of Porto, R. Dr. Roberto Frias, 4200-465 Porto, Portugal; pmpp@fe.up.pt (P.P.); fbf@fe.up.pt (F.F.); lcosta@fe.up.pt (M.d.L.L.)
* Correspondence: cvieira@fe.up.pt

Received: 13 March 2020; Accepted: 4 May 2020; Published: 8 May 2020

Abstract: In recent years, environmental concerns related to the overexploitation of natural resources and the need to manage large amounts of wastes arising from construction activities have intensified the pressure on the civil engineering industry to adopt sustainable waste recycling and valorisation measures. The use of recycled construction and demolition (C&D) wastes as alternative backfill for geosynthetic-reinforced structures may significantly contribute towards sustainable civil infrastructure development. This paper presents a laboratory study carried out to characterise the interaction between a fine-grained C&D material and two different geogrids (a polyester (PET) geogrid and an extruded uniaxial high-density polyethylene (HDPE) geogrid) through a series of large-scale pullout tests. The effects of the geogrid specimen size, displacement rate and vertical confining pressure on the pullout resistance of the geogrids are evaluated and discussed, aiming to assess whether they are in line with the current knowledge about the pullout resistance of geogrids embedded in soils. Test results have shown that the measured peak pullout resistance of the geogrid increases with the specimen size, imposed displacement rate and confining pressure. However, the pullout interaction coefficient has exhibited the opposite trend with the specimen size and confining pressure. The pullout interaction coefficients ranged from 0.79 and 1.57 and were generally greater than or equal to the values reported in the literature for soil-geogrid and recycled material-geogrid interfaces.

Keywords: sustainability in geotechnics; recycled construction and demolition materials; geogrids; pullout behaviour; pullout test parameters

---

1. Introduction

Recent years have witnessed an increasing environmental awareness and the recognition of the importance of reducing the production of wastes and the exploitation of non-renewable natural resources in order to foster sustainable development. The civil engineering industry is among the major contributors to the worldwide consumption of natural resources (such as sand, gravel and stone reserves), being responsible for about 50% of all the materials extracted from the earth's crust [1]. On the other hand, construction and demolition (C&D) waste is one of the heaviest and most voluminous waste streams generated in the European Union (EU), representing approximately 25–30% of all waste generated in the EU [2].Billions of tons of construction and demolition (C&D) wastes are produced every year from different activities, including the construction, maintenance and demolition of buildings and civil infrastructure, which raises severe environmental concerns and intensifies the need for more efficient waste management in the construction sector. In particular, the large volumes of C&D waste generated across the EU and their high valorisation potential have led the European Commission to classify these materials as a priority waste stream [3].

Geotechnical design and construction, which is often placed early in a civil engineering project, can significantly contribute to enhance the overall sustainable development by incorporating sustainable practices, among which is the use of alternative, environment friendly materials and the reuse of waste materials, such as the C&D wastes [4–6]. In Europe about 40% of the natural aggregates are consumed in unbound layers of transportation infrastructures [7]. This suggests that the reliance on natural aggregates in geotechnical applications is high and the inclusion of recycled aggregates can contribute significantly to preserve the environment. In view of the above, several studies have recently been conducted to evaluate the feasibility of using recycled C&D wastes in diverse geotechnical applications, such as ground improvement works [8,9], pipe bedding and backfilling [10,11], construction of paved and unpaved roads [12–18] and backfilling of geosynthetic-reinforced structures [19–22].

Most of the studies carried out on recycled aggregates from C&D waste are related to recycled concrete aggregates [9,10,14,23] or reclaimed asphalt pavement materials [16,17]. However, particularly in Southern European Countries, C&D wastes sent out at the recycling plants are mainly mixed wastes (comprising concrete, mortars, stones, ceramics,). The recycled aggregates coming from mixed C&D waste have limited market acceptance, particularly to concrete production and base layers of roadways. Coarse recycled aggregates are sometimes applied as aggregates in sub-base layers of transportation infrastructures [13] or unpaved roads [15,24].

During the recycling process of C&D waste, particularly in Portuguese recycling plants, a fine-grain recycled material (0–10 mm) is produced. This fine grain fraction has reduced market acceptance and the recycling operators have difficulties in commercialize it. Rodrigues et al. [25] evaluated physical and chemical properties relevant for the incorporation in concrete of 10 samples of fine aggregates from C&D waste obtained from seven Portuguese recycling plants and concluded that none of the samples has all characteristics within the limits imposed in the European Standards to allow its incorporation in concrete. Based on these evidences, a research study has been carried out to evaluate the feasibility of using these recycled materials in the construction of structural embankments, in particular geosynthetic reinforced embankments [11,20,21].

The interaction between the geosynthetic reinforcement and the backfill material is of critical importance for the safe design and adequate performance of geosynthetic-reinforced structures, such as walls, slopes and bridge abutments [26,27]. Various test methods have been used by numerous researchers over the last decades to characterise soil-geosynthetic interaction, such as the direct shear test [28–31], inclined plane test [32–35], pullout test [36–40] and in-soil tensile test [41], each of which allows simulating a different type of deformation at the backfill-reinforcement interface. For instance, the direct shear test is commonly used to analyse soil-reinforcement interaction when sliding of the backfill on the geosynthetic surface is anticipated, whereas the pullout test simulates the interaction between the backfill and the reinforcement in the anchorage zone of geosynthetic-reinforced soil walls and slopes (i.e., beyond the hypothetical failure surface). In fact, a condition for verification of internal stability of these structures is that the pullout capacity of the geosynthetic in the anchorage zone should not be lower than the tensile force acting on the reinforcement. The pullout resistance of the geosynthetic is therefore an important parameter required by design codes for geosynthetic-reinforced structures [42–45].

The pullout resistance of the geogrids is developed primarily by the combination of the passive resistance mobilised against their transverse members and the skin friction at both sides of the reinforcement [27]. While the latter mechanism depends mainly on the type of backfill material and geogrid surface roughness, the contribution of the passive resistance mechanism to the overall pullout resistance is dependent upon several factors, including the confining pressure, geogrid geometry and ratio of the mean grain size of the backfill material to the geogrid opening size.

The feasibility of using C&D recycled materials in the construction of geogrid-reinforced structures has been studied in recent years by some research groups [19–21,46,47]. These researches have been focused mainly on full-scale testing [19] or on the study of interfaces behaviour through direct shear [20,47–49] or pullout tests [20,22,49]. Nevertheless, the studies on the pullout resistance of geogrids embedded in C&D recycled materials are limited and based on valid assumptions for soils.

Vieira et al. [20] carried out pullout tests on three geosynthetics for soil reinforcement (a uniaxial HDPE geogrid, a uniaxial PET geogrid and a high-strength composite geotextile) embedded in a fine grain recycled C&DW obtained from a Portuguese recycling plant. The tests procedures were defined in accordance with the European standard for determination of pullout resistance in soil [50], being the tests were carried out with a constant displacement rate of 2 mm/min and under normal stress of approximately 31 kPa at interface level. Vieira et al. [22] report an experimental study carried out to assess the pullout behaviour of two geosynthetics (a uniaxial geocomposite reinforcement and an extruded HDPE geogrid) embedded in a recycled C&D material under cyclic and post-cyclic loading conditions. Soleimanbeigi et al. [49] performed pullout tests on a recycled concrete aggregate reinforced with a woven geotextile or a uniaxial geogrid following the procedures outlined in the ASTM standard for measuring geosynthetic pullout resistance in soil [51]. walls. A pullout displacement rate of 1.0 mm/min was used and the tests were performed under 20, 30, 50, 100 and 200 kPa normal stress.

The aim of the present study is assessing whether the effects of the different parameters with influence on the pullout resistance of geogrids (namely, the geogrid specimen size, the displacement rate and the normal stress) when they are embedded in a C&D recycled material are in line with the current knowledge about the pullout resistance of geogrids in soils. Previous studies on the pullout behaviour of interfaces between geosynthetics and C&D recycled materials have been performed following the guidance for common backfill materials (cohesionless soil soils) [20,22,49]. Thus, it is of great importance to evaluate whether some assumptions are still valid for alternative backfill materials. To this end, a series of large-scale laboratory pullout tests were carried out using a compacted C&D recycled material and two distinct uniaxial geogrids: a laid and welded geogrid consisting of extruded polyester (PET) bars and an extruded uniaxial high-density polyethylene (HDPE) geogrid. The effects of the geogrid specimen size, displacement rate and vertical confining pressure on the measured pullout response of the reinforcements are assessed. The pullout interaction coefficients for the studied interfaces are then derived, discussed and compared with the values typically reported in the literature for soil-geogrid interfaces. The main conclusions and the implications to the design of geogrid-reinforced structures are also depicted.

2. Materials and Methods

2.1. General Overview

Figure 1 presents a flowchart of the experimental study for a better understanding of this research. In this section the materials are characterized, pullout test apparatus and procedures are described, the test programme is summarised and the processing of the pullout test results is introduced.

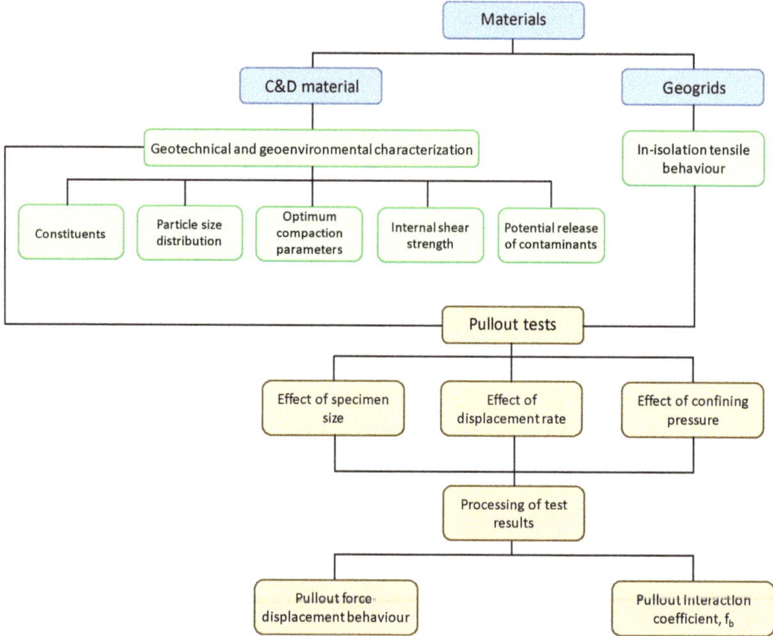

Figure 1. Flowchart of the experimental study.

2.2. Construction and Demolition (C&D) Material

A fine-grained C&D recycled material was collected at a Portuguese recycling plant and used throughout the current study. This fine grain fraction is produced during the recycling process of C&D wastes and has little market acceptance due, mainly, to the likely high soil content and heterogeneity. Table 1 lists the respective constituents determined on the basis of the European Standard [52]. This standard refers to the constituents of coarse recycled aggregates, so some adjustments have been implemented, namely to estimate the "soil" constituent. The presented results clearly show that this particular C&D material was essentially composed of concrete and mortar products, unbound aggregates, masonries and soil.

Table 1. Constituents of the recycled C&D material [22].

Constituents	Values
Concrete, concrete products, mortar, concrete masonry units [%]	40.0
Unbound aggregates, natural stone, aggregates treated with hydraulic binders [%]	36.5
Clay and calcium silicate masonry units, aerated non-floating concrete [%]	10.8
Bituminous materials [%]	0.5
Glass [%]	1.3
Soils [%]	10.8
Other materials [%]	0.1
Floating particles [cm^3/kg]	10.0

The gradation was evaluated by sieving and sedimentation, in accordance with the Standards [53,54], respectively (Figure 2). It is worth noting that this material complies with the gradation requirements of the Federal Highway Administration (FHWA) [43] for reinforced soil slopes and of the National Concrete Masonry Association (NCMA) [44] for segmental retaining walls, but does not fulfil the criteria established by the FHWA for mechanically stabilised earth walls. Following the

FHWA criteria, this C&D material can be used as backfill material for geosyntetic-reinforced steep slopes (face inclinations of less than 70 degrees).

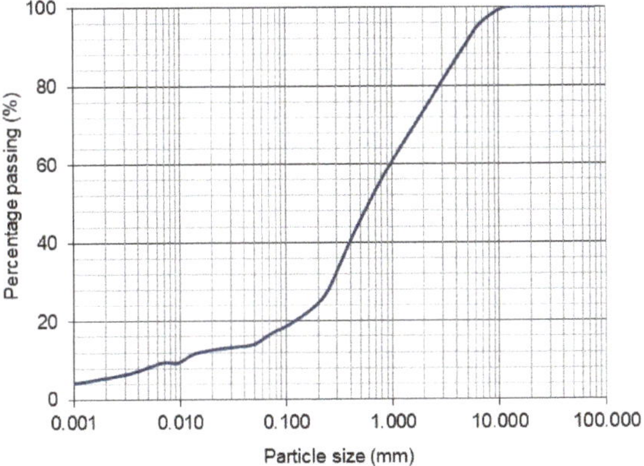

Figure 2. Particle size distribution curve of the C&D material.

The relevant geotechnical properties of this C&D material are summarised in Table 2. The optimum compaction parameters (i.e., maximum dry unit weight, $\gamma_{d,max}$ = 20.1 kN/m^3 and optimum moisture content, w_{opt} = 9%) were estimated using the modified Proctor test, according to [55]. The quality of fines was assessed using the methylene blue test, following the standard [56]. The methylene blue value, MB (expressed in grams of dye per kilogram of the 0–2 mm size fraction) was determined as 3.2 g/kg.

Table 2. Geotechnical properties of the C&D material.

Properties	Values
D_{10} [mm]	0.01
D_{50} [mm]	0.65
D_{60} [mm]	1.03
Fines fraction (No. 200 sieve) [%]	16.9
Minimum void ratio, e_{min}	0.434
Maximum void ratio, e_{max}	0.877
Particles density, G_s	2.58
Maximum dry unit weight, $\gamma_{d,max}$ [kN/m^3]	20.1
Optimum moisture content, w_{opt} [%]	9.0
Methylene blue value, MB [g/kg]	3.2
Peak friction angle, φ [°]	37.6
Cohesion, c [kPa]	16.3

The internal shear strength of the material was estimated by large-scale direct shear tests, using a prototype facility described in previous publications [28,29]. The C&D material was compacted inside the direct shear box (600 mm × 300 mm in plan) at the dry unit weight, γ_d = 16.1 kN/m^3 (corresponding to 80% of its maximum dry density) and at the optimum moisture content (w_{opt} = 9%) and then subjected to shearing under the normal stresses of 25, 50, 100 and 150 kPa. Based on the Mohr-Coulomb failure criterion, a peak friction angle (φ) of 37.6° and cohesion (c) of 16.3 kPa were obtained (Table 2).

The use of C&D materials in geotechnical applications may lead to environmental issues as far as groundwater contamination is concerned. When the rainfall percolates a solid material (in the case, solid waste or recycled aggregates) produces a leachate which, in case it contains hazardous substances, can contaminate groundwater. This issue is even more relevant when C&D recycled materials are placed in direct contact with the ground, as it is generally the case of geotechnical applications.

To evaluate the potential short-term release of contaminants, laboratory leaching tests were carried out on the recycled C&D material, as per the standard [57]. The obtained results are given in Table 3, which also shows the acceptance criteria of maximum leached concentration for inert landfill according to the European Council Decision 2003/33/EC [58]. As can be seen from the analysis of Table 3 only the value of sulphate, SO_4, is above the threshold value.

Table 3. Leaching test results and acceptance criteria according to Council Decision 2003/33/EC [58].

Parameter	Value [mg/kg Dry Matter]	Acceptance Criteria [mg/kg Dry Matter]
Arsenic, As	0.021	0.5
Lead, Pb	<0.01	0.5
Cadmium, Cd	<0.003	0.04
Chromium, Cr	0.012	0.5
Copper, Cu	0.10	2
Nickel, Ni	0.011	0.4
Mercury, Hg	<0.002	0.01
Zinc, Zn	<0.1	4
Barium, Ba	0.11	20
Molybdenum, Mo	0.018	0.5
Antimony, Sb	<0.01	0.06
Selenium, Se	<0.02	0.1
Chloride, Cl	300	800
Fluoride, F	6.1	10
Sulphate, SO_4	3200	1000
Phenol index	<0.05	1
Dissolved Organic Carbon, DOC	220	500
pH	8.2	-

The source of sulphates in C&D recycled materials is commonly attribute to the gypsum drywall [59], as well as, to concrete and mortar, natural aggregates and ceramic components [60]. Table 2 does not provide evidence of high gypsum content, but concrete and mortar, natural aggregates and ceramic components are the predominant components of the recycled material.

It should be noted however that, according to the above-mentioned Decision [58], if the waste material does not comply with the value for sulphate, it can still be considered as meeting the acceptance criteria if the leaching (estimated either by a batch leaching test or by a percolation test under conditions approaching local equilibrium) does not exceed 6000 mg/kg at a liquid to solid ratio of 10 L/kg (L/S = 10). Based on that it can be concluded that this recycled material meet the acceptance criteria set out by the European legislation for inert materials.

The Federal Highway Administration [43] recommends the use of fill materials with pH values above 3 and in the range of 3–9 for mechanically stabilized earth structures involving geosynthetic reinforcements manufactured from polyolefin (PP and HDPE) and polyester, respectively. The pH value of this C&D material (pH = 8.2) fulfils the FHWA specified requirements.

The content of water soluble sulphates was determined by spectrophotometry on the basis of Section 10 of the standard [61]. Firstly, the specimens of C&D material were sieved through the 4 mm sieve. The retained particles were crushed so as to pass the same sieve. Then, the specimens were mixed with hot water to extract water-soluble sulphate ions and barium chloride was added so that sulphate ions precipitate as barium sulphate. The content of water soluble sulphates obtained by weighting and expressed as a percentage of sulphate ions by mass of C&D material was about 0.14%.

2.3. Geogrids

Two commercially available geogrids were tested (Figure 3): a laid uniaxial geogrid consisting of extruded polyester (PET) bars with welded rigid junctions (GGR1) and an extruded uniaxial geogrid manufactured from high-density polyethylene, HDPE (GGR2). These geogrids in particular were selected as they are widely used in Portugal, they are produced with different polymers and manufacturing processes, they have nominal strength as close as possible and they are suitable for medium-high reinforced structures (i.e., neither with very low nor very high strength).

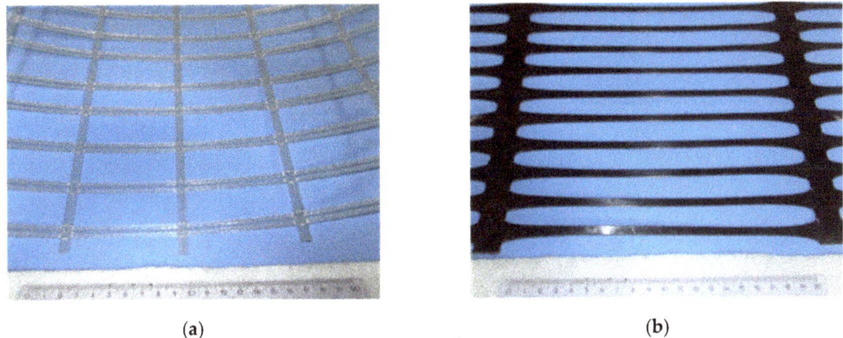

(a)   (b)

Figure 3. Photographic views of the geogrids (ruler in centimetres): (**a**) GGR1 (**b**) GGR2.

The tensile load-strain behaviour of the geogrids was assessed by in-isolation tensile tests carried out on a Universal Testing Machine (LR50K) following the standard [62]. The main physical and mechanical properties of the geogrids can be found in Table 4.

Table 4. Main physical and mechanical properties of the geogrids.

Property	GGR1	GGR2
Raw material	PET	HDPE
Mass per unit area (g/m^2)	380	450
Aperture dimensions (mm)	30×73	16×219
With of longitudinal members (mm)	10	6
With of transverse members (mm)	7	16
Thickness of longitudinal members (mm)	1.0	1.1
Thickness of transverse members (mm)	1.0	2.5 to 2.7
Mean value of the tensile strength [1] (kN/m)	80	68
Mean value of the tensile strength [2] (kN/m)	92.2	60.3
Elongation at maximum load [1] (%)	≤8	11 ± 3
Elongation at maximum load [2] (%)	5.6	10.1
Secant stiffness at 5% strain [2] (kN/m)	1640	718

[1] As per the manufacturer specifications (machine direction). [2] As per the laboratory tensile tests performed in this study (machine direction).

2.4. Pullout Test Apparatus and Test Procedures

Figure 4a presents an overall view of the large-scale pullout test apparatus used in this study. The pullout box consists of a modular structure with internal dimensions of 1.00 m wide, 1.53 m long and 0.80 m deep. A 0.20 m long sleeve is fixed to the front wall to minimise the frictional effects of this rigid boundary during testing (Figure 4b).

The C&D material was compacted inside the pullout box in several 0.15 m thick layers at the optimum moisture content (w_{opt} = 9%) and at the dry density of 16.1 kN/m^3 (i.e., 80% of its maximum dry density, based on the modified Proctor test), using an electric vibratory hammer. Once the first two layers were compacted, the geosynthetic specimen was clamped and laid over the C&D material, with the machine direction members oriented parallel to the pullout direction. For the geogrid GGR1 (Figure 4c), preliminary testing showed that the inextensible wires used to monitor the displacements throughout the length of the reinforcement had the potential to damage the geogrid ribs and lead to rupture of the junctions during testing. Hence, to avoid any reduction of the measured pullout resistance due to the presence of the inextensible wires, they were not used in these tests. In the tests involving the geogrid GGR2, a set of inextensible wires were attached to the specimen (Figure 4d) and connected to linear potentiometers placed at the back of the pullout box. Two additional layers of C&D material were then placed and compacted over the geosynthetic specimen, resulting in a total height of filling material of 0.60 m. To reduce the influence of the top boundary rigidity on the test results and attain more uniform distribution of the vertical load, a neoprene slab was placed on the top of the C&D material and beneath a wooden loading plate. The vertical load was applied by ten hydraulic jacks and monitored by a load cell.

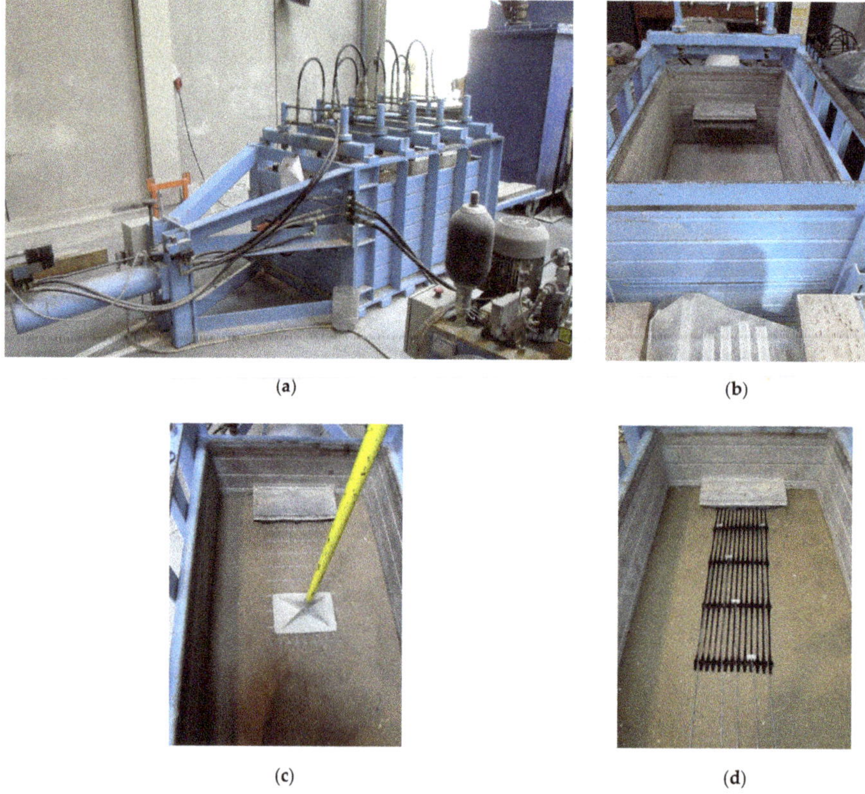

(a)

(b)

(c)

(d)

Figure 4. Large-scale pullout test apparatus at the University of Porto. (**a**) Overall view. (**b**) Detail of the pullout box and steel sleeve. (**c**) GGR1 specimen. (**d**) Inextensible wires fixed along a GGR2 specimen.

The pullout force was applied to the geogrid specimen through a hydraulic system under displacement-controlled conditions (i.e., by adopting a constant displacement rate throughout the test) and the associated front displacement was measured by a linear potentiometer. During the tests, relevant parameters, such as the imposed vertical pressure, pullout load, front displacement of the geogrid specimen (i.e., displacement of the clamped end) and displacements along the length of the reinforcement (for GGR2) were continuously monitored, using an automatic data acquisition system. A more comprehensive description of the pullout test apparatus can be found elsewhere [36,39].

2.5. Test Programme

A summary of the test conditions adopted in the current study is presented in Table 5. As mentioned before, two different geogrids were tested (GGR1 and GGR2). To investigate the possible effect of specimen size on the measured data, different specimen dimensions were considered for both reinforcements (200 mm × 600 mm versus 250 mm × 750 mm for GGR1, and 200 mm × 600 mm versus 300 mm × 900 mm for GGR2), while keeping the ratio of the specimen confined length to width equal to three, as recommended by the European Standard [50]. The role of the vertical confining pressure (σ_v) on the GGR1 pullout behaviour was evaluated by imposing vertical stresses at the interface level of 10, 25 and 50 kPa. According to the European Standard [50], geosynthetic pullout tests involving free draining soils, where excess pore water pressures are not anticipated should be conducted at a constant displacement rate of 2 mm/min (±10%). However, the American Standard [51] suggests that the pullout force be applied at a slower rate of 1 mm/min (±10%). In order to determine whether the displacement rate affects the pullout test results, different displacement rates (d_r) ranging from 1 to 4 mm/min were adopted in the tests involving the geogrid GGR2. To analyse the repeatability of results, the initial six tests (T1–T6) were carried out three times under identical conditions. Therefore, 22 pullout tests were performed in this study.

Table 5. Test programme.

Test	Geogrid	Specimen Size [mm]	Displacement Rate [mm/min]	Confining Pressure [kPa]	Number of Specimens
T1	GGR1	200 × 600	2	10	3
T2	GGR1	200 × 600	2	25	3
T3	GGR1	200 × 600	2	50	3
T4	GGR1	250 × 750	2	10	3
T5	GGR1	250 × 750	2	25	3
T6	GGR1	250 × 750	2	50	3
T7	GGR2	200 × 600	1	10	1
T8	GGR2	200 × 600	2	10	1
T9	GGR2	200 × 600	4	10	1
T10	GGR2	300 × 900	2	10	1

2.6. Processing of the Test Results

Based on the pullout load continuously monitored by the data acquisition system, the pullout force per unit width of the geogrid is evaluated and plotted against the front displacement of the geogrid specimen. The displacements over the geogrid length (only for GGR2 as mentioned in Section 2.4) recorded through linear potentiometers can be plotted for any value of the pullout force or front displacement. Graphs of all the relevant parameters (imposed vertical pressure, pullout load, front displacement of the geogrid specimen and displacements along the geogrid length) as a function of time can also be prepared.

In the design of geosynthetic-reinforced structures, such as retaining walls and slopes, one of the essential parameters is the geosynthetic-backfill material pullout interaction coefficient. The pullout interaction coefficient (f_b) can be estimated as the ratio of the maximum shear stress mobilised at the backfill-geosynthetic interface during the pullout test (τ_p) to the direct shear strength of the backfill material (τ_s), under the same confining pressure:

$$f_b = \frac{\tau_p}{\tau_s} \tag{1}$$

The maximum shear stress developed at the interface during the pullout test can be determined as:

$$\tau_p = \frac{P_R}{2\,L_R} \tag{2}$$

where P_R is the pullout resistance (i.e., maximum pullout force per unit width of reinforcement) and L_R is the confined length of the reinforcement at maximum pullout force.

The shear strength of the backfill material (τ_s) was evaluated through direct shear tests carried out on a large scale prototype. Details on these tests can be found in a previous publication [48].

3. Results and Discussion

In this section the results are presented and discussed, firstly, in terms of pullout force-displacement behaviour (Sections 3.1–3.3) and subsequently, regarding the values of the pullout interaction coefficient—Section 3.4 (see Figure 1).

3.1. Effect of Specimen Size

Figure 5 shows the effect of specimen geometry on the variation of the pullout force with the frontal displacement for geogrid GGR1. Here, the three graphs illustrate the representative curves corresponding to different vertical confining pressures imposed at the interface level: 10 kPa (Figure 5a), 25 kPa (Figure 5b) and 50 kPa (Figure 5c).

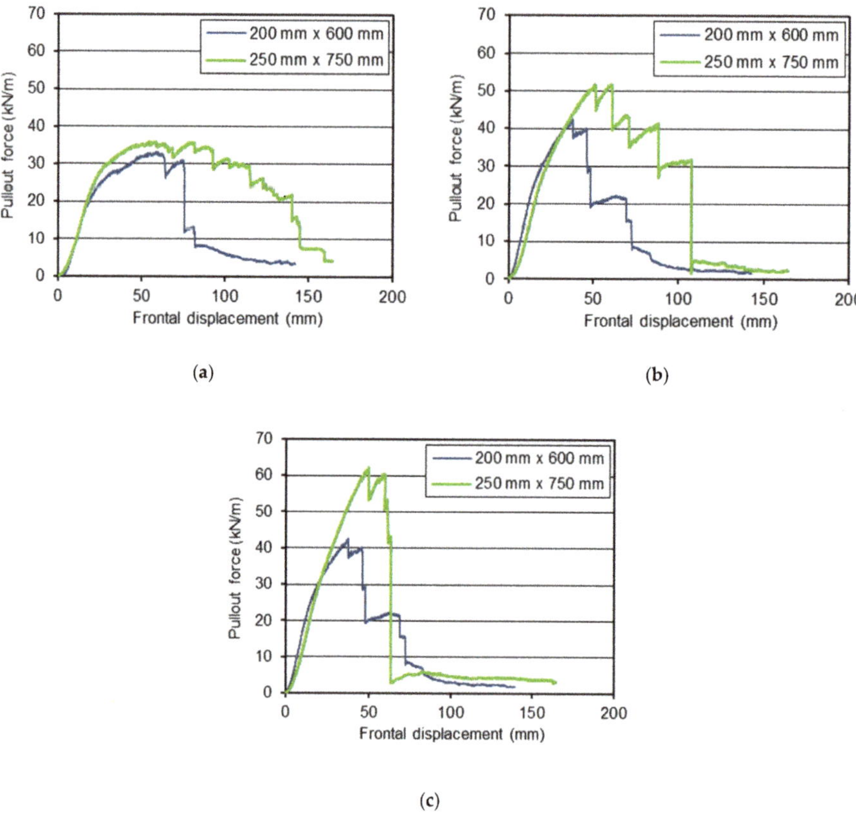

Figure 5. Effect of specimen size on the pullout behaviour of GGR1 (d_r = 2 mm/min): (**a**) σ_v = 10 kPa; (**b**) σ_v = 25 kPa; (**c**) σ_v = 50 kPa.

The plotted data show that, regardless of the confining pressure applied in the tests, the increase in specimen size from 200 mm × 600 mm to 250 mm × 750 mm led to an increment in the peak pullout resistance of the geogrid. Moreover, the influence of specimen dimensions on the peak pullout resistance appears to be more significant at higher confining pressures. Under the lowest vertical pressure (10 kPa), the peak pullout resistance of the geogrid increased about 10% (on average) with the increase in specimen size, and the frontal displacement at peak load remained nearly constant (Figure 5a). However, for the vertical pressures of 25 kPa and 50 kPa, the increase in pullout resistance (associated with the increase in specimen size) was more pronounced (21% and 17%, respectively) and higher frontal displacements were required for mobilisation of the peak load when larger specimens were employed (Figure 5b,c).

It is well known that the pullout resistance of the geogrid is developed primarily by the combination of the passive resistance mobilised against the transverse members and the skin friction at both sides of the reinforcement. Therefore, the increase in the peak pullout resistance of the geogrid achieved when the longer specimens were used is associated with an increase in the surface area available for mobilisation of skin friction along the reinforcement, as well as an increment in the passive resistance provided by the recycled C&D material due to the presence of two additional transverse members contributing to the mobilised forces (the shorter specimens had 7 transverse members, whereas the longer specimens had 9 transverse members). Since the increase in the confining pressure leads to the increase in the passive resistance of recycled C&D material mobilised against the geogrid transverse members during pullout movement, greater increments in the pullout resistance were attained when the longer specimens were subjected to higher confining pressures of 25 and 50 kPa (Figure 5b,c).

The influence of specimen size on the pullout response of GGR2 was evaluated through pullout tests carried out under a vertical pressure of 10 kPa. The results presented in Figure 6a indicate that, similar to the trend observed for GGR1, the increase in specimen size resulted in an increment in the peak pullout resistance of the geogrid. This is mainly related to the fact that the number of transverse members (i.e., bearing members) increased when the specimen length changed from 600 to 900 mm (from 3 to 4 bars). Due to the importance of the passive resistance developed against those bars to the overall pullout resistance of the reinforcement, significantly greater (17%) peak pullout resistance was achieved when the 900 mm long specimen was tested. However, the frontal displacement at peak pullout load remained nearly constant in both tests.

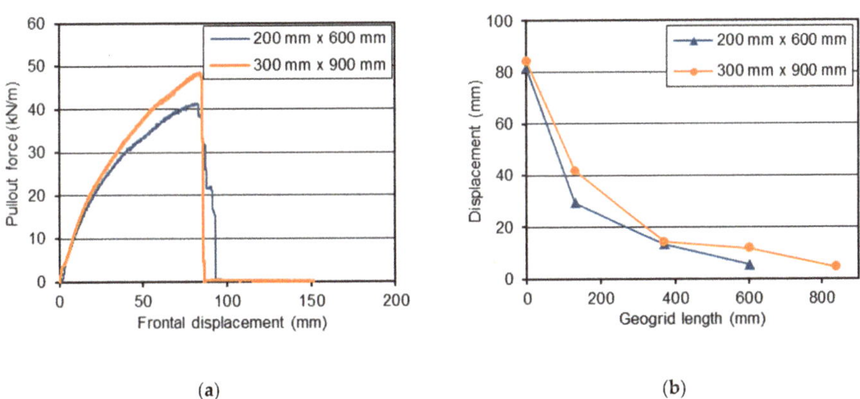

(a) (b)

Figure 6. Effect of specimen size on the pullout behaviour of GGR2 (σ_v = 10 kPa, d_r = 2 mm/min): (a) Pullout force plotted against frontal displacement; (b) Displacements over the geogrid length at maximum pullout force.

The displacements recorded by the potentiometers throughout the length of the specimens at maximum pullout force are shown in Figure 6b. It can be noted that the total displacements measured at the front and rear ends of the reinforcement were rather similar for both specimens. Regardless of the specimen dimensions, the full length of the reinforcement was mobilised during testing, thus contributing to the mobilised forces. For the shorter specimen, the deformations (i.e., elongations) decreased progressively along the reinforcement length. For the longer specimen, the deformations at the front half of the reinforcement (i.e., the two sections located closer to the clamp system) were significantly greater than those at the back half of the geogrid.

3.2. Effect of Displacement Rate

As mentioned previously, the influence of the displacement rate on the pullout test results involving the geogrid GGR2 was investigated by adopting clamp displacement rates of 1, 2 and 4 mm/min (tests T7 to T9). These tests were carried out under a vertical pressure of 10 kPa (at the interface level) and using 200 mm × 600 mm specimens (Table 5). The obtained pullout force—displacement curves and the profiles of the displacements recorded over the length of the geogrid at maximum pullout force are presented in Figure 7a,b, respectively. Regardless of the displacement rate adopted in the tests, all three specimens failed in tension. Nevertheless, the displacement rate affected the measured pullout resistance of the reinforcement (Figure 7a). In fact, for the conditions investigated, the peak pullout resistance increased progressively and up to 16.2% with increasing displacement rate. The variation of peak pullout resistance attributed to the increment in the displacement rate was more pronounced under lower displacement rates (i.e., 1–2 mm/min). Furthermore, the frontal displacement at peak increased considerably when the rate of displacement varied from 1 to 2 mm/min, but did not significantly change upon a further increase in the displacement rate (i.e., from 2 to 4 mm/min).

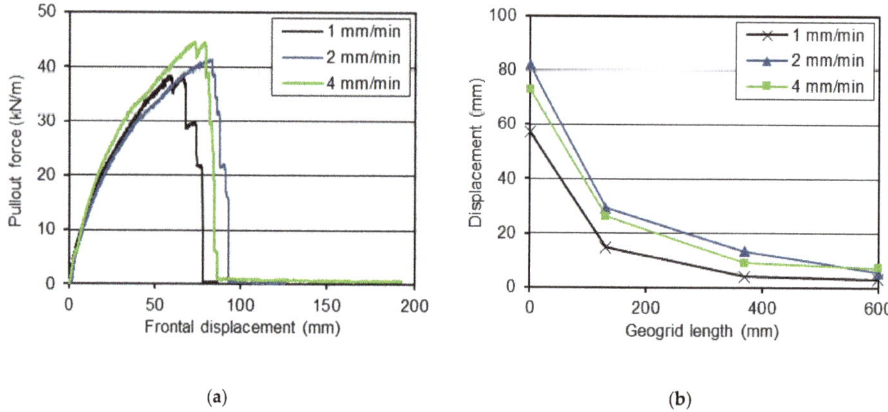

Figure 7. Effect of displacement rate on the pullout behaviour of GGR2 (σ_v = 10 kPa, specimen size = 200 mm × 600 mm): (**a**) Pullout force plotted against frontal displacement; (**b**) Displacements over the geogrid length at maximum pullout force.

The displacements recorded by the potentiometers along the reinforcement length at peak pullout force (Figure 7b) indicate that the deformations induced by pullout loading were more significant along the first confined section of the geogrid and reduced progressively with increasing distance to the point of application of the pullout load (front geogrid end). As expected, the higher the frontal displacement at peak load, the greater the displacements mobilised over the geogrid length.

Previous studies have shown that, because of the intrinsic viscous, time-dependent response of polymeric geosynthetic reinforcements, the peak strength is sensitive to rate of loading and generally decreases with a reduction in the strain rate at failure [36,63]. Lopes and Ladeira [36] evaluated the influence of the displacement rate on the pullout test results of a uniaxial HDPE geogrid embedded in a gravelly sand. The authors observed a progressive increment in the measured pullout resistance of the geogrid with increasing displacement rate from 1.8 to 22 mm/min. Over the range investigated, the geogrid pullout resistance increased by up to 30%, whereas the displacements along the reinforcement resulting from elongation tended to decrease. Hirakawa et al. [63] investigated the loading rate effects on the tensile load-strain behaviour of different geosynthetics and found out that the rupture strength increases linearly with the logarithm of the strain rate at rupture. The results obtained in this study are in good agreement with these previous findings, showing that the strain rate influences the peak load measured in pullout tests, as well as the deformations along the reinforcement at maximum load.

3.3. Effect of Confining Pressure

The effect of the vertical confining pressure on the pullout response of the geogrid GGR1 is shown in Figure 8. Figure 8a presents the pullout force—displacement curves from three representative specimens tested with initial dimensions of 200 mm × 600 mm and subjected to different vertical stresses (tests T1 to T3), whereas the results from tests carried out using 250 mm × 750 mm specimens (tests T4 to T6) are plotted in Figure 8b. These tests were conducted at a constant displacement rate of 2 mm/min (Table 5). As expected, the pullout resistance increased significantly with the confining pressure, regardless of the specimen dimensions. Indeed, when the confining pressure varied from 10 to 50 kPa, the pullout resistance of the reinforcement increased by 57% and 67% for 200 mm × 600 mm and 250 mm × 750 mm specimens, respectively. Moreover, the increment in pullout resistance associated with the confining pressure increase was more pronounced at lower confining pressures (i.e., 10 to 25 kPa).

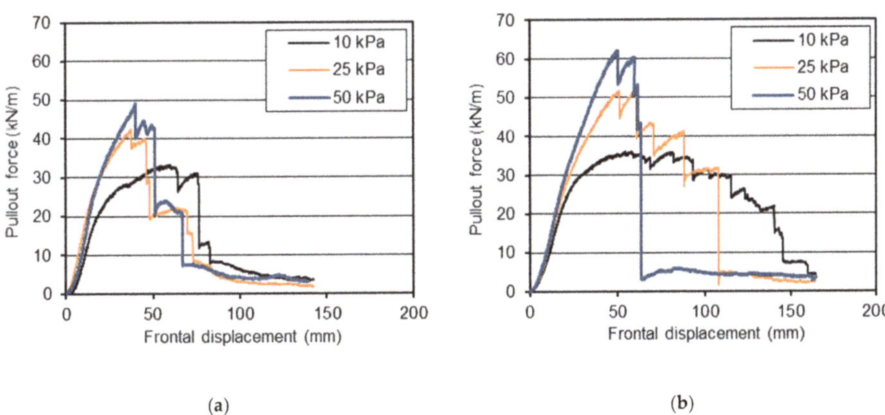

Figure 8. Effect of vertical confining pressure on the pullout behaviour of GGR1: (**a**) Specimen size of 200 mm × 600 mm; (**b**) Specimen size of 250 mm × 750 mm.

Despite the fact that the increase in the confining pressure from 10 to 50 kPa may have reduced the tendency of the recycled C&D material to dilate, the increase in the passive resistance of the C&D material mobilised against the geogrid transverse bars (resulting from the confining pressure increase) led to substantially greater pullout resistance of the reinforcement.

3.4. Discussion

The pullout interaction coefficients for the interfaces studied herein were obtained based on Equations (1) and (2), using the results from pullout tests and those from large-scale direct shear tests on the recycled C&D material.

Table 6 summarises the pullout test results for each individual test specimen, including the pullout resistance (P_R) and corresponding frontal displacement (d_{PR}), the failure mode (i.e., pullout or tensile failure) observed in the test, the pullout interaction coefficient (f_b), as well as the values of τ_p and τ_s used in its calculation. Also shown in this table are the average values of P_R, d_{PR} and f_b from the repeatability tests. Values in brackets in the test designation refers to the sample number. The analysis of the results summarised in Table 6 should be simultaneously performed with Table 5.

Table 6. Summary of results.

Test	P_R [kN/m]	P_R—Average [kN/m]	d_{PR} [mm]	d_{PR}—Average [mm]	Failure Mode	τ_p [kPa]	τ_s [kPa]	f_b	f_b Average
T1 (1)	30.84		56.78		Pullout	28.39	24.01	1.18	
T1 (2)	33.23	33.27	59.81	55.75	Pullout	30.76	24.01	1.28	1.27
T1 (3)	35.73		50.67		Pullout	32.52	24.01	1.35	
T2 (1)	41.71		43.18		Pullout	37.45	35.92	1.04	
T2 (2)	42.48	43.32	37.31	47.54	Pullout	37.74	35.92	1.05	1.09
T2 (3)	45.76		62.14		Pullout	42.54	35.92	1.18	
T3 (1)	49.23		40.05		Pullout	43.95	52.21	0.84	
T3 (2)	46.45	52.12	40.15	42.76	Tensile	41.48	52.21	0.79	0.89
T3 (3)	60.69		48.07		Tensile	54.98	52.21	1.05	
T4 (1)	40.83		54.52		Pullout	29.35	24.01	1.22	
T4 (2)	33.19	36.67	49.45	54.05	Pullout	23.69	24.01	0.99	1.10
T4 (3)	36.00		58.19		Pullout	26.02	24.01	1.08	
T5 (1)	51.97		60.51		Pullout	37.69	35.92	1.05	
T5 (2)	54.62	52.42	61.46	57.62	Pullout	39.66	35.92	1.10	1.05
T5 (3)	50.67		50.88		Pullout	36.24	35.92	1.01	
T6 (1)	64.41		61.77		Tensile	46.79	52.21	0.90	
T6 (2)	56.91	61.17	58.81	56.87	Tensile	41.17	52.21	0.79	0.85
T6 (3)	62.20		50.04		Tensile	44.43	52.21	0.85	
T7	38.38	-	57.26	-	Tensile	32.15	24.01	1.34	-
T8	41.37	-	81.94	-	Tensile	34.80	24.01	1.45	-
T9	44.58	-	73.00	-	Tensile	37.60	24.01	1.57	-
T10	48.56	-	84.46	-	Tensile	27.12	24.01	1.13	-

It can be observed that the interaction coefficients for the interface involving the laid and welded PET geogrid (GGR1) ranged from 0.79 to 1.35, whereas the values varied from 1.13 to 1.57 for the interface involving the HDPE geogrid (GGR2).

On average the values of f_b decreased with increasing confining pressure, regardless of specimen dimensions, which is consistent with the findings of the study by Mohiuddin [64] involving cohesive soil-geosynthetic interfaces.

The increase in the specimen size resulted in an increment in the pullout resistance, P_R, for both geogrids (Table 6). However, it should be noted that the confinement length of the geogrid at maximum pullout force, L_R, depends on the geogrid length, L, and frontal displacement, d_{PR}, ($L_R = L\text{-}d_{PR}$) being higher when the samples are longer. This leads to maximum shear stress (τ_p) variations and, on average, to its decrease as the sample size increases. According to Equation (1), the decrease of the maximum shear stress mobilised at the interface (τ_p) induces the reduction of the pullout coefficient, f_b. Thus, on average the values of f_b decreased when the geogrid specimen size increased.

Comparing the results of tests T7, T8 and T9 carried out on GGR2 samples with dimensions of 200 mm × 600 mm, one can conclude that the pullout resistance (P_R) increased with increasing displacement rate. Similar trend was observed for the pullout interaction coefficient (f_b). These results follow the know trend for soil-geogrid interfaces [36,63].

The values of f_b obtained in this study are generally greater than or equal to the values usually reported in the literature for soil-geosynthetic and recycled material-geosynthetic interfaces. Vieira et al. [20] reported values in the range 0.58–0.63 for interfaces between two geogrids and a fine-grain C&D recycled material under normal stress of approximately 31 kPa. Soleimanbeigi et al. [49] found that the interaction coefficient for reinforced recycled concrete aggregate decreases with increasing normal stress and is lower than 0.5.

Goodhue et al. [65] obtained f_b values ranging between 0.25 and 1.4 for different soil-geosynthetic interfaces involving a uniformly-graded quartz sand. Mohiuddin [64] reported values in the range of 0.44–1.04 for a variety of geosynthetics tested in a cohesive soil. Tang et al. [66] investigated the interaction between different geogrids and dense-graded crushed stone and obtained pullout interaction coefficients in the range of 0.62–1.00. Hsieh et al. [67] reported values of f_b ranging from 0.18 and 1.25 derived from pullout tests of geosynthetics embedded in granular soils.

It is worth noting that most of the f_b values attained in this study exceed the values typically assumed in the design of geosynthetic-reinforced soil structures in the absence of test data.

4. Conclusions

An experimental study was carried out to assess the effects of different parameters with influence on the pullout resistance of geogrids (namely, the geogrid specimen size, the displacement rate and the normal stress) when they are embedded in a C&D recycled material. Previous studies on the pullout behaviour of interfaces between geosynthetics and C&D recycled materials have been performed following the guidance for common backfill materials so, it is important to evaluate whether some assumptions are still valid for alternative backfills. The most relevant findings of the study are listed below.

Regardless of the vertical confining pressure imposed in the tests (10, 25 or 50 kPa), the peak pullout resistance of the geogrids increased with the specimen size. This increment was more pronounced at higher confining pressures (i.e., 25 and 50 kPa).

The displacement rate adopted in the pullout tests affected the measured pullout resistance of the HDPE geogrid (GGR2). This is related to the intrinsic viscous response of polymeric geosynthetics under tensile loads (particularly for geosynthetics manufactured from HDPE). For the conditions investigated in this study, the peak pullout resistance of GGR2 increased by up to 16% with increasing displacement rate (from 1 mm/min to 4 mm/min).

The pullout resistance of the geogrid is positively correlated to the confining pressure acting at the interface level. When the confining pressure varied from 10 to 50 kPa, the peak pullout resistance of the geogrid GGR1 increased by 57% (for 200 mm × 600 mm specimens) and 67% (for 250 mm × 750 mm specimens), which is associated with the increase in the passive resistance of the recycled C&D material developed against the geogrid transverse members.

The values of pullout interaction coefficient, f_b, obtained in this study are generally greater than the values usually reported in the literature for soil-geosynthetic and recycled material-geosynthetic interfaces. The pullout interaction coefficients, f_b, for the interface involving the PET geogrid (GGR1) ranged from 0.79 to 1.35. For the HDPE geogrid (GGR2), the values of f_b were generally higher, ranging from 1.13 to 1.57.

Although the pullout resistance of the geogrids has increased with the specimen size, the pullout interaction coefficient, f_b, showed in general the opposite trend.

The increase in the displacement rate led to both an increase in the geogrids pullout resistance and an increase in the pullout interaction coefficient, f_b.

Albeit the pullout resistance of the geogrids is positively correlated to the confining pressure, on average and regardless of specimen size the pullout interaction coefficient, f_b, decreased with this parameter.

This study has confirmed the expected trends regarding the pullout resistance: higher pullout resistance achieved in laboratory pullout tests for higher confining pressure, displacement rate and specimen size. However, the conclusions concerning pullout interaction coefficient, f_b, are very interesting and should be properly considered. Even keeping the ratio of the specimen confined length to width equal to three, as recommended by the European Standard (ASTM standard recommends two), the pullout results are dependent upon geogrid length. It is important to point out that higher pullout resistance (expected in longer geogrids) does not imply higher pullout interaction coefficient. The increase in the confining pressure also led to the decrease of f_b.

In the design of geosynthetic-reinforced structures one of the essential parameters is the pullout interaction coefficient, f_b. The f_b values attained in this study exceed the values typically assumed in the design in the absence of test data (0.5–0.7), which allows to follow the usual practices when the conventional backfill materials (soils) are used.

It should be noted that while the results of this study support the feasibility of using C&D recycled materials in the construction of geogrid-reinforced structures, with obvious benefits in terms of environmental protection and sustainability, the conclusions are limited to the utilized materials and procedures followed.

Author Contributions: C.S.V.: Conceptualization; Funding acquisition; Methodology; Project administration; Supervision; Writing—review & editing. P.P.: Methodology; Data collection; Data curation. F.F.: Data curation; Formal analysis; Writing—original draft preparation. M.d.L.L.: Funding acquisition; Methodology. All authors have read and agreed to the published version of the manuscript.

Funding: This work was financially supported by: Project PTDC/ECI-EGC/30452/2017—POCI-01-0145-FEDER-0304 52—funded by FEDER funds through COMPETE2020—Programa Operacional Competitividade e Internacionalização (POCI) and by national funds (PIDDAC) through FCT/MCTES.

Acknowledgments: The authors would like to acknowledge the contribution of the company RCD for the supply of the C&D materials and Naue and Tensar International for providing geogrid samples.

References

1. European Commission, Competitiveness of the Construction Industry. *A Report drawn up by the Working Group for Sustainable Construction with Participants from the European Commission, Member States and Industry*; European Commission: Brussels, Belgium, 2001.

2. EC DGE. *European Commission Directorate-General for Environment, Service Contract on Management of Construction and Demolition Waste SR1 Final Report. Task 2-Study*; EU publications: Brussels, Belgium, 2011.

3. European Commission. EU Construction & Demolition Waste Management Protocol. Available online: https://ec.europa.eu/growth/content/eu-construction-and-demolition-waste-protocol-0_en (accessed on 4 April 2020).

4. Basu, D.; Misra, A.; Puppala, A.J. Sustainability and geotechnical engineering: Perspectives and review. *Can. Geotech. J.* **2014**, *52*, 96–113. [CrossRef]

5. Vieira, C.S.; Pereira, P.M. Use of recycled construction and demolition materials in geotechnical applications: A review. *Resour. Conserv. Recycl.* **2015**, *103*, 192–204. [CrossRef]

6. Gomes Correia, A.; Winter, M.G.; Puppala, A.J. A review of sustainable approaches in transport infrastructure geotechnics. *Transp. Geotech.* **2016**, *7*, 21–28. [CrossRef]

7. Dhir, R.K.; Brito, J.; Silva, R.V.; Lye, C.Q. *Sustainable Construction Materials*; Woodhead Publishing: Cambridge, UK, 2019; p. 652.

8. Henzinger, C.; Heyer, D. Soil improvement using recycled aggregates from demolition waste. *Proc. Inst. Civ. Eng: Ground Improv.* **2018**, *171171*, 74–81. [CrossRef]

9. Kianimehr, M.; Shourijeh, P.T.; Binesh, S.M.; Mohammadinia, A.; Arulrajah, A. Utilization of recycled concrete aggregates for light-stabilization of clay soils. *Constr. Build. Mater.* **2019**, *227*, 116792. [CrossRef]

10. Rahman, M.A.; Imteaz, M.; Arulrajah, A.; Disfani, M.M. Suitability of recycled construction and demolition aggregates as alternative pipe backfilling materials. *J. Clean. Prod.* **2014**, *66*, 75–84. [CrossRef]

11. Vieira, C.S.; Lopes, M.L.; Cristelo, N. Geotechnical Characterization of Recycled C&D Wastes for Use as Trenches Backfilling. In *Proceedings of the International Conference WASTES: Solutions, Treatments and Opportunities*; CRC Press: Boca Raton, FL, USA, 2018; pp. 175–182.

12. Del Rey, I.; Ayuso, J.; Barbudo, A.; Galvín, A.P.; Agrela, F.; Brito, J. Feasibility study of cement-treated 0–8 mm recycled aggregates from construction and demolition waste as road base layer. *Road Mater. Pavement Des.* **2016**, *17*, 678–692. [CrossRef]

13. Tavira, J.; Jiménez, J.R.; Enrique, F.; Ledesma, E.F.; López-Uceda, A.; Ayuso, J. Real-scale study of a heavy traffic road built with in situ recycled demolition waste. *J. Clean. Prod.* **2020**, *248*, 119219. [CrossRef]

14. Yaghoubi, E.; Arulrajah, A.; Choy, W.Y.-C.; Horpibulsuk, S. Stiffness properties of recycled concrete aggregate with polyethylene plastic granules in unbound pavement applications. *J. Mater. Civ. Eng.* **2017**, *29*, 04016271. [CrossRef]

15. Pereira, P.M.; Ferreira, F.B.; Vieira, C.S.; Lopes, M.L. Use of Recycled C&D Wastes in Unpaved Rural and Forest Roads—Feasibility Analysis. In *Proceedings of the International Conference WASTES: Solutions, Treatments and Opportunities*; CRC Press: Boca Raton, FL, USA, 2019; pp. 161–167.

16. Freire, A.C.; Neves, J.M.C.; Roque, A.J.; Martins, I.M.; Antunes, M.L. Feasibility study of milled and crushed reclaimed asphalt pavement for application in unbound granular layers. *Road Mater. Pavement Des.* **2019**, 1–21. [CrossRef]

17. Plati, C.; Cliatt, B. A Sustainability Perspective for Unbound Reclaimed Asphalt Pavement (RAP) as a Pavement Base Material. *Sustainability* **2019**, *11*, 78. [CrossRef]

18. Zhang, J.; Ding, L.; Li, F.; Peng, J. Recycled aggregates from construction and demolition wastes as alternative filling materials for highway subgrades in China. *J. Clean. Prod.* **2020**, *255*, 120223. [CrossRef]

19. Santos, E.C.; Palmeira, E.M.; Bathurst, R.J. Performance of two geosynthetic reinforced walls with recycled construction waste backfill and constructed on collapsible ground. *Geosynth. Int.* **2014**, *21*, 256–269. [CrossRef]

20. Vieira, C.S.; Pereira, P.M.; Lopes, M.L. Recycled construction and demolition wastes as filling material for geosynthetic reinforced structures. Interface properties. *J. Clean. Prod.* **2016**, *124*, 299–311. [CrossRef]

21. Vieira, C.S.; Pereira, P.M. Use of mixed construction and demolition recycled materials in geosynthetic reinforced embankments. *Indian Geotech. J.* **2018**, *48*, 279–292. [CrossRef]

22. Vieira, C.S.; Ferreira, F.B.; Pereira, P.M.; Lopes, M.L. Pullout behaviour of geosynthetics in a recycled construction and demolition material – Effects of cyclic loading. *Transp. Geotech.* **2020**, *23*, 100346. [CrossRef]

23. Poon, C.S.; Chan, D. Feasible use of recycled concrete aggregates and crushed clay brick as unbound road sub-base. *Constr. Build. Mater.* **2006**, *20*, 578–585. [CrossRef]

24. Jiménez, J.R.; Ayuso, J.; Galvín, A.P.; López, M.; Agrela, F. Use of mixed recycled aggregates with a low embodied energy from non-selected CDW in unpaved rural roads. *Constr. Build. Mater.* **2012**, *34*, 34–43. [CrossRef]

25. Rodrigues, F.; Carvalho, M.T.; Evangelista, L.; Brito, J. Physical-chemical and mineralogical characterization of fine aggregates from construction and demolition waste recycling plants. *J. Clean. Prod.* **2013**, *52*, 438–445. [CrossRef]

26. Palmeira, E.M. Soil-geosynthetic interaction: Modelling and analysis. *Geotext. Geomembr.* **2009**, *27*, 368–390. [CrossRef]

27. Lopes, M.L. Soil-Geosynthetic Interaction. In *Handbook of Geosynthetic Engineering*; Shukla, S.K., Ed.; ICE Publishing: London, UK, 2012; pp. 45–66.

28. Vieira, C.S.; Lopes, M.L.; Caldeira, L.M. Sand-geotextile interface characterisation through monotonic and cyclic direct shear tests. *Geosynth. Int.* **2013**, *20*, 26–38. [CrossRef]

29. Ferreira, F.B.; Vieira, C.S.; Lopes, M.L. Direct shear behaviour of residual soil–geosynthetic interfaces—Influence of soil moisture content, soil density and geosynthetic type. *Geosynth. Int.* **2015**, *22*, 257–272. [CrossRef]

30. Khoury, C.N.; Miller, G.A.; Hatami, K. Unsaturated soil-geotextile interface behavior. *Geotext. Geomembr.* **2011**, *29*, 17–28. [CrossRef]

31. Liu, C.-N.; Zornberg, J.G.; Chen, T.-C.; Ho, Y.-H.; Lin, B.-H. Behavior of geogrid-sand interface in direct shear mode. *J. Geotech. Geoenvironmental Eng.* **2009**, *135*, 1863–1871. [CrossRef]

32. Lopes, M.L.; Ferreira, F.B.; Carneiro, J.R.; Vieira, C.S. Soil-geosynthetic inclined plane shear behavior: Influence of soil moisture content and geosynthetic type. *Int. J. Geotech. Eng.* **2014**, *8*, 335–342. [CrossRef]

33. Ferreira, F.B.; Vieira, C.S.; Lopes, M.L. Soil-Geosynthetic Interface Strength Properties from Inclined Plane and Direct Shear Tests—A Comparative Analysis. In *Proceedings of GA 2016-6th Asian Regional Conference on Geosynthetics: Geosynthetics for Infrastructure Development*; Indian Chapter of International Geosynthetics Society: New Delhi, India, 2016.

34. Pitanga, H.N.; Gourc, J.P.; Vilar, O.M. Enhanced measurement of geosynthetic interface shear strength using a modified inclined plane device. *Geotech. Test. J.* **2011**, *34*, 643–652.

35. Briançon, L.; Girard, H.; Gourc, J.P. A new procedure for measuring geosynthetic friction with an inclined plane. *Geotext. Geomembr.* **2011**, *29*, 472–482. [CrossRef]

36. Lopes, M.L.; Ladeira, M. Influence of the confinement, soil density and displacement rate on soil-geogrid interaction. *Geotext. Geomembr.* **1996**, *14*, 543–554. [CrossRef]

37. Raju, D.M.; Fannin, R.J. Load-strain-displacement response of geosynthetics in monotonic and cyclic pullout. *Can. Geotech. J.* **1998**, *35*, 183–193. [CrossRef]

38. Moraci, N.; Recalcati, P. Factors affecting the pullout behaviour of extruded geogrids embedded in a compacted granular soil. *Geotext. Geomembr.* **2006**, *24*, 220–242. [CrossRef]

39. Ferreira, F.B.; Vieira, C.S.; Lopes, M.L.; Carlos, D.M. Experimental investigation on the pullout behaviour of geosynthetics embedded in a granite residual soil. *Eur. J. Environ. Civ. Eng.* **2016**, *20*, 1147–1180. [CrossRef]

40. Ferreira, F.; Vieira, C.; Lopes, M. Pullout Behavior of Different Geosynthetics—Influence of Soil Density and Moisture Content. *Front. Built Environ.* **2020**, *6*, 12. [CrossRef]

41. Mendes, M.J.; Palmeira, E.M.; Matheus, E. Some factors affecting the in-soil load-strain behaviour of virgin and damaged nonwoven geotextiles. *Geosynth. Int.* **2007**, *14*, 39–50. [CrossRef]

42. BS 8006. Code of Practice for Strengthened/Reinforced Soils and other Fills. In *British Standard Institution*; British Standards Institution (BSI): London, UK, 2010; p. 260.

43. FHWA. Design and Construction of Mechanically Stabilized Earth Walls and Reinforced Soil Slopes. In *FHWA-NHI-10-024*; Berg, R.R., Christopher, B.R., Samtani, N.C., Eds.; National Highway Institute: Washington, DC, USA, 2010.

44. NCMA. Design Manual for Segmental Retaining Walls. In *National Concrete Masonry Association*, 3rd ed.; National Concrete Masonry Association: Herndon, VA, USA, 2010; p. 206.

45. AASHTO. LRFD Bridge. Design Specifications. In *American Association of State Highway and Transportation Officials*, 8th ed.; American Association of State Highway and Transportation Officials: Washington, DC, USA, 2017.

46. Vieira, C.S. Valorization of Fine-Grain Construction and Demolition (C&D) Waste in Geosynthetic Reinforced Structures. *Waste Biomass Valorization* **2020**, *11*, 1615–1626.

47. Arulrajah, A.; Rahman, M.A.; Piratheepan, J.; Bo, M.W.; Imteaz, M.A. Evaluation of Interface Shear Strength Properties of Geogrid-Reinforced Construction and Demolition Materials using a Modified Large Scale Direct Shear Testing Apparatus. *J. Mater. Civ. Eng.* **2014**, *26*, 974–982. [CrossRef]

48. Vieira, C.S.; Pereira, P.M. Interface shear properties of geosynthetics and construction and demolition waste from large-scale direct shear tests. *Geosynth. Int.* **2016**, *23*, 62–70. [CrossRef]

49. Soleimanbeigi, A.; Tanyu, B.F.; Aydilek, A.H.; Florio, P.; Abbaspour, A.; Dayioglu, A.Y.; Likos, W.J. Evaluation of recycled concrete aggregate backfill for geosynthetic-reinforced MSE walls. *Geosynth. Int.* **2019**, *26*, 396–412. [CrossRef]

50. *EN 13738: 2004, Geotextiles and Geotextile-Related Products-Determination of Pullout Resistance in Soil*; European Committee for Standardization: Brussels, Belgium, 2004.

51. *ASTM D6706-01: 2013, Standard Test. Method for Measuring Geosynthetic Pullout Resistance in Soil*; ASTM International: West Conshohocken, PA, USA, 2013.

52. *EN 933-11:2009, Tests for Geometrical Properties of Aggregates—Part. 11: Classification Test for the Constituents of Coarse Recycled Aggregate*; CEN: Brussels, Belgium, 2009; p. 16.

53. *EN 933-1: 2012, Tests for Geometrical Properties of Aggregates—Part. 1: Determination of Particle Size Distribution–Sieving Method*; CEN: Brussels, Belgium, 2012.

54. *CEN ISO/TS 17892-4: 2004, Geotechnical Investigation and Testing-Laboratory Testing of Soil—Part. 4: Determination of Particle Size Distribution*; CEN: Brussels, Belgium, 2004.

55. EN 13286-2: 2002, Unbound and hydraulically bound mixtures—Part. 2: Test. methods for laboratory reference density and water content-Proctor compaction. *Ger. Version EN* **2004**, *34*.

56. EN 933-9: 2009, Tests for geometrical properties of aggregates. Assessment of fines. Methylene blue test. *Assess. Fines. Methylene Blue Test* **2009**.

57. EN 12457-4: 2002, Characterisation of waste–Leaching–Compliance test for leaching of granular waste material and sludges. Part 4: One stage batch test at a liquid to solid ratio of 10 l/kg for materials with particle size below 10 mm (without or with size reduction). *Part* **2002**, *2*, 30.

58. EC. Council Decision 2003/33/EC establishing criteria and procedures for the acceptance of waste at landfills pursuant to Article 16 of and Annex II to Directive 1999/31/EC. *Off. J. Eur. Union* **2003**, *11*, 27–39.

59. Jang, Y.-C.; Townsend, T. Sulfate leaching from recovered construction and demolition debris fines. *Adv. Environ. Res.* **2001**, *5*, 203–217. [CrossRef]

60. Barbudo, A.; Galvín, A.P.; Agrela, F.; Ayuso, J.; Jiménez, J.R. Correlation analysis between sulphate content and leaching of sulphates in recycled aggregates from construction and demolition wastes. *Waste Manag.* **2012**, *32*, 1229–1235. [CrossRef]

61. *EN 1744-1: Tests for Chemical Properties of Aggregates—Part. 1: Chemical Analysis*; CEN: Brussels, Belgium, 2009.

62. *EN ISO 10319: 2008, Wide-Width Tensile Tests*; European Committee for Standardization: Brussels, Belgium, 2008.

63. Hirakawa, D.; Kongkitkul, W.; Tatsuoka, F.; Uchimura, T. Time-dependent stress-strain behaviour due to viscous properties of geogrid reinforcement. *Geosynth. Int.* **2003**, *10*, 176–199. [CrossRef]

64. Mohiuddin, A. Analysis of Laboratory and Field Pull-Out Tests of Geosynthetics in Clayey Soils. Master's Thesis, Faculty of the Louisiana State University and Agricultural and Mechanical College, Baton Rouge, LA, USA, 2003.

65. Goodhue, M.J.; Edil, T.B.; Benson, C.H. Interaction of foundry sands with geosynthetics. *J. Geotech. Geoenvironmental Eng.* **2001**, *127*, 353–362. [CrossRef]

66. Tang, X.; Chehab, G.R.; Palomino, A. Evaluation of geogrids for stabilising weak pavement subgrade. *Int. J. Pavement Eng.* **2008**, *9*, 413–429. [CrossRef]

67. Hsieh, C.W.; Chen, G.H.; Wu, J.H. The shear behavior obtained from the direct shear and pullout tests for different poor graded soil-geosynthetic systems. *J. Geoengin.* **2011**, *6*, 15–26.

sustainability

Article

Evaluation of Changes in the Permeability Characteristics of a Geotextile–Polynorbornene Liner for the Prevention of Pollutant Diffusion in Oil-Contaminated Soils

Jeongjun Park

Incheon Disaster Prevention Research Center, Incheon National University, Incheon 22012, Korea; jjpark72@inu.ac.kr

Abstract: In this study, changes in the permeability characteristics of a geotextile–polynorbornene liner at different oil pollutant contact times were evaluated. Experiments and numerical analyses were performed, and ASTM D5887 and ASTM D6766 were applied as test methods. The test results show that, when the pollutant contact time and pressure head were 4 h and 75 kPa, the reaction between the geotextile–polynorbornene liner and the pollutant was almost complete. Moreover, a numerical analysis was used to measure the ratio of the concentration of the pollutant that permeated through the geotextile–polynorbornene liner to the initial pollutant concentration at different pollutant contact times. The ratio was between 70 and 83% after a pollutant contact time of 0.5 h and between 0.1 and 1.0% after 4 h. The test and numerical analysis results confirm that, as a reactive medium, the geotextile–polynorbornene liner can effectively prevent the diffusion of oil pollutants by changing its permeability characteristics.

Keywords: oil-contaminated soils; geotextile–polynorbornene liner; pollutant adsorption; diffusion; permeability alteration

Citation: Park, J. Evaluation of Changes in the Permeability Characteristics of a Geotextile–Polynorbornene Liner for the Prevention of Pollutant Diffusion in Oil-Contaminated Soils. *Sustainability* 2021, *13*, 4797. https://doi.org/10.3390/su13094797

Academic Editors: Castorina Silva Vieira and Chunjiang An

Received: 3 April 2021
Accepted: 23 April 2021
Published: 24 April 2021

Publisher's Note: MDPI stays neutral with regard to jurisdictional claims in published maps and institutional affiliations.

1. Introduction

The rapid growth of the urban population has resulted in higher population densities, enhanced urbanization and industrialization, and increased anthropogenic inputs in the environment, which may threaten sustainable development and worsen several environmental problems, including groundwater and soil pollution [1–3]. Specifically, as a result of increased industrialization in South Korea, higher concentrations of chemicals and increased waste generation from industries have become national concerns because of their negative effects on the soil matrix. Thus, to protect the population from exposure to soil contaminants, several countries have established soil quality standards (SQS) and environmental impact assessments (EIA) for evaluating and monitoring SOC development projects. In particular, some of the most disastrous effects on the soil matrix are the result of oil and chemical pollution, as these pollutants have short- and long-term consequences on the ecosystem and soil makeup. The significant growth in oil and chemical consumption has resulted in increased concentrations of these pollutants in the soil due to leaks and spills from oil storage tanks in gas stations and chemical storage facilities, as well as pipeline ruptures, well blowouts, anthropogenic inputs, and transport accidents. Specifically, total petroleum hydrocarbon (TPH) is frequently released to the environment through accidents in commercial and private facilities and from storage facilities in military bases and industrial complexes. Ławniczak et al. [4] reported that crude oil-based hydrocarbons constitute the largest class of environmental pollutants in the world. With damage at such large scales, many remediation methods, treatment plans, and control strategies are costly and difficult to implement.

In the past years, various remediation techniques have been used to restore contaminated soils using eco-friendly approaches at a relatively low cost. These methods are

309

divided into ex situ (presence of excavation) and in situ (absence of excavation) treatments, depending on the characteristics of the location, the nature of the pollutants, the degree of pollution, and the types and characteristics of the pollutants in contaminated soils. Ex situ treatment is a remediation strategy that involves the physical removal of certain sites of contamination to another area, preferably within the same location. On the other hand, in situ treatment methods remediate contaminated soils at the original location without excavation [5,6]. These remediation technologies restore contaminated soils. However, it takes considerable time to identify contaminated soil after a polluting event. By the time it is identified, extensive damage has already occurred due to the diffusion of pollutants. Therefore, rather than applying remediation techniques after contamination, the application of proactive treatments at sites where the leakage of pollutants can be reasonably anticipated (gas stations, oil storage facilities, and industrial complexes) can significantly prevent the diffusion of pollutants and reduce the scale of damage. Therefore, researching technology that can prevent the diffusion of pollutants and restore contaminated areas is essential.

Many studies have been conducted on the remediation of oil-contaminated soils. Jeong et al. [7] artificially contaminated soils with different amounts of oil and used TPH analysis to evaluate the effects on the soil composition. On the other hand, Lee et al. [8] applied land farming and high-temperature thermal desorption as a remediation method for petroleum hydrocarbon-contaminated soils and evaluated the pollutant removal efficiency. In addition, Cho et al. [9] researched a mechanism of pollutant removal from TPH-contaminated soils using microwave heating. Sayed et al. [10] reviewed several previous studies on the application of bioremediation to environments contaminated with crude oil, TPH, and related petroleum products. Han [11] evaluated the removal efficiency of a biopile when it was used to restore soils that had been contaminated with low-concentration TPH for 100 days.

The main methods used to remediate oil-contaminated soils are chemical oxidation, which oxidizes pollutants into water and carbon dioxide using an oxidizing agent, and soil washing, which removes pollutants through contact between an aqueous solution containing a cleaning agent and contaminated soils (Feng et al. [12]). Lee et al. [13] used soil washing as a method to reduce the pollutant concentration in oil-contaminated soils and evaluated its efficiency in removing TPH from diesel-contaminated sand. Previous studies related to the remediation of oil-contaminated soils using soil washing have mainly used nonionic and anionic surfactants as cleaning agents. Khalladi et al. [14] reported that surfactants were effective in removing TPH adsorbed on the surface of soil particles by reducing the interfacial tension between the soil particles and oil. Vreysen and Maes [15] used sandy loam that was artificially contaminated with diesel and reported that nonionic surfactants had a removal efficiency of 50%. In addition, Hernández-Espriú et al. [16] reported that soil washing could achieve a TPH removal rate of 60%. Jang et al. [17] used plasma blasting for the remediation of contaminated soils and demonstrated the applicability of the technique by evaluating the fluid diffusion effect, the improved permeability of the contaminated soils, and the purification efficiency.

The diffusion of pollutants is caused by concentration changes that occur in the liquid state. Previous studies have suggested that this phenomenon can be prevented by applying a reactive medium that reduces the concentrations of solid (e.g., heavy metals) and liquid pollutants (e.g., oil) in groundwater. In particular, contaminated groundwater can be remediated using permeable reactive barriers (PRBs), which utilize effective, eco-friendly, and cost-efficient reactive media, as well as appropriate construction methods for the site [18–21]. PRBs are generally installed in the ground where contaminant plumes exist and then use their hydraulic flow. They remove pollutants by inducing physicochemical and biological reactions between reactive media and pollutants [18,22].

Moreover, many studies have been conducted on liner systems, PRBs, reactive barriers, and reactive media to prevent the diffusion of pollutants. In particular, geosynthetic clay liners (GCLs), in which the bentonite layer is surrounded by geotextiles or geomembranes, have been widely distributed to prevent the diffusion of pollutants in fluids. GCLs have been applied to many geotechnical fields, including landfills, dams, artificial lakes, sewage treatment ponds, storage tanks, and contaminated soils. The popularity of PRBs is due to the variety of advantages that they offer, such as low permeability coefficients, low hydraulic conductivity, high mechanical stability, and simple and rapid on-site installation [23–27]. Xue et al. [28] conducted permeability tests on GCLs soaked in various types of solutions with different concentrations and analyzed the relationship between the expansion and permeability coefficient of GCLs.

Kim and Lee [29] evaluated the treatment efficiency of groundwater contaminated with heavy metals by applying zero-valent iron, steel slag, activated carbon, and tree bark to PRBs as reactive media. Ji and Cheong [30] applied PRBs as a method for remediating contaminated leachate from mines on-site and recommended the application of organic carbon mixtures as reactive media to remove high concentrations of aluminum. Furthermore, Chung and Lee [31] evaluated the suitability and limitations of Moringa oleifera mass bentonite (MOM-bentonite) as the reactive media of reactive barriers for treating aquifers contaminated by the movement of PCE-contaminated groundwater on site. Guerin et al. [32] evaluated the applicability of PRBs for the remediation of groundwater contaminated with petroleum hydrocarbons. Moreover, Cho et al. [33] evaluated the applicability of pyrophyllite as a reactive medium for PRBs to prevent the diffusion of pollutants in contaminated groundwater and reduce the environmental pollution of soils. Kim et al. [34] evaluated the concentration of solidifying agents containing fly ash and lime as well as the optimum water content for the formation of mixed barriers to purify contaminated groundwater in soils classified as SW-SC. They also evaluated the performance of a mixed liner and cover materials containing solidifying agents. In addition, Yun et al. [35] conducted a compaction test on calcium bentonite–sand mixtures to determine the optimum water content. They also evaluated the permeability characteristics of the mixed liner and cover materials by conducting variable-head permeability tests according to the mixing ratio of calcium bentonite.

Incineration or recovery methods applied after using various oil-absorbing materials are widely known treatments for oil-contaminated soils [36–38]. To prevent soil contamination from oil spills, fabric-based oil absorbents have been primarily used for ground surfaces. Non-woven fabrics that use hydrophobic hydrocarbon-based fibers have been most frequently utilized [39,40]. In recent years, oil-absorbing resins that use various adsorption or gelation-type polymers have been increasingly studied [41–44]. Jeong [45] evaluated the oil adsorption characteristics of polypropylene (PP) materials treated with a lipophilic acrylic resin. Gelling agents have a high rate of reaction with oil, and the substances generated in the reaction can be easily recovered. Therefore, Yun et al. [46] applied a mixture of calcium bentonite and a gelling agent as a liner and cover material and evaluated its permeability characteristics after it reacted with trichloroethylene (TCE)—a dense non-aqueous phase liquid (DNAPL) substance—in contaminated soil. Nguyen et al. [47] (2021) evaluated adsorption materials containing polydimethylsiloxane (PDMS) for the removal of oil, and Taylor et al. [48] evaluated the performance of materials that can adsorb pollutants from hydrocarbon-contaminated groundwater.

As mentioned above, many studies have been conducted to prevent the diffusion of pollutants in oil-contaminated soils or to remove such pollutants. However, most of the studied technologies are applied after the occurrence of pollution accidents. Therefore, in this study, a geotextile–polynorbornene liner was used as the oil-absorbing material in reactive barriers to instantly prevent the diffusion of oil pollutants in soils in the event of an oil spill in facilities where such accidents may occur. As the permeability performance is the most important factor in preventing the diffusion of pollutants, the applicability of the geotextile–polynorbornene liner and cover material was evaluated based on changes in its permeability characteristics when it contacts the oil pollutant. The applicability was evaluated using experimental and numerical analyses.

2. Overview of the Geotextile–Polynorbornene Liner

In general, bentonite minerals that constitute GCLs selectively adsorb moisture and swell. When bentonite particles that exhibit swelling behavior above a certain level are constrained using upper and lower fabric layers, the GCLs become impermeable. In other words, if a synthetic resin that adsorbs oil replaces bentonite to prevent the diffusion of oil pollutants, it exhibits the same behavior as that of GCLs. This concept is illustrated in Figure 1. The barrier is formed by using upper and lower geosynthetic layers to constraining an oil-absorbing synthetic resin that reacts only with oil; therefore, before an oil spill, the groundwater flows normally because water does not react with the synthetic resin. However, in the event of an oil spill on the ground, the absorption, swelling, and gelation of the oil-absorbing synthetic resin occur when it contacts the oil. In addition, because an impermeable layer is formed by the chemical reaction of the synthetic resin, it is possible to prevent the diffusion of oil. In this study, polynorbornene, which has excellent gelation properties, was applied as a reactive material to oil, and a geotextile was applied as a geosynthetic liner that constrains polynorbornene. Therefore, the oil-absorbing material was named the geotextile–polynorbornene liner.

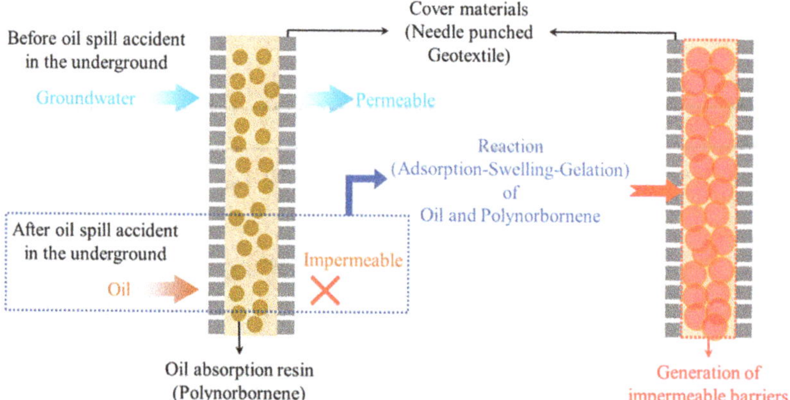

Figure 1. The generation of impermeable barriers of the geotextile–polynorbornene liner.

If the geotextile–polynorbornene liner is applied as a reactive medium to form reactive barriers such as PRBs. Then, in the event of an oil spill, the reaction of the material can prevent the diffusion of oil pollutants. Figure 2 shows a conceptual diagram of the prevention of oil pollutant diffusion.

Figure 3 shows the morphology of polynorbornene powder at 100× magnification. Polynorbornene powder particles have a very irregular geometry. To examine the degree of adsorption and swelling of the polynorbornene powder, a simple test on the change in state was conducted, as shown in Figure 4. It can be inferred from the figure that 24 h after mixing polynorbornene powder with diesel, the weight and volume increased as the powder reacted with and absorbed the oil. This indicates that polynorbornene powder can be utilized as an impermeable material through its gelation.

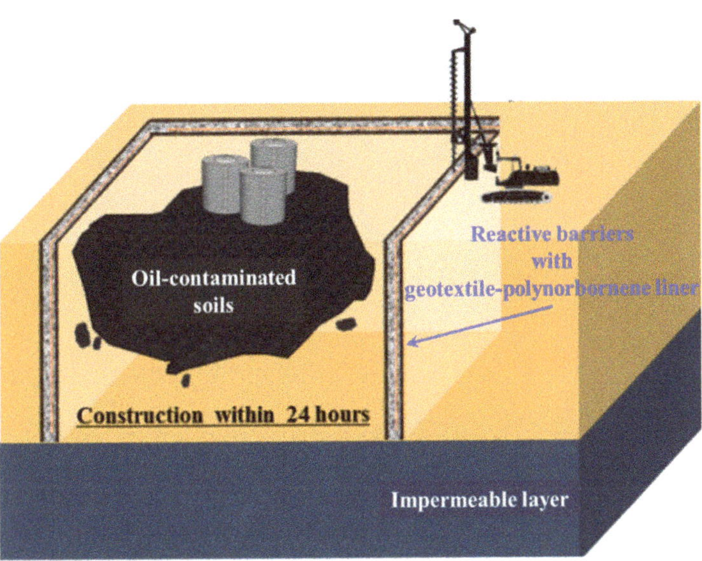

Figure 2. Conceptual diagram of reactive barriers with the geotextile–polynorbornene liner.

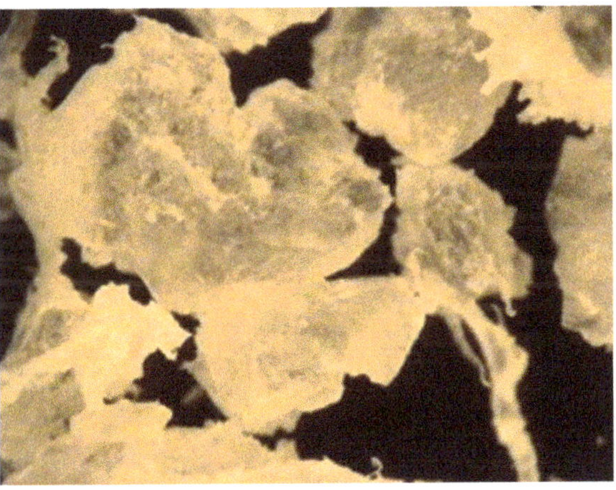

Figure 3. Morphology of polynorbornene powder at $100\times$ magnification.

The impermeability performance of polynorbornene after completing gelation must be confirmed before applying it as a liner. From the perspective of geoenvironmental engineering, the impermeability performance of a material is evaluated using its permeability coefficient, which is defined as 10^{-7} cm/s or less for a typical impermeable layer. As it is necessary to evaluate the permeability coefficient of the geotextile–polynorbornene liner over time, experiments were conducted in this study to measure changes in its permeability characteristics over time.

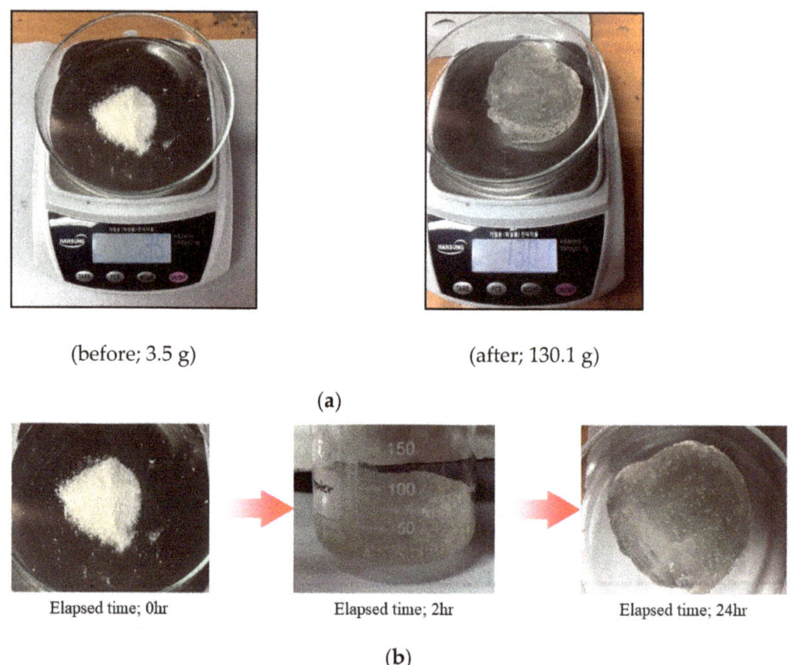

(before; 3.5 g)   (after; 130.1 g)

(a)

Elapsed time; 0hr   Elapsed time; 2hr   Elapsed time; 24hr

(b)

Figure 4. Adsorption and expansion of polynorbornene powder: (**a**) weight change after reaction with diesel; (**b**) gelation over time.

3. Materials and Methods

3.1. Test Apparatus and Materials

To evaluate the permeability of the geotextile–polynorbornene liner, a permeability coefficient similar to that of typical soils in the absence of pollutants was employed with a range of 10^{-2} to 10^{-4} cm/s. Soil that comes into contact with pollutants has a typical impermeability of 10^{-7} cm/s or less.

There are several test methods used to evaluate the permeability performance of materials, but methods based on reactions between pollutants and the liner and cover materials are limited. Conventionally, permeability tests for typical liners and cover materials such as GCLs have been conducted using water after swelling for $\geq$48 h. Furthermore, the impermeability performance of the materials against oil pollutants with varying concentrations must also be assessed. Therefore, test methods that consider these conditions were used in this study. Two American Society for Testing and Materials (ASTM) International methods for evaluating the permeability and impermeability performances of liners and cover materials were adopted: ASTM D5887 is the standard test method for measuring the index flux through saturated geosynthetic clay liner specimens using a flexible wall permeameter, and ASTM D6766 is the standard test method for evaluating the hydraulic properties of geosynthetic clay liners permeated with potentially incompatible aqueous solutions. In addition, a test apparatus that can evaluate changes in the permeability characteristics of the geotextile–polynorbornene liner according to its contact time with the oil pollutant was also adopted. As shown in Figure 5, the test apparatus consists of a water and air controller, a pressure controller, and an upper/lower pressure cell controller.

Figure 5. Test apparatus used for permeability performance evaluation.

Figure 6 shows the cross-section of the geotextile–polynorbornene liner used in the test. This material has a non-woven fabric made of PP and polyethylene (PET) at the top and a woven fabric made of PP at the bottom. In addition, polynorbornene powder was located between the non-woven fabric (top) and woven fabric (bottom). Lastly, diesel was used as an oil pollutant to induce a reaction with the geotextile–polynorbornene liner.

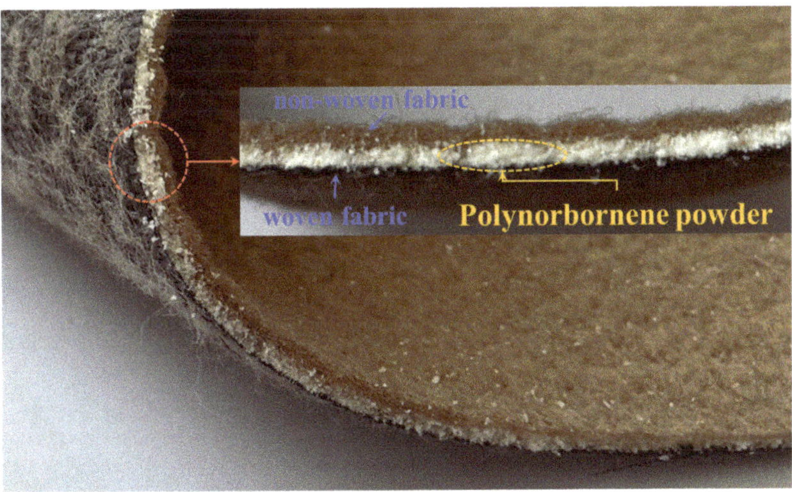

Figure 6. Cross-section of the geotextile–polynorbornene liner.

3.2. Test Procedure

Changes in the permeability characteristics of the geotextile–polynorbornene liner were tested using the following procedure, in accordance with the methods of ASTM D5887 and ASTM D6766: (1) 100 mm diameter circular samples (geotextile–polynorbornene liner) were prepared; (2) the sample holder was installed at the bottom inside the cell, the lower porous plate was installed, the sample was placed in the holder, and the upper porous plate was installed; (3) the membrane and O-ring were installed to prevent the leakage of the oil pollutant (diesel) during the reaction; (4) the cell pressure (35 kPa) and upper/lower back pressure (7–14 kPa) were established; (5) the cell and back pressures were increased every ten minutes; (6) the final pressures (cell pressure = 550 kPa, back pressure = 515 kPa) were established; (7) the sample was stabilized under pressure for 40 h; (8) a pressure head of 15 kPa was generated after setting the lower back pressure to 530 kPa to cause the upward penetration of the pollutant; (9) the burette reading was recorded over time after inducing the permeation of the oil through the sample; and (10) the flux and permeability coefficient were calculated. The procedure is summarized in Figure 7.

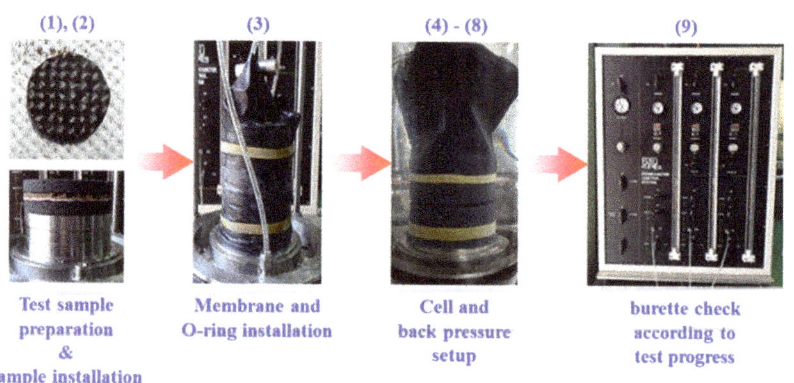

Figure 7. Test procedure.

The flux was obtained to calculate the permeability coefficient. The flux is the laminar flow per unit volume, that is, the flow of water that passes through the cross-section of the sample per unit time, and it is expressed in units of velocity. The flux calculation results were used to determine the permeability coefficient, which is defined as the laminar flow that passes through the cross-section per unit time under the influence of a hydraulic gradient and is expressed in units of velocity. Therefore, the permeability coefficient can be calculated based on the flux using Equations (1)–(4).

$$F = \frac{Q}{A} \tag{1}$$

$$V = ki \tag{2}$$

$$i = \frac{\Delta h}{L} \tag{3}$$

$$\Delta h = \frac{\Delta P}{\rho g} \tag{4}$$

where F is the flux (m³/m²·s), Q is the flow rate (m³/s), A is the cross-sectional area (m²), V is the discharge velocity (cm/s), k is the permeability coefficient (cm/s), i is the hydraulic gradient, L is the specimen length (m), ΔP is the pressure head (kPa), Δh is the total head (m), ρ is the density of water (ton/m³), and g is the gravitational acceleration.

Table 1 lists the test conditions used in the study. The pollutant contact time started at 0 h (before contact with the pollutant) and was monitored 0.5, 1, 2, 4, 16, and 24 h after initial contact with the pollutant. The pressure head ranged between 15 and 45 kPa (three times), 75 kPa (five times), and 105 kPa (seven times) to analyze the discharge time. In addition, the discharge time was set according to the pressure head to ensure a constant flow for each pollutant contact time.

Table 1. Test conditions.

Contact Time of Oil Pollutant	Pressure Head (ΔP, kPa)	Total Head (Δh, m)	Discharge Time (t, s)	Flow Rate (Q, cm^3/s)	Specimen Area (A, cm^2)	Specimen Length (L, cm)
0 h	15	1.53	8.2	40	50.27	2
	45	4.59	3.8			
	75	7.65	2.8			
	105	10.71	2.3			
0.5 h	15	1.53	16			
	45	4.59	13			
	75	7.65	12			
	105	10.71	9			
1 h	15	1.53	540			
	45	4.59	94			
	75	7.65	34			
	105	10.71	15			
2 h	15	1.53	780			
	45	4.59	110			
	75	7.65	57			
	105	10.71	29	0.5		
4 h	15	1.53	8242			
	45	4.59	2438			
	75	7.65	967			
	105	10.71	195			
16 h	15	1.53	8402			
	45	4.59	2631			
	75	7.65	990			
	105	10.71	210			
24 h	15	1.53	8420			
	45	4.59	2638			
	75	7.65	970			
	105	10.71	195			

4. Results and Discussion

4.1. Changes in the Permeability Characteristics of the Geotextile–Polynorbornene Liner over Time after Contacting the Pollutant

Table 2 reports the test results, and Figure 8 shows the corresponding graphs. The test results for each test condition in Table 2 are the mean values of three experiments. In each test condition, the three experimental results had different values. However, the error was ignored because the difference between the values was outside the range of significant figures. When there was no contact with the oil pollutant (pollutant contact time = 0 h), the permeability coefficient ranged from 10^{-3} to 10^{-4} cm/s depending on the size of the pressure head, which is similar to the flow velocity of groundwater in the weathered granite soils in Korea. Hence, the flow of groundwater was not affected by contact with the geotextile–polynorbornene liner.

Table 2. Test results.

Contact Time of Oil Pollutant	Pressure Head (ΔP, kPa)	Hydraulic Gradient	Flux (F, cm^3/cm$^2\cdot$s)	Permeability Coefficient (k, cm/s)
0 h	15	77	9.68×10^{-2}	1.26×10^{-3}
	45	230	2.11×10^{-1}	9.17×10^{-4}
	75	383	2.88×10^{-1}	7.53×10^{-4}
	105	536	3.46×10^{-1}	6.46×10^{-4}
0.5 h	15	77	6.22×10^{-4}	8.12×10^{-6}
	45	230	7.65×10^{-4}	3.33×10^{-6}
	75	383	8.29×10^{-4}	2.17×10^{-6}
	105	536	1.11×10^{-3}	2.06×10^{-6}
1 h	15	77	1.84×10^{-5}	1.24×10^{-6}
	45	230	1.06×10^{-4}	7.64×10^{-7}
	75	383	2.93×10^{-4}	4.61×10^{-7}
	105	536	6.63×10^{-4}	2.41×10^{-7}
2 h	15	77	1.28×10^{-5}	6.40×10^{-7}
	45	301	9.04×10^{-5}	4.56×10^{-7}
	75	383	1.74×10^{-4}	3.01×10^{-7}
	105	536	3.43×10^{-4}	1.67×10^{-7}
4 h	15	77	1.21×10^{-6}	9.52×10^{-8}
	45	230	4.08×10^{-6}	2.69×10^{-8}
	75	383	1.03×10^{-5}	1.78×10^{-8}
	105	536	5.10×10^{-5}	1.58×10^{-8}
16 h	15	77	1.18×10^{-6}	8.84×10^{-8}
	45	230	3.78×10^{-6}	2.63×10^{-8}
	75	383	1.00×10^{-5}	1.65×10^{-8}
	105	536	4.74×10^{-5}	1.55×10^{-8}
24 h	15	77	1.18×10^{-6}	9.52×10^{-8}
	45	230	3.77×10^{-6}	2.68×10^{-8}
	75	383	1.03×10^{-5}	1.64×10^{-8}
	105	536	5.10×10^{-5}	1.54×10^{-8}

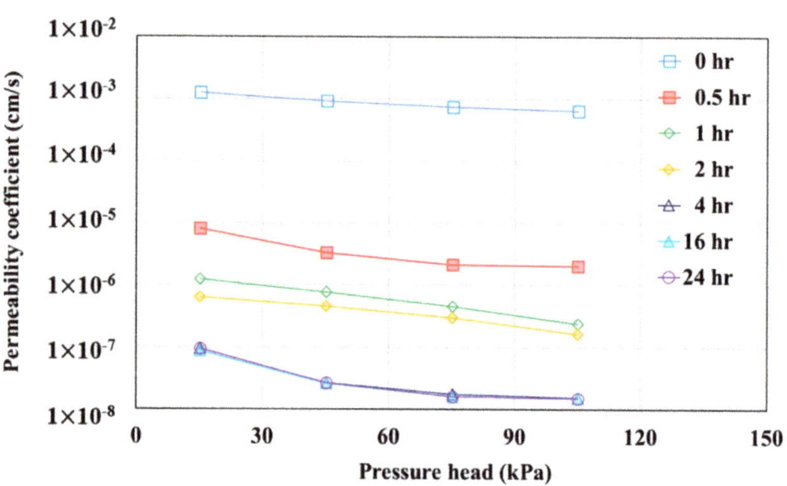

Figure 8. Permeability coefficient according to the pressure head.

As shown in Figure 8, changes in the permeability coefficient according to the pressure head were measured at different pollutant contact times. First, regardless of the pollutant contact time, the permeability coefficient decreased as the pressure head increased (the hydraulic gradient increased). When the pollutant contact time was 4 h or longer, the permeability coefficient was 10^{-7} cm/s or less, which is defined as almost an impermeable layer. The values showed a tendency to slowly converge when the pressure head was 75 kPa or higher. In other words, the geotextile–polynorbornene liner was likely to further react with the pollutant when the contact time was shorter than 4 h and the pressure head was lower than 75 kPa. However, when the contact time was longer and the pressure head was higher, the reaction between the geotextile–polynorbornene liner and the pollutant was almost complete.

Figure 9 shows the permeability coefficient over time under different pressure head conditions. The permeability coefficient was high when there was no contact with the pollutant, but it sharply decreased at a pollutant contact time of 0.5 h. In addition, regardless of the pressure head, the permeability coefficient did not substantially change once a pollutant contact time of 4 h was reached. Thus, the reaction between the geotextile–polynorbornene liner and the pollutant was completed after 4 h of contact. This result indicates that it is possible to form an impervious layer that can block pollutants.

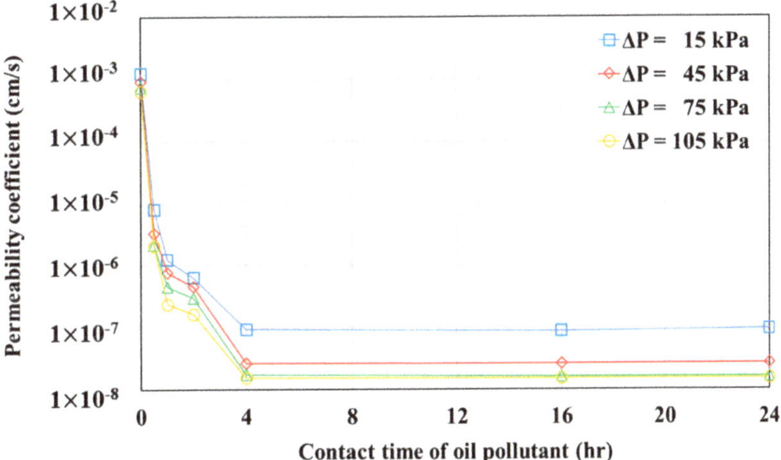

Figure 9. Permeability coefficient according to contact time.

Changes in the permeability characteristics of the geotextile–polynorbornene liner mentioned in this section were measured by applying diesel as an oil pollutant. As oil spills on the ground occur at various concentrations, changes in the permeability characteristics of the reactive material must be analyzed at different oil concentrations to evaluate its applicability. Therefore, a three-dimensional (3D) numerical analysis was conducted to simulate different oil concentrations, and the results are presented in Section 4.2.

4.2. Numerical Analysis

4.2.1. Finite Difference Analysis (FDA)

In this study, FDA was conducted using the well-known environmental simulation software MT3D (Visual MODFLOW; USGS, Denver, CO, USA) to analyze changes in the permeability characteristics of the geotextile–polynorbornene liner resulting from different concentrations of the oil pollutant. MT3D facilitates the 3D FDA of a hydraulic model for solute movement in a complicated hydrogeological structure. This software has been widely used for pollutant diffusion analysis because it can account for the steady-state flow, transient flow, anisotropic dispersion, first-order decay, chemical reactions between solutes, and linear and nonlinear adsorption.

Figure 10 shows the analysis model implemented in 3D and its plane view. The analysis model was composed of an oil tank that can generate the pressure head of the oil pollutant, soils with a permeability coefficient of 10^{-4} cm/s, and the geotextile–polynorbornene liner in the soils. For the mesh in the analysis, a square of 0.1 m was used for the oil tank and soils. The thickness of the geotextile–polynorbornene liner was 0.06 m. In addition, pollutant monitoring wells at four locations were simulated to examine the concentration of pollutants that passed through the geotextile–polynorbornene liner. The dimensions of the oil tank and soils were set to 0.24 × 0.5 × 1.2 m and 1.0 × 0.5 × 0.5 m (L × W × H). It is worth noting that a soil box can be used in further research.

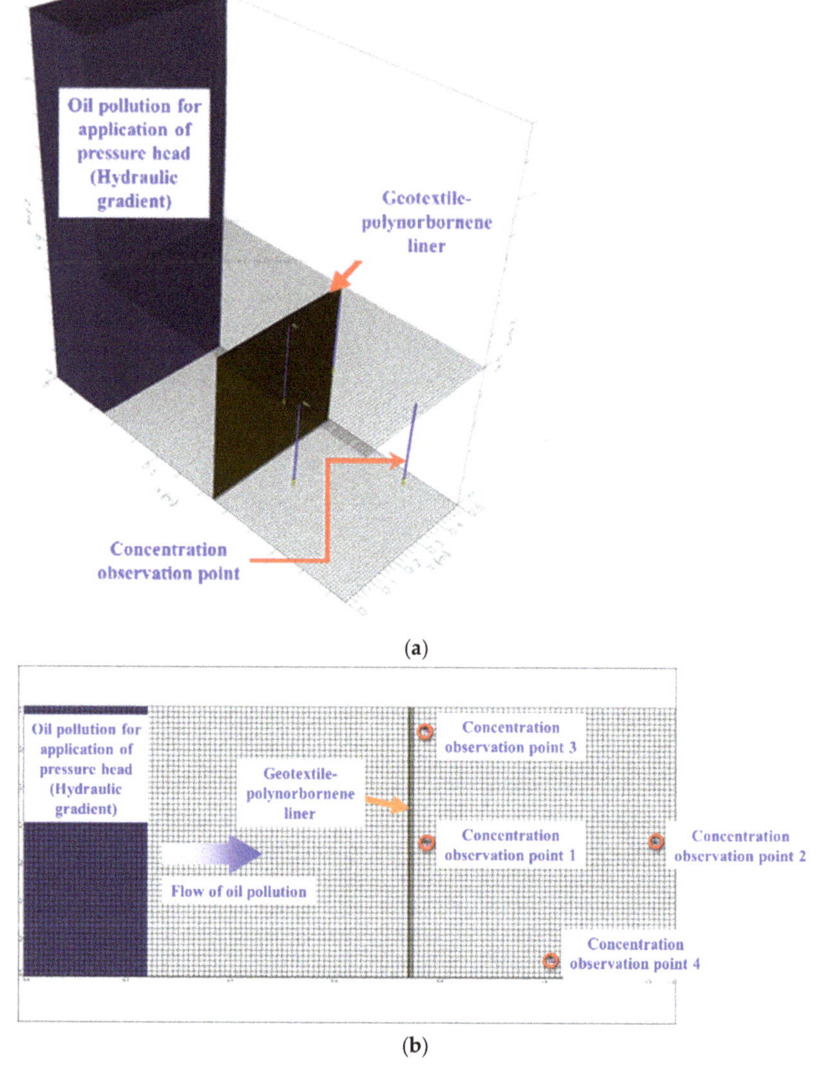

Figure 10. The 3D FDA model: (**a**) 3D FDA model; (**b**) plane view of the analysis model.

The hydraulic conductivity represents the degree of smoothness of the liquid material flow in the soil. A higher hydraulic conductivity indicates a smoother flow of the liquid material. Therefore, the hydraulic conductivity of the oil tank was set to 1000 cm/s so that the oil could be smoothly introduced to the simulated soils. In addition, 10^{-4} cm/s was applied as the permeability coefficient of the soils. Table 3 lists the conditions for the analysis model.

Table 3. FDA model conditions.

Classification	Oil Tank	Soils
Porosity	0.9	0.25
Horizontal permeability coefficient (cm/s)	1	10^{-4}
Vertical permeability coefficient (cm/s)	1	10^{-4}
Specific storativity (m^{-1})	10^{-5}	10^{-5}
Specific yield	0.9	0.15
Contact time of oil pollutant (h)	96	

In general, the processes that govern the transport of pollutants include groundwater flow, pollutant adsorption, advection, diffusion, dispersion, and biodegradation. The purpose of this study, however, was to examine the impermeability performance of the geotextile–polynorbornene liner when oil pollutants with different concentrations are released in soils. Therefore, only the influences of advection, diffusion, and dispersion were considered for the prediction of pollutant movement by FDA. TPH, which can simulate diesel, was applied as the pollutant type. In addition, because the coefficient results show that permeability changed to impermeability as the pollutant contact time increased from 0.5 to 4 h, the permeability coefficients obtained when the geotextile–polynorbornene liner was in contact with the pollutant for 0.5 and 4 h were applied in the FDA. Preliminary analysis confirmed that a pressure head of 15 kPa was too small to affect the FDA results. Therefore, 45, 75, and 105 kPa were applied as pressure head conditions. Table 4 lists the FDA cases.

Table 4. FDA cases.

Analysis Cases	Pollutant (TPH) Concentration (ppm)	Contact Time of Pollutant (h)	Pressure Head (ΔP, kPa)	Permeability Coefficient of Geotextile–Polynorbornene Liner (cm/s)
Case HC-1			45	3.33×10^{-6}
Case HC-2		0.5	75	2.17×10^{-6}
Case HC-3	6000		105	2.06×10^{-6}
Case HC-4			45	2.69×10^{-8}
Case HC-5		4	75	1.78×10^{-8}
Case HC-6			105	1.58×10^{-8}
Case MC-1			45	3.33×10^{-6}
Case MC-2		0.5	75	2.17×10^{-6}
Case MC-3	2000		105	2.06×10^{-6}
Case MC-4			45	2.69×10^{-8}
Case MC-5		4	75	1.78×10^{-8}
Case MC-6			105	1.58×10^{-8}
Case LC-1			45	3.33×10^{-6}
Case LC-2		0.5	75	2.17×10^{-6}
Case LC-3	500		105	2.06×10^{-6}
Case LC-4			45	2.69×10^{-8}
Case LC-5		4	75	1.78×10^{-8}
Case LC-6			105	1.58×10^{-8}

4.2.2. Changes in the Permeability Characteristics of the Geotextile–Polynorbornene Liner According to the Concentration of the Oil Pollutant

Figures 11–13 show the concentration of the pollutant at each observation point over time after the pollutant passed through the geotextile–polynorbornene liner at different TPH concentrations.

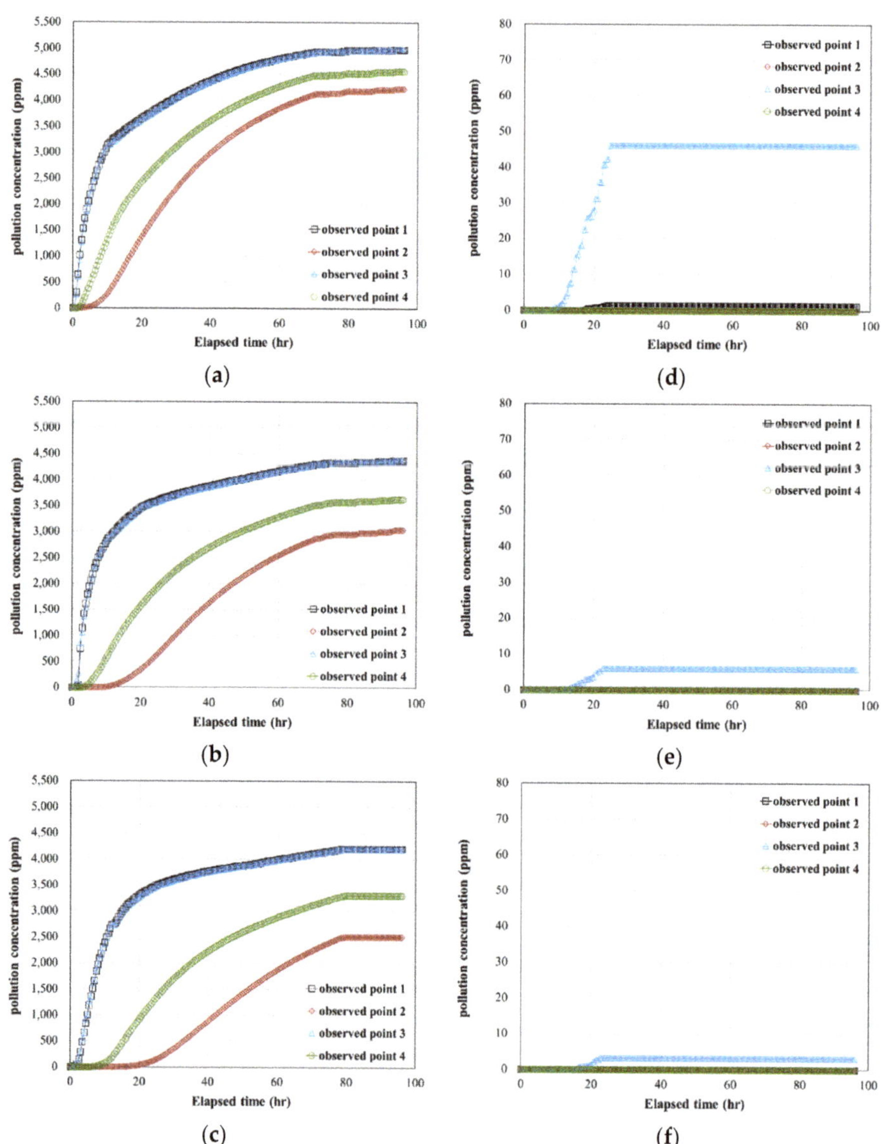

Figure 11. FDA results for the high concentration (6000 ppm) condition: (**a**) HC-1; (**b**) HC-2; (**c**) HC-3; (**d**) HC-4; (**e**) HC-5; (**f**) HC-6.

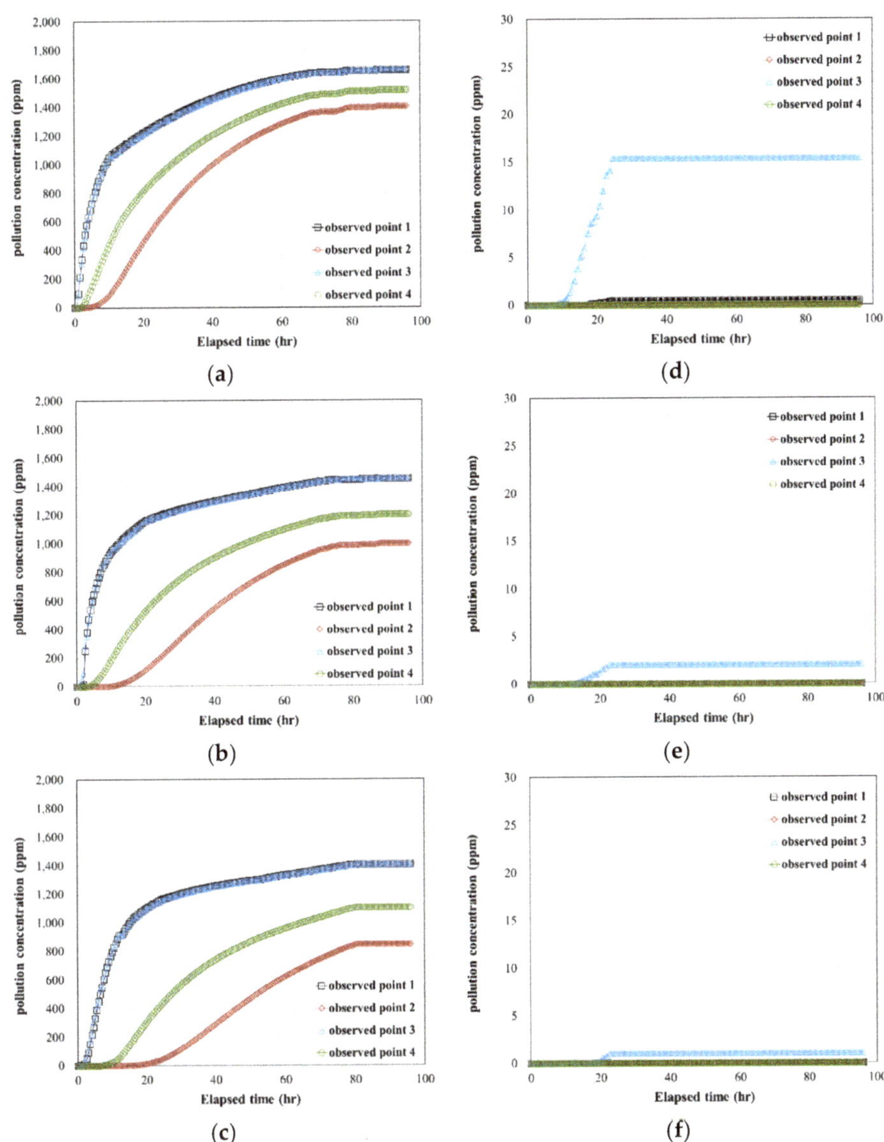

Figure 12. FDA results for the moderate concentration (2000 ppm) condition: (**a**) MC-1; (**b**) MC-2; (**c**) MC-3; (**d**) MC-4; (**e**) MC-5; (**f**) MC-6.

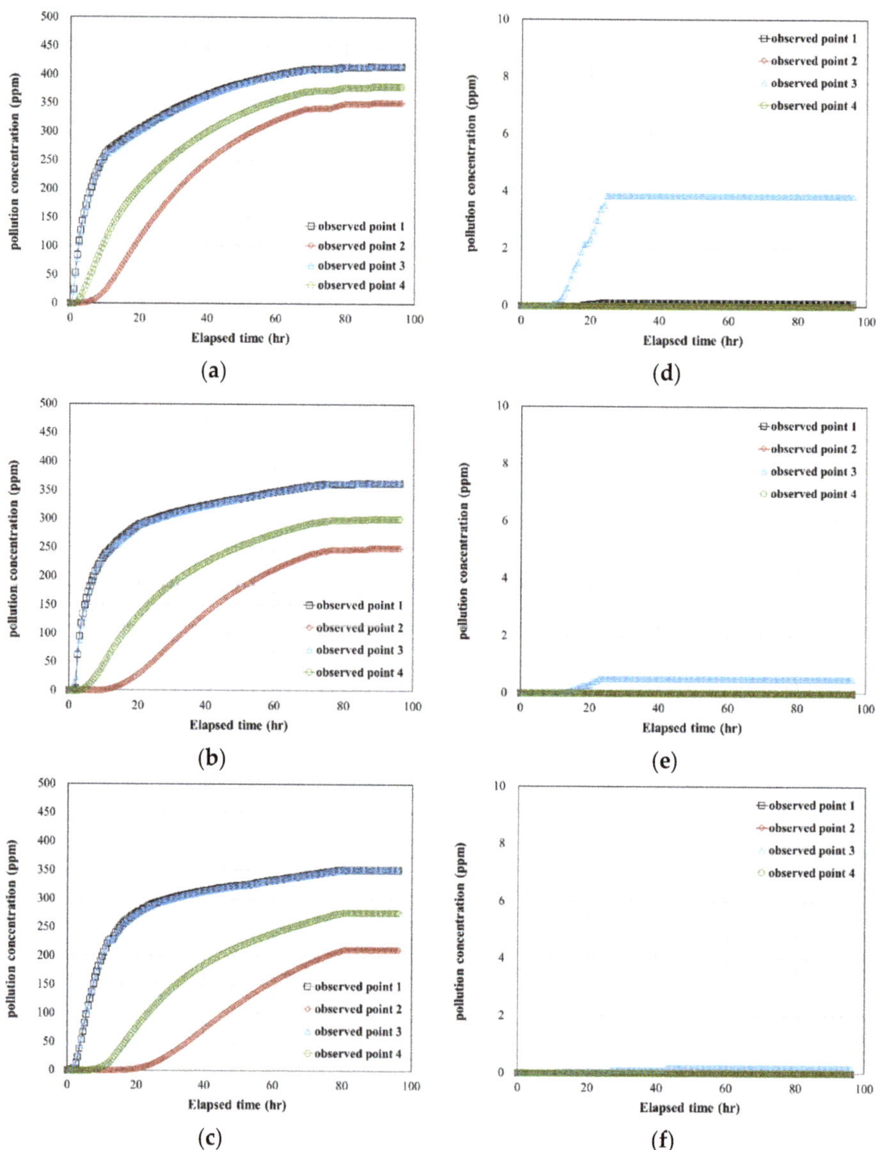

Figure 13. FDA results for the low concentration (500 ppm) condition: (**a**) LC-1; (**b**) LC-2; (**c**) LC-3; (**d**) LC-4; (**e**) LC-5; (**f**) LC-6.

Figure 11a–c (cases HC-1–HC-3) shows the concentration change over time for 0.5 h of contact between the pollutant and the geotextile–polynorbornene liner. As the permeability coefficient of the geotextile–polynorbornene liner decreased, the concentration of the pollutant tended to decrease at each observation point, but the concentration continued to increase over time. In addition, for each permeability coefficient, a higher concentration of the pollutant was released at the observation points adjacent to the geotextile–polynorbornene liner, and the pollutant concentration significantly increased within a short period of time. As the distance from the geotextile–polynorbornene liner increased, however, the pollutant concentration decreased, and the rate of pollutant increase was lower than that in the adjacent observation points.

Figure 11d–f (cases HC-4–HC-6) shows the concentration change over time for a contact time of 4 h between the pollutant and the geotextile–polynorbornene liner. As the permeability coefficient of the geotextile–polynorbornene liner decreased, the concentration of the pollutant tended to decrease at each observation point. In addition, the pollutant concentrations at the observation points adjacent to the geotextile–polynorbornene liner were higher than those at the observation points far away from it. However, the change in concentration with the pollutant contact time significantly decreased for the same permeability coefficient and observation point. Furthermore, at most observation points, the pollutant concentration did not substantially change over time. This is the result of the impermeability effect of the geotextile–polynorbornene liner as well as the influence of the distances of the observation points.

The same tendencies of the analysis results mentioned above were observed at moderate (Figure 12) and low concentrations (Figure 13). Under each analysis condition, the maximum pollutant concentration was the highest at the observation point that was closest to the geotextile–polynorbornene liner, as shown in Table 5.

Table 5. Maximum pollutant concentration at the observation points adjacent to the geotextile–polynorbornene liner.

Analysis Cases	Pollutant (TPH) Concentration (ppm)	Contact Time of Pollutant (h)	Pressure Head (ΔP, kPa)	Maximum Concentration of Observed Point 1 (ppm)
Case HC-1			45	4985.8
Case HC-2		0.5	75	4379.6
Case HC-3	6000		105	4200.9
Case HC-4			45	46.1
Case HC-5		4	75	6.0
Case HC-6			105	3.2
Case MC-1			45	1660.5
Case MC-2		0.5	75	1453.2
Case MC-3	2000		105	1405.3
Case MC-4			45	15.4
Case MC-5		4	75	2.0
Case MC-6			105	1.0
Case LC-1			45	415.13
Case LC-2		0.5	75	363.31
Case LC-3	500		105	351.32
Case LC-4			45	3.8
Case LC-5		4	75	0.5
Case LC-6			105	0.2

At an initial pollutant concentration of 6000 ppm and a contact time of 0.5 h, the ratio of the concentration of the pollutant that permeated through the geotextile–polynorbornene liner to the initial pollutant concentration ranged from 70.02 to 83.1%. For a pollutant contact time of 4 h, however, the concentration ranged from 0.08 to 0.92% compared to that at 0.5 h, and the ratio ranged from 0.05 to 0.77%. When the initial pollutant concentrations were 2000 and 500 ppm, the concentrations of the pollutant that permeated through the geotextile–polynorbornene liner over time were similar to the results for 6000 ppm. In other words, the numerical analysis results show that the geotextile–polynorbornene liner has a pollutant blocking effect over time.

5. Conclusions

In this study, changes in the permeability characteristics of an oil-absorbing medium were tested, and experiments and numerical analysis were used to evaluate a geotextile–

polynorbornene liner for its ability to prevent the diffusion of pollutants. The results are as follows:

1. When changes in the permeability coefficient were examined at different pressure heads and different pollutant contact times, the permeability coefficient decreased as the pressure head increased (the hydraulic gradient increased) regardless of the pollutant contact time. In addition, when the pollutant contact time was 4 h or longer, the permeability coefficient of the geotextile–polynorbornene liner was 10^{-7} cm/s or less, which is defined as almost an impermeable layer.
2. Changes in the permeability coefficient were examined over time under different pressure head conditions. There was almost no change in the permeability coefficient starting from the pollutant contact time of 4 h. Thus, when the pollutant contact time reaches 4 h or more, the geotextile–polynorbornene liner has an impermeable layer that can block pollutants.
3. The results of the 3D pollutant diffusion analysis showed that, for a pollutant contact time of 4 h, the maximum concentration of the pollutant that permeated through the geotextile–polynorbornene liner was less than approximately 0.8% compared to the initial pollutant concentration. Therefore, the numerical analysis results confirm that the geotextile–polynorbornene liner has a pollutant blocking effect over time.
4. The test and numerical analysis results confirm the impermeability performance of the geotextile–polynorbornene liner against oil pollutants. Therefore, it has potential as an application for the prevention of pollutant diffusion.

Funding: This research received no external funding.

Institutional Review Board Statement: Not applicable.

Informed Consent Statement: Not applicable.

Data Availability Statement: The data presented in this study are available on request to the corresponding author. The data are not publicly available as they form part of an ongoing study.

Conflicts of Interest: The author declares no conflict of interest.

References

1. *Environment, Purification of Pollution Guideline*; Environment of Korea: Sejong-si, Korea, 2007.
2. Awad, Y.M.; Kim, S.C.; Abd EI-Azeem, S.A.M.; Kim, K.H.; Kim, K.R.; Kim, K.J.; Jeon, C.; Lee, S.S. Veterinary antibiotics contamination in water, sediment, and soil near a swine manure composting facility. *Environ. Earth Sci.* **2014**, *71*, 1433–1440. [CrossRef]
3. Tang, Z.; Zhang, L.; Huang, Q.; Yang, Y.; Nie, Z.; Cheng, J.; Yang, J.; Wang, Y.; Chao, M. Contamination and risk of heavy metals in soils and sediments from a typical plastic waste recycling area in North China. *Ecotoxicol. Environ. Saf.* **2015**, *112*, 343–351. [CrossRef]
4. Ławniczak, Ł.; Woźniak-Karczewska, M.; Loibner, A.P.; Heipieper, H.J.; Chrzanowski, Ł. Microbial Degradation of Hydrocarbons—Basic Principles for Bioremediation: A Review. *Molecules* **2020**, *25*, 856. [CrossRef]
5. Teefy, D.A. Remediation technologies screening matrix and reference guide: Version III. *Remediat. J.* **1997**, *8*, 115–121. [CrossRef]
6. Ministry of Environment, Guidelines for Cleaning Contaminated. 2007. (In Korean). Available online: http://www.me.go.kr/home/web/policy_data/read.do;jsessionid=U8n8W4UDhpcnTttwLejCuHvj.mehome1?pagerOffset=2990&maxPageItems=10&maxIndexPages=10&searchKey=&searchValue=&menuId=10261&orgCd=&condition.deleteYn=N&seq=3297 (accessed on 13 April 2007).
7. Jeong, J.; Kim, H.; Lee, S.; Jeong, S.W. Effects of Diesel Dose and Soil Texture on Variation in the Concentration of Total Petroleum Hydrocarbon in the Diesel-Contaminated Soil. *J. Korean Soc. Environ. Eng.* **2017**, *37*, 69–72. [CrossRef]
8. Lee, S.H.; Lee, J.H.; Jung, W.C.; Park, M.; Kim, M.S.; Lee, S.J.; Park, H. Changes in Soil Health with Remediation of Petroleum Hydrocarbon Contaminated Soils Using Two Different Remediation Technologies. *Sustainability* **2020**, *12*, 10078. [CrossRef]
9. Cho, K.; Myung, E.; Kim, H.; Purev, O.; Park, C.; Choi, N. Removal of Total Petroleum Hydrocarbons from Contaminated Soil through Microwave Irradiation. *Int. J. Environ. Res. Public Health* **2020**, *17*, 5952. [CrossRef]
10. Sayed, K.; Baloo, L.; Sharma, N.K. Bioremediation of Total Petroleum Hydrocarbons (TPH) by Bioaugmentation and Biostimulation in Water with Floating Oil Spill Containment Booms as Bioreactor Basin. *Int. J. Environ. Res. Public Health* **2021**, *18*, 2226. [CrossRef]
11. Han, S. Remediation Study of Total Petroleum Hydrocarbons Contaminated Soil Using Biopile. Master's Thesis, Hankuk University of Foreign Studies, Seoul, Korea, 2017.

12. Feng, D.; Lorenzen, L.; Aldrich, C.; Mar, P.W. Exsitu diesel contaminated soil washing with mechanical methods. *Miner. Eng.* **2001**, *14*, 1093–1100. [CrossRef]

13. Lee, C.D.; Yoo, J.C.; Yang, J.S.; Kong, J.; Baek, K. Extraction of Total Petroleum Hydrocarbons from Petroleum Oil-Contaminated Sandy Soil by Soil Washing. *J. Soil Groundw. Environ.* **2013**, *18*, 18–24. [CrossRef]

14. Khalladi, R.; Benhabiles, O.; Bentahar, F.; Moulai-Mostefa, N. Surfactant remediation of diesel fuel polluted soil. *J. Hazard. Mater.* **2009**, *164*, 1179–1184. [CrossRef] [PubMed]

15. Vreysen, S.; Maes, A. Remediation of Diesel Contaminated, Sandy-Loam Soil Using Low Concentrated Surfactant Solutions. *J. Soils Sediments* **2005**, *5*, 240–244. [CrossRef]

16. Hernández-Espriú, A.; Sánchez-León, E.; Martnez-Santos, P.; Torres, L.G. Remediation of a diesel-contaminated soil from a pipeline accidental spill: Enhanced biodegradation and soil washing processes using natural gums and surfactants. *J. Soils Sediments* **2013**, *13*, 152–165. [CrossRef]

17. Jang, H.S.; Kim, K.J.; Song, J.Y.; An, S.G.; Jang, B.A. An Experimental Study to Improve Permeability and Cleaning Efficiency of Oil Contaminated Soil by Plasma Blasting. *J. Eng. Geol.* **2020**, *30*, 557–575.

18. Blowes, D.W.; Ptacek, C.J.; Jambor, J.L. In-Situ remediation of chromate contaminated groundwater using permeable reactive walls. *Environ. Sci. Technol.* **1997**, *31*, 3348–3357. [CrossRef]

19. USEPA. *In-Situ Remediation Technology Status Report, Treatment Walls*; 542-K-94-004; USEPA: Washington, DC, USA, 1995.

20. USEPA. *Permeable Reactive Barrier Technologies for Contaminant Remediation*; 600-R-98-125; USEPA: Washington, DC, USA, 1998.

21. USEPA. *Cost Analyses for Selected Groundwater Cleanup Projects; Pump and Treat System and Permeable Reactive Barriers, Solid Waste and Emergency Response (5102G)*; 542-R-00-013; USEPA: Washington, DC, USA, 2001.

22. Boni, M.R.; Sbaffoni, S. The potential of compost-based biobarriers for Cr(VI) removal from contaminated groundwater, Column test. *J. Hazard. Mater.* **2009**, *166*, 1087–1095. [CrossRef] [PubMed]

23. Kong, D.J.; Wu, H.N.; Chai, J.C.; Arulrajah, A. State-Of-The-Art Review of Geosynthetic Clay Liners. *Sustainability* **2017**, *9*, 2110. [CrossRef]

24. Shackelford, C.D.; Meier, A.; Sample-Lord, K. Limiting membrane and diffusion behavior of a geosynthetic clay liner. *Geotext. Geomembr.* **2016**, *44*, 707–718. [CrossRef]

25. Liu, Y.; Bouazza, A.; Gates, W.P.; Rowe, R.K. Hydraulic performance of geosynthetic clay liners to sulfuric acid solutions. *Geotext. Geomembr.* **2015**, *43*, 14–23. [CrossRef]

26. Wu, H.N.; Shen, S.L.; Liao, S.M.; Yin, Z.Y. Longitudinal structural modelling of shield tunnels considering shearing dislocation between segmental rings. *Tunn. Undergr. Space Technol.* **2015**, *50*, 317–323. [CrossRef]

27. Wu, H.N.; Shen, S.L.; Yang, J. Identification of tunnel settlement caused by land subsidence in soft deposit of Shanghai. *J. Perform. Constr. Facil.* **2017**, *31*, 4017092. [CrossRef]

28. Xue, Q.; Zhang, Q.; Liu, L. Impact of High Concentration Solutions on Hydraulic Properties of Geosynthetic Clay Liner Materials. *Materials* **2012**, *5*, 2326–2341. [CrossRef]

29. Kim, J.B.; Lee, J.Y. A Feasibility Study on the Complex Media Permeable Reactive Barrier using Waste Resources. *J. Korea Soc. Waste Manag.* **2017**, *34*, 159–167. [CrossRef]

30. Ji, S.W.; Cheong, Y.W. Experiment of Reactive Media Selection for the Permeable Reactive Barrier Treating Groundwater contaminated by Acid Mine Drainage. *Econ. Environ. Geol.* **2005**, *38*, 237–245.

31. Chung, S.L.; Lee, D.H. Remediation of PCE-contaminated Groundwater Using Permeable Reactive Barrier System with M0M-Bentonite. *J. Soil Groundw. Environ.* **2012**, *17*, 73–80. [CrossRef]

32. Guerin, T.F.; Hornerb, S.; McGovernb, T.; Davey, B. An application of permeable reactive barrier technology to petroleum hydrocarbon contaminated groundwater. *Water Res.* **2002**, *36*, 15–24. [CrossRef]

33. Cho, K.; Kim, H.; Choi, N.C.; Park, C.Y. The Characterization of Pyrophyllite Based Ceramic Reactive Media for Permeable Reactive Barriers. *J. Mineral. Soc. Korea* **2019**, *31*, 227–234. [CrossRef]

34. Kim, Y.H.; Lim, D.H.; Lee, J.Y. A Feasibility Study on the Deep Soil Mixing Barrier to Control Contaminated Groundwater. *J. Soil Groundw. Environ.* **2001**, *6*, 53–59.

35. Yun, S.Y.; An, H.K.; Oh, M.; Lee, J.Y. A study on the Evaluation of Permeability and Structure for Calcium Bentonite-Sand Mixtures. *J. Korean Geosynth. Soc.* **2019**, *18*, 1–10.

36. Lee, S.K.; Kim, S.H.; Lee, K.J.; Shin, K.I. Functional Treatment of Recycled Ultrafine Fibrous Nonwovens. *Text. Sci. Eng.* **2005**, *42*, 370–375.

37. Lee, Y.H.; Kim, J.S.; Kim, D.H.; Shin, M.S.; Jung, Y.J.; Lee, D.J.; Kim, H.D. Effect of Blend Ratio of PP/kapok Blend Nonwoven Fabrics on Oil Sorption Capacity. *Environ. Technol.* **2013**, *34*, 3169–3175. [CrossRef] [PubMed]

38. Wang, J.; Zheng, Y.; Wang, A. Effect of Kapok Fiber Treated with Various Solvents on Oil Absorbency. *Ind. Crop. Prod.* **2012**, *40*, 178–184. [CrossRef]

39. Shin, H.S.; Yoo, J.H.; Jin, L. A Study on Oil Absorption Rate and Oil Absorbency of Melt-blown Nonwoven. *Text. Color. Finish.* **2010**, *22*, 257–263. [CrossRef]

40. Rengasamy, R.S.; Das, D.; Karan, C.P. Study of Oil Sorption Behavior of Filled and Structured Fiber Assemblies Made from Polypropylene, Kapok and Milkweed Fibers. *J. Hazard. Mater.* **2011**, *186*, 526–532. [CrossRef]

41. Lelaune, R.D.; Lindau, C.W.; Jugsujinda, A. Effectiveness of Nochar Solidifier Polymer in Removing Oil from Open Water in Coastral Wetlands. *Spill Sci. Technol. Bull.* **1999**, *5*, 357–359.

42. Atta, A.; Arndt, K.F. Swelling and Network Parameters of High Oil Absorptive Network Based on 1-Octene and Isodecyl Acrylate Copolymers. *J. Appl. Polym. Sci.* **2005**, *97*, 80–91. [CrossRef]
43. Shin, Y.S.; Cha, H.Y.; Park, S.S.; Woo, J.W.; Choi, J.S. Oil-absorption Capacity of poly(2-ethylhexylacrylate) Polymer in the Presence of Ultrasonic Wave. *J. Korean Ind. Eng. Chem.* **2002**, *13*, 326–329.
44. Chang, S.C.; Chun, S.H.; Lee, M.S.; Lee, K.B.; Choi, J.S. Oil-absorption Behaviors and Kinetics of Poly(dodecyl acrylate). *Appl. Chem.* **1999**, *3*, 141–143.
45. Jeong, H.J. Oil Absorptive Properties of Polypropylene Knit Fabric Treated with Oleophilic Acrylic Resin. *J. Korea Acad. Ind. Coop. Soc.* **2016**, *17*, 528–535.
46. Yun, S.Y.; Choi, J.W.; Oh, M.; Lee, J.Y. Characteristic of Permeability with the Sand, Calcium Bentonite and Solidifier Mixtures according to Selective Reaction of TCE. *J. Korean Geosynth. Soc.* **2020**, *19*, 25–33.
47. Nguyen, D.C.; Bui, T.T.; Cho, Y.B.; Kim, Y.S. Highly Hydrophobic Polydimethylsiloxane-Coated Expanded Vermiculite Sorbents for Selective Oil Removal from Water. *Nanomaterials* **2021**, *11*, 367. [CrossRef] [PubMed]
48. Taylor, N.M.; Toth, C.R.A.; Collins, V.; Mussone, P.; Gieg, L.M. The Effect of an Adsorbent Matrix on Recovery of Microorganisms from Hydrocarbon-Contaminated Groundwater. *Microorganisms* **2021**, *9*, 90. [CrossRef] [PubMed]

Article

HDPE Geomembranes for Environmental Protection: Two Case Studies

Fernando Luiz Lavoie [1,2,*], Clever Aparecido Valentin [2], Marcelo Kobelnik [2], Jefferson Lins da Silva [2] and Maria de Lurdes Lopes [3]

1 Department of Civil Engineering, Mauá Institute of Technology, 09580-900 São Caetano do Sul, Brazil
2 São Carlos School of Engineering, University of São Paulo-USP, 05508-220 São Paulo, Brazil; cclever@sc.usp.br (C.A.V.); mkobelnik@gmail.com (M.K.); jefferson@sc.usp.br (J.L.d.S.)
3 Department of Civil Engineering, University of Porto, 4099-002 Porto, Portugal; lcosta@fe.up.pt
* Correspondence: fernando.lavoie@maua.br; Tel.: +55-119-8105-8718

Received: 3 September 2020; Accepted: 12 October 2020; Published: 20 October 2020

Abstract: High-density polyethylene (HDPE) geomembranes have been used for different applications in engineering including sanitation, such as landfills and waste liquid ponds. For these applications, the material can be exposed to aging mechanisms as thermal and chemical degradation, even to UV radiation and biological contact, which can degrade the geomembrane and decrease the material's durability. This paper aims to present an experimental evaluation of two exhumed HDPE geomembranes, the first was used for 2.75 years in a sewage treatment aeration pond (LTE sample) and another was used for 5.17 years in a municipal landfill leachate pond (LCH sample). Physical and thermal analyses were used such as thermogravimetry (TG), differential thermal analysis (DTA), differential scanning calorimetry (DSC) and dynamic mechanic analysis (DMA). The thermogravimetric analyses showed significant changes in the LCH sample's thermal decomposition probably caused by the interaction reactions between the polymer and the leachate. For the DSC analyses, the behavior seen in the LTE sample was not observed in the LCH sample. In the DMA analyses, the behavior of the LTE sample storage module shows which LCH sample is less brittle. The LTE sample presented low stress cracking resistance and low tensile elongation at break, following the DMA results.

Keywords: geomembrane; HDPE; durability; thermal analysis; sewage; leachate

1. Introduction

A geomembrane is a product of the geosynthetics family used as a liner in the environmental protection system. This polymeric product can be manufactured by the industry in different polymers and it is installed in the field. The high-density polyethylene (HDPE) geomembrane is the most used type of geomembrane in the world, especially for landfills and waste liquid ponds. The high chemical and mechanical resistance of HDPE associated with a low permeability coefficient and low cost of production are the advantages of this product [1–5].

For landfill liner applications, the HDPE geomembrane can be exposed to aging mechanisms such as thermal and chemical degradations. Moreover, for slopes, in the installation time, UV radiation exposition occurs. For leachate ponds and liquid waste treatment applications, the product can be exposed to UV radiation and high temperatures, as well as the chemical and biological contact. This set of aging mechanisms can degrade the geomembrane and decrease the material's durability [6–9].

The synergic effects of aging mechanisms can significantly reduce the lifetime of the product. A package of additives is incorporated into the resin to protect the polymer and guarantee the long-term service life [10,11]. In general, carbon black (2–3%) is incorporated into the polymeric resin as UV protection and antioxidants and thermostabilizers (0.5–1.0%) as thermal and oxidative protection [12–14].

Islam and Rowe [15] carried out a study with high-density polyethylene geomembranes with nominal thicknesses of 1.5, 2.0 and 2.5 mm immersed in synthetic leachate at four temperatures (22, 55, 70 and 85 °C). The authors used the standard oxidative induction time (OIT) test to evaluate the antioxidant depletion of the samples. They understood which samples with thicker thicknesses demand more time to consume the additives. They concluded that the thicker geomembranes have longer service lives.

Research conducted by Rowe et al. [16] studied an HDPE geomembrane sample (1.5 mm of thickness) in contact with four synthetic leachates with different combinations of volatile fatty acids, inorganic nutrients, trace metal solution and surfactant at four temperatures (the lowest was 22 °C and the highest was 85 °C). The Arrhenius modeling was used to analyze the antioxidant depletion of the geomembrane. The four leachates examined were similar in terms of the antioxidant depletion rate, but the faster depletion occurred in acidic and basic leachates.

Rowe et al. [17] evaluated a 2.0 mm thick HDPE geomembrane for 8 to 10 years exposed to synthetic leachate, water and air in some temperature incubations. The sample contained approximately 97% polyethylene, 2.5% carbon black, and trace amounts of antioxidants and heat stabilizers. The synthetic leachate compound was based in the Keele Valley Landfill leachate in Canada. Several properties of the product were investigated, including the antioxidant depletion, stress cracking resistance, melt flow index and surface analysis. Using the Arrhenius Method to predict the durability of the sample, the results showed that the service life of the product can reach more than 50 years at 50 °C, about 300 years at 35 °C and more than 700 years immersed in leachate at 20 °C.

Lodi and Bueno [18] studied HDPE geomembrane samples with thicknesses of 0.8 and 2.5 mm immersed in synthetic leachate and exposed to weathering using the thermogravimetric analysis (TG). The time in both leachate and weathering expositions was 30 months. It was observed that for the 0.8 mm sample, the thermal stability temperature was higher for the exposed samples compared with the fresh (virgin) sample. The 2.5 mm sample was observed at the same behavior, but the difference among the temperatures was lower than in the 0.8 mm sample. It was observed that the carbon black content of these samples was very low compared to the specifications. The authors commented that the use of multiple analysis such as TG, melt flow index (MFI) and oxidative induction time (OIT) can support the degradation analysis for HDPE geomembranes.

Research was conducted by Ewais et al. [19] over 17 years on an HDPE geomembrane immersed in synthetic leachate, water and air at some temperatures. A stress cracking test, OIT test, MFI test and tensile test were used in the analysis. The authors concluded that the losses in the tensile properties and the stress cracking resistance confirm the oxidative degradation in the polymer. The predicted nominal failure estimated was about 18 years at 60 °C for water and about 13 years at 60 °C for leachate.

Reis et al. [20] carried out a huge field study about HDPE geomembranes exposed by the weather in eight different parts of Portugal. The authors noted that for the higher UV indexes regions, the consequences were higher for tensile properties of the samples. Moreover, the exposed geomembranes had changes in other properties, especially in the OIT results.

Antioxidant depletion analysis was carried out by Safari et al. [21] in an exhumed HDPE geomembrane that was 1.5 mm thick after 25 years of operation in a hazardous waste landfill in Canada. The authors used modern HDPE geomembranes to compare it with the exhumed geomembrane. The results showed that the leachate exposure condition could significantly influence the antioxidant depletion because the bottom samples presented higher OIT depletion than the wall samples. For this study, the sample location and, consequently, the exposure condition differences can represent different behaviors of the samples. The OIT (standard) and some HP-OIT (HP—high pressure) values were found to be significantly lower than those of modern virgin geomembranes.

Some studies evaluated the behavior of high-density polyethylene geomembranes immersed in chlorinated solutions showing a huge polymer degradation capacity of this chemical solution. Abdelaal and Rowe [22] immersed an HDPE geomembrane without HALS (hindered amine light stabilizers) in chlorinated water solutions (0.5, 1.0, 2.5 and 5.0 ppm of free chlorine) at 25, 40, 65, 75 and 85 °C for over 3 years. The authors observed that the antioxidant depletion and the stress cracking resistance in a solution of 5.0 ppm were much faster than synthetic leachate. Abdelaal et al. [23] analyzed the behavior of a high-density polyethylene geomembrane with HALS immersed in four chlorinated solutions with concentrations of 0.5–5.0 ppm at different temperatures for over 5 years. According to the authors, the sample degradation occurred quickly after immersion in all concentrations and incubation temperatures, except for the lower temperature (25 °C). Properties such as tensile and stress cracking decreased when the free chlorine concentration was increased. Finally, the authors noted that the behavior of geomembrane samples with and without HALS showed a huge difference for the chlorinated water solution incubation, but the same difference was not noted for other solution incubations.

Therefore, this work evaluated the final conditions of two exhumed geomembrane samples in sanitation applications using thermoanalytical, physical and mechanical analyses to contribute with the knowledge of HDPE geomembrane behavior applied in environmental facilities.

2. Materials and Methods

2.1. Materials

Two different high-density polyethylene geomembrane samples were evaluated exhumed from sanitation construction works. The first sample was exhumed from a sewage treatment aeration pond (called LTE) after 2.75 years of operation. This geomembrane was damaged during the operation, entailing the pond liner exchange. Table 1 presents the typical characteristics of the sewage. Figure 1 shows the sewage pond which presented an HDPE geomembrane with 1.0 mm of nominal thickness. Another sample was exhumed from a municipal landfill leachate pond (called LCH) after 5.17 years of operation. The exhumation of the geomembrane occurred because the site was used for the landfill expansion. Table 2 presents the chemical characteristics of the municipal landfill leachate. Figure 2 shows the landfill leachate pond that presented an HDPE geomembrane with 2.0 mm of nominal thickness. Both construction works are located in Brazil.

Table 1. Typical characteristics of the sewage [24].

Characteristic	Unit	Result
Fixed Suspension Solids	mg/L	80
Volatile Suspended Solids	mg/L	320
Total Suspended Solids	mg/L	350
Fixed Dissolved Solids	mg/L	400
Volatile Dissolved Solids	mg/L	300
Total Dissolved Solids	mg/L	700
Sedimentable Solids	mg/L	15
Total solids	mg/L	1100
pH	-	6.5–7.5
BOD	mg/L	100–400

Figure 1. Sewage treatment aeration pond under operation.

Table 2. Chemical characteristics of the municipal landfill leachate [25].

Characteristic	Unit	Result
Total alkalinity	mg CaCO$_3$/L	6912
Calcium	mg/L	366
Cadmium	mg/L	Not Detected
Lead	mg/L	Not Detected
Chloride	mg Cl-/L	3502
Total Coliforms	NMP/100 mL	8.3
Conductivity	μS/cm	23,210
BOD	mg O$_2$/L	775
Iron	mg/L	11.4
Magnesium	mg/L	0.165
Mercury	mg/L	Not Detected
Nickel	mg/L	0.230
pH	-	8.19
Mineral Oils and Greases	mg/L	Not Detected

Figure 2. Municipal landfill leachate pond under demobilization.

2.2. Physical Properties

The thickness [26] was determined by measuring the difference between the dead-weight loading gauge and the geomembrane specimen thickness with 0.001 mm precision, applying a pressure force of 200 ± 0.2 kPa. The carbon black content (CBC) [27] was determined using a muffle furnace at $605 \pm 5\ °C$

for 3 min in an aluminum dish by pyrolysis. It had been used for 1 ± 0.1 g of each geomembrane specimen. The masses before and after being burned were determined using an analytical balance with 0.0001 g precision. The measure of density [28] was performed in isopropyl alcohol at $21 \pm 0.1\,°C$, mass sample $1.0\,g \pm 0.1$ g in apparatus that included an analytical balance with 0.0001 g precision and an immersion vessel and a beaker. The melt flow index (MFI) [29] of the studied samples were obtained using a plastometer with a smooth bore 2.095 ± 0.005 mm in diameter and 8000 ± 0.025 mm long. The polymer was extruded with $190 \pm 0.08\,°C$ with a deadweight load of 5.0 kg and its mass was measured for 10 min using an analytical balance with 0.0001 g precision.

2.3. Mechanical Properties

Tensile and tear tests were performed using an EMIC Universal Machine, model DL 3000, manufactured by EMIC at São José dos Pinhais, Brazil, with pneumatic grips and a 2-kN load cell. The tensile test was performed using the type IV dog bone specimen with a test speed of 50 mm min^{-1}, which is the speed test indicated for HDPE geomembranes [30]. The tear test was performed using a test specimen which produces tearing in a small area of stress concentration at rates far below those usually encountered in the field. This test uses a test speed of 51 mm min^{-1} with an initial jaw separation of 25.4 mm [31].

2.4. Stress Cracking (SC) and Oxidative Induction Time (OIT) Properties

The stress cracking test was performed using an equipment manufactured by WT Indústria at São Carlos, Brazil, with a capacity for 20 specimens simultaneously. The test used was the NCTL-SP (notched constant load test-single point) [32]. Moreover, 30% of yield tensile stress was applied at the specimen using a deadweight with a 10 g precision. The specimen was immersed in a solution of 10% Igepal CO 630 and 90% of water at $50 \pm 1\,°C$. A notch of 20% of the specimen thickness with 0.001 mm of precision was taken in each specimen. The rupture sample time (1 s precision) was measured using an electronic device for each specimen.

For measurements of high-pressure oxidative induction time (OIT-HP) [33], DSC equipment, model Q20, manufactured by TA Instruments at New Castle, United States of America, with high-pressure cell Q series DSC Pressure Cell with a sample mass of 5.0 mg was used. The measurements were proceeded in an open aluminum crucible, under a heating rate of $20\,°C$ min^{-1}, with nitrogen gas purge at a constant pressure of 5 Psi, from ambient temperature to $150\,°C$. After this stage, nitrogen was exchanged for oxygen at a constant pressure of 500 Psi, maintaining the isotherm of $150\,°C$ until the complete sample oxidation.

2.5. Thermal Analysis Methodology

The thermal analysis was used mainly to evaluate the samples' conditions and observe the possible changes that occurred. The TG/DTG and DTA curves evaluate the thermal stability temperatures. Furthermore, the activation energy of each sample can be obtained. DSC curves can evaluate the glass transition temperature (T_g), crystallization temperature ($T_{crist.}$) and the melting point of each sample. In addition, the DMA curves evaluate the effects precisely of applied stresses and temperatures under the sample, which allows better precision in the data on molecular relaxation. The knowledge of molecular relaxation helps in the analysis of physical changes in the material, providing information on the protection of the polymer, as molecular relaxation makes the material susceptible to a short lifetime.

The samples were evaluated on dynamic mechanic analysis (DMA) equipment, in a flexural mode, using a DMA thermal analyzer, model Q800, manufactured by TA Instruments at New Castle, United States of America. The samples had a dimension of 35×13 mm and were performed under a heating rate of $10\,°C$ min^{-1} with nitrogen purge gas (flow of 50 mL min^{-1}). The oscillation amplitude of 20 mm was used, with a frequency of 1 Hz and a temperature range from -80 to $125\,°C$.

The thermogravimetry (TG/DTG) and differential thermal analysis (DTA) were performed at an SDT 2960 (TA Instruments, USA) with heating rates of 5, 10, 20 and $30\,°C$ min^{-1} under carbonic gas and synthetic air purge gases, with a flow of 100 mL min^{-1}. These polymers were evaluated in an α-alumina crucible in a temperature range of 30 to 600 °C. The activation energy was obtained by the Flynn-Wall-Ozawa method using the DTG curves to obtain the data [34–36].

In this study, DSC was used to measure the changes in both materials. Measurements were conducted using a DSC1 Stare, manufactured by Mettler Toledo at Columbus, United States of America, with samples of a diameter of 3 mm^2 and masses around 7.5 mg to LTE and 15.5 mg to LCH. These samples were performed in the aluminum crucible in the temperature range of −80 to 200 °C. The first step was cooling from 25 to −80 °C, followed by the second step, where the samples were heated from −80 to 200 °C; in the third step, the samples cooled again from 200 to −80 °C; in the fourth step, the sample was heating from −80 to 200 °C and finally, the fifth step, the sample was cooling from 200 to 25 °C. The heating and cooling rate was 30 °C min^{-1} under nitrogen purge gas (flow of 50 mL min^{-1}).

3. Results and Discussion

3.1. Physical Evaluations

Table 3 shows the results of exhumed samples' physical tests. The average values of the thickness are following the minimum values of the GRI-GM13 [37]. The carbon black content values obtained are also according to the GRI-GM13 [37], which determines values between 2–3%. The obtained density values follow the GRI-GM13 [37], which requires a minimum density value of 0.940 g/cm^3. According to Telles et al. [38], low MFI values exhibit good stress cracking environmental resistance for HDPE geomembranes. The obtained results showed low MFI values for both tested samples.

Table 3. Physical test results of exhumed samples.

Sample	Thickness/(mm)	CBC/(%)	Density/(g/cm^3)	MFI/(g/10 min)
LTE	1.001 (±0.038)	2.49 (±0.47)	0.959 (±0.001)	0.4555 (±0.0061)
LCH	2.075 (±0.036)	2.36 (±0.11)	0.946 (±0.002)	0.5008 (±0.0072)

The standard deviations are shown between brackets. CBC = carbon black content.

3.2. Mechanical Evaluations

Table 4 shows the results of the exhumed samples' mechanical tests. The tests were conducted only for the machine direction. The samples' results are according to the GRI-GM13 [37] concerning the minimum values of the tensile break (27 kN m^{-1} for 1.0 mm of thickness and 53 kN m^{-1} for 2.0 mm of thickness) and tear resistance (125 N for 1.0 mm of thickness and 249 N for 2.0 mm of thickness). For the analyzed samples, only the LCH sample presented tensile elongation at breaks higher than 700%, which is the minimum value prescribed by GRI-GM13 [37].

Table 4. Mechanical test results of exhumed samples.

Sample	Tens. Break Resist./(kN m^{-1})	Tens. Break Elong./(%)	Tear Resist./(N)
LTE	27.12 (±1.30)	679.33 (±27.53)	170.13 (±2.05)
LCH	60.40 (±7.66)	752.60 (±81.38)	321.80 (±8.92)

The standard deviations are shown between brackets.

3.3. Stress Cracking (SC) and Oxidative Induction Time (OIT) Evaluations

Table 5 shows the results of the stress cracking test (Notched Constant Tensile Load Test–Single Point-NCTL-SP) and the high-pressure oxidative induction time test (OIT-HP). According to the GRI-GM13 [37], the minimum required value for the stress cracking test (NCTL-SP) is 500 h. The LCH sample performed an average value higher than 500 h but presented a high variation in each specimen value tested, two of the five specimens tested obtained values higher than 1000 h, but the other two of the five specimens obtained values less than 200 h. The mean stress cracking value is in agreement with the tensile behavior of this sample. For the other exhumed sample tested, the results showed a low value of stress cracking resistance. The LTE sample, despite having good tensile behavior and an adequate melt flow index, presented unexpected stress cracking results.

The minimum required value for the OIT High-Pressure test, for the GRI-GM13 [37] is 400 min. According to Mueller and Jakob [39], the main function of antioxidants is to prevent initiation of oxidation chain reactions. Antioxidants are more effective over a certain range of temperatures. As an instance, phosphites are more effective at higher temperatures whereas hindered amine light stabilizers (HALS) are effective at ambient temperature. Hindered phenols, however, are used as long-term stabilizers since they are effective over a wide range of temperatures. None of the samples presented OIT-HP values equal to or higher than 400 min. The presence of HALS in the additive package increases the results of the OIT-HP, as this test is performed at 150 °C. Probably none of the samples presented HALS in the additive package. Both samples presented result values lower than 400 min. The LCH sample obtained the highest OIT-HP value of the exhumed samples tested.

Table 5. Tests conducted to the SC and high-pressure oxidative induction time (OIT-HP) of the exhumed high-density polyethylene (HDPE) geomembrane samples.

Sample	SC (NCTL-SP) (hours)	OIT-High-Pressure (min)
LTE	30.89 (±12.31)	180.0 (±1.41)
LCH	542.15 (±508.17)	231.50 (±2.12)

The standard deviations are shown between brackets.

3.4. Thermal Analysis Evaluations

3.4.1. Thermogravimetry (TG) and Differential Thermal Analysis (DTA)

An experiment series was systematically carried out using TG-DTA analysis simultaneously, with heating rates of 5, 10, 20 and 30 °C min^{-1}. The obtained results from both HDPE geomembrane samples are shown in Figures 3–6.

Figure 3 shows the TG, DTG and DTA curves of the LTE sample in synthetic air with different heating rates. It was observed that the thermal stability of the LTE sample (TG curves in Figure 3A), which was in contact with the sewage, reached 238 °C for the heating rate of 5 °C and gradually increased to 262 °C for the heating rate of 30 °C. The first sample's mass variation for the heating rate of 5 °C occurs in one stage only, in the range of 248–369 °C, as seen in the TG curve, whereas in the DTG curve (Figure 3B) only a small deviation from the baseline is seen. The second stage, attributed to the material's thermal decomposition process, occurs with overlapping reactions, which have presented variations in the intensity of the reactions for the different heating rates, as seen in the TG curves. Furthermore, in the DTG curves it can be observed that with the increase in the heating rate, there is a widening in the decomposition events. For the verification of the mass variations and the respective temperature ranges, the values of each stage are shown in Table 1. For both LTE and LCH samples, at the end of the analyses, the ash formation was observed, which was easily removed from the crucible by a breath. From the DTA curves (Figure 3C), endothermic peaks can be observed at 127, 130, 135 and 140 °C, respectively, for the heating rates of 5, 10, 20 and 30 °C min^{-1}, which were attributed to material melting. This melting point temperature variation is due to the speed of the sample heat absorption, which occurs for different heating rates. Other exothermic peaks are linked to the different stages of the material's thermal decomposition and, as an effect of the speeds of the different heating rates, the peaks widened.

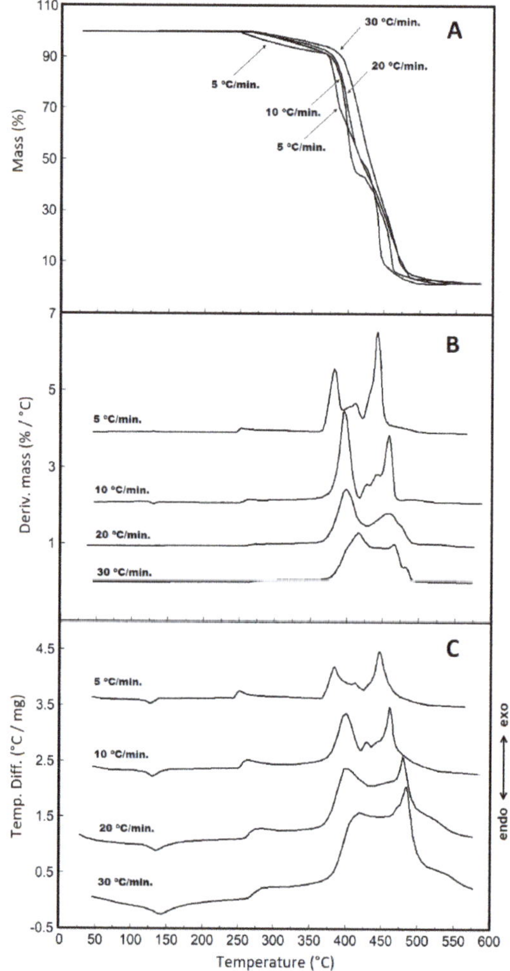

Figure 3. (**A**) TG curves for LTE sample in the synthetic gas purge, with mass samples around 7.50 mg; (**B**) DTG curves for the LTE sample in the synthetic gas purge, with mass samples around 7.50 mg; (**C**) DTA curves for the LTE sample in the synthetic gas purge, with mass samples around 7.50 mg. (all analyses conducted with heating rates of 5, 10, 20 and 30 °C min^{-1} with the flow of 110 mL min^{-1} in α-alumina crucible).

In the LTE sample analysis, there were no changes in the thermal behavior of the geomembrane, despite the contact of the sample with the sewage for 2.75 years, compared with the results of virgin HDPE geomembrane samples studied by Valentin et al. [40]. However, for the LCH sample evaluation which had been in contact with the leachate for 5.17 years, it was observed that the thermogravimetric behavior presented significant changes in the sample's thermal decomposition. To ensure this result, two other TG curves were performed in each heating rate to verify the thermal behavior observed once again. Indeed, the other analyses showed that there had been undoubtedly a change in the sample's thermal behavior due to contact with the leachate. The leachate is a highly concentrated organic substance and this fact added to the 5.17 years of sample exposition can explain the changes in the sample's thermal behavior. Thus, as seen in the TG curves (Figure 4A), the organic material absorption by the geomembrane probably changed the material's behavior, causing interaction reactions between the polymer and the leachate. Figure 4A shows that the heating rate of 10 °C min^{-1} is different from the other curves. In this specific curve, the first mass variation occurs between 259–406 °C and the second mass variation occurs from 406 °C, in two stages, which can be seen in the DTG curve (Figure 4B). The 5 °C heating rate curve shows three stages of mass variation, in the temperature ranges of 248-369, 369–392 and 392–459 °C. For the heating rates of 20 and 30 °C from the second mass variation, the DTG curves are wider, which indicates that the decomposition occurs with overlapping reactions. The mass variation data for each thermal decomposition stage are shown in Table 6.

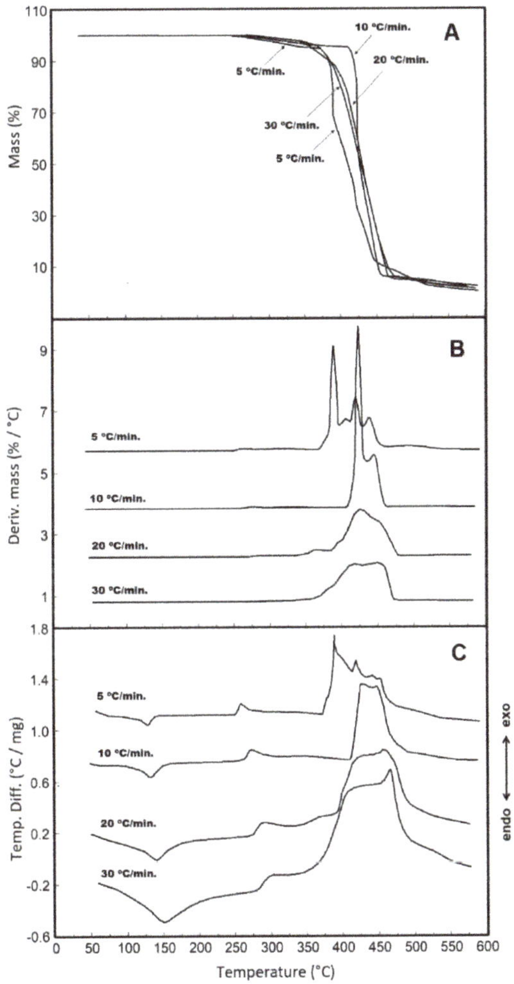

Figure 4. (**A**) TG curves for LCH sample in synthetic gas purge, with mass samples around 15.50 mg; (**B**) DTG curves for the LCH sample in synthetic gas purge, with mass samples around 15.50 mg; (**C**) DTA curves for the LCH sample in synthetic gas purge, with mass samples around 15.50 mg (all analyses conducted with heating rates of 5, 10, 20 and 30 °C min^{-1} with the flow of 110 mL min^{-1} in α-alumina crucible).

The obtained results from the DTA curves are shown in Figure 4C for the LCH sample. It can be observed that the first event is an endothermic reaction (without mass variation in the TG/DTG curves), which occurs at temperatures of 127, 132, 138 and 150 °C, for the four heating rates, respectively, representing the samples' melting point. As both samples (LTE and LCH) were produced with the same type of polymer, the melting point values differ slightly from each other probably due to their different uses of the conditions. The following stages of thermal decomposition are exothermic reactions for the four heating rates. The heating rate of 5 °C min^{-1} showed a sharp peak in the second stage of thermal decomposition, which is attributed to a sample combustion reaction. In addition, as with the LTE sample, it can be observed that with the increase in the heating rate, there is a widening effect of the exothermic reactions due to the overlapping reactions.

Table 6. Temperature intervals (°C) obtained from TG/DTG curves for the thermal decomposition stages in synthetic air and carbonic gas, with heat flow rates of 5, 10, 20 and 30 °C min^{-1}.

Sample	5 °C min^{-1}	10 °C min^{-1}	20 °C min^{-1}	30 °C min^{-1}
	248–369 °C	259–406 °C	264–352 °C	264–358 °C
	5.33%	4.64%	3.47%	2.95%
	369–392 °C	406–465 °C	352–392 °C	358–578 °C
	29.15%	89.52%	8.68%	91.89%
LCH Synthetic air	392–459 °C	465–600 °C	392–484 °C	578–600 °C
	55.77 °C	4.85%	82.48%	2.72%
	459–600 ºC	——	484–600 °C	——
	9.51%		3.35%	
	Residue	Residue	Residue	Residue
	0.24%	0.99%	2.95%	2.44%
	238–363 °C	241–364 °C	249–362 °C	262-372 °C
	7.89%	8.40%	6.33%	6.01%
	363–422 °C	364–417 °C	362–426 °C	372–500 °C
	44.09%	48.08%	49.29%	90.57%
LTE Synthetic air	422–468 °C	417–475 °C	426–497 °C	500–580 °C
	42.77%	38.92%	39.65%	1.34%
	468–600 °C	475–592 °C	497–586 °C	——
	3.77%	3.46%	2.85%	——
	Residue	Residue	Residue	Residue
	1.48%	1.14%	1.88%	2.08%
	378–498 °C	383–435 °C	387–443 °C	401–451 °C
	96.38%	3.66%	1.33%	3.00%
LCH Carbonic air	——	435–508 °C	443–523 °C	451–530 °C
	——	92.93%	94.89%	93.48%
	Residue	Residue	Residue	Residue
	3.62%	3.41%	3.78%	3.52%
	376–496 °C	381–507 °C	405–518 °C	410–525 °C
	94.04%	94.28%	94.97%	96.15%
LTE Carbonic air	Residue	Residue	Residue	Residue
	5.96%	5.72%	5.03%	3.85%

The TG/DTG curves in a non-isothermal condition, shown in Figures 5 and 6, in carbon dioxide purge gas, with the analyses at different heating rates, were used to obtain the kinetic data of the LTE and LCH samples, respectively.

For the LTE sample, as seen in Figure 5, there was no change in mass between the initial temperature and 376 °C for the analysis in the heating rate of 5 °C, while for the other heating rates, the values were: 381 °C (10 °C min^{-1}), 405 °C (20 °C min^{-1}) and 410 °C (30 °C min^{-1}). The mass variation values are shown in Table 1. It can also be observed that the TG curves show similarities since the beginning of the thermal decomposition behavior, with the presence of a shoulder at the beginning of the decomposition reaction and later homogeneity during the thermal decomposition.

For the LCH sample, as seen in Figure 6, the geomembrane thermal stability occurs up to the following temperatures and respective heating rates: 378 °C (5 °C min^{-1}), 383 °C (10 °C min^{-1}), 387 °C (20 °C min^{-1}) and 401 °C (30 °C min^{-1}). The thermal decomposition process showed that at the beginning of the reaction, a small shoulder was formed and then a thermal decomposition occurred. During the main decomposition process, a shoulder formation also occurred in the DTG curves, except for the heating rate of 5 °C min^{-1} which was attributed to overlapping decomposition reactions.

At the end of the thermal decomposition reaction, for the analysis of both samples, the presence of carbonaceous material impregnated in the crucible was observed, which was removed.

The activation energy values are shown in Table 2, which presents the intervals used to obtain the kinetic values for the analyses in carbonic gas and synthetic air, except for the LCH sample. This sample presented an altered decomposition behavior, that is, the heating rates did not show homogeneous displacement among them, as seen in the other analyses. Therefore, this sample was not analyzed in the synthetic purge gas.

Figure 7 shows the behavior of the activation energy during thermal decomposition. For the analysis of the LTE sample in synthetic air, it can be observed that the activation energy has a lower value than the analysis carried out on carbon dioxide and shows that there is a gradual decrease. This shows that the thermal decomposition of the material occurred using the purge gas during the reaction, that is, an oxidation reaction occurs between the oxygen present in the synthetic air and the polymer. For the LTE sample reaction in carbonic gas, it can be observed that the initial value of its activation energy is close to the synthetic air analysis. However, there is a gradual increase in the activation energy. This gradual increase indicates which decomposition occurs due to the increase in temperature, considering that carbonic gas is inert. Likewise, for the LCH sample, in addition to having initial activation energy values higher than those of the LTE sample, the same behavior of a gradual increase in activation energy also occurs. The HDPE activation energy values under a nitrogen atmosphere were reported by Valentin et al. [40]. These authors showed that the activation energies values were similar to those obtained in this work, however, the authors did not proceed with any evaluation in another atmosphere.

Additionally, the correlation coefficient values (Table 7) show a linear pattern for the analyzed samples, which shows that the kinetic data follow the same behavior trend.

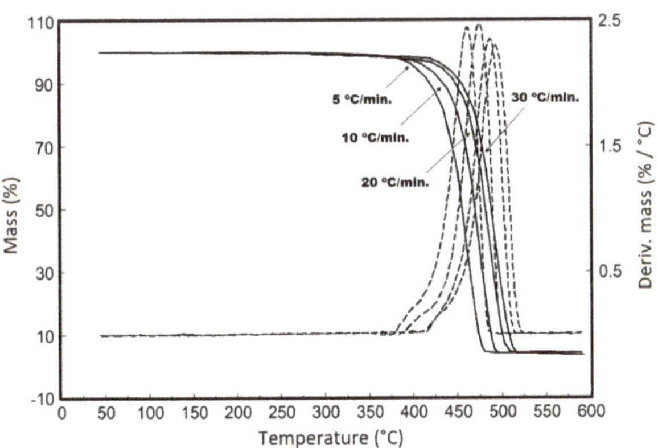

Figure 5. LTE sample TG/DTG curves in carbonic purge gas, with mass samples around 7.50 mg, with analyses conducted with heating rates of 5, 10, 20 and 30 °C min^{-1} with the flow of 110 mL min^{-1} in α-alumina crucible.

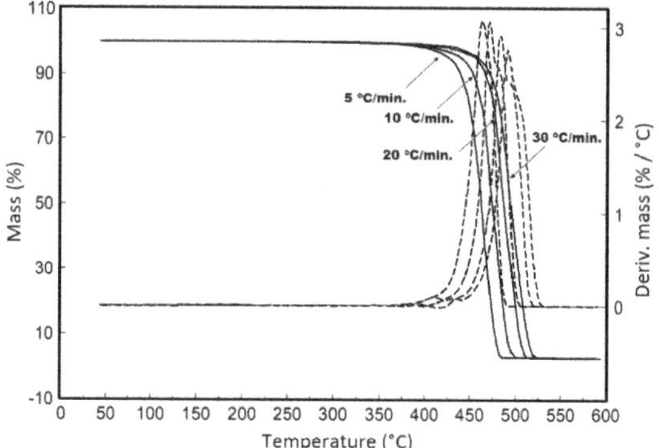

Figure 6. LCH sample TG/DTG curves in carbonic purge gas, with mass samples around 15.50 mg, with analyses conducted with heating rates of 5, 10, 20 and 30 °C min^{-1} with the flow of 110 mL min^{-1} in an α-alumina crucible.

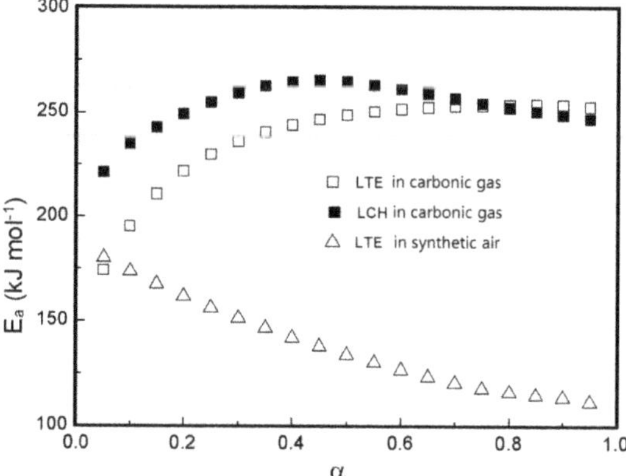

Figure 7. Activation energy versus degree conversion for LTE and LCH samples.

Table 7. Temperature intervals used for the kinetic analysis, activation energy and correlation coefficient values.

Purge Gas and Sample	Temperature Ranges for Kinetic Evaluation (DTG Curves)	E_a/kJ mol^{-1}	R
Synthetic air LTE	(5 °C) 243–281 °C (10 °C) 250–287 °C (20 °C) 258–294 °C (30 °C) 262–300 °C	137.94 ($\pm$ 0.15)	0.99357
Carbonic gas LTE	(5 °C) 376–490 °C (10 °C) 386–503 °C (20 °C) 406–516 °C (30 °C) 418–525 °C	237.83 ($\pm$ 0.09)	0.99634
Carbonic gas LCH	(5 °C) 390–492 °C (10 °C) 405–507 °C (20 °C) 425–524 °C (30 °C) 441–533 °C	253.07 ($\pm$ 0.04)	0.99945

The standard deviations are shown between brackets.

3.4.2. Differential Scanning Calorimetry (DSC)

Figures 8 and 9 show, respectively, the LTE and LCH samples analyses under heating and cooling conditions performed to verify the materials' transitions.

Figure 8A shows that during the first cooling there was a decrease in the LTE sample heat flow, which coincides with the baseline of the second cooling between 5 to −28 °C. The first heating of the sample (Figure 8B) shows that there had been a change in the baseline of the curve between −55 to −25 °C. For the second cooling, there was a heat flow decrease in the range of −28 to −40 °C, which is attributed to a difference in the heat capacity of the sample. However, in the second and third heating, there was an inverse behavior, that is, there was an increase in the material's heat flow. These changes in the sample's heat flow occurred due to the behavior of the material's molecules, which during cooling, the polymer molecules undergo contraction due to low temperature, getting closer to one another, and consequently changing the sample dimensions. During heating, the molecules tend to distance themselves from each other, and thus, the heat capacity changes, causing the accumulated energy to be released [41]. In contrast, in the first heating, the material melted and then there was a crystallization, where there is a molecular reorganization and a temperature decrease. Then, the molecular structure experiences a decrease in the heat flow between −25 and −39 °C, which shows an energy loss that corresponds to a material's molecular approximation. As the material was heated and recrystallized again, there was an even more random molecular reorganization, which causes the change in the heat flow. When reheating this sample again, in the second heating (Figure 8B) there is an increase in the sample's heat flow, and then the new fusion, which has a wider peak area than the first fusion. When recrystallizing again, as seen in the third cooling, the peak is also smaller and wider. However, as seen, there was a small change in the baseline between −4 to −19 °C. Finally, in the third heating, there was an overlap with the event observed in the second cooling, attributed to the molecular distance.

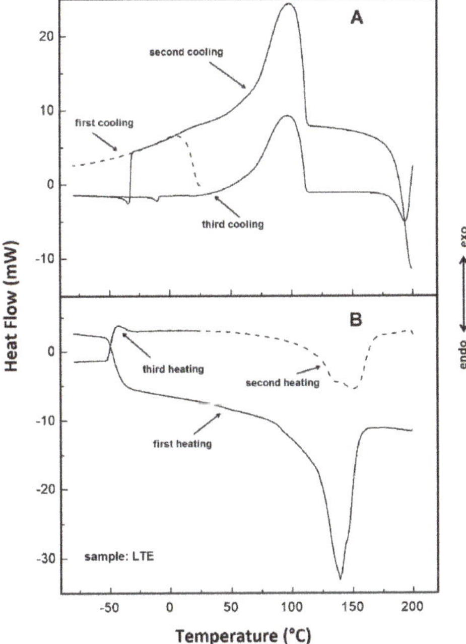

Figure 8. LTE sample DSC curves with a heating rate of 30 °C min^{-1} under nitrogen gas purge with the flow of 50 mL min^{-1} in an aluminum crucible with sample masses around 3.50 mg: (**A**) heating and (**B**) cooling.

For the LCH sample, the behavior seen in the LTE sample was not observed. It indicates that there was probably an effect of leachate in the sample, causing a change in the heat flow. As seen in Figure 9A, the first cooling and the first heating have the same behavior observed for the LTE sample. After the melting and the first crystallization, a slight change in the baseline is seen in the second cooling (17 to 8 °C) and the third cooling (7 to −2 °C), both in agreement with what is seen in the third cooling of the LTE sample. It is important to note that during the heating of this sample, there was no change in the material's baseline between (−54 to −16 °C). Thus, it can be reaffirmed that the effect caused on the DSC curve for the LCH sample is attributed to the presence of leachate molecules, which altered the material's behavior after the melting process.

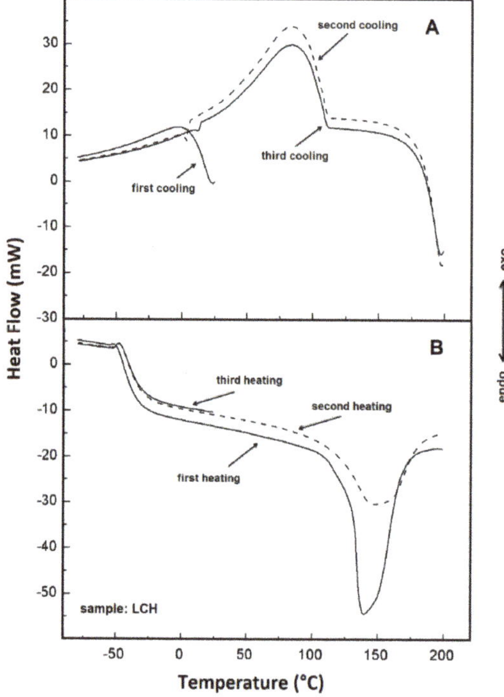

Figure 9. LCH sample DSC curves with a heating rate of 30 °C min^{-1} under nitrogen gas purge with the flow of 50 mL min^{-1} in an aluminum crucible with sample masses around 3.50 mg: (**A**) heating and (**B**) cooling.

3.4.3. Dynamic Mechanical Analysis (DMA)

DMA was used to track changes in the molecular relaxation of the materials, to learn whether these changes were a function of the time in operation and an indication of the material's degradation, and to learn about the effects of leachate and sewage on the material. DMA analyses usually have different modules, which allow a better characterization of the material, that is, the material's capacity to store energy (storage modulus), its capacity to lose energy (loss modulus) and the proportion of these effects (tan-δ), which is called damping (damping factor) [42].

Figure 10 shows the DMA analysis under nitrogen purge gas, at a range temperature of −80 to 120 °C for both samples, where it is possible to verify that both samples have different behavior between each other. The temperature range utilized in this analysis is situated above the glass transition temperature, which is a region of high hardness and therefore, the information obtained from the temperature of −80 °C refers to the transition region performed in this work [43]. The behavior of the

LTE sample storage module (Figure 10A) shows that this sample has a higher value than the LCH sample, which means that the LCH sample is less brittle, that is, more elastic. The basic definition of the storage module is given as a measure of the mechanical energy that the material is capable of storing, in the form of potential or elastic energy [44]. This result shows that the potential energy of the LTE sample decreases gradually until the temperature of 32 °C, and after this temperature, this sample increases its potential energy again until the temperature of 78 °C. This effect is attributed to the simultaneous effect of raising the temperature and the interaction of the geomembrane polymer with molecules from the sewage which were impregnated in the geomembrane, which causes an increase in stiffness (becomes less elastic). The same effect is observed for the LCH sample, but to a lesser degree. The expected trend would be an increase in the elasticity of the material and a consequent decrease in stiffness.

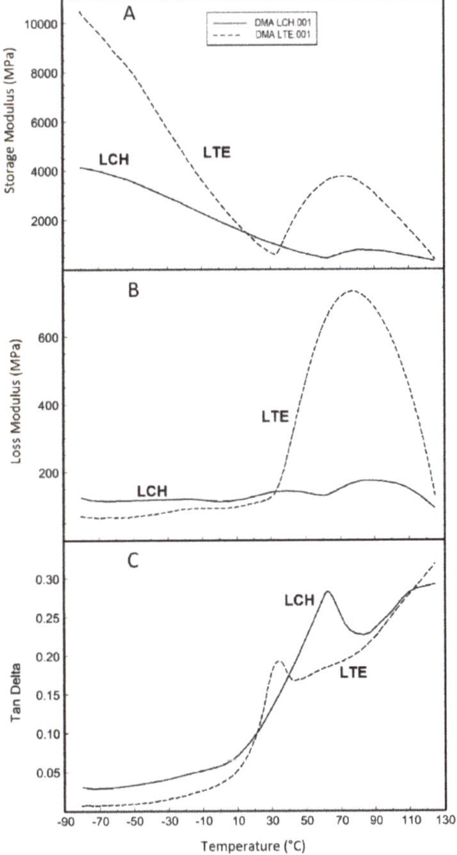

Figure 10. DMA curves of LCH and LTE samples with a heating rate of 5 °C min^{-1} under nitrogen gas purge with a flow of 50 mL min^{-1}: (**A**) storage modulus, (**B**) loss modulus and (**C**) tan delta.

These results show that the geomembranes' contact with the sewage and leachate caused the alteration in elastic potential, which is attributed to the molecular interaction in the structure of the polymeric matrix. Furthermore, from temperatures of 72 °C for the LTE sample and 85 °C for the LCH sample, both samples tend to increase viscosity since the increase in temperature leads to the melting stage, that is, the polymer chains start to have a greater movement due to the increase in the temperature with loss of stored energy, until the consequent fusion.

The loss module (E″) is shown in Figure 10B, which shows that the LTE sample has a much higher curve behavior than the LCH sample curve in the same temperature range (32 to 125 °C). This information indicates that the LTE sample, when dissipating the energy stored in this interval, shows that its elasticity decreases again, while the LCH sample has a constant behavior, showing variations in different temperature ranges [45].

In Figure 10C, tan-δ is shown, which is a relationship between the material's elastic/stiffness behavior [42,43]. As both materials are stiffer at low temperatures, tan-δ tends to increase with the temperature increasing. Thus, for the LTE sample at the temperature of 32 °C a peak of tan-δ, there was a peak in the decrease in the material's stiffness and after this temperature there was a decrease in the tan-δ, indicating an increase in stiffness. For the LCH sample there was the same behavior but with the tan-δ peak at 63 °C, which corresponds to the temperature at which is the value of the lowest potential energy of this sample (Figure 10A), that is, after this temperature, this sample increases its potential energy up to 82 °C. As with the LTE sample, there was an increase in the LCH stiffness, and therefore, the tan-δ peak (63 °C) corresponds to this situation. Tan-δ decreases with increasing stiffness, that is, if the material is at a constant stiffness, tan-δ will be constant (will have a zero value) [46,47].

4. Conclusions

This paper analyzed two HDPE geomembrane samples exposed to different sanitation environments. The interaction and the aging of the studied samples may be different, not only in relation to the different fluids but also in relation to the different geomembranes' thicknesses and to the different exposed periods of time. The samples presented some similarities and some differences in their behaviors. The thermogravimetric analyses showed significant changes in the LCH sample's thermal decomposition. The organic material absorption by the geomembrane probably changed the material's behavior, causing interaction reactions between the polymer and the leachate. In the LTE sample analysis, no changes were observed in the geomembrane thermal behavior.

In the kinetic evaluation, for the analysis of the LTE sample, with a synthetic air purge gas, it can be observed that the activation energy has a lower value than the analysis carried out in carbon dioxide. For the LTE sample reaction in carbonic gas, it can be observed that the initial value of its activation energy is close to the synthetic air analysis. However, there is a gradual increase in the activation energy. This gradual increase indicates which decomposition occurs due to the increase in temperature, considering that carbonic gas is inert. The LCH sample presented an altered decomposition behavior, that is, the heating rates did not show homogeneous displacement among them, as seen in the other analyses.

For the DSC analyses, the behavior seen in the LTE sample was not observed in the LCH sample. It indicates that probably there was an effect of leachate in the sample, causing a change in the heat flow. For the LCH sample, after the melting and the first crystallization, a slight change in the baseline is seen in the second cooling and the third cooling, both in agreement with what is seen in the third cooling of the LTE sample. It is important to note that during the heating of this sample, there was no change in the material's baseline. Thus, it can be reaffirmed that the effect caused on the DSC curve for the LCH sample is attributed to the presence of leachate molecules, which altered the material's behavior after the melting process.

The DMA analyses showed different behaviors between the samples. The behavior of the LTE sample storage module shows that this sample has a higher value than for the LCH sample, which means that the LCH sample is less brittle, that is, more elastic. These results follow the physical results, because the LTE sample presented low stress cracking resistance and low tensile elongation at break. By analyzing the loss module, can be observed that the LTE sample, when dissipating the energy stored in this interval, shows that its elasticity decreases again, while the LCH sample has a constant behavior, showing variations in different temperature ranges. Both samples presented the similar behavior in the tan-δ analysis, but the temperature at which is the value of the lowest potential energy was different for the samples. All polymeric materials, with rare exceptions, are elastic. The molecular interaction will determine the elasticity of the material, that is, the greater the molecular disorganization, the more amorphous the material will be.

For future studies, we recommended obtaining and characterizing the virgin sample and monitoring the exposed samples during the exposition at predetermined times. Besides, to characterize the fluids in contact with the geomembranes and to monitor their properties along the exposed time is recommended.

Author Contributions: Conceptualization, F.L.L., C.A.V., J.L.d.S. and M.d.L.L.; methodology, M.K.; validation, F.L.L., C.A.V. and M.K.; formal analysis, C.A.V. and M.K.; investigation, F.L.L., C.A.V. and M.K.; resources, F.L.L. and M.K.; writing—original draft preparation, F.L.L., C.A.V. and M.K.; writing—review and editing, F.L.L., C.A.V., J.L.d.S. and M.d.L.L.; visualization, F.L.L. and C.A.V.; supervision, F.L.L.; project administration, J.L.d.S. All authors have read and agreed to the published version of the manuscript.

Funding: This research received no external funding.

References

1. Rollin, A.R.; Rigo, J.M. *Geomembranes—Identification and Performance Testing*, 1st ed.; Chapman and Hall: London, UK, 1991.
2. Palmeira, E.M. *Geossintéticos em Geotecnia e Meio Ambiente*, 1st ed.; Oficina de Textos: São Paulo, Brazil, 2018.
3. Koerner, R.M. *Designing with Geosynthetics*, 5th ed.; Prentice Hall: Upper Saddle River, NJ, USA, 2005.
4. Scheirs, J. *A Guide to Polymeric Geomembranes: A Practical Approach*, 1st ed.; Wiley: London, UK, 2009.
5. Vertematti, J.C. *Manual Brasileiro de Geossintéticos*, 2nd ed.; Blücher: São Paulo, Brazil, 2015.
6. Rowe, R.K.; Sangam, H.P. Durability of HDPE Geomembranes. *Geotext Geomembr.* **2002**, *20*, 77–95. [CrossRef]
7. Kay, D.; Blond, E.; Mlynarek, J. Geosynthetics durability: A polymer chemistry issue. In Proceedings of the 57th Canadian Geotechnical Conference, Quebec, QC, Canada, 24–26 October 2004.
8. Rowe, R.K.; Abdelaal, F.B.; Brachman, R.W.I. Antioxidant depletion of HDPE geomembrane with sand protection layer. *Geosynth. Int.* **2013**, *20*, 73–89. [CrossRef]
9. Koerner, G.R.; Hsuan, Y.G.; Koerner, R.M. The durability of geosynthetics. In *Geosynthetics in Civil Engineering*, 1st ed.; Sarsby, R.W., Ed.; Woodhead Published Limited: Cambridge, UK, 2007; pp. 36–65.
10. Hsuan, Y.G.; Koerner, R.M. The single point-notched constant tension load test: A quality control test for assessing stress crack resistance. *Geosynth. Int.* **1995**, *2*, 831–843. [CrossRef]
11. Hsuan, Y.G.; Koerner, R.M. Antioxidant depletion lifetime in high density polyethylene geomembranes. *J. Geotech. Geoenviron.* **1998**, *124*, 532–541. [CrossRef]
12. Ewais, A.M.R.; Rowe, R.K.; Scheirs, J. Degradation behaviour of HDPE geomembranes with high and low initial high-pressure oxidative induction time. *Geotext. Geomembr.* **2014**, *42*, 111–126. [CrossRef]
13. Lodi, P.C.; Bueno, B.S.; Vilar, O.M. The Effects of Weathering Exposure on the Physical, Mechanical, and Thermal Properties of High-density Polyethylene and Poly (Vinyl Chloride). *Mater. Res.* **2013**, *16*, 1331–1335. [CrossRef]
14. Rowe, R.K.; Ewais, A.M.R. Ageing of exposed geomembranes at locations with different climatological conditions. *Can. Geotech. J.* **2015**, *52*, 326–343. [CrossRef]
15. Islam, M.Z.; Rowe, R.K. Effect of HDPE geomembrane thickness on the depletion of antioxidants. In Proceedings of the 60th Canadian Geotechnical Conference and the 8th Joint CGS/IAH-CNC Groundwater Conference, Ottawa, ON, Canada, 21–24 October 2007.
16. Rowe, R.K.; Islam, M.Z.; Hsuan, Y.G. Leachate chemical composition effects on OIT depletion in an HDPE geomembrane. *Geosynth. Int.* **2008**, *15*, 136–151. [CrossRef]
17. Rowe, R.K.; Rimal, S.; Sangam, H. Ageing of HDPE geomembrane exposed to air, water and leachate at different temperatures. *Geotext. Geomembr.* **2009**, *27*, 137–151. [CrossRef]
18. Lodi, P.C.; Bueno, B.S. Thermo-gravimetric Analysis (TGA) after Different Exposures of High Density Polyethylene (HDPE) and Poly Vinyl Chloride (PVC) Geomembranes. *Electron. J. Geotech. Eng.* **2012**, *17*, 3339–3349.

19. Ewais, A.M.R.; Rowe, R.K.; Rimal, S.; Sangam, H.P. 17-year elevated temperature study of HDPE geomembrane longevity in air, water and leachate. *Geosynth. Int.* **2018**, *25*, 525–544. [CrossRef]

20. Reis, R.K.; Barroso, M.; Lopes, M.G. Evolução de cinco geomembranas expostas a condições climáticas em Portugal durante 12 anos. *GEOTECNIA* **2017**, *141*, 41–58. [CrossRef]

21. Safari, E.; Rowe, R.K.; Markle, J. Antioxidants in an HDPE geomembrane used in a bottom liner and cover in a PCB containment landfill for 25 years. In Proceedings of the Pan Am CGS Geotechnical Conference, Toronto, ON, Canada, 2–6 October 2011.

22. Abdelaal, F.B.; Rowe, R.K. Degradation of an HDPE geomembrane without HALS in chlorinated water. *Geosynth. Int.* **2019**, *26*, 354–370. [CrossRef]

23. Abdelaal, F.B.; Morsy, M.S.; Rowe, R.K. Long-term performance of a HDPE geomembrane stabilized with HALS in chlorinated water. *Geotext Geomembr.* **2019**, *47*, 815–830. [CrossRef]

24. Leite, A.F.R.; Ligeiro, L.P.M. Características, Tratamento e Potencial Utilização de Esgoto Produzido em Shopping Centers: Estudo de caso do Catarina Fashion Outlet. Master's Thesis, University of São Paulo, São Paulo, Brazil, 2017.

25. Química Pura Laboratório de Análises Químicas. Relatório de Ensaios n° 49808/18. Giruá, Rio Grande do Sul, Brazil. 2018.

26. ASTM (American Society for Testing and Materials). *ASTM D 5199 Standard Test Methods for Measuring the Nominal Thickness of Geosynthetics*; ASTM: West Conshohocken, PA, USA, 2012; p. 4.

27. ASTM (American Society for Testing and Materials). *ASTM D 4218 Standard Test Method for Determination of Carbon Black Content in Polyethylene Compounds by the Muffle-Furnace Technique*; ASTM: West Conshohocken, PA, USA, 2020; p. 4.

28. ASTM (American Society for Testing and Materials). *ASTM D 792 Standard Test Methods for Density and Specific Gravity (Relative Density) of Plastics by Displacement*; ASTM: West Conshohocken, PA, USA, 2013; p. 6.

29. ASTM (American Society for Testing and Materials). *ASTM D 1238 Standard Test Methods for Melt Flow Rates of Thermoplastics by Extrusion Plastometer*; ASTM: West Conshohocken, PA, USA, 2013; p. 16.

30. ASTM (American Society for Testing and Materials). *ASTM D 6693 Standard Test Methods for Determining Tensile Properties of Nonreinforced Polyethylene and Nonreinforced Flexible Polypropylene Geomembranes*; ASTM: West Conshohocken, PA, USA, 2015; p. 5.

31. ASTM (American Society for Testing and Materials). *ASTM D 1004 Standard Test Methods for Tear Resistance (Graves Tear) of Plastic Film and Sheeting*; ASTM: West Conshohocken, PA, USA, 2013; p. 4.

32. ASTM (American Society for Testing and Materials). *ASTM D 5397 Standard Test Method for Evaluation of Stress Crack Resistance of Polyolefin Geomembranes Using Notched Constant Tensile Load Test*; ASTM: West Conshohocken, PA, USA, 2020; p. 7.

33. ASTM (American Society for Testing and Materials). *ASTM D 5885 Standard Test Method for Oxidative Induction Time of Polyolefin Geosynthetics by High-Pressure Differential Scanning Calorimetry*; ASTM: West Conshohocken, PA, USA, 2017; p. 5.

34. Dias, D.S.; Crespi, M.S.; Ribeiro, C.A.; Kobelnik, M. Evaluation by thermogravimetry of the interaction of the poly(ethylene terephthalate) with oil-based paint. *Eclét. Quím.* **2015**, *40*, 77–85. [CrossRef]

35. Dias, D.S.; Crespi, M.S.; Kobelnik, M.; Ribeiro, C.A. Calorimetric and SEM studies of polymeric blends of PHB-PET. *J. Anal. Calorim.* **2009**, *97*, 581–584. [CrossRef]

36. Lima, J.S.; Kobelnik, M.; Ribeiro, C.A.; Capela, J.M.V.; Crespi, M.S. Kinetic study of crystallization of PHB in presence of hydroxy acids. *J. Anal. Calorim.* **2009**, *97*, 525–528. [CrossRef]

37. GRI (Geosynthetic Research Institute). *GRI—GM13 Test Methods, Test Properties and Testing Frequency for High Density Polyethylene (HDPE) Smooth and Textured Geomembranes*; GRI: Folsom, PA, USA, 2019; p. 11.

38. Telles, R.W.; Lubowitz, H.R.; Unger, S.L. *Assessment of Environmental Stress Corrosion of Polyethylene Liners in Landfills and Impoundments*, 1st ed.; U.S. EPA: Cincinnati, OH, USA, 1984.

39. Mueller, W.; Jakob, I. Oxidative resistance of high density polyethylene geomembranes. *Polym. Degrad. Stab.* **2003**, *79*, 161–172. [CrossRef]

40. Valentin, C.A.; Silva, J.L.; Kobelnik, M.; Ribeiro, C.A. Thermoanalytical and dynamic mechanical analysis of commercial geomembranes used for fluid retention of leaching in sanitary landfills. *J. Anal. Calorim.* **2018**, *136*, 471–481. [CrossRef]

41. Höhne, G.W.H.; Hemminger, W.F.; Flammersheim, H.J. *Differential Scanning Calorimetry*, 2nd ed.; Springer: Berlin/Heidelberg, Germany, 2003.

42. Gabbott, P. *Principles and Applications of Thermal Analysis*, 1st ed.; Blackwell Publishing Ltd.: Oxford, UK, 2008.

43. Menard, K.P. *Dynamic Mechanical Analysis: A Practical Introduction*, 2nd ed.; CRC Press Taylor & Francis Group: Boca Raton, FL, USA, 2008.

44. Cassu, S.N.; Felisberti, M.I. Comportamento dinâmico-mecânico e relaxações em polímeros e blendas poliméricas. *Quim. Nova* **2005**, *28*, 255–263. [CrossRef]

45. Hatakeyama, T.; Quinn, F.X. *Thermal Analysis: Fundamentals and Applications to Polymer Science*, 2nd ed.; John Wiley & Sons Ltd.: Baffins Lane Chichester, UK, 1999.

46. Krongauz, V.V. Diffusion in polymers dependence on crosslink density. Eyring approach to mechanism. *J. Anal. Calorim.* **2010**, *102*, 435–445. [CrossRef]

47. Suceska, M.; Liu, Z.Y.; Sanja Matecic Musanic, S.M.; Fiamengo, I. Numerical modelling of sample–furnace thermal lag in dynamic mechanical analyser. *J. Anal. Calorim.* **2010**, *100*, 337–345. [CrossRef]

Article

Geotextile Tube Dewatering Performance Assessment: An Experimental Study of Sludge Dewatering Generated at a Water Treatment Plant

Maria Alejandra Aparicio Ardila [1,*], Samira Tessarolli de Souza [1], Jefferson Lins da Silva [1], Clever Aparecido Valentin [1] and Angela Di Bernardo Dantas [2]

[1] São Carlos School of Engineering (EESC), University of São Paulo (USP), São Carlos, São Paulo 13566-590, Brazil; samira.tessarolli.souza@usp.br (S.T.d.S.); jefferson@sc.usp.br (J.L.d.S.); cclever@sc.usp.br (C.A.V.)

[2] Hidrosan Engenharia, São Carlos, São Paulo 13560-900, Brazil; angela@hidrosanengenharia.com.br

[*] Correspondence: maparicio@usp.br; Tel.: +55-16-3373-8220

Received: 12 August 2020; Accepted: 22 September 2020; Published: 2 October 2020

Abstract: Using geotextile tubes as dewatering technology may significantly contribute to sustainable treatment of sludge generated in different industries, such as the water industry. This is an economical alternative for dewatering sludge from a Water Treatment Plant (WTP), which prevents sludge from being directly deposited in water bodies and makes it possible to then transfer the sludge to landfills. This paper presents a laboratory study and a statistical analysis, carried out to evaluate the geotextile tube dewatering of sludge from a WTP, discussing the relation between the independent variables (initial Total Solids (TS) of the sludge and polymer dosing) and dependent variables (performance indices used in the literature) evaluated using semi-performance tests. Sludge from a WTP and three different types of geotextiles bags were used. Changes in the geotextiles' characteristics after dewatering were also evaluated, quantitatively using permittivity tests and qualitatively by Scanning Electron Microscopy (SEM). The results indicated turbidity of effluent that met the Brazilian regulations for the discharge of effluents into Class 2 water bodies, as well as higher percent-solids than those obtained with mechanical dewatering technologies. This study underscores the importance of semi-performance tests to understand dewatering in geotextile tubes.

Keywords: geosynthetics; geotextile tubes; sludge; dewatering; total solids; polymer dosing; response surface

---

1. Introduction

Surface water sources have been increasingly mistreated by releasing debris, which is a result of population growth, industrial activities, and the disorderly occupation of protected areas [1]. Due to the low-quality conditions of water bodies, increasing quantities of chemical products need to be used to treat the water, thus increasing the generation of sludge. A Water Treatment Plant (WTP) sludge is a high-water content material, with a granulometric distribution of fine sediments. It originates mainly in decanters and filter washing, and its characteristics depend on different factors, such as the type and quality of crude water, chemical products used in treatment systems, and the operational conditions of the WTP [1–3].

Most Brazilian WTPs dispose of their sludge into watercourses, contradicting current legislation and causing environmental impacts [4], although Brazilian legislation (Law 9.605/98 and Law 12.305/10) establishes that the release of effluents into water bodies, when not approved by environmental agencies, is considered an environmental crime [5,6].

As environmental awareness is raised and more stringent regulations related to the treatment of WTP sludge emerge, technologies that aim to dewater sludge to facilitate its treatment and disposal become increasingly more important. Approximately two decades ago, sludge dewatering was carried out almost exclusively with conventional technologies such as settling ponds, mechanical presses, and centrifuges [7]. Despite the various alternatives and technologies available on the market, the main obstacles to WTP sludge dewatering are the high cost and operational complexity [3]. In this context, geotextile tube technology emerged. It was used for the first time in the 1990s by Fowler et al. [8] for dewatering sludge from a sewage treatment plant. It led to the initial understanding of how geotextile tubes can be used in dewatering applications [9]. Thus, showing the importance and relevance of the environment segment [10].

Geotextile tubes foster the natural physical separation between the solid and liquid fraction of the sludge, in addition to possibly containing contaminants present in the sludge [9], showing, in some cases, a better performance compared to conventional dewatering technologies. The solid fraction can be transferred directly to sanitary landfills and the liquid fraction (effluent) can be returned to the interior of the system, or sent directly to water bodies, as long as it complies with environmental regulations, which, if not met, will require a secondary treatment [11].

Chemical conditioning of the sludge and the filter cake formation is a fundamental aspect to be considered in geotextile tubes dewatering. Adding polymers, particularly polyacrylamide-based ones, to the sludge has become an essential component for the dewatering process in geotextile tubes, where the polymers act as flocculants, improving dewatering characteristics, increased dewatering rate, particle retention, and reducing the risk of clogging [9,12–19].

Another important aspect is the filter cake, which is a type of clogging inherent to the dewatering process, which occurs due to the suffusion of the fine particles of sludge, which form a layer on the inner surface of the tube. This layer is usually formed after the first filling cycle of the system, and substantially reduces the drainage capacity of the geotextile, governing the filtration [20–28]. Weggel and Ward [29] developed a numerical model of the formation of the filter cake in the geotextile tube during the dewatering process where the size distribution of the particles in various layers within the filter cake can be determined from the model. The model was verified by Weggel and Dortch [30] through tests with two low permittivity woven geotextiles and three types of sediments, where they compared the flow rate through the experiments with the rate predicted by the theory, obtaining satisfactory results in the prediction of the filter cake accumulation on geotextiles.

In order to improve the dewatering performance in geotextile tubes, several works have already been developed, using different treatments, test methodologies, residues or sludge, and polymers. Bourgès-Gastaud et al. [26] evaluated the dewatering of residues with different clay content using nonwoven geotextiles, showing the feasibility of using this type of geotextile in dewatering residues with fine granulometric characteristics. They also observed that the samples of residues with less than 25% of silt in the composition obtained less dewatering efficiency than the others, indicating that the sludge composition, and not the geotextile characteristics, determines the system's dewatering efficiency, confirming the statement by Christopher and Fischer [31].

Compared with other natural technologies of dewatering, geotextile tubes show a lower dependence on meteorological conditions, as there is a lower input of rainwater through the geotextile [28]. These systems can be manufactured in different sizes, are simple to transport and use, and are significantly more economical [9]. Geotextile tubes in the national and international panorama present great potential for application, making them an efficient and viable solution from a technical and economical point of view.

The filtration criteria of geotextiles have limited applicability in geotextile tubes, due to the fact that the properties of the sludge are the dominant control factors in the filtration process [21,25]. However, knowledge of its filtration, operation, and improvement of design procedures is essential. The success of this application, and the duration of the dewatering and consolidation, depends on the filtration compatibility between the sludge and the geotextiles used to make the tubes [21,32]. Therefore, making preliminary performance tests is fundamental to assess the design conditions before installing the technology. Researchers and professionals have used several test methods alike as a means of evaluating the dewatering performance. These procedures comprise laboratory or field tests [33]. These methods include bench-scale tests (not standardized) such as the cone test (e.g., [28]), Falling Head Test (e.g., [34,35]), Pressure Filtration Test (e.g., [14,22,25,36,37]), and the Pressurized 2-Dimensional Dewatering Test (e.g., [33,38]). Moreover, midscale tests or semi-performance tests such as the Hanging Bag Test (HBT) formalized as a standard by the Geosynthetics Research Institute—GT14 [39] and the Geotextile Tube Dewatering Test (GDT) formalized as a standard by the Geosynthetics Research Institute—GT15 and the American Society for Testing and Materials (ASTM) D7880 [40,41]. Finally, full performance tests have limited use due to their complexity [9,25].

In studies carried out to compare the two methods of semi-performance tests, Koerner and Koerner [15] concluded that the GDT has more advantages compared to the HBT as it is user friendly and it evaluates more parameters such as the filling pressure. Therefore, carrying out this type of semi-performance test together with bench-scale tests is recommended when a full-scale test cannot be done.

The present study aims to contribute to the discussion and knowledge on geotextile tube technology used in WTP sludge dewatering. The implementation of this dewatering technology in the WTP could reduce the environmental impacts related to the disposal of WTP discharge. In addition, the present study seeks to evaluate the viability of using nonwoven geotextiles, a material with better filtration characteristics than the commonly used materials (woven geotextiles) that can be produced at a lower cost [42] and its use can economically benefit the installation of the systems. For this purpose, tests were carried out concurrently under the same conditions on two nonwoven geotextiles and a woven geotextile commonly used for this application, in order to have a performance reference. Thus, the study presents an evaluation of the dewatering of WTP sludge through GDT (semi-performance test) using sludge from a Brazilian WTP and bags made from geotextiles commercially used for dewatering application. Besides, a series of permittivity tests and SEM were carried out to analyze the geotextile characteristic changes due to sludge dewatering. Moreover, response surfaces graphs that establish the relationship between the dependent and independent variables that influence the dewatering performance are presented.

2. Materials and Methods

2.1. Sludge From WTP

The sludge used in the research was from a conventional WTP located in Nova Odessa, state of São Paulo, Brazil. This WTP treats an average of 15.5 million liters of water per day and uses poly aluminum chloride in the coagulation process. In order to carry out the tests, a single sludge sample was collected at the outlet of the decanter while washing one of the four decanters (Figure 1a,b) in a 1000-L reservoir (Figure 1c). The determination of total, fixed, and volatile solids was carried out in the WTP laboratory, according to test procedures of the Standard Methods for Examination of Water and Wastewater [43]. The results obtained in the tests of the solids contents are presented in Table 1.

Table 1. Total, fixed, and volatile solids of the sludge.

Parameter	%
Total Solids (TS)	3.80
Fixed Solids (FS)	41.20
Volatile Solids (VS)	58.80

Figure 1. Sludge from a Water Treatment Plant (WTP): (**a**) WTP decanters. (**b**) Washing the decanter. (**c**) Sludge sample collected.

The specific gravity of the grains test and granulometric analysis was carried out in the Soil Mechanics Laboratory at the University of São Paulo (USP) in São Carlos, state of São Paulo, Brazil. These tests were carried out according to the procedures of NBR 7181 and NBR 6458 [44,45], where the weight of the samples had to be adapted. The sample sludge presented specific gravity of the grains of 2.4 g/cm^3. Figure 2 shows the granulometric distribution of the sludge. The granulometric curve of the sludge shows that the grain size distribution comprises 0.5% sand, 16.5% silt, and 83.0% clay.

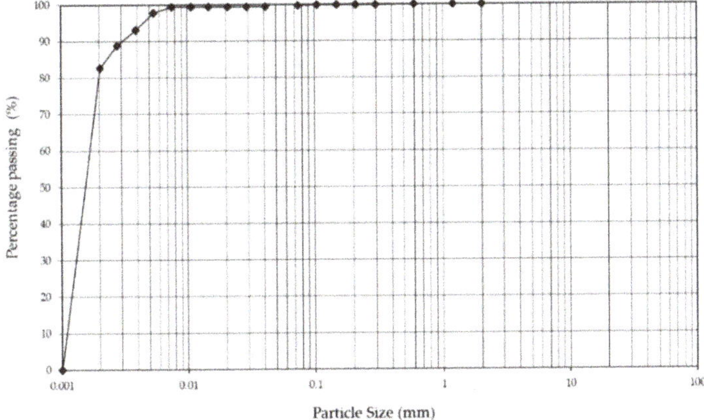

Figure 2. Particle size distribution curve of the WTP sludge.

2.2. Geotextiles

Three types of geotextiles were chosen that are commercially available in Brazil for dewatering applications. They consisted of two nonwoven (NW1 and NW2) and one woven geotextile (W1). The properties of the geotextiles are presented in Table 2. All the geotextile tests were carried out in the Geosynthetic Laboratory at the University of São Paulo (USP) in São Carlos, state of São Paulo, Brazil.

Table 2. Geotextile properties.

Properties	Test Method	NW1	NW2	W1
Structure, polymer type [1]	–	NW, PET	NW, PET	W, PP
Mass per unit area (g/m^2)	ABNT NBR ISO 9864 [46]	612	895	414
Thickness (mm)	ABNT NBR ISO 9863-1 [47]	3.96	4.87	2.81
Apparent opening size (μm)	ABNT NBR ISO 12956 [48]	52	44	200
Permittivity (s^{-1})	ASTM D4491 [49]	1.36	0.80	0.84
Tensile strength per unit width MD × CD [2] (kN/m)	ABNT NBR ISO 10319 [50]	35 × 28	53 × 40	109 × 106

[1] NW, nonwoven; W, woven; PET, polyester; PP, polyester. [2] MD, machine direction; CD, cross direction.

2.3. Polymer Floculant

The polymer was selected using jar test and cone tests. These tests were carried out in the WTP laboratory, aiming to identify the optimum polymer (flocculant) and dosage to increase the dewatering rate and minimize the effluent turbidity in the geotextile tubes. Ten polyacrylamide-derived polymers (anionic, cationic, and nonionic) were tested and the polymer that presented the best performance was the cationic polymer C8396. In recent years, many Brazilian studies have been carried out using the polymer C8396 (e.g., [28,51]).

2.4. Performance Index

The most common indexes for evaluating the dewatering performance are the Filtration Efficiency—FE (retention index) and Dewatering Efficiency—DE (dewatering index) [25]. Moo-Young

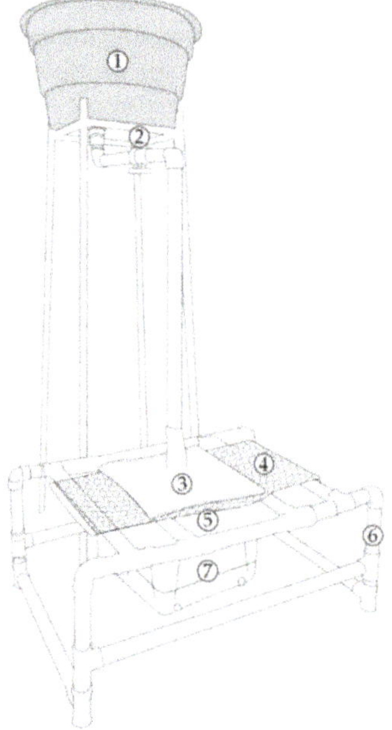

Figure 3. Geotextile tube dewatering test scheme.

In order to observe the changes (degradation) in the geotextiles after dewatering in the different scenarios analyzed, samples were collected from the bottom face of each bag to measure the permittivity property and for the SEM analyses. From each bag, five specimens were collected for permittivity tests according to the ASTM D4491 [49] standard. The exhumation was performed one month after each test, during which the bags remained closed. The SEM photomicrographs were obtained from the Chemical and Instrumental Analysis Center at the São Carlos Institute of Chemistry (CAQI/IQSC/USP) using ZEISS LEO 440 equipment (Cambridge, England) with an OXFORD detector (model 7060), operating with electron beam 15 kV, 2.82 A current, and 200pA I probe. The samples were covered with 6nm gold in a Coating System metallizer BAL-TEC MED 020 (BAL-TEC, Liechtenstein) and kept in a desiccator until the moment of analysis. The samples were covered with 6nm gold, in a Coating System metallizer BAL-TEC MED 020 (BAL-TEC, Liechtenstein) and kept in a desiccator until being analyzed.

3. Results and Discussion

In this section, the results of the dependent variables evaluated are presented and discussed: performance indices adopted in this work (effluent turbidity and PD), the dewatering rate, and the percent-solids observed at the end of the monitoring (Sections 3.1–3.3). The permittivity test and SEM analyses after dewatering are also shown (Sections 3.4 and 3.5).

3.1. Effluent Turbidity

The results of turbidity obtained for each type of geotextile are analyzed in two steps, the initial step equal to 5 min after the start of the GDT and the second step 25 min after. The results of turbidity were presented in the initial periods evaluated of 5 and 25 min in Nephelometric Turbidity Unit (NTU), because data on the turbidity were missing in some scenarios in the periods of 1 h, 2 h, and 24 h due to the fact that the percolated volume was not sufficient to measure the turbidity. Considering that the bags were filled in a single cycle, a tendency to decrease turbidity was observed in the scenarios where it was possible to monitor the turbidity evolution during the 24 h. The second-order polynomial equations associated to the effluent turbidity results in each step for experimental design (geotextile type) are:

$$Turbidity\ at\ 5\ min\ in\ NW1\ (NTU) = 31.31 - 34.01\ x_2 + 80.45\ x_2^2 - 44.20\ x_1\ x_2 \tag{4}$$

$$Turbidity\ at\ 5\ min\ in\ NW2\ (NTU) = 24.55 + 14.31\ x_1 + 33.73\ x_2^2 + 16.01\ x_1\ x_2 \tag{5}$$

$$Turbidity\ at\ 5\ min\ in\ W1\ (NTU) = 114.49 + 63.88\ x_1 - 130.23\ x_2 + 144.84\ x_2^2 - 161.76\ x_1\ x_2 \tag{6}$$

$$Turbidity\ at\ 25\ min\ in\ NW1\ (NTU) = 7.22 + 15.27\ x_1 + 13.62\ x_1^2 + 7.80\ x_1\ x_2 \tag{7}$$

$$Turbidity\ at\ 25\ min\ in\ NW2\ (NTU) = 9.75 + 19.81\ x_1 + 17.58\ x_1^2 + 8.63\ x_1\ x_2 \tag{8}$$

$$Turbidity\ at\ 25\ min\ in\ W1\ (NTU) = 25.48 + 23.31\ x_1 - 12.36\ x_2 - 15.82\ x_1\ x_2 \tag{9}$$

The mathematical models of the turbidity evaluated in the two times, showed that the two variables analyzed x_1 (initial ST) and x_2 (polymer dosing) presented statistical significance. The mathematical models were subjected to ANOVA and later the response surface graphs were generated. Figures 4–6 show the response surfaces of turbidity in the GDT carried out with a bag made in NW1, NW2, and W1, respectively. The determination coefficients obtained for each response (Figures 4–6) show that the regression models fit the experimental data, considering the inherent variability of the sludge. The values of determination coefficients (R^2) were between 0.79 (Figure 5a) and 0.98 (Figure 6a).

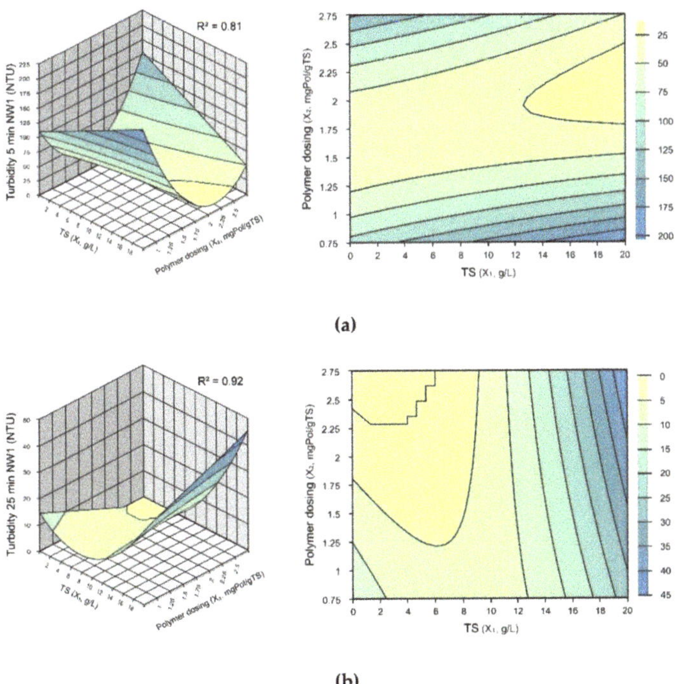

Figure 4. Turbidity response surfaces in NW1: (**a**) at 5 min. (**b**) At 25 min.

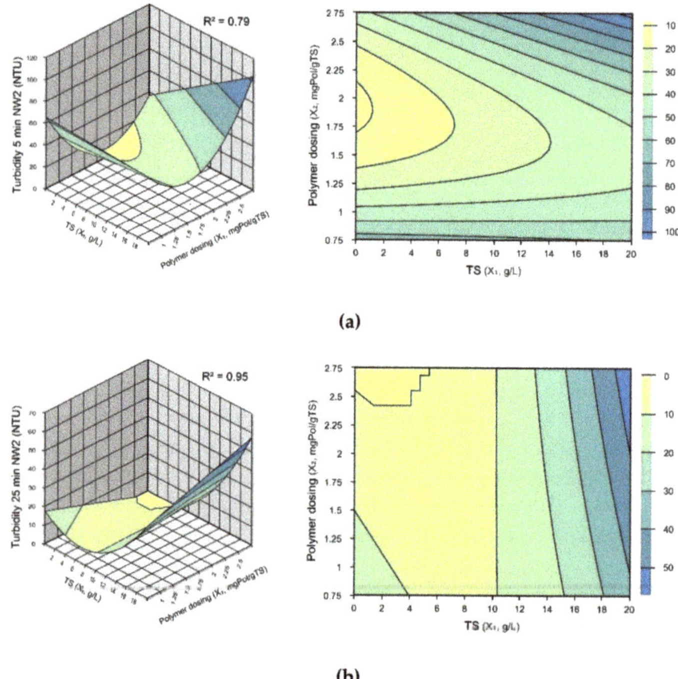

(a)

(b)

Figure 5. Turbidity response surfaces in NW2: (**a**) at 5 min. (**b**) At 25 min.

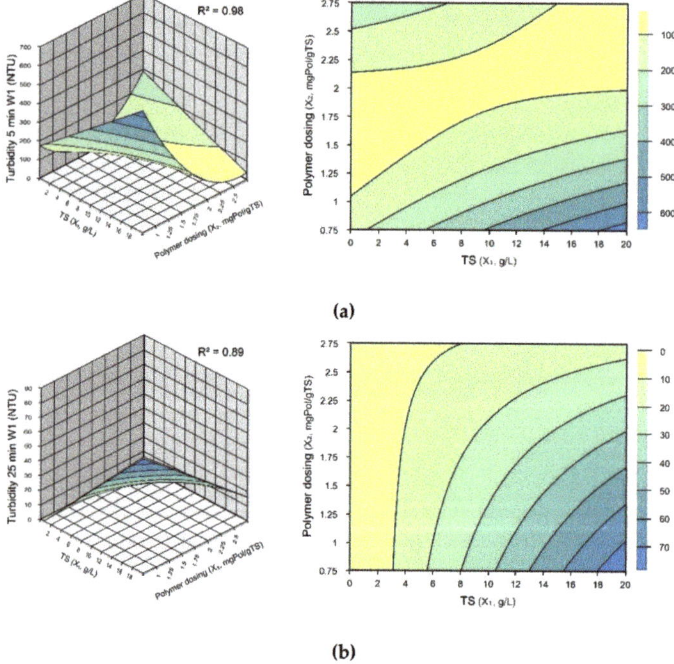

(a)

(b)

Figure 6. Turbidity response surfaces in W1: (**a**) at 5 min. (**b**) At 25 min.

After 25 minutes of testing (Figure 4b, Figure 5b, and Figure 6b) it was observed that in the three types of geotextiles evaluated there was a reduction in turbidity values, indicating that the filter cake was formed between the first 5 and 25 min of testing. Figures 4b and 5b show that after 25 min the influence of the polymer dosing on the turbidity was weak for NW1 and NW2 (nonwoven geotextiles). Turbidity, in addition to being considered a retention index, is a parameter that indicates water quality. Analyzing turbidity as a quality parameter, turbidity values below 100 NTU were observed in all the experimental designs, indicating that after 25 min of test time, the effluent from the bags can be released into Class 2 water bodies, complying with Brazilian legislation [54,55]. It should be mentioned that turbidity is only one of the different parameters that evaluates the quality of the effluent and other physical–chemical parameters of the effluent established in the Resolution No. 430 [55] must be verified.

Regions that indicate a tendency to remove turbidity in the tests performed with NW1 and NW2 were observed (Figures 4b and 5b). In the NW1 (Figure 4b), this region is delimited by the initial TS concentration below 5.75 g/L and polymer dosing above 2.33 mgPol/gTS. On the other hand, the turbidity removal effect in the NW2 (Figure 5b) was observed for sludge with initial TS also below 5.75 g/L and polymer dosing above 2.46 mgPol/gTS.

For Class 2 water bodies, the Brazilian National Environment Council (CONAMA) Resolution No.357 and No.430 [54,55] establish the effluent characteristics to discharge. However, in Brazil, there are no specific rules or legislation for the recirculation of effluents in the WTP [37,53]. As a result, internationally recommended values are used, such as the recommendation for recirculation of filtered washing water in the United Kingdom, which establishes a maximum value of turbidity of 5 NTU [56]. In compliance with the recommendation and considering the possible recirculation of the effluent generated in the dewatering of sludge, the response surfaces of the bags manufactured in each geotextile type were analyzed. In the bags made in NW1 (Figure 4b), the region that presents turbidity below 5 NTU is delimited by the initial TS concentration below 9.15 g/L and polymer dosing greater than 1.2 mgPol/gTS. In the bags made in NW2 (Figure 5b), this region is bounded by the initial TS concentration between 1.7 and 6.95 g/L and polymer dosing greater than 1.55 mgPol/gTS. Finally, in the bags made in W1 (Figure 6b), the region is delimited by the initial TS concentration below 1.85 g/L and polymer dosage below 2.50 mgPol/gTS. It is observed that for the three types of geotextile, the region that presented turbidity below 5 NTU is related to low initial TS concentrations.

3.2. Dewatering Rate

The rate of sludge dewatering in the bags manufactured in different types of geotextiles was evaluated in the initial and final monitoring stages. The rate of the initial stage was calculated with the volume of effluent collected during the first 5 minutes of testing, and the final rate calculated with the accumulated volume of effluent collected during the 24 hours of testing. The mathematical models of initial dewatering rate in each geotextile type are:

$$\text{Initial dewatering rate in NW1 (cm}^3\text{/s}) = 57.97 - 20.01\,x_1 + 17.12\,x_1{}^2 + 4.95\,x_2 \tag{10}$$

$$\text{Initial dewatering rate in NW2 (cm}^3\text{/s}) = 56.68 - 24.04\,x_1 + 13.84\,x_1{}^2 + 11.93\,x_2 \tag{11}$$

$$\text{Initial dewatering rate in W1 (cm}^3\text{/s}) = 59.47 - 23.08\,x_1 + 21.49\,x_1{}^2 + 6.44\,x_2 - 9.11\,x_2{}^2 + 8.04\,x_1\,x_2 \tag{12}$$

In all mathematical models (Equations (10)–(12)), the independent variables analyzed (x_1 and x_2) were statistically significant. The initial dewatering rate in the nonwoven geotextiles (Equations (10) and (11)) presented a second-order polynomial equation according to the same terms (x_1, $x_1{}^2$, x_2), and the woven geotextile (Equation (12)) presented an equation according to all possible terms in a second-order polynomial equation.

The response surfaces for each type of geotextile and the determination coefficients are presented in Figure 7. It can be concluded that the models adjusted well to the experimental data. In fact, the minimum R^2 value obtained in the response surfaces of the initial dewatering rate is 88% (Figure 7b), indicating that 12% of the total variation is not explained by the model.

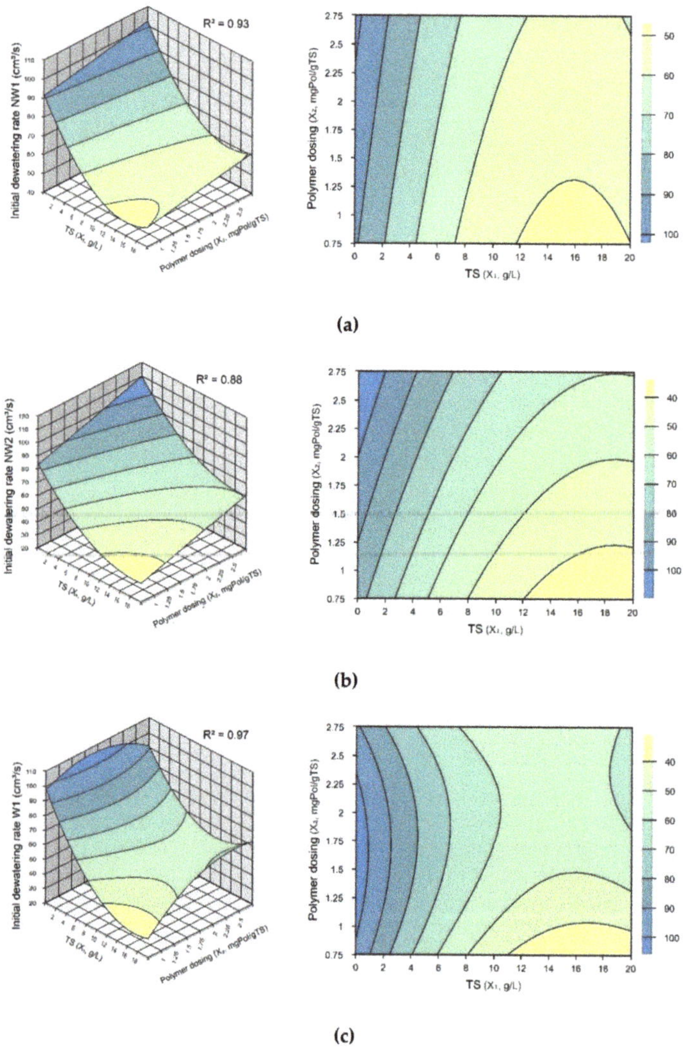

Figure 7. Initial dewatering rate response surfaces: (**a**) in NW1. (**b**) In NW2. (**c**) In W1.

Analyzing the response surfaces of the initial dewatering rate in the three geotextiles (Figure 7), it is observed that the dewatering responses in bags manufactured from NW1 geotextile (Figure 7a), presented higher dewatering rates compared to those observed in the other response surfaces (Figure 7b,c). The surface response (Figure 7a) shows dewatering rates greater than 50 cm^3/s, with the exception of the sludge with TS concentration greater than 11.90 g/L and polymer dosage between 0.80 mgPol/gTS and 1.32 mgPol/gTS.

On the other hand, the mathematical models of the final dewatering rate presented the same equation (Equation (13)) in the different types of geotextile evaluated.

$$\textit{Final dewatering rate (cm}^3\textit{/s)} = 0.30 - 0.03\,x_1 \tag{13}$$

The equation indicates that only the linear term of the variable x_1 (initial TS) have statistical significance in the model. The response surfaces of the final dewatering rate of each geotextile and the values of R^2 are presented in the Figure 8. The decrease in the final dewatering rate may be related to evaporation, due to the fact that this rate is determined in function to the volume of effluent collected. It is noteworthy that the effluent was stored in the reservoir during the 24 h of testing and exposed to environmental conditions.

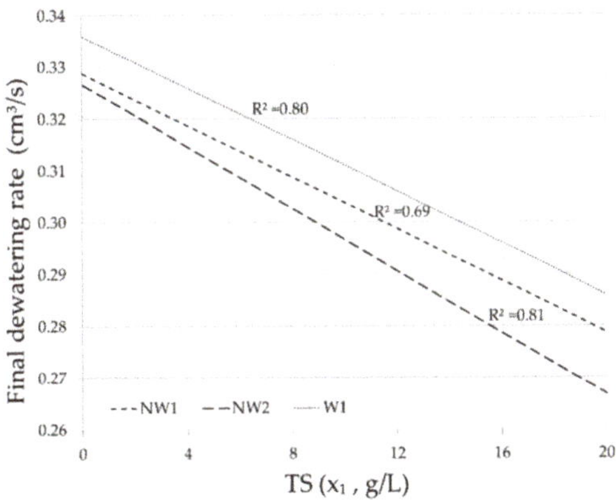

Figure 8. Final dewatering rate response surfaces.

3.3. Percent-Solids and PD

The results of the percent-solids presented, referred to the end of the monitoring (168 h after testing). To calculate the dewatering index PD using Equation (3), the final water content of the sludge was the retained in the bag 168 h after testing. The mathematical models of percent-solids and PD are:

$$Percent\text{-}solids\ in\ NW1\ (\%) = 58.66 + 17.83\ x_1 - 40.82\ x_1{}^2 \tag{14}$$

$$Percent\text{-}solids\ in\ NW2\ (\%) = 41.37 + 13.27\ x_1 - 28.10\ x_1{}^2 \tag{15}$$

$$Percent\text{-}solids\ in\ W1\ (\%) = 46.11 + 15.06\ x_1 - 31.04\ x_1{}^2 \tag{16}$$

$$PD\ in\ NW1\ (\%) = 99.05 + 48.16\ x_1 - 50.90\ x_1{}^2 \tag{17}$$

$$PD\ in\ NW2\ (\%) = 98.44 + 47.13\ x_1 - 51.31\ x_1{}^2 \tag{18}$$

$$PD\ in\ W1\ (\%) = 98.51 + 47.63\ x_1 - 50.88\ x_1{}^2 \tag{19}$$

As in the mathematical model of the final dewatering rate, the models of percent-solids and PD have statistical significance only of the variable initial TS (x_1). The equations are presented in the function of the linear and quadratic terms of the variable x_1. Figure 9 shows the response surfaces of percent-solids and PD generated after the ANOVA.

The values of determination coefficients of percent-solids were between 0.70 and 0.88 (Figure 9a). It is observed that the maximum percent-solid point for the three types of geotextiles is obtained when the sludge has an initial TS concentration of 12.1 g/L. The maximum values of final percent-solids for NW1, NW2, and W1 are 60.59%, 42.9%, and 47.9%, respectively.

The determination coefficients obtained for the PD responses show that the regression model fits the experimental data correctly. In fact, R^2 in PD was 99.90% in the three response surface models. It is observed that the maximum PD point is obtained at the initial TS concentration of 10.13 g/L (level 0 of the variable x_1) for the three types of geotextiles. It is noteworthy that the PD response surface shown in Figure 9b shows values greater than 100%, because in the analysis of variance that influences the creation of the response surface, the maximum value is not established. However, the maximum values of the final PD in the experimental design for the geotextiles NW1, NW2, and W1 were 99.05%, 98.44%, and 98.51%, respectively, as observed in the mathematical models (Equations (17)–(19)) when $x_1 = 0$.

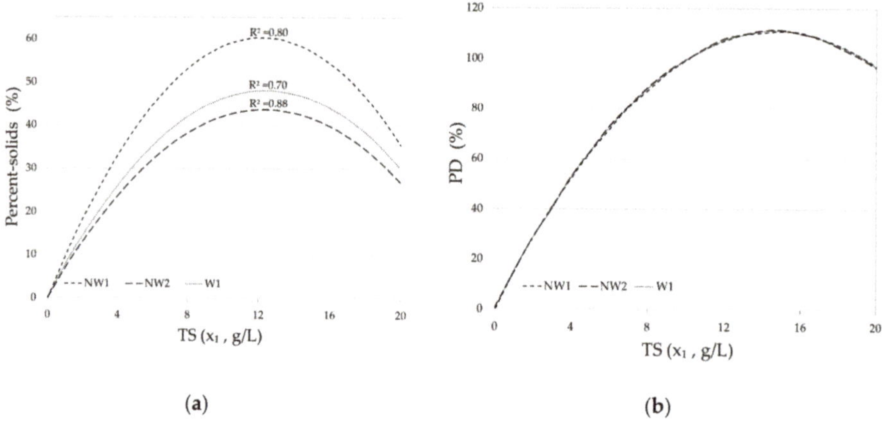

Figure 9. Response surfaces of the geotextiles: (**a**) Percent-solids. (**b**) Percent Dewatered (PD).

It is observed that the GDT carried out in bags made in NW1 reached the highest final percent-solids and PD. The values of percent-solids obtained from tests with bags manufactured in W1 are within the range of values obtained by Guimarães [53] after 168 h of testing (from 40% to 60%). It should be stressed that the author followed the evolution of the content of solids inside the bags until reaching values above 90% (for approximately 40 days). Such results show the efficiency of sludge dewatering in the three types of geotextiles evaluated when compared with the literature that recommends solids content inside geotextile tubes above 20% [53], to a subsequent transfer to sanitary landfills.

Analyzing the evolution of percent-solids inside the bags during the test period in the three geotextiles tested (Figure 10), evaporation influences in the dewatering performance are clear by observing the increase of the percent-solids. According to Müller and Vidal [57], water loss occurs most significantly at the beginning of the dewatering process, through drainage, representing a large initial volume reduction, and subsequently, the evaporation represents the main water exit path of the geotextile tubes. Unfortunately, the evaporation of dewatering performance could not quantify in the GDT. It should underscore that the bags remained closed during the monitoring (168 h) and its feeding sleeve opened quickly only to collect samples of the sludge cake. In addition, the bags did not filter a significant volume of effluent after 24 h of testing. It was not possible to present the evolution of the percent-solids of the tests with an initial TS of 0.25 g/L (scenarios −1A, −1B, and −1C), as there was not enough volume of sludge kept inside the bag to determine the percent-solids. Figure 10a shows the evolution of the percent-solids for sludge with an initial TS of 10.13 g/L (scenarios 0A, 0B*-the average value of the central points of each experimental design, and 0C) and Figure 10b the evolution of the percent-solids for sludge with an initial TS of 20 g/L (scenarios 1A, 1B, and 1C). In general, it is observed that the GDT with initial TS of 10.13 g/L presented a final percent-solids (after 168 hours) higher than the GTD with an initial TS of 20 g/L. Comparing the test scenarios, the scenario with the highest final percent-solids was 0B* (with polymer dosing of 1,7 mgPol/gTS) in the GDT with bags elaborated in NW1 (Figure 10a), followed by the GDT in scenario 0A (with polymer dosing of 0.8 mgPol/gTS) where the evolution of the percent-solids in the GDT in the three types of geotextiles showed similar behavior (Figure 10a).

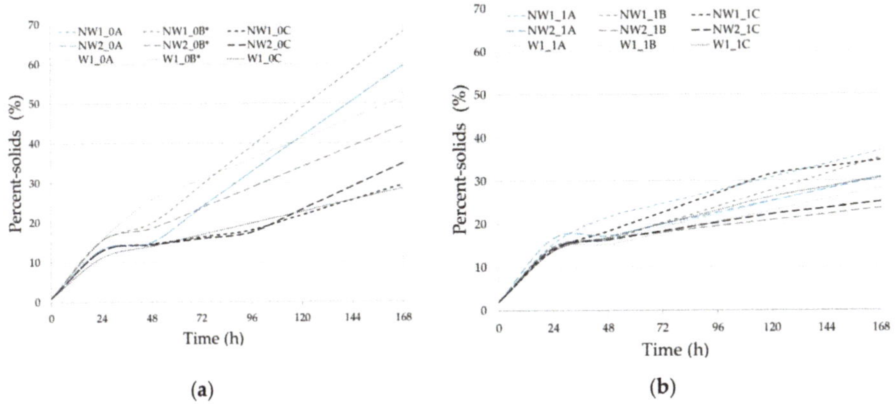

Figure 10. Evolution of the percent-solids in the Geotextile Tube Dewatering Test (GDT) scenarios with: (a) initial TS of 10.13 g/L (0A, 0B* and 0C) and (b) initial TS of 20.00 g/L (1A, 1B and 1C).

3.4. Geotextiles Permittivity

Table 5 shows the average values of permittivity carried out according to the ASTM D4491 [49], with the confidence interval (in parentheses) in each evaluated scenario where the reference is the permittivity value of the virgin sample of the geotextiles, and 0B* is the average permittivity value of the central points of each experimental design (average value of scenarios 0B1, 0B2, and 0B3).

Table 5. Permittivity properties of the geotextiles before and after dewatering of sludge.

Scenario	Permittivity (s^{-1})		
	NW1	NW2	W1
Reference	1.36 (±0.25)	0.80 (±0.12)	0.84 (±0.07)
−1A	0.93 (±0.23)	0.60 (±0.07)	0.48 (±0.05)
0A	1.02 (±0.13)	0.52 (±0.08)	0.51 (±0.05)
1A	0.77 (±0.16)	0.55 (±0.13)	0.46 (±0.02)
−1B	0.92 (±0.13)	0.57 (±0.05)	0.44 (±0.02)
0B*	0.87 (±0.19)	0.46 (±0.03)	0.50 (±0.01)
1B	1.02 (±0.09)	0.62 (±0.13)	0.50 (±0.04)
−1C	0.89 (±0.10)	0.40 (±0.07)	0.50 (± 0.04)
0C	0.98 (±0.16)	0.61 (±0.12)	0.56 (±0.02)
1C	0.88 (±0.18)	0.65 (±0.13)	0.46 (±0.03)

Intervals with a 98% confidence level in parentheses; 0B* (average value of triplicate).

Figure 11 shows the average permittivity values of the three types of geotextiles after dewatering in the nine scenarios (scenarios 0B1, 0B2, and 0B3 = scenario 0B*) resulting from the experimental design and the permittivity values of the virgin samples (Reference). In each bar (scenario), the sample variation is illustrated. After dewatering, it was observed that there was a decrease in permittivity in all scenarios evaluated in all the geotextiles (NW1, NW2, and W1) in relation to the average permittivity value of the virgin sample (Reference), highlighting that the reduction of this property over time is considered for the assessment of clogging in geotextiles [58–60]. It was found that the dewatering of the sludge generated a drastic reduction in the permittivity values, which affects the permeability, the decrease of which is related to the formation of the filter cake, which according to Moo-Young et al. [21], after its formation improves the characteristics of the effluent, but reduces the permeability of the system.

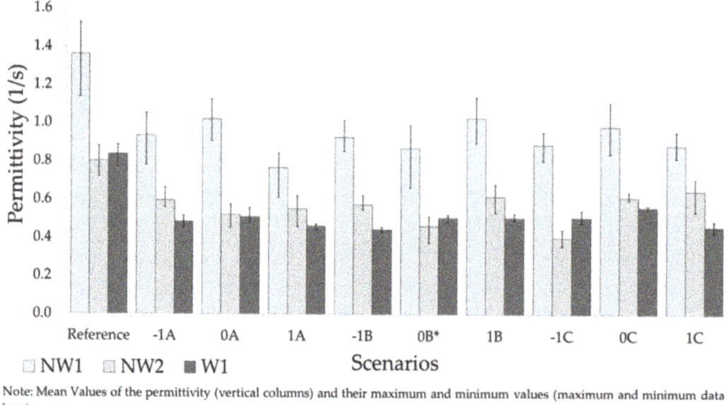

Note: Mean Values of the permittivity (vertical columns) and their maximum and minimum values (maximum and minimum data bars)

Figure 11. Geotextiles permittivity—effects of sludge dewatering.

In the general analysis, the percentages of permittivity reduction varied from 19.27% (scenario 1C—NW2) to 50.04% (scenario −1C—NW2), in which NW2 was the geotextile with the greatest variation. Comparing the geotextiles in the scenarios with the same initial TS (Table 5 and Figure 11), it is observed that for the initial ST of 0.25 g/L (level −1) in nonwoven geotextiles (NW1 and NW2), the permittivity decreased with the increase in the polymer dosing. As for the initial TS of 10.13 g/L (level 0) in the three types of geotextile evaluated (NW1, NW2, and W1), the 0B* scenario with a dosage of 1.7 mgPol/gTS was the one that showed the greatest decrease in permittivity. In scenarios with initial TS of 20 g/L (level 1) in the nonwoven geotextiles (NW1 and NW2), scenario 1A with a dosage of 0.8 mgPol/gTS showed the greatest decrease in permittivity in relation to the virgin sample, and in the woven geotextile (W1) the scenarios 1A and 1C showed the same decrease in permittivity (45.20%). It was observed that the formation of the filter cake reflects in the decrease of the permeability [21], in this case in the permittivity that is directly proportional to the permeability.

3.5. SEM

Using SEM photomicrographs to evaluate geotextiles characteristics has been considered in the scientific literature, particularly for estimating the pore size distribution of nonwoven geotextiles used for filtration [61]. The SEM analyses were used to evaluate qualitatively the changes in the geotextiles, and highlight details of the disposition of sludge particles in the geotextiles fibers, complementing the results obtained in permittivity tests. SEM images are presented of the scenarios where there was a greater and lesser reduction in the permittivity. Figures 12 and 13 show the nonwoven geotextiles SEM images that are presented with magnification 500×.

Figures 12a and 13a show the virgin samples of the NW1 and NW2 geotextiles (both of polyester) where tiny portions of material adhered to the filaments and cracks in the fibers of the geotextiles are noted, which, according to the supplier, are due to the oil content used in the manufacturing process to reduce the friction between the polyester fibers during the needling process. This technique is commonly used in the manufacturing of nonwoven geotextiles. It noted that these characteristics do not compromise the filtration function of the geotextiles. The SEM analyses showed similar fiber sizes between the geotextiles, but the spacing is greater between the NW2 geotextile fibers, which represent its larger filtration opening.

Figure 12b,c show the images of the scenarios with the lowest (scenario 1B) and the largest (scenario 1A) reduction in permittivity in relation to the virgin sample in the geotextile NW1. In the scenario 1A (Figure 12c) the presence of larger sludge particles adhered to the geotextile fibers is high compared with the presented in the scenario 1B (Figure 12b). This particle adhesion can cause blinding, explaining the reduction of permittivity in the NW1.

Figure 12. NW1 SEM (magnification: 500×). (**a**) Virgin sample. (**b**) Sample with less permittivity reduction (scenario 1B). (**c**) Sample with high permittivity reduction (scenario 1A).

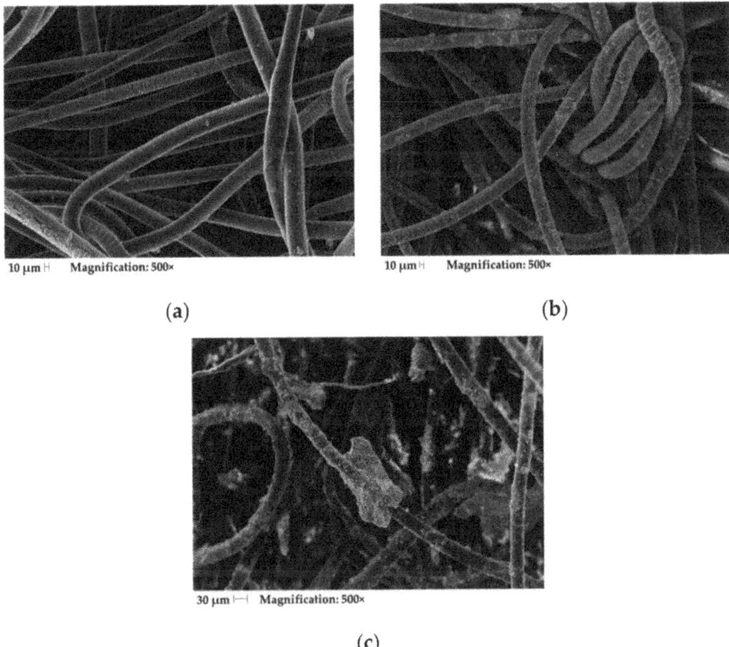

Figure 13. NW2 SEM (magnification: 500×). (**a**) Virgin sample. (**b**) Sample with less permittivity reduction (scenario 1C). (**c**) Sample with high permittivity reduction (scenario −1C).

The NW2 geotextile is the material with the highest mass per unit area (895 g/m^2) used in the study and the geotextile that presented the scenario with the greatest reduction in permittivity between the three types of geotextiles evaluated. NW2 underwent a permittivity reduction of 50% in the 1C scenario (Figure 13c), as a result of the partial closure of the pores of the geotextile due to the amount of sludge particles adhered to the fibers. In the NW2 geotextile as well, it is noted by SEM analysis that in scenario 1C (Figure 13b) there were smaller particles adhered to the geotextile filaments, which influenced to a lesser extent the decrease in permittivity.

In the virgin sample of the W1 presented in Figure 13a, there are tiny portions of adhered material resulting from the manufacturing process. The images of this type of material were obtained with a magnification of up to 75× (Figure 14), as it was not necessary to have more detail, due to the fact that the fibers are organized and of larger size when compared to nonwoven geotextiles.

(a)   (b)

(c)

Figure 14. W1 SEM (magnification: 60× and 75×). (**a**) Virgin sample. (**b**) Sample with less permittivity reduction (scenario 0C). (**c**) Sample with high permittivity reduction (scenario −1B).

When evaluating the image of geotextile W1 in the scenarios with less and greater reduction in permittivity (Figure 14b,c), it was noted that the images do not provide much information, as only some sludge particles were deposited at the intersection of the fibers in both scenarios, showing the importance of carrying out an analysis of the cross section of the geotextile as performed by Mlynarek and Rollin [62], which in this case was not possible due to the difficulty of obtaining a sample of the unformed cross section. The importance of the representativeness of the samples is highlighted, which could be improved with the comparison of samples collected at different points of the same material.

The fact of not observing differences between the scenarios can also be related to the sample chosen to perform the SEM, considering the sample size is approximately 1 cm^2 and considering that there could be places with greater representativeness, which can be a disadvantage in the time to evaluate the results obtained.

4. Conclusions

The present experimental study aims to contribute to the knowledge and the dissemination of the geotextile tubes technology used in the WTP sludge dewatering. Based on the results, the following conclusions are presented below:

- The use of the Faced Centered Design of the Response Surface Methodology was presented as an efficient tool for the experimental planning of the tests, optimizing the results obtained and facilitating their interpretation. The levels adopted for the analyzed independent variables (initial sludge TS and polymer dosing) were effective, since these were statistically significant.

- Despite the fact that the independent variable polymer dosing did not have statistical significance in all the dependent variables analyzed, such as the final dewatering rate, the percent-solids, and the PD. The result is debatable as several related studies and preliminary cone tests carried out in this work have shown the importance of the chemical conditioning of the sludge as an accelerator in the dewatering. It is noteworthy that the results analyzed for percent-solids and PD were the final results obtained 168 h after the beginning of the test, and the final dewatering rate obtained 24 h after the beginning of the test, which could indicate that the polymer dosing influences the initial dewatering stage.

- The GDT carried out proved to be adequate for the achievement of the proposed objectives, showing the importance of carrying out tests to evaluate the dewatering in geotextile tubes.

- After dewatering, there was a decrease in permittivity in the three types of geotextile in all the evaluated scenarios, thus verifying the formation of the filter cake in the geotextile bags that was corroborated by the SEM analyses. It was also verified that the formation of the filter cake contributed to the improvement of the effluent quality.

- In order for the effluent, resulting from sludge dewatering, to be directly deposited in water bodies or recirculated in the same WTP, it is proposed that the effluent collected at the beginning of the dewatering (before the formation of the filter cake) should be recirculated inside the geotextile tube so that it can be filtered again, and therefore improve its quality.

- Analyzing the performance indices, the quality of the effluent, and the final percent-solids, this latter a fundamental parameter in the routine of treatment of WTP sludge. It is concluded that under the same test conditions, the dewatering performance was better in the bags fabricated in NW1 (nonwoven geotextile), followed by the bags fabricated in W1 (woven geotextile commonly used), indicating the feasibility of using nonwoven geotextile tubes for the dewatering of WTP sludge.

- It is recommended to quantify the influence of evaporation on the dewatering of geotextile tubes in future research, due to this form of water loss is the one that runs the dewatering in the post-drainage stage. It is also interesting that full performance tests (real scale) are performed to verify the scale-effect and the influence of the dependent variables evaluated in the present work (initial TS and polymer dosing) in the dewatering performance.

Author Contributions: Conceptualization, investigation, and resources, M.A.A.A. and S.T.d.S.; methodology, M.A.A.A. and C.A.V.; software, A.D.B.D. and M.A.A.A.; validation, formal analysis, data curation, visualization and writing—original draft preparation, M.A.A.A.; writing—review and editing, J.L.d.S. and C.A.V.; supervision, J.L.d.S. and A.D.B.D.; project administration and funding acquisition, J.L.d.S. All authors have read and agreed to the published version of the manuscript.

Funding: This work was financially supported by the Laboratory of Geosynthetics of the University of São Paulo (USP) and the Coordenação de Aperfeiçoamento de Pessoal de Nível Superior (CAPES; 001).

Acknowledgments: The authors are grateful for the financial support received from the CAPES, Ober and Huesker for providing geotextile tubes and the CODEN for providing the sludge and the physical space in which conduct the tests. Also to, SEM facilities were provided by CAQI/IQSC/USP.

Conflicts of Interest: The authors declare no conflict of interest. The funders had no role in the design of the study; in the collection, analyses, or interpretation of data; in the writing of the manuscript, or in the decision to publish the results.

References

1. Cordeiro, J.S. Processamento de lodos de Estações de Tratamento de Água (ETAs). In *Resíduos Sólidos do Saneamento: Processamento, Reciclagem e Diposição Final*; Projeto PROSAB: Rio de Janeiro, Brazil, 2001; pp. 121–142.
2. Reali, M.A.P. Principais características quantitativas e qualitativas do lodo de ETAs. In *Noções Gerais de Tratamento e Disposição Final de Lodos de ETA*; ABES/PROSAB: Rio de Janeiro, Brazil, 1999; pp. 21–39.
3. Dantas, A.D.B.; Voltan, P.E.N. *Métodos e Técnicas de Tratamento de Água*, 3rd ed.; LDiBe: São Carlos, Brazil, 2017; p. 1246.
4. Achon, C.L.; Barroso, M.M.; Cordeiro, J.S. Resíduos de estações de tratamento de água e a ISO 24512: Desafio do saneamento brasileiro. *Eng. Sanit. e Ambient.* **2013**, *18*, 115–122. [CrossRef]
5. LAW, Brazilian Environmental Crimes; Brazilian Law 9605/1998. 1998. Available online: http://www.planalto.gov.br/ccivil_03/leis/l9605.htm#:~{}:text=LEI%20N%C2%BA%209.605%2C%20DE%2012%20DE%20FEVEREIRO%20DE%201998.&text=Disp%C3%B5e%20sobre%20as%20san%C3%A7%C3%B5es%20penais,ambiente%2C%20e%20d%C3%A1%20outras%20provid%C3%AAncias (accessed on 8 September 2020).
6. LAW, National Policy on Solid Waste, Alters Law 9605/1998 and Makes Other Provisions; Brazilian Law 12305/2010. 2010. Available online: http://www.planalto.gov.br/ccivil_03/_ato2007-2010/2010/lei/l12305.htm (accessed on 8 September 2020).
7. Grzelak, M.D.; Maurer, B.W.; Pullen, T.S.; Bhatia, S.K.; RamaRao, B. A Comparison of Test Methods Adopted for Assessing Geotextile Tube Dewatering Performance. *Geo-Front. 2011 Adv. Geotech. Eng.* **2011**, 2141–2151. [CrossRef]
8. Fowler, J.; Bagby, R.M.; Trainer, E. Dewatering sewage sludge with geotextile tubes. In Proceedings of the 49th Canadian Geotechnical Conference, St. John's, NL, Canada, 23–25 September 1996; pp. 1–31.
9. Lawson, C. Geotextile containment for hydraulic and environmental engineering. *Geosynth. Int.* **2008**, *15*, 384–427. [CrossRef]
10. Guimarães, M.G.A.; Urashima, D.d.C. Dewatering sludge in geotextile closed systems: Brazilian experiences. *Soils Rocks* **2013**, *36*, 251–263.
11. Muthukumaran, A.; Ilamparuthi, K. Laboratory studies on geotextile filters as used in geotextile tube dewatering. *Geotext. Geomembr.* **2006**, *24*, 210–219. [CrossRef]
12. Fowler, J.; Duke, M.; Schmidt, M.L.; Crabtree, B.; Bagby, R.M.; Trainer, E. Dewatering sewage sludge and hazardous sludge with geotextile tubes. In Proceedings of the 7th International Conference on Geosynthetics, Nice, France, 22–27 September 2002; Delmas, P., Gourc, H., Girard, G.P., Eds.; Swets & Zeitlinger: Lisse, The Netherlands, 2002; pp. 1007–1012.
13. Koerner, G.; Koerner, R. Geotextile tube assessment using a hanging bag test. *Geotext. Geomembr.* **2006**, *24*, 129–137. [CrossRef]
14. Satyamurthy, R.; Bhatia, S.K. Effect of polymer conditioning on dewatering characteristics of fine sediment slurry using geotextiles. *Geosynth. Int.* **2009**, *16*, 83–96. [CrossRef]
15. Suits, L.D.; Sheahan, T.C.; Koerner, R.M.; Koerner, G.R. Performance Tests for the Selection of Fabrics and Additives When Used as Geotextile Bags, Containers, and Tubes. *Geotech. Test. J.* **2010**, *33*, 236–242. [CrossRef]
16. Weggel, J.R.; Dortch, J.; Gaffney, D. Analysis of fluid discharge from a hanging geotextile bag. *Geotext. Geomembr.* **2011**, *29*, 65–73. [CrossRef]
17. Maurer, B.W.; Gustafson, A.; Bhatia, S.; Palomino, A. Geotextile dewatering of flocculated, fiber reinforced fly-ash slurry. *Fuel* **2012**, *97*, 411–417. [CrossRef]
18. Khachan, M.; Bhatia, S.K.; Bader, R.A.; Cetin, D.; RamaRao, B. Cationic starch flocculants as an alternative to synthetic polymers in geotextile tube dewatering. *Geosynth. Int.* **2014**, *21*, 119–136. [CrossRef]
19. Kang, J.; McLaughlin, R. Simple systems for treating pumped, turbid water with flocculants and a geotextile dewatering bag. *J. Environ. Manag.* **2016**, *182*, 208–213. [CrossRef] [PubMed]
20. Pilarczyk, K.W. BOOK REVIEW/CRITIQUE DE LIVRE Geosynthetics and geosystems in hydraulic and coastal engineering. *Can. J. Civ. Eng.* **2001**, *28*, 878. [CrossRef]
21. Moo-Young, H.K.; A Gaffney, D.; Mo, X. Testing procedures to assess the viability of dewatering with geotextile tubes. *Geotext. Geomembr.* **2002**, *20*, 289–303. [CrossRef]

22. Liao, K.; Bhatia, S.K. Geotextile Tube: Filtration Performace of woven geotextiles under pressure. North American Geosynthetic Conference. Presented at NAGS 2005/GRI 19 Cooperative Conference, Las Vegas, NV, USA, 14–16 December 2005; pp. 1–15.

23. Cantré, S.; Saathoff, F. Design parameters for geosynthetic dewatering tubes derived from pressure filtration tests. *Geosynth. Int.* **2011**, *18*, 90–103. [CrossRef]

24. Yee, T.; Lawson, C.; Wang, Z.; Ding, L.; Liu, Y. Geotextile tube dewatering of contaminated sediments, Tianjin Eco-City, China. *Geotext. Geomembr.* **2012**, *31*, 39–50. [CrossRef]

25. Bhatia, S.K.; Maurer, B.W.; Khachan, M.M.; Grzelak, M.D.; Pullen, T.S. Performance Indices for Unidirectional Flow Conditions Considering Woven Geotextiles and Sediment Slurries. *Found. Eng. Face Uncertain.* **2013**, 318–332. [CrossRef]

26. Bourgès-Gastaud, S.; Stoltz, G.; Sidjui, F.; Touze-Foltz, N. Nonwoven geotextiles to filter clayey sludge: An experimental study. *Geotext. Geomembr.* **2014**, *42*, 214–223. [CrossRef]

27. Khachan, M.; Bhatia, S. The efficacy and use of small centrifuge for evaluating geotextile tube dewatering performance. *Geotext. Geomembr.* **2017**, *45*, 280–293. [CrossRef]

28. Guimarães, M.; Urashima, D.D.C.; Vidal, D.D.M. Dewatering of sludge from a water treatment plant in geotextile closed systems. *Geosynth. Int.* **2014**, *21*, 310–320. [CrossRef]

29. Weggel, J.R.; Ward, N.D. A model for filter cake formation on geotextiles: Theory. *Geotext. Geomembr.* **2012**, *31*, 51–61. [CrossRef]

30. Weggel, J.R.; Dortch, J. A model for filter cake formation on geotextiles: Experiments. *Geotext. Geomembr.* **2012**, *31*, 62–68. [CrossRef]

31. Christopher, B.; Fischer, G. Geotextile Filtration Principles, Practices and Problems. *Geosynth. Filtr. Drain. Eros. Control* **1992**, 1–17. [CrossRef]

32. Shin, E.C.; Oh, Y.I. Analysis of geotextile tube behaviour by large-scale field model tests. *Geosynth. Int.* **2003**, *10*, 134–141. [CrossRef]

33. Driscoll, J.; Rupakheti, P.; Bhatia, S.K.; Khachan, M.M. Comparison of 1-D and 2-D Tests in Geotextile Dewatering Applications. *Int. J. Geosynth. Ground Eng.* **2016**, *2*, 2. [CrossRef]

34. Huang, C.-C.; Luo, S.-Y. Dewatering of reservoir sediment slurry using woven geotextiles. Part I: Experimental results. *Geosynth. Int.* **2007**, *14*, 253–263. [CrossRef]

35. Huang, C.-C.; Jatta, M.; Chuang, C.-C. Dewatering of reservoir sediment slurry using woven geotextiles. Part II: Analytical results. *Geosynth. Int.* **2012**, *19*, 93–105. [CrossRef]

36. Moo-Young, H.K.; Tucker, W.R. Evaluation of vacuum filtration testing for geotextile tubes. *Geotext. Geomembr.* **2002**, *20*, 191–212. [CrossRef]

37. Queiroz, S.C.B. Influência das Características da Água Bruta no Desaguamento de Resíduos Gerados no Tratamento de Água Pela Filtração em Tubo Geotêxtil. Ph.D. Thesis, Universidade de Ribeirão Preto, Ribeirão Preto, SP, Brazil, 2019.

38. Ratnayesuraj, C.R.; Bhatia, S.K. Testing and analytical modeling of two-dimensional geotextile tube dewatering process. *Geosynth. Int.* **2018**, *25*, 132–149. [CrossRef]

39. GRI Test Method GT14:2004. *Standard Test Method for Hanging Bag Test for Field Assessment of Fabrics Used for Geotextile Tubes and Containers*; Geosynthetics Research Institute: Folsom, CA, USA, 2004.

40. GRI Test Method GT15:2009. *Standard Test Method for the Pillow Test for Test Field Assessment of Fabrics/Additives Used for Geotextile Bags, Containers and Tubes*; Geosynthetics Research Institute: Folsom, CA, USA, 2009.

41. ASTM D7880:2013. *Standard Test Method for Determining Flow Rate of Water and Suspended Solids Retention from a Closed Geosynthetic Bag*; ASTM International: West Conshohocken, PA, USA, 2013.

42. Rawal, A.; Anandjiwala, R. Comparative study between needlepunched nonwoven geotextile structures made from flax and polyester fibres. *Geotext. Geomembr.* **2007**, *25*, 61–65. [CrossRef]

43. Clesceri, L.S. *Standard Methods for the Examination of Water and Wastewater*, 22nd ed.; Rice, E.W., Baird, R.B., Eaton, A.D., Clesceri, L.S., Eds.; American Public Health Association: Washington, DC, USA, 2012; p. 724.

44. *Solo-Análise Granulométrica*; Associação Brasileira de Normas Técnicas: Rio de Janeiro, Brazil, 1984; ABNT NBR 7181.

45. *Grãos de Pedregulho Retidos na Peneira de Abertura 4,8 mm—Determinação da Massa Específica, da Massa Específica Aparente e da Absorção de Água*; Associação Brasileira de Normas Técnicas: Rio de Janeiro, RJ, Brazil, 2017; ABNT NBR 6458.

46. *Geossintéticos: Método de Ensaio Para Determinação da Massa Por Unidade de Área de Geotêxteis e Produtos Correlatos*; Associação Brasileira de Normas Técnicas: Rio de Janeiro, RJ, Brazil, 2013; ABNT NBR ISO 9864:2013.

47. *Geossintéticos: Determinação da Espessura a Pressões Especificadas Parte 1: Camada Única*; Associação Brasileira de Normas Técnicas: Rio de Janeiro, RJ, Brazil, 2013; ABNT NBR ISO 9863-1:2013.

48. *Geotêxteis e Produtos Correlatos: Determinação da Abertura de Filtração Característica*; Associação Brasileira de Normas Técnicas: Rio de Janeiro, RJ, Brazil, 2013; ABNT NBR ISO 12956:2013.

49. *Standard Test Methods for Water Permeability of Geotextiles by Permittivity*; ASTM International: West Conshohocken, PA, USA, 2015; ASTM D4491:2015.

50. *Geossintéticos: Ensaio de Tração de Faixa Larga*; Associação Brasileira de Normas Técnicas: Rio de Janeiro, RJ, Brazil, 2013; ABNT NBR ISO 10319:2013.

51. Guanaes, E.A.; Guimarães, M.G.A.; de Carvalho Urashima, D.; Pontes, P.P. Análise laboratorial do desaguamento do lodo residual de estação de tratamento de água por meio de geossintéticos. *Educ. Tecnol.* **2009**, *14*, 33–39.

52. Rodrigues, M.I.; Iemma, A.F. *Experimental Design and Process Optimization*; CRC Press: New York, NY, USA, 2014.

53. Guimarães, G. Eficiência do Desaguamento de Resíduo de Estação de Tratamento de Água em Tubo Geotêxtil. Ph.D. Thesis, Universidade de Ribeirão Preto, Ribeirão Preto, SP, Brazil, 2019.

54. Resolution, Brazilian National Council of Environment, Establishes Provisions for the Classification of Water Bodies as Well as Environmental Directives for Their Framework, Establishes Conditions and Standards for Effluent Releases and Makes Other Provisions; CONAMA Resolution 357/05. Available online: http://www2.mma.gov.br/port/conama/legiabre.cfm?codlegi=459 (accessed on 8 September 2020).

55. Resolution, Brazilian National Council of Environment, Provisions the Conditions and Standards of Effluents and Complements and Changes Resolution 357/2005; CONAMA Resolution 430/2011. 2011. Available online: http://www2.mma.gov.br/port/conama/legiabre.cfm?codlegi=646 (accessed on 8 September 2020).

56. UKWIR. *Guidance Manual Supporting the Water Treatment Recommendations from the Badenoch Group of Experts on Cryptosporidium*; Water Industry Research Limited: London, UK, 1998.

57. Müller, M.; Vidal, D. Comparison between Open and Closed System for Dewatering with Geotextile: Field and Comparative Study. *Int. J. Civ. Environ. Eng.* **2019**, *13*, 634–639.

58. Junqueira, F.F.; Silva, A.R.; Palmeira, E.M. Performance of drainage systems incorporating geosynthetics and their effect on leachate properties. *Geotext. Geomembr.* **2006**, *24*, 311–324. [CrossRef]

59. Yaman, C.; Martin, J.P.; Korkut, N.E. Effects of wastewater filtration on geotextile permeability. *Geosynth. Int.* **2006**, *13*, 87–97. [CrossRef]

60. Palmeira, E.M.; Remigio, A.F.; Ramos, M.L.; Bernardes, R.S. A study on biological clogging of nonwoven geotextiles under leachate flow. *Geotext. Geomembr.* **2008**, *26*, 205–219. [CrossRef]

61. Silva, R.A.E.; Negri, R.G.; Vidal, D.D.M. A new image-based technique for measuring pore size distribution of nonwoven geotextiles. *Geosynth. Int.* **2019**, *26*, 261–272. [CrossRef]

62. Mlynarek, J.; Rollin, A.L. Bacterial clogging of geotextiles: Overcoming engineering concerns. In Proceedings of the Geosynthetics '95 Conference, Nashville, TN, USA, 21–23 February 1995; pp. 177–188.

Article

Use of Incinerator Bottom Ash as a Recycled Aggregate in Contact with Nonwoven Geotextiles: Evaluation of Mechanical Damage Upon Installation

Filipe Almeida *, José Ricardo Carneiro and Maria de Lurdes Lopes

Construct-Geo, Faculty of Engineering, University of Porto, Rua Dr. Roberto Frias, 4200-465 Porto, Portugal; rcarneir@fe.up.pt (J.R.C.); lcosta@fe.up.pt (M.d.L.L.)
* Correspondence: filipe.almeida@fe.up.pt

Received: 2 October 2020; Accepted: 27 October 2020; Published: 3 November 2020

Abstract: The recycling and reuse of materials is crucial to reducing the amount of generated waste and the exploitation of natural resources, contributing to achieving environmental sustainability. During the incineration process of municipal solid waste, a residue known as incinerator bottom ash is generated in considerable amounts, being important the development of solutions for its valorization. In this work, three nonwoven geotextiles were submitted to mechanical damage under repeated loading tests with incinerator bottom ash and, for comparison purposes, with three natural aggregates (sand 0/4, gravel 4/8 and *tout-venant*) and a standard aggregate (*corundum*). Damage assessment was carried out by monitoring the changes that occurred in the short-term tensile and puncture behaviors of the geotextiles. Results showed that the damage induced by incinerator bottom ash on the short-term mechanical behavior of the geotextiles tended to be lower than the damage induced by the natural aggregates or by the standard aggregate. Therefore, concerning the mechanical damage caused on geotextiles, there are good prospects for the use of incinerator bottom ash as a filling material in contact with those construction materials, thereby promoting its valorization.

Keywords: incinerator bottom ash; geotextiles; mechanical damage; sustainable engineering; waste valorization

---

1. Introduction

Environment, economy and society are the three dimensions that make up the universal purpose known as Sustainable Development. The quality of life of current and upcoming generations depends on how consistent the measures that are being implemented in terms of environmental policies are. In the European Union, Directive 2008/98/EC [1] establishes key measures regarding the protection of the environment and human health. The success of environmental policies cannot be assigned to a particular person or institution but to several agents that comprise our society, particularly legislators, researchers, and manufacturers, who should be responsible for involving the population in the overall process.

Actions should be carried out to avoid unreasonable exploitation of natural resources and consumption of energy, which are overwhelming the sustainability and health of planet Earth. Circular economy emerged as a model that aims to extend the lifetime of products, materials and resources, and to reduce the generation of waste. For that purpose, solutions should be developed to promote the reuse and recycling of products and materials, which lead to the reduction, for instance, of incineration and/or landfilling. In addition, efforts should be carried out to understand how residues that are generated can be introduced into valorization chains, in which they could be labelled as raw materials.

Incinerator bottom ash (IBA) is a residue that results from the incineration of municipal solid waste. According to Blasenbauer et al. [2], there are 463 municipal solid waste incineration plants currently functioning in the European Union, Norway, and Switzerland, leading to the generation of 17.6 megatons of IBA per year. Considering the large amount of IBA that is produced, and in order to prevent its landfilling, there is a need to develop innovative solutions in which IBA plays the role of a noble raw material. The academic community has started on this path, since different investigations have been carried out to find useful roles for IBA, namely: (1) its use in cementitious materials, (2) as a recycled aggregate, or (3) in geotechnical projects. Due to its chemical composition, IBA has been the subject of research aiming to develop alternative cementitious materials to ordinary Portland cement [3–6]. In the domain of concrete, IBA showed potential to be used as a recycled aggregate to replace natural aggregates [7–9]. Regarding the geotechnics field, IBA was evaluated as a raw material in road construction, where it was used in asphalt concrete or in cement-bound mixtures [10], and it was also studied as an aggregate in road pavements [11–13]. In the latter example, IBA may have contact with geosynthetics. Despite the encouraging findings of the previous works, the environmental behavior of IBA must be deeply studied in order to understand its suitability for being introduced into innovative solutions. Indeed, it is crucial to ensure that IBA does not contain hazardous substances that can be released into soil or water.

Geosynthetics are construction materials that can perform many different functions, e.g., reinforcement, protection, separation, drainage, filtration, erosion control or fluid barrier. This high versatility, associated with their high efficiency, low cost and ease of installation, provides the possibility of using these materials in a wide range of engineering projects such as embankments, roadways, railways, landfill sites or coastal protection structures. Within the framework of geotechnical engineering structures, in which IBA may be used as a filling material (for example, in roadways or railways), geotextiles (one of the main types of geosynthetics) can be suitable construction materials to perform the functions of protection, filtration or separation.

Installation on site may cause damage to geosynthetics, resulting in undesirable changes to their mechanical and/or hydraulic properties. The damage that occurs during the installation process (for example, cuts in components, tears, formation of holes or abrasion) is essentially caused by handling the geosynthetics and by the placement and compaction of filling materials over them. In some cases, the stresses induced during the installation process can be higher than those considered in the design for service conditions [14].

Mechanical damage is often associated with the installation process of geosynthetics. In order to induce mechanical damage to geosynthetics, the European Committee for Standardization developed a standard method (EN ISO 10722 [15]). This method has been used by some authors to evaluate the damage that occurs during the installation of geosynthetics [16,17], while others tried to correlate it with field installation conditions [18,19]. The evaluation of installation damage can also be carried out by field tests, which provide more reliable data about the behavior of geosynthetics. However, these tests are more expensive and time-consuming and require the use of heavy equipment and skilled workers, making them unsuitable for routine analysis. Damage assessment is usually carried out by monitoring changes in the mechanical behavior of geosynthetics [20–22]. Some authors have also monitored changes in their hydraulic behavior [23–25].

Filling materials (which can come into contact with geosynthetics in many applications) are often natural and may, in some cases, be replaced by recycled aggregates. When considering the use of recycled aggregates in applications where they will be in contact with geosynthetics, it is expected that the recycled aggregates do not cause higher mechanical damage to these construction materials, compared with the natural aggregates commonly used. Existing studies about the use of recycled aggregates in contact with geosynthetics are relatively few. Vieira and Pereira [26] and Vieira et al. [27] developed investigations in which the interface properties between recycled aggregates resulting from construction and demolition waste and geosynthetics were studied. The mechanical damage induced to geosynthetics by recycled aggregates resulting from the previously mentioned waste stream was also addressed, namely by Vieira and Pereira [28] and Carlos et al. [29], whose findings offered positive perspectives when it came to exploiting the use of recycled aggregates in applications involving contact with geosynthetics.

Similar to other recycled aggregates, one relevant aspect that has to be evaluated when considering the possibility of using IBA as a filling material in contact with geosynthetics is the degree of mechanical damage induced by IBA to those construction materials. If the outcomes are positive, a door is opened to using IBA in this particular application, which would contribute to promoting more environmentally friendly practices towards a circular economy in the domain of geotechnical engineering.

In this work, three nonwoven geotextiles were submitted to laboratory mechanical damage under repeated loading tests with IBA and also, for comparison purposes, three natural aggregates (sand 0/4, gravel 4/8 and *tout-venant*) and a standard aggregate (*corundum*). The changes that occurred in the short-term tensile and puncture behaviors of the geotextiles were monitored. The main goals of the work included: (1) evaluation of the effect of IBA on geotextiles, (2) comparison between the damage induced by IBA and by the natural and standard aggregates, and (3) evaluation of whether IBA can be a viable alternative as a filling material in replacement of natural aggregates (in terms of mechanical damage induced on the geotextiles).

2. Materials and Methods

2.1. Geotextiles

The experimental campaign conducted in this work included the use of three nonwoven polypropylene geotextiles, which were designated by GT100, GT300 and GT450 (the number following the abbreviation 'GT' (geotextile) corresponds to the mass per unit area of the geotextile, as defined by its manufacturer). Depending on the requirements defined by the designers, GT100, GT300 and GT450 may be adequate to perform functions like separation or filtration. GT300 and GT450 may also be suitable to accomplish the function of protection. The physical properties (mass per unit area and thickness) of the geotextiles can be found in Table 1.

Table 1. Physical properties of the geotextiles.

Geotextile	Mass per Unit Area [1] (g.m^{-2})	Thickness [2] (mm)
GT100	117 ($\pm$ 6)	0.96 ($\pm$ 0.06)
GT300	322 ($\pm$ 15)	3.83 ($\pm$ 0.11)
GT450	457 ($\pm$ 14)	4.54 ($\pm$ 0.17)

[1] Determined according to EN ISO 9864 [30]. [2] Determined according to EN ISO 9863-1 [31]. (95% confidence intervals in brackets).

2.2. Mechanical Damage Under Repeated Loading Tests

2.2.1. Equipment and Test Method

The mechanical damage (MD) under repeated loading tests (hereinafter designated by MD tests) were performed according to the procedures described in EN ISO 10722 [15] (with an exception for the use of different aggregates other than *corundum*).

The equipment used in the MD tests was a laboratory prototype developed at the Faculty of Engineering of the University of Porto in accordance with the specifications of EN ISO 10722 [15]. The equipment included a test-container (a rigid metal box where the geotextiles and aggregates were placed), a loading plate (with dimensions of 200 mm × 100 mm) and a compression machine. The test-container, having a square base with a side of 300 mm, was divided into two parts: a lower box and an upper box, each with a height of 87.5 mm. A schematic representation of the equipment is illustrated in Figure 1. In this outline, the upper box is represented in a distinct plane to provide a better understanding of the elements of the equipment.

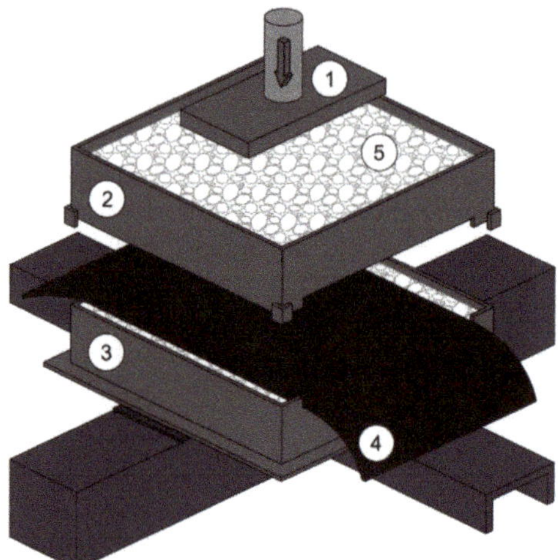

Figure 1. Schematic representation of equipment used for the MD tests: (1) loading plate, (2) upper box, (3) lower box, (4) test-specimen, (5) aggregate.

The MD tests comprised of several steps. First, a sublayer of aggregate (five different aggregates were used) with a height of 37.5 mm was placed in the lower box and submitted to compaction. Then, the remaining half of the box was filled with another sublayer of aggregate (with a height of 37.5 mm), followed again by compaction. Each compaction process consisted of placing a flat plate over the sublayers of aggregate and applying a pressure of (200 ± 2) kPa for 60 s (pressure was applied over the whole area of the box).

The next step was the placement of a test-specimen (with a width and length of 250 and 500 mm, respectively) over the compacted layer of aggregate, followed by the installation of the upper box. A single layer of loose aggregate with a height of 75 mm was afterwards introduced into the upper box (it is important to stress that this layer was not submitted to a compaction process). Damaging actions were induced in the test-specimen by applying a vertical dynamic loading between (5.0 ± 0.5) kPa and (500 ± 10) kPa at a frequency of 1 Hz for 200 cycles. At the end of the MD test, the test-specimen was carefully removed, avoiding additional damage.

A total of 50 test-specimens of each geotextile were submitted to MD tests, in accordance with the following plan: 10 test-specimens for each aggregate, of which 5 were further characterized by tensile tests and 5 by static puncture tests.

2.2.2. Aggregates

The aggregates used in the MD tests included IBA and, for comparison purposes, *corundum* (a synthetic aggregate of aluminum oxide used in the procedure described in EN ISO 10722 [15]) and three natural aggregates: sand 0/4 (river sand), gravel 4/8 and *tout-venant* (well graded untreated mixed aggregate) (Figure 2). IBA was supplied by a Portuguese incineration plant (Lipor II-Maia) and resulted from the incineration of municipal solid waste. This recycled aggregate was used as provided by the supplier, without further treatment other than drying. Both IBA and the other aggregates were dried to constant mass in a ventilated oven at 110 °C.

Figure 2. Aggregates used in the MD tests: (**a**) IBA, (**b**) sand 0/4, (**c**) gravel 4/8, (**d**) *tout-venant*, (**e**) *corundum*.

The particle size distribution of the aggregates was determined by sieving (tests carried out according to EN 933-1 [32]) and can be found in Figure 3. The main parameters for the characterization of the particle size distributions (D_{10}–effective 10% particle size, D_X–particle size corresponding to X% passing, and D_{Max}–maximum particle size) are summarized in Table 2.

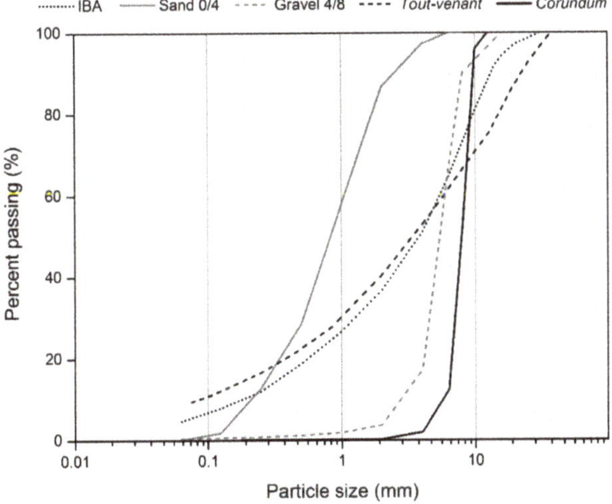

Figure 3. Particle size distribution of the aggregates.

Table 2. Characterization of the particle size distribution of the aggregates.

Aggregate	%<0.063 mm	D_{10} (mm)	D_{30} (mm)	D_{50} (mm)	D_{60} (mm)	D_{Max} (mm)
IBA	4.7	0.19	1.33	3.83	5.48	14.0
Sand 0/4	0.2	0.20	0.52	0.86	1.07	4.0
Gravel 4/8	0.4	2.92	4.37	5.79	5.97	8.0
Tout-venant	8.0	0.08	1.04	3.67	6.07	31.5
Corundum	0.1	5.77	7.05	7.91	8.36	10.0

2.3. Evaluation of the Damage Suffered by the Geotextiles

The damage suffered by the geotextiles (in the MD tests) was evaluated qualitatively by visual inspection and quantitatively by monitoring the changes that occurred in their tensile and puncture properties.

Tensile tests were performed in the machine direction of production of the geotextiles according to EN ISO 10319 [33], under displacement control at a constant rate of 20 mm.min^{-1} (the test-specimens had a length of 100 mm (between grips) and a width of 200 mm). Elongation was determined by expressing the relative displacement of the grips as percentage of the original length (100 mm).

Static puncture tests, which were conducted under displacement control at a constant rate of 50 mm.min^{-1}, followed the guidelines of EN ISO 12236 [34]. In these tests, a stainless steel plunger (a cylinder with a diameter of 50 mm and a leading edge with a radius of 2.5 mm) was pushed-through the test-specimens, which had a diameter of 150 mm between the clamping rings.

Tensile and static puncture tests were performed using a Lloyd Instruments LR10K Plus testing machine (Bognor Regis, UK) fitted with a load cell of 10 kN. The mechanical parameters resulting from the tensile tests included tensile strength (T, in kN.m^{-1}) and elongation at maximum load (E_{ML}, in %), while puncture strength (F_P, in kN) and push-trough displacement (displacement at maximum force) (h_P, in mm) were the parameters obtained from the static puncture tests. The results for the tensile and puncture properties of the geotextiles, which correspond to the mean values of 5 specimens, are presented with 95% confidence intervals determined according to Montgomery and Runger [35].

The changes that occurred in T and F_P are also presented in terms of residual strengths. The residual tensile strength ($T_{Residual}$, in %) was obtained by dividing the T of the damaged samples by the respective strength of the reference samples (undamaged). Residual puncture strength ($F_{P\ Residual}$, in %) was determined as $T_{Residual}$, taking into account the F_P of damaged and undamaged samples.

3. Results and Discussion

3.1. Geotextile GT100

The MD tests affected GT100 distinctively, depending on the aggregate. Tests with sand 0/4, *tout-venant* and IBA induced practically no visible damage to the nonwoven structure, only some minor abrasion on the contact surfaces between the geotextile and the aggregates. On the other hand, gravel 4/8 and *corundum* provoked visible damage in GT100 like some small cuts, punctures and abrasion (the damage caused by *corundum* appeared to have been slightly higher than that induced by gravel 4/8). In all cases, it was possible to find some fine particles (constituent particles of the aggregates or particles resulting from their fragmentation during the MD tests) imprisoned in the nonwoven structure. The mechanical properties of GT100, before and after the MD tests, can be found in Table 3.

Table 3. Tensile and puncture properties of geotextile GT100 before and after the MD tests.

Mechanical Damage Test	T (kN.m^{-1})	E_{ML} (%)	F_P (kN)	h_P (mm)
Undamaged	7.92 ($\pm$ 0.34)	48.4 ($\pm$ 5.6)	1.58 ($\pm$ 0.07)	50.1 ($\pm$ 5.7)
MD test with IBA	6.12 ($\pm$ 0.52)	36.2 ($\pm$ 5.6)	1.14 ($\pm$ 0.16)	42.0 ($\pm$ 4.3)
MD test with sand 0/4	6.36 ($\pm$ 0.58)	36.8 ($\pm$ 3.7)	1.17 ($\pm$ 0.09)	45.1 ($\pm$ 6.6)
MD test with gravel 4/8	4.79 ($\pm$ 0.52)	29.1 ($\pm$ 1.6)	0.74 ($\pm$ 0.17)	39.7 ($\pm$ 5.6)
MD test with *tout-venant*	5.88 ($\pm$ 1.32)	34.2 ($\pm$ 6.0)	1.11 ($\pm$ 0.15)	42.8 ($\pm$ 5.5)
MD test with *corundum*	3.94 ($\pm$ 0.56)	28.6 ($\pm$ 2.7)	0.68 ($\pm$ 0.09)	40.9 ($\pm$ 4.3)

(95% confidence intervals in brackets).

Contrary to what was expected, the MD tests with sand 0/4, *tout-venant* and IBA (aggregates that apparently did not induce relevant damage) caused reductions (between 19.7% and 25.8%) in the T of GT100 (the decrease was slightly more pronounced after the MD tests with *tout-venant*). These results indicated that, although not visibly detectable, the MD tests induced some physical damage on the nonwoven structure.

Like the previous aggregates, gravel 4/8 and *corundum* also provoked reductions in T, but these were much more pronounced (39.5% and 50.3%, respectively). These reductions were, once again, much more significant than those expected, taking into account the damage visibly found in GT100. Yet, as expected from visual analysis, the MD tests with gravel 4/8 were less damaging than the MD tests with *corundum*.

Similar to T, E_{ML} also suffered some changes (Table 3). The highest reductions in E_{ML} occurred after the MD tests with gravel 4/8 and *corundum* (reductions from 48.4% to 29.1% and to 28.6%, respectively). Like the losses in T, the reductions in E_{ML} after the MD tests with sand 0/4, *tout-venant* and IBA were relatively similar and lower than the reductions induced by gravel 4/8 and *corundum*.

The puncture properties of GT100 also suffered relevant changes after the MD tests (Table 3). These changes depended, once again, on the characteristics of the aggregates. Like T, F_P experienced pronounced losses after the MD tests with gravel 4/8 and *corundum* (53.2% and 57.0%, respectively). The reductions provoked by sand 0/4, *tout-venant* and IBA were identical (between 25.9% and 29.7%) and significantly lower than those caused by gravel 4/8 and *corundum*. Besides the losses in F_P, reductions also occurred in h_P (reduction trend not as evident as that found for F_P). It should be highlighted that the reductions in F_P were slightly more pronounced than those that occurred in T.

3.2. Geotextile GT300

The tests with sand 0/4, *tout-venant* and IBA did not lead to visible damage in GT300. On the other hand, the tests with gravel 4/8 and *corundum* induced minor visible damage, namely small punctures and some abrasion. Like for GT100, the damage caused by gravel 4/8 seemed to be slightly less than the damage provoked by *corundum*. In all cases, and like before, fine particles were found imprisoned in the nonwoven structure. The tensile and puncture properties of GT300 can be found in Table 4.

Table 4. Tensile and puncture properties of geotextile GT300 before and after the MD tests.

Mechanical Damage Test	T (kN.m^{-1})	E_{ML} (%)	F_P (kN)	h_P (mm)
Undamaged	22.43 ($\pm$ 1.03)	135.3 ($\pm$ 12.3)	4.25 ($\pm$ 0.21)	76.2 ($\pm$ 1.6)
MD test with IBA	18.89 ($\pm$ 0.87)	93.7 ($\pm$ 7.3)	3.81 ($\pm$ 0.42)	58.7 ($\pm$ 3.8)
MD test with sand 0/4	18.31 ($\pm$ 1.03)	85.1 ($\pm$ 5.0)	3.57 ($\pm$ 0.30)	58.3 ($\pm$ 7.6)
MD test with gravel 4/8	14.99 ($\pm$ 1.06)	75.5 ($\pm$ 2.1)	2.82 ($\pm$ 0.21)	60.6 ($\pm$ 1.5)
MD test with *tout-venant*	17.36 ($\pm$ 1.45)	88.5 ($\pm$ 2.6)	3.47 ($\pm$ 0.21)	65.3 ($\pm$ 1.9)
MD test with *corundum*	12.08 ($\pm$ 1.18)	68.6 ($\pm$ 9.3)	2.19 ($\pm$ 0.24)	54.5 ($\pm$ 3.7)

(95% confidence intervals in brackets).

As observed in GT100, the MD tests with sand 0/4, *tout-venant* and IBA also induced reductions in the T of GT300. These reductions were relatively similar (between 15.8% and 22.6%) but more

377

pronounced than those expected from the apparent non-existence of relevant physical damage on the geotextile. The losses provoked by the MD tests with gravel 4/8 and *corundum* (33.6% and 48.5%, respectively) were significantly more pronounced than those caused by sand 0/4, *tout-venant* and IBA. The trend observed in the variation of E_{ML} was, in general, identical to the trend found in T (Table 4).

Losses that occurred in F_P were relatively identical to those observed in T (comparing the same MD tests). For example, the MD tests with IBA induced, respectively, reductions of 15.8% and 10.4% in the T and F_P of GT300. The same trend was observed for gravel 4/8 (there were reductions of 33.2% and 33.6%, respectively) and *corundum* (with reductions of 46.1% and 48.5%, respectively). The h_P also decreased after the MD tests (again, the reduction trend was not as evident as that found for F_P).

Resistance losses that occurred after the MD tests were more pronounced for GT100 than for GT300, particularly regarding F_P. This indicates that the mass per unit area influences the degree of degradation of the geotextiles in MD tests (there was better survivability for the geotextile with a higher mass per unit area).

3.3. Geotextile GT450

Similar to what was observed for GT300, only minor defects (such as small punctures and abrasion) were detected in GT450 after MD tests with gravel 4/8 and *corundum*. The tests with IBA, sand 0/4 and *tout-venant* did not cause detectable damage. The results obtained for the tensile and puncture properties of GT450 are exhibited in Table 5.

Table 5. Tensile and puncture properties of geotextile GT450 before and after the MD tests.

Mechanical Damage Test	T (kN.m^{-1})	E_{ML} (%)	F_P (kN)	h_P (mm)
Undamaged	36.16 (± 2.01)	133.3 (± 17.6)	7.17 (± 0.47)	74.3 (± 3.6)
MD test with IBA	32.42 (± 2.14)	100.0 (± 6.5)	6.52 (± 0.73)	58.9 (± 3.6)
MD test with sand 0/4	29.30 (± 1.37)	97.4 (± 10.8)	5.96 (± 0.71)	54.8 (± 6.2)
MD test with gravel 4/8	27.64 (± 1.60)	94.0 (± 7.9)	5.84 (± 0.79)	54.7 (± 3.4)
MD test with *tout-venant*	29.46 (± 2.35)	97.3 (± 11.5)	6.37 (± 0.54)	60.0 (± 4.3)
MD test with *corundum*	25.38 (± 1.79)	82.5 (± 8.2)	5.29 (± 0.41)	53.5 (± 2.9)

(95% confidence intervals in brackets).

The behavior of GT450 after the MD tests was relatively identical to the behavior of GT100 and GT300, leading to analogous conclusions. However, reductions that occurred in its T and F_P tended to be less pronounced than those observed in GT100 and GT300. For example, the MD tests with IBA led to reductions in T of 22.7%, 15.8% and 10.3% in GT100, GT300 and GT450, respectively. It is worth mentioning that the losses that occurred in the T of GT450 tended to be slightly higher than those found in F_P. The higher resistance of GT450 (compared with the other geotextiles) shows, once again, that the increase of mass per unit area resulted in a better performance with regard to the MD tests.

3.4. Comparison of the Effect of the Aggregates

The effect of the different aggregates was compared by the impact that they caused on the tensile and puncture properties of the geotextiles. Comparison of the $T_{Residual}$ of the geotextiles after MD tests can be found in Figure 4. Figure 5 illustrates a similar comparison for $F_{P\ Residual}$. It is worth mentioning that the points representing the residual strengths of the geotextiles in Figures 4 and 5 were joined by lines in order to highlight the effect of mass per unit area on the resistance of the geotextiles against mechanical damage.

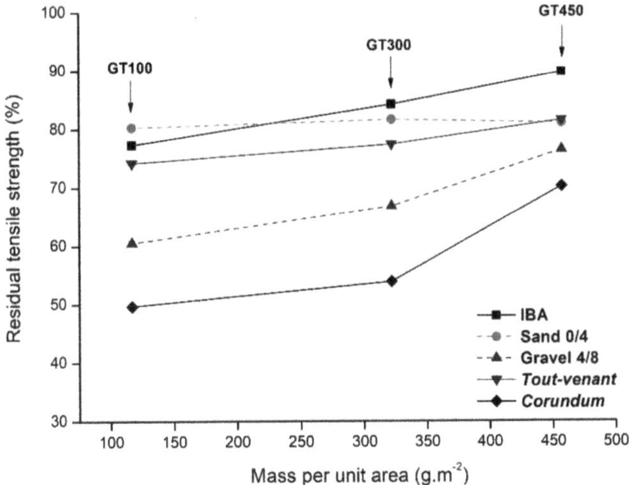

Figure 4. Comparison of the residual tensile strengths of the geotextiles after the MD tests.

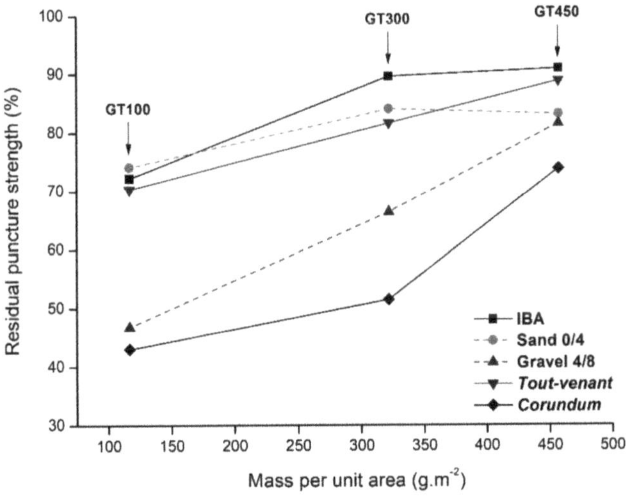

Figure 5. Comparison of the residual puncture strengths of the geotextiles after the MD tests.

The MD tests with sand 0/4 led to some reductions in the T of the geotextiles (the $T_{Residual}$ was between 80.3% and 81.6%). Regarding the effect on F_P, GT100 was slightly more affected (there was a reduction of 25.9%) than GT300 and GT450 (with reductions of 16.0% and 16.9%, respectively). Compared with the other aggregates, sand 0/4 was one of the less damaging. The geometry and dimensions of this aggregate (D_{10}, D_{50} and D_{60} of, respectively, 0.20, 0.86 and 1.07 mm) fostered the formation of a plane and regular surface after compaction. Therefore, there was a high contact surface area between sand 0/4 and the geotextiles, which could have contributed to a good distribution of the applied loads (and, thus, would have minimized the occurrence of damage). However, as a consequence of the non-cohesive nature of sand 0/4, considerable settlements occurred during the MD tests, which affected the aforementioned good distribution of the applied loads. Even though the particles were of low dimensions, they had a rough texture, which may have induced some damage (not very pronounced nor visibly detected) on the nonwoven structures (which explains the relatively minor deterioration of the mechanical properties of the geotextiles).

As sand 0/4, *tout-venant* also did not provoke extensive degradation in the geotextiles. The lower degradation caused by *tout-venant* (when comparing with gravel 4/8 or *corundum*) may be related to its well graded classification. Indeed, despite having the highest D_{Max} (31.5 mm), *tout-venant* had a relatively high percentage of fine particles (8.0% of the particles had a particle size lower than 0.063 mm) and a low quantity of larger particles compared with other aggregates (for example, the D_{50} of *tout-venant* was lower than the D_{50} of gravel 4/8 or *corundum*). Hence, *tout-venant* originated in a fairly flat and smooth surface when compacted (the particles of lower dimensions were surrounding their counterparts of higher dimensions and thereby fulfilling the voids), thus creating a large contact area with the geotextiles and allowing for better distribution of the applied loads. Yet, the particles of higher dimensions with angular shape were capable of inducing some damage to the geotextiles.

The MD tests with gravel 4/8 provoked pronounced losses in the mechanical resistance of the geotextiles. Gravel 4/8 was a poorly graded aggregate (D_{10}, D_{50} and D_{Max} of, respectively, 2.92, 5.79 and 8.0 mm) formed by rough particles with angular shape, which explains it being the natural aggregate that caused the highest reductions in the T and F_P of the geotextiles.

The reductions in resistance imposed by the natural aggregates were lower compared with *corundum*, which was a poorly graded synthetic aggregate (D_{10}, D_{50} and D_{Max} of, respectively, 5.77, 7.91 and 10.0 mm). In addition, the particles forming *corundum* had an angular shape (which can promote cuts in the nonwoven structures), were rough and had a high abrasive effect (the abrasive effect of *corundum* was noticeably higher than the other aggregates). Therefore, the use of *corundum* in EN ISO 10722 [15] seems to establish a conservative approach to evaluating the mechanical damage suffered by geotextiles. Indeed, all natural aggregates induced lower reductions in the resistance of the geotextiles than those caused by *corundum*. The execution of field tests (with the same natural aggregates and geotextiles), and further comparison with laboratory results, may allow reliable conclusions about the suitability of *corundum* (and the MD test described in EN ISO 10722 [15]) in simulating field installation conditions.

Finally, the effects of IBA were identical to the effects of sand 0/4 and slightly less pronounced compared with *tout-venant*. The particle size distribution of IBA and *tout-venant* were much identical (Figure 3), which explains the similarities between the effects of these aggregates on the mechanical properties of the geotextiles. Indeed, despite having different D_{Max} (higher for *tout-venant*), the D_{30}, D_{50} and D_{60} of IBA and *tout-venant* were not significantly different. As *tout-venant*, IBA also had a relatively high number of fine particles (when compared with sand 0/4, gravel 4/8 or *corundum*), which surrounded the larger particles and originated a relatively plane and regular surface for the transference of stresses during the MD tests. Although IBA had a higher D_{Max} than gravel 4/8 and *corundum* (14.0 mm for IBA and, respectively, 8.0 and 10.0 mm for gravel 4/8 and *corundum*), D_{60}, D_{50}, D_{30} and D_{10} were lower compared with those aggregates. This circumstance also contributes to the lower degradation induced by IBA to the geotextiles when compared with gravel 4/8 or *corundum*. The damaging effect of IBA may be ascribed to the presence of some cutting materials in its composition (such as glass, metals and ceramic waste), which may have induced some cuts (not detected by naked eye) on the geotextiles.

The behavior of IBA was very promising for the possible use of this recycled aggregate as a filling material that comes into contact with geotextiles. For instance, in embankment projects, the substitution of local soils with soils with proper properties is a common practice when they are not in compliance with demanded requirements. This procedure implies removing soils from their original location, leading to negative environmental impacts. Another solution is the potential use of IBA as a replacement of *tout-venant* in the construction of transport infrastructures, since they have similar particle size distributions. If IBA meets the requirements to be considered as a recycled aggregate that can be used in place of natural aggregates in engineering projects, steps are being taken towards more environmentally friendly practices within this domain.

4. Conclusions

MD tests with IBA caused some deterioration of the tensile and puncture properties of three nonwoven geotextiles. However, the damage caused by IBA was not that different (in some cases, even lower) to the damage induced by sand 0/4 or by *tout-venant* (two natural aggregates). When compared with gravel 4/8 (another natural aggregate) and *corundum* (a synthetic aggregate used in the standard method for inducing mechanical damage on geosynthetics), the effect of IBA on the mechanical behavior of the geotextiles was significantly less pronounced.

Regarding the possibility of using IBA as a filling material in contact with geotextiles (and/or other geosynthetics), this research does not allow definitive conclusions, since it only evaluated the mechanical damage induced by that recycled aggregate on geotextiles. Still, and as previously stated, the effect of IBA on the short-term mechanical behavior of the geotextiles tended to be less pronounced than (or, at least, identical to) the effects of natural aggregates or the effects of *corundum*. These results open good potential for using IBA as a filling material that comes into contact with geotextiles. However, before assigning this role to IBA, field damage tests should be carried out to verify if the conclusions obtained in the laboratory effectively represent the behavior of the geotextiles under real conditions.

Even if IBA shows positive behavior in terms of the mechanical damage induced to the geotextiles, additional studies are required before its application as a filling material is carried out. Indeed, it is also important to evaluate the effects of IBA on the long-term behavior of geotextiles. The use of alternative filling materials, like IBA, may also address environmental concerns (for example, the possible contamination of soils and water). Therefore, the characterization of the chemical composition of IBA and the performance of leachate tests may help in assessing the potential amounts of hazardous substances that could be released into the environment. It is also important to evaluate if IBA meets the requirements established for the use of recycled aggregates in the domain of geotechnical applications.

Current circumstances require the development of solutions including the reuse and recycling of residues. Considering the interesting performance exhibited by IBA in this research, it seems reasonable to recognize its potential use as a filling material that comes into contact with geotextiles, allowing for its valorization and thereby contributing to more sustainable solutions within engineering applications.

Author Contributions: Conceptualization, J.R.C. and M.d.L.L.; methodology, F.A., J.R.C. and M.d.L.L.; validation, F.A., J.R.C. and M.d.L.L.; formal analysis, F.A. and J.R.C.; investigation, F.A.; writing—original draft preparation, F.A. and J.R.C.; writing—review and editing, F.A., J.R.C. and M.d.L.L.; project administration, J.R.C. and M.d.L.L.; funding acquisition, J.R.C. and M.d.L.L. All authors have read and agreed to the published version of the manuscript.

Funding: This work was financially supported by: (1) project PTDC/ECI-EGC/28862/2017—POCI-01-0145-FEDER-028862, funded by FEDER funds through COMPETE 2020—Programa Operacional Competitividade e Internacionalização (POCI) and by national funds (PIDDAC) through FCT/MCTES; (2) Base Funding—UIDB/04708/2020 of the CONSTRUCT—Instituto de I&D em Estruturas e Construções—funded by national funds through the FCT/MCTES (PIDDAC).

Acknowledgments: The authors would like to thank Lipor II (Maia, Portugal) for supplying the incinerator bottom ash.

Conflicts of Interest: The authors declare no conflict of interest.

References

1. European Commission. Directive 2008/98/EC of the European Parliament and of the Council of 19 November 2008 on waste. *Off. J. Eur. Union* **2008**, *51*, 3–30.
2. Blasenbauer, D.; Huber, F.; Lederer, J.; Quina, M.J.; Blanc-Biscarat, D.; Bogush, A.; Bontempi, E.; Blondeau, J.; Chimenos, J.M.; Dahlbo, H.; et al. Legal situation and current practice of waste incineration bottom ash utilisation in Europe. *Waste Manag.* **2020**, *102*, 868–883. [CrossRef] [PubMed]

3. Qiao, X.C.; Tyrer, M.; Poon, C.S.; Cheeseman, C.R. Novel cementitious materials produced from incinerator bottom ash. *Resour. Conserv. Recycl.* **2008**, *52*, 496–510. [CrossRef]

4. Lancellotti, I.; Cannio, M.; Bollino, F.; Catauro, M.; Barbieri, L.; Leonelli, C. Geopolymers: An option for the valorization of incinerator bottom ash derived "end of waste". *Ceram. Int.* **2015**, *41*, 2116–2123. [CrossRef]

5. Garcia-Lodeiro, I.; Carcelen-Taboada, V.; Fernández-Jiménez, A.; Palomo, A. Manufacture of hybrid cements with fly ash and bottom ash from a municipal solid waste incinerator. *Constr. Build. Mater.* **2016**, *105*, 218–226. [CrossRef]

6. Wongsa, A.; Boonserm, K.; Waisurasingha, C.; Sata, V.; Chindaprasirt, P. Use of municipal solid waste incinerator (MSWI) bottom ash in high calcium fly ash geopolymer matrix. *J. Clean. Prod.* **2017**, *148*, 49–59. [CrossRef]

7. Pera, J.; Coutaz, L.; Ambroise, J.; Chababbet, M. Use of incinerator bottom ash in concrete. *Cem. Concr. Res.* **1997**, *27*, 1–5. [CrossRef]

8. Müller, U.; Rübner, K. The microstructure of concrete made with municipal waste incinerator bottom ash as an aggregate component. *Cem. Concr. Res.* **2006**, *36*, 1434–1443. [CrossRef]

9. Kuo, W.-T.; Liu, C.-C.; Su, D.-S. Use of washed municipal solid waste incinerator bottom ash in pervious concrete. *Cem. Concr. Compos.* **2013**, *37*, 328–335. [CrossRef]

10. Toraldo, E.; Saponaro, S.; Careghini, A.; Mariani, E. Use of stabilized bottom ash for bound layers of road pavements. *J. Environ. Manag.* **2013**, *121*, 117–123. [CrossRef] [PubMed]

11. Forteza, R.; Far, M.; Seguí, C.; Cerdá, V. Characterization of bottom ash in municipal solid waste incinerators for its use in road base. *Waste Manag.* **2004**, *24*, 899–909. [CrossRef] [PubMed]

12. Le, N.H.; Abriak, N.E.; Binetruy, C.; Benzerzour, M.; Nguyen, S.-T. Mechanical behavior of municipal solid waste incinerator bottom ash: Results from triaxial tests. *Waste Manag.* **2017**, *65*, 37–46. [CrossRef]

13. Lynn, C.J.; Ghataora, G.S.; Dhir Obe, R.K. Municipal incinerated bottom ash (MIBA) characteristics and potential for use in road pavements. *Int. J. Pavement Res. Technol.* **2017**, *10*, 185–201. [CrossRef]

14. Shukla, S.K.; Yin, J.-H. *Fundamentals of Geosynthetic Engineering*, 1st ed.; Taylor & Francis/Balkema: Leide, The Netherlands, 2006.

15. CEN. *Geosynthetics-Index Test Procedure for the Evaluation of Mechanical Damage under Repeated Loading-Damage Caused by Granular Material*; CEN-European Committee for Standardization: Brussels, Belgium, 2007; EN ISO 10722.

16. Huang, C.-C. Laboratory simulation of installation damage of a geogrid. *Geosynth. Int.* **2006**, *13*, 120–132. [CrossRef]

17. Huang, C.-C.; Chiou, S.-L. Investigation of installation damage of some geogrids using laboratory tests. *Geosynth. Int.* **2006**, *13*, 23–35. [CrossRef]

18. Huang, C.-C.; Wang, Z.-H. Installation damage of geogrids: Influence of load intensity. *Geosynth. Int.* **2007**, *14*, 65–75. [CrossRef]

19. Pinho-Lopes, M.; Lopes, M.L. Tensile properties of geosynthetics after installation damage. *Environ. Geotech.* **2014**, *1*, 161–178. [CrossRef]

20. Hufenus, R.; Rüegger, R.; Flum, D.; Sterba, I.J. Strength reduction factors due to installation damage of reinforcing geosynthetics. *Geotext. Geomembr.* **2005**, *23*, 401–424. [CrossRef]

21. Carlos, D.M.; Carneiro, J.R.; Pinho-Lopes, M.; Lopes, M.L. Effect of Soil Grain Size Distribution on the Mechanical Damage of Nonwoven Geotextiles Under Repeated Loading. *Int. J. Geosynth. Ground Eng.* **2015**, *1*, 9. [CrossRef]

22. Dias, M.; Carneiro, J.R.; Lopes, M.L. Resistance of a nonwoven geotextile against mechanical damage and abrasion. *Ciênc. Tecnol. Mater.* **2017**, *29*, 177–181. [CrossRef]

23. Carneiro, J.R.; Morais, L.M.; Moreira, S.P.; Lopes, M.L. Evaluation of the Damages Occurred During the Installation of Non-Woven Geotextiles. *Mater. Sci. Forum* **2013**, *730*, 439–444. [CrossRef]

24. Cheah, C.; Gallage, C.; Dawes, L.; Kendall, P. Measuring hydraulic properties of geotextiles after installation damage. *Geotext. Geomembr.* **2017**, *45*, 462–470. [CrossRef]

25. Carlos, D.M.; Carneiro, J.R.; Lopes, M.L. Effect of Different Aggregates on the Mechanical Damage Suffered by Geotextiles. *Materials* **2019**, *12*, 15. [CrossRef]

26. Vieira, C.S.; Pereira, P.M. Interface shear properties of geosynthetics and construction and demolition waste from large-scale direct shear tests. *Geosynth. Int.* **2016**, *23*, 62–70. [CrossRef]

27. Vieira, C.S.; Pereira, P.M.; Lopes, M.L. Recycled Construction and Demolition Wastes as filling material for geosynthetic reinforced structures. Interface properties. *J. Clean. Prod.* **2016**, *124*, 299–311. [CrossRef]

28. Vieira, C.S.; Pereira, P.M. Damage induced by recycled Construction and Demolition Wastes on the short-term tensile behaviour of two geosynthetics. *Transp. Geotech.* **2015**, *4*, 64–75. [CrossRef]

29. Carlos, D.M.; Carneiro, J.R.; Lopes, M.L. Mechanical damage of geotextiles caused by recycled c&dw and other aggregates. In Proceedings of the Wastes-Solutions, Treatments and Opportunities III, Lisbon, Portugal, 4–6 September 2019; pp. 250–256.

30. CEN. *Geosynthetics-Test Method for the Determination of Mass Per Unit Area of Geotextiles and Geotextile-Related Products*; CEN-European Committee for Standardization: Brussels, Belgium, 2005; EN ISO 9864.

31. CEN. *Geosynthetics-Determination of Thickness at Specified Pressures-Part 1: Single Layers*; CEN-European Committee for Standardization: Brussels, Belgium, 2016; EN ISO 9863-1.

32. CEN. *Tests for Geometrical Properties of Aggregates-Part 1: Determination of Particle Size Distribution-Sieving Method*; CEN-European Committee for Standardization: Brussels, Belgium, 2012; EN 933-1.

33. CEN. *Geosynthetics-Wide-Width Tensile Test*; CEN-European Committee for Standardization: Brussels, Belgium, 2015; EN ISO 10319.

34. CEN. *Geosynthetics-Static Puncture Test (CBR Test)*; CEN-European Committee for Standardization: Brussels, Belgium, 2006; EN ISO 12236.

35. Montgomery, D.C.; Runger, G.C. *Applied Statistics and Probability for Engineers*, 5th ed.; John Wiley & Sons, Inc.: New York, NY, USA, 2010; p. 784.

Publisher's Note: MDPI stays neutral with regard to jurisdictional claims in published maps and institutional affiliations.

 *sustainability*

Article

Stabilization/Solidification of Zinc- and Lead-Contaminated Soil Using Limestone Calcined Clay Cement (LC^3): An Environmentally Friendly Alternative

Vemula Anand Reddy [1,*], **Chandresh H. Solanki** [1], **Shailendra Kumar** [1], **Krishna R. Reddy** [2,*] **and Yan-Jun Du** [3]

[1] Civil Engineering Department, Sardar Vallabhbhai National Institute of Technology, Surat 395007, India; chs@amd.svnit.ac.in (C.H.S.); skumar@amd.svnit.ac.in (S.K.)
[2] Department of Civil and Materials Engineering, University of Illinois at Chicago, Chicago, IL 60607, USA
[3] Jiangsu Key Laboratory of Urban Underground Engineering & Environmental Safety, Institute of Geotechnical Engineering, Southeast University, Nanjing 210096, China; duyanjun1972@163.com
* Correspondence: d16am001@amd.svnit.ac.in (V.A.R.); kreddy@uic.edu (K.R.R.)

Received: 10 April 2020; Accepted: 28 April 2020; Published: 4 May 2020

Abstract: Due to increased carbon emissions, the use of low-carbon and low-cost cementitious materials that are sustainable and effective are gaining considerable attention recently for the stabilization/solidification (S/S) of contaminated soils. The current study presents the laboratory investigation of low-carbon/cost cementitious material known as limestone-calcined clay cement (LC^3) for the potential S/S of Zn- and Pb-contaminated soils. The S/S performance of the LC^3 binder on Zn- and Pb-contaminated soil was determined via pH, compressive strength, toxicity leaching, chemical speciation, and X-ray powder diffraction (XRPD) analyses. The results indicate that immobilization efficiency of Zn and Pb was solely dependent on the pH of the soil. In fact, with the increase in the pH values after 14 days, the compressive strength was increased to 2.5–3 times compared to untreated soil. The S/S efficiency was approximately 88% and 99%, with increase in the residual phases up to 67% and 58% for Zn and Pb, respectively, after 28 days of curing. The increase in the immobilization efficiency and strength was supported by the XRPD analysis in forming insoluble metals hydroxides such as zincwoodwardite, shannonite, portlandite, haturite, anorthite, ettringite (Aft), and calcite. Therefore, LC^3 was shown to offer green and sustainable remediation of Zn- and Pb-contaminated soils, while the treated soil can also be used as safe and environmentally friendly construction material.

Keywords: low carbon materials; heavy metal immobilization; sustainable remediation; environmentally friendly materials

---

1. Introduction

Soil contaminated with heavy metals is a serious threat to the sustainable development and global food security [1–4]. In contrast to water and air pollution, heavy metal pollution in soils is an invisible and unseen problem [1,5–9]. Many of the world's contaminated sites have become the dump sites of various industrial by-products that contain inorganic pollutants such as heavy metals. As these heavy metals come in contact with water, the human health and environment within the ecosystem become potentially at risk [10,11]. Among the hazardous heavy metals zinc (Zn) and lead (Pb) are considered as the harmful pollutants that exist at elevated levels in most of the contaminated sites around the world [12]. Further, Zn and Pb are not only harmful to human health and environment, but also lead to mechanical–chemical degradation of contaminated soils, which in turn results in unfavorable conditions for the redevelopment of contaminated sites. It is therefore imperative to identify a time- and cost-effective remediation method for the treatment of heavy metal-contaminated soils, consequently the treated soils can be reused as safe and environmentally friendly construction materials.

Stabilization/solidification (S/S) is considered to be most appropriate method for immobilization of heavy metal-contaminated soils due to its ease and workability among the available effective remediation methods [13–16]. Besides, the United States Environmental Protection Agency (USEPA) recognizes S/S as the best demonstrated available technology (BDAT) for treating hazardous metals [7,17,18]. The mechanisms involved in S/S treatment is as follows: stabilization refers to reducing the hazard potential by converting contaminants into their least soluble/toxic form [19,20], whereas solidification is the encapsulation of waste in a monolith mass of high structural integrity that involves both mechanical binding and chemical interaction between solidifying agents such as cementitious materials, which further restricts the movement of heavy metals by isolating them into less/insoluble crystalline phases [14]. The performance of S/S depends on the nature of the contaminants (organic/inorganic) and binders used. Inorganic heavy metals are commonly immobilized via chemical reaction and physical encapsulation by forming barely insoluble metal hydroxides. Thus, the binder plays a key role in the S/S process, and the development of novel binders has gained special attention recently, specifically low carbon/cost binders. In a previous study, the authors revealed that partial replacement of calcined clay (CC) and limestone (LS) with ordinary Portland cement (OPC) has better immobilization efficiency for Zn-contaminated soils (Reddy et al. [21]). In addition, various hydration products, such as portlandite, ettringite, tri-calcium silicate, and wulfingite, were found to be responsible for the immobilization of Zn-contaminated soils. In addition, Wang et al. [6] reported that supplementary cementitious materials (SCMs) such as CC and LS have improved immobilization efficiency in treating both oxy-anionic As- and cationic Pb-contaminated soils. The leachability efficiency of CC and LS was approximately 96% and 99% for As and Pb, respectively. Further, addition of LS to CC promotes the transformation of metastable hydroxyl-rich Afm to stable carbonate rich Afm, which increases the degree of polymerization in calcined clay hydrates, resulting in enhancement of mechanical properties. Therefore, replacing SCMs with the conventional cement binders has a better performance in treating heavy metal-contaminated soils.

Recently, a new ternary blend known as limestone-calcined clay cement LC^3 was successfully demonstrated in the authors' previous study on Zn-contaminated soils [21]. LC^3 is known as a low-carbon and low-cost binder since the production process involves replacement of low grade calcined clays (CC) and limestone (LS). Typically, LC^3 is a ternary blend of 30% CC, 15% LS, and 5% gypsum replaced with 50% cement clinker. The replacement of low-grade limestone and low-grade kaolinitic clay with (kaolinite content > 40%) when calcined at 750 °C undergoes hydroxylation to form CC/metakaolin (MK) [6] as presented in Equation (1), which possess high pozzolanic reactivity due to the presence of alumina- and silicate-rich phases. Further, when LS reacts with CC it produces carboaluminosilicates-rich mineral phases that are responsible for the formation of primary and secondary hydration products such as calcium silicate hydrate (C-S-H), calcium hydrate (C-H), calcium aluminate silicate hydrate (C-A-S-H), and calcium aluminate hydrate (C-A-H) [14–16]. Furthermore, the production of 1 ton cement produces 0.82 ton CO_2 whereas 1 ton CC produces 0.175 ton CO_2 emissions. Therefore, replacing 50% OPC with CC and LS reduces the carbon footprint up to 40% [22–24], which makes the binder low-carbon/cost and also an environmentally friendly alternative material [24–28]. Although the influence of LC^3 is validated for Zn alone, its effectiveness and mechanism involved for the immobilization of Zn and Pb when they co-exist are unknown and need additional investigations.

$$Al_2O_3(SiO_2)_2 \cdot (H_2O)_2 \; (Kaolinite) \rightarrow Al_2O_3(SiO_2)_2 \cdot (H_2O)_x (calcined\ clay) + (2-x)H_2O. \qquad (1)$$

The objective of the study was to evaluate the feasibility of the LC^3 binder upon S/S of Zn- and Pb-contaminated soils individually as well as combined at elevated levels in terms of strength, toxicity leaching, chemical speciation, and XRPD analysis. The research aimed to provide scientific insights on environmentally friendly alternative LC^3, such as: (1) to investigate the immobilization mechanisms involved in the soils treated with Zn and Pb; (2) to study the effect of curing time and binder dosage on physical strength and pH; and (3) to elucidate the hydration products responsible for the S/S of treated soils. This study provides the feasibility of using a sustainable binder LC^3 for the treatment of contaminated soils, while the treated soil can be reused as safe and environmentally friendly construction material.

2. Materials and Methods

2.1. Materials

Clean soil used in this study was collected from the nearby open area at Sardar Vallabhbhai National Institute of Technology, Surat, India. Approximately 250 kg of the soil sample was collected from the 0.5–1.0 m depth. Later, soil was homogenized, then air dried and passed into a 2 mm screen before use. The soil was classified as CH as per the Unified Soil Classification System based on ASTM D2487 [29], where initial water content, specific gravity, and pH were 19.7, 2.59, and 6.8, respectively, and the detailed chemical composition of the clean soil is shown in Table 1. Further, the binder LC^3 was procured from Technological Action and Rural Advancement (TARA), Delhi, India and the major oxides present were CaO, SiO_2, Al_2O_3, and Fe_2O_3 of 61.4%, 24.38%, 6.52%, and 4.31%. In addition, the remaining physicochemical parameters of soil and chemical composition of LC^3 binder used in the study can be found in authors' previous study [21].

Table 1. Chemical composition of clean soil used in the study.

Oxide	Value (%) [a]
Silicon oxide (SiO_2)	54.26
Aluminium oxide (Al_2O_3)	17.86
Ferric oxide (Fe_2O_3)	12.17
Calcium oxide (CaO)	4.24
Magnesium oxide (MgO)	7.17
Potassium oxide (K_2O)	0.06
Titanium oxide (TiO_2)	ND [c]
Sulphur trioxide (SO_3)	1.28
Loss on ignition [b]	2.67

[a] Analyzed using Rigaku WD-XRF machine. [b] Value of loss on ignition is referred to as 950 °C. [c] Not Detected.

2.2. Artificially Contaminated Soil and S/S Samples Preparation

The target metals used in the study were lead (Pb) and zinc (Zn) as they are considered as the most commonly encountered heavy metals at contaminated sites worldwide [15,18]. The analytical grade zinc nitrate hexa-hydrate Zn $(NO_3)_2 \cdot 6H_2O$ and lead nitrate hexa-hydrate Pb $(NO_3)_2.6H_2O$ were used and nitrate anion was chosen because it is inert and also eliminates unexpected precipitates with other ions during hydration and pozzolanic reaction [5,15]. Further, the essential volume of stock solutions was added to the air dried soil until the stock solution content reached to 29%, i.e., the optimum moisture content (OMC) of the soil, and then stayed untouched for 14 days to ensure the necessary contact between soil and heavy metals Pb and Zn. Similar procedure for the preparation of artificially contaminated soil was reported by Du et al. [18,30]. Further, the concentrations of 5000 mg/kg and 10,000 for both Zn and Pb and the combination of both at 10,000 mg/kg were used to represent typical field concentration levels. In addition, for comparison purposes, the untreated soil concentration was maintained at 10,000 mg/kg. The samples were designated as ZnU, PbU (untreated Zn and Pb), and Zn 0.5, Zn 1.0, Pb 0.5, Pb 1.0, and ZnPb 1.0 in the study. Furthermore, the binder LC^3 was added to artificially contaminated soil on predetermined dry soil weight basis at 8%. The soil–binder mixture was thoroughly mixed using an electronic mixer for 5–10 min in order to obtain a homogenous mix, until the water content reached a predetermined OMC and maximum dry density (MDD), which are shown in Table 2. The mixture was compacted in three layers of 5 cm-diameter and 10 cm-height PVC molds using a hydraulic jack until it reached the MDD. The molds were carefully sealed in a polythene bags and demolded after curing periods of 3,7,14, 28, and 56 days. The mixing, curing, and compaction procedures were followed as per ASTM C192 [31] to ensure the similarity among all the samples.

Table 2. Maximum dry density (MDD) and optimum moisture content (OMC) of untreated and 8% limestone-calcined clay cement (LC^3)-treated Zn- and Pb-contaminated soils.

Sample Designation	OMC (%)	MDD (kg/m^3)
ZnU	29	1.47×10^3
PbU	29	1.46×10^3
Zn 0.5	32	1.42×10^3
Zn 1.0	33	1.43×10^3
Pb 0.5	32	1.43×10^3
Pb 1.0	33	1.43×10^3
ZnPb 1.0	33	1.44×10^3

2.3. Testing Methods

The primary objective of the study was to determine the mechanical strength, leaching, chemical speciation, and mineralogy of the untreated and LC^3-treated specimens. Physical strength was determined using an unconfined compression strength (UCS) test as per ASTM D2166 [32] with a controlled strain rate of 1%/minute. Later, the crushed samples were taken for determination of leaching, chemical speciation, and mineralogy tests. In addition, pH values were measured in the leachate by using HANNA waterproof tester, as per ASTM standard [33].

Toxicity leaching was performed as per standard toxicity characteristic leaching protocol (TCLP) EPA method 1311 [34]. In total, 10 g of the soil was mixed in TCLP fluid#1, i.e., CH$_3$COOH and NaOH mixture at pH 4.93 ± 0.05, and the soil solution mixture was rotated for 18 ± 2 h at 30 rpm using an end-to-end shaker. Later, the leachant solution was separated using centrifuging/decantation at 3000 rpm for 8–10 min. Finally, the leachate was subjected to pH analysis and acidified using HNO$_3$ at (pH ≤ 2) before proceeding to heavy metal analysis. All the samples were tested in triplicates/quadruplicates for ensuring the repeatability of results, and average values of the results are reported in the study. In addition, to understand the leaching performance, S/S efficiencies of heavy metals [35] were determined using Equations (2) and (3), where S represents S/S efficiency and L is the leaching factor, which is defined by heavy metal concentration in leachate divided by the initial soil contamination condition.

$$S = 100 - L \tag{2}$$

$$L = \frac{Mass\ of\ contamination\ in\ leachate\ (mg)}{Initial\ mass\ of\ soil\ contamination\ (mg)} * 100. \tag{3}$$

Further, the chemical speciation analysis was performed using a modified Community Bureau of Reference three step sequential extraction procedure (BCR-SEP) [36–38]. The test method was comprised of four phases P1 = acid soluble phase, extraction in 0.11 mol/L CH$_3$COOH at pH 2.8; P2 = reducible phase, extraction in 0.5 mol/L NH$_2$OH·HCl at pH 1.5; P3 = oxidizable phase, oxidation in acid using 30% H$_2$O$_2$ and extraction in 1 mol/L in CH$_3$COONH$_4$, both at pH 2; and P4 = residual phase, extraction with total digestion using mixture of (3:1) concentrated 70% HNO$_3$ and 30% HCl using 11,466 protocols [39]. Furthermore, the P1 phase was comprised of heavy metals that were precipitated and co-precipitated in a carbonate phase, which were present in a bioavailable form. The P2 phase was made-up of iron (Fe) and manganese (Mn) oxides that can be activated in high pH conditions (acidic). The P3 phase was incorporated into a stable organic matter and sulfides that were mobilizable and not bioavailable during oxidation. The P4 phase contained primary and secondary minerals, which could hold the heavy metals within the crystal lattices [2,38]. The P4 phase was expected to remain for longer periods in the contaminated soil and very difficult to release, even at high pH conditions. In addition, to measure the reliability of sequential extraction procedure, metal recovery rate (MRR), given in Table 4, is the most commonly used parameter [38,40]. It is defined as the sum of four phases divided by total concentration obtained from the complete digestion as presented in Equation (4).

$$MRR\ (\%) = \frac{P1 + P2 + P3 + P4}{\text{Total Concentration}} \times (100). \tag{4}$$

Moreover, XRPD analysis was performed after hydration stoppage in the treated samples. The samples were powdered using 75 μm mesh and tested using a Rigaku X-ray diffractometer with (Cu-Kα) radiation λ = 1.540538 Å at the 2θ (Theta) of 10°–70° with the step size of 0.02° under room temperature conditions. The system was operated at 45 kV and 30 mA, with a scan time of 20 s at each step in the step scan mode, and XRPD results were analyzed using PANanlytical Xpert High Scomᵢre plus software V.3e (Malvern Panalytical, Worcestershire, United Kingdom) [41] for attaining phase identification of minerals.

3. Results and Discussion

3.1. Leachate pH

Figure 1 represents the leachate pH of Zn- and Pb-treated soil with varying curing periods. It can be seen that LC^3 treatment significantly increased by approximately 1.3–2.4 and 1.1–1.4 units, respectively, for Zn and Pb at 7 days of curing as compared to untreated soil. However, Pb failed to reach the remediation goal compared to Zn. This may be due to the molar concentration of Pb, which is much lower than that of Zn, which induces more significant retardant effect on hydration in the system. For instance, at 28 days, mean pH values increased from 8.33 to 9.12 and 7.22 to 7.94, i.e., 2.66–3.45 and 1.98–2.7 units for Zn and Pb, respectively. Further, it can be noted that the pH values of Pb 1.0 and ZnPb 1.0 showed similar trends, which indicates Pb at higher concentrations retards the hydration mechanism in the system, which agrees well with Wang et al. and Xia et al. [17,42]. Furthermore, the increase in the pH values was noted at 56 days curing time by 8.71–9.36 and 7.37–8.38, which is 3.06–3.71 and 3.17–2.16 units. The increase in the pH values is attributed to the chemical composition of the binder, which facilitates the release of OH^-, Ca^+, and Al^+ ions in the pore water, creating an alkaline environment in the system [43,44]. Further, the dissolution of aluminates and silicates expedite the pozzolanic reaction over time, which are responsible for the binding of heavy metals to form insoluble metal hydroxides [24,45,46]. Therefore, the increase in pH over time in the treated system supports the formation of various hydration products, such as $[ZnAl(OH)_2(SO)_2]$, $[Pb_2O(CO_3)]$, $[Ca(OH)_2]$, Ca_3O_5Si, $[CaAl_2Si_2O_8]$, (Aft) $[Ca_6Al_2(SO4)_3(OH)_{12}.26H_2O]$, and $[CaCO_3]$, which is also validated by XRPD analysis Section 3.5 in the study.

3.2. Zn and Pb Leachability

Figure 2 shows the leached Zn and Pb concentrations in the TCLP test. It was observed that leached Zn and Pb concentrations exceeded Hazardous Waste Management (HWM) rules [47] limits of Zn (250 mg/L) and Pb (5 mg/L), respectively, suggesting that the soil is toxic and requires remediation. The average leached concentrations of LC^3-stabilized soils decreased with the rising curing time. Besides, the leached Zn concentrations were below regulatory limit after 14 days of curing. Whereas, Pb reached the regulatory limit only after 28 days of curing, showing the solidification efficiency of approximately 88% and 99% with leaching factors 11.89 and 1.22 for Zn and Pb. The decrease in the leached concentrations could be due to increased pH values in the previous Section 3.1 and formation of metal hydroxides in the presence of freely available Ca $(OH)_2$ and Ca^+ ions in the binder [43]. Moreover, LS, when added with CC, reacts to form carbo-aluminates, which produces C-S-H and C-A-H-based hydration products that have a tendency to arrest the heavy metals in forming insoluble metal hydroxides that increase the immobilization efficiency and reduce leaching. Overall, it can be concluded that LC^3 stabilization promotes the immobilization of Zn- and Pb-contaminated soils, which further reduces leachability.

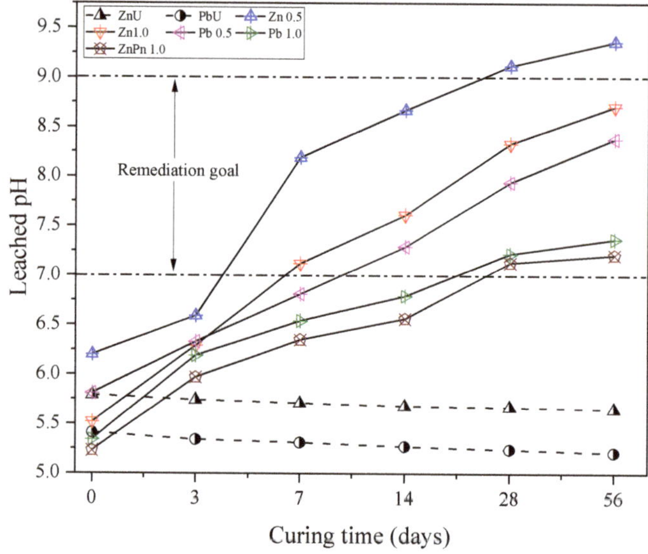

Figure 1. pH of the untreated and LC^3-treated soils with varying curing periods.

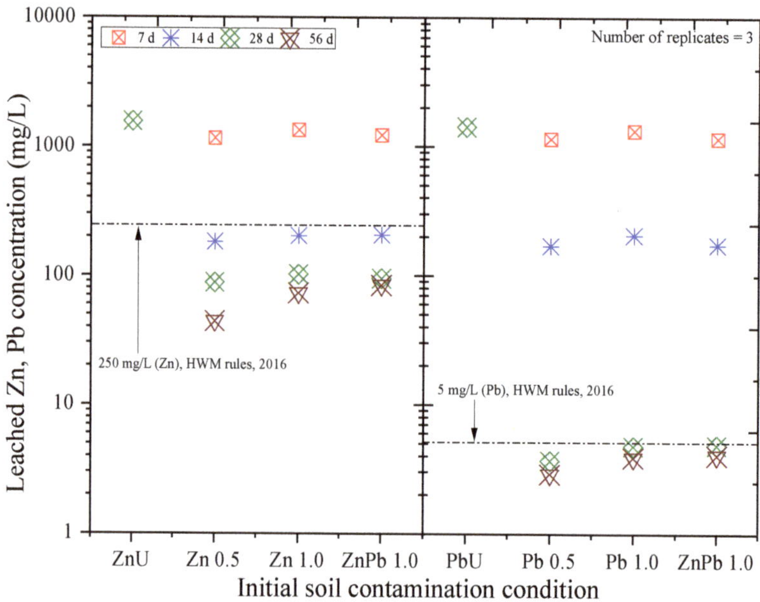

Figure 2. Toxicity characteristic leaching protocol (TCLP) of leached Zn and Pb concentrations with varying curing times.

3.3. Unconfined Compressive Strength

Figure 3 represents the unconfined compressive strength behavior of LC^3-treated samples with varying curing time. In all the treated samples, increasing compression strength was observed over untreated soil with curing time. However, Pb has the retardation effect on the treated soil at higher concentrations, which ultimately decreases the strength values. For instance, the increase in the strength values of LC^3-treated soil was approximately 1.50–1.62 and 2.42–2.57 for Zn, whereas 1.46–1.69 and 2.46–2.94 for Pb at 7 and 28 days of curing, and the additional values can be found in Table 3. Moreover, it was seen that increases in the strength values were observed even after 28 days of curing, which is because LC^3 can improve strength even up to 365 days of curing [48] as a consequence of hydration reactions [45,49]. In addition, binder LC^3 includes partial replacement of cement with calcined clay and limestone, and during the hydration process the combination of Ca (OH)$_2$ and CC increases the pozzolanic reactivity [24,28,41,43,50], which produces more binding phases, resulting in improved density and reduced pore spaces that ultimately lead to increased compressive strength. Therefore, the results demonstrate that Zn and Pb concentrations at higher levels have synergetic effects that could favorably affect the strength behavior of contaminated soils.

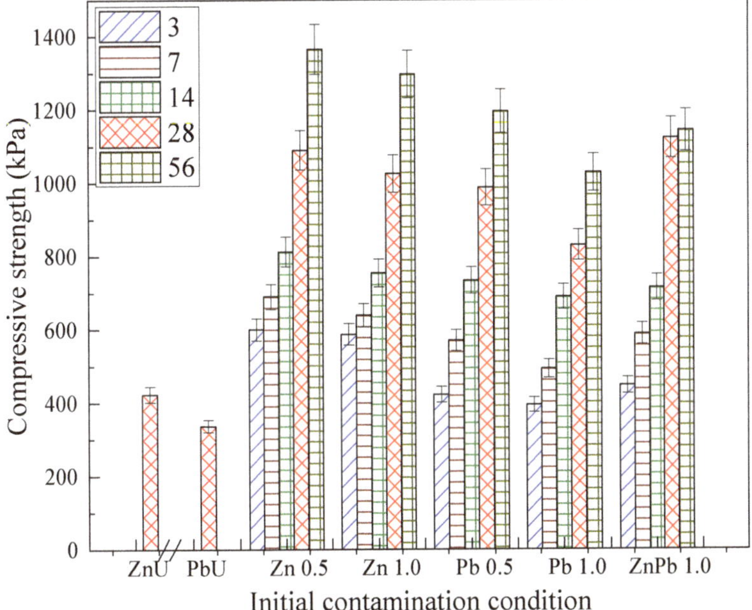

Figure 3. Compressive strength results of the untreated and LC^3-treated soil with varying curing period.

Table 3. Unconfined compressive strength values of initial soil contamination condition and curing time.

Initial Soil Contamination Condition Curing (Days)	ZnU	PbU	Zn 0.5	Zn 1.0	Pb 0.5	Pb 1.0	ZnPb 1.0
3	-	-	599.65	586.35	422.54	394.22	447.35
7	-	-	689.67	638.37	569.54	492.33	587.26
14	-	-	812.34	754.32	733.51	688.34	713.35
28	423.35	336.84	1089.38	1025.37	986.82	829.64	1121.67
56	-	-	1364.38	1296.35	1195.24	1027.33	1142.54

Note: Units for compressive strength values are (kPa).

3.4. Chemical Speciation of Heavy Metals

Typically, four phases of soil sample are analyzed to recognize the environmental activity and bioavailability of heavy metals, namely acid soluble (P1), reducible (P2), oxidizable (P3), and residual phases (P4). Commonly, P1 and P2 phases are considered to be bioavailable in nature due to their weak binding capacity in the acidic/low pH environment [38]. The higher the proportion of P1 and P2 in an active fraction, the greater the heavy metal's ion mobility [36,51,52]. Therefore, the phases P1 and P2, particularly the P1 phase, are not stable and impose environmental risks resulting from leached heavy metals in the environment. Further, to assess the reliability of the phase extractions, the metal recovery rate (MRR) is given in Table 4. Figure 4a,b shows the histograms of Zn and Pb metal distribution after LC^3 treatment at 28 and 56 days of curing. As shown in Figure 4a, the P1 of Zn in LC^3-stabilized soil was 21–32% lower than 61% of ZnU (untreated Zn) and approximately 37–51% of Zn in the stabilized soil was bound to the P4 phase. Besides, the P1 phase of Pb in the LC^3-stabilized soil ranged from 33–46% lower than 58% of PbU (untreated Pb), and 36–44% of Pb in the stabilized soil was bound to the P4 phases after 28 days of curing. While at 56 days of curing, as shown in Figure 4b, the increase in the P4 phase in the stabilized soil ranged from 53–67% and 41–58% for Zn and Pb, respectively. The increase in the P4 phase thus promotes the development of highly insoluble and immobile complexes that are responsible for making the heavy metals less bioavailable in nature under low pH/acidic environmental conditions. It can be concluded that LC^3 stabilization/solidification results in the transformation of (acid-soluble) P1 phases of Zn- and Pb-contaminated soil into more insoluble (residual) P4 phases.

Table 4. Metal recovery rate (MRR) after modified BCR sequential extraction test.

Sample Designation	ZnU	PbU	Zn 0.5	Zn 1.0	Pb 0.5	Pb 1.0	ZnPb 1.0
Metal recovery rate (MRR%)	91	82	86	93	88	94	95

3.5. XRPD Analysis

The mineralogical analysis was conducted after 28 days of curing for both untreated and LC^3-treated samples to examine the effect of hydration and pozzolanic reaction on various phases of the contaminants, i.e., Pb and Zn in a stabilized matrix, as shown in Figure 5. The results show quartz [SiO_2], muscovite [$(KF)_2(Al_2O_3)_3(SiO_2)_6(H_2O)$], albite [$NaAlSi_3O_8$], wulfingite [$Zn(OH)_2$], zinc silicate [$Zn_2SiO_4$], and lead silicate [$Pb_2SiO_3$] were the primary minerals in the untreated Zn and Pb soil. Whereas hydration products such as zincwoodwardite [$ZnAl(OH)_2(SO)_2$], shannonite [$Pb_2O(CO_3)$], portlandite [$Ca(OH)_2$], haturite Ca_3O_5Si, anorthite [$CaAl_2Si_2O_8$], ettringite (Aft) [$Ca_6Al_2(SO4)_3(OH)_{12}.26H_2O$], and calcite [$CaCO_3$] were the major cementitious products responsible for stabilization in the treated samples and these hydrated products controlled the heavy metal migration in the LC^3 samples, which agrees well with Wang et al. [48]. The formation of aluminates and silicates-based products after LC^3 treatment was also noticeable, which was due to the availability of carbo-aluminate phases in the hydroxyl-rich calcined clay and limestone [24,28,46]. The changes in the structural and crystalline phases were due to lime, which is effectively activated by calcined clay, which further enhances metal hydrates and hydroxide phase formation. Accordingly, these products are normally insoluble, promoting Zn and Pb immobilization, which improves soil stabilization and reduction in the leaching of contaminated soils.

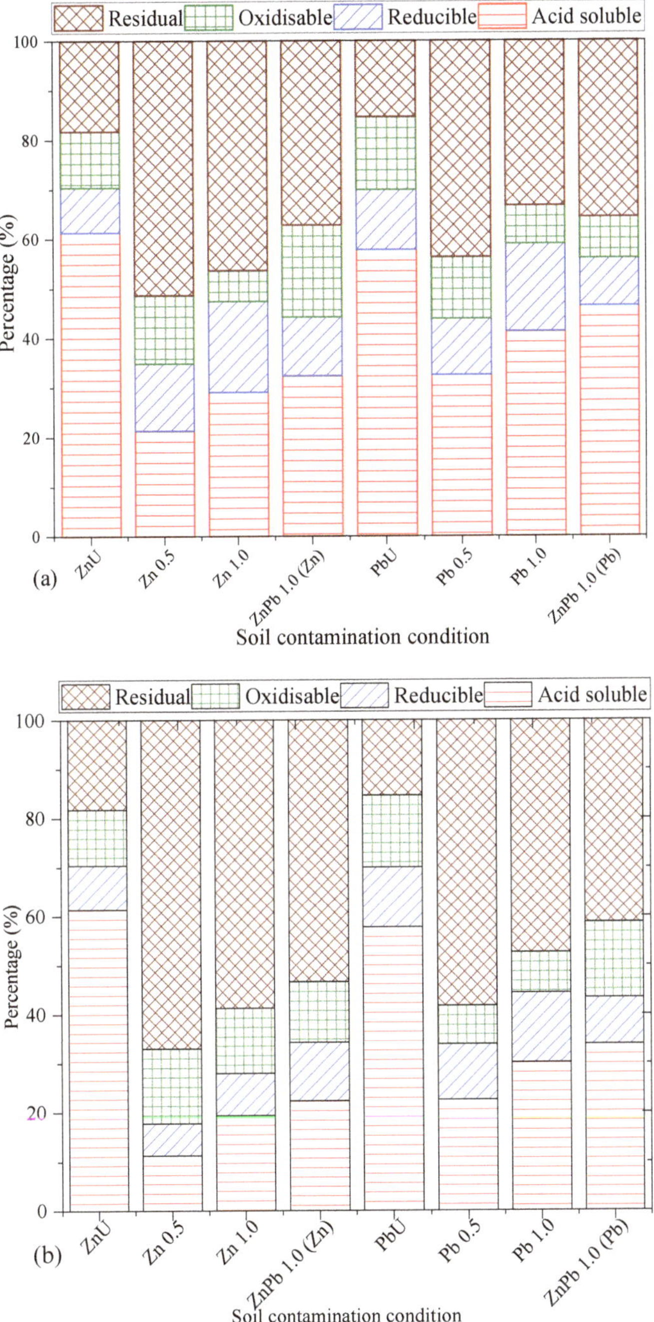

Figure 4. Distribution of heavy metals in the untreated and LC^3 treated soils after (**a**) 28 days (**b**) 56 days of curing.

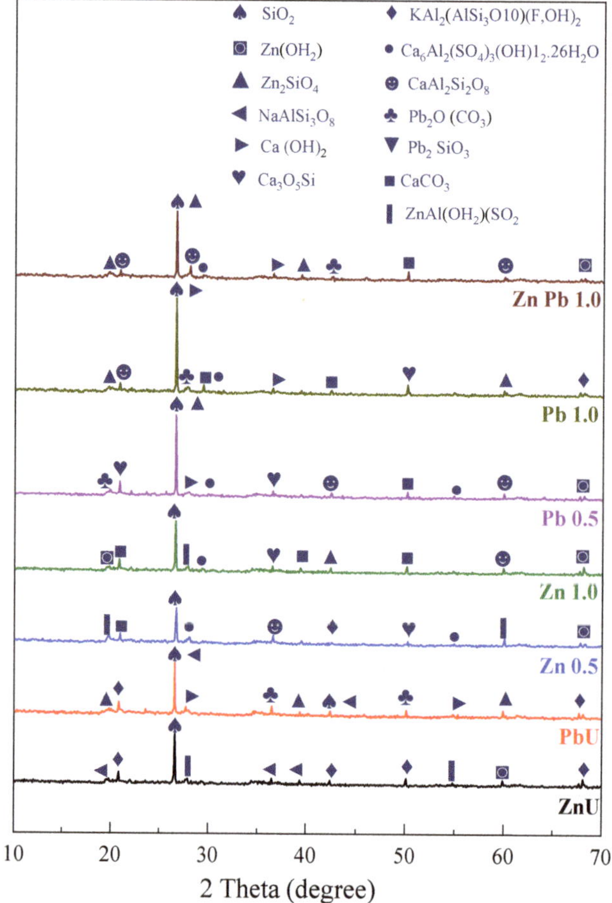

Figure 5. X-ray powder diffraction (XRPD) analysis of untreated and LC^3-treated soils after 28 days of curing.

4. Conclusions

The present study investigated the role of LC^3 on Zn- and Pb-contaminated soils and evaluated the S/S performance through analyses of unconfined compressive strength, chemical speciation, leaching, and XRD. XRD results showed that Zn and Pb had become an integral part of the crystalline phase physically and chemically in forming Si- and Al-based carbo-alumino-silicate products such as $Ca(OH)_2$, Ca_3O_5Si, $CaAl_2Si_2O_8$, and $Ca_6Al_2(SO_4)_3(OH)_{12}.26H_2O$ (Aft), which were responsible for the immobilization of heavy metals. The compressive strength results indicated that the addition of the LC^3 binder at 8% could improve the strength values up to three times compared to untreated Zn- and Pb-contaminated soils. The pH transformation of acidic to alkaline nature after a 14 day curing period allowed the adsorption of heavy metals in forming various insoluble metal hydroxides. Leachability of Zn-contaminated soil reached the regulatory limit after 14 days of curing, whereas Pb-contaminated and mixed contaminated soil, ZnPb, reached the limit only after 28 days of curing. Chemical speciation results indicated that reduction in the acid-soluble phases and increased residual phases significantly supported the formation of insoluble hydration products, which were responsible for increased immobilization efficiency and strength. The results illustrated that LC^3 is a promising binder in solidifying/stabilizing the contaminated soils and the treated soils can be reused as safe and sustainable construction materials.

Author Contributions: V.A.R., laboratory investigation, result analysis, writing—original manuscript; C.H.S. and S.K., supervision, writing—review and editing; K.R.R., conceptualization, supervision, writing—review and editing; Y.-J.D. writing—review and editing. All authors have read and agreed to the current version of the manuscript.

Funding: This research received no external funding.

Conflicts of Interest: The authors declare no conflict of interest.

References

1. Xia, W.Y.; Du, Y.J.; Li, F.S.; Guo, G.L.; Yan, X.L.; Li, C.P.; Arulrajah, A.; Wang, F.; Wang, S. Field evaluation of a new hydroxyapatite based binder for ex-situ solidification/stabilization of a heavy metal contaminated site soil around a Pb-Zn smelter. *Constr. Build. Mater.* **2019**, *210*, 278–288. [CrossRef]

2. Xia, W.Y.; Feng, Y.S.; Jin, F.; Zhang, L.M.; Du, Y.J. Stabilization and solidification of a heavy metal contaminated site soil using a hydroxyapatite based binder. *Constr. Build. Mater.* **2017**, *156*, 199–207. [CrossRef]

3. Li, J.S.; Wang, L.; Cui, J.L.; Poon, C.S.; Beiyuan, J.; Tsang, D.C.W.; Li, X.D. Effects of low-alkalinity binders on stabilization/solidification of geogenic As-containing soils: Spectroscopic investigation and leaching tests. *Sci. Total Environ.* **2018**, *631*, 1486–1494. [CrossRef] [PubMed]

4. Du, Y.J.; Wu, J.; Bo, Y.L.; Jiang, N.J. Effects of acid rain on physical, mechanical and chemical properties of GGBS–MgO-solidified/stabilized Pb-contaminated clayey soil. *Acta Geotech.* **2020**, *15*, 923–932. [CrossRef]

5. Wang, F.; Shen, Z.; Liu, R.; Zhang, Y.; Xu, J.; Al-Tabbaa, A. GMCs stabilized/solidified Pb/Zn contaminated soil under different curing temperature: Physical and microstructural properties. *Chemosphere* **2020**, *239*, 124738. [CrossRef] [PubMed]

6. Wang, L.; Cho, D.; Tsang, D.C.W.; Cao, X.; Hou, D.; Shen, Z.; Alessi, D.S.; Sik, Y.; Sun, C. Green remediation of As and Pb contaminated soil using cement-free clay- based stabilization/solidi fi cation. *Environ. Int.* **2019**, *126*, 336–345. [CrossRef]

7. Xia, W.Y.; Du, Y.J.; Li, F.S.; Li, C.P.; Yan, X.L.; Arulrajah, A.; Wang, F.; Song, D.J. In-situ solidification/stabilization of heavy metals contaminated site soil using a dry jet mixing method and new hydroxyapatite based binder. *J. Hazard. Mater.* **2019**, *369*, 353–361. [CrossRef]

8. Devi, P.; Kothari, P.; Dalai, A.K. Stabilization and solidification of arsenic and iron contaminated canola meal biochar using chemically modified phosphate binders. *J. Hazard. Mater.* **2020**, *385*, 121559. [CrossRef]

9. Wu, H.L.; Du, Y.J.; Yu, J.; Yang, Y.L.; Li, V.C. Hydraulic conductivity and self-healing performance of Engineered Cementitious Composites exposed to Acid Mine Drainage. *Sci. Total Environ.* **2020**, *716*, 137095. [CrossRef]

10. Capasso, I.; Lirer, S.; Flora, A.; Ferone, C.; Cioffi, R.; Caputo, D.; Liguori, B. Reuse of mining waste as aggregates in fly ash-based geopolymers. *J. Clean. Prod.* **2019**, *220*, 65–73. [CrossRef]

11. Ferone, C.; Capasso, I.; Bonati, A.; Roviello, G.; Montagnaro, F.; Santoro, L.; Turco, R.; Cioffi, R. Sustainable management of water potabilization sludge by means of geopolymers production. *J. Clean. Prod.* **2019**, *229*, 1–9. [CrossRef]

12. El-Eswed, B.I.; Aldagag, O.M.; Khalili, F.I. Efficiency and mechanism of stabilization/solidification of Pb(II), Cd(II), Cu(II), Th(IV) and U(VI) in metakaolin based geopolymers. *Appl. Clay Sci.* **2017**, *140*, 148–156. [CrossRef]

13. Pantazopoulou, E.; Ntinoudi, E.; Zouboulis, A.; Mitrakas, M.; Yiannoulakis, H.; Zampetakis, T. Heavy metal stabilization of industrial solid wastes using low-grade magnesia, Portland and magnesia cements. *J. Mater. Cycles Waste Manag.* **2020**, 1–11. [CrossRef]

14. Sharma, H.D.; Reddy, K.R. *Geoenvironmental Engineering: Site Remediation, Waste Containment, and Emerging Waste Management Technologies*; Wiley: Hoboken, NJ, USA, 2004; ISBN 978-0-471-21599-8.

15. Du, Y.-J.; Jiang, N.-J.; Liu, S.-Y.; Jin, F.; Singh, D.N.; Puppala, A.J. Engineering properties and microstructural characteristics of cement-stabilized zinc-contaminated kaolin. *Can. Geotech. J.* **2014**, *51*, 289–302. [CrossRef]

16. Wu, H.L.; Jin, F.; Bo, Y.L.; Du, Y.J.; Zheng, J.X. Leaching and microstructural properties of lead contaminated kaolin stabilized by GGBS-MgO in semi-dynamic leaching tests. *Constr. Build. Mater.* **2018**, *172*, 626–634. [CrossRef]

17. Xia, W.-Y.; Feng, Y.-S.; Du, Y.-J.; Reddy, K.R.; Wei, M.-L. Solidification and Stabilization of Heavy Metal–Contaminated Industrial Site Soil Using KMP Binder. *J. Mater. Civ. Eng.* **2018**, *30*, 04018080. [CrossRef]

18. Du, Y.-J.; Wei, M.-L.; Reddy, K.R.; Jin, F.; Wu, H.-L.; Liu, Z.-B. New phosphate-based binder for stabilization of soils contaminated with heavy metals: Leaching, strength and microstructure characterization. *J. Environ. Manag.* **2014**, *146*, 179–188. [CrossRef]

19. Patel, H.; Pandey, S. Evaluation of physical stability and leachability of Portland Pozzolona Cement (PPC) solidified chemical sludge generated from textile wastewater treatment plants. *J. Hazard. Mater.* **2012**, *207*, 56–64. [CrossRef]

20. Soundararajan, R. An overview of present day immobilization technologies. *J. Hazard. Mater.* **1990**, *24*, 199–212. [CrossRef]

21. Reddy, V.A.; Solanki, C.H.; Kumar, S.; Reddy, K.R.; Du, Y.J. New ternary blend limestone calcined clay cement for solidification/stabilization of zinc contaminated soil. *Chemosphere* **2019**, *235*, 308–315. [CrossRef]

22. Tironi, A.; Trezza, M.A.; Scian, A.N.; Irassar, E.F. Cement & Concrete Composites Assessment of pozzolanic activity of different calcined clays. *Cem. Concr. Compos.* **2013**, *37*, 319–327.

23. Fernandez, R.; Martirena, F.; Scrivener, K.L. Cement and Concrete Research The origin of the pozzolanic activity of calcined clay minerals: A comparison between kaolinite, illite and montmorillonite. *Cem. Concr. Res.* **2011**, *41*, 113–122. [CrossRef]

24. Cancio, Y.; Heierli, U.; Favier, A.R.; Machado, R.S.; Scrivener, K.L.; Fernando, J.; Hernández, M.; Habert, G. Limestone calcined clay cement as a low-carbon solution to meet expanding cement demand in emerging economies. *Dev. Eng.* **2017**, *2*, 82–91. [CrossRef]

25. Kavitha, O.R.; Shanthi, V.M.; Arulraj, G.P.; Sivakumar, V.R. Microstructural studies on eco-friendly and durable Self-compacting concrete blended with metakaolin. *Appl. Clay Sci.* **2016**, *124*, 143–149. [CrossRef]

26. Pan, S.Y.; Shah, K.J.; Chen, Y.H.; Wang, M.H.; Chiang, P.C. Deployment of accelerated carbonation using alkaline solid wastes for carbon mineralization and utilization toward a circular economy. *ACS Sustain. Chem. Eng.* **2017**, *5*, 6429–6437. [CrossRef]

27. Antoni, M.; Rossen, J.; Martirena, F.; Scrivener, K. Cement substitution by a combination of metakaolin and limestone. *Cem. Concr. Res.* **2012**, *42*, 1579–1589. [CrossRef]

28. Scrivener, K.; Martirena, F.; Bishnoi, S.; Maity, S. Calcined clay limestone cements (LC3). *Cem. Concr. Res.* **2018**, *114*, 49–56. [CrossRef]

29. Standard Practice for Classification of Soils for Engineering Purposes (Unified Soil Classification System). Available online: https://compass.astm.org/EDIT/html_annot.cgi?D2487+17e1 (accessed on 19 December 2016).

30. Du, Y.J.; Wei, M.L.; Reddy, K.R.; Wu, H. Effect of carbonation on leachability, strength and microstructural characteristics of KMP binder stabilized Zn and Pb contaminated soils. *Chemosphere* **2016**, *144*, 1033–1042. [CrossRef]

31. ASTM C192/C192M Standard Practice for Making and Curing Concrete Test Specimens in the Laboratory. 2016, pp. 1–8. Available online: https://compass.astm.org/EDIT/html_annot.cgi?C192+19 (accessed on 19 December 2016).

32. ASTM Standard D2166 Standard Test Method for Unconfined Compressive Strength of cohesive soil. 2008. Available online: https://compass.astm.org/EDIT/html_annot.cgi?D2166+16 (accessed on 19 December 2016).

33. ASTM Standard D4972 Test Method for pH of Soils. 2007. Available online: https://compass.astm.org/EDIT/html_annot.cgi?D4972+19 (accessed on 19 December 2016).

34. USEPA. USEPA Method 1311 Toxicity Characteristic Leaching Procedure (TCLP). 1992. Available online: https://www.epa.gov/hw-sw846/sw-846-test-method-1311-toxicity-characteristic-leaching-procedure (accessed on 23 August 2016).

35. Muhammad, F.; Huang, X.; Li, S.; Xia, M.; Zhang, M.; Liu, Q. Strength evaluation by using polycarboxylate superplasticizer and solidi fi cation ef fi ciency of Cr 6 þ, Pb 2 þ and Cd 2 þ in composite based geopolymer. *J. Clean. Prod.* **2018**, *188*, 807–815. [CrossRef]

36. Ščančar, J.; Strazar, M.; Burica, O. Total metal concentrations and partitioning of Cd, Cr, Cu, Fe, Ni and Zn in sewage sludge. *Sci. Total Environ.* **2000**, *250*, 9–19. [CrossRef]

37. Tessier, A.; Campbell, P.G.; Bisson, M. Sequential Extraction Procedure for the speciation of particulate trace metals. *Anal. Chem.* **1979**, *51*, 844–851. [CrossRef]

38. Dai, Z.; Wang, L.; Tang, H.; Sun, Z.; Liu, W.; Sun, Y.; Su, S.; Hu, S.; Wang, Y.; Xu, K.; et al. Speciation analysis and leaching behaviors of selected trace elements in spent SCR catalyst. *Chemosphere* **2018**, *207*, 440–448. [CrossRef] [PubMed]

39. ISO. Determination of Elements in Aqua Regia and Nitric Acid Digests by Flame Atomic Absorption Spectrometry. 2002, pp. 1–17. Available online: https://horizontal.ecn.nl/docs/society/horizontal/STD6241_AAS-Flame.pdf (accessed on 1 January 2017).

40. Wang, F.; Zhang, F.; Chen, Y.; Gao, J.; Zhao, B. A comparative study on the heavy metal solidification/stabilization performance of four chemical solidifying agents in municipal solid waste incineration fly ash. *J. Hazard. Mater.* **2015**, *300*, 451–458. [CrossRef] [PubMed]

41. Dhandapani, Y.; Santhanam, M. Assessment of pore structure evolution in the limestone calcined clay cementitious system and its implications for performance. *Cem. Concr. Compos.* **2017**, *84*, 36–47. [CrossRef]

42. Wang, Y.S.; Dai, J.G.; Wang, L.; Tsang, D.C.W.; Poon, C.S. Influence of lead on stabilization/solidification by ordinary Portland cement and magnesium phosphate cement. *Chemosphere* **2018**, *190*, 90–96. [CrossRef] [PubMed]

43. Marangu, J.M. Physico-chemical properties of Kenyan made calcined Clay -Limestone cement (LC3). *Case Stud. Constr. Mater.* **2020**, *12*, e00333. [CrossRef]

44. Danner, T.; Norden, G.; Justnes, H. Characterisation of calcined raw clays suitable as supplementary cementitious materials. *Appl. Clay Sci.* **2018**, *162*, 391–402. [CrossRef]

45. Lothenbach, B.; Le Saout, G.; Gallucci, E.; Scrivener, K. Influence of limestone on the hydration of Portland cements. *Cem. Concr. Res.* **2008**, *38*, 848–860. [CrossRef]

46. Dhandapani, Y.; Vignesh, K.; Raja, T.; Santhanam, M. Development of the Microstructure in LC3 Systems and Its Effect on Concrete Properties. In *Calcined Clays for Sustainable Concrete*; Scrivener, K., Favier, A., Eds.; Springer: Dordrecht, The Netherlands, 2018; Volume 10, pp. 131–140. ISBN 978-94-017-9938-6.

47. Ministry of Environment and Forests. MOEF Hazardous and Other Wastes-Ministry of Environment and Forests. *Gazzate India* **2016**, *1981*, 1–68.

48. Gu, Y.C.; Li, J.L.; Peng, J.K.; Xing, F.; Long, W.J.; Khayat, K.H. Immobilization of hazardous ferronickel slag treated using ternary limestone calcined clay cement. *Constr. Build. Mater.* **2020**, *250*, 118837. [CrossRef]

49. Matschei, T.; Lothenbach, B.; Glasser, F.P. The role of calcium carbonate in cement hydration. *Cem. Concr. Res.* **2007**, *37*, 551–558. [CrossRef]

50. Krishnan, S.; Emmanuel, A.C.; Shah, V.; Parashar, A.; Mishra, G.; Maity, S.; Bishnoi, S. Industrial production of limestone calcined clay cement: Experience and insights. *Green Mater.* **2019**, *7*, 15–27. [CrossRef]

51. Feng, Y.S.; Du, Y.J.; Reddy, K.R.; Xia, W.Y. Performance of two novel binders to stabilize field soil with zinc and chloride: Mechanical properties, leachability and mechanisms assessment. *Constr. Build. Mater.* **2018**, *189*, 1191–1199. [CrossRef]

52. Liu, G.; Tao, L.; Liu, X.; Hou, J.; Wang, A.; Li, R. Heavy metal speciation and pollution of agricultural soils along Jishui River in non-ferrous metal mine area in Jiangxi Province, China. *J. Geochem. Explor.* **2013**, *132*, 156–163. [CrossRef]

 *sustainability*

Article

Efficacy of Enzymatically Induced Calcium Carbonate Precipitation in the Retention of Heavy Metal Ions

Arif Ali Baig Moghal [1], Mohammed Abdul Lateef [2], Syed Abu Sayeed Mohammed [2], Kehinde Lemboye [3], Bhaskar C. S. Chittoori [4] and Abdullah Almajed [3,*]

[1] Department of Civil Engineering, National Institute of Technology, Warangal 506004, Telangana State, India; baig@nitw.ac.in or reach2arif@gmail.com (A.A.B.M.)
[2] Department of Civil Engineering, HKBK College of Engineering, Bengaluru 560045, India; abdulmohammed040@gmail.com (M.A.L.); abubms@gmail.com (S.A.S.M.)
[3] Department of Civil Engineering, College of Engineering, King Saud University, P.O. Box 800, Riyadh 11421, Saudi Arabia; 438105781@student.ksu.edu.sa (K.L.)
[4] Department of Civil Engineering, Boise State University, Boise, ID 83725, USA; bhaskarchittoori@boisestate.edu (B.C.S.C.)
* Correspondence: alabduallah@ksu.edu.sa

Received: 30 July 2020; Accepted: 25 August 2020; Published: 28 August 2020

Abstract: This study evaluated the efficacy of enzyme induced calcite precipitation (EICP) in restricting the mobility of heavy metals in soils. EICP is an environmentally friendly method that has wide ranging applications in the sustainable development of civil infrastructure. The study examined the desorption of three heavy metals from treated and untreated soils using ethylene diamine tetra-acetic acid (EDTA) and citric acid ($C_6H_8O_7$) extractants under harsh conditions. Two natural soils spiked with cadmium (Cd), nickel (Ni), and lead (Pb) were studied in this research. The soils were treated with three types of enzyme solutions (ESs) to achieve EICP. A combination of urea of one molarity (M), 0.67 M calcium chloride, and urease enzyme (3 g/L) was mixed in deionized (DI) water to prepare enzyme solution 1 (ES1); non-fat milk powder (4 g/L) was added to ES1 to prepare enzyme solution 2 (ES2); and 0.37 M urea, 0.25 M calcium chloride, 0.85 g/L urease enzyme, and 4 g/L non-fat milk powder were mixed in DI water to prepare enzyme solution 3 (ES3). Ni, Cd, and Pb were added with load ratios of 50 and 100 mg/kg to both untreated and treated soils to study the effect of EICP on desorption rates of the heavy metals from soil. Desorption studies were performed after a curing period of 40 days. The curing period started after the soil samples were spiked with heavy metals. Soils treated with ESs were spiked with heavy metals after a curing period of 21 days and then further cured for 40 days. The amount of $CaCO_3$ precipitated in the soil by the ESs was quantified using a gravimetric acid digestion test, which related the desorption of heavy metals to the amount of precipitated $CaCO_3$. The order of desorption was as follows: Cd > Ni > Pb. It was observed that the average maximum removal efficiency of the untreated soil samples (irrespective of the load ratio and contaminants) was approximately 48% when extracted by EDTA and 46% when extracted by citric acid. The soil samples treated with ES2 exhibited average maximum removal efficiencies of 19% and 10% when extracted by EDTA and citric acid, respectively. It was observed that ES2 precipitated a maximum amount of calcium carbonate ($CaCO_3$) when compared to ES1 and ES3 and retained the maximum amount of heavy metals in the soil by forming a $CaCO_3$ shield on the heavy metals, thus decreasing their mobility. An approximate improvement of 30% in the retention of heavy metal ions was observed in soils treated with ESs when compared to untreated soil samples. Therefore, the study suggests that ESs can be an effective alternative in the remediation of soils contaminated with heavy metal ions.

Keywords: heavy metals; soil; enzyme solutions; desorption; extractant

---

1. Introduction

Increase in population and attempts to satisfy the ever-growing demands of the same have led to industrialization, widespread construction activity, and extensive mining. Industrial effluents contaminate land and thereby pose a threat to the environment [1,2]. The number of contaminated sites identified in the United States of America alone until 2004 was 294,000 [3], which indicates an alarming need to implement remediation and decontamination methods to maintain a balance in nature. Remediation of the sites is costly because conventional remediation methods, such as excavation and dumping on unused land, are outdated, and contaminant removal by physical methods is difficult. As a result, contaminated soils arrive in engineered landfills. Engineered landfills, which are identified as the most viable means of landfilling solid municipal wastes, have also become sources of leachates rich with high levels of toxicity, fluorides, nitrates, and heavy metals [4–6]. Hence, attempts to mitigate the adverse effects of hazardous wastes present in landfills has attracted significant research attention. Urbanization also poses a worldwide challenge for landfill management. Developing an environmentally friendly method to decontaminate soils is a priority research topic.

One of the methods adopted to decrease landfill hazards involves adding liners to landfills that act as barriers relative to leachate infiltration from the landfills to the soil beneath [7–9]. A liner is an essential part of an engineered landfill and should be durable and properly designed for the safety of the circumferential environment [10]. The use of locally available, fine, and cohesive soil as liner material is an easy and convenient option [11]. However, not all locally available material is suitable for such uses. In such cases, suitable material is imported, which adds to the overall costs of the project. Another method of decontaminating soils is by washing the soils with suitable chemicals to neutralize the contaminants. In general, these methods for decontaminating soils are costly owing to difficulties involved with physical separation or soil washing. Furthermore, due to the heterogeneous soil texture, removal of soil contaminants by soil washing solutions becomes difficult [3].

Another method to strengthen the landfill liner is to stabilize the soil using lime or cement. Stabilization of soil by cement and lime have been vastly used [12–15] in landfill liners, and these methods contribute to the emission of greenhouse gases, leading to large scale global warming. Production of cement and lime result in about 800–900 and 600–700 kg CO_2 per ton, respectively [16]. The cement industry contributes about 5% of CO_2 emissions globally [17,18]. With such an alarming situation, researchers have searched for sustainable stabilization methods for soil, wherein partial or total replacement of the cement or lime binders are tested for their reliability. These stabilizers include palm oil fuel ash [19], flyash [20], rice husk ash [21,22], residue of calcium carbide [23], alkali-activated agro-waste [24], and so forth.

As an attempt to contribute to the field of soil stabilization, bio cementation by $CaCO_3$ is also employed as a sustainable method of soil stabilization [25–29], which includes improvement of geotechnical properties of soil as well as contaminant remediation. This study evaluates the use of one such method, known as enzyme induced calcite precipitation (EICP), to immobilize heavy metals as a method of decontaminating soil. The main objective is to study the ability of the EICP method to retain heavy metal contaminants such as nickel, cadmium, and lead in soil and to determine the dosage of Enzyme Solution (ES) that leads to the maximum retention of heavy metals. Another aim is to identify the percentage of heavy metal contaminants retained in the soil as a result of EICP.

2. Background

The dumping of contaminated soils into landfills has made landfill management a major challenge, as these sites become major sources of leachates rich in heavy metals, fluorides, and other contaminations. Furthermore, the intrusion of leachates from landfills has been found to introduce large quantities of heavy metals, fluorides, and nitrates to ground water used for irrigation, industrial purposes, and drinking [30]. High concentrations of metals in a low-pH soil increase soil acidity, thereby making the soil vulnerable to contamination [6]. The presence of less than 1000 mg/kg of heavy metals in soil is rarely toxic, although human activities involving the disposal of them into the soil above these levels poses a threat to the flora and fauna. Heavy metals that are naturally present in the soil cause less damage than those that accumulate due to human activity [31]. Specifically, Cd is naturally available in earth's crust with a concentration of 0.1–0.5 ppm associated with zinc, copper, and lead ores [32]. Industries play an important role in increasing heavy metal concentrations in surface soils above permissible levels by releasing toxic fumes into the atmosphere, because traces of fumes find their way to the soil surface [33,34].

Soil in its natural form exhibits different components, such as phyllosilicates, humic substances, and carbonates, which contribute to the sorption process of heavy metals [35]. Concentrations of metals in the soil are governed by anthropogenic effects, pedogenic processes, and parent material [36]. Immobilization [37], phytoremediation [38,39], and soil washing [40] methods are adopted to achieve remediation of heavy metals in soils [7,41,42]. Remediation of contaminated soils can be performed using in situ or ex situ methods based on site conditions [4,43]. Landfill management continues to constitute a complex issue and should be examined further [39].

Industries also emit toxic elements into water bodies. These elements are then absorbed by aquatic plants, making them unsafe for consumption. Sahu et al. [44] examined concentrations of different heavy metals in macrophytes and observed an average of 13 mg/kg of Pb in seven aquatic plants from the Kharun River in India. Metal contaminants also pose a threat to terrestrial plants because of the increase in anthropogenic activities, which eventually leads to their intrusion in the food chain [45]. Heavy metals that are most harmful to human health include Pb, Cd, As, Zn, Cu, Hg, Cr, and Ni. Heavy metals happen to be the most commonly found carcinogens among the pollutants; for example, Hg leads to mutations and genetic damage, and Cu and Pb can affect the brain and bones [46]. Heavy metals are generally removed from the soil via precipitation–dissolution, oxidation-reduction, and adsorption–desorption processes, among which adsorption–desorption is observed as the most effective geochemical process for contaminant remediation [47]. The removal of metals can also be performed by chemical precipitation, bio-precipitation, ion exchange, adsorption, biosorption, physical separation, electrochemical separation, solvent extraction, flotation, and cementation [48,49]. Another technique adopted to decrease the hazardous effects of heavy metals involves retaining contaminants in soil via encapsulation. Methods adopted for retaining heavy metals by encapsulation decrease the mobility of heavy metal ions in the soil. The use of nano calcium silicate in retaining Cd, Ni, and Pb was observed as an effective [50,51] approach towards contaminant remediation. However, the costs associated with production of nano compounds on a macro scale is not economically feasible.

Among the approaches to encapsulate heavy metals in soils, the use of microorganisms is also adopted by researchers. This method, popularly known as bioremediation, involves processes such as microbial induced calcite precipitation (MICP) to encapsulate heavy metals inside the precipitated calcium carbonate. The MICP technique fosters metabolic activity in certain types of soil bacteria (*Sporosarcina pasteurii*), which results in the formation of inorganic compounds (such as $CaCO_3$) outside the cellular structure; these compounds can bind soil particles together and stabilize the soil. In addition, it is possible to encapsulate heavy metals present in the soil inside the $CaCO_3$ crystals. Another branch of bioremediation, known as enzyme induced calcite precipitation (EICP), uses plant-based urease enzymes to precipitate calcium carbonate. The use of enzymes is advantageous, because they are non-toxic and ecofriendly [52].

Nathan et al. [53] found the enzyme treatment effective in reducing the heavy metals in paper pulp. EICP is a biologically inspired soil improvement process designed to initiate urea hydrolysis using urease enzyme extracted from jack beans [25,54–57] or watermelon seeds [58]. The precipitation of $CaCO_3$ through enzyme activity is obtained by mixing urea, calcium chloride, and urease enzyme with deionized water. This solution can then be mixed with soil to prepare soil samples for testing [59]. Additionally, MICP is a process in which $CaCO_3$ is precipitated by the activity of bacteria, such as *Sporosarcina pasteurii*, to improve the geotechnical properties of soil [60]. Calcium carbonate precipitates are observed to fill the voids between soil grains, thereby reducing the permeability of soil and improving unconfined compressive strength (UCS) [61,62] and shear strength of the soil [63].

Furthermore, the precipitates mitigate liquefaction below the existing structures, stabilize roadways, control the flow of groundwater, immobilize contaminants [37], and remediate hazardous trace metals in soils [64]. The urease enzyme used in the study was crystallized from jack beans (*Canavalia ensiformis*) and formed the main source of urea hydrolysis [65]. Plant-based urease is optimal if obtained from jack bean and is identified as the first crystallized enzyme as well as the first nickel metalloenzyme [66]. Concerning remediation of soil contaminants, $CaCO_3$ precipitates are relatively easy to implement, useful in retaining the contaminants, and economic. They are readily accepted in society, based on an approximate scoring proposed by Dejong et al. [67].

3. Materials and Methods

3.1. Soil

The present study was performed on two soils, namely black cotton soil collected from Yadgir district (16°45′20.556″ N, 77°9′4.5072″ E) and red soil from Bangalore district (13°2′14.1396″ N, 77°37′11.928″ E) in Karnataka, India. The dominant mineral in black cotton soil was identified as montmorillonite (henceforth referred to as 'Soil M') and kaolinite in red soil (referred to as 'Soil K'). Basic tests of these soils were conducted and are listed in Table 1. Soil M was classified as clay with high plasticity (CH), whereas Soil K was classified as clay with low plasticity (CL) according to the Unified Soil Classification System.

Table 1. Engineering properties of tested soils.

Property	Soil K	Soil M
Color	Red	Black
Specific gravity	2.6	2.5
Liquid limit (%)	30	54
Plastic limit (%)	17	27
Plasticity index (%)	13	27
Classification (USCS)	CL	CH
pH	5.7	8.3
Organic content %	0.87	0.77
Optimum water content (%)	16.8	15.2
Maximum dry density (kN/m^3)	17.9	15.8
Unconfined compressive strength (MPa)	0.07	0.013

3.2. Heavy Metal Contaminants

Three heavy metals, namely cadmium (Cd), nickel (Ni), and lead (Pb), were spiked in the soils by preparing a stock solution with predetermined load ratios to obtain the target concentrations of contaminants in the soil as described by Mohammed and Moghal [2].

Analytical reagent grade (AR) nitrates of nickel ($Ni(NO_3)_2$), cadmium ($Cd(NO_3)_2$), and lead ($Pb(NO_3)_2$) were used to prepare the stock solutions. The target load ratio was identified as 50 and 100 mg/kg of heavy metal contaminants. Salt solutions were prepared in containers of borosilicate glass to maintain a sufficiently wet consistency to spike soils with contaminants. Oven-dried soil samples of a predetermined quantity were placed in glassware and properly washed with deionized double-distilled water (DI), and the heavy metal stock solution was added to the soil samples and mixed thoroughly to achieve the target Ni, Cd, and Pb load ratios [68]. Soil containers were covered with thin sheets of aluminum with minute perforations to ensure free airflow and to avoid dust intrusion and cross-contamination [51]. Subsequently, the samples were placed on a flat dry platform in a temperature- and humidity-controlled room where the temperature was maintained at 23–27 °C with proper ventilation and average relative humidity of 45%. The soil samples were left to cure for 40 days such that the probability of contaminants desorbing from the soil in the initial stages was minimized. After the soil samples were cured for 40 days, acid digestion was conducted, and the actual amount of heavy metal concentration in the soils was established using an atomic absorption spectrophotometer (AAS) (PerkinElmer Model A-Analyst 400). The results are tabulated in Table 2.

Table 2. Determination of actual load ratio of metal ions via acid digestion method.

Load Ratio Available with 50 mg/kg Initial Load Ratio							
Metal Ion	Soil K	Soil K + ES1	Soil K + ES2	Soil K + ES3	Soil M	Soil M + ES1	Soil M + ES2	Soil M + ES3
Ni	49.37	48.67	46.19	47.69	48.37	47.65	45.61	46.97
Cd	48.97	47.68	44.93	46.37	48.69	47.86	45.57	46.59
Pb	49.01	48.79	46.82	48.03	48.04	47.63	45.39	46.55

Load Ratio Available with 100 mg/kg Initial Load Ratio							
Metal Ion	Soil K	Soil K + ES1	Soil K + ES2	Soil K + ES3	Soil M	Soil M + ES1	Soil M + ES2	Soil M + ES3
Ni	99.09	98.67	95.98	97.64	99.12	98.36	95.63	97.86
Cd	98.39	98.49	96.15	97.35	98.87	98.65	95.58	97.55
Pb	99.2	98.52	96.28	97.36	99.24	99.83	96.05	97.64

The same procedure was repeated for the soils treated with ESs, and the concentrations of the contaminants in the soils treated with the ESs were determined for comparison with those of untreated soils.

3.3. Enzyme Solutions

Three types of enzyme solutions were prepared to treat the soil using Analytical Reagent (AR) grade materials. Specifically, the following materials were used to prepare the ESs:

- Urea (CH_4N_2O)
- Calcium chloride ($CaCl_2·2H_2O$)
- Urease enzyme (Canavalia ensiformis (jack bean) Type III, powder, 15,000–50,000 units/g solid)
- Non-fat milk powder

The chemicals were procured from Winlab Chemical, Market Harborough, United Kingdom, and the urease enzyme was procured from Sigma-Aldrich, St. Louis, MO, USA. The composition of each of the ESs is presented in Table 3. It can be noted here that ES1 and ES2 have similar compositions, except for the use of non-fat milk powder, which aided in developing nucleation sites in the soil masses, thereby leading to the precipitation of $CaCO_3$ at the contact points [25]. Similarly, ES2 and ES3 differ by the concentration of the urease enzyme, used as per Almajed et al. [25].

Table 3. Composition of ESs.

Solution	Concentration of Component 1	Concentration of Component 2	Concentration of Component 3	Concentration of Component 4
	Urea (CH_4N_2O)	Calcium Chloride ($CaCl_2·2H_2O$)	Urease Enzyme	Non-Fat Milk Powder
ES1	1.00 M	0.67 M	3.00 g/L	-
ES2	1.00 M	0.67 M	3.00 g/L	4.00 g/L
ES3	0.37 M	0.25 M	0.85 g/L	4.00 g/L

3.4. EICP Treatments

Enzyme-treated soil samples were prepared by compacting soil passing through a 425-μ sieve and mixed with an ES to obtain a moisture content equal to the optimum moisture content of the soil by weight. The soil was mixed with ES in a 5-cm diameter mold with a depth of 10 cm in three layers to ensure maximum density and minimal voids. Subsequently, the soil samples were cured in a desiccator for 21 days after they were sealed in an airtight seal pack to ensure proper calcite precipitation.

3.4.1. UCS Tests

UCS tests were also performed for the enzyme-treated soil samples as per ASTM D2166 [69] to gain a better understanding of the effect of ESs on the strength properties of soils. Figure 1 shows the UCS results obtained for soil samples using ESs cured for 7, 14, and 21 days. It was observed that the UCS values of the soil samples increased to the maximum level when the samples were treated with ES2 and cured for 21 days. Non-fat milk powder in ES2 is the main strength booster, because it creates sites of nucleation in the soil, enhancing the precipitation of $CaCO_3$. However, non-fat milk powder is also used in ES3, and the strength gain is less than that of the ES2, because the amount of urease enzyme used in ES3 is 0.85 g/L, which is less than that of ES2 [25]. The same reason potentially holds in terms of understanding the amount of $CaCO_3$ precipitated in the soil when ES3 is used, which constitutes a reduction in $CaCO_3$ precipitated when compared to precipitation given the use of ES2.

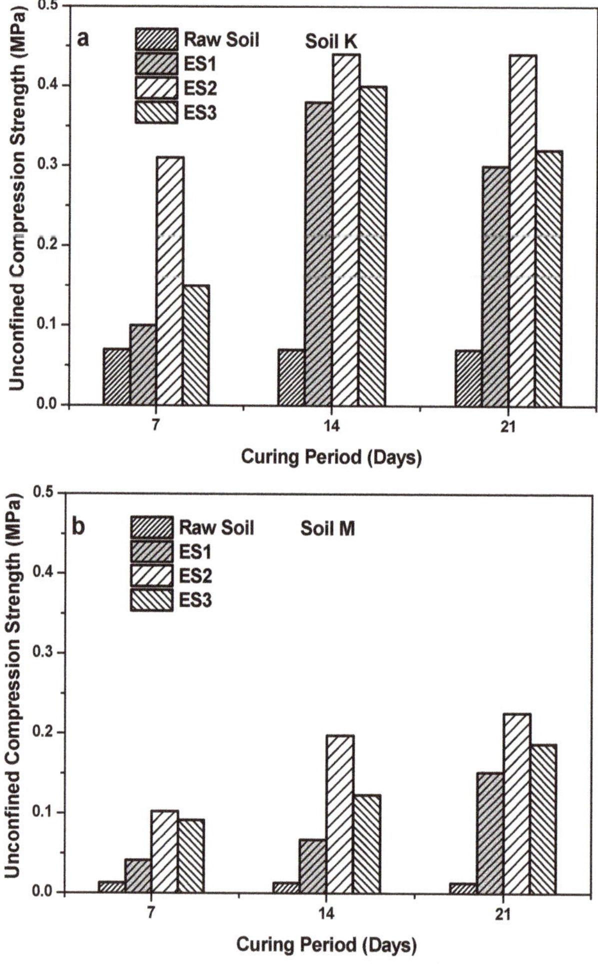

Figure 1. UCS results of EICP treated soils with different enzyme solutions: (**a**) Soil K (**b**) Soil M.

3.4.2. Measurement of Calcium Carbonate Precipitation

The ES-treated soil samples were kept in a desiccator after they were sealed in air-tight packs and were tested to identify the amount of $CaCO_3$ precipitated by gravimetric acid digestion. 40 to 50 g of soil specimens was soaked in 1 M hydrochloric acid for an hour until the disappearance of the effervescence was observed after soaking (due to the dissolution of $CaCO_3$ in the soil). Subsequently, the samples were rinsed and placed in an oven for drying at a temperature of 105 °C. The difference in the masses of the soil samples before and after soaking in 1.0 M hydrochloric acid was determined to calculate the percentage of $CaCO_3$ precipitated in soil. The soil samples were tested for the amount of $CaCO_3$ precipitated over 7, 14, and 21 days to understand the precipitation trend. These results are presented in Table 4.

Table 4. Calcium Carbonate Precipitation (%) as obtained by gravimetric acid digestion test.

Curing Period (Days)	Soil K			Soil M		
	ES1	ES2	ES3	ES1	ES2	ES3
7	1.315	1.387	1.367	1.581	2.631	1.671
14	1.769	2.286	1.933	2.463	4.722	2.632
21	1.689	2.043	1.811	2.026	4.582	2.328

3.5. Extractants

The extractants used for the desorption tests included ethylene diamine tetra-acetic acid (EDTA) and citric acid in three molar concentrations (i.e., 0.1, 0.25, and 0.5 M) for the removal of heavy metals retained in the soil by acid digestion. A solid-to-liquid ratio of 1:20 was maintained. Both EDTA [70] and citric acid [71] were proven to be effective in extracting heavy metals from soil surfaces. Gu and Yeung [71] noted that citric acid dominant industrial wastewater is effective in desorbing Cd from soil surfaces in the pH range of 4 to 8.

3.6. Desorption Tests

A desorption test was conducted on the soil samples spiked with contaminants after a curing period of 40 days. Five grams of contaminated soil samples was placed in 0.1, 0.25, and 0.5 M extractants in 50-mL polytetrafluoroethylene bottles and shaken as per ASTM D3987 [72] in a mechanical shaker at 30 rpm for 24 h to ensure that the soil particles were uniformly distributed in the solution, and thereby releasing heavy metal contaminants. Subsequently, the slurry was filtered through filter papers (Whatman No. 42) to obtain solutions that were tested in AAS to identify their metal contamination levels. Each set of tests was performed with three samples, and an average of the three results was considered as the final value. The pH values of the solutions tested in AAS were determined, and their removal efficiencies were calculated as follows [73].

$$\text{Removal Efficiency } (\%) = \frac{\text{Contaminant mass from desorption}}{\text{Initial contaminant mass in soil}} \tag{1}$$

4. Results and Discussions

The data obtained from the desorption experiments were plotted (Figures 2–4) with the molar concentrations of the extractants on the abscissa and heavy metal removal efficiencies (%) on the ordinate. A lower removal efficiency percentage is favorable, as it means that heavy metal contaminants are immobilized in the soil and cannot be extracted from it. Among the three heavy metals, it was observed that the removal efficiency of Cd was minimal when compared to that of Ni or Pb.

The formation of $CaCO_3$ was initiated in the soil by the reaction between urea and calcium chloride; this occurs when the calcium ions and carbonate ions combine [74]. The process of precipitation of calcium carbonate is accelerated by the urease enzyme [75]. The recipe of the ESs adopted provides an environment for efficient utilization of enzyme for $CaCO_3$ precipitation [76]. $CaCO_3$ is precipitated along the nucleation sites with ES2 due to the use of non-fat milk powder [25]. The main process that leads to the formation of calcite is hydrolysis of urea by the urease enzyme into CO_2 and NH_3, and the speciation of NH_3 leads to the development of NH^{4+} ions, which creates a suitable environment for precipitation of $CaCO_3$ in calcium rich solution of $CaCl_2$ [26]. It is also proposed that the heavy metal ions with ion radii close to that of Ca^{2+} (e.g., Pb^{2+}, Cu^{2+}, Cd^{2+}, and Sr^{2+}) are incorporated in $CaCO_3$ crystal lattice by replacing Ca^{2+} ions or by creating defects on the calcium carbonate crystals, or even by penetrating the $CaCO_3$ interstice. This indicates that the EICP technique can be used for the remediation of heavy metals [77]. The formation of carbonates of heavy metals occurs in the microenvironment of the mineral carbonates, and the process is even applicable for radionuclides such as strontium forming strontium carbonate ($SrCO_3$) [74]. Metal ions in soil tend to cluster with carbonates that are already present in the soil, and heavy metal retention is expected due to the precipitation of heavy metal carbonates [78,79]. Enzymatic mechanism of bioremediation is found to be effective in issues concerning heavy metal bioaccumulation in paper pulp [53].

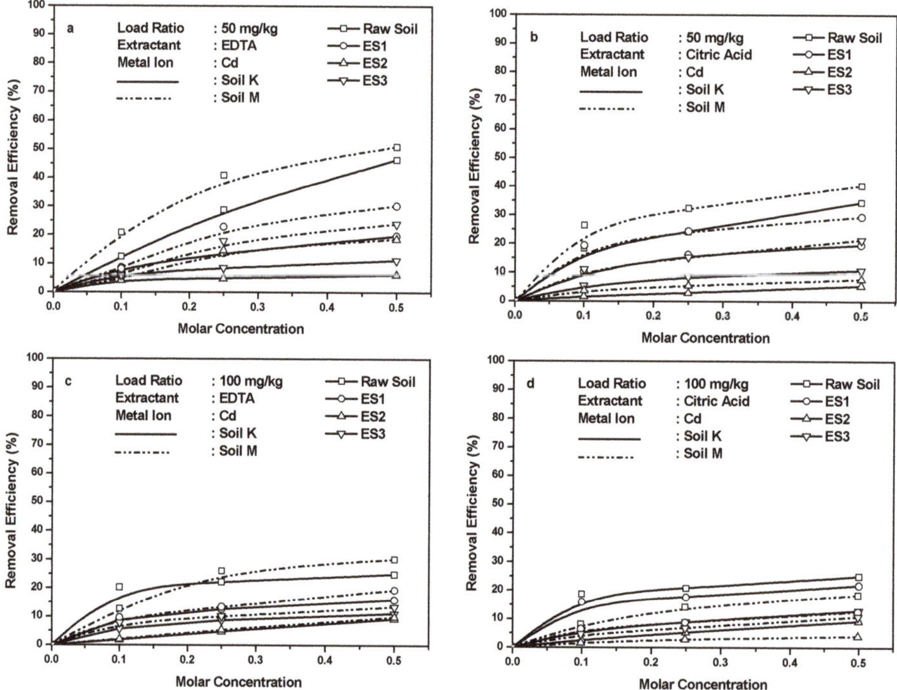

Figure 2. Effect of extractants on 'Cd' removal efficiencies of soils treated with different enzymatic solutions with (**a**) EDTA extractant and load ratio of 50 mg/kg, (**b**) Citric acid extractant and load ratio of 50 mg/kg, (**c**) EDTA extractant and load ratio of 100 mg/kg, (**d**) Citric acid extractant and load ratio of 100 mg/kg.

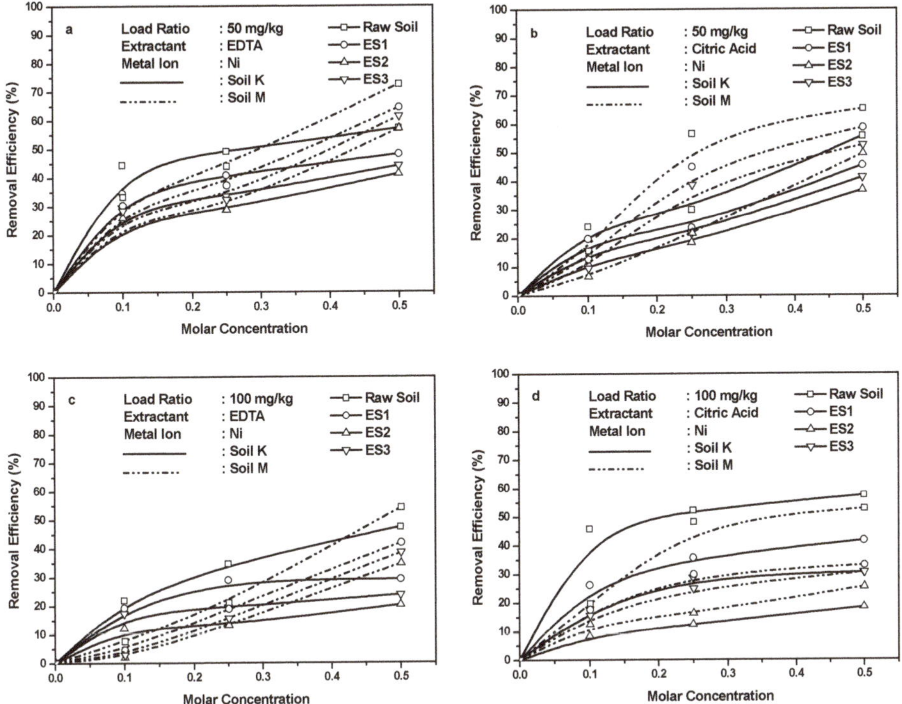

Figure 3. Effect of extractants on 'Ni' removal efficiencies of soils treated with different enzymatic solutions with (**a**) EDTA extractant and load ratio of 50 mg/kg, (**b**) Citric acid extractant and load ratio of 50 mg/kg, (**c**) EDTA extractant and load ratio of 100 mg/kg, (**d**) Citric acid extractant and load ratio of 100 mg/kg.

4.1. Cd Retention in Soils

The highest removal efficiency of Cd was observed as 50.39% for untreated Soil M and 46.37% for untreated Soil K (see Figure 2). The lowest removal efficiency for Cd when extracted by 0.5 M citric acid was observed as 3.97% for Soil M and 5.39% for Soil K treated with ES2 (see Figure 2). An increase in clay or silt particles in the soil decreases the release of Cd due to strong deposits of heavy metal traces that occur in the finer fractions of clay [80]. Hence, it was observed that higher silt or clay content expanded soil capacity for heavy metal retention [81,82]. Our results are in line with these observations.

The Cd removal efficiency was also low due to the precipitation of a comparatively large amount of CaCO$_3$ in soil treated with ES2 (see Table 4). Wang et al. [83] found that the presence of CaCO$_3$ played a major role in immobilizing heavy metals, as observed in their study, when treating soils with CaCO$_3$. The EICP method used in this study for the retention of heavy metals is comparatively more effective in immobilizing Cd. Therefore, the EICP treatment can be termed an adsorbent selective to Cd contamination.

The role of pH is also important in the retention of Cd in the soil, and the desorption of Cd is observed to increase when pH decreases. Additionally, increase in the pH of soil results in immobilization of Cd ions [32]. Citric acid is observed to desorb Cd to a lesser extent at lower molar concentrations and to a greater extent at higher molar concentrations (see Figure 2). This can be attributed to the formation of Cd-citric acid complexes in an aqueous state, which detach from soil surfaces. EDTA is observed to retain greater amounts of Cd, even at a pH of 5.05 [84]. The range of pH was between 5–8 with citric acid as chelant with 0.1 M for extraction of Cd. This pH reduced (within range of 3–6) at 0.5 M molarity, leading to comparatively higher removal of Cd. The removal efficiencies and associated standard deviation values are provided in the Supplementary Materials (Tables S1 and S2 respectively).

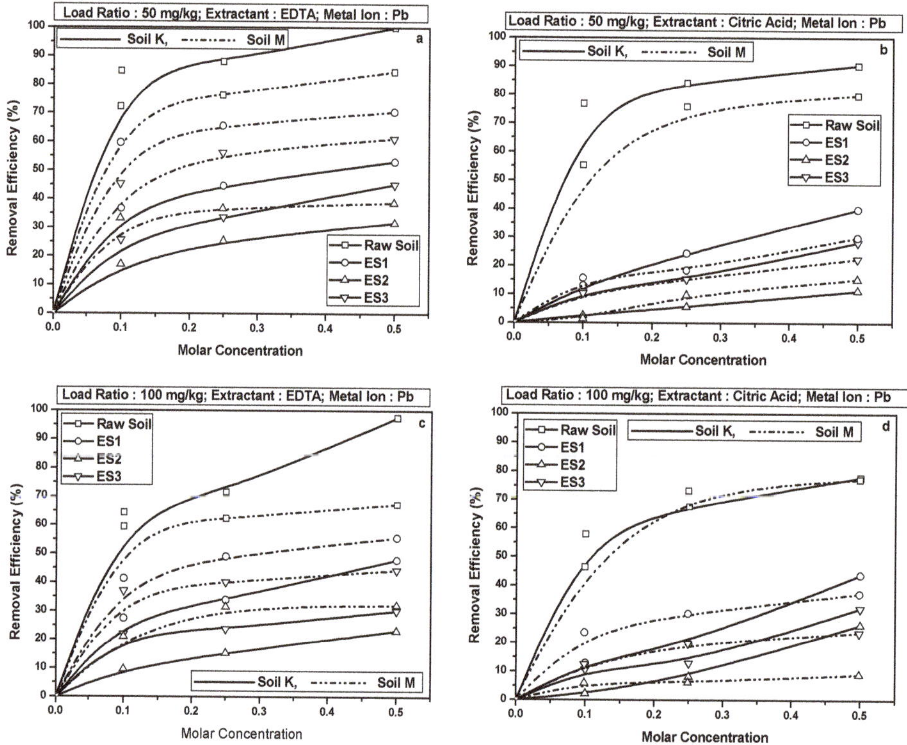

Figure 4. Effect of extractants on Pb removal efficiencies of soils treated with different enzymatic solutions with (**a**) EDTA extractant and load ratio of 50 mg/kg, (**b**) Citric acid extractant and load ratio of 50 mg/kg, (**c**) EDTA extractant and load ratio of 100 mg/kg, (**d**) citric acid extractant and load ratio of 100 mg/kg.

4.2. Ni Retention in Soils

Figure 3 shows the plots for Ni extracted from both soils using EDTA and citric acid. Figure 3a,b shows the removal efficiency for load ratios of 50 mg/kg. Figure 3c,d shows the removal efficiency for load ratios of 100 mg/kg. The plots indicate that the encapsulation of Ni in the soils with ES2 exceeded that in the soils with ES1 and ES3. The difference in removal efficiency was not significant between the two load ratios (50 and 100 mg/kg). Both load ratios yielded removal efficiency values of approximately 15–20%.

The maximum range of Ni desorption was observed as approximately 72.36–57.2% for soils M and K, respectively, and the minimum amounts of Ni desorption were 34.48% for Soil M and 20.26% for Soil K when extracted by 0.5 M EDTA. Nickel is one of the most commonly found contaminants in the brownfields, and nickel contamination originates from the discharge of industries involving metal plating, nickel refinement, and mining sites. Nickel precipitates into a stable compound in the form of nickel hydroxide [Ni(OH)$_2$] in slightly alkaline and neutral solutions. Nickel can be effectively removed by citric acid and EDTA chelating agents [70], and nickel retention can be understood from the relationship between the sorptive surface and ion concentration that decreases metal ion removal.

When metal ion concentrations are low, the number of binding sites for heavy metals is initially high and results in the immobilization of heavy metals [85]. The mechanism of Ni retention can also be expressed as a mechanism in which metal binds to a carboxylic group, given the competition between protons, metals, and ion exchange interactions in the solution, and leads to the immobility of Ni ions [86]. Precipitation of heavy metals occurs when a solution is saturated with a specific element through a homogeneous or heterogeneous aggregation processes. The former aggregation process is an outcome of nucleation of the supersaturated phase in the soil solution, whereas the latter process involves precipitation formed by the nucleation of other materials (i.e., soil particles), which hold metals as sorbents on the surface [87]. The urease enzyme used in the study is reportedly active and stable when the EDTA solution exhibits a neutral pH. The enzymatic activity plays a major role in already present nickel ions in the urease enzyme. It is also established that nickel ions are active in strengthening the enzyme activity completely; thus, attempts to remove nickel from the urease enzyme were not successful [66]. Thus, it can be assumed that preexisting nickel in urease and added nickel potentially cluster and this behavior leads to the retention of Ni in the soil. This can explain the results shown in Figure 3.

4.3. Pb Retention in Soils

Figure 4 shows the Pb removal efficiency. As shown in the figure, Pb retention was minimal when compared to Ni and Cd retention. Raw soil K contaminated with Pb demonstrated very high removal efficiency, which ranges from 99.96% to 97.52% for soil treated with 0.5 M EDTA extractant and with load ratios of 50 and 100 mg/kg, respectively. The removal efficiencies after acid digestion by citric acid decreased to 11.2% and 26.26% for the same soil with 50 and 100 mg/kg load ratios, respectively, when Soil K was treated with ES2.

Pb and Cd get adsorbed on $CaCO_3$ precipitates, limiting their mobility. The retention of Pb and Cd was observed as predominant by polydopamine $CaCO_3$ when compared to natural $CaCO_3$ [88]. Thus, it was inferred that decreases in Cd and Pb removal efficiency observed in the present study occurred due to the presence of $CaCO_3$ in the soil. Pb retention can also occur via diffusion of solid-state and precipitation reactions resulting in $PbSO_4$ and $PbCO_3$ precipitates when the metal contamination levels exceed solubility levels of the carbonates and hydroxides at a given pH [89,90]. Kumpiene et al. [91] performed Pb immobilization studies via phosphorus-containing chemicals, such as apatites and hydroxyapatites, in synthetic and natural forms. This is because precipitation and ionic exchange of pyromorphite-type minerals decrease Pb mobility, and because compounds of Ca are generally efficient in performing Pb immobilization as soil pH increases within the range 8–9, thereby leading to the retention of Pb. Almost 99% of Pb retention was obtained by Wang et al. [83] using reagent-grade stabilizers, namely $Ca(H_2PO_4)_2$ and $CaCO_3$ in a dumpsite in Taiwan.

The heavy metal desorption process is affected by the pH of the pore fluid solution [42]. Even waste soils with neutral pH values exhibit significant levels of heavy metal sorption [92]. Harter [93] examined the effects of soil pH on heavy metal adsorption and stated that the degree to which metal ions hydrolyze at a specific extractant pH (leading to their release from the host soil) is unknown. In a previous study, Harter [93] suggested that retention of univalent Pb hydroxides (when compared to divalent ions) can increase by 60% when pH increases from 6 to 8 and forms precipitates of metals at high pH levels. The same study revealed that Ni was retained due to a reaction that formed precipitates at high pH levels and concluded that the validity of pH, as a deciding factor in ascertaining the quantity of heavy metal that can be safely retained in the soil, is uncertain, although the pH of the extractants used for Pb retention was within the range 3–5.

Gupta and Lataye [94] indicated that the pH of a solution affects the charge and process of ionization of the sorbent in the solution. Therefore, it is presumed that pH increases the duration for which metal contaminants are released by extractants. The process of heavy metal removal was performed in the study by conducting separate desorption tests on soils spiked with individual contaminants, because the simultaneous removal of heavy metals using a single extraction method is difficult [95].

5. Conclusions

The aim of this study was to examine the efficacy of EICP in fixing Cd, Ni, and Pb that were spiked in Soil M and Soil K in addition to improving their strength characteristics. The following conclusions were made from the study:

- Soil grains adhered to each other due to $CaCO_3$ precipitation initiated by the urease enzyme, and there exists a high probability that metal ions are encapsulated between the soil grains and $CaCO_3$ precipitates. This leads to the effective retention of heavy metals in the soil matrix.
- Heavy metal retention in the soil occurred in the following order: Cd > Ni > Pb. The EICP-treated soil retained the maximum quantity of Cd among all the heavy metals. Additionally, Cd retention exceeded Ni or Pb retention even after treatment with chelants EDTA and citric acid. This was potentially due to the formation of $CdCO_3$ in the soil matrix.
- EICP treatment using ES2 was observed to be better in terms of retaining heavy metals in the soil when compared to ES1 and ES3.
- The use of non-fat milk powder in the preparation of ES2 played a major role in boosting the UCS strength of the soil and also in retaining heavy metals in the soil due to the precipitation of $CaCO_3$. Overall, the effects of $CaCO_3$ precipitation due to the ESs consisted of improvements in heavy metal retention and UCS strength when compared to those of raw soils. Hence, it was concluded that the precipitates of $CaCO_3$ hold heavy metals and improve UCS strength irrespective of whether the quantity of precipitation varies for different ESs.
- Based on these results, it is clear that EICP has sufficient ability to immobilize heavy metals in contaminated soil. This study portrayed appreciable outcomes for Cd when compared to Ni and Pb; this technique can be employed to immobilize specific contaminants by identifying its effectiveness on other heavy metals also. However, further testing in the field is necessary before drawing this conclusion.

Supplementary Materials: The following are available online at http://www.mdpi.com/2071-1050/12/17/7019/s1, Table S1: Removal efficiencies (%) for different contaminants for EDTA and Citric Acid extractants, Table S2: Standard deviation values obtained for the removal efficiencies.

Author Contributions: The authors confirm their contributions to the present work as follows: A.A.B.M. and S.A.S.M. came up with the concept; methodology was proposed by A.A.B.M., S.A.S.M. and B.C.S.C.; validation and approval of the work was carried out by A.A.B.M., A.A., B.C.S.C. and K.L.; formal analysis was done by A.A.B.M., A.A., M.A.L. and K.L.; original draft preparation was carried out by A.A.B.M., M.A.L. and K.L.; and further review and editing was done by A.A.B.M., A.A., B.C.S.C. and S.A.S.M.; Supplementary Materials containing the removal efficiency of the heavy metals from treated and untreated soil samples are provided by A.A.B.M., M.A.L. and S.A.S.M.; All authors have read and agreed to the published version of the manuscript.

Funding: Deanship of Scientific Research at King Saud University grant number: RG-1440-073.

Acknowledgments: The authors extend their appreciation to the Deanship of Scientific Research at King Saud University for funding the study through Research Group No. RG- 1440-073.

Conflicts of Interest: The authors declare no conflict of interest.

References

1. Chalermyanont, T.; Arrykul, S.; Charoenthaisong, N. Potential use of lateritic and marine clay soils as landfill liners to retain heavy metals. *Waste Manag.* **2009**, *29*, 117–127. [CrossRef] [PubMed]
2. Mohammed, S.A.S.; Moghal, A.A.B. Soils Amended with Admixtures as Stabilizing Agent to Retain Heavy Metals. In Proceedings of the Geo-Congress, Atlanta, Georgia, 23–26 February 2014; pp. 2216–2225. [CrossRef]
3. Tsang, D.C.W.; Lo, I.M.C.; Surampalli, R.Y. *Chelating Agents for Land Decontamination Technologies*; American Society of Civil Engineers: Reston, VA, USA, 2012; ISBN 978-0-7844-1218-3.

4. Reddy, K.R.; Kumar, G.; Giri, R.K. System Effects on Bioreactor Landfill Performance Based on Coupled Hydro-Bio-Mechanical Modeling. *J. Hazard. Toxic Radioact. Waste* **2018**, *22*, 04017024. [CrossRef]

5. Zhang, Q.; Lu, H.; Liu, J.; Wang, W.; Zhang, X. Hydraulic and mechanical behavior of landfill clay liner containing SSA in contact with leachate. *Environ. Technol.* **2018**, *39*, 1307–1315. [CrossRef] [PubMed]

6. Srivastava, S.K.; Ramanathan, A.L. Geochemical assessment of groundwater quality in vicinity of Bhalswa landfill, Delhi, India, using graphical and multivariate statistical methods. *Environ. Geol.* **2008**, *53*, 1509–1528. [CrossRef]

7. Wuana, R.A.; Okieimen, F.E. Heavy Metals in Contaminated Soils: A Review of Sources, Chemistry, Risks and Best Available Strategies for Remediation. *ISRN Ecol.* **2011**, *2011*. [CrossRef]

8. de Galvão, T.C.B.; Kaya, A.; Ören, A.H.; Yükselen, Y. Geomechanics of Landfills—Innovative Technology for Liners. *Soil Sediment Contam. Int. J.* **2008**, *17*, 411–424. [CrossRef]

9. Lo, I.M.-C. Innovative Waste Containment Barriers for Subsurface Pollution Control. *Pract. Period. Hazard. Toxic Radioact. Waste Manag.* **2003**, *7*, 37–45. [CrossRef]

10. Tuncan, A.; Tuncan, M.; Koyuncu, H.; Guney, Y. Use of natural zeolites as a landfill liner. *Waste Manag. Res. J. Int. Solid Wastes Public Clean. Assoc. ISWA* **2003**, *21*, 54–61. [CrossRef]

11. Naeini, S.A.; Gholampoor, N.; Jahanfar, M.A. Effect of leachate's components on undrained shear strength of clay-bentonite liners. *Eur. J. Environ. Civ. Eng.* **2019**, *23*, 395–408. [CrossRef]

12. Sariosseiri, F.; Muhunthan, B. Effect of cement treatment on geotechnical properties of some Washington State soils. *Eng. Geol.* **2009**, *104*, 119–125. [CrossRef]

13. Celaya, M.; Veisi, M.; Nazarian, S.; Puppala, A. Accelerated Design Process of Lime-Stabilized Clays. In Proceedings of the Geo-Frontiers, Dallas, TX, USA, 13–16 March 2011; pp. 4468–4478. [CrossRef]

14. Horpibulsuk, S.; Rachan, R.; Suddeepong, A. Assessment of strength development in blended cement admixed Bangkok clay. *Constr. Build. Mater.* **2011**, *25*, 1521–1531. [CrossRef]

15. Dash, S.K.; Hussain, M. Lime Stabilization of Soils: Reappraisal. *J. Mater. Civ. Eng.* **2012**, *24*, 707–714. [CrossRef]

16. Fasihnikoutalab, M.H.; Asadi, A.; Kim Huat, B.; Westgate, P.; Ball, R.J.; Pourakbar, S. Laboratory-scale model of carbon dioxide deposition for soil stabilisation. *J. Rock Mech. Geotech. Eng.* **2016**, *8*, 178–186. [CrossRef]

17. Sabine, C.L.; Feely, R.A.; Gruber, N.; Key, R.M.; Lee, K.; Bullister, J.L.; Wanninkhof, R.; Wong, C.S.; Wallace, D.W.R.; Tilbrook, B.; et al. The Oceanic Sink for Anthropogenic CO_2. *Science* **2004**, *305*, 367–371. [CrossRef]

18. Feely, R.A.; Sabine, C.L.; Lee, K.; Berelson, W.; Kleypas, J.; Fabry, V.J.; Millero, F.J. Impact of Anthropogenic CO_2 on the $CaCO_3$ System in the Oceans. *Science* **2004**, *305*, 362–366. [CrossRef]

19. Yin, C.Y.; Wan Ali, W.S.; Lim, Y.P. Oil palm ash as partial replacement of cement for solidification/stabilization of nickel hydroxide sludge. *J. Hazard. Mater.* **2008**, *150*, 413–418. [CrossRef]

20. Horpibulsuk, S.; Rachan, R.; Raksachon, Y. Role of Fly Ash on Strength and Microstructure Development in Blended Cement Stabilized Silty Clay. *Soils Found.* **2009**, *49*, 85–98. [CrossRef]

21. Anupam, A.K.; Kumar, P.; Ransingchung, R.N.G.D. Performance evaluation of structural properties for soil stabilised using rice husk ash. *Road Mater. Pavement Des.* **2014**, *15*, 539–553. [CrossRef]

22. Ali, F.H.; Adnan, A.; Choy, C.K. Geotechnical properties of a chemically stabilized soil from Malaysia with rice husk ash as an additive. *Geotech. Geol. Eng.* **1992**, *10*, 117–134. [CrossRef]

23. Kampala, A.; Horpibulsuk, S. Engineering Properties of Silty Clay Stabilized with Calcium Carbide Residue. *J. Mater. Civ. Eng.* **2013**, *25*, 632–644. [CrossRef]

24. Pourakbar, S.; Asadi, A.; Huat, B.B.K.; Fasihnikoutalab, M.H. Soil stabilisation with alkali-activated agro-waste. *Environ. Geotech.* **2015**, *2*, 359–370. [CrossRef]

25. Almajed, A.; Tirkolaei, H.K.; Kavazanjian, E.; Hamdan, N. Enzyme Induced Biocementated Sand with High Strength at Low Carbonate Content. *Sci. Rep.* **2019**, *9*, 1135. [CrossRef] [PubMed]

26. Kavazanjian, E.; Almajed, A.; Hamdan, N. Bio-Inspired Soil Improvement Using EICP Soil Columns and Soil Nails. In Proceedings of the Grouting, Honolulu, HI, USA, 9–12 July 2017; pp. 13–22. [CrossRef]

27. Arab, M.; Omar, M.; Aljassmi, R.; Nasef, R.; Nassar, L.; Miro, S. EICP Cemented Sand Modified with Biopolymer. In *International Congress and Exhibition"Sustainable Civil Infrastructures"*; Springer: Cham, Switzerland, 2019; pp. 74–85.

28. Almajed, A.; Abbas, H.; Arab, M.; Alsabhan, A.; Hamid, W.; Al-Salloum, Y. Enzyme-Induced Carbonate Precipitation (EICP)-Based methods for ecofriendly stabilization of different types of natural sands. *J. Clean. Prod.* **2020**, *274*, 122627. [CrossRef]

29. Zhao, Z.; Hamdan, N.; Shen, L.; Nan, H.; Almajed, A.; Kavazanjian, E.; He, X. Biomimetic Hydrogel Composites for Soil Stabilization and Contaminant Mitigation. *Environ. Sci. Technol.* **2016**, *50*, 12401–12410. [CrossRef]

30. Singh, R.K.; Datta, M.; Nema, A.K.; Pérez, I.V. Evaluating Groundwater Contamination Hazard Rating of Municipal Solid Waste Landfills in India and Europe Using a New System. *J. Hazard. Toxic Radioact. Waste* **2013**, *17*, 62–73. [CrossRef]

31. Kumar, M.; Gogoi, A.; Kumari, D.; Borah, R.; Das, P.; Mazumder, P.; Tyagi, V.K. Review of Perspective, Problems, Challenges, and Future Scenario of Metal Contamination in the Urban Environment. *J. Hazard. Toxic Radioact. Waste* **2017**, *21*, 04017007. [CrossRef]

32. Mohammed, S.A.S.; Sanaulla, P.F.; Moghal, A.A.B. Sustainable Use of Locally Available Red Earth and Black Cotton Soils in Retaining Cd2+ and Ni2+ from Aqueous Solutions. *Int. J. Civ. Eng.* **2016**, *14*, 491–505. [CrossRef]

33. Steinnes, E.; Lierhagen, S. Geographical distribution of trace elements in natural surface soils: Atmospheric influence from natural and anthropogenic sources. *Appl. Geochem.* **2018**, *88*, 2–9. [CrossRef]

34. Srinivasa Gowd, S.; Ramakrishna Reddy, M.; Govil, P.K. Assessment of heavy metal contamination in soils at Jajmau (Kanpur) and Unnao industrial areas of the Ganga Plain, Uttar Pradesh, India. *J. Hazard. Mater.* **2010**, *174*, 113–121. [CrossRef]

35. Violante, A.; Pigna, M. Sorption-Desorption Processes of Metals and Metalloids in Soil Environments. *Rev. Cienc. Suelo Nutr. Veg.* **2008**, *8*, 95–101. [CrossRef]

36. dos Santos, N.M.; do Nascimento, C.W.A.; de Aguiar Accioly, A.M. Guideline Values and Metal Contamination in Soils of an Environmentally Impacted Bay. *Water Air Soil Pollut.* **2017**, *228*, 88. [CrossRef]

37. DeJong, J.T.; Kavazanjian, E. Bio-mediated and Bio-inspired Geotechnics. In *Geotechnical Fundamentals for Addressing New World Challenges*; Lu, N., Mitchell, J.K., Eds.; Springer Series in Geomechanics and Geoengineering; Springer International Publishing: Cham, Switzerland, 2019; pp. 193–207, ISBN 978-3-030-06249-1.

38. Salt, D.E.; Blaylock, M.; Kumar, N.P.B.A.; Dushenkov, V.; Ensley, B.D.; Chet, I.; Raskin, I. Phytoremediation: A Novel Strategy for the Removal of Toxic Metals from the Environment Using Plants. *Bio/Technology* **1995**, *13*, 468–474. [CrossRef] [PubMed]

39. Ciumasu, I.M.; Costica, M.; Costica, N.; Neamtu, M.; Dirtu, A.C.; de Alencastro, L.F.; Buzdugan, L.; Andriesa, R.; Iconomu, L.; Stratu, A.; et al. Complex Risks from Old Urban Waste Landfills: Sustainability Perspective from Iasi, Romania. *J. Hazard. Toxic Radioact. Waste* **2012**, *16*, 158–168. [CrossRef]

40. Rojas, L.A.; Yáñez, C.; González, M.; Lobos, S.; Smalla, K.; Seeger, M. Characterization of the metabolically modified heavy metal-resistant Cupriavidus metallidurans strain MSR33 generated for mercury bioremediation. *PLoS ONE* **2011**, *6*, e17555. [CrossRef]

41. Mulligan, C.N.; Yong, R.N.; Gibbs, B.F. Remediation technologies for metal-contaminated soils and groundwater: An evaluation. *Eng. Geol.* **2001**, *60*, 193–207. [CrossRef]

42. Moghal, A.A.B.; Mohammed, S.A.S.; Almajed, A.; Al-Shamrani, M.A. Desorption of Heavy Metals from Lime-Stabilized Arid-Soils using Different Extractants. *Int. J. Civ. Eng.* **2020**, *18*, 449–461. [CrossRef]

43. Dermont, G.; Bergeron, M.; Mercier, G.; Richer-Laflèche, M. Metal-Contaminated Soils: Remediation Practices and Treatment Technologies. *Pract. Period. Hazard. Toxic Radioact. Waste Manag.* **2008**, *12*, 188–209. [CrossRef]

44. Sahu, Y.K.; Deb, M.K.; Patel, K.S.; Martín-Ramos, P.; Towett, E.K.; Tarkowska-Kukuryk, M. Bioaccumulation of Nutrients and Toxic Elements with Macrophytes. *J. Hazard. Toxic Radioact. Waste* **2020**, *24*, 05019007. [CrossRef]

45. Muhammad, D.; Chen, F.; Zhao, J.; Zhang, G.; Wu, F. Comparison of EDTA- and citric acid-enhanced phytoextraction of heavy metals in artificially metal contaminated soil by Typha angustifolia. *Int. J. Phytoremediat.* **2009**, *11*, 558–574. [CrossRef]

46. Mishra, S.P. Adsorption–desorption of heavy metal ions. *Curr. Sci.* **2014**, *107*, 12.

47. Goldberg, S.; Criscenti, L.J.; Turner, D.R.; Davis, J.; Cantrell, K.J. Adsorption-Desorption Processes in Subsurface Reactive Transport Modeling. *Vadose Zone J.* **2007**, *6*, 407–435. [CrossRef]

48. Blais, J.F.; Djedidi, Z.; Cheikh, R.B.; Tyagi, R.D.; Mercier, G. Metals Precipitation from Effluents: Review. *Pract. Period. Hazard. Toxic Radioact. Waste Manag.* **2008**, *12*, 135–149. [CrossRef]

49. Moghal, A.A.B.; Reddy, K.R.; Mohammed, S.A.S.; Al-Shamrani, M.A.; Zahid, W.M. Lime-Amended Semi-arid Soils in Retaining Copper, Lead, and Zinc from Aqueous Solutions. *Water Air Soil Pollut.* **2016**, *227*, 372. [CrossRef]

50. Mohammed, S.A.S.; Moghal, A.A.B. Efficacy of nano calcium silicate (NCS) treatment on tropical soils in encapsulating heavy metal ions: Leaching studies validation. *Innov. Infrastruct. Solut.* **2016**, *1*, 21. [CrossRef]

51. Moghal, A.A.B.; Reddy, K.R.; Mohammed, S.A.S.; Al-Shamrani, M.A.; Zahid, W.M. Sorptive Response of Chromium (Cr^{+6}) and Mercury (Hg^{+2}) From Aqueous Solutions Using Chemically Modified Soils. *J. Test. Eval.* **2017**, *45*, 105–119. [CrossRef]

52. Madhu, A.; Chakraborty, J.N. Developments in application of enzymes for textile processing. *J. Clean. Prod.* **2017**, *145*, 114–133. [CrossRef]

53. Nathan, V.K.; Rani, M.E.; Gunaseeli, R.; Kannan, N.D. Enhanced biobleaching efficacy and heavy metal remediation through enzyme mediated lab-scale paper pulp deinking process. *J. Clean. Prod.* **2018**, *203*, 926–932. [CrossRef]

54. Almajed, A.A. Enzyme Induced Carbonate Precipitation (EICP) for Soil Improvement. Ph.D. Thesis, Arizona State University, Tempe, AZ, USA, 2017.

55. Almajed, A. Enzyme induced cementation of biochar-intercalated soil: Fabrication and characterization. *Arab. J. Geosci.* **2019**, *12*, 403. [CrossRef]

56. Almajed, A.; Khodadadi, H.; Kavazanjian, E. Sisal Fiber Reinforcement of EICP-Treated Soil. In Proceedings of the IFCEE, Orlando, FL, USA, 5–10 March 2018; pp. 29–36. [CrossRef]

57. Pasillas, J.N.; Khodadadi, H.; Martin, K.; Bandini, P.; Newtson, C.M.; Kavazanjian, E. Viscosity-Enhanced EICP Treatment of Soil. In Proceedings of the IFCEE, Orlando, FL, USA, 5–10 March 2018; pp. 145–154. [CrossRef]

58. Javadi, N.; Khodadadi, H.; Hamdan, N.; Kavazanjian, E. EICP Treatment of Soil by Using Urease Enzyme Extracted from Watermelon Seeds. In Proceedings of the IFCEE, Orlando, FL, USA, 5–10 March 2018; pp. 115–124. [CrossRef]

59. Oliveira, P.J.V.; Freitas, L.D.; Carmona, J.P.S.F. Effect of Soil Type on the Enzymatic Calcium Carbonate Precipitation Process Used for Soil Improvement. *J. Mater. Civ. Eng.* **2017**, *29*, 04016263. [CrossRef]

60. Zhao, Q.; Li, L.; Li, C.; Li, M.; Amini, F.; Zhang, H. Factors Affecting Improvement of Engineering Properties of MICP-Treated Soil Catalyzed by Bacteria and Urease. *J. Mater. Civ. Eng.* **2014**, *26*, 04014094. [CrossRef]

61. Venda Oliveira, P.J.; da Costa, M.S.; Costa, J.N.P.; Nobre, M.F. Comparison of the Ability of Two Bacteria to Improve the Behavior of Sandy Soil. *J. Mater. Civ. Eng.* **2015**, *27*, 06014025. [CrossRef]

62. Hommel, J.; Akyel, A.; Frieling, Z.; Phillips, A.J.; Gerlach, R.; Cunningham, A.B.; Class, H. A Numerical Model for Enzymatically Induced Calcium Carbonate Precipitation. *Appl. Sci.* **2020**, *10*, 4538. [CrossRef]

63. Chandra, A.; Ravi, K. Application of Enzyme Induced Carbonate Precipitation (EICP) to improve the shear strength of different type of soils. In Proceedings of the Indian Geotechnical Conference, Bengaluru, India, 13–15 December 2018; p. 8.

64. Handley-Sidhu, S.; Sham, E.; Cuthbert, M.O.; Nougarol, S.; Mantle, M.; Johns, M.L.; Macaskie, L.E.; Renshaw, J.C. Kinetics of urease mediated calcite precipitation and permeability reduction of porous media evidenced by magnetic resonance imaging. *Int. J. Environ. Sci. Technol.* **2013**, *10*, 881–890. [CrossRef]

65. Dharmakeerthi, R.S.; Thenabadu, M.W. Urease activity in soils: A review. *J. Natl. Sci. Found. Sri Lanka* **1996**, *24*, 159–195. [CrossRef]

66. Blakeley, R.L.; Zerner, B. Jack bean urease: The first nickel enzyme. *J. Mol. Catal.* **1984**, *23*, 263–292. [CrossRef]

67. Dejong, J.T.; Soga, K.; Kavazanjian, E.; Burns, S.; Van Paassen, L.A.; Al Qabany, A.; Aydilek, A.; Bang, S.S.; Burbank, M.; Caslake, L.F.; et al. Biogeochemical processes and geotechnical applications: Progress, opportunities and challenges. *Géotechnique* **2013**, *63*, 287–301. [CrossRef]

68. Mohammed, S.A.S.; Moghal, A.A.B.; Sanaulla, P.F.; Kotresha, K.; Reddy, H.P. Cadmium Fixation Studies on Contaminated Soils Using Nano Calcium Silicate—Treatment Strategy. In Proceedings of the Geotechnical Frontiers, Orlando, FL, USA, 12–15 March 2017; pp. 434–442. [CrossRef]

69. ASTM D2166/D2166M. *Test Method for Unconfined Compressive Strength of Cohesive Soil*; ASTM International: West Conshohocken, PA, USA, 2016.

70. Khodadoust, A.P.; Reddy, K.R.; Maturi, K. Removal of Nickel and Phenanthrene from Kaolin Soil Using Different Extractants. *Environ. Eng. Sci.* **2004**, *21*, 691–704. [CrossRef]

71. Gu, Y.-Y.; Yeung, A.T. Desorption of cadmium from a natural Shanghai clay using citric acid industrial wastewater. *J. Hazard. Mater.* **2011**, *191*, 144–149. [CrossRef]

72. ASTM D3987. *Standard Test Method for Shake Extraction of Solid Waste with Water*; ASTM International: West Conshohocken, PA, USA, 2012.

73. Moghal, A.A.B.; Ashfaq, M.; Al-Shamrani, M.A.; Al-Mahbashi, A. Effect of Heavy Metal Contamination on the Compressibility and Strength Characteristics of Chemically Modified Semiarid Soils. *J. Hazard. Toxic Radioact. Waste* **2020**, *24*, 04020029. [CrossRef]

74. Dhami, N.K.; Reddy, M.S.; Mukherjee, A. Biomineralization of calcium carbonates and their engineered applications: A review. *Front. Microbiol.* **2013**, *4*, 314. [CrossRef]

75. Krajewska, B. Urease-aided calcium carbonate mineralization for engineering applications: A review. *J. Adv. Res.* **2018**, *13*, 59–67. [CrossRef] [PubMed]

76. Almajed, A.; Khodadadi Tirkolaei, H.; Kavazanjian, E. Baseline Investigation on Enzyme-Induced Calcium Carbonate Precipitation. *J. Geotech. Geoenviron. Eng.* **2018**, *144*, 04018081. [CrossRef]

77. Ran, D.; Kawasaki, S. Effective Use of Plant-Derived Urease in the Field of Geoenvironmental/Geotechnical Engineering. *J. Civ. Environ. Eng.* **2016**, *6*, 1–13. [CrossRef]

78. Madrid, L.; Diaz-Barrientos, E. Influence of carbonate on the reaction of heavy metals in soils. *J. Soil Sci.* **1992**, *43*, 709–721. [CrossRef]

79. Torres-Aravena, Á.E.; Duarte-Nass, C.; Azócar, L.; Mella-Herrera, R.; Rivas, M.; Jeison, D. Can Microbially Induced Calcite Precipitation (MICP) through a Ureolytic Pathway Be Successfully Applied for Removing Heavy Metals from Wastewaters? *Crystals* **2018**, *8*, 438. [CrossRef]

80. Vega, F.A.; Covelo, E.F.; Andrade, M.L. A versatile parameter for comparing the capacities of soils for sorption and retention of heavy metals dumped individually or together: Results for cadmium, copper and lead in twenty soil horizons. *J. Colloid Interface Sci.* **2008**, *327*, 275–286. [CrossRef] [PubMed]

81. Acosta, J.A.; Jansen, B.; Kalbitz, K.; Faz, A.; Martínez-Martínez, S. Salinity increases mobility of heavy metals in soils. *Chemosphere* **2011**, *85*, 1318–1324. [CrossRef]

82. González Costa, J.J.; Reigosa, M.J.; Matías, J.M.; Covelo, E.F. Soil Cd, Cr, Cu, Ni, Pb and Zn sorption and retention models using SVM: Variable selection and competitive model. *Sci. Total Environ.* **2017**, *593–594*, 508–522. [CrossRef]

83. Wang, Y.M.; Chen, T.C.; Yeh, K.J.; Shue, M.F. Stabilization of an elevated heavy metal contaminated site. *J. Hazard. Mater.* **2001**, *88*, 63–74. [CrossRef]

84. Zhou, D.M.; Wang, S.Q.; Chen, H.M. Interaction of Cd and citric acid, EDTA in red soil. *J. Environ. Sci.* **2001**, *13*, 153–156.

85. Manjeet, B.; Diwan, S.; Garg, V.K.; Pawan, R. Use of Agricultural Waste for the Removal of Nickel Ions from Aqueous Solutions: Equilibrium and Kinetics Studies. *Int. J. Environ. Sci. Eng.* **2009**, *1*, 108–114. [CrossRef]

86. Attar, K.; Demey, H.; Bouazza, D.; Sastre, A.M. Sorption and Desorption Studies of Pb(II) and Ni(II) from Aqueous Solutions by a New Composite Based on Alginate and Magadiite Materials. *Polymers* **2019**, *11*, 340. [CrossRef] [PubMed]

87. Karna, R.R.; Luxton, T.; Bronstein, K.E.; Redmon, J.H.; Scheckel, K.G. State of the science review: Potential for beneficial use of waste by-products for in situ remediation of metal-contaminated soil and sediment. *Crit. Rev. Environ. Sci. Technol.* **2017**, *47*, 65–129. [CrossRef] [PubMed]

88. Li, C.; Qian, Z.; Zhou, C.; Su, W.; Hong, P.; Liu, S.; He, L.; Chen, Z.; Ji, H. Mussel-inspired synthesis of polydopamine-functionalized calcium carbonate as reusable adsorbents for heavy metal ions. *RSC Adv.* **2014**, *4*, 47848–47852. [CrossRef]

89. Peters, R.W. Chelant extraction of heavy metals from contaminated soils. *J. Hazard. Mater.* **1999**, *66*, 151–210. [CrossRef]

90. Cline, S.R.; Reed, B.E. Lead Removal from Soils via Bench-Scale Soil Washing Techniques. *J. Environ. Eng.* **1995**, *121*, 700–705. [CrossRef]

91. Kumpiene, J.; Lagerkvist, A.; Maurice, C. Stabilization of As, Cr, Cu, Pb and Zn in soil using amendments— A review. *Waste Manag.* **2008**, *28*, 215–225. [CrossRef]

92. Tian, K.; Benson, C.H.; Tinjum, J.M. Chemical Characteristics of Leachate in Low-Level Radioactive Waste Disposal Facilities. *J. Hazard. Toxic Radioact. Waste* **2017**, *21*, 04017010. [CrossRef]

93. Harter, R.D. Effect of Soil pH on Adsorption of Lead, Copper, Zinc, and Nickel. *Soil Sci. Soc. Am. J.* **1983**, *47*, 47–51. [CrossRef]

94. Gupta, T.B.; Lataye, D.H. Adsorption of Indigo Carmine Dye onto *Acacia Nilotica* (Babool) Sawdust Activated Carbon. *J. Hazard. Toxic Radioact. Waste* **2017**, *21*, 04017013. [CrossRef]
95. Maturi, K.; Reddy, K.R. Extractants for the Removal of Mixed Contaminants from Soils. *Soil Sediment Contam. Int. J.* **2008**, *17*, 586–608. [CrossRef]

Article

Bioengineering Techniques Adopted for Controlling Riverbanks' Superficial Erosion of the Simplício Hydroelectric Power Plant, Brazil

Vinicius F. Vianna [1], Mateus P. Fleury [2,*], Gustavo B. Menezes [3], Arnaldo T. Coelho [4], Cecília Bueno [5], Jefferson Lins da Silva [2] and Marta P. Luz [1,6]

[1] Eletrobras Furnas, Rio de Janeiro 22281-900, Brazil; viniciusvian@gmail.com (V.F.V.); martaluz@furnas.com.br or marta.eng@pucgoias.edu.br (M.P.L.)

[2] São Carlos School of Engineering (EESC), University of São Paulo (USP), São Carlos 13566-590, Brazil; jefferson@sc.usp.br

[3] Department of Civil Engineering, California State University, Los Angeles, CA 90032, USA; gmeneze@calstatela.edu

[4] Ingá Engenharia e Consultoria, Belo Horizonte 30320-130, Brazil; arnaldo@ingaengenharia.com.br

[5] Biological Science and Professional Master in Environmental Sciences, Universidade Veiga de Almeida (UVA), Rio de Janeiro 20271-020, Brazil; cecilia.bueno@uva.br

[6] Industrial and Systems Engineering Postgraduate Program (MEPROS), Pontifical Catholic University of Goiás, Goiânia 74605-010, Brazil

* Correspondence: mateusfleury@usp.br; Tel.: +55-16-3373-8220

Received: 13 July 2020; Accepted: 17 September 2020; Published: 24 September 2020

Abstract: Controlling and preventing soil erosion on slope surfaces is a pressing concern worldwide, and at the same time, there is a growing need to incorporate sustainability into our engineering works. This study evaluates the efficiency of bioengineering techniques in the development of vegetation in soil slopes located near a hydroelectric power plant in Brazil. For this purpose, twelve different bioengineering techniques were evaluated, in isolation and in combination, in the slopes (10 m high) of two experimental units (approximately 70 m long each) located next to the Paraíba do Sul riverbanks, in Brazil. High-resolution images of the slopes' frontal view were taken in 15-day interval visits in all units for the first 90 days after implantation, followed by monthly visits up to 27 months after the works were finished. The images were treated and analyzed in a computer algorithm that, based on three-color bands (red–green–blue scale), helps to assess the temporal evolution of the vegetative cover index for each technique adopted. The results showed that most of the solutions showed a deficiency in vegetation establishment and were sensitive to climatological conditions, which induced changes in the vegetation phytosanitary aspects. Techniques which provided a satisfactory vegetative cover index throughout the investigated period are pointed out.

Keywords: bioengineering techniques; vegetative cover index; slope's superficial erosion; phytosanitary aspects; climatological conditions

---

1. Introduction

Soil erosion in slopes is a natural process that involves several processes such as landslides and detachment, dissolution and/or wear of soil particles, followed by their transport and deposition caused by the action of an erosive agent (e.g., water, wind, and/or gravity [1]). Human activities aggravate this process by removing existing vegetation, influencing agriculture and overstocked pasturing, and causing change in natural slopes by cutting operations routinely used in transportation works. This accelerated erosion shows relevant environmental, social, and economic impacts around the world, reducing soil fertility and promoting soil sedimentation in river flows [2]. Therefore, erosion control is a present concern and different techniques have been proposed over the last decades to address this issue, from hard engineering to more sustainable solutions, such as bioengineering [3,4].

Bioengineering is an ancient technique that has recently regained interest and popularity for use in erosion control [5]. It combines live vegetation with or without inert components (e.g., rocks, wood, metal, geosynthetics, etc.) to reinforce soil, prevent and stabilize the erosion process, decrease surface runoff, and increase water infiltration, promoting ecosystem restoration [6–8]. The key aspect of this technique, according to Bischetti et al. [5], is that it is intentionally used considering an environmental and landscape perspective.

In a context with an increasing need to account for sustainability in engineering projects, bioengineering systems have clear advantages, because they show lower environmental impact compared to conventional stabilization methods that rely on hard structures such as retaining walls [2,9]. In this context, in Europe and North America, soil bioengineering has been widely used [10].

Research on soil bioengineering has experienced a significant increase over the last decades, leading to major advances over the past 20 years according to Stokes et al. [11], who highlight the relevance and interest of the scientific community [9]. Focus was given to the history and types of bioengineering techniques and applications [5,10,12–14], their technical, economic, environmental, and ecological benefits [15–19], the characterization of relevant physical and mechanical attributes of plant species application in such systems [20–22], and some successful case studies in transportation, urban settings, or riverbank restoration in the European Alps [23,24], North America [6,9,25], Central America [26], and South Asia [15,27,28]. The life cycle assessment (LCA) model for soil bioengineering constructions was also proposed [29].

In Brazil, the favorable climate to plant growth (especially the tropical one—characterized by a dry winter and rainy summer) supports the use of bioengineering techniques. However, few studies investigated the use of bioengineering systems in the Brazilian context, especially in large field test studies such as the one presented in this paper. Nonetheless, soil degradation is a major problem throughout the entire country [30]. Sattler et al. [31] highlight that the use of soil bioengineering in southeast Brazil (the economic pole of the country) for restoration and rehabilitation of degraded areas is still incipient, and when applied, relies mostly on the use of non-native plant species that have been shown to work in other tropical regions. After a 15-month monitoring period, the authors [31] reported successful riverbank slope protection when using hedgerow terraces (which work as living fences) with rooted or unrooted live cuttings of five native plant species (shrub or tree) associated with a biodegradable polymer.

Outside the Brazilian context, several studies reported the use of soil bioengineering techniques. Ansted et al. [29] monitored two willow spilling projects (used to prevent riverbank erosion) during the first year after installation to assess their biological and geomorphological function. Rey and Labonne [32] observed a 45% survival rate of willow (Salix) cuttings (after four growing seasons) on bioengineering structures installed in the Francon Catchment (Southern French Alps). Brush layer inclusions were used to stabilize steep slopes along a roadway in Massachusetts [6]. Petrone and Preti [16] assessed which native species were most suited (survival rate higher than 60%) for soil bio-engineering purposes in Nicaragua. Dhital and Tang [15] reported the effectiveness of vegetative check dams and wire net check dams along with vegetation in the stabilization of riverbanks located in Nepal. Furthermore, a laboratory study was carried out by Muhammed et al. [33].

Monitoring restored sites is as important as the planning stage of the restoration program. It can significantly influence the success of the solution used and provides invaluable data for future bioengineering projects, such as the lifetime and efficacy of the bioengineering system on slope stability and erosion control in different regions [2,34]. Nonetheless, studies on the ecological efficiency of slope restoration techniques are still scarce [15].

This study aims to assess the performance of 12 soil bioengineering techniques (systems) in erosion prone slopes of the Simplício Hydroelectric Power Plant-FURNAS in Brazil. We evaluated their viability and performance over 27 months after implantation. The latter was evaluated by periodic qualitative observations (integrity, anchoring, soil stability, germination, phytosanitary aspect, pests' occurrence, nutritional status, and other general aspects) in site visits, and, most importantly by a quantitative evaluation of the evolution and success of vegetative cover index throughout three years to access the long term erosion protection via permanent vegetation establishment. In this way, we intend to contribute to the knowledge building on using soil bioengineering techniques in Brazilian territory, possibly helping to broaden its use in the country.

2. Methodology

This study reports the application and monitoring of different bioengineering techniques for superficial slope erosion control and analyses the evolution of the vegetation developed in the slope surfaces evaluated. It is worth mentioning that the experimental units have been subjected only to the erosion process caused by the rainfall, which means the river was not in contact with any section of both experimental units (even during the flood seasons). The bioengineering techniques adopted, their application and monitoring, and procedures for determining the vegetative cover index are detailed as follows.

2.1. Experimental Units

Two slopes located on the Paraiba do Sul riverbanks, close to the Simplício Hydroeletric Power Plant (Chiador-MG in Brazil; Figure 1) were selected as experimental units in this study. In terms of the geology context, the experimental units are located on a Proerozoic gneiss-magmatic terrain (part of Juiz de Fora and Paraíba do Sul geological complexes), grouped in two lithological types: high-grade paragneisses and orthogneisses (dominant in the area of experiment).

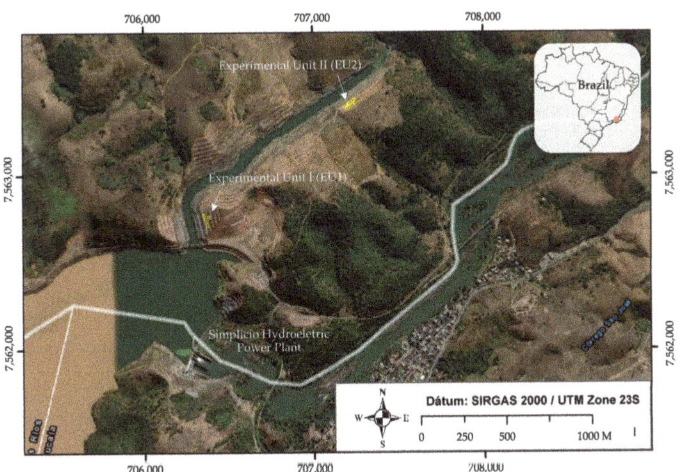

Figure 1. Location of the experimental units I (EU1; 22°1′40.06″ S; 43°0′1.49″ W) and II (EU2; 22°1′16.49″ S; 42°59′30.36″ W).

Experimental unit I (EU1) consists of two 90-m-long slopes (west facing slope) of approximately 9.5 m height (H) and a slope angle (α) close to 32°, as indicated in Figure 2a. Similarly, the experimental unit II (EU2) consists of two 70-m-long slopes (northwest facing slope): the bottom one with 8.5 m height (H) and α equal to 35°, and the upper one with H = 9.5 m and α = 32° (Figure 2b). To avoid people traffic and animal trampling, the units were demarcated with fences (appropriate materials and equipment were used). The soil samples were collected 300 mm deep from the slopes' surfaces (15 m distant apart from each other) along both units (EU1 and EU2) to provide a representative characterization of its fertility. A total of 10 soil samples was tested for the unit EU1, whereas for the

EU2, 6 soil samples were collected and tested. Following the test procedures recommended by the Brazilian Agricultural Research Corporation (EMBRAPA [35]), the soil fertility test results (Table 1; classification according to CFSEMG (1999)) indicate that the experimental units show similar fertility, with a low content for all parameters evaluated and medium acidity.

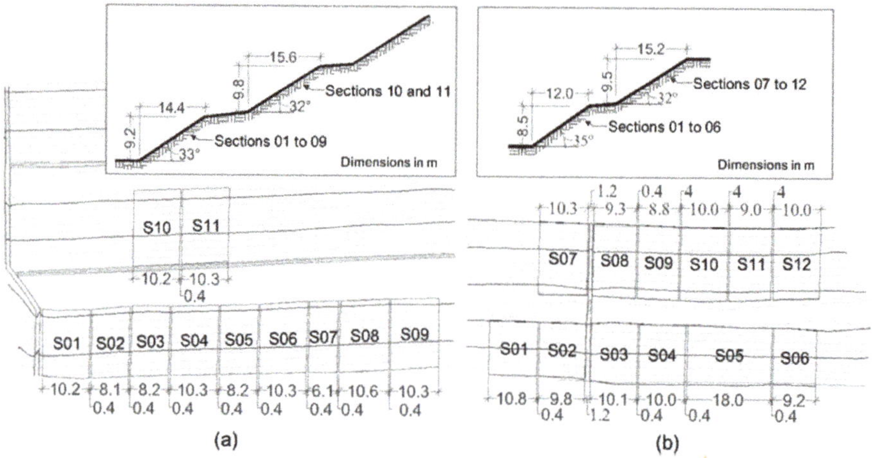

Figure 2. Topographic and division of the experimental units (**a**) I (EU1) and (**b**) II (EU2) into sections. [Note: Representative cross-sections are shown in the figure's upper region].

Additionally, Table 2 summarizes characterization and soil strength test results for the experimental units, performed based on the Brazilian Standard (NBR) proposed by the Brazilian Association of Technical Standards (ABNT). The laboratory tests consisted of specific gravity of soil solids (NBR 6458 [36]), grain size distribution (NBR 7181 [37]; classification according to American Society for Testing and Materials (ASTM; D 2487-06 [38]), Atterberg limits (NBR 6459 [39]; NBR 7180 [40]), standard Proctor compaction (NBR 7182 [41]), and a direct shear test (unsaturated soil samples compacted at similar density to the soil from the slopes; ASTM D 3080-98 [42]).

Prior to following the bioengineering techniques, the superficial soil of the slopes of the experimental units was prepared to provide an adequate slope (inclination) and condition for the execution process. The preparation included cleaning, removal of bulges and poorly consolidated soil masses, and manual conformation. Additionally, linear erosions (gullies and deep grooves) close to the experimental units that could impair this study were recovered using organic sediment.

Table 1. Soil fertility test results for the experimental units' slope surface (mean values from samples taken 30 cm deep from soil surface at 15-m linear intervals).

Parameter	Unit	EU1 [a]		EU2 [b]	
		Value	Classification [c]	Value	Classification [c]
pH	-	5.8 (0.2)	Medium acidity	5.4 (0.2)	Medium acidity
Hydrogen (H) [d]	meq/100 cm^3	1.7 (0.5)	-	1.7 (0.3)	-
Aluminum (Al) [d]	meq/100 cm^3	0.2 (0.2)	Low	0.6 (0.4)	Medium
Calcium (Ca) [d]	meq/100 cm^3	1.1 (0.8)	Low	1.0 (0.5)	Low
Magnesium (Mg) [d]	meq/100 cm^3	0.4 (0.5)	Low	0.2 (0.2)	Low
Phosphorus (P) [e]	ppm	2.3 (2.5)	Low	2.2 (0.8)	Low
Potassium (k) [e]	ppm	37.7 (55.3)	Low	19.7 (17.3)	Low
Organic matter (O.M.)	%	0.10 (0.14)	Low	0.07 (0.07)	Low

Note: standard deviation values are shown in parenthesis; [a] ten soil samples tested; [b] six soil samples tested; [c] classification according to CFSEMG [43]; [d] exchangeable content; and [e] available content.

Table 2. Geotechnical characteristics of the experimental units.

Characteristic	Value	Standard
Specific gravity of soil solids (ρ_s)	2.657 g/cm^3	NBR 6458 [36]
Soil classification	Sandy silt (ML)	NBR 7181 [37] ASTM D 2487-06 [38]
Liquid limit (LL)	44%	NBR 6459 [39]
Plastic limit (PL)	27%	NBR 7180 [40]
Plasticity index (PI)	17%	-
Maximum dry unit weight (γ_{d_m})	16.57 kN/m^3	NBR 7182 [41]
Optimum water content (w_{op})	17.8%	NBR 7182 [41]
Friction angle (φ)	33.9°	ASTM D 3080-98 [a] [42]
Cohesion (c)	16.54 kPa	ASTM D 3080-98 [a] [42]

Note: [a] Direct shear tests performed in unsaturated soil samples compacted at similar density to the soil from the slopes (normal stresses of 12.5, 25, 50, and 100 kPa).

Retainers (OSRs; 0.20-m or 0.40-m width). OSRs consists of polypropylene screen coating filled with herbaceous straw or coconut fiber, prepared to become a suitable substrate for developing vegetation.

2.2. Bioengineering Techniques

This study assessed 12 different techniques: 6 isolated bioengineering techniques (IBT) and 6 mixed bioengineering techniques (MBT, more than one technique), both summarized in Table 3. The techniques were chosen based on bibliographic research and consulting to specialized bioengineering companies (information of most common bioengineering techniques adopted in Brazil). The experimental units were divided into sections (EU1: S01 to S11; EU2: S01 to S14; dimensions indicated in Figure 2), and the bioengineering techniques were executed contiguously (randomly) along the experimental units (Figure 3) according to Table 3. A brief description of each technique is shown in the following table. Manual seeding was adopted as the control technique to evaluate the efficiency of the other techniques adopted herein. Manual seeding consists of the manual launch of seeds (species and quantity described in Table 4), and it was adopted in isolation conditions in EU1-S08 and EU2-S07 and in combination with organic sediment retainers (OSRs) in EU1-S06, EU1-S11, and EU2-S12. Manual seeding attached with organic material (cellulose mulch) was installed in EU1-S09 and in EU1-S10 (combined with OSRs) as an alternative to help germination and vegetation. The specimens presented in Table 4 were chosen based on bibliographic research, an indication of specialized Brazilian companies (including nurseries) and professors from Spellman College, California State University, Universidade Federal de Viçosa, and Universidade Federal de Minas Gerais.

Table 3. Bioengineering techniques evaluated and respective installation site—Experimental Unit (EU) and Section (S).

Code	Bioengineering Technique	Section Installed	
		EU1	EU2
Isolated bioengineering techniques (IBT)			
IBT-1	Manual seeding	S08	S07
IBT-2	Live stakes	-	S01
IBT-3	Live Organic Sediment Retainer (L-OSR)	-	S06
IBT-4	Live Rolled Erosion Control Products (L-RECPs)	S01	S02; S03
IBT-5	Rolled Erosion Control Products (RECP-01)	S02	
	Rolled Erosion Control Products (RECP-02)		S11
	Rolled Erosion Control Products (RECP-03)		S09
IBT-6	Geocellular containment system (GCSs)	S07	S05

Table 3. *Cont.*

Code	Bioengineering Technique	Section Installed	
		EU1	**EU2**
Mixed bioengineering techniques (MBT)			
MBT-1	Manual seeding + Cellulose mulch	S09	-
MBT-2	Manual seeding + Organic Sediment Retainer (OSR)	S06; S11	S12
MBT-3	Manual seeding + Cellulose mulch + Organic Sediment Retainer (OSR)	S10	-
MBT-4	Live stakes + Live Organic Sediment Retainer (L-OSRs)	-	S04
MBT-5	Live stakes + Live Organic Sediment Retainer (L-OSRs) + Manual seeding	S03; S04	-
MBT-6	Rolled Erosion Control Products (RECP-02) + Organic Sediment Retainer (OSR)	S05	S08;
	Rolled Erosion Control Products (RECP-03) + Organic Sediment Retainer (OSR)	-	S10

Figure 3. Contiguous execution of the bioengineering techniques with the experimental unit.

Table 4. Vegetative species (scientific and common name) adopted in bioengineering techniques.

Scientific Name	Common Name [a]	Quantity (g/m^2)
Alternanthera ficoidea [b]	White carpet	2.0
Avena strigose [c]	Black oats	4.0
Brachiaria humidicola [b]	Koronivia grass	2.5
Brachiaria decumbens [b]	Palisade grass	2.5
Hyparrhenia rufa [b]	Jaragua grass	2.5
Lablab purpureus [b]	Lab lab bean	3.0
Calopogonium mucunoides [d]	Calopo	1.5
Melinis minutiflora [b]	Molasses grass	1.5
Mucuna aterrima [b]	Florida beans	3.0
Cajanus cajan [b]	Pigean bean	3.0

Notes: [a] The common names may depend on the region/country; [b] introduced species; [c] non-native species; and [d] native species.

Live and rootable vegetative cuttings, herein called "live stakes", were installed (tamped) into the ground of EU1-S01 in isolation conditions and in EU1-S03, EU1-S04, and EU2-S04 combined with organic sediment retainer (OSR). The live stakes were cultivated in greenhouses and treated with fungicidal solution (Benomyl, 0.03%), nutritive fertilizer (Nitrogen (N), Phosphorus (P) and Potassium (K) (NPK) 20:20:20; 10 g/L), and rooting inductors (indole-butyric acid; 100 ppm). Live stakes were comprised by non-native (*Eritrina mulungu, Croton urucurana, Hibiscus tiliaceus, Morus alba, Psidium guajava, Mimosa* sp.*, Ficus gameleira, Joanesia princeps, Psidium cattleianum* and *Chorisia speciosa*) and native (*Hymenea corbaril*, *Caesalpinia leiostachya* and *Inga* sp.) species. Images of the root system and installation procedures are available in Appendix A (Figure A1).

This study also investigated the adoption of flexible organic or synthetic rolls manufactured in greenhouses (live rolled erosion control products; L-RECPs). Structural geogrids (three-dimensional UV-stabilized polypropylene mesh) were layered in a levelled watertight surface followed by a layer of organic RECP. Over the RECPs surface, a mix of fast-growing herbaceous and legume seeds (indicated in Table 4, 10 g/m^2), chemical fertilizer (5 g/m^2), organic compounds (500 g/m^2), and hydrogel (5 g/m^2) was spread. Finally, irrigation (2–5 L/m^2) was conducted at regular intervals. Images of L-RECPs storing and installation procedures are available in Appendix A (Figure A2). These elements were installed in isolation at EU1-S01, EU2-S02, and EU2-S03.

Live elements associated with organic sediment retainers (L-OSRs) were cultivated in greenhouses, with dimensions of 0.20 × 2.00 m and 0.40 × 1.60 m (diameter × length). They were installed in isolation (EU2-S06) and associated with live stakes (EU1-S03, EU1-S04, and EU2-S04). The seedlings were planted directly inside the OSR after making beds filled with 100 g of vermiculite associated with 2 g of hydrogel and 20 g of NPK 20:20:20 fertilizer powder. Images from the cultivation up to its installation are available in Appendix A (Figure A3).

This study also investigated the adoption of erosion control products commercially available in the Brazilian market. Three rolled erosion control products (RECPs) were adopted. The first one (RECP-01) is a coconut fiber blanket (mass per unit area 450 g/m^2) composed of a high flexible, UV-stabilized, and non-degradable three-dimensional polypropylene woven matrix, coupled with high resistance metallic hexagonal reinforcement mesh (installed in isolation in EU1-S02). The second RECP (RECP-02) is a two-dimensional erosion control blanket consisting of coconut fibers (mass per unit area 400 g/m^2) interlaced and incorporated in photodegradable polypropylene nets (installed in EU1-S05 and EU2-S08 in combination with OSR and in isolation in EU2-S11). Finally, the third RECP (RECP-03) is a straw erosion control blanket consisting of vegetable fibers (agricultural straw; mass per unit area 600 g/m^2) interlaced and incorporated in photodegradable polypropylene nets and a third UV-resistant net in the upper face (installed in isolation in EU2-S09 and combined with OSRs in EU2-S10).

An organic sediment retainer (OSR) manufactured using dehydrated vegetable fibers was combined with manual seeding in EU1-S06, EU1-S11, and EU2-S12 and further combined with cellulose mulch in EU1-S10. The experimental sections EU1-S05, EU1-S08, and EU2-S10 received a combination of OSR and rolled erosion control products (RECP). Additionally, a geocellular containment system (GCS) was adopted in sections EU1-S07 and EU2-S05. The GCSs are geobags (manufactured with a geosynthetic mesh of high-density polyethylene (HDPE) combined with an organic geotextile of coconut fiber) filled with herbaceous straw, organic matter, seeds, and fertilizers.

2.3. Monitoring the Experimental Units

Monitoring of the experimental unit sections started in December 2016 after completing the bioengineering techniques' implantation works. The monitoring occurred periodically, with 15-day interval visits in all units (i.e., EU1-S01 to EU1-S11 and EU2-S01 to EU2-S12) for the first 90 days after the implantation, followed by monthly visits up to 27 months after the works were finished. In this period, the sections were visited and detailed inspections were carried out using high-quality images taken from the front view of each section. During the visits, preventive and corrective actions were used to take care of leaf-cutting ants, leaf chlorosis, and fence conditions.

Furthermore, local qualitative analyses were conducted with evaluation cards (presented in Appendix B) to register integrity, anchoring, soil stability, germination, phytosanitary aspects, pests' occurrence, nutritional status, and other general aspects of the vegetation and/or bioengineering techniques in the experimental units' sections. Due to the existence of diagnostic (observer) bias in the set of characteristics evaluated, these results require a detailed analysis that may include the (observer) bias error and are not in the scope of this study. These qualitative results are aimed to be made available in future publications.

2.4. Vegetative Cover Index Determination

The evolution of vegetative cover for each bioengineering technique adopted was evaluated by means of the vegetative cover index, calculated through a computer code in the MATLAB software (2015). The images captured (on a smartphone with wide-angle lens) during the experimental units' monitoring period (27 months; e.g., Figure 4a) were treated and analyzed with a MATLAB script (or algorithm) developed using the method proposed by Woebbecke et al. [44]. This semi-quantitative analysis is proposed as an effective approach to assess the vegetation performance under climatological changes (dry and wet spell cycles).

Prior to its analysis, all images were treated by trimming out the regions outside the analyzed section area (Figure 4b). The MATLAB algorithm splits the image into three color bands (red, green, and blue—RGB scale). The value of each pixel was used to determine the normalized ratio of each color in the band by dividing it by 255, which is the maximum value for each color, as shown in the following Equations (1)–(3):

$$R^* = R/255 \tag{1}$$

$$G^* = G/255 \tag{2}$$

$$B^* = B/255 \tag{3}$$

where R^*, G^*, and B^* are the normalized ratio of red, green, and blue colors, respectively, and R, G, and B are the color value of each pixel, which ranges from 0 to 255, respectively.

The normalized colors were used to calculate the ratio of each color, r, g, and b, with respect to all three colors in the pixel, as shown in the Equations below (4)–(6):

$$r = R^*/(R^* + G^* + B^*) \tag{4}$$

$$g = G^*/(R^* + G^* + B^*) \tag{5}$$

$$b = B^*/(R^* + G^* + B^*) \tag{6}$$

Finally, the excess green (*ExG*) value is calculated for each pixel using the Woebbecke et al. [45] excess green Equation (7):

$$ExG = 2g - r - b \tag{7}$$

The *ExG* value for each pixel was compared to a minimum threshold value that was determined for each image. Thus, if the pixel's *ExG* value is greater than this threshold value, the pixel is set to represent vegetative covered areas (that means a vegetation pixel). The vegetative cover index is determined for each image by the ratio between the number of green pixels and the total number of pixels in the treated image (white pixels are not considered). In addition, the vegetation pixels were superimposed in the treated image for a visual comparison (Figure 4c).

(a)　　　　　　　　(b)　　　　　　　　(c)

Figure 4. Image analysis processes: (**a**) original image; (**b**) treated image submitted to the computer algorithm; and (**c**) comparison between the vegetation pixels provided by the computer algorithm and the treated image.

The vegetative cover index analysis was performed for all images taken while monitoring the experimental units (total of 601) for all sections, and bioengineering techniques were applied. The results were plotted in function of the time after the implanting the bioengineering works to evaluate the evolution of this parameter during the investigated period. In this study, vegetative cover index values higher than 70% are considered to represent a satisfactory superficial slope cover condition [45], in other words, to ensure the protection/control against erosion of the slope superficial soil.

2.5. Climatological Data

To compare the evolution of the vegetative cover index with the regional seasons, climatological data was obtained from the closest working weather station, 31 km from the experimental units. The precipitation data (in mm), obtained with a rain gauge, are summarized in Figure 5 for the inspection period adopted herein. In summary, the local climatological condition is comprised of a rainy (wet) period during the spring and summer seasons and a dry period during the autumn and winter seasons.

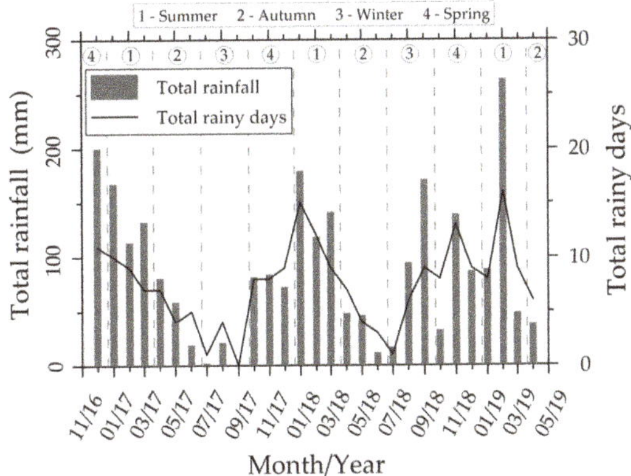

Figure 5. Total rainfall and number of rainy days obtained in a weather station located 31 km distant from the experimental units' location.

3. Results and Discussion

Figure 6 shows the vegetative cover index evolution for the bioengineering techniques installed in isolation at the experimental units. The vegetative cover index evolutions can be divided into five different groups. The first one exhibits vegetative cover index values lower than 40% (non-satisfactory values) during the whole period investigated. The index experienced small increases during the wet periods and remained or decreased during the dry ones. The manual seeding (IBT-1; EU1-S08 and EU2-S07), live stakes (IBT-2; EU2-S01), and live organic sediment retainer (L-OSRs; IBT-3; EU2-S06) techniques comprise this group (Figure 6a). The second group presented satisfactory vegetative cover index values (higher than 70%) 25 days after it was implemented and small reductions at the end of the wet periods (especially the third period). Two techniques installed in the first experimental unit compose this group: live rolled erosion control products (L-RECPs; IBT-4; EU1-S01) and rolled erosion control product (RECP-01; IBT-5; EU1-S02).

The third group identified in Figure 6 comprises L-RECPs installed at the second experimental unit (IBT-4; EU2-S02 and EU2-S03) and the geocellular containment system (GCS) installed in EU1-S07. Similar to the second group, this group exhibited satisfactory vegetative cover index values very soon (25 days after its implementation). However, there was a significant reduction in the index values during the first dry period (softer reduction for the GCS-EU1-S07) followed by an increase in the following two periods, and a small reduction in the index values at the third wet period. RECPs installed in the second experimental unit (RECP-2 in EU2-S09 and RECP-03 in EU2-S11) are considered the fourth group and showed a high variation in the vegetative cover index values during the period investigated (increases in the wet periods and sharp decreases in the dry ones). A particular case in Figure 6 (the fifth group) is the GCS installed at EU2-S05 that started a significant increase in the vegetative cover index values at the middle of the first dry period and maintained a value close to 60% until the end of the investigated period.

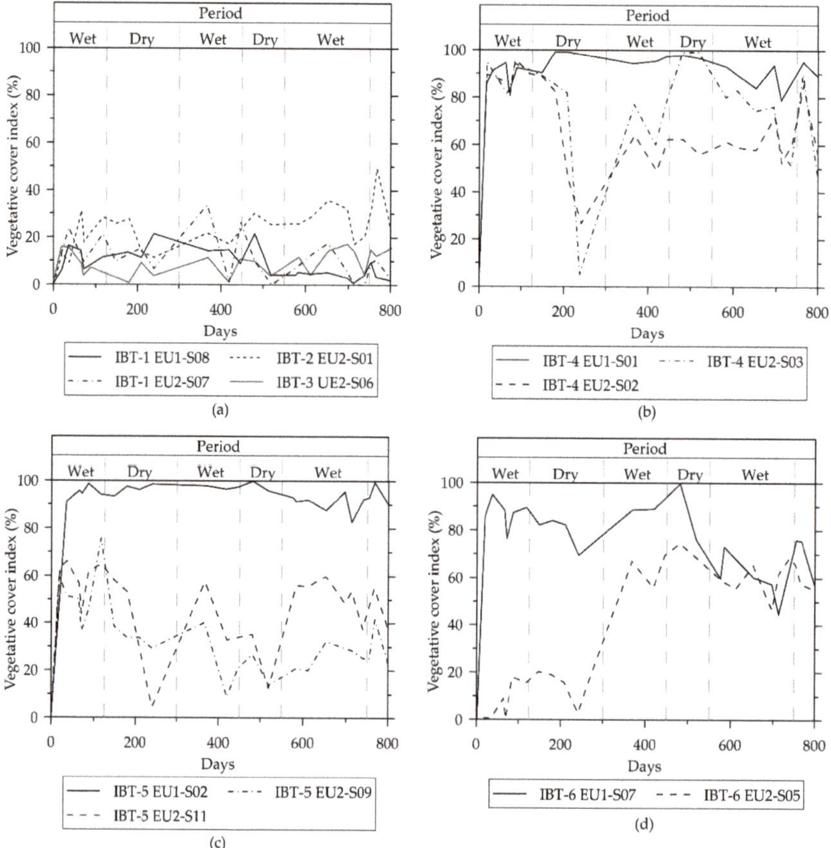

Figure 6. Vegetative cover index evolution for different in-isolation bioengineering techniques: (**a**) manual seeding (IBT-1; EU1-S08 and EU2-S07), live stakes (IBT-2; EU2-S01), and live organic sediment retainers (L-OSRs; IBT-3; EU2-S06); (**b**) live rolled erosion control products (L-RECPs; IBT-4; EU1-S01, EU2-S02, and EU2-S03); (**c**) rolled erosion control products (RECPs; IBT-5; EU1-S02, EU2-S09, and EU2-S110); and (**d**) geocellular containment system (GCSs; IBT-6; EU1-S07 and EU2-S05).

The results presented in Figure 6a indicate that the adoption of the in-isolation manual seeding technique (IBT-1) can be insufficient to prevent and control slope superficial soil erosion. Considering the vegetation establishment, difficulties can be observed since the first wet period, and it proves to be highly influenced by the changes in the climatological condition (especially during the dry periods). In addition, widespread laminar erosion and small grooves were reported in EU1-S08 at the end of the inspection period, proving the inefficiency of this technique.

Adopting live stakes in an isolated condition (IBT-2; EU2-S01) presented minimal effectiveness for the soil conditions and vegetation species used in this study. Three main factors impair its application: the values of the vegetative cover index for IBT-2 were slightly higher than the ones obtained with IBT-1

(Figure 6a), the time consumed for the cultivation of the stakes, and the costs involved in the processes. Similarly, adopting live organic sediment retainers (L-OSRs) in isolation conditions (IBT-3; EU2-S06) proves to be inefficient. In fact, this technique (IBT-3) provided values of vegetative cover index smaller than the ones obtained with IBT-1 (Figure 6a).

Live rolled erosion control products (L-RECPs) installed in isolation (IBT-4; EU1-S01, EU2-S02 and EU2-S03) proved to be an attractive technique to prevent slope superficial erosion. However, for the sections of the second experimental unit (EU-2), sharp drops (dry periods or beginning of the wet periods) and sensitive reductions (especially during wet periods) characterize the evolution of vegetative cover index for this technique (Figure 6b).

Changes in the vegetation phytosanitary aspects (chlorosis and phytopathogen conditions) explain the sharp drops observed. The predecessor dry period could change the phytosanitary aspects of the vegetation, leading to an intensive decay of its green colors (becoming faded). Considering the computer code adopted in this study, when the treated image (containing faded vegetation) is subjected to the RGB scale, a decrease may occur in the pixel's green value, culminating in its identification as a pixel that does not represent vegetation, decreasing the vegetative cover index.

Considering the sensitive reductions, they can be attributed to the predominant vegetation composition (type). Some vegetative species may not adapt to the soil and/or climatological conditions (especially the non-native species), leading to changes in the vegetation phytosanitary aspects (initially) and/or the vegetation death. Further studies are required to validate this assumption and identify which species are unsuitable.

Three types of rolled erosion control products (RECPs) commercially available on the Brazilian market were installed in isolated conditions (IBT-5). RECP-01 (EU1-S02) exhibited a quick establishment of the vegetation—less than 30 days after it was implanted—and it has not been significantly influenced by climate changes (Figure 6c). Similar results were reported by Álvarez-Mozos et al. [46]. RECP-02 (EU2-S11) and RECP-03 (EU2-S09) also presented a quick establishment of the vegetation, but sharp drops and sensitive reductions in the vegetative cover index values are clear (Figure 6c).

Despite the similar behavior of the vegetative cover index evolution, RECP-02 exhibited a better re-establishment of the vegetation in the third wet period. Thus, one must consider that RECPs comprising coconut fibers could be more susceptible to vegetation development than the ones comprising vegetable fiber (straw). It is worth mentioning that these conclusions are specific for the conditions of soil and vegetation species used in this study. Furthermore, these products are indicated for temporary slope protection, as high degrees of deterioration were evident at the end of the inspection period—similar results were reported by Vishnudas [47].

Considering the last in-isolation technique investigated (IBT-6), the geocellular containment system (GCS) installed in EU1-S07 and EU2-S05 exhibited similar vegetation evolutions (Figure 6d). The non-satisfactory vegetative cover index values reported are associated with the difficulty of seeds' germination and establishment. The geotextiles act as a barrier between the seeds and the soil [44,48] and between the sunlight and soil, leading to low germination rates [49]. However, the IBT-6 technique proved to avoid slope superficial erosion processes, as no erosive processes were reported during the inspections performed in the investigated period.

One must be aware of the shortcomings of comparing the performance of the in-isolation techniques in the experimental units one (EU-1) and two (EU-2). As shown by the results of the manual seeding (IBT-1; Figure 6a), L-RECPs (IBT-4; Figure 6b), and GCS (IBT-6; Figure 6d) techniques, differences in the vegetation development (and establishment) between sections installed in EU-1 and EU-2 were noted (especially in the dry periods). Overall, these results show that the sections of the second experimental unit (EU2) were more sensitive to climatological changes. In fact, because of the experimental units' different orientation (slope aspects stated in topic 2.1), the sections of EU2 face sunlight exposure for a longer period compared to the sections of EU1. This difference hampers the comparison between the results obtained with the RECPs' isolated technique (IBT-5, Figure 6c). It is not ideal to compare the results obtained with RECP-01 with RECP-02 and/or RECP-03, because they were installed under the different experimental units and a different type of RECPs was adopted in each case.

Figure 7 shows the vegetative cover index evolution for the mixed bioengineering techniques installed at the experimental units. The vegetative cover index evolutions can be divided into three different groups. The first group exhibits vegetative cover index evolution similar to the first group of the in-isolation bioengineering techniques: unsatisfactory vegetative cover index values throughout the whole investigated period and increases during wet periods and maintenance or decrease of the index values during the dry periods. The manual seeding and cellulose mulch (MBT-1; EU1-S09), manual seeding and organic sediment retainers (OSRs; MBT-2; EU1-S06), and live stakes and live organic sediment retainer (L-OSRs; MBT-4; EU2-S04) techniques comprise this group. Manual seeding and ORSs (MBT-2; EU2-S12); live stakes, L-ORSs, and manual seeding (MBT-5; EU1-S04); and RECPs and OSRs (MBT-6; EU2-S10) techniques set the second group. This group is characterized by vegetative cover index values between 20 and 70%, sharp drops in the index values at the end of the dry periods, and its increase during the wet ones.

Finally, the third group of mixed bioengineering techniques is characterized by a satisfactory vegetative cover index 30 days after the techniques were implemented, followed by variations (increases and decreases) in the index values for the whole period investigated (decreases occurred especially at the end of the periods—regardless of wet or dry ones). Despite this variability, the sections of this group exhibited a satisfactory vegetative cover index at the end of the investigation period. This group is comprised by manual seeding and ORSs (MBT-2; EU1-S11); manual seeding, cellulose mulch and OSRs (MBT-3; EU1-S10); live stakes, L ORSs and manual seeding (MBT-5; EU1-S03); and rolled erosion control products (RECPs) and OSRs (MBT-6; EU1-S05 and EU2-S08).

The results presented in Figure 7a revealed that adopting organic material (cellulose mulch) attached with manual seeding (MBT-1) did not provide any improvement in the vegetation establishment compared with the in-isolation manual seeding technique (IBT-1; Figure 6a). In addition, widespread laminar erosion and small grooves were reported in the section (EU1-S09) at the end of the inspection period. Thus, adding organic material attached to the manual seeding technique does not seem to be an effective technique for the soil conditions studied in this paper. Further studies are required to assess other types of organic material attached to the manual seeding technique.

In the case of the combined adoption of organic sediment retainers (OSRs) and manual seeding technique (MBT-02), better establishment of the vegetation (Figure 7a) compared with IBT-1 (Figure 6a) is evident. However, MBT-2 exhibited high variability in the vegetative cover index evolution between the sections installed (EU1-S06, EU1-S11 and EU2-S12)—a notable difference occurs between EU1-S06 and EU1-S11, installed at the same experimental unit (EU1). Furthermore, this technique proved to be vulnerable to climatologic changes (dry–wet cycles) for the particular vegetation species used. Despite these issues, this mixed bioengineering technique prevented the occurrence of slope superficial erosion, as no erosion process was encountered in the sections up to the end of the investigation period. This evidence was expected, as the OSR decreases runoff on the slopes, retains humidity and nutrients in the system, and promotes vegetation establishment [50].

Despite the variability encountered for the aforementioned technique (MBT-2), the OSRs provided an excellent vegetation development when added to the manual seeding technique attached to organic material—cellulose mulch (MBT-3, EU1-S10; Figure 7a). MBT-3 proves to be an effective technique as slope superficial soil erosion protection.

The combination of live stakes (IBT-02) and L-OSRs (IBT-03) techniques (MBT-4; EU2-S04; Figure 7b) did not provide an improvement in the vegetation development compared to its in-isolation techniques (IBT-02 and IBT-03; Figure 6a). Despite the slight influence of the climatological condition in the vegetative cover index values, difficulties in the vegetation establishment occurred even during the wet period. Moreover, laminar erosion and grouting processes were reported during the inspections.

On the other hand, the combination of live stakes, live organic sediment retainers (L-OSRs), and manual seeding (MBT-5; EU1-S03 and EU1-S04) exhibited an improvement of the vegetation establishment (Figure 7) compared to its applications in isolation conditions (IBT-1, IBT-2, and IBT-3; Figure 6a). However, this technique exhibits an unexpected variability (MBT-5): vegetative cover index evolution between sections EU1-S03 and EU1-S04 is very different. As both sections are located side-by-side (Figure 2) and under identical climatological conditions, the difference in the vegetation evolution may be caused due to intrinsic variability of the technique—more than one material can contribute with its own variability.

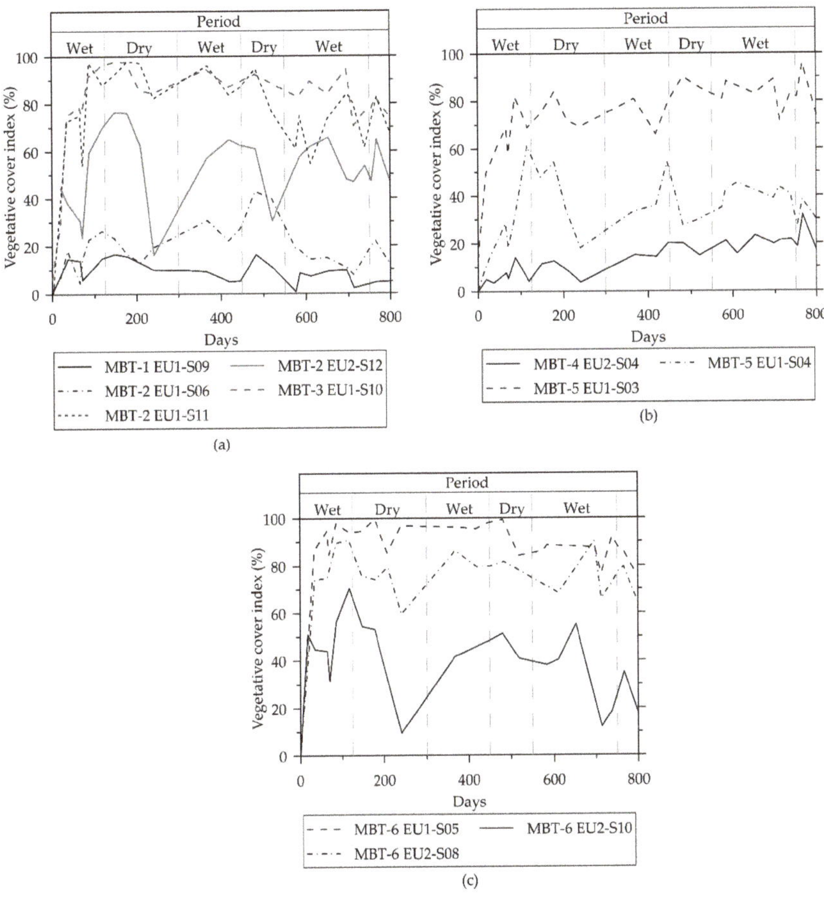

Figure 7. Vegetative cover index evolution for different mixed bioengineering techniques: (**a**) manual seeding and cellulose mulch (MBT-1; EU1-S09); manual seeding and organic sediment retainers (OSRs; MBT-2; EU1-S06, EU1-S11 and EU2-S12); and manual seeding, cellulose mulch and OSRs (MBT-3, EU1-S10); (**b**) live stakes and live organic sediment retainers (L-OSRs; MBT-4; EU2-S04) and live stakes, L-OSRs and manual seeding (MBT-5; EU1-S03 and EU1-S04); and (**c**) rolled erosion control products (RECPs; MBT-6; EU1-S05, EU2-S08, and EU2-S10).

Finally, the last mixed bioengineering technique (MBT-6) evaluated comprised the combination of RECPs and OSRs (Figure 7c). RECP-2 installed with OSRs (EU1-S05 and EU2-S08) exhibited better vegetation establishment, compared to RECP-3 installed with OSRs, during the whole period investigated (EU2-S10). Once again, the RECPs comprised by coconut fibers proves to be more susceptible for vegetation development than the ones comprised of vegetable fiber (straw) for similar conditions (soil and vegetation species used), as considered in this study.

Differences between the vegetative cover index evolution between the same technique installed in both experimental unit sections were also evident in MBT-3 and MBT-6 (with RECP-2). These differences are related to the experimental units' different orientation (stated in topic 2.1) as previously discussed.

Despite the benefits of using geosynthetics (polymeric based materials) in geotechnical works, one must be aware of the long-term environmental impacts caused by using these materials. The small particles resulting from the geosynthetic degradation (micro plastic) may enter the environment, resulting in soil pollution and spoiling the fauna [48,51–53].

4. Conclusions

In this study, different types of bioengineering techniques were evaluated, in isolation and in combined conditions, as systems to prevent/control superficial erosion processes in the slopes of Paraíba do Sul-MG (Brazil). High-quality images taken from periodical visits conducted over the course of 27 months were submitted to a computer code to assess the percentage of vegetation developed in the slopes' superficial soil through time. Based on the results obtained in this study, the following conclusions can be drawn:

- Despite the similar characteristics of the soil of the two selected experimental units, the effectiveness of the same technique applied to both (EU1 and EU2) seems to be influenced by the differences in their climatological conditions. As the experimental units are only 1.2 km apart, differences in their climatological conditions can hardly be attributed to differences in rainfall events. Thus, due to the experimental units' different orientation, experimental unit two (EU2) was exposed to sunlight incidence for longer periods compared to experimental unit one (EU1). In this case, EU2 experienced significant changes in the vegetation's phytosanitary aspects, impairing the vegetative cover index's determination. This hypothesis is supported by the sharp drops in the vegetative cover index for most techniques at the end or after the dry periods in the EU2.

- Among the six in-isolation bioengineering techniques evaluated in this study, adopting manual seeding, live stakes, and live organic sediment retainers (L-OSRs) exhibited vegetative cover index values smaller than 40%, with difficulties in vegetation establishment over the 27-month investigated period (fluctuation in its values were also present). These techniques have not proven to be effective to prevent/control slopes' superficial erosion. Despite the high values of vegetative cover index observed when geocellular containment systems (GCSs) were applied, this technique exhibited fluctuation throughout the period investigated, which indicates vegetation establishment deficiency.

- Live rolled erosion control products (L-RECPs) and a highly flexible, UV-stabilized, and non-degradable three-dimensional matrix rolled erosion control product (RECP) investigated in isolation conditions exhibited a high tendency for vegetation development. L-RECPs led to a swift establishment of the vegetation. Adopting different types of RECPs available in the Brazilian market has shown that RECPs comprised of coconut fibers were more susceptible for vegetation development than the ones comprised of vegetable fiber (straw) for the specific soil conditions and vegetation species used herein. However, one must consider that for these techniques (L-RECPs and RECPs), the development of vegetation seems to be highly susceptible to harsh climatological conditions (especially the dry periods).

- Incorporating organic material (cellulose mulch) to the manual seeding technique did not improve the vegetation development. The inclusion of organic sediment retainers (OSR) to the manual seeding technique exhibited an expressive variability on the vegetative cover index, with a high influence of the dry periods in the vegetation establishment. However, the combination of manual seeding, cellulose mulch, and organic sediment retainers (OSR) proved to be an excellent technique, inducing a quick establishment of vegetation with satisfactory vegetative cover index and only slight fluctuations during the period investigated. Live stakes combined with organic sediment retainers (OSR) and the manual seeding technique exhibited high variability and non-satisfactory vegetative cover index. OSRs combined with commercially available RECPs exhibited a significant variability and proved to be sensitive in locations with a harsh climatological condition (especially the dry periods).

- The sharp drop and sensitive reductions in the vegetative cover index can be attributed to different factors. The sharp drops, especially at the end or after the exposure to dry periods, possibly occurred due to significant changes in the vegetation's phytosanitary aspect, in other words, an intensive decay on the vegetation's green color. Regarding the sensitive reductions in the vegetative cover index, especially in wet periods, they can be attributed to difficulties in some vegetative species (possibly the non-native ones) to adapt to the soil and/or climatological conditions of the area, resulting in a similar change in the vegetative phytosanitary aspects or vegetation death. In both cases, the reductions experienced in the vegetative cover index derive from the methodology adopted in this study, which did not include faded vegetation in the vegetative cover index. It should be noted that this conservative procedure aims to neglect the vegetation without adequate phytosanitary aspects, considering that they are not able to prevent the surface erosion of the slopes.

Finally, the conclusions presented in this study are restricted to the specific site-conditions investigated herein: the given slope, a sandy silt soil with low fertility and medium acidity in a tropical climate, and for the specific selected vegetation species used in the treatments. Therefore, caution should be exercised when using these conclusions for different site-conditions.

Author Contributions: The individual contributions of each authors are highlighted as follows: conceptualization: M.P.L.; data curation: M.P.F. and A.T.V.; formal analysis: M.P.F.; funding acquisition: M.P.L.; investigation: V.F.V. and G.B.M.; methodology: G.B.M.; project administration: M.P.L.; software: G.B.M. and A.T.C.; supervision: J.L.d.S.; validation: V.F.V.; visualization: C.B.; writing—original draft: M.P.F.; writing—review & editing: C.B. and J.L.d.S. All authors have read and agreed to the published version of the manuscript.

Funding: This research was funded by the Agência Nacional de Energia Elétrica (ANEEL; PD-0394-1603/2016) and the Coordenação de Aperfeiçoamento de Pessoal de Nível Superior (CAPES; 001).

Acknowledgments: The authors would like to thank the University of São Paulo, California State University Los Angeles, and all of the Eletrobras FURNAS community for the support provided to the research activities reported in this paper and Agência Nacional de Energia Elétrica (ANEEL; National Agency of Electric Energy) for promoting this research to Bioengineering (PD-ANEEL number 0394-1603/2016). This study was financed in part (support granted to the second author) by the Brazilian Federal Agency for Support and Evaluation of Graduate Education (Coordenação de Aperfeiçoamento de Pessoal de Nível; CAPES)-Finance Code 001.

Conflicts of Interest: The authors declare that they have no known competing financial interests or personal relationships that could have appeared to influence the work reported in this paper. Thus, the authors declare no conflict of interest. Moreover, the funders had no role in the design of the study; in the collection, analyses, or interpretation of data; in the writing of the manuscript; or in the decision to publish the results.

Appendix A

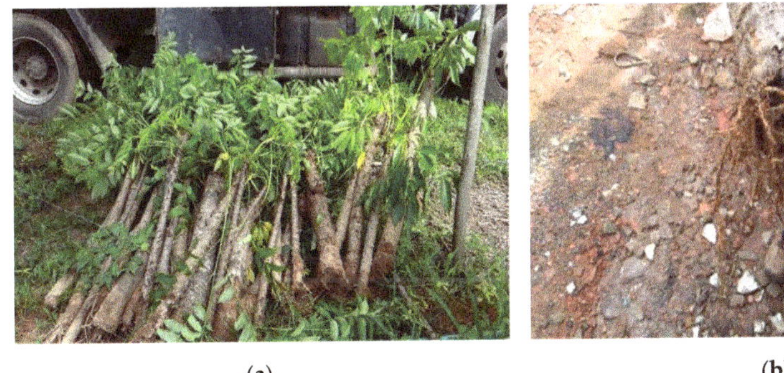

(a)   (b)

Figure A1. *Cont.*

(c)   (d)

Figure A1. Live stakes: (**a**) storage after transport in an air-conditioned environment, (**b**) root system detail, (**c**) installation procedure in manually excavated cavities, and (**d**) substrate material application.

(a)   (b)

(c)   (d)

Figure A2. Live rolled erosion control products (L-RECPs): (**a**) cultivation at greenhouses, (**b**) coil-shaped collection, (**c**) transportation, and (**d**) installation of the elements in the experimental units slopes.

(a)   (b)

Figure A3. Organic sediment retainers associated with live elements (L-OSRs): (**a**) storage in the experimental units; and (**b**) installation of the elements in manually excavated trenches.

Appendix B

Table A1. Checklist for the visual inspections performed during visits carried out during 27 months in the experimental units' sections area.

Inspection Check List of the Sections	
Local: *Simplício* UHE	Experimental Unit: "Number of the EU"
Section: "Number of the section"	Bioengineering technique: "Name of the bioengineering technique"
Data: "Data of the inspection/visit"	Time: "Time of the inspection/visit"
Climatological Condition: "Sunny, cloudy, partially cloudy, rainy"	Data of the end of execution: "Data of the conclusion of the bioengineering technique's execution process in the section"
General aspects	Good (Satisfactory general aspects) Medium (Non-satisfactory factors present which does not compromise the integrity and treatment efficiency) Poor (Presence of non-satisfactory factors that may impair the integrity and treatment efficiency)
General aspects of the vegetation/structure	
Structural integrity	Good (Absence of apparent damage) Medium (Presence of less significant damage) Poor (Presence of damage that impairs the integrity and treatment efficiency)
Anchoring/Stapling	Great (Efficient stapling, absence of loose or uprooted staples) Good (Efficient stapling, presence of loose or uprooted staples in a rate up to 0.1 staples/m^2) Medium (Efficient stapling, presence of loose or uprooted staples in a rate of 0.1–0.5 staples/m^2) Poor (Efficient stapling, presence of loose or uprooted staples in a rate higher than 0.5 staples/m^2)
Soil stability/Erosion	Good (general aspects of the section are satisfactory: without sediment mobilization points) Medium (section with the presence of sediment mobilization points); Poor (presence of linear erosion)
Germination/Vegetation stakes setting (%)	Good (Germination of 25% up to 1 month after cultivation or germination of 50% for 1–3 months after cultivation or germination of 70% after 3 months of cultivation) Medium (Germination of 15% up to 1 month after cultivation or germination of 30% for 1–3 months after cultivation or germination of 50% after 3 months of cultivation) Good (Germination less than 25% up to 1 month after cultivation or germination less than 30% for 1–3 months after cultivation or germination less than 50% after 3 months of cultivation)

Table A1. *Cont.*

Predominant species	Mix of 10 herbaceous species (attached)
Phytosanitary aspect	Good (Absence of phytopathogen) Medium (Presence of chlorosis and phytopathogen up to 20% of the total vegetation) Poor (Presence of chlorosis and phytopathogen higher than 20% of the total vegetation)
Pest occurrence	Presence or absence of pests
Nutritional status	Good (Absence of chlorosis in vegetation) Medium (Occurrence of chlorosis up to 20% in the total vegetation) Poor (Occurrence of chlorosis higher than 20% in the total vegetation)
Fencing condition	Good (Absence of fence breakage) Poor (Presence of fence breakage)
Additional aspects	Other aspects that must be highlighted
Photographic report	
Photo 01	Photo 02
Photo 03	Photo 04

References

1. Ellison, W.D. Soil Erosion. *Soil Sci. Soc. Am. J.* **1948**, *12*, 479–484. [CrossRef]
2. Stokes, A.; Douglas, G.B.; Fourcaud, T.; Giadrossich, F.; Gillies, C.; Hubble, T.; Kim, J.H.; Loades, K.W.; Mao, Z.; McIvor, I.R.; et al. Ecological mitigation of hillslope instability: Ten key issues facing researchers and practitioners. *Plant Soil* **2014**, *377*, 1–23. Available online: http://link.springer.com/10.1007/s11104-014-2044-6 (accessed on 18 September 2020). [CrossRef]
3. Koerner, R.M. Emerging and Future Developments of Selected Geosynthetic Applications. *J. Geotech. Geoenviron. Eng.* **2000**, *126*, 293–306. [CrossRef]
4. Theisen, M. The role of geosynthetics in erosion and sediment control: An overview. *Geotext. Geomembr.* **1992**, *11*, 535–550. [CrossRef]
5. Bischetti, G.B.; Dio, M.D.F.; Florineth, F. On the Origin of Soil Bioengineering. *Landsc. Res.* **2012**, *39*, 583–595. [CrossRef]
6. Gray, D.H.; Sotir, R.B. Biotechnical stabilization of steepened slopes. *Transp. Res. Rec.* **1995**, *1474*, 23–29. Available online: http://onlinepubs.trb.org/Onlinepubs/trr/1995/1474/1474-003.pdf (accessed on 18 September 2020).
7. Wu, K.; Austin, D. Three-dimensional polyethylene geocells for erosion control and channel linings. *Geotext. Geomembr.* **1992**, *11*, 611–620. [CrossRef]
8. Lewis, L.; Salisbury, S.L.; Hagen, S. *Soil Bioengineering for Upland Slope Stabilization*; Report WA-RD491.1; Washington State Transportation Center (TRAC): Washington, DC, USA, 2001.
9. Simon, K.; Steinemann, A. Soil Bioengineering: Challenges for Planning and Engineering. *J. Urban Plan. Dev.* **2000**, *126*, 89–102. [CrossRef]
10. Evette, A.; LaBonne, S.; Rey, F.; Liébault, F.; Jancke, O.; Girel, J. History of bioengineering techniques for erosion control in rivers in western europe. *Environ. Manag.* **2009**, *43*, 972–984. [CrossRef] [PubMed]
11. Stokes, A.; Raymond, P.; Polster, D.; Mitchell, S.J. Engineering the ecological mitigation of hillslope stability research into the scientific literature. *Ecol. Eng.* **2013**, *61*, 615–620. [CrossRef]
12. Stokes, A.; Sotir, R.; Chen, W.; Ghestem, M. Soil bio- and eco-engineering in China: Past experience and future priorities. *Ecol. Eng.* **2010**, *36*, 247–257. [CrossRef]
13. Rey, F.; Bifulco, C.; Bischetti, G.B.; Bourrier, F.; De Cesare, G.; Florineth, F.; Graf, F.; Marden, M.; Mickovski, S.; Phillips, C.; et al. Soil and water bioengineering: Practice and research needs for reconciling natural hazard control and ecological restoration. *Sci. Total Environ.* **2019**, *648*, 1210–1218. [CrossRef]
14. Lewis, L. *Soil Bioengineering: An Alternative for Roadside A Practical Guide*; San Dimas Technology & Development Center: San Dimas, CA, USA, 2000.
15. Dhital, Y.P.; Tang, Q. Soil bioengineering application for flood hazard minimization in the foothills of Siwaliks, Nepal. *Ecol. Eng.* **2015**, *74*, 458–462. [CrossRef]

16. Petrone, A.; Preti, F. Suitability of soil bioengineering techniques in Central America: A case study in Nicaragua. *Hydrol. Earth Syst. Sci.* **2008**, *12*, 1241–1248. [CrossRef]

17. Rauch, H.; Sutili, F.J.; Hörbinger, S. Installation of a Riparian Forest by Means of Soil Bio Engineering Techniques—Monitoring Results from a River Restoration Work in Southern Brazil. *Open J. For.* **2014**, *4*, 161–169. [CrossRef]

18. Woolsey, S.; Capelli, F.; Gonser, T.; Hoehn, E.; Hostmann, M.; Junker, B.; Paetzold, A.; Roulier, C.; Schweizer, S.; Tiegs, S.D.; et al. A strategy to assess river restoration success. *Freshw. Biol.* **2007**, *52*, 752–769. [CrossRef]

19. Hagen, S.; Salisbury, S.; Wierenga, M.; Xu, G.; Lewis, L. Soil Bioengineering as an Alternative for Roadside Management Benefit—Cost Analysis Case Study. *Transp. Res. Rec.* **2001**, *1794*, 97–104. [CrossRef]

20. Ghestem, M.; Cao, K.-F.; Ma, W.; Rowe, N.; Leclerc, R.; Gadenne, C.; Stokes, A. A Framework for Identifying Plant Species to Be Used as 'Ecological Engineers' for Fixing Soil on Unstable Slopes. *PLoS ONE* **2014**, *9*, e95876. [CrossRef]

21. Cazzuffi, D.; Cardile, G.; Gioffre, D. Geosynthetic Engineering and Vegetation Growth in Soil Reinforcement Applications. *Transp. Infrastruct. Geotechnol.* **2014**, *1*, 262–300. [CrossRef]

22. Stokes, A.; Atger, C.; Bengough, A.G.; Fourcaud, T.; Sidle, R.C.; Bengough, A.G. Desirable plant root traits for protecting natural and engineered slopes against landslides. *Plant Soil* **2009**, *324*, 1–30. [CrossRef]

23. Cavaillé, P.; Dommanget, F.; Daumergue, N.; Loucougaray, G.; Spiegelberger, T.; Tabacchi, E.; Evette, A. Biodiversity assessment following a naturality gradient of riverbank protection structures in French prealps rivers. *Ecol. Eng.* **2013**, *53*, 23–30. [CrossRef]

24. Tisserant, M.; Janssen, P.; Evette, A.; González, E.; Cavaillé, P.; Poulin, M. Diversity and succession of riparian plant communities along riverbanks bioengineered for erosion control: A case study in the foothills of the Alps and the Jura Mountains. *Ecol. Eng.* **2020**, *152*, 105880. [CrossRef]

25. Sotir, R.B. Integration of Soil Bioengineering Techniques for Watershed Management. In *Wetlands Engineering and River Restoration Conference 2001*; Intergovernmental Panel on Climate Change, Ed.; American Society of Civil Engineers (ASCE): Reston, VA, USA, 2001; pp. 1–8. Available online: https://www.cambridge.org/core/product/identifier/CBO9781107415324A009/type/book_part (accessed on 18 September 2020).

26. Petrone, A.; Preti, F. Soil bioengineering for risk mitigation and environmental restoration in a humid tropical area. *Hydrol. Earth Syst. Sci.* **2010**, *14*, 239–250. [CrossRef]

27. Li, X.; Zhang, L.; Zhang, Z. Soil bioengineering and the ecological restoration of riverbanks at the Airport Town, Shanghai, China. *Ecol. Eng.* **2006**, *26*, 304–314. [CrossRef]

28. Zhang, H.-L.; Zhao, Z.; Ma, G.; Sun, L. Quantitative evaluation of soil anti-erodibility in riverbank slope remediated with nature-based soil bioengineering in Liaohe River, Northeast China. *Ecol. Eng.* **2020**, *151*, 105840. [CrossRef]

29. Von Der Thannen, M.; Hoerbinger, S.; Paratscha, R.; Smutny, R.; Lampalzer, T.; Strauss, A.; Rauch, H.P. Development of an environmental life cycle assessment model for soil bioengineering constructions. *Eur. J. Environ. Civ. Eng.* **2017**, *24*, 141–155. [CrossRef]

30. Sattler, D.; Seliger, R.; Nehren, U.; De Torres, F.N.; Da Silva, A.S.; Raedig, C.; Hissa, H.R.; Heinrich, J.; Filho, W.L.; De Freitas, L.E. Pasture Degradation in South East Brazil: Status, Drivers and Options for Sustainable Land Use under Climate Change. In *Climate Change Adaptation in Latin America Managing Vulnerability, Fostering Resilience*; Filho, W.L., de Freitas, L.E., Eds.; Springer: Berlin/Heidelberg, Germany, 2017; pp. 3–17.

31. Sattler, D.; Raedig, C.; Hebner, A.; Wesenberg, J. Use of Native Plant Species for Ecological Restoration and Rehabilitation Measures in Southeast Brazil. In *Strategies and Tools for a Sustainable Rural Rio de Janeiro*; Springer: Berlin/Heidelberg, Germany, 2018; pp. 191–204.

32. Rey, F.; LaBonne, S. Resprout and Survival of Willow (Salix) Cuttings on Bioengineering Structures in Actively Eroding Gullies in Marls in a Mountainous Mediterranean Climate: A Large-Scale Experiment in the Francon Catchment (Southern Alps, France). *Environ. Manag.* **2015**, *56*, 971–983. [CrossRef]

33. Muhammad, M.M.; Alias, M.N.; Yusouf, K.W.; Mustafa, M.U.; Ghani, A.A. Suitability of bioengineering channels in erosion control: Application to urban stormwater drainage systems. *Adv. Appl. Fluid Mech.* **2016**, *19*, 765–785. [CrossRef]

34. Giupponi, L.; Borgonovo, G.; Giorgi, A.; Bischetti, G.B. How to renew soil bioengineering for slope stabilization: Some proposals. *Landsc. Ecol. Eng.* **2018**, *15*, 37–50. [CrossRef]

35. Donagema, G.K.; Campos, D.V.B.; Calderano, S.B.; Teixeira, W.G.; Viana, J.H.M. *Soil Analysis Methods Manual*, 2nd ed.; Brazilian Agricultural Research Corporation (EMBRAPA): Rio de Janeiro, Brazil, 2011; p. 230. Available online: https://www.embrapa.br/busca-de-publicacoes/-/publicacao/990374/manual-de-metodos-de-analise-de-solo (accessed on 18 September 2020).

36. Brazilian Association of Technical Standards. *NBR 6458: Gravel Grains Retained on the 4,8 mm Mesh Sieve—Determination of the Bulk Specific Gravity, of the Apparent Specific Gravity and of Water Absorption*; ABNT: Rio de Janeiro, Brazil, 2016; p. 14.

37. Brazilian Association of Technical Standards. *NBR 7181: Soil—Grain Size Analysis*; ABNT: Rio de Janeiro, Brazil, 2016; p. 16.

38. American Society for Testing and Materials. *ASTM D 2487-06: Standard Practice for Classification of Soils for Engineering Purposes (Unified Soil Classification System)*; ASTM International: West Conshohocken, PA, USA, 2006; p. 12.

39. Brazilian Association of Technical Standards. *NBR 6459: Soil—Liquid Limit Determination*; ABNT: Rio de Janeiro, Brazil, 2016; p. 9.

40. Brazilian Association of Technical Standards. *NBR 7180: Soil—Plasticity Limit Determination*; ABNT: Rio de Janeiro, Brazil, 2016; p. 7.

41. Brazilian Association of Technical Standards. *NBR 7182: Soil—Compaction Test*; ABNT: Rio de Janeiro, Brazil, 2016; p. 13.

42. American Society for Testing and Materials. *ASTM D 3080-98: Standard Test Method for Direct Shear Test of Soils under Consolidated Drained Conditions*; ASTM International: West Conshohocken, PA, USA, 1998; p. 6.

43. Soil Fertility Commission of the State of Minas Gerais. *Recommendations for Use of Fertilizer Correctives in Minas Gerais*, 5th ed.; Ribeiro, A.H.A., Guimarães, P.T.G., Alvarez, V.H., Eds.; CFSEMG: Viçosa, MG, Brazil, 1999; p. 359.

44. Woebbecke, D.M.; Meyer, G.E.; Von Bargen, K.; Mortensen, D.A. Color Indices for Weed Identification Under Various Soil, Residue, and Lighting Conditions. *Trans. ASAE* **1995**, *38*, 259–269. [CrossRef]

45. Coppin, N.J.; Richards, I.G. *Use of Vegetation in Civil Engineering*; Coppin, N.J., Richards, I.G., Eds.; Construction Industry Research and Information Association (CIRIA): London, UK, 1990; p. 40.

46. Álvarez-Mozos, J.; Abad, E.; Goñi, M.; Giménez, R.; Campo, M.A.; Díez, J.; Casalí, J.; Arive, M.; Diego, I. Catena Evaluation of erosion control geotextiles on steep slopes. Part 2: In fl uence on the establishment and growth of vegetation. *Catena* **2014**, *121*, 195–203. [CrossRef]

47. Vishnudas, S.; Savenije, H.; Van Der Zaag, P.; Anil, K.R.; Balan, K. The protective and attractive covering of a vegetated embankment using coir geotextiles. *Hydrol. Earth Syst. Sci.* **2006**, *10*, 565–574. [CrossRef]

48. Wiewel, B.V.; Lamoree, M. Geotextile composition, application and ecotoxicology—A review. *J. Hazard. Mater.* **2016**, *317*, 640–655. [CrossRef] [PubMed]

49. Chen, S.-C.; Chang, K.-T.; Wang, S.-H.; Lin, J.-Y. The efficiency of artificial materials used for erosion control on steep slopes. *Environ. Earth Sci.* **2010**, *62*, 197–206. [CrossRef]

50. Olsen, M.J.; Rikli, A.M. Investigation of Straw Wattle Influence on Surficial Slope Stability. In Proceedings of the Transportation Research Board 91st Annual Meeting, Washington, DC, USA, 22–26 January 2012; p. 16.

51. Hsuan, Y.G.; Schroeder, H.F.; Rowe, K.; Muller, W.; Greenwood, J.; Cazzuffi, D.; Koerner, R.M. Long-term performance and lifetime predition of geosynthetics. In Proceedings of the 4th European Conference on Geosynthetics, Edinburgh, Scotland, UK, 7–10 September 2008; p. 40.

52. Prambauer, M.; Wendeler, C.; Weitzenböck, J.; Burgstaller, C. Geotextiles and Geomembranes Biodegradable geotextiles—An overview of existing and potential materials. Geotext Geomembranes. *Geotext. Geomembr.* **2019**, *47*, 48–59. [CrossRef]

53. Browne, M.A.; Galloway, T.; Thompson, R.; Chapman, P.M. Microplastic—An emergig contaminant of potential concern? *Integr. Environ. Assess Manag.* **2007**, *3*, 559–561. [CrossRef]

MDPI
St. Alban-Anlage 66
4052 Basel
Switzerland
Tel. +41 61 683 77 34
Fax +41 61 302 89 18
www.mdpi.com

Sustainability Editorial Office
E-mail: sustainability@mdpi.com
www.mdpi.com/journal/sustainability